CONCISE ENCYCLOPEDIA OF
COMPOSITE MATERIALS
REVISED EDITION

ADVANCES IN MATERIALS SCIENCE AND ENGINEERING

This is a series of Pergamon scientific reference works, each volume providing comprehensive, self-contained and up-to-date coverage of a selected area in the field of materials science and engineering. The series has been developed primarily from the highly acclaimed *Encyclopedia of Materials Science and Engineering*, published in 1986. Other titles in the series are listed below.

BEVER† (ed.)
Concise Encyclopedia of Materials Economics, Policy & Management

BROOK (ed.)
Concise Encyclopedia of Advanced Ceramic Materials

CAHN & LIFSHIN (eds.)
Concise Encyclopedia of Materials Characterization

CARR & HERZ (eds.)
Concise Encyclopedia of Mineral Resources

CORISH (ed.)
Concise Encyclopedia of Polymer Processing & Applications

EVETTS (ed.)
Concise Encyclopedia of Magnetic & Superconducting Materials

MAHAJAN & KIMERLING (eds.)
Concise Encyclopedia of Semiconducting Materials & Related Technologies

MOAVENZADEH (ed.)
Concise Encyclopedia of Building & Construction Materials

SCHNIEWIND (ed.)
Concise Encyclopedia of Wood & Wood-Based Materials

WILLIAMS (ed.)
Concise Encyclopedia of Medical & Dental Materials

CONCISE ENCYCLOPEDIA OF

COMPOSITE MATERIALS

REVISED EDITION

Editor

ANTHONY KELLY

Extraordinary Fellow, Churchill College
Cambridge, UK

Executive Editor

ROBERT W CAHN

University of Cambridge, UK

Senior Advisory Editor

MICHAEL B BEVER†

MIT, Cambridge, MA, USA

PERGAMON

UK	Elsevier Science Ltd, The Boulevard, Langford Lane, Kidlington, Oxford OX5 1GB, England
USA	Elsevier Science Inc, 660 White Plains Road, Tarrytown, New York 10591-5133, USA
JAPAN	Elsevier Science Japan, Tsunashima Building Annex, 3-20-12 Yushima, Bunkyo-ku, Tokyo 113, Japan

---

Revised edition 1994
Reprinted 1995

Library of Congress Cataloging in Publication Data
Concise encyclopedia of composite materials / editor, Anthony Kelly; executive editor, Robert W. Cahn; senior advisory editor, Michael B. Bever. — Rev. ed.
p. cm. — (Advances in materials science and engineering)
Includes index.
1. Composite materials—Encyclopedias. I. Kelly, A. (Anthony) II. Cahn, R. W. (Robert W.), 1924– . III. Series.
TA418.9.C6C635 1994
620.1'18'03—dc20 93–48542

British Library Cataloguing in Publication Data
A catalogue record for this book is available from the Brirish Library

ISBN 0–08–042300–0

(∞)™ The paper used in this publication meets the minimum requirements of the American National Standard for Information Sciences—Permanence of Paper for Printed Library Materials, ANSI Z39.48–1984.

Printed and bound in Great Britain by BPC Wheatons Ltd, Exeter.

CONTENTS

HONORARY EDITORIAL ADVISORY BOARD

FOREWORD

In the time since its publication, the *Encyclopedia of Materials Science and Engineering* has been accepted throughout the world as the standard reference about all aspects of materials. This is a well-deserved tribute to the scholarship and dedication of the late Editor-in-Chief, Professor Michael Bever, the Subject Editors and the numerous contributors.

During its preparation, it soon became clear that change in some areas is so rapid that publication would have to be a continuing activity if the Encyclopedia were to retain its position as an authoritative and up-to-date systematic compilation of our knowledge and understanding of materials in all their diversity and complexity. Thus, the need for some form of supplementary publication was recognized at the outset. The Publisher has met this challenge most handsomely: both a continuing series of Supplementary Volumes to the main work and a number of smaller encyclopedias, each covering a selected area of materials science and engineering, have now been published.

Professor Robert Cahn, the Executive Editor, was previously the editor of an important subject area of the main work and many other people associated with the Encyclopedia have contributed to its Supplementary Volumes and derived Concise Encyclopedias. Thus, continuity of style and respect for the high standards set by the *Encyclopedia of Materials Science and Engineering* are assured. They have been joined by some new editors and contributors with knowledge and experience of important subject areas of particular interest at the present time. Thus, the Advisory Board is confident that the new publications will significantly add to the understanding of emerging topics wherever they may appear in the vast tapestry of knowledge about materials.

Walter S Owen
Chairman
Honorary Editorial Advisory Board

EXECUTIVE EDITOR'S PREFACE

As the publication of the *Encyclopedia of Materials Science and Engineering* approached, Pergamon resolved to build upon the immense volume of work which had gone into its creation by embarking on a follow-up project. This project had two components. The first was the creation of a series of Supplementary Volumes to the Encyclopedia itself. The second component of the new project was the creation of a series of Concise Encyclopedias on individual subject areas included in the Main Encyclopedia to be called *Advances in Materials Science and Engineering*.

These Concise Encyclopedias are intended, as their name implies, to be compact and relatively inexpensive volumes (typically 400–600 pages in length) based on the relevant articles in the Encyclopedia (revised where need be) together with some newly commissioned articles, including appropriate ones from the Supplementary Volumes. Some Concise Encyclopedias offer combined treatments of two subject fields which were the responsibility of separate Subject Editors during the preparation of the parent Encyclopedia (e.g., dental and medical materials).

Eleven Concise Encyclopedias have been published. These and their editors are listed below.

Concise Encyclopedia of Advanced Ceramic Materials	Prof. Richard J Brook
Concise Encyclopedia of Building & Construction Materials	Prof. Fred Moavenzadeh
Concise Encyclopedia of Composite Materials	Prof. Anthony Kelly
Concise Encyclopedia of Magnetic & Superconducting Materials	Dr Jan Evetts
Concise Encyclopedia of Materials Characterization	Prof. Robert W Cahn FRS & Dr Eric Lifshin
Concise Encyclopedia of Materials Economics, Policy & Management	Prof. Michael B Bever †
Concise Encyclopedia of Medical & Dental Materials	Prof. David F Williams
Concise Encyclopedia of Mineral Resources	Dr Donald D Carr & Prof. Norman Herz
Concise Encyclopedia of Polymer Processing & Applications	Mr P J Corish
Concise Encyclopedia of Semiconducting Materials & Related Technologies	Prof. Subhash Mahajan & Dr Lionel C Kimerling
Concise Encyclopedia of Wood & Wood-Based Materials	Prof. Arno P Schniewind

Many of the new or substantially revised articles in the Concise Encyclopedias have been published in one or other of the Supplementary Volumes, which are designed to be used in conjunction with the Main Encyclopedia. The Concise Encyclopedias, however, are "free-standing" and are designed to be used without necessary reference to the parent Encyclopedia.

The Executive Editor is personally responsible for the selection of topics and authors of articles for the Supplementary Volumes. In this task, he has the benefit of the advice of the Senior Advisory Editor and of other members of the Honorary Editorial Advisory Board, who also exercise general supervision of the entire project. The Executive Editor is responsible for appointing the Editors of the various Concise Encyclopedias and for supervising the progress of these volumes.

Robert W Cahn FRS
Executive Editor

ACKNOWLEDGEMENTS FOR THE FIRST EDITION

I am very grateful to all the authors who responded to my invitation to write for this Concise Encyclopedia. That a single editor has been able to put this work together is a tribute to the authors and to my secretary, Mrs Anne Roberts, who gives so generously of herself in helping me in so many things. I am also grateful to my colleagues at Surrey University, particularly to Dr P J Mills and Dr P A Smith, who stepped in to help me where others failed.

Anthony Kelly
Editor

ACKNOWLEDGEMENTS FOR THE REVISED EDITION

I am indeed grateful to the authors of the first edition for having produced a work which has established itself so quickly. This has warranted this revised edition. I am also very grateful to the new contributors who have produced their articles so well and speedily.

Anthony Kelly
Editor

GUIDE TO USE OF THE ENCYCLOPEDIA

This Concise Encyclopedia is a comprehensive reference work covering all aspects of composite materials. Information is presented in a series of alphabetically arranged articles which deal concisely with individual topics in a self-contained manner. This guide outlines the main features and organization of the Encyclopedia, and is intended to help the reader to locate the maximum amount of information on a given topic.

Accessibility of material is of vital importance in a reference work of this kind and article titles have therefore been selected, not only on the basis of article content, but also with the most probable needs of the reader in mind. On this basis, articles dealing with the properties of composite materials generally are to be found under the property (thus *Strength of Composites* and *Nonmechanical Properties of Composites*). An alphabetical list of all the articles contained in this Encyclopedia is to be found on p. xv.

Articles are linked by an extensive cross-referencing system. Cross-references to other articles in the Encyclopedia are of two types: in-text and end-of-text. Those in the body of the text are designed to refer the reader to articles that present in greater detail material on the specific topic under discussion. They generally take one of the following forms:

> ...as fully described in the article *Paper and Paperboard: An Overview*.

> ...an immersion test method (see *Nondestructive Evaluation of Composites*).

The cross-references listed at the end of an article serve to identify broad background reading and to direct the reader to articles that cover different aspects of the same topic.

The nature of an encyclopedia demands a higher degree of uniformity in terminology and notation than many other scientific works. The widespread use of the International System of Units has determined that such units be used in this Encyclopedia. It has been recognized, however, that in some fields Imperial units are more generally used. Where this is the case, Imperial units are given with their SI equivalent quantity and unit following in parentheses. Where possible the symbols defined in *Quantities, Units, and Symbols*, published by the Royal Society of London, have been used.

All articles in the Encyclopedia include a bibliography giving sources of further information. Each bibliography consists of general items for further reading and/or references which cover specific aspects of the text. Where appropriate, authors are cited in the text using a name/date system as follows:

> ...as was recently reported (Smith 1991).

> Jones (1987) describes...

The contributor's name and the organization to which they are affiliated appear at the end of each article. All contributors can be found in the alphabetical List of Contributors, along with their full postal address and the titles of the articles of which they are authors or co-authors.

The Introduction to this Concise Encyclopedia provides an overview of the field of composite materials and directs the reader to many of the key articles in this work.

The most important information source for locating a particular topic in the Encyclopedia is the multilevel Subject Index, which has been made as complete and fully self-consistent as possible.

ALPHABETICAL LIST OF ARTICLES

AN INTRODUCTION TO COMPOSITE MATERIALS

by Anthony Kelly

The term composite originally arose in engineering when two or more materials were combined in order to rectify some shortcoming of a particularly useful component. For example, cannons which had barrels made of wood were bound with brass because a hollow cylinder of wood bursts easily under internal pressure. The early clipper sailing ships, which were said to be of composite construction, consisted of wood planking on iron frames, with the wood covered by copper plates to counter the attack of marine organisms on the wood.

For the purposes of this Encyclopedia, a composite material can be defined as a heterogeneous mixture of two or more homogeneous phases which have been bonded together. Provided the existence of the two phases is not easily distinguished with the naked eye, the resulting composite can itself be regarded as a homogeneous material. Such materials are familiar: many natural materials are composites, such as wood; so are automobile tires, glass-fiber-reinforced plastics (GRP), the cemented carbides used as cutting tools, and paper—a composite consisting of cellulose fibers (sometimes with a filler, often clay). Paper is essentially a mat of fibers, with interfiber bonding being provided by hydrogen bonds where the fibers touch one another.

It is sometimes a little difficult to draw a distinction between a composite material and an engineered structure, which contains more than one material and is designed to perform a particular function. The combination is usually spoken of as a composite material provided that it has its own distinctive properties, such as being much tougher than any of the constituent materials alone, having a negative thermal expansion coefficient, or having some other property not clearly shown by any of the component materials.

In many cases, the dimensions of one of the phases of a composite material are small, say between 10 nm and a few micrometers, and under these conditions that particular phase has physical properties rather different from that of the same material in the bulk form; such a material is sometimes referred to as a nanocomposite.

The breaking strength of fibers of glass, graphite, boron and pure silica, and of many whisker crystals, is much greater than that of bulk pieces of the same material. In fact, it may be that all materials are at their strongest when in fiber form. In order to utilize the strength of such strong fibers, they must be stuck together in some way: for example, a rope is appropriate for fibers of hemp or flax, and indeed carbon and glass fibers are often twisted into tows. However, for maximum utilization, a matrix in which to embed such strong fibers is required, in order to provide a strong and stiff solid for engineering purposes. The properties of the matrix are usually chosen to be complementary to the properties of the fibers: for example, great toughness in a matrix complements the tensile strength of the fibers. The resulting combination may then achieve high strength and stiffness (due to the fibers), and resistance to crack propagation (due to the interaction between fibers and matrix).

Nowadays, the term advanced composite means specifically this combination of very strong and stiff fibers within a matrix designed to hold the fibers together. This type of composite combines the extreme strength and stiffness of the fibers and, due to the presence of the matrix, shows much greater toughness than would otherwise be obtainable.

1. Prediction of Physical Properties

An important question, both for engineering design of a composite material and for scientific understanding of its properties, is that of how the overall properties of the composite depend upon those of the individual constituents.

Properties of composite materials can be considered under two headings: (a) those that depend solely upon the geometrical arrangement of the phases and their respective volume fractions, and not at all upon the dimensions of the components. For this to be so, the smallest dimension of each phase must usually be greater than 10 nm, and sometimes much greater than this; (b) those that depend on structural factors such as periodicity of arrangement or the sizes of the pieces of the two or more component phases.

1.1 Properties Determined by Geometry—Additive Properties

The geometrical arrangement of the phases in a composite material can often be described in simple terms (Fig. 1).

The simplest physical property of a composite—namely its density ρ_c—is given by the volume-weighted average of the densities of the components:

$$\rho_c = \rho_1 V_1 + \rho_2 V_2 \tag{1}$$

where V is the volume fraction and subscripts 1 and 2 refer to the components. If there are no voids, $V_1 + V_2 = 1$. More complicated physical properties, e.g., those described by a second-rank tensor, relate two vectors: either a solenoidal vector and an irrotational vector (as with magnetic or electric susceptibility), or else a flux vector and the gradient of a scalar function (as with diffusivity, and electric and thermal conductivity). The relations derived for these properties for a composite material in terms of the same

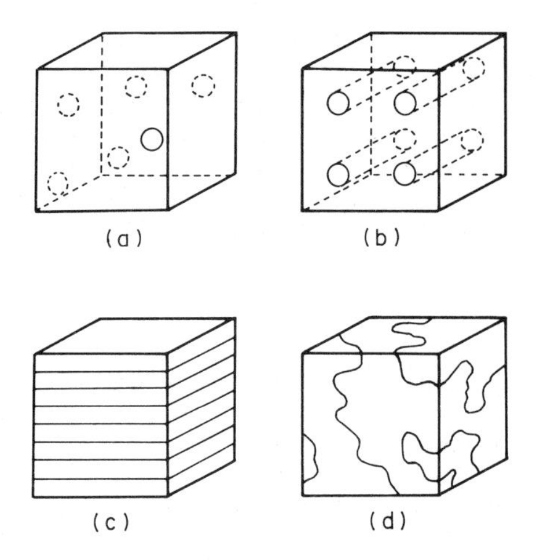

Figure 1
Composite geometries: (a) random dispersion of spheres in a continuous matrix; (b) regular array of aligned filaments; (c) continuous laminae; and (d) irregular geometry (after Hale 1976)

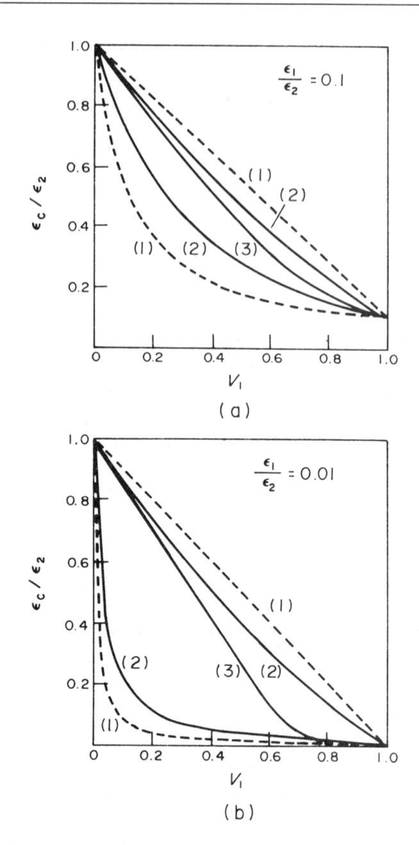

Figure 2
Dielectric constant of composite materials: (a) $\varepsilon_1/\varepsilon_2=0.1$ and (b) $\varepsilon_1/\varepsilon_2=0.01$. (1) Bounds from Eqns. (1) and (3); (2) Hashin and Shtrikman bounds for arbitrary geometry (Eqn. (2)); (3) Self-consistent approximation for spheres in a continuous matrix (after Hale 1976)

property in each of the submaterials or components are all formally identical. They have usually been discussed in terms of the dielectric constant ε and, because of this, the general attempt to predict the overall physical properties of a multiphase composite from the same properties of the individual phases is sometimes called the dielectric problem. Attempts to devise exact, generally applicable theoretical expressions for the dielectric constants of mixtures of unknown phase geometry are futile. However, the dielectric constant of a mixture of two phases, ε_c, must lie between certain limits, whatever the geometry. If some knowledge of the phase geometry is available, still closer bounds can be set on the value of ε_c; the same is true of other properties. There is, therefore, a good deal of utility in the search for bounds on the properties. For a few special geometries, exact solutions can be obtained. Approximate solutions are also often obtained for two-phase composites when the volume fraction of one of the components is low. Such solutions are often extended to nondilute situations by use of a self-consistent scheme, in which each particle of the minor component is assumed to be surrounded not by the other component but by the average composite material.

For example, as given by Hashin and Shtrikman (1962), the effective dielectric constant of an isotropic composite must lie in the range

$$\varepsilon_1+\frac{V_2}{[(\varepsilon_2-\varepsilon_1)^{-1}+(V_1/3\varepsilon_1)]}\leqslant\varepsilon_c\leqslant\varepsilon_2+\frac{V_1}{[(\varepsilon_1-\varepsilon_2)^{-1}+(V_2/3\varepsilon_2)]}\qquad(2)$$

where $\varepsilon_2>\varepsilon_1$. These are the best possible bounds obtainable for the dielectric constant of an isotropic two-phase material if no structural information, apart from the volume fractions, is available.

Figure 2 compares the predictions of those bounds for the cases $\varepsilon_1/\varepsilon_2=0.1$ and $\varepsilon_1/\varepsilon_2=0.01$ with the most elementary bounds which can be rigorously derived for an isotropic composite. These elementary bounds correspond to simple Voigt (polarization in parallel) and Reuss (polarization in series) estimates. The Voigt estimate is of the form of Eqn. (1), while the Reuss estimate gives the composite dielectric constant as

$$1/\varepsilon_c=V_1/\varepsilon_1+V_2/\varepsilon_2\qquad(3)$$

If ε_1 and ε_2 do not differ greatly, the bounds given by Eqn. (2) are very close together, but for $\varepsilon_1/\varepsilon_2=0.01$ they are far apart (such as in Fig. 2); therefore, in many practical cases, better bounds are required.

In the foregoing discussion, the properties of the individual components or submaterials within the composite are assumed to be identical with the same properties measured in a piece of the submaterial outside the composite. In addition, the composite is taken to be statistically homogeneous in the sense that, if we extracted small elements of the material, these would have the same physical properties as the whole sample. In addition, the implicit assumptions are made that there are no voids, that space-charge effects and polarization are absent (in the case of electrical conductivity) and that there is no discontinuity in temperature at the interface (in the case of heat flux).

The prediction of the elastic properties of particulate composites is discussed in the article *Particulate Composites*; for fiber composites see *Fibrous Composites: Thermomechanical Properties*.

Thermal expansion coefficients of composites involve both the elastic constants and the thermal expansion coefficients of the individual phases. An important function of reinforcing fillers and fibers in plastics is the reduction and control of thermal expansion. For example, with dental filling materials, a difference in thermal expansion between the filling material and the tooth substance can lead to a marginal gap, and hence composite filling materials are designed to have a thermal expansion coefficient very close to that of the tooth substance.

With some components, especially at low fiber volume fractions, the transverse thermal expansion coefficient of, for example, a glass-fiber–epoxy-resin composite, can be greater than that of the matrix. This effect is particularly noticeable with fibers of high modulus and low expansion coefficient (e.g., boron or carbon) in a low-modulus matrix.

Fibers are often used in laminated arrangements because properties parallel to the fibers are very different from properties perpendicular to the fibers. The effective in-plane thermal expansion coefficients for angle-ply laminates (in which the fibers are arranged at plus and minus an angle ϕ to a particular direction) show that in such laminates a scissoring or lazy-tongs type of action can occur, and, with appropriate values of ϕ, can lead to a zero or even negative thermal expansion coefficient in one direction.

1.2 Special Property Combinations—Product Properties

Composite materials can, in principle, be thought of as materials which produce properties unobtainable in a single material: for example, by combing a piezoelectric material with a material showing magnetostriction, the composite should show a magnetoelectric effect—that is, an applied magnetic field would induce an electric dipole moment; or a material could be produced in which an applied magnetic field produced optical birefringence, by coupling a material which shows strain-induced birefringence with one showing magnetostriction.

This idea lead van Suchtelen (1972) to classify such effects as product properties of composites and so now there are considered to be two different types of physical property of composite materials.

The first is the type discussed so far in Sect. 1.1. These are sum or additive properties where the composite property is related to that same property of each of the components and so depends on the geometry of arrangment of the two components. The geometry of arrangement includes of course the volume fraction. Examples are elastic stiffness, relating applied stress and measured strain, or electrical resistance, relating applied electric field and measured current density, or the simple example of mass density. The value of the physical property of the composite in general lies between those of the components.

There is a subclass of additive properties in which the value of the property of the composite can lie well outside the bounds set by the values of the property of the components; examples are Poisson's ratio involving the ratio of two compliance coefficients under the action of a single applied stress, or acoustic wave velocity, which depends on the ratio of elastic modulus to density. Here, because the elastic modulus and the density can follow quite different variations with volume fraction, e.g., elastic modulus following Eqn. (3) and density following Eqn. (1), the acoustic wave velocity of the composite can lie well outside the value for either component.

A further, but less obvious, example is the thermal expansion coefficient, which depends on the thermal expansion coefficients and the elastic constants of the two phases. Newnham (see *Nonmechanical Properties of Composites*) distinguishes this last set of properties and calls them combination properties (distinct from sum or additive properties) when the composite property lies outside the bounds of the same property of the two or more constituents. However, since a value lying outside the range of the constituents' properties can in principle occur for all composite properties (even mass density) it seems best to view all cases in which the same property of the composite and of the components is considered as cases of additive or sum properties.

Product properties are those cited at the beginning of this section: the property of the composite depends specifically on the interaction between its components. Any physical property can be considered as the action of a physical quantity X resulting in physical quantity Y giving the X–Y effect. Product properties are those in which an X–Y effect in submaterial 1 produces a Y–Z effect in submaterial 2, producing in the composite an X–Z effect. A good example is the one cited of a magnetoelectric effect in a composite material having one magnetostrictive and one piezoelectric phase. Application of a magnetic field produces a change in shape in the magnetostrictive phase which then stresses the piezoelectric phase and hence generates an electric field. In this case the coupling is mechanical but the coupling could also be electrical, optical, magnetic, thermal or chemical (van Suchtelen 1972).

Viewed in this way, sum properties are those in which an $X-Y$ effect in submaterial 1 and the same $X-Y$ effect in submaterial 2 combine to give an $X-Y$ effect in the composite.

Table 1 classifies some physical properties or phenomena according to the input–output parameters X and Y. A small selection of possible product properties is given in Table 2. Practical examples of these effects are given in *Nonmechanical Properties of Composites*.

1.3 Properties Dependent on Phase Dimensions and Structural Periodicity

In Sects. 1.1 and 1.2 the physical properties of the components or submaterials have been assumed to be unaltered by the incorporation of the components into the composite. This is usually not the case. The effects of phase changes or chemical reactions during fabrication are of importance in a wide range of composites. Excessive shrinkage of one component can result in high internal stresses, which may lead to premature failure or even preclude successful fabrication. Matrix shrinkage also has an important indirect effect on the mechanical behavior of fiber composites, since the resulting internal stresses can determine the frictional forces at the fiber–matrix interface.

The dimensions and periodicity of a composite structure also have an important effect on the properties when they become comparable with, for example, the wavelength of incident radiation, the size of a magnetic domain or the thickness of the space-charge layer at an interface. The dimensions are also particularly important when the properties of a material depend on the presence of defects of a particular size (e.g., cracks). The strength of a brittle solid containing a crack depends on the square root of the length of the crack. Since very small particles cannot contain long cracks, and because the surface region of most solids behaves differently from the interior (a small fiber or sphere contains proportionately more surface material than a large piece), the breaking strength depends on the dimensions of the piece tested. This is particularly important when considering mechanical strength and toughness of composites.

Many of the optical effects which depend on structural dimensions also depend in a very complex way on other factors, and optical effects produced by dispersion of a second phase within a material are often used to determine the distribution of that phase (e.g., determination of the molecular weights of polymers or of the size of crystal nuclei in glasses), rather than regarding the material as a composite with special optical properties. However, a composite material containing aligned elongated particles of an optically isotropic material in an optically isotropic matrix may exhibit double refraction as a straightforward consequence of the relationship between dielectric constant and refractive index.

In a ferromagnetic material at temperatures below the Curie point, the electronic magnetic moments are aligned within small contiguous regions—the magnetic domains—within each of which the local magnetization is saturated. Magnets of very high coercivity can be obtained by using very fine particles: a particle with a diameter of between 10^{-1} and 10^{-3} μm would be magnetized almost to saturation, since the formation of a flux-closure configuration would be energetically unfavorable. Magnetization reversal cannot take place in a sufficiently small single-domain particle by boundary displacement; it must occur by a single jump against the magnetocrystalline anisotropy and the shape factor. Fine-particle permanent magnets of rare-earth alloys such as $SmCo_5$, enveloped in an inert matrix such as tin, have been developed which have superior properties; for example, the energy product can be up to 30 MGOe ($1\ GOe = 79.6 \times 10^{-4}\ T\,A\,m^{-1} = 8\ J\,m^{-3}$). The preparation of these fine-particle magnets shows that attention to small size alone is not enough, and that care must be taken to prevent domain-wall nucleation at the surfaces of the particles.

Considerable attention has also been paid to rod-type eutectics as permanent magnetic materials. If the permanent magnet is long and thin, the demagnetization energy can be very large as a consequence of the shape anisotropy, even in the absence of high magnetocrystalline anisotropy. This is the concept of the elongated single domain. In the Bi–Mn, Bi system, very high coercive forces (e.g., $24\ kOe = 1.9 \times 10^6\ A\,m^{-1}$) can be obtained but, because the volume fraction of magnetic material is low, so is the energy product.

Composite principles are also used in the construction of powerful electromagnets containing filamentary superconducting composites: for example, high-performance electromagnets producing field densities of 16 T with a Nb–Sn (intermetallic) winding, or 8 T with Nb–Ti. One of the problems here is instability in the interaction between the superconductor and the magnetic field, which causes quenching of superconductivity. When this occurs, a very large current must be carried by other means, and one solution to this problem is to surround the conductor with, say, ten times its volume in copper. The copper can then cool the superconductor and return it to the superconducting state. The energy dissipated is proportional to r^2, where r is the radius of the wire, and the cooling is proportional to r. The ratio of surface area to cross-sectional area per unit length varies as $1/r$ and therefore small wires are more effectively cooled.

The theory of how instability is produced also shows that it can be almost eliminated by ensuring that the diameter of the individual superconducting filaments is reduced below approximately 20–50 μm. This size effect arises because instability occurs whenever a small rise in temperature reduces the shielding current (proportional to the superconducting current),

Table 1
Matrix classification of some physical properties or phenomena in materials according to the type of input and ouput parameters (after van Suchtelen 1972)

	Input parameter (X)					
Output parameter (Y)	Mechanical (force/ deformation)(1)	Magnetic (field/ polarization)(2)	Electrical (field/ polarization, current)(3)	Optical and particle radiation (light or particle flux)(4)	Thermal (temperature, temperature gradient, heat current)(5)	Chemical (chemical composition, chemical composition gradient)(6)
Mechanical (force/ deformation) (1)	elasticity	magnetostriction magnetoviscosity (suspension)	electrostriction Kirkendall effect electroviscosity (suspension) indirect: thermal expansion		thermal expansion	osmotic pressure
Magnetic (field/polarization) (2)	piezomagnetism	magnetic susceptibility	superconductors galvanic deposition of ferromagnetic layer direct generagtion of magnetic field	photomagnetic effect	thermomagnetism ferromagnetic material at $T \simeq T_c$ (+ magnetic field)[a]	dependence of T_c on ferromagnetic composition
Electrical (field/ polarization, current) (3)	piezoelectricity piezoresistivity	magnetoresistance (+electric current)[a] Hall effect (+electric current)[a] ac resonance induction of voltage	dielectric constant, dielectric polarization Hall effect (+magnetic field)[a]	photoconductivity photoemission photoelectromagnetic effect (+magnetic field)[a] ionization	thermoelectricity ferroelectrics at $T \simeq T_c$ (+electric field)[a] temperature-dependent resistivity (+electric current)[a]	dependence of T_c on ferroelectric composition
Optical and particle radiation (light or particle flux) (4)	stress birefringence triboluminescence	Faraday effect magnetooptic Kerr effect deflection of charged particles	electroluminescence laser junctions refractive index Kerr effect absorption by galvanic deposits cold emission of electrons	refractive index fluorescence scintillation color-center activation	thermoluminescence	chemoluminescence
Thermal (temperature, temperature gradient, heat current) (5)	heat of transition of pressure-induced phase transition piezoresistivity and Joule heating	adiabatic demagnetization Nernst-Ettingshausen temperature gradient effect (+ electric current)[a] magnetoresistnace effect + Joule heating (+electric field)[a]	dissipation in resistance Peltier effect Nernst-Ettingshausen temperature gradient effect (+ magnetic field)[a]	absorption	thermal conductivity	reaction heat
Chemical (chemical composition, chemical composition gradient) (6)	pressure-induced phase transition		electromigration galvanic deposition	light or particle stimulated reactions (photosensitive layers)	Soret effect (temperature gradient)[a] phase transition change of chemical equilibrium	

[a] Indicates the parameter in parenthesis is essential as a second input

Table 2
Product properties of composite materials (after van Suchtelen 1972)

X–Y–Z (Table 1)	Property of phase 1 (*X–Y*)	Property of phase 2 (*Y–Z*)	Product property (*X–Z*)
1–2–3	Piezomagnetism	Magnetoresistance	Piezoresistance Phonon drag
1–2–4	Piezomagnetism	Faraday effect	Rotation of polarization by mechanical deformation
1–3–4	Piezoelectricity	Electroluminescence	Piezoluminescence
1–3–4	Piezoelectricity	Kerr effect	Rotation of polarization by mechanical deformation
2–1–3	Magnetostriction	Piezoelectricity	Magnetoelectric effect
2–1–3	Magnetostriction	Piezoresistance	Magnetoresistance Spin-wave interaction
2–5–3	Nernst–Ettingshausen effect	Seebeck effect	Quasi-Hall effect
2–1–4	Magnetostriction	Stress-induced birefringence	Magnetically induced birefringence
3–1–2	Electrostiction	Piezomagnetism	Electromagnetic effect
3–1–3	Electrostriction	Piezoresistivity	Coupling between resistivity and electric field (negative differential resistance, quasi-Gunn effect)
3–4–3	Electroluminescence	Photoconductivity	
3–1–4	Electrostriction	Stress-induced birefringence	Electrically induced birefringence light modulation
4–2–1	Photomagnetic effect	Magnetostriction	Photostriction
4–3–1	Photoconductivity	Electrostriction	Photostriction
4–3–4	Photoconductivity	Electroluminescence	Wavelength changer (e.g., infrared-visible)
4–4–3	Scintillation	Photoconductivity	Radiation-induced conductivity (detectors)
4–4–4	Scintillation, fluorescence	Fluorescence	Radiation detectors, two-stage fluorescence

causing a flux change which in turn leads to heating, and so on. The energy liberated per flux jump is smaller if the superconducting element is narrower, and the time to cool the superconductor from outside is smaller if the diameter is smaller. Both effects produce an upper limit to the diameter of the superconductor for effective stabilization. For ac applications it is also necessary to reduce the eddy-current losses in the copper. This can be done by using a material with a high electrical resistivity to decouple the wires.

To produce the very highest magnetic fields attainable in the laboratory (50–100 T)pulsed-field systems are used. These require the conductor to show high mechanical strength as well as very good electrical conductivity; a composite has the ideal geometry for producing this combination of properties.

2. *Advanced Fiber Composites*

The best known and most striking example of a modern composite material is as we have said, the fiber composite, designed for high strength, high stiffness and low weight.

The breaking strength and the stiffness of typical specimens of a number of modern fibers are shown in Fig. 3. Steel wire and glass provide the strongest materials and pitch-based carbon fibers the stiffest.

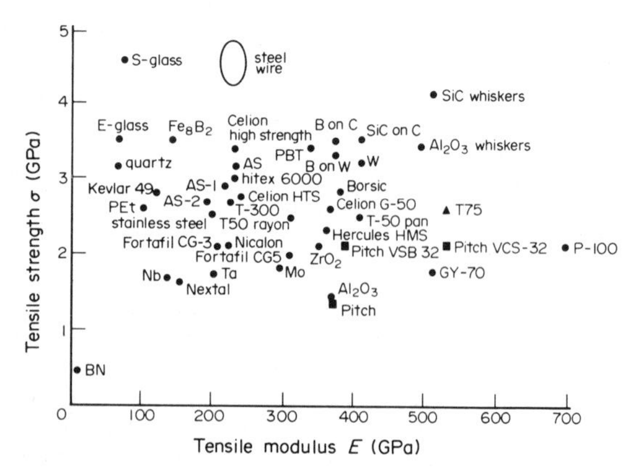

Figure 3
Tensile strengths and stiffness of a variety of fibers

The strength and stiffness of the same fibers divided by their density and by the acceleration due to gravity is shown in Fig. 4. The resulting physical unit is a length and, as far as the strengths are concerned, represents (intriguingly) the greatest length of a uniform rod which could in principle be picked up from one end on the surface of the Earth without breaking.

Quantities such as specific strength and stiffness are often quoted in units such as strength divided by

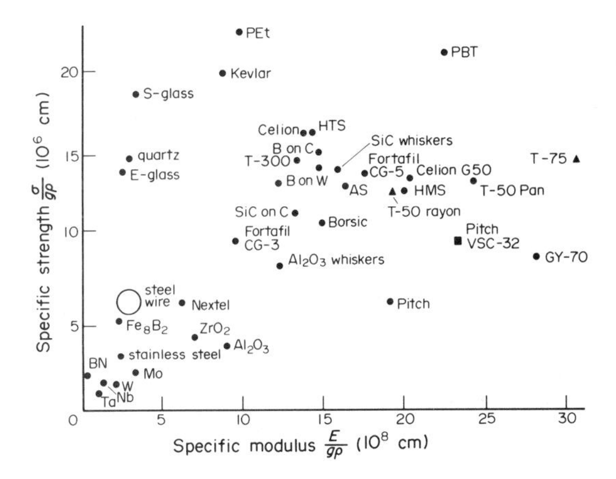

Figure 4
Specific strength and stiffness of a variety of fibers. The specific value is the value of the property divided by $g\rho$ where ρ is the density and g the acceleration due to terrestrial gravity

specific gravity, in which case units of stress are used, or else as stress divided by density, the resulting unit being a velocity squared. In the case of stiffness the velocity represents that of longitudinal sound waves in the material. In Fig. 4 a unit of specific stiffness, $E/\rho g$ (or of specific strength, $\sigma/\rho g$) of 10^7 cm is equivalent to a value of stiffness, E/ρ, (or of strength, σ/ρ) of $9.8 \times 10^9\ \mathrm{cm}^2\ \mathrm{s}^{-2}$ (E is Young's modulus and σ is tensile strength).

Figure 3 shows that the stiffest materials are ceramics composed of elements from the first rows of the periodic table (B, C, SiC, Al_2O_3, etc.). This is because these elements contain the highest possible packing of strong covalent bonds directed in three dimensions. Such materials also each have a high melting point, a low coefficient of thermal expansion and a low density. All of these are very desirable engineering properties. The strongest materials are the glasses and the metals.

In all structural applications, the weight of the structure necessary to bear a given load and to do so with a minimum elastic deflection is important, and so strong fibers are compared on the basis of strength divided by density. The fibers shown in Fig. 3 are compared in this way in Fig. 4. When this is done the stiffest materials on a weight-for-weight basis remain the ceramics but the strongest are the plastics (PEt, PBT, Kevlar) and the glasses. The metals make a very poor showing on either stiffness or strength. However, fibers cannot be used directly and must be bound together within a matrix.

A set of aligned fibers in a pliant matrix is only useful as an engineering material under direct tensile forces parallel to the fibers. The salient properties are represented by those of a pocket handkerchief which is stiff in the direction of the warp and weft (parallel to the fibers) but shears easily parallel to these. Hence the properties of fiber composites are very directional (see Fig. 5), being very strong and stiff parallel to the fibers, but rather weak in shear parallel to the fibers (because this property depends principally upon the shear properties of the matrix), and very weak indeed in tension perpendicular to the fibers. In order to overcome this, the fibers are arranged in laminae, each containing parallel fibers, and these are stuck together (Fig. 6) so as to provide a more isotropic material with a high volume loading of fibers. Figure 6 shows the specific strengths and stiffness values of laminated forms of fibrous composites made of some of the fibers whose properties are shown in Figs. 3 and 4. The advantage over conventional isotropic constructional materials represented by the metal alloys is seen to be considerable but is much less marked than in Figs. 3 and 4.

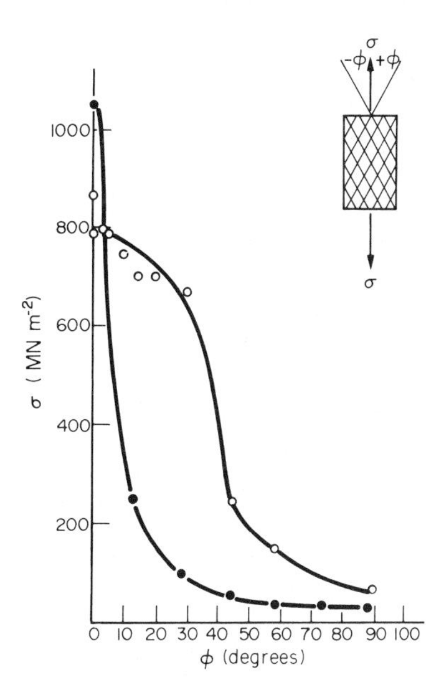

Figure 5
Measured variation of tensile strength with angle ϕ for specimens consisting of a number of alternate layers of fibers. The fibers in each layer are parallel and continuous. Alternate layers are at $+\phi$ and at $-\phi$ to the tensile axis. Open circles represent data for a volume fraction of 40% of silica fibers in aluminum. Full circles represent data for a volume fraction of 66% of E-glass fiber in an epoxy resin

Alternatively, the fibers may be randomly arranged in a plane or in three dimensions; such arrangements limit the obtainable fiber packing density. Fibers are also often woven into mats before incorporation into the composite because this aids the handling of fiber arrays.

When carbon- or other fiber-reinforced plastics are made with woven fabric rather than nonwoven material, distortion of the load-carrying fibers parallel to the applied stress reduces the tensile strength further

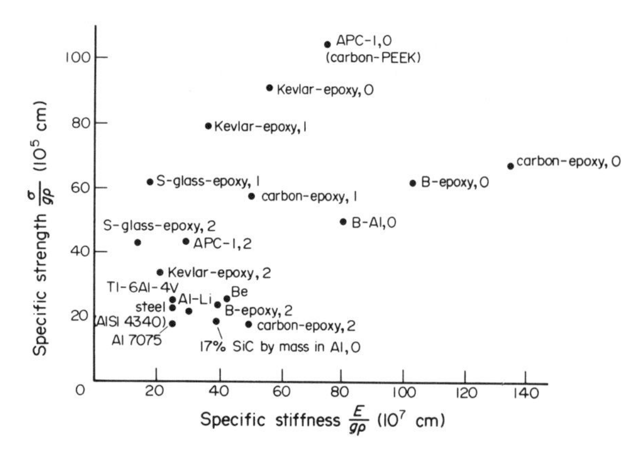

Figure 6
Specific strength and stiffness of some isotropic materials and of fiber composites. The designation 1 on the composites means the following arrangements of fibers: 50% at 0°; 40% at ±45° and 10% at 90° to the stress; 2 denotes balanced laminates with equal proportions at 45°, 90° and 135°; O indicates aligned fibers in the specified matrix. The volume fraction of fibers in the various composites are not the same in the different systems. They vary between 40 and 60%. The metal alloys are those without such designation

and reduces stiffness and toughness. When, however, woven fabric is oriented at 45° to the load direction these properties compare favorably with those for nonwoven ±45° material. Indeed, the residual strengths after impact can be greater. At present between one third and one half of the market of carbon-fiber-reinforced plastic (CFRP) consists of woven fabric; the rest consists of aligned material.

The anisotropy of the properties of advanced fiber composites represents a completely new feature in engineering design. The nonisotropy can be varied; in fact it is possible now to place the fibers precisely where they are needed in a structure to bear the loads. In fact it is also possible, for instance, to vary the amount and stacking of the fibers in various directions along the length of a beam so as to vary the stiffness and torsional rigidity of the material.

This is precisely what is needed in helicopter rotor blades, and within a few years' time all rotor blades except perhaps those for heavy-lift helicopters will be made from GRP or combinations of GRP and CFRP, despite the fact that such composites were not applied to helicopter blades until the mid-1960s.

The AV 8B, the latest version of the Harrier vertical takeoff and landing aircraft, employs composites in torque boxes, auxiliary flaps, the forward fuselage and cone structure, the horizontal stabilizers and in the ammunition and gun pods.

It would be impossible to make use of the inherent advantages of anisotropic materials such as CFRP without the invention of computer-aided methods. The reiterations and the number of variables involved require computer programmes both to relate the properties of the individual laminae to the properties of the fibers and to take into account the interactions in terms of bending–twisting coupling between the various laminae. Initially, of course, these complications were viewed as a great disadvantage of the material, since the engineer was not happy with anisotropy. Now the problem of how to use it and make a virtue, not so much out of necessity, but of a supposed vice has been solved. The latest Grumman X29 aircraft makes use of bending–twisting coupling so that, as the loads on an aircraft wing increase, the structure can deform but remain tuned to the aerodynamic requirements. The material possesses almost a form of "gearing," so that it changes its shape automatically without the need for sensors and levers.

3. *Applications of Composites*

High-performance composites use fibers in order to attain the inherent properties of the fiber, coupling this with a judicious choice of matrix so that toughness and impact damage are not lost.

Strong fibers, whether they be stiff or not, have the great advantage of restraining cracking in what are called brittle matrices. It is this use of fibers which is often referred to when showing how the ancients employed composite materials. Quotations from the Bible are made, such as that from the Book of Exodus, Chapter V, verses 6 *et seq.*, which refer to the difficulty of making of bricks without straw. Fibers restrain cracking because they bridge the cracks. The principle is very simple. It has been used in asbestos-reinforced cement and the principle of reinforced concrete is not too far from the same idea. The principle is illustrated in Fig. 7. A crack passing through a brittle material may not enter the fibers but leaves a crack straddled by fibers so that a material which would normally have broken with a single crack now has to be cracked in a

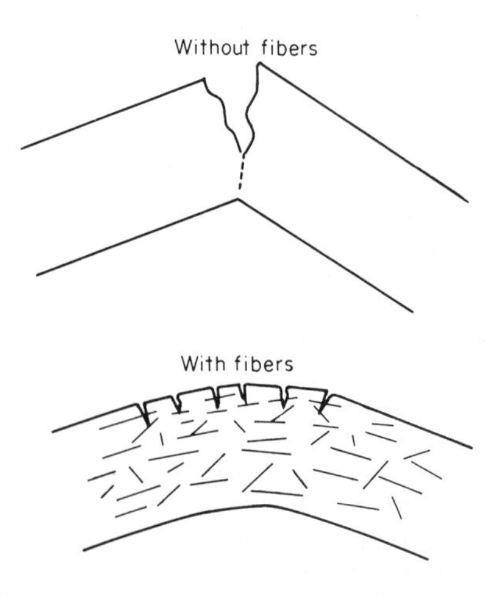

Figure 7
Demonstration of how cracks are prevented from running in a brittle material because of the fibers in their path

large number of places before the fibers themselves fail or pull out. This is utilized in fiber-reinforced cement, which is based on the same principles as reinforced concrete, but since the fibers are very much finer than reinforcing bars and are often not visible to the naked eye, a homogeneous material is effectively produced which can show quasiplastic deformation.

This judicious use of fibers enables us to think nowadays in terms of using brittle materials for construction purposes, because if they are cracked the fibers will render the cracks harmless. Building panels and pipes, of course, are made of such material. Glass-reinforced cement is becoming commonplace. However, a much greater prize appears ahead. Metals have limited high-temperature capability. The ceramic materials described above as providing the best fibers also provide the materials of highest melting point. For the construction of prime movers at high temperature these will have to replace metals. They cannot do this as monolithic pieces because, as we have seen, these would be easily broken. However, they can resist cracking if they contain fibers. The composite principle then leads to suggesting that even fibers of the same material as the matrix may give marked resistance to cracking and fracturing. Such is true of carbon–carbon composites which are used in brakes for Concorde and in certain ballistic missile and reentry vehicle applications. Of course carbon oxidizes and so would have to be protected if exposed to hot air. In contrast, therefore, a good deal of interest centers on the use of refractory oxide glasses such as cordierite or a material of almost amorphous silicon nitride or silicon carbide for use at high temperature, containing fibers again of silicon nitride or silicon carbide. These will restrain cracking and give an engineering material usable for thousands of hours at temperatures in excess of 1100 °C. As yet these are not commercially available but aircraft engine manufacturers are making and testing pieces of ceramics, sometimes unreinforced, for aerospace applications; but the main thrust of research is towards reinforced materials.

Ceramic materials are of widespread abundance and are all essentially cheap. If they can be fashioned by simple chemical means, such as reacting acids with alkalis, they are classed as phosphate-bonded materials. These again have great potential for replacing, say, sheet steel in normal household use, again provided cracking can be restrained. Fibers will do this and, if the material is not exposed to high temperature, glass fibers or even textile fibers may be employed.

4. *Arrangement of the Encyclopedia*

With this introduction the reader will now hopefully be able to understand the appearance of the various topics within this Encyclopedia. It is arranged, of course, with the articles (all by world experts) in alphabetical order to facilitate ease of access by the reader. The content of the book follows a pattern perhaps not apparent without explanation.

It is the existence of manufactured strong stiff fibers which has given rise to the utility of composite materials. Fibers, not necessarily very strong and stiff, are used alone for many purposes: in textiles, hence in clothing and apparel; in industrial belting in gaskets; in automobile tires; and in many other uses. Nowadays both man-made and natural fibers are used. The newer, strongest and stiffest artificial fibers have extended many of the conventional uses of textiles. The uses of fibers in textiles are reviewed in the article *Fibers and Textiles: An Overview.*

5. *Fibers*

The very strong and stiff fibers comprise the following, each with a separate section in this Encyclopedia: asbestos (a naturally occurring fiber), boron and carbon of varying degrees of graphitization, made in principle from a number of starting materials but in practice mainly two: PAN (polyacrylonitrile) and pitch. Inorganic glasses based on silica yield many fibers. Those made in the form of continuous filaments are most important for composite materials. Many linear polymers such as polyparabenzamide or simple polyethylene can be processed to form very stiff and strong fibers. The ways of doing so are very varied. They are all described in *High-Modulus High-Strength Organic Fibers.* Fibers of Al_2O_3 are important for high-temperature use in filtration and other applications because of their chemical inertness. These, and fibers based on silicon carbide and nitride, again made via various routes, are described in *Oxide Inorganic Fibers*, *Silicon Carbide Fibers* and *Silicon Nitride Fibers*, respectively. Whiskers or tiny (~ 1 μm diameter) fibers are known of most materials. They are often very strong and stiff and possibly will have uses as agents producing toughness, stiffness or wear resistance when introduced into other materials. Their genesis is described in *Whiskers*.

Historically, the first materials designed specifically by chemists to hold fibers in a composite material were the organic thermosetting resins. The modern varieties of these resins are described in *Thermosetting Resin Matrices*.

6. *Examples of Composite Materials*

Many natural products such as wood, shells and manufactured derivatives such as paper and plywood are of course composites, and the understanding of the physical properties of these materials both illuminates our understanding of the properties of wholly synthetic composites and, more frequently, is itself enhanced from our knowledge of the behavior of the modern composite materials. The sections in this Encyclopedia on the principal composite material systems describe both. Hence there are sections on natural composite materials, on the breaking strength of woods, on the production and properties of paper and paperboard, on plywood, and on molded fiber products (where articles are molded using essentially only water as the dispersing agent, which is subsequently removed) and on wood–polymer composites. To these must be added a description of composites based on natural fibers other than wood, such as sisal, sunhemp, coir and jute. The incorporation of these into matrices such as polyesters is described in *Natural-Fiber-Based Composites.*

The principal composite systems in use commercially are of course described. Glass-reinforced plastic (GRP) is by far the largest in commercial volume. The fabrication and properties of glass-reinforced plastics based on thermosetting resins are described in *Glass-Reinforced Plastics: Thermosetting Resins*, and the properties of GRP based on the thermoplastic resins in *Glass-Reinforced Plastics: Thermoplastic Resins.* Carbon-fiber-reinforced plastic (CFRP) has been developed rapidly within the last ten years or so; based initially on thermosetting resins, they have recently employed thermoplastics. The methods of fabrication and some of the properties are described in *Carbon-Fiber-Reinforced Plastics.* The newer thermoplastic polymers such as poly(ether ether ketone), polysulfone or the modified polyamides have advantages over the thermosetting resins: they can withstand higher strain before failure, they have lower moisture absorption and they can be shaped when heated and repaired without requiring a lengthy and intricate cycle of cure. They also have, in effect, an infinite storage life at room temperature. The newer thermoplastic polymers are, however, expensive. Their properties when containing glass and the stiffer fibers are described in *High-Performance Composites with Thermoplastic Matrices.*

The automobile tire is made of a composite material based on an elastomeric matrix. Its properties are better appreciated than analytically understood. An analysis of its performance in terms usually used in the description of modern composites is given in *Automobile Tires.*

There is no reason why a fibrous composite need contain fibers of only one chemical type. Hybrids are possible containing more than one type of fiber and the advantage of these is described in *Hybrid Fiber–Resin Composites.* Nor is it necessary that fiber and matrix be chemically different: carbon–carbon composites, in which strong stiff carbon fibers are incorporated into a matrix of quasiamorphous carbon produced by pyrolysis, chemical vapor deposition or by other means, have a number of important uses and great potential for use at higher temperatures (see *Carbon–Carbon Composities*). It is the potential of ceramics to displace metals for use at high temperatures (> 1000 °C), because of the ceramics' greater oxidation resistance and lower density, that primarily causes interest in the use of these materials in composites containing fibers. Ceramic materials used alone are brittle and hence produce fragile components. One way, perhaps the only way, to produce very high fracture toughness in them is to incorporate fibers. Fiber-reinforced ceramics possess great potential but have few uses at present. However, the incorporation of fibers to increase toughness has been commercially realized (see *Fiber-Reinforced Cements*). In order to provide adequate toughness, the fibers must be spatially oriented in three dimensions. The modern three-dimensional arrangements of fibers are described in the article *Three-Dimensional Fabrics for Composites.*

Under some conditions metals such as aluminum or magnesium may be better vehicles for carrying fibers, and hence for making a useful fiber composite, than polymer matrices. These metals have been reinforced with a large number of fibers because as a matrix they show advantages over all thermosetting resins and some thermoplastic resins; they have high thermal conductivity, little hydrothermal degradation and possess dimensional stability, and they are not susceptible to radiation damage or low-temperature brittleness. Their higher melting point can be an advantage. The article *Metal-Matrix Composites* describes these and also the use of fibers in order to ameliorate the properties of a particular metal, e.g., lead or copper. Some metals, e.g., tungsten, are extremely stiff (tungsten has a Young's modulus of 350 GPa) and possess very high melting points besides being well known as fibers. If they can be protected from oxidation, they offer the possibility of being used as a reinforcement of another metal, e.g., nickel-base alloy, or of a ceramic for high-temperature use.

The idea of making a composite for engineering use has arisen in recent years from the desire to utilize very stiff and strong fibers. The fibers cannot be used alone, so they require a matrix. The need to manufacture the two component materials separately could be avoided if fibers were made in situ in a matrix. Here ingenuity is at a premium since the resulting composite, if it has adequate properties, is likely to be more cheaply produced than if each component is made separately. It may well also be that the two components in the final product are so chemically related that the stability of each in the presence of the other is ensured. *In Situ Composites: Fabrication* describes the processes used with metals and inorganic materials; processes such as plastic deformation, eutectic solidification, precipitation and the recently discovered method of great promise, melt oxidation, which can form ceramic fibers or particles either alone or within a metal matrix.

The corresponding processes in organic-polymer systems are based on copolymerisation and extrusion, and the drawing of multicomponent mixtures containing polymers of different melting points are described in *In Situ Polymer Composites* and *Polymer–Polymer Composites.*

Although the great majority of composites referred

to are advanced composites, the increasing use of stiff fibers and the interest in these enhances interest in particulate composites where a stiff and thermally resistant phase of nonfibrous form is used in the composite. Some of these composites have been known for years and have found use as cutting tools, wear resistant parts, etc. They are described in *Particulate Composites*.

7. *Properties of Composites*

The science of the physical properties of composite materials relates the properties of the composite to those of the individual constituents and of the interface between them. Most of the simple physical properties such as density, elastic modulus and thermal expansion coefficient can be calculated and, noting the remarks in Sect 1.1, these calculations and their comparison with experiment are dealt with in *Fibrous Composites: Thermomechanical Properties* essentially for unidirectional fibrous composites. The elastic properties of laminates which consist of thin layers (lamellae) stacked upon one another, the fibers being parallel within each layer, is an important topic for design and details of how these calculations can be made appear in *Laminates:Elastic Properties*. Another form of arrangement of nonparallel fibers is that of the woven fabric and the elastic properties of these are dealt with in *Woven-Fabric Composites:Properties*.

An engineer may wish to choose a particular arrangement of fibers of given properties within a given matrix in order to attain or to approach a particular set of properties within the composite. For the simpler properties, where there is confidence in the prediction of a property from consideration of the properties of the constituents, such as for elastic modulus or for thermal expansion coefficient, attainable combinations of properties can be displayed graphically, leading to the concept of a structure–performance map, described in *Structure–Performance Maps*.

Properties such as strength and toughness of composite materials are not as well understood as the simpler elastic properties because in many cases the modes of failure under a given system of external load are not predictable in advance. An added complication is the fact that the breaking strength of individual fibers within a population shows a spread of values. This variation is described generally by what are called Weibull statistics. Since the breaking strength of a composite will depend not only on the highest breaking strength of the individual fibers but also on how the individual breaks are arranged in space within the composite, the accurate prediction of the breaking strength requires a complicated statistical theory even when time-dependent effects are ignored. In the article *Strength of Composites: Statistical Theories*, a simplified account of the breaking strenght of unidirectional compositics is given. Despite the complications of arriving at an exact prediction of the strength, advanced composites are used nonetheless and their strength measured. Because of the anisotropy of elasticity and strength, these quantities must be measured in special ways. Methods of achieving reliable measurements of strength and what simple rules there are for variation of strength with orientation, fiber length, and so on are dealt with authoritatively in *Strength of Composites*.

In engineering practice, the static strength of a composite in the presence of notches and stress concentrations due to geometric form is at least as important as that of simple test pieces used in the laboratory; a very short account of some important features is given in *Failure of Composites: Stress Concentrations, Cracks and Notches*. In practice, composite materials must be joined to one another and to other structural members. The peculiar methods of joining employed are dealt with in *Joining of Composites*. Here every effort is made to minimize stress-concentrating effects such as abrupt changes of elastic moduli and of geometrical section or form.

The ability of a material to remain serviceable when containing cracks, or when cracks arise during service, depends upon its toughness. It is by no means clear that measures of toughness or crack resistance, such as the concept of a critical stress intensity factor derived from linearly elastic mechanics, are applicable to composite materials, and some aspects of the problem are dealt with in *Failure of Composites:Stress Concentrations, Cracks and Notches*. Some progress can be made in elucidating the primary mechanisms responsible for the energy required to break a composite material. These are the sliding friction between fibers and matrix if broken fibers pull past one another, the deformation within the fiber and matrix, and the formation of subsiduary cracks. These energy-dissipating mechanisms depend on the variables of volume fraction, diameter of fibers, etc., and so representative diagrams indicating the relative importance of these can be constructed, as in the article *Toughness of Fibrous Composites*.

This description of the mechanical properties of composite materials has so far made no mention of time-dependent properties; in particular, strength has been described almost as a static property. The strength under oscillating stress is described in *Fatigue of Composites*, and the variation of deformation of the composite with time is described in *Creep of Composites*. Both are aspects of a general investigative analysis of the failure of composite materials which has been called damage mechanics. In *Fatigue of Composites*, it is shown that emphasis on the strain range to which a given load subjects a composite can simplify greatly the interpretation of the results. During cyclic testing of a composite material, cracks in one of the constituents occur usually within the matrix. The same occurs in a static test where it is called multiple fracture. It is necessary then to understand the physical properties of a material containing a large number of cracks.

The creep deformation, though not the creep rupture, of aligned fibrous composites with continuous fibers stressed parallel to these cracks is, on the other hand, relatively well understood and can be modelled in terms of the constitutive equations of deformation for the fibers and matrix deforming in parallel. Discontinuous-fiber composites can also be dealt with to some extent and hence deformation normal to the fibers at least partly understood.

The long-term stability of composite materials under load depends on chemical factors, oxidative stability of the two components, and so on, particularly at high temperatures. For most composites based on resin matrices, water absorption at room and slightly elevated temperatures is of great importance. Epoxy resins, for example, absorb water and this alters their glass transition temperature. If the fibers are susceptible to water attack, as glass is, the ingress of water leads to attack on the reinforcing fibers (see *Long-Term Degradation of Polymer-Matrix Composites*.

Before composite materials enter service, and while they are in engineering service, their properties must be evaluated in a nondestructive manner. In the article *Nondestructive Evaluation of Composites*, this is dealt with from firsthand experience. The principal objective of nondestructive evaluation is to provide assurance on the quality and structural integrity of a particular component. This can be achieved directly by using a nondestructive evaluation technique (NDE) or indirectly by monitoring or controlling the fabrication process. The latter is really only an extension of the usual procedures of process control necessary to achieve a consistent product. But NDE can also assist in optimizing fabrication procedure by, for example, monitoring the local state of cure. Since composite materials are essentially arrays of fibers assembled into place, the possibility arises of using the fibers themselves as monitors of their performance in service or alternatively doing this by incorporation of other types of sensors. The monitoring of fracture within a glass fiber or of the failure of resin adhesion to the surface of a fiber of glass is quite possible using visible light, and other fibers may be interrogated using other wavelengths of radiation.

These ideas raise intriguing possibilities of monitoring performances in situ by NDE so that the material itself becomes perceptive and says "how it feels." These ideas may be of great importance for the use of advanced materials generally, not just for composites (Kelly 1988). They have developed greatly recently and are described in detail in the article *Smart Composite Materials Systems*.

A section is of course included on nonmechanical properties. In *Nonmechanical Properties of Composites*, composite structures such as multilayer capacitors, piezoelectric transducers, varistors and sensors of a variety of types are dealt with.

Composites have become such an exciting part of materials science and engineering that other structures, many of which have only recently been recognized as composites, are now being described in the same terms as are composites (see *Nanocomposites*).

8. Applications: Use of Composite Materials in Engineering

It is already apparent that the applications of composite materials, even if mention is not made of those based on naturally occurring composites such as wood, are very wide, ranging from a simple glass reinforced plastic (GRP) tank or boat hull to a sophisticated aeroplane wing aeroelastically designed, or to another primary aircraft structure such as a helicopter rotor blade.

In *Applications of Composites: An Overview*, applications of GRP in the building industry, in marine applications (boats and minehunters), in building, transport and the electrical industry are covered. In many of these applications it has been corrosion resistance which has determined the success in substituting for a traditional material. It is mainly in leisure goods, medical materials and in aerospace, where performance requirements overcome the considerations of cost, that the newer advanced composites are gaining ground. Consideration of manufacturing methods (see *Manufacturing Methods for Composites: An Overview*) will often dominate which composite is to be used, because an important principle of composite materials is that almost any desired combination of physical properties can, within certain limits, be obtained.

The manufacturing methods depend, of course, upon the feedstock, and again the glass-reinforced-plastics industry has developed most of the methods employed. Glass is available in many forms, as rovings, chopped fibers or fabrics. The fibers may be laid up and then impregnated with resin, or guns may be used to spray liquid resin or fibers onto a suitably shaped core. Alternatively, preimpregnated arrays of fibers (prepregs) are used which in their simplest form consist of parallel tows, rovings or aligned individual fibers spread out to produce a uniform distribution of fibers within the thickness of the sheet, which is impregnated with resin to produce a material of controlled volume fraction. The resin is partially cured (B-staged) to a slightly tacky condition. Sometimes this must be refrigerated until used. Sheet, dough and bulk molding compounds are variants of this, generally using discontinuous fibers.

All manufacturing methods aim to avoid the entrapment of air or the formation of voids since these represent gross defects. Articles may be made by hand or spray placement, by press molding using heated matched male/female tools under pressures of

3–7 MPa, or by vacuum molding using a flexible membrane to obviate the need for a press. If higher density and lower void content are required, molding is done in an autoclave using pressures of, say, 1–2 MPa at an elevated temperature, ~ 100°C or so.

Alternatively, fibers may be enclosed in a mold and the resin injected, the resin being precatalyzed so that it cures. In reaction injection molding, two fast-reacting components (usually based on urethanes) of initial low viscosity are pumped into the mold. This method gives low cycle times (1–2 minutes).

In the process of pultrusion, continuous fibers are passed through a bath of resin and the impregnated fibers passed through a heated die so that the resin cures. This process is good for the production of rod, sheet, tube and bar forms. The most accurate positioning of fibers necessary for some high-performace applications is attained by filament winding, in which fibers impregnated with resin are wrapped onto a former mandrel which is withdrawn after the resin is cured. Modern filament winding machinery can produce complicated nonaxisymmetric shapes as a result of the application of robotics. The process is admirably suited to computer control and this will become more widespread in the future.

The extrusion process is described in *Fiber-Reinforced Polymer Systems: Extrusion*. This is essentially a form of polymer processing involving the incorporation of discontinuous reinforcing fibers such as chopped glass-fiber or natural cellulose.

All of these more conventional methods are used with high-performance composites but the increasing interest, and some use, of higher melting point thermoplastic matrices (such as poly(ether ether ketone)) containing carbon fibers, is leading to the development of forming methods in the solid state by pressing, drawing and extrusion comparable to conventional metal-forming methods.

The application of the new composites in a variety of industrial sectors and with diverse applications are also described in this Encyclopedia. (See *Aircraft and Aerospace Applications of Composites*; *Artificial Bone*; *Automotive Components: Fabrication*; *Composite Armor*; *Dental Composites*; *Friction and Wear Applications of Composites*; *Helicopter Applications of Composites*; *Solid Fiber Composites as Biomedical Materials*).

The introduction into use and the commercial prospects of all materials are greatly affected by the possibilities for extensive recycling and so an article on this topic has been included (see *Recycling of Polymer–Matrix Composites*).

Bibliography

Bunsell A R (ed.) 1988 *Fibre Reinforcements for Composite Materials*. Elsevier, Amsterdam

Bunsell A R, Kelly A, Massiah A (eds.) 1993 Developments in the science and technology of composite materials. *Proc. 6th European Conf. Composite Materials*. Woodhead, Cambridge

Chou T W 1992 *Microstructural Design of Fiber Composites*. Cambridge University Press, Cambridge

Christensen R M 1979 *Mechanics of Composite Materials*. Wiley, New York

Clyne T W, Withers P J 1993 *Introduction to Metal Matrix Composites*. Cambridge University Press, Cambridge

Cogswell F N 1992 *Thermoplastic Aromatic Polymer Composites*. Butterworth-Heinemann, Oxford

Dhingra A K, Lauterbach H G 1986 Fibers, Engineering. In: Marle H F, Bikales N M, Overberger C G, Menges G 1986 *Encyclopedia of Polymer Science and Engineering*, Vol. 6, 2nd ed. Wiley, New York, pp. 756–802

Gerstle F P Jr 1986 Composites In: Marle H F, Bikales N M, Overberger C G, Menges G 1986 *Encyclopedia of Polymer Science and Engineering*, Vol. 3, 2nd edn. Wiley, New York, pp. 776–820

Hale D K 1976 The physical properties of composite materials. *J. Mater. Sci.* 11: 2105–41

Hannant D J 1978 *Fiber Cements and Concretes*. Wiley, New York

Hashin Z, Shtrikman S 1962 A variational approach to the theory of the effective magnetic permeability of multiphase materials. *J. Appl. Phys.* 33: 3125–31

Hull D 1981 *An Introduction to Composite Materials*. Cambridge University Press, Cambridge

Kelly A 1988 Advanced new materials; substitution via enhanced mechanical properties. In: *Advancing with Composites, International Conference on Composite Materials*. CUEN, Naples, pp. 15–23

Kelly A, Macmillan N H 1986 *Strong Solids*. 3rd edn. Clarendon, Oxford

Kelly A, Rabotnov Y N (eds.) 1983 *Handbook of Composites* (4 Volumes). North Holland, Amsterdam

Maddock B J 1969 Superconductive composites. *Composites* 1: 104–11

Naslain R, Lamon J, Doumeingts D (eds.) 1993 High temperature ceramic matrix composites. *Proc. 6th European Conf. Composite Materials*. Woodhead, Cambridge, UK

Suchtelen J van 1972 Product properties: A new application of composite materials. *Phillips Res. Rept.* 27: 28–37

Suchtelen J van 1980 Non-structural applications of composite materials. *Ann. Chim. Fr.* 5: 139–53

A

Aircraft and Aerospace Applications of Composites

Fibrous polymer composites have found applications in aircraft from the first flight of the Wright Brothers' *Flyer 1*, in North Carolina on December 17th 1903, to the plethora of uses now enjoyed by them on both military and civil aircraft, in addition to the more exotic applications on space launcher vehicles and satellites. Their growing use has arisen from their high specific strength and stiffness, when compared to the more conventional existing and developing materials, and the ability to shape and tailor their structure to produce more aerodynamically efficient structures.

While fibrous polymer composites can, and will in the future, contribute up to 40–50% of the structural mass of an aircraft, the development of conventional alloys, aluminum–lithium alloys and the emerging metal and ceramic matrix composites will find increasing use, the latter in the more aggressive environments to which reentry vehicles and vertical take-off aircraft will be subjected.

1. Composites For Aircraft Applications

Glass-fiber-reinforced composites have been used on military aircraft dating from 1940, but their poor relative specific stiffness has prevented them from extending the foothold they have found on fairings, doors etc. to the primary structural applications of wings, stabilizers and major fuselage sections. Aramid fibers introduced in the 1960s found parallel applications with glass fibers, but their lack of specific stiffness and poor compressive strength limited their use, despite the tolerance to damage that composites utilizing these fibers can afford.

The adoption of composite materials as a major contribution to aircraft structures followed on from the discovery of carbon fiber at the Royal Aircraft Establishment at Farnborough, UK, in 1964. However, not until the late 1960s did these new composites start to be applied, on a demonstrator basis, to military aircraft. Examples of such demonstrators were trim tabs, spoilers, rudders and doors. With increasing application and experience of their use came improved fibers and matrix materials resulting in composites with improved properties, allowing them to displace the more conventional materials—aluminum and titanium alloys—from primary structures.

1.1 Structural Properties of Composites

Whereas composite strength is primarily a function of fiber properties, the ability of the matrix to both support the fibers and provide out-of-plane strength is, in many load situations, equally important. The aim of the material supplier is to provide a system with a balanced set of properties. While improvements in fiber and matrix properties can lead to improved laminate properties, the all-important field of fiber–matrix interface must not be neglected.

All laminate properties, with the exception of those relying on interlaminar strength and stiffness, are almost directly proportional to the basic strength of the fiber. As laminate stiffness also follows this relationship then clearly any improvement in fiber properties will almost inevitably lead to a product having improved laminate properties with a consequential improvement in their application to aerospace structures.

(*a*) *Effects of fiber properties.* The early carbon fibers, following on from the Farnborough discovery, exhibited properties moderate by today's standards and improvements have occurred both in stiffness, strength and a combination of the two. Figure 1 shows a carpet plot of fiber strength against modulus with an indication of the years in which fibers were developed. While it is difficult to be specific as to the direction in which developments should best take place, the trend of development required for military aircraft would be one in which both properties were increased simultaneously along the "ideal" direction indicated. Satellite applications, in contrast, benefit from the use of high fiber modulus, improving stability and stiffness for reflector dishes, antennas and their supporting structure.

(*b*) *Effects of matrix properties.* In order to produce a laminated structural element the fibers have to be bonded one to the other and while thermoplastic materials are becoming available, the more conventional matrix material is a thermosetting epoxy. The matrix material is the Achilles' heel of the system and limits the fiber from exhibiting its full potential in terms of laminate properties. The matrix performs a number of functions, amongst which are: stabilizing the fiber in compression, translating the fiber properties into the laminate, minimizing damage due to impact by exhibiting plastic deformation and providing out-of-plane properties to the laminate.

Matrix-dominated properties are reduced when the glass transition temperature is exceeded and whereas with a dry laminate this is close to the cure temperature, the inevitable absorption of moisture reduces this temperature and hence limits the application of most high-temperature-cure thermoset epoxy composites to less than 120 °C.

The first generation of aerospace composites introduced in the 1960s and 1970s utilized brittle-matrix

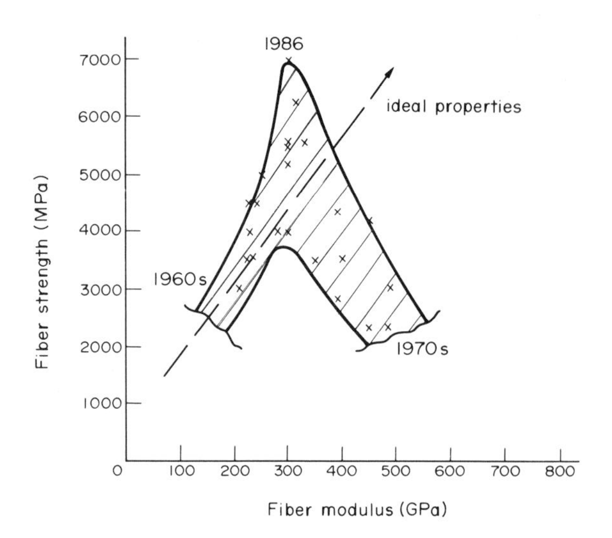

Figure 1
Tensile strength versus tensile modulus for carbon fiber

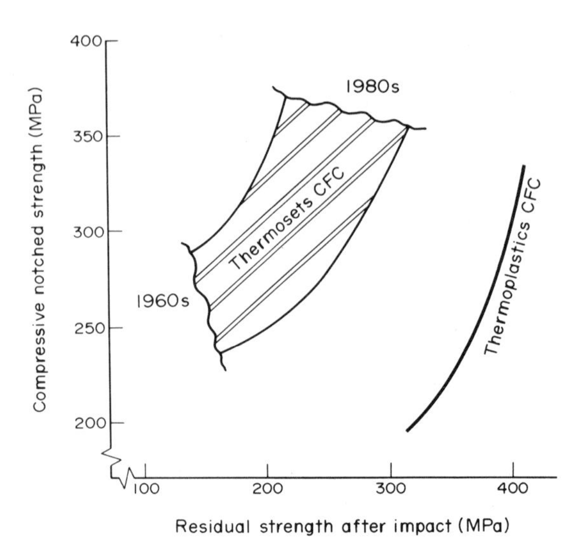

Figure 2
Comparison of notched compression and residual strength after impact of carbon-fiber composite (CFC)

systems leading to laminates with a poor tolerance to low-energy impact. These impacts can be caused by runway debris thrown up by the aircraft wheels or the impacts occurring during manufacture and subsequent servicing operation. Although the emerging toughened epoxy systems provide improvements in this respect, they fall far short of the damage tolerance provided by the new solvent-resistant thermoplastic materials. A measure of damage tolerance is the laminate compression strength after impact and this property is plotted on the abscissa of Fig. 2 with the laminate notched compression strength on the ordinate. The ideal solution is to provide a material exhibiting equal properties and it can be seen that whereas the thermoplastic systems are tougher they have not capitalized on this by yielding higher notched compression properties.

(c) *The fiber–matrix interface.* This, the third factor governing strength, is considered by many to be crucial in maximizing the degree to which the fiber properties can be translated into the laminate, be the matrix either thermosetting or thermoplastic. The interface between the fiber and matrix may be less than one micrometer yet excessive bonding between the fiber and matrix will result in poor notched tensile strength while an inadequate bond will result in poor interlaminar properties. Conflict therefore exists and the designer must select the material most nearly meeting his requirements.

2. *Alternative Fabrication Materials*

Of two materials with identical mechanical properties, the one with the lowest density will produce the minimum mass structure if all or part of the material is stability designed. As a consequence, carbon-fiber composites have an advantage over the more conventional materials used in aircraft fabrication. Composite materials are displacing the 2000 and 7000 series aluminum alloys from their dominant position in aerospace but emerging metallic materials have the potential to challenge the growing composite applications. Aluminum–lithium alloys, again a development from the Royal Aircraft Establishment, Farnborough, UK, are offering improvements in strength with a parallel reduction in density. These, in their own right, offer a challenge but when reinforced by ceramic particulates, such as silicon carbide, yield specific properties approaching those of today's composites. However, these are prohibitively expensive and difficult to manipulate and will only find application in niche areas in the near future. A comparison has been made, and shown on Fig. 3, of the specific notched tensile strength of various aerospace materials including both thermoset and thermoplastic materials.

3. *Implications of Composites on Design*

Aircraft design from the 1940s has been based primarily on the use of aluminum alloys and as such a plethora of data and experience exists to facilitate the design process. With the advent of laminated composites exhibiting anisotropic properties the methodology of design had to be reviewed and, in many areas, replaced. It is accepted that designs in composites should not merely replace the metallic alloy but should take advantage of exceptional composite properties if the most efficient designs are to evolve.

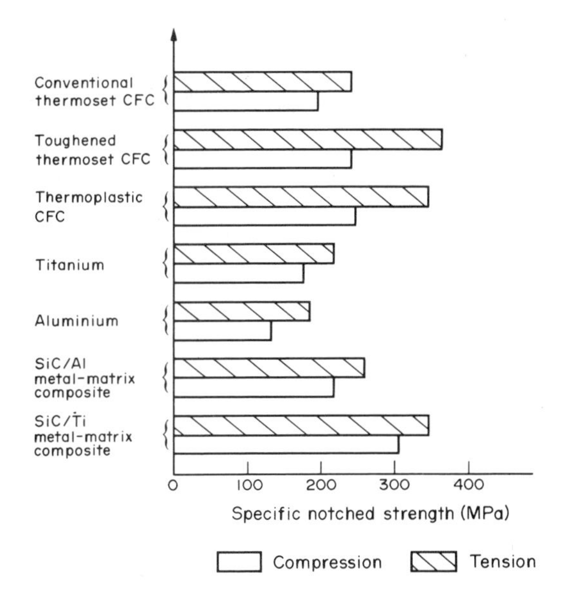

Figure 3
Structural material strength comparison

Hitherto the material properties for direct use by the designer had been available and his function was, based on a limited number of properties, to ensure that the structure offered sufficient strength and stiffness. With composite materials, however, the extra function of designing the material from a set of laminae properties has to be undertaken. Although in many instances laminae properties are available and documented for strength, stiffness and stability, stacking sequences have to be compromised to ensure continuity and compatibility of structure. An example of the complexity of the process is the derivation of the bending stiffness of an anisotropic laminate which involves the manipulation of a 6×6 matrix, a task which can only be performed by computer. Although these complexities lengthen the design process, they are more than compensated for by the mass savings and improvements in aerodynamic efficiency that result.

3.1 Aerodynamic Improvements

Because of the limitation of metal-manipulation techniques, the majority of aircraft control–lift surfaces produced have a single degree of curvature. Improvements in aerodynamic efficiency can be obtained by moving to double curvature allowing, for example, the production of variable camber, twisted wings. Composites allow the shape to be tailored to meet the required performance targets at various points in the maneuver envelope.

A further and equally important benefit is the ability to tailor the aeroelastics of the surface to further improve the aerodynamic performance. This tailoring can involve adopting laminate configurations which allow the cross-coupling of flexure and torsion such that wing twist can result from bending and vice versa. Modern computational techniques of structural optimization, using finite-element analysis techniques, allow this process of aeroelastic tailoring, along with strength and dynamic stiffness (flutter) requirements, to be performed automatically with a minimum of postanalysis engineering yielding a minimum mass solution.

3.2 Airworthiness Procedures

High material variability and the deleterious effects of temperature and moisture preclude the use of the conventional airworthiness route required for the clearance of aircraft structures. The composite design allowables, particularly in compression, can be as low as 60% of the mean test result obtained from a test conducted at room temperature. As both material variability and degradation are failure-mode dependent, a room-temperature test cannot practically demonstrate the true strength of a structure. This is exacerbated if the structure is hybrid, containing metallic components, as the variability and thermal and moisture degradation of these materials can be negligible.

As the most critical condition, the production of minimum design values, generally takes place at elevated temperature when the laminate has absorbed moisture then the obvious method would be to undertake the testing under these conditions. However, testing in this manner on full-scale flight vehicles can be prohibitive in terms of both cost and timescales and, as a consequence, alternative routes have been found to ensure the competence of the structure.

These routes have as their foundation the collection of data from many thousands of specimens, giving strength for the particular property along with variability and environmental degradation factors: from this, design data are produced. Verification of these data on elements containing structural features is then obtained by undertaking some of the tests under ambient conditions, and other elements are degraded to confirm the degradation factor. Paralleling this is the verification of strain distribution by comparing strain measurements during the test with those obtained from a finite-element analysis of the specimen. The full-scale test is then performed at ambient conditions with each of the critical cases taken to the ultimate design load with the results from the extensively instrumented structure extrapolated to verify that, had the test proceeded to higher loads, both variability and degradation would have been accommodated. Should failure not have occurred at ultimate load then the test would be taken to failure, with this occurring in the area with lowest variability and degradation effects. A metallic component usually shows this property. Figure 4 has been constructed in an attempt to demonstrate graphically the way in which a full-scale static test undertaken at ambient

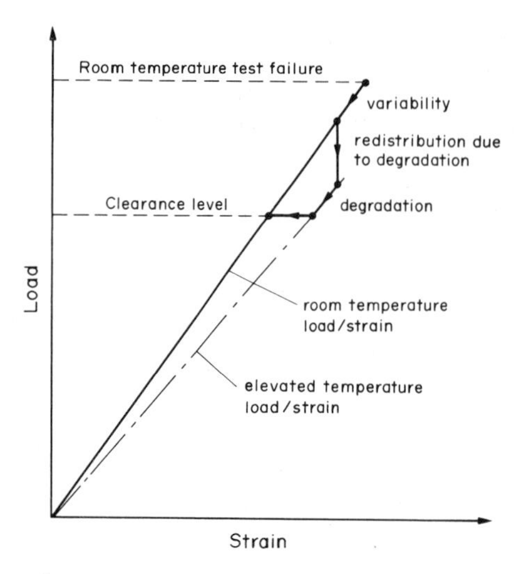

Figure 4
Airworthiness clearance of composite structures

conditions can be read across to provide the effective ambient condition design allowable.

3.3 Test Instrumentation

Instrumentation, particularly strain measurement, is essential to the airworthiness clearance of composite structures. The two major agents for understanding the strain distribution are electrical resistance strain gauges and photoelastic coatings, the former giving field strains while the latter are more useful for indicating the distribution around stress concentrations.

Work is ongoing, taking advantage of the laminated form of composite structures, to embed strain-measuring fiber optics into the laminate during manufacture, producing a matrix of multipoint data measurements leading to an "intelligent" structure. Currently, however, only single-point strain measurement along a fiber is possible.

More direct methods of failure prediction have been investigated by many researchers and prominent amongst these methods is acoustic emission (AE). This relies on the ultrasonic emission from structures caused by matrix cracking, fiber failure, joint slip etc. AE has been used as an inspection tool for such items as pressure vessels where a characteristic pressure–emission profile can be obtained from acceptable cylinders and then used to clear production components. However, on structures where this characteristic has not been determined, the current capability to predict failure is limited. Emission characterization is being developed to identify the impending failure mode and it is considered that this may lead to a more accurate prediction capability.

3.4 Electromagnetic Compatibility

All aircraft contain electronic equipment and with modern "fly-by-wire" aircraft the use of such equipment is increasing rapidly in the critical flight safety area of the systems. On metallic aircraft these delicate systems are in the main adequately protected from electromagentic interference by the inherent shielding provided by the fuselage skins. Fibrous polymer composites unfortunately do not provide this protection and, as a consequence, special action has to be taken to ensure that this deficiency is overcome.

Electromagnetic interference can result from natural phenomena: lightning strike, radio transmitters, etc., or in the case of military aircraft from malicious intent. The effect of the traditional aluminum structure can be replicated by lining the composite structure with foil and ensuring electrical continuity. The alternative is to provide adequate shielding to all the equipment and the interconnecting cabling. The use of optical fibers for communication will minimize the mass increase associated with this shielding requirement.

While a lightning strike can produce electromagnetic pulses affecting the electronic equipment, its direct effect can be catastrophic, producing significant structural damage. Arcing in fuel tanks, for example, can cause explosive disintegration of these tanks with potential loss of the vehicle. The inherent low conductivity of composite structures leads to the poor dissipation of arc current, producing hot spots, mechanical damage and preferential attachment to metallic items, leading to the internal arcing discussed above. Protection against the direct effects of lightning strike are therefore aimed at limiting the physical damage to an acceptable level, particularly over fuel.

The above-mentioned internal foil is inadequate for system protection, and external foils or meshes have to be provided on laminates in the more critical zones of the aircraft. Meshes can be cocured along with the laminate, as can foils, but an attractive protection capable of easy repair is that of flame or plasma spraying metallic coatings.

A side issue that must be addressed is the conflicting requirements of adequate electrical continuity and corrosion. In the presence of aluminum, carbon-fiber composites act as noble metals with the former corroding in their presence. Careful design is therefore essential to prevent the creation of galvanic cells and the resulting corrosion.

4. Manufacture

The aerospace industry has, over the past decades, geared itself to the batch manufacture of metallic parts for aircraft. This has lead to sophisticated methods of manipulating these metals with numerically controlled machines for metal removal, routing, forming etc. Further improvements, with the introduction of flexible manufacturing systems, ensure the optimum availability of part, tool and machine, producing a system capable of manufacturing a number of different components with minimal interference to efficiency.

With the extensive introduction of fibrous composites in the 1970s, new skills had to be developed in both the production engineering of the design and the subsequent manufacture of the component.

Mold tools capable of operating at the cure and postcure temperatures (200 °C) while remaining stable and durable had to be designed and, while this can be readily effected in steel, the mismatch in coefficient of thermal expansion between the two materials requires careful consideration. Early composite mold tools were neither durable nor stable but modern tooling materials, usually in preimpregnated form, are proving more reliable. Although the steel thermal mismatch can be overcome on laminate molding tools, the resolution of the problem is more difficult if this material is to be used in assembly tools requiring thermal cycling.

4.1 Laminate Manufacture

The largest proportion of carbon-fiber composites used on primary class-one structures is fabricated by placing layer upon layer of unidirectional material to the designers requirement in terms of ply profile and fiber orientation. On less critical items, woven fabrics very often replace the prime unidirectional form.

A number of techniques have been developed for the accurate placement of the material, ranging from labor intensive hand layup techniques to those requiring high capital investment in automated tape layers. These latter machines can require, in certain instances, investments of up to US$3 million (at 1988 prices).

Processes adopted by the aerospace industry range from hand layup using foil-transfer techniques, through the use of numerically controlled ply profiling machines to numerically controlled tape laying or filament winding. The first of these is an obsolescent process and has, in general, been replaced by the latter two. It involves the hand profiling of plies and their subsequent location on the mold tool, again by hand, using draughting foils for their location.

Use has been made in both the tailoring and shoe industries of automated profiling equipment and out of these has arisen the broadgoods system in which numerically controlled machines under two-axis control profile the plies to the appropriate shape. Three cutting media have been used on these machines; high pressure water, reciprocating knife cutting and laser cutting, with reciprocating knife becoming prominent because of the moisture uptake with the former and charring of ply ends with the latter. Although the use of broadgoods systems reduces the effort in profiling the plies, their manual laying remains a costly aspect of the laminate production. A number of attempts have been made to automate this ply-placement process, with success only being found on small components with limited complexity and curvature.

An obvious development is to combine the ply profiling and laying into one operation and this has been accomplished with limited success in automatic tape-laying machines operating under numerical control. These machines are currently limited in production applications to flat layup, and significant effort is being directed by machine manufacturers at overcoming the many problems associated with laying on contoured surfaces.

Cost reductions in the manufacture of carbon-fiber composite components have been obtained by the progression from hand layup to the use of tape-laying machines. Table 1 shows the corresponding increase in productivity.

4.2 Assembly of Aerospace Components

Early composite designs were replicas of those which employed metallic materials and, as a consequence, the high material cost and man-hour-intensive laminate production jeopardized their acceptance. This was compounded by the increase in assembly costs due to the initial difficulties of machining and hole production. Evidence exists to support the argument that cost is directly proportional to the number of parts in the assembly and, as a consequence, design and manufacturing techniques had to be modified to integrate parts, thereby reducing the number of associated fasteners.

A number of avenues are available for reducing the parts count, amongst which are the use of honeycomb sandwich panels—sometimes frowned upon in metallic structures due to corrosion—the use of integrally stiffened structures and, more ambitiously, the cocuring or cobonding of substructures onto lift surfaces such as wings and stabilizers. Figure 5 has been constructed to demonstrate the cost of manufacture of aircraft structure in terms of units of cost per kilogram of structure using the conventional aluminum structure as the datum. Hand layup techniques and conventional assembly results in manufacturing costs 60% higher than the datum and only with the progressive introduction of automated layup and advanced assembly techniques can composites compete with their metallic predecessors.

Table 1
CFC layup rate comparison

Method of manufacture	Deposition rate ($kg\,hr^{-1}$)
Foil transfer	0.1–0.25
Broadgoods	0.7–1.25
Tape layer	5
Filament winding	20

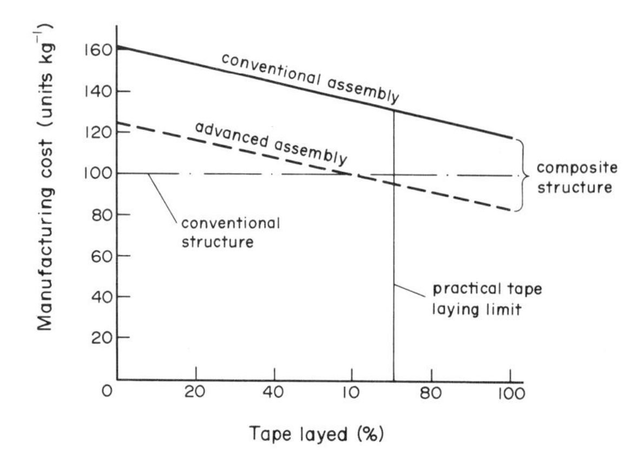

Figure 5
Manufacturing cost of conventional and composite structures

4.3 Nondestructive Examination

All aerospace materials are quality assured to a high standard with the bulk of metallic materials being examined by the suppliers using ultrasonics, x-ray and other appropriate processes. The carbon-fiber pre-impregnated material supplier quality-assures his product by various techniques, but the verification of a sound cured structural element is now the responsibility of the aircraft fabricator.

Composites attract a high level of both conventional inspection and the more sophisticated nondestructive examination (NDE) and these combined can account for up to 25% of the direct production effort. Of this figure, the proportion required for NDE is by far the largest. The main tool to ensure that any defects above those defined by the designer as tolerable are detected is ultrasonic examination, invariably on 100% of the structure. This, in many instances, can be supplemented by 100% radiographic examination.

A number of avenues are open for reducing this burden, by relaxing standards, based on defect investigation programs, by zoning of components through correlating allowable defect to stress–strain levels, or by developing existing and new techniques. Current techniques based on single probes or multiprobes are time consuming and attention is being directed at phased arrays, thermography, etc. (See *Nondestructive Evaluation of Composites.*)

5. *Aircraft and Aerospace Applications*

From the cautious application of composites in the early 1970s has developed the more confident use of these materials in the primary load-carrying structures of both military and civil aircraft. Although military applications are far in excess of civil aircraft applications, the latter are growing rapidly.

5.1 Civil Aircraft Applications

These have concentrated on replacing the secondary-structure, bounding major structural elements, with fibrous composites where the reinforcing media has either been carbon, glass, Kevlar, or hybrids of these. The matrix material, a thermosetting epoxy system, is either a 125 °C or 175 °C curing system with the latter becoming dominant because of its greater tolerance to environmental degradation. Typical examples of the extensive application of composites in this manner are the later aircraft from Boeing (the 757 and 767) and from Europe, the aircraft produced by Airbus Industrie. This consortium of aircraft companies is currently producing the A320 which, like the later versions of the A310, carries a vertical stabilizer, a primary aerodynamic and structural member, fabricated almost in its entirety from carbon composite. In addition, the A320 has extended the use of composites to the horizontal stabilizer in addition to the plethora of panels and secondary control surfaces.

In two small transport aircraft attempts have been made to apply composite materials to almost the entire aircraft; these are the Lear Fan fabricated in Belfast, Northern Ireland, and the US Beech Starship. Of these ambitious projects the former failed for technical and commercial reasons although the latter is showing every prospect of entering service.

5.2 Military Aircraft Applications

These applications throughout the world, initially funded from national research and development budgets, have enabled composite technology to be developed to its current level. This includes material manufacture, design techniques and product manufacture as well as in-service experience and maintenance.

Without exception all agile fighter aircraft currently being designed throughout the world contain in the region of 40% of composites in the structural mass, covering some 80% of the surface area of the aircraft. Without this degree of compositization the essential agility of the aircraft would be lost because of the consequential mass increase.

The US advanced tactical fighter (ATF) and the European Fighter Aircraft (EFA) are examples of the proposed degree of compositization for aircraft entering service in the 1990s. Examples of current applications in which the structural mass of composites is 25% of the total can be found in the McDonnell–Douglas AV8B (Harrier Mk. II) and the European Experimental Aircraft Programme (EAP).

5.3 Space Applications

Satellites and space platforms clearly require ultralight structures to minimize the launch costs and, to this end, exotic materials have been developed. These include ultrathin lamina and high-modulus materials, the former to enable precise tailoring to the structural requirement and the latter to provide stability for the antennas and reflectors. With the demand for observation and communication satellites ever increasing, the developments for this niche market will inevitably continue.

Current launch-vehicle availability in terms of tonnes per annum is incapable of meeting demand and the available vehicles, the US Space Shuttle and the European Arianne rocket, are prohibitively expensive. Worldwide studies are determining the feasibility of fully reusable launch vehicles requiring minimum turnaround activity, thereby reducing launch costs significantly. Amongst these are the US National Aerospaceplane (NASP) and the British Horizontal Take-Off Launch Vehicle (HOTOL). The latter experiences structural temperatures ranging from that of liquid hydrogen (20 K) to those caused by reentry (1750 K). Although materials are available that will perform structurally at these temperatures, their performance, particularly towards the upper temperature limit, is inadequate to meet the demanding specific-strength requirements. These inadequacies in available materials provide the next major challenge to material suppliers, designers and fabricators involved in the introduction of fibrous composites with polymer, ceramic or metal matrices.

See also: Applications of Composites: An Overview; Nondestructive Evaluation of Composites

Bibliography

Ashton J E, Halpin J C, Petit P H 1969 *Primer on Composite Materials Analysis.* Technomic, Stanford, Connecticut
Curtiss P T (ed.) 1984 Crag test methods for the measurement of the engineering properties of fiber reinforced composites, Royal Aircraft Establishment Report RAE TR84102. RAE, Farnborough, UK
Haresceugh R I 1987 Composites—the way ahead. *Proc.* 6th Int. Conf. on Composite Materials, 2nd *European Conf. on Composite Materials*, Vol. 5. Elsevier, London
Jones R M 1975 *Mechanics of Composite Materials.* McGraw–Hill, New York
Lechnitski S G 1963 *Theory of Elasticity of an Anisotropic Elastic Body.* Holden–Day, San Francisco
Royal Aeronautical Society 1986 *Materials In Aerospace*, Vols. 1, 2. RAS, London

R. I. Haresceugh
[British Aerospace Military Aircraft Division, Preston, UK]

Applications of Composites: An Overview

Composite materials now exert an influence on the daily lives of most people in industrialized societies. Much personal transport, and indeed safety, depends upon the cord or fiber reinforced elastomer tire; many commercial airliners fly with some parts of the primary structure and much of the interior furnishings and trim made from composites; many types of sporting and leisure goods, such as boats, gliders, sailboards, skis and racquets, make extensive use of composite materials in their construction; and in the home, most plastic-bodied appliances incorporate reinforcement in the form of short, chopped fibers.

The term composite material can be defined in a number of ways. The most commonly held view is of a material in which strong, high modulus man-made fibers, with diameters of the order 10–100 μm, are embedded in a matrix material which may be a polymer, a metal, a glass or a ceramic. The objective is to enhance the mechanical or physical properties of the host material. Alternatively, from a mechanical standpoint, the matrix may be regarded simply as a binder whose role is to transfer stress to the reinforcing fibers and ensure their cooperative interaction. The matrix may also fulfill other functions, such as protecting the fine reinforcing filaments from corrosion, oxidation, or other forms of environmental degradation.

Reinforcing agents are not necessarily confined to man-made fibers, although continuous filaments of glass, carbon, ceramics and high-modulus polymers are the most widely used at present. However, vegetable fibers, inorganic whiskers, metal wires and particles of refractory solids all find a place in the spectrum of reinforcing materials. On a somewhat different scale, concrete reinforced with steel bars can be regarded as a composite with mechanical behavior analogous to that of fiber composites, but where application in tonnage terms dwarfs that of other constructional materials. However, this article is concerned with the families of materials known as fiber-reinforced plastics (FRPs), metal-matrix composites (MMCs) and ceramic-matrix composites (CMCs) in which at least two of the three dimensions of the reinforcement are typically of the order 100 μm or less.

Of these materials, the group that has reached the most advanced stage of both technical development and market penetration is fiber-reinforced plastics. One of their principal attributes is low density and to indicate the scale of their application, world consumption of glass-fiber-reinforced plastics (GRP) is compared in Fig. 1 to that of aluminum, a material of roughly comparable cost and density, but established in the market place some 50 years earlier. The data for aluminum are plotted from the time (approximately 1890–1900) when commercial quantities became available following the development of the Hall process for the electrolysis of fused cryolite. Figures for GRP date from the mid-1940s when significant quantities of glass fiber first became commercially available. Both curves reflect the recession in world trade during the 1980s, and the aluminum plot also illustrates the boost and subsequent fall in demand associated with World War II, and the influence of the 1973 oil crisis.

Although present consumption of GRP is small compared to that of aluminum, it is apparent from Fig. 1 that the rate of growth of consumption over the

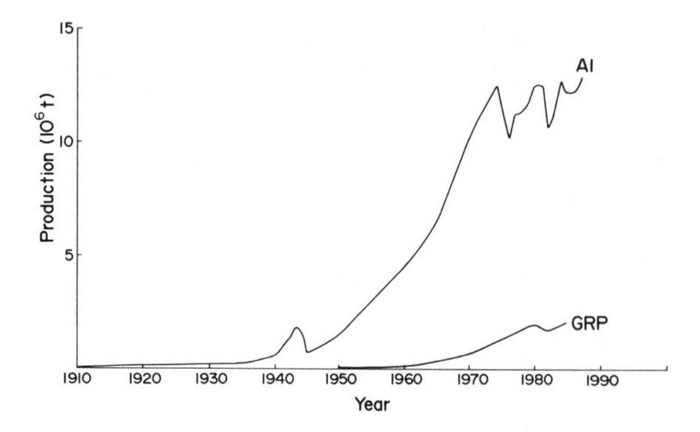

Figure 1
Comparative growth of aluminum and GRP consumption

first 30 years of composites technology, averaging 10–15% per annum, has been considerably greater than in the first 30 years of aluminum technology. It is probably not meaningful to pursue the comparison further in view of the general increase in the pace of technological development that has occurred over the past few decades, except to note that at least part of the growth for GRP has resulted from metals substitution, and that most market projections are for continuing growth in the applications of composites.

Glass-fiber-reinforced plastics, in volume terms, represent the major part of the reinforced plastics market, mainly on account of their relatively low cost, good corrosion resistance, good specific strength and modulus, and the ease with which complex shapes can be molded or fabricated by a number of well-established processes. The same general advantages apply to plastics reinforced with even stronger or stiffer fibers, notably of carbon or of highly oriented polymers such as the aramids, and although the resulting composites are considerably more expensive than GRP, they can nevertheless be highly cost-effective in applications where weight reduction gives rise to operating economies, as in aerospace. For example, one source estimates the acceptable cost of weight saving in helicopter construction as US\$250 kg^{-1}, and in a satellite as US\$650 kg^{-1} (Jardon and Costes 1987).

One of the major limitations of polymer-based composites is the decrease in their mechanical properties with increased temperature. A large number of polymer formulations, both thermosetting and thermoplastic, are used as matrices for composites and although a few, presently expensive, materials are suitable for extended service above 300 °C, most reinforced plastics are limited to temperatures below ~200 °C for continuous operation. It is necessary to employ more refractory reinforcements with metal or ceramic matrices in order to exploit the advantages of composite materials for high temperature applications, such as in heat engines. Table 1, from Hancox and Phillips (1985), indicates maximum temperatures for continuous operation of the three principal types of composite material considered in this article.

Reinforcements currently being explored with metal and ceramic-matrix composites include continuous fibers of carbon, silicon carbide, boron, aluminum oxide and whiskers of silicon nitride and silicon carbide. Most of the resulting composites are still at the research and development stage (see, e.g., Anderson et al. 1988), and to give some indication of the relative scale of applications for polymer, metal- and ceramic-matrix composites, Table 2 lists some estimates of approximate consumption levels for the principal reinforcing materials. Despite the fact that carbon fibers are used to reinforce some low-melting-point metals and alloys, it can be assumed that virtually all the glass, carbon and aramid fibers listed in Table 2 are used for plastics reinforcement at the present time. Although the quantities of carbon, aramid and ceramic fibers used in composites are small compared to glass fiber consumption, most market analysts predict growth rates up to ~20% per annum for high-performance polymer-matrix composites, based largely

Table 1
Upper temperature limits for continuous operation of composite materials

Composite	Maximum operating temperature (°C)
Polymer matrix	400
Metal matrix	580
Ceramic matrix	1000

Table 2
Approximate production levels for the principal reinforcements for composite materials

Reinforcement	Production capacity/consumption (tonnes per annum)
Continuous fibers	
glass	[a]1280×10^3
aramid	3600[a]
carbon	3300[a]
ceramics (B, SiC, Al_2O_3)	few tonnes[b]
Short ceramic fibers	~40[b]
Whiskers (SiC, Si_3N_4)	~100[b]

a Brehm and Sprenger 1985 b Feest et al. 1986

on assessments of the continuing adoption of these materials by the aerospace industry.

In discussing applications for composite materials it is important to recognize that they are generally more expensive than other construction materials. Table 3 gives broad ranges of cost for reinforcements, matrix resins and some conventional construction materials. Exact costs will depend on the form in which the basic materials are to be supplied.

In the case of reinforced plastics, for example, cost will be determined by whether the fibers are in the form of rovings or woven fabric, whether preimpregnated with resin, and on the tow size or areal weight of the fabric. The costs of manufactured items will also depend on the nature of the production process and on the complexity of the component. For GRP, the cost of a finished component may be typically two or three times the materials cost, and the added value may be significantly higher if there is a substantial degree of proof testing and quality assurance involved in the manufacturing process, as with many aerospace parts.

1. Applications of Fiber-Reinforced Plastics

In view of the difference between market sizes for GRP and for composites based on carbon and aramid fibers, it is convenient, for a broad description of applications, to discuss the two areas separately. There is, however, no clear-cut definition of what constitutes a high-performance composite in terms of its composition. Thus, although some types of GRP fulfill relatively undemanding uses such as decorative panels or cable ducting, other types are used in primary structures of some aircraft designs (Lubin and Donohue 1980) and in the construction of deep-sea submersibles (Oliver 1980).

1.1 Applications of GRP

Table 4 shows a breakdown of the European market in terms of applications for GRP based mainly on the relatively low-priced polyester resins (Chevalier 1987). Some attempt is made below to give an indication of the types of application in some of the larger sectors.

Table 3
Approximate price ranges for reinforcements, thermosetting resins and some traditional materials of construction

Material	Approximate price, 1988 ($£\ 10^3\ t^{-1}$)
Continuous fibers	
glass	1–2
aramid	20–75
carbon	25–100
SiC	200–400
Thermosetting resins	1–10
Whiskers	50–100
Unreinforced materials	
metals (e.g., steel, aluminum)	0.1–1
timber	0.3–1.5
engineering thermoplastics	2–10

Table 4
European market for glass-resin composites in 1986, breakdown by application (estimated volume 860 000 t)

Application	Distribution (%)
Transport	21
Electrical industry	19
Industrial and agricultural equipment	20
Building and public works	16
Sports and leisure	6
Consumer goods	6
Others	12

(*a*) *Transport.* In Western Europe, some 170 000 t of GRP in its various forms is used in land transportation applications, and in the USA, total consumption in this sector is substantially greater. The incentive to introduce reinforced plastics into land-based transport is mainly to take advantage of their corrosion resistance and of manufacturing economies. Weight reduction, although leading to improved fuel economy or performance, is unlikely to be an important factor in new designs unless it can be achieved without a cost penalty.

The use of reinforced plastic panels for car bodies is at present mainly restricted to small-volume production of specialist vehicles for which the versatility of GRP fabrication methods enables design changes to be made fairly easily, and where the high capital cost of large metal press tools is not justified. The largest production run to date is probably the General Motors' Fiero car which employs body panels produced by press-molding a sheet molding compound (SMC) and by the reinforced reaction injection molding (RRIM) process, and these are attached to a steel structural frame (Ferrarini et al. 1984). Similar schemes have been used for many years for the production of cabs for heavy commercial vehicles (Seamark 1981).

The adoption of molded reinforced plastics for a wide variety of panels, bumpers and interior fittings appears to be growing steadily, at least in terms of the average quantity per car (Waterman 1986). In the USA, reinforced plastic front and rear ends have almost completely superseded die-cast zinc because a single molding can replace an assembly of many individually cast items. In Europe, Renault were producing 960C GRP bumpers per day in 1983, and the Citroen BX features an SMC bonnet (hood) and tailgate, produced at a rate of the order of 1000 per day (Buisson 1983). Attention has also been directed towards underbonnet components, of which one of the most ambitious has been the molding, by a lost-wax-type process, of a dough molding compound (DMC) inlet manifold (Suthurst and Rowbotham 1980).

The use of composites for structural automotive components has not been overlooked, and there are numerous references in the literature to the development of stressed components such as drive shafts and GRP springs. Leaf springs make use of the excellent fatigue properties of fiber composites and can give extended lives in comparison to steel springs, particularly for commercial vehicles (Lea and Dimmock 1984, Jardon and Costes 1987). Figure 2 shows a GRP spring undergoing fatigue testing. Such springs provide a substantial reduction in mass over steel springs, as well as providing an improved ride and a decrease in the noise transmitted to the driving compartment.

The use of composites is not confined to road transport, however. Applications in rolling stock on European railways have been reviewed (Anon 1980) and the construction of the cabs on British Rail's Intercity vehicles has been described by Gotch and Plowman (1978). In the USA a rail-borne grain transporter, 15 m long and 4.6 m in diameter, has been produced by filament winding (Ruhmann et al. 1982).

(*b*) *Electrical.* The most common type of glass fiber used for reinforcement purposes is E-glass, which was originally developed for electrical insulation purposes, so it is not surprising that, when combined with organic resins, a class of materials results which can readily be molded into complex shapes possessing excellent insulating properties and a high dielectric strength. Thus switch casings, junction boxes, cable and distribution cabinets, relay components and lamp housings, to name just a few examples, are produced from a variety of glass-fiber molding compounds. Cable ducts and standard cross-sectional shapes for transformer insulation are produced by the pultrusion process, and reinforcing bands for rapidly rotating electrical machinery are often filament wound. Mathweb, a proprietary process for winding lattice-type

Figure 2
Fatigue test of a GRP automotive leaf spring (Photograph by courtesy of GKN Technology Ltd.)

structural beams (Preedy and Kelly 1982), is now used by British Rail to provide emergency support to overhead power lines because the light weight simplifies replacement of damaged gantries. Pultruded composites are also used in the construction of pantograph arms for current pickup on electric traction vehicles.

(*c*) *Industrial and Agricultural.* Generally speaking, applications in agriculture, in chemical process plants and in tanks and pipes for industrial purposes rely upon the combination of low density and good corrosion resistance of GRP. One of the advantages of the material is the relative ease with which large and complex-shaped components can be fabricated on a one-off basis. Typical applications include vats and silos for the storage of aggressive chemicals, pipelines for the transport of water and sewerage, water storage tanks, wine vats and the construction of certain types of chemical process plants as, for example, in the manufacture of chlorine. A particular advantage of the low density is that the cost of transporting components is minimized and this particularly facilitates the installation of pipelines in remote or rugged terrain. It is also possible to transport and handle longer lengths of pipe, thus minimizing the number of joints and hence the assembly costs. Figure 3 illustrates a PVC-lined GRP storage vessel for hydrochloric acid, of dimensions 10.5 m in length and 3.5 m in diameter.

(*d*) *Building.* The corrosion resistance and lightness, and hence ease of transport and installation of GRP, coupled with the ability to mold surfaces with decorative textures and color, has led to extensive use of this material as cladding for prestige buildings (Jaafari et al. 1976). Considerable quantities of GRP in the form of flat and corrugated translucent sheeting are also used for roofing purposes, particularly in outdoor canopies of large structures such as stadia, where the reduced weight of the roofing panels enables economies to be made in the design of the roof support structure. In the USA, the development of prefabricated bathroom units comprising sanitary ware, furniture and partitioning has generated a significant market (Seymour and Tompkins 1975).

Figure 3
GRP tank lined with PVC for storage of HCl (Photograph by courtesy of Plastics Design and Engineering Ltd.)

(*e*) *Marine.* This area provides a good example of how a number of attributes of GRP in combination have resulted in almost complete substitution for traditional materials, particularly timber. Low density, corrosion resistance and the ability to produce complete hull and deck moldings repetitively from a single set of molds have revolutionized the leisure and small workboat building industries, and reduced maintenance costs have played a large part in gaining widespread consumer acceptance of GRP. Of less general applicability, but of particular importance to naval mine-clearance operations, is the nonmagnetic character of GRP. The UK pioneered the introduction of GRP minesweepers with the launch, in 1972, of HMS Wilton with an overall length of 46 m (Dixon et al. 1973), and vessels up to 60 m in length and with displacements of ~600 t are now being built in a number of countries (Anon 1987).

1.2 Applications of High-Performance Composites

As indicated earlier, the term high-performance composites is somewhat loosely employed to describe composites based primarily on the more expensive carbon and aramid reinforcements, although it also reflects the improved performance to be gained in the two major market sectors to have developed thus far—aerospace and sports goods.

(*a*) *Aerospace.* The use of lighter materials of construction means either that a greater payload can be carried or that operating costs can be reduced for the same payload. The outstanding example of the cost-effectiveness of lightweight composites is in spacecraft technology where the cost of launching may often exceed the cost of designing and building the spacecraft (see, e.g., Zweben 1981, Burke 1986). Consequently, carbon- and aramid-fiber-reinforced plastics are frequently used for such items as the basic structure of a spacecraft, as supports for solar arrays, for the construction of dish antennas, in pressure vessels for propellant gases and in apogee motor cases and thrust cones. In addition to weight savings, the low axial thermal expansivity of these materials enables very high dimensional stability to be achieved in antennas, microwave resonant cavities and filters and in telescope support structures (see, e.g., Reibaldi 1985). Such components may experience temperature excursions in orbit within the range −100 °C to approximately 100 °C depending on whether the component is exposed to full sunlight or is in the shadow of the spacecraft.

In military aircraft, composites offer improved performance, and a number of aircraft design and construction programmes are in progress involving major use of advanced composites (Hadcock 1986). For example, approximately 35% of the structural weight of the projected European Fighter Aircraft (EFA) is likely to be built from composite materials, involving the main wing, the forward fuselage and the fin and rudder. Figure 4 illustrates the use of composites in the construction of the AV-8B Mk. 5 Harrier II airframe giving rise to a 25% weight saving in the airframe compared to all-metal construction (Riley 1986). In commercial aircraft, composites have for some time been used to reduce weight in internal fittings such as cargo floors, galley furniture, luggage lockers and for internal trim. Seat frames are now beginning to be built in carbon-fiber-reinforced plastics (CFRP) since they not only lead to weight savings but occupy less space. With regard to the use of composites in primary aircraft structures, considerable caution has been, and still is, exercised in proving such materials before flight certification is granted, but nevertheless major structural items are now being flown on passenger aircraft. In the A310 Airbus, for example, the vertical stabilizer makes substantial use of composites construction involving glass, carbon and aramid fibers, with a total weight saving of 397 kg (Pinzelli 1983). The A320 Airbus will use CFRP in the fin and tailplane leading to a weight saving of 800 kg over aluminum-alloy skin construction (Anon 1986). As an indication of the benefit of such weight savings, it has been estimated that a 1 kg weight reduction on a DC-18 saves over 2900 l of fuel per year (Zweben 1981).

On a smaller scale, a number of designs of light aircraft and gliders (Riddell 1985), built primarily of GRP, have been flying for a number of years and attention is now being directed to the development of all-composite aircraft in the executive jet category. In this class, two prototypes of the Lear Fan 2100 aircraft with 70% of structural weight in CFRP were flown before the company went into liquidation for reasons unconnected with the use of composites (Noyes 1983). The Beech Starship, currently under development, is also constructed largely from composite materials and one phase of the programme involves a comparison between hand-lay and filament-winding techniques for the construction of the fuselage (Wood 1986).

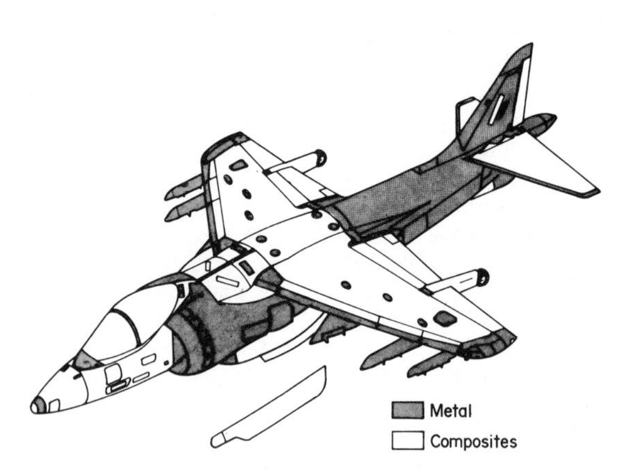

Figure 4
The extent of composites construction in the AV-8B Mk. 5 Harrier II aircraft (From Riley 1986)

As frequently happens when an existing technology is threatened by innovation, the aluminum industry has responded to the introduction of composites with the development of alloys, such as the Al–Li series, with improved strength and stiffness (Clementson 1985). Although their specific properties do not match those of the advanced composites, they possess the advantage that established metal skin design and fabrication procedures can be applied, obviating the need for investment in new skills to handle design with anisotropic materials and the very different manufacturing techniques involved with composites. Nevertheless, whatever the outcome of the trade-off between metal or composite properties, one important economic advantage of composites remains, namely the ability to fabricate large and complex parts as a single molding, as opposed to assembly of a number of individual metal pressings.

Thus far, little has been said of the excellent fatigue behavior of composites in fiber-dominated directions. An outstanding example of a component whose life is governed by fatigue damage is the helicopter rotor blade which experiences complex dynamic stresses including torsion, together with bending in, and perpendicular to, the plane of the rotor (Goddard 1985). Composite blades are now fitted to most major helicopter designs on the basis that they give virtually unlimited fatigue life in relation to the lifetime of the aircraft, reduce manufacturing costs and enable novel and more efficient aerofoil structures to be created. Most helicopter manufacturers also plan to use composites extensively in future airframes. For example, the Boeing 360 demonstrator has an all-composite fuselage consisting of CFRP frames to which are bolted and bonded honeycomb sandwich panels with aramid-fiber-reinforced skins. This form of construction gives a 25% weight saving in the fuselage and substantial cost reductions related to parts integration and the shorter time required for assembly (Anon 1987a). As with helicopter rotor blades, aircraft propeller blades made from composites show excellent fatigue and corrosion resistance compared to metals, and at reduced weight. McCarthy (1986) has described the design and construction of composite blades and has illustrated the weight savings compared to metal blades.

(*b*) *Leisure and sporting goods.* Generally speaking, the production of aerospace structures and components is a low-volume manufacturing operation measured typically in terms of a few tens or hundreds of parts per year. The second largest market sector for composites based on carbon and aramid fibers, and to some extent glass fibers, has in recent years been the leisure goods industry, where production rates have been measured in thousands or tens of thousands of units per year. However, in items like tennis and squash racquets, fishing rods, skis and ski poles, golf shafts, windsurfers and other small sailing craft, the quantities of expensive reinforcements are small, but aggregate to approximately one-third of present carbon and aramid fiber consumption.

Reference to the use of composites in sailplanes and small boats has already been made in this article. A feature of the use of such materials in high-performance sailing craft is that the advanced materials and techniques developed for the aerospace industry are being adapted to produce racing vessels with strong rigid hulls constructed from honeycomb sandwich panels, giving substantial reductions in weight, and increases in performance. For similar reasons, many racing cars now employ a composite chassis consisting of CFRP skins bonded to a honeycomb core (Clarke 1985).

Figure 5
Toyota diesel engine piston. The metal-matrix composite insert can be seen as the darker region incorporating the upper piston ring groove

2. *Metal-Matrix Composites*

Among the attributes of fiber or whisker reinforced MMCs, in addition to improved strength and stiffness, are increased creep and fatigue resistance, giving rise to higher operating temperatures than those of the unreinforced metal (Feest 1986). The presence of ceramic reinforcements, whether in fibrous or particulate form, can also provide increased hardness, wear and abrasion resistance. These properties are potentially exploitable in many aspects of pump and engine technology including compressor bodies, vanes and rotors, piston sleeves and inserts, connecting rods, cylinder heads and clutch components. Many such components are under development (Feest et al. 1986) but few have yet entered commercial production. One exception is the use by the Toyota Motor Corporation (Donomoto et al. 1983) of aluminum-alloy diesel engine pistons incorporating a ceramic-fiber-reinforced top land and piston ring groove (Fig. 5). Honda have also engaged in a limited production run of aluminum-alloy connecting rods reinforced with stainless steel wire for their 1.2-liter City vehicle in Japan. On a much smaller scale of application, boron-fiber-reinforced aluminum-alloy struts are used in the midsection framework of the US space shuttle fuselages (Irving 1983). An application that exploits the thermal properties of aluminum reinforced with silicon carbide particles is the packaging of microwave circuits (Thaw et al. 1987). In this respect, the presence of the silicon carbide reduces the thermal expansion coefficient of the packaging material, providing a better match to the ceramic circuit substrate without significantly decreasing the high thermal conductivity required for good heat dissipation from the device.

3. *Ceramic-Matrix Composites*

The present picture for CMCs is similar in many ways to that of MMCs in that considerable research and development activity exists (Phillips 1985) but there are few established applications. Potential applications for CMCs include high-temperature seals, bearings, blades and rotors, particularly in gas turbine engines, piston crowns in reciprocating engines, and wear-resistant duties such as valve bodies, cutting and forming tools. In the latter context, can-forming punches made from alumina reinforced with silicon carbide whiskers are reported to show better performance than tungsten carbide punches in machine that produce 400 cans per minute round the clock (Anon 1988).

The availability of strong, tough, lightweight materials with good thermal shock resistance, capable of continuous operation in air at temperatures above 1200 °C, would make possible significant increases in the efficiency of gas turbine engines. Phillips (1987) and Jamet (1987) have reviewed the prospects of developing ceramic-matrix composites to meet this type of requirement. Existing ceramic reinforcements suffer degradation in strength or undergo creep in the range 1000–1200 °C and the only material available in the late 1980s with attractive mechanical properties above 1200 °C is carbon-fiber-reinforced carbon, operating in a nonoxidizing atmosphere. (Hill et al. 1974). Carbon–carbon materials are currently used for ablative purposes on those surfaces of spacecraft and reentry vehicles subject to severe aerodynamic heating, and considerable research is being devoted to identifying ways of increasing oxidation resistance, particularly for reusable space vehicles.

At lower temperatures and for short periods of time at temperature, carbon–carbon materials have proved suitable in disk brakes for aircraft and racing cars, where sufficiently large amounts of energy are generated in emergency braking to cause conventional steel disk brakes to fade. For a review of carbon–carbon technology, see Fitzer (1987).

See also: Aircraft and Aerospace Applications of Composites; Automobile Tires; Automotive Components: Fabrication; Helicopter Applications of Composites

Bibliography

Anderson L I, Lilholt H, Pederson O B (eds.) 1988 Mechanical and physical behaviour of metallic and ceramic composites. *Proc. 9th Risø Int. Symp. on Metallurgy and Materials Science* Risø National Laboratory, Denmark

Anon 1980 GRP for rolling stock on European railways. *Vetrotex Fibreworld* 9: 11–18

Anon 1986 A320—Fly-by-wire airliner. *Flight Int.* 130(4026): 86–94

Anon 1987a Boeing 360—Helicopter hi-tech. *Flight Int.* 131(4058): 22–27

Anon 1987b Growth of composites in military naval construction: Slowly but surely. *Vetrotex Fibreworld* 24: 6–7

Anon 1988 ACMC's ceramic composites—Tooling up for a first in cans. *Mater. Edge* 7:10

Brehm B, Sprenger K H 1985 Modern fibre reinforced materials. *Sprechsaal* 118 (3): 253

Buisson M 1983 SMC and the car. *Vetrotex Fibreworld* 17:10–12

Burke W R (ed.) 1986 Composites design for space applications. *Proc. Workshop ESA SP*-243. European Space Agency Publications Division, Noordwijk, The Netherlands

Chevalier A 1987 Glass/resin composite market. *Vetrotex Fibreworld* 24:8

Clarke G P 1985 The use of composite materials in racing car design. *Proc. 3rd Int. Conf. on Carbon Fibres—Uses and Prospects*, Paper 18. Plastics and Rubber Institute, London

Clementson A 1985 Materials and manufacturing in aerospace. *Proc. 2nd Conf. on Materials Engineering.* Institute of Mechanical Engineers, London, pp. 189–94

Dixon R H, Ramsey B W, Usher P J 1973 Design and build of the GRP hull of HMS Wilton. *Proc. Symp. on GRP Ship Construction.* Royal Institute of Naval Architects, London. pp. 1–32

Donomoto T, Funatani K, Miura N, Miyake N 1983 *Ceramic Fiber Reinforced Piston for High Performance Diesel Engines*, SAE technical paper series no. 830252. Society of Automotive Engineers, Warrendale, Pennsylvania

Feest E A 1986 Metal matrix composites for industrial application. *Mater. Des.* 7(2):58–64

Feest E A, Ball M J, Begg A R, Biggs D A 1986 *Metal Matrix Composites Development in Japan*, Report on OSTEM visit to Japan, Oct. 1986. Harwell Laboratory, Didcot, UK

Ferrarini L J, Spence D H, Walker M G 1984 Broadening the limits of reinforced polyurethane RIM. *Proc 1st Int. Conf. on Fibre Reinforced Composites*, Paper 25. Plastics and Rubber Institute, London

Fitzer E 1987 The future of carbon–carbon composites. *Carbon* 25 (2):163–90

Goddard P N 1985 The use of new materials in helicopter load bearing structures—Present trends and future predictions. *Proc. 2nd Conf. on Materials Engineering.* Institute of Mechanical Engineers, London, pp. 243–51

Gotch T M, Plowman P E R 1978 Improved production processes for manufacture of GRP on British Rail. *Proc. Reinforced Plastics Congr.* Paper 4. British Plastics Federation, London

Hadcock R N 1986 Design of advanced composite aircraft structures. In: Dorgham M A (ed.) 1986 *Designing with Plastics and Advanced Plastic Composites. Proc. Int. Association for Vehicle Design.* Interscience, Geneva

Hancox N L, Phillips D C 1985 Fibre composites for intermediate and high temperature applications. *Proc. 2nd Conf. on Materials Engineering.* Institute of Mechanical Engineers, London, pp. 139–44

Hill J, Thomas C R, Walker E J 1974 Advanced carbon–carbon composites for structural applications. *Proc. Conf. on Carbon Fibers, Their Place in Modern Technology*, Paper 19. Plastics Institute, London

Irving R R 1983 Metal matrix composites pose a big challenge to conventional alloys. *Iron Age* 226(2):35–39

Jaafari A, Holloway L, Burstell M L 1976 Analysis of the use of glass reinforced plastics in the construction industry. *Proc. Reinforced Plastics Congress*, Paper 27. British Plastics Federation, London

Jamet J F 1987 *Ceramic—Ceramic Composites for Use at High Temperature, New Materials and Their Applications*, Inst. Phys. Conf. Ser. No. 89. Institute of Physics, Bristol, pp. 63–75

Jardon A, Costes M 1987 Mass production composites, *Proc. 6th Int. Conf. on Composite Materials*, Vol. 1. Elsevier, London, pp. 1. 1–1.4

Lea M, Dimmock J 1984 The development of leaf springs for commercial vehicle applications. *Proc. 1st Int. Conf. on Fibre Reinforced Composites*, Paper 10. Plastics and Rubber Institute, London

Lubin G, Donohue P 1980 Real life ageing properties of composites. *Proc. 35th Annu. Tech. Conf. New Orleans*, Paper 17-E. Society for Plastics Industry, New York

McCarthy R 1986 *Manufacture of Composite Propellor Blades for Commuter Aircraft*, SAE Trans. 850875. Society of Automotive Engineers, Warrendale, Pennsylvania, pp. 4. 606–4.613

Noyes J V 1983 Composites in the construction of the Lear Fan 2100 aircraft. *Composites* 14(2):129–39

Oliver P C 1980 Applications for composites in the offshore environment. *Proc. 3rd Int. Conf. on Composite Materials.* Pergamon, Oxford, p. 2395

Phillips D C 1985 Fibre reinforced ceramics. In: Davidge R W (ed.) *Ceramic Composites for High Temperature Engineering Applications.* CEC Publication EUR 9565EN. Commission of the European Communities, Luxembourg

Phillips D C 1987 High temperature fibre composites. In: Matthews F L, Buskell N C R, Hodgkinson J M, Morton J (eds.) 1987 *Proc. 6th Int. Conf. on Composite Materials*, Vol. 2. Elsevier Applied Science, London, pp. 1–32

Pinzelli R 1983 Concept, benefits and applications of aramid fiber in hybrid composites. *Proc. 4th Int. Conf. SAMPE European Chapter.* Comité Pour la Développement des Matériaux Composites, Bordeaux, pp. 63–77

Preedy J, Kelly J F 1982 An application for Mathweb structures—Development of a temporary overhead electrical gantry by British Rail. *Proc. Reinforced Plastics Congress*, Paper 23. British Plastics Federation, London
Riebaldi G G 1985 Dimensional stability of CFRP tubes for space structures. *Proc. Workshop on Composites Design for Space applications*, ESA-SP243. European Space Agency Publications Division, Noordwijk, The Netherlands
Riddell J C 1985 Composite materials and the sailplane market. *Proc. 2nd Conf. on Materials Engineering*. Institute of Mechanical Engineers, London, pp. 253–58
Riley B L 1986 AV-8B/GR Mk. 5 airframe composite applications. *Proc. Inst. Mech. Eng.* 200(50):1–17
Ruhmann D C, Mundlock J D, Britton R A 1982 Glass-hopper—The fiberglass reinforced polyester covered hopper car. *Proc. 37th Annu. Conf. SPI*, Paper 4-E. Society for Plastics Industry, New York
Seamark M J 1981 Facelifting the world's first all-SMC clad truck cab after 5 years—A unique case study. *Proc 36th Annu. Conf. SPI*, Paper 11-C. Society for Plastics Industry, New York
Seymour M W, Tompkins D D 1975 Design approaches to the fibrous glass reinforced polyester bathroom as related to market needs. *Proc. 30th Ann. Tech. Conf. SPI*, Paper 3-A. Society for Plastics Industry, New York
Suthurst G D, Rowbotham E M 1980 Achieving the impossible—Plastic intake manifold. *Proc. Reinforced Plastics Congress*, Paper 33. British Plastics Federation, London
Thaw C, Minet R, Zemany J, Zweben C 1987 Metal matrix composite microwave packaging components. *SAMPE J.* 23(6):40–43
Waterman N A 1986 The economic case for plastics. In: Dorgham M A (ed.) 1986 *Designing with Plastics and Advanced Plastic Composites*, Proc. Int. Assn. for Vehicle Design. Interscience, Geneva, pp. 1–15
Wood A S 1986 The majors are taking over in advanced composites. *Mod. Plast. Int.* 16(4):40–43
Zweben C 1981 Advanced composites for aerospace applications. *Composites* 12(4):235–40

D. H. Bowen
[AEA Technology, Didcot, UK]

Artificial Bone

Synthetic materials have been applied for some years as replacements for natural bone in orthopedic surgery, with outstanding clinical success achieved in operations ranging from bone grafting to the total replacement of arthrotic joints, using prostheses based on conventional engineering materials, such as stainless steel, cobalt–chromium alloys, titanium alloys, alumina and polyethylene (Bonfield 1987a). However, it has become apparent that the lifetime of these procedures is often limited, not by a failure of the implant materials, but by a progressive loss of bone mass from the surrounding tissue leading to loosening of the prostheses. This effect is a consequence of the continual remodelling of bone in response to physiological loading. If such loading is altered irreversibly from the normal levels required for bone to remain in an equilibrium state by the introduction of a different material and/or structure, then bone resorption will occur. The kinetics of this process depend on the age and activity: it is sufficiently slow to allow 12–15 years lifetime for a total hip arthroplasty in a patient over the age of 65, but too rapid to permit an identical operation in a patient under the age of 45. The recognition of this circuit between the implant material and the natural tissue provided the stimulus for the development of artificial bone, i.e., a synthetic analogue of bone which is both biocompatible (i.e., does not produce an adverse tissue reaction) and mechanically compatible (i.e., has similar deformation behavior) and hence, in association with the natural tissue, produces bone stability or augmentation rather than bone resorption. With such a material, the prospect of prostheses with an infinite lifetime becomes attainable.

1. Structure of Natural Bone

The starting point for the development of a bone analogue material is a definition of the mechanical behavior of the natural tissue in the context of its complex structure, in which several levels of organization from macroscale to microscale can be identified (Bonfield 1984). In general, the major support bones consist of an outer load-bearing shell of cortical (or compact) bone with a medullary cavity containing cancellous (trabecular or spongy) bone towards the bone ends. Cortical bone itself is anisotropic with osteons (about 100–300 μm in diameter) in a preferred orientation parallel to the long axis of the bone, interspersed in regions of nonoriented bone. Each osteon has a central Haversian canal ($\sim$ 20–40 μm in diameter) containing a blood vessel, which supplies the elements required in bone remodelling. The osteons are composed of concentric lamellae ($\sim$ 5 μm in thickness) consisting of two major components, collagen and hydroxyapatite ($(Ca_{10}(PO_4)_6OH)_2$). In mature bone, hydroxyapatite occupies $\sim$0.5 volume fraction, is mainly crystalline (hexagonal unit cell) with a rod-like (or sometimes plate-like) habit and dimensions of $\sim 5 \times 5 \times 50$ nm. The hydroxyapatite crystal lattice has recently been imaged in situ in bone with high resolution electron microscopy, and dislocations resolved. The precise microstructural organization varies between different bones, between different locations in the same bone and as a function of age, with the result that cortical bone has a range of associated properties rather than an unique set of values. In materials science terms, two levels of composite structure can be identified: first, the hydroxyapatite-reinforced collagen composite making up individual lamella (on the nanometer to micrometer scale), and second, osteon (considered as hollow hexagons) reinforced interstitial bone (on the micrometer to millimeter scale). Indeed, composite structures are a feature of biological materials in general as is discussed in the article *Natural Composites*.

2. Mechanical Properties of Cortical Bone and Comparisons with Current Implant Materials

An assessment of the mechanical behavior of bone may be made on whole bones *in vivo* (but because of their irregular shape it is difficult to interpret) or on sections prepared with careful machining into standard specimens, such as those used for conventional materials testing, which are tested *in vitro*. The conditions required to prepare and test dead bone specimens so as to give meaningful data representative of living bone are now well established and it is particularly important to maintain the water content of the bone. A standard tensile test of wet cortical bone at ambient temperature at a quasistatic strain rate gives an almost linear stress–strain curve, with a small viscoelastic component, and culminates in brittle fracture at a total strain of only 0.5–3.0%. This mechanical behavior is as expected from a composite of a linear elastic ceramic reinforcement (hydroxyapatite) and a compliant, ductile polymer matrix (collagen). The values of stiffness of cortical bone, as represented by the Young's modulus E, range between 7 and 30 GPa (depending on orientation, location and age). These values demonstrate that bone is significantly less stiff than the various alloys and ceramics currently utilized as prosthetic materials, but is stiffer than polymer implants, as shown in Table 1. The actual Young's modulus values for bone can be modelled to within a factor of 2 by a rule-of-mixtures calculation on the basis of a 0.5 volume fraction hydroxyapatite-reinforced collagen composite.

As the fracture of cortical bone occurs in a brittle mode, the values determined for ultimate tensile strength ($\sim$50–150 MPa) are less critical than a measure of its fracture toughness, which has recently been established, with a critical stress intensity factor (K_{IC}) of $\sim$2–12 MN m$^{-3/2}$ and a critical strain energy release rate (G_{IC}) of $\sim$500–6000 J m^{-2} (Bonfield 1987b). These values of fracture toughness for bone are considerably lower than those of the alloy implant materials, somewhat lower than that of polyethylene and marginally superior to those of alumina and poly (methyl methacrylate) (PMMA) bone cement, as shown in Table 1.

Such a comparison of stiffness and fracture toughness allows a definition of the approximate mechanical compatibility required of artificial bone in an exact structural replacement of bone or to stabilize a bone–implant interface. (A precise matching requires a comparison of all the elastic stiffness coefficients.)

3. Development of a Cortical Bone Analogue

From Table 1 it can be seen that a possible approach to the development of a mechanically compatible artificial bone material is by stiffening one of the polymers already used as an implant (i.e., polyethylene or PMMA bone cement) with a suitable second-phase addition. The general effect has been well demonstrated for polymers in that a progressive increase in the volume fraction of the second phase will correspondingly increase the Young's modulus but decrease the fracture toughness. Hence, starting with a ductile polymer, an eventual transition into brittle behavior is observed. As PMMA is brittle, while polyethylene is ductile, with an elongation to fracture of $\sim$100%, polyethylene provides the more suitable starting matrix material and has comparable mechanical behavior to collagen, the natural bone matrix. Various materials could be considered as additions to polyethylene which would stiffen the material at the expense of fracture toughness, but the selection of hydroxyapatite (Bonfield et al. 1981, 1984a) which is the

Table 1
Comparison of some mechanical properties of current implant materials with those of cortical bone

Materials	Young's modulus (GPa)	Ultimate tensile strength (MPa)	Critical stress intensity factor K_{IC} (MN m$^{-3/2}$)	Critical strain energy release rate G_{IC} (J m^{-2})
Alumina	365	6–55	$\sim$3	$\sim$40
Cobalt–chromium alloys	230	900–1540	$\sim$100	$\sim$50000
Austenitic stainless steel	200	540–1000	$\sim$100	$\sim$50000
Ti–6Wt% Al–4Wt%V	106	900	$\sim$80	$\sim$60000
Cortical bone	7–30	50–150	2–12	$\sim$600–5000
PMMA bone cement	3.5	70	1.5	$\sim$400
Polyethylene	1	30		$\sim$8000

reinforcing constituent of natural bone and hence biocompatible, proved particularly appropriate in the development of artificial bone. Hydroxyapatite, in the form of calcined bone, is in fact already a significant tonnage engineering material as a constituent of bone china and alternatively can be readily prepared in synthetic form in the laboratory.

Starting with polyethylene granules (of average molecular weight of ~400 000) and hydroxyapatite particles (~0.5–20 μm), composites of hydroxyapatite-reinforced polyethylene are prepared to various volume fractions by a compounding and molding technique designed to minimize polymer degradation (Abram et al. 1984). As no coupling agents are used, all the components of the composite are biocompatible. The composite artifacts produced are pore free with a homogeneous distribution of hydroxyapatite particles, with various volume fractions from 0.1 to 0.5. It was demonstrated (Bonfield et al. 1984b) that the Young's modulus, as predicted, increases progressively with an increase in volume fraction from a starting value of ~1 GPa to ~9 GPa at the 0.5 level. This increase in modulus is accompanied by a decrease in percentage elongation-to-fracture. However, the composite remains unusually ductile, and hence fracture tough, up to a hydroxyapatite volume fraction of 0.4, when the elongation to fracture is still ~30% (and hence fracture mechanics analysis is inappropriate). From 0.4 to 0.5 volume fracture, the elongation to fracture decreases markedly, with brittle fracture observed at the 0.45 and 0.5 volume fractions. At these volume fractions, values of K_{IC} of ~3–8 $MN\,m^{-3/2}$ were established.

Hence in terms of cortical bone replacement, a polyethylene–hydroxyapatite composite with 0.5 volume fraction hydroxyapatite has a Young's modulus within the lower band of the values for bone and a comparable fracture toughness while a polyethylene–hydroxyapatite composite with 0.4 volume fraction hydroxyapatite is less stiff ($E \sim 5$ GPa), but has a superior fracture toughness. Composites with both volume fractions, or intermediate levels, would have applications as artificial bone, depending on the nature of the bone being replaced and the applied physiological loading.

For application as a prosthetic material it is also essential to demonstrate that the hydroxyapatite-reinforced polyethylene composites are biocompatible with bone. Following sterilization by γ irradiation, a satisfactory zero cytotoxic response of fibroblast cells placed on the composite surface in a cell culture was demonstrated. In addition the presence of bone-making cells (osteoblasts) in the cell culture was found to promote osteoid (the precursor of bone) formation at the composite surface, which indicates a favorable response of the natural bone to hydroxyapatite-reinforced polyethylene. This result was confirmed *in vivo* by implantation of sterilized machined composite pins, 2.5 mm in diameter, in the lateral femoral condyle of adult rabbits, for periods up to 6 months. It was established by histological sectioning and microprobe analysis, for both 0.4 and 0.5 hydroxyapatite volume fraction composites, that an initial fibrous encapsulation was succeeded by localized bone apposition at the implant surface to create a secure bond between the natural bone and the implant (Bonfield et al. 1986). As the limb was not immobilized, this favorable bone response occurred during physiological loading and demonstrates the absence of significant relative movement at the bone–implant interface resulting from the mechanical compatibility of the natural and artificial bone. The stable interface and the presence of hydroxyapatite provide the necessary factors for a favorable condition of bioactivity in which bone growth around the implant is encouraged.

Further refinements to the hydroxyapatite-reinforced polyethylene composite are in progress, particularly with respect to modifying the scale and shape of the synthetic hydroxyapatite particles to match natural hydroxyapatite more closely. However, the concept of artificial bone based on hydroxyapatite-reinforced polyethylene is now well-established, clinical trials are in progress and there is every prospect that this composite will contribute significantly to advances in orthopedic practice.

Bibliography

Abram J, Bowman J, Behiri JC, Bonfield W 1984 The influence of compounding route on the mechanical properties of highly loaded particulate filled polyethylene composites. *Plast. Rubber Process. Appl.* 4: 261–69

Bonfield W 1984 Elasticity and viscoelasticity of cortical bone. In: Hastings DW, Ducheyne P (eds.) 1984 *Natural and Living Materials.* CRC Press, Boca Raton, Florida, pp. 43–60

Bonfield W 1987a Materials for the replacement of osteoarthrotic hip joints. *Met. Mater.* 3: 712–16

Bonfield W 1987b Advances in the fracture mechanics of cortical bone. *J. Biomech.* 20: 1071–81

Bonfield W, Behiri JC, Doyle C, Bowman J, Abram J 1984 Hydroxyapatite reinforced polyethylene composites for bone replacement. In: Ducheyne P, van der Perre G, Aubert AE (eds.) 1984 *Biomaterials and Biomechanics* 1983. Elsevier, Amsterdam, pp. 421–26

Bonfield W, Doyle C, Tanner KE 1986 In vivo evaluation of hydroxyapatite reinforced polyethylene composites. In: Christel P, Meunier A, Lee AJC (eds.) 1986 *Biological and Biomechanical Performance of Biomaterials.* Elsevier, Amsterdam, pp. 153–58

Bonfield W, Grynpas MD, Bowman J 1984 A prosthesis comprising composite material. UK Patent GB 2,085,461B

Bonfield W, Grynpas MD, Tully AE, Bowman J, Abram J 1981 Hydroxyapatite reinforced polyethylene—A mechanically compatible implant material for bone replacement. *Biomaterials* 2: 185–86

W. Bonfield
[Queen Mary and Westfield College,
London, UK]

Asbestos Fibers

Asbestos is a generic name applied to certain specific minerals which occur in nature in fibrous form. The common features which make these minerals unique are their flexibility, high tensile strength and value as reinforcing fibers in composite materials.

1. Types of Asbestos

Chrysotile is the most abundant type of asbestos and accounts for 95% of total world availability of asbestos fiber. It is one of the serpentine group of minerals, and the only one of such to occur in fibrous form.

The other types of asbestos belong to the amphibole group of minerals which occur widely in many igneous and metamorphic rocks. Asbestiform amphiboles, however, have a restricted occurrence. The principal commercial varieties, crocidolite and amosite, are mined only in the Precambrian ironstone formations in South Africa.

Three other asbestiform amphiboles—anthophyllite, tremolite and actinolite—have little commercial importance nowadays.

World production of asbestos by country is given in Table 1; the total may be a slight overestimate. US Bureau of Mines condensed data released in 1987 indicates an estimated annual world production of 4.1 Mt for 1985 and 1986. Various other sources suggest that output from the USSR will continue to exceed the deficits suffered by Canada, the USA and Southern Africa. Forecasts for an increase in asbestos consumption over the next decade range from 3 to 5% per annum.

Table 1
Asbestos: world production 1984

Country	Production (t)
Australia	10 000
Brazil	160 000
Canada (shipments)	922 000
China	160 000
Cyprus	16 000
Greece	110 000
India	25 000
Indonesia	25 000
Italy	140 000
Korea	15 000
South Africa	170 000
Swaziland	30 000
USSR	2 300 000
USA	57 422
Yugoslavia	10 400
Zimbabwe	165 000
Others, including Argentina, Bulgaria, Colombia, Egypt, Japan, Mozambique, Taiwan, Turkey	22 475
Total (approx)	4 338 000

Source: R A Clifton, *Asbestos*, US Bureau of Mines Minerals Yearbook 1984

Some years ago, it was estimated that there were 3000 applications for asbestos fibers, but environmental restrictions in many countries have severely reduced the range of use mainly to those products where the fiber is said to be "locked in." About 75% of all asbestos produced goes into asbestos-cement products and the remainder is spread over such applications as friction materials, papers and felts, textiles, vinyl floor tiles and asphaltic coatings. Despite the loss of many applications, not least among these a wide variety of insulation materials, the overall consumption of asbestos has diminished only by a limited extent, principally because of the increase in recent years of the world capacity for asbestos-cement manufacture.

2. Properties of Asbestos

The structure, composition, chemistry and form of chrysotile on the one hand and amphibole fibers on the other differ considerably. Only in terms of physical properties are there close similarities between these important varieties of asbestos fiber.

Chrysotile is a hydrated magnesium silicate. It has a layered lattice structure, consisting of a sheet of Mg–OH units superimposed on a sheet of Si_2O_5 units. These layers are curved in one direction, and in chrysotile they form scrolls or concentric tubes of up to 20 turns. A chrysotile fiber consists of a number of such fibrils in parallel orientation, together with interstitial material of similar composition. The fibers possess characteristic flexibility.

The basic structure of the fibrous amphiboles consists of two parallel chains of Si_4O_{11} units linked by a ribbon of cations of Fe^{3+}, Fe^{2+}, Mg, Ca and Na, the distribution of which determines the type of amphibole. These fibers may be likened to a uniform stack of laths of the basic structure which are cross-linked to each other.

Thus, in contrast to chrysotile asbestos, amphibole fibers are considerably less flexible and are generally referred to as harsh fibers.

In the period 1970 to 1982 investigative work on the microstructures of asbestos fibers at levels of atomic resolution enabled knowledge of chrysotile structures to be expanded considerably, and revealed in the amphibole fibers the nature of fibrosity. The collective term, biopyribole, has been reintroduced to explain structural relations and alteration mechanisms between sheet silicates, pyroxenes and amphiboles, notably between talc, chrysotile of vermiculite and tremolite or anthophyllite (Hodgson 1986).

The idealized chemical formulae of the three main types of asbestos are given in Table 2. Chrysotile has a predominantly basic composition, whereas the composition of the amphiboles is balanced in favor of the acidic (Si_4O_{11}) component. This has an important bearing on the chemical properties of asbestos. Whereas the amphibole varieties show considerable resistance to decomposition by acids, chrysotile is easily decomposed by loss of its MgO and OH components (Table 3). Indeed, part of the Mg^{2+} content of chrysotile can be leached out by water alone.

Degradation by heating is also distinctive for the different types of asbestos. Chrysotile loses water gradually from 100°C upwards until at about 650 °C all 13% of its water content is lost. At about 800 °C, the residual material recrystallizes to forsterite and free silica. All the amphiboles lose up to 2% of water but at widely varying temperatures. The ferrous iron content of crocidolite oxidizes at 400 °C, that of amosite at about 650 °C, and these temperatures are coincident with dehydroxylation. Ultimate decomposition occurs at 900 °C for crocidolite and 600 °C for amosite.

Table 3 gives the main physical properties of asbestos insofar as they are relevant to composite materials. Fiber strength and modulus are high and are substantially greater than most comparable man-made fibers. The information on specific gravity and magnetic susceptibility is important for specific applications in fiber-reinforced resin systems. Table 3 also shows the percentage loss in tensile strength of asbestos fiber as it is heated up to 600 °C. Chrysotile resists changes in tensile strength as temperature increases and even gains at 500 °C, because of slight dimensional shrinkage. Thereafter, it shows catastrophic loss in strength. Crocidolite holds up well to 300 °C but both crocidolite and amosite lose a substantial proportion of strength at and above 400 °C.

Table 2
Idealized chemical formulae of the main types of asbestos

Name	Formula
Chrysotile	$Mg_3Si_2O_5\,(OH)_4$
Crocidolite	$Na_2Fe_2^{3+}(Fe^{2+}Mg)_3(Si_4O_{11})_2(OH)_4$
Amosite	$(Fe^{2+}Mg)_7(Si_4O_{11})_2(OH)_4$

3. Reinforcing Properties of Asbestos

It is both a theoretical and observed fact that the best dimensional form of asbestos fibers in reinforcement terms corresponds to a critical length between 1 and 2 mm and an aspect ratio between 150 and 250. Asbestos fiber of a quality within these narrow limits could give reinforcing strengths in asbestos cement, for example, some four to five times greater than actually observed for this product.

In commercially produced asbestos there is, however, a very wide spectrum of fiber lengths and diameters and it is generally the preponderance of length within certain broad limits which determines the grade of asbestos. Observations of fracture failure in asbestos-reinforced composite materials indicate that long, thick fibers, and short fibers below the critical length, pull out of the matrix whereas all thin fibers, as long as they exceed critical length, are most likely to break at the plane of fracture. Since the tensile properties of the fiber contribute more than do pull out properties to the reinforcement strength of a composite material, the

Table 3
Some physical and chemical properties of asbestos

Property	Chrysotile	Crocidolite	Amosite
Tensile strength[a] ($MN\,m^{-2}$)	3100	3500	2500
Elastic modulus[a] ($GN\,m^{-2}$)	160	190	160
Specific gravity	2.55	3.43	3.37
Magnetic susceptibility (mean χ at 10 kOe)	5.3×10^{-6}	78.7×10^{-6}	60.9×10^{-6}
% loss in tensile strength[a]			
at 300 °C	0	13	37
400 °C	(2.7)[b]	63	61
500 °C	(13.5)[b]	78	80
600 °C	84	83	96
% weight loss by acid decomposition[c] in period			
0.5 h	60	6	8
2 h		7	15
4 h		7.5	22
8 h		8.5	30

a Measurements made on micro fibers at 4 mm length between jaws b Gain in tensile strength c Under conditions of boiling refluxed 4 M hydrochloric acid aim in preparing asbestos fiber for use in many applications is to improve the fineness of the fiber without appreciable loss of length.

A variety of both wet- and dry-milling methods are used in the preparation of asbestos fiber and their effectiveness may well vary with the type of fiber being processed. Chrysotiles in general withstand vigorous milling but amosite requires careful preparation to avoid length reduction. Crocidolite asbestos responds in a way intermediate between chrysotile and amosite.

The most important tests for asbestos fiber concern length, diameter (as expressed by apparent surface area) and reinforcement properties. Some abridged but typical properties of asbestos fiber used in asbestos cement, and based upon the above tests, are given in Table 4.

4. *Applications of Asbestos*

The manufacture of asbestos-cement products accounts for an annual turnover of some 3 million tonnes of asbestos. Such products contain 10–17% of asbestos, mainly chrysotile. Chrysotile gives plasticity to a wet asbestos-cement system, essential to product molding. Crocidolite and amosite provide ease of drainage in the manufacturing process, and crocidolite, in conjunction with medium-length chrysotiles, ensures the required circumferential strength in pressure pipes.

Asbestos is compatible with calcium silicate matrices formed by reaction between lime and silica in high-pressure steam autoclaves. The asbestos is unaffected by the conditions of the reaction. Amosite asbestos is used in such a matrix in the manufacture of fireproof insulation boards, although this industry has diminished in favor of products based on mineral wools and cellulose.

The use of asbestos in components based on reinforced plastics is small, because of the preference for glass-reinforced plastic products. Although the strongly polar nature of chrysotile interferes to some extent with its bonding with resins, this type of asbestos is a crucial requirement in friction materials which are basically asbestos-reinforced phenolic resins, with modifiers to control friction and wear characteristics. Both types of amphibole asbestos have been used to advantage in reinforced thermosetting and thermoplastic resins, but in composites embodying thermoplastics it is necessary to use complex stabilizing systems to prevent higher-temperature depolymerization. The ferrous iron content of the asbestos is the active agent in depolymerization.

Vinyl floor tiles account for the use of a limited quantity of the less expensive short-fiber chrysotiles. The fiber is an extender in the system and also provides a measure of resilience to the product. Chrysotile asbestos is also formed into papers and felts which may be used in flooring materials and in the manufacture of electrical components based on synthetic resins.

Asbestos textiles based on chrysotile continue to be in demand, particularly for fire protection purposes and in making heavy-duty brake linings. Modern wet yarn-drawing processes have partly replaced conventional carding and spinning methods in this industry. Some asbestos cloth, as well as ordinary chrysotile fiber, is widely used in the manufacture of heat resistant glands, packings and seals.

5. *Health and Environmental Hazards Associated with Asbestos*

In the course of processing asbestos at any stage from mining and milling through to industrial processes, exceedingly fine fibers are generated to form airborne dust clouds of concentrations up to 1000 fibers per cm^3 of air. Inhalation of asbestos dust is a hazard to health and measures of environmental control are necessary to ensure minimum risk to those who work with asbestos.

Safety in the asbestos industry nowadays is governed by control measures, coupled with current dust-control technology, and by an international policy of information and awareness concerning the hazards.

Table 4
Typical properties of asbestos fiber used in asbestos-cement manufacture

	Chrysotile (medium length)	Chrysotile (short)	Crocidolite (medium length)	Amosite (medium length)
Length distribution fraction (%)				
> 5 mm	10	1	40	15
5–1 mm	20	4	20	20
1–0.1 mm	30	35	5	25
< 0.1 mm	40	60	35	40
Apparent surface area ($m^2\ kg^{-1}$)	1800	1800	1100	900
MR[a] ($MN\ m^{-2}$)	32	20	36	25

a Modulus of rupture of asbestos-cement test pieces with 10% fiber content and density 1.6 $g\,cm^{-3}$

Stringent regulations in many countries are designed to protect workers in the industry and to ensure that dust levels in the workplace are maintained at 0.25 fiber per ml of air. Many dusty processes have been eliminated, such as spray insulation and asbestos-based thermal insulation and fireproof boards. The uses for asbestos today are those in which the fiber is locked in and in which the manufacturing process can be kept in tight control.

At the same time, there has been considerable activity in recent years towards the development and application on alternatives to asbestos in many product lines. While there has been success in some areas, there are undeniable constraints to the use of alternative raw materials in terms of costs and availabilities, technical acceptance, and even possible new health hazards.

Bibliography

American Society for Testing and Materials *Annual Books of ASTM Standards.* ASTM, Philadelphia, Pennsylvania

Asbestos International Association *Health and Safety Publications; Proceedings of Biennial Conferences.* AIA, London

Deer W A, Howie R A, Zussman J 1962 *Rock Forming Minerals*, Vols. 2, 3. Longman, London

Hodgson A A 1987 *Alternatives to Asbestos and Asbestos Products*, 2nd ed. Anjalena, Crowthorne, UK

Hodgson A A 1986 *Scientific Advances in Asbestos 1967 to 1985.* Anjalena, Crowthorne, UK

Michaels L, Chissick S S (eds.) 1979 *Asbestos: Properties Applications and Hazards.* Wiley, Chichester

Roskill Information Services 1986 *The Economics of Asbestos*, 5th edn. Roskill Information Services, London

World Health Organisation 1986 *Environmental Health Criteria, 53: Asbestos and other Mineral Fibers.* WHO, Geneva

A. A. Hodgson
[Crowthorne, UK]

Automobile Tires

Although automobile and truck tires are manufactured in vast quantities—over 200 000 000 per year—it is not generally recognized that the tire is a highly sophisticated composite structure. Layers of rubber and rubber-impregnated sheets of parallel fiber bundles (cords) are combined to make what is basically a toroidal laminate with different properties in each of its three principal directions. Indeed, it is this carefully chosen anisotropy that enables the tire to perform diverse and often conflicting functions. Tire mechanics, outlined here, deals with the appropriate choice of materials and designs in order to achieve five main goals:

(a) high sliding friction in wet and dry conditions, over a wide range of temperatures and with varied road surfaces, to provide adequate acceleration, braking and steering;

(b) low rolling resistance under the same circumstances;

(c) a low vertical stiffness to cushion the ride;

(d) high longitudinal and lateral stiffnesses to minimize sliding motions in the contact patch (see Sect. 3); and

(e) resistance to cutting, puncturing, abrasion, etc.

How these diverse requirements are met is discussed below, after a description of the materials of which the tire is composed.

1. Tire Materials

R. W. Thomson, in 1845, was the first to employ air to provide pneumatic cushioning in a rubber tire, but the first practical use was developed in 1888 by Dr J. B. Dunlop. In modern tires, loss of air is minimized by a special rubber inner lining, usually of butyl rubber, chosen for its low permeability (Fig. 1).

The tire must be reinforced by layers of relatively inextensible cords to contain the air pressure and to restrict longitudinal and lateral deformations and growth of the tire in service. Such cord layers have been made of a variety of materials over the years: cotton, rayon, nylon, polyester, glass, polyaramide and steel, the latter two materials being the dominant ones at the present time. The choice of cord material is based on the need for high stiffness, good resistance to repeated flexing, high strength-to-weight ratio, good adhesion to rubber and low cost.

Rubber sidewalls of the tire (Fig. 1) must resist scraping, flexing, and the attack of ozone in the air. Rubber treads must primarily resist abrasive wear. Indeed, this is one of the most severe applications of

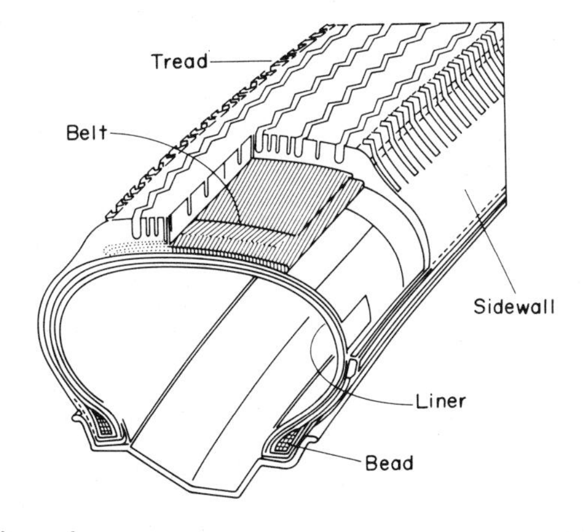

Figure 1
Structure of a radial tire

rubber, and the continuous improvement in wear resistance by improved formulations has been a remarkable technological achievement. Typical formulations for tire sidewalls and treads are given in Table 1.

It must be emphasized, however, that rubber formulation is still more art than science. For example, the reason why a tread recipe based on polybutadiene is superior in abrasion resistance to one based on polyisoprene or on a butadiene–styrene copolymer is not known. It should also be pointed out that strong rubber compounds are themselves composite in nature. Typically, about 30% by volume of a fine particulate solid filler, usually carbon black, is included in the formulation. Without this ingredient, the abrasion resistance of rubber would be much lower. Indeed, the introduction of carbon black reinforcement in rubber about 1910 was a major factor in prolonging tire life, from about 8000 km to about 16 000 km. But, again, the mechanism of reinforcement of rubber by fillers is still obscure.

2. *Tire Structure*

The most important feature of tire structures is the arrangement of the reinforcing cords, which determines the degree of anisotropy. At first they were placed to minimize distortion. For a straight tube, alternate layers would then be arranged at a cord (crown) angle to the tube axis of $\alpha = 57°$ (Fig. 2) so that the line tensions along and across the tube, given by $nt \cos^2 \alpha$ and $nt \sin^2 \alpha$, would meet the inflation requirements, $PR/2$ and PR, respectively, where P is the inflating pressure, R is the tube radius, t is the cord tension and n is the number of cords per unit of width normal to the cord direction. In the early 1950s, however, Michelin introduced the radial tire (patented by Gray and Sloper in 1913), in which one set of cords is arranged to lie almost parallel to the tire circumference, forming a circumferential belt (Fig. 1). In practice, because the tire is a toroid rather than a cylindrical tube, the belt cords are made to lie at a crown angle of about 11° rather than at 0°. This arrangement maximizes the line tension in the circumferential direction for a given value of the inflation pressure, and hence gives greater resistance to lateral distortion of the tire on cornering. This feature provides improved steering and wear resistance (see Sect. 4).

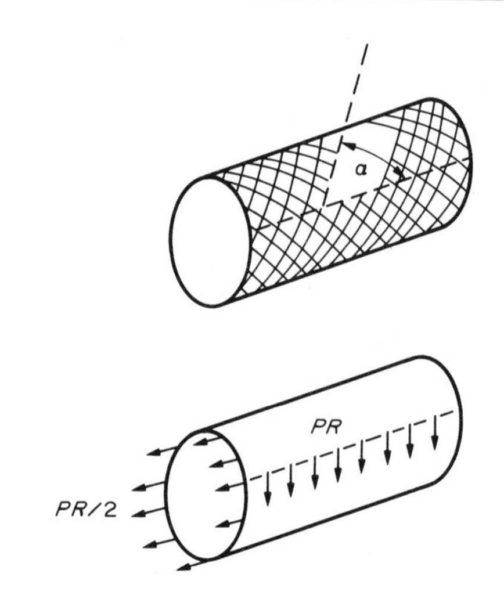

Figure 2
Cord angle α

Table 1
Representative formulations for passenger car tires in parts by weight. The mixes are converted into strong elastic solids by chemically interlinking the rubber molecules ("vulcanization"), on heating for about 20 minutes at a temperature of about 150 °C in a tire mold

Ingredient	Compound	
	Sidewall	Tread
Natural rubber	50	–
Butadiene–styrene copolymer (75/25)	–	65
Cis-1,4 polybutadiene	50	35
Heavy aromatic oil	–	25
Carbon black (HAF)	50	65
Processing oil	10	15
Zinc oxide	4	4
Stearic acid	1	1
Sulfur	2	2
Vulcanization accelerator[a]	1	1
Antioxidant[b]	3	2
Protective wax	2	2

a Example: *N*-cyclohexyl-2-benzothiazolesulfenamide b Example: *N*-(1,3 dimethylbutyl)-*N*′-phenyl-paraphenylenediamine

3. *Tire Traction*

Consider a steadily applied traction or vehicle-maneuvering force, as in acceleration, braking or cornering. The tire footprint takes up a steady-state condition, in which tread rubber, entering and advancing through the footprint, is gripped fixedly by the ground while the interior of the tire body moves or "slips" with respect to the ground. Thus the tread is progressively strained under the ground and wheel reactions. Figure 3 illustrates vehicle cornering, where the tread is displaced laterally from the plane of wheel rotation, set (by the driver) at a "slip" angle θ to the direction of vehicle motion. A point is reached, usually near the exit from the footprint, where the ground can no longer exert a sufficiently large stress to maintain the deformation because the local friction has been exceeded. The deformation of the tire then recovers by sliding of the surface over the ground, back to the plane of the wheel.

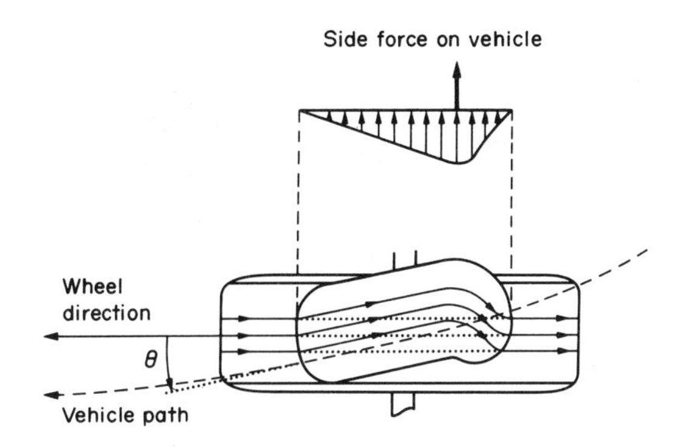

Figure 3
Tire deformation and ground stresses set up in cornering

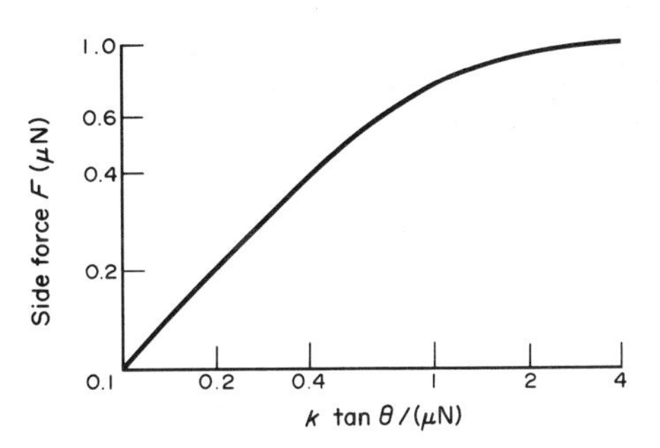

Figure 4
Side force F vs slip angle θ. N is the normal load on tire, μ the coefficient of friction, and k the lateral stiffness × the half-length of contact patch

The total lateral force exerted on the wheel is the summation of the ground stress distribution over the footprint. It varies with the slip angle θ typically as shown in Fig. 4. The initial slope is related to the lateral stiffness of the tire, and the maximum value is governed by the coefficient of friction of tread rubber on the road, reached at a slip angle where the ground stress is everywhere equal to the friction limit. Above this value, the tire skids.

An analogous situation holds under longitudinal forces of braking and acceleration where the tire slips by "windup," producing a strain gradient in the wheel plane and a longitudinal stress distribution similar to that shown in Fig. 3. Here the slip is the difference between vehicle and tire surface speeds, relative to vehicle speed. However, in braking, the tire gets so hot that the effective coefficient of friction of rubber against the road decreases, and Fig. 4 becomes modified as shown in Fig. 5. A so-called friction peak is encountered, which antiskid devices try to maintain.

The initial slopes of curves like those shown in Figs. 4 and 5 are steeper for the stiffer radial construction so that more rapid responses and higher levels of tractive force are obtained for the same steering or braking (or accelerative) inputs.

Because the resultant of lateral and longitudinal forces cannot exceed the coefficient of friction times the weight on the tire, when both act together, as in a braked turn, the resistance to skidding is diminished.

Traction is maximized by the choice of tread materials having high hysteresis or energy dissipation, which raises the coefficient of friction of rubber. Addition of carbon black and the use of high styrene–butadiene copolymers have this effect.

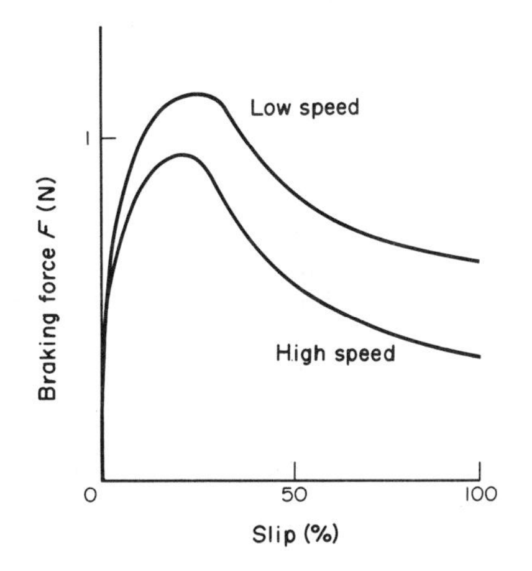

Figure 5
Braking force F vs slip (windup), expressed as the slip velocity relative to the circumferential velocity of the wheel

4. Tire Wear

Life expectancy of tires has increased to approach 150 000 km because of advances in rubber compounding and tire design. Use of improved vulcanization systems and blending of elastomers, including the abrasion resistant *cis*-polybutadiene, has greatly improved the wear resistance of the tread. The almost universal adoption of the superior radial construction is another important factor.

However, radial tires are more prone than bias tires to uneven wear, in which parts of the tread pattern tend to wear away at a greater rate than their surroundings. When such concentrated wear reaches the vicinity of the belt, the tire must be replaced even though most of the tread remains. Uneven wear is minimized by empirical design of tread elements and by ensuring that the vehicle suspension does not impose severe camber or slip angles on the tire.

Wear proceeds by the mechanism mentioned in Sect. 3. At some point in the footprint the cumulative stress on the tread reaches the friction limit and the deformed tire slides back to its undistorted shape, scrubbing the tread against the road. When wear occurs evenly over the tread face, the rate R ($m^3\ m^{-1}$) of material abraded from the tread in unit distance travelled is

$$R = \rho\gamma F^2/k$$

where ρ is the tire resilience or fraction of energy not dissipated in a loading–unloading cycle and therefore available to produce abrasion; γ is the abradability ($m^3\ J^{-1}$), defined as the amount of material abraded for a unit of energy expended in sliding; F (N) is the root mean square value of the distortion force over the tire run, arising from accelerations imposed by the road configuration, by the necessity to stop and start, by the vehicle speed and driver habit; and k is the appropriate stiffness coefficient in the ground plane, defined as the product of the half-length of the contact patch and the lateral or longitudinal force (depending on whether F refers to wear in cornering, or in braking and acceleration) for a unit of displacement of the contact patch with respect to the wheel rim.

Higher k means smaller excursions of the tread in slip and therefore shorter sliding paths and reduced wear. This explains the enhanced tread life of the stiffer radial over the softer bias construction.

Bibliography

Clark S K (ed.) 1981 *Mechanics of Pneumatic Tires*, 2nd edn. US Government Printing Office, Washington, DC

Eirich F R (ed.) 1978 *Science and Technology of Rubber*. Academic Press, New York

Hays D F, Browne A L (eds.) 1974 *The Physics of Tire Traction*. Plenum, New York

Pearson H C 1906 *Rubber Tires and All About Them*. India Rubber Publishing Company, New York

Tompkins E 1981 *The History of the Pneumatic Tire*. Lavenham Press. Lavenham, UK

A. N. Gent
[University of Akron, Akron, Ohio, USA]

D. I. Livingston
[Livingston Associates, Akron, Ohio, USA]

Automotive Components: Fabrication

There is a fundamental difference in the strategy of application of composites between the aerospace industry and the automobile industry. This is primarily due to the volume requirements of the two businesses. In aerospace and defense, the design of the structure is optimized to provide the required functionality and performance, and the manufacturing process (and associated cost) is subsequently selected on the basis that the process is capable of achieving the desired design. In direct contrast, in high-volume production industries such as the automotive industry, the rate of manufacture is critical to satisfying the economics of this consumer industry. Thus, manufacturing processes which are capable of satisfying production output are the primary consideration, and design of a component or structure must be within the boundary constraints of the selected fabrication process. It is extremely important to appreciate this philosophical difference of approach to understand the difficulty of translating the extensive design and fabrication experience in aerospace and defense into the automotive industry. Perhaps the clearest example is the use of prepreg materials and hand layup procedures, common in aerospace and amenable to optimal design, which is clearly unacceptable in an industry requiring high manufacturing output but with a quality level of the same degree.

The use of plastic-based composite materials in automotive applications has gradually evolved during the 1970s and 1980s. Virtually all uses of these materials in high-volume vehicles are limited to decorative or semistructural applications. Sheet-molding component (SMC) materials are the highest-performance composites in general automotive use today, and the most widely used SMC materials consist of approximately 25 wt% chopped glass fibers in a polyester matrix and so cannot really be classified as high-performance composites. Typically, SMC materials are used for grille opening panels on many car lines, and closures panels (hoods, deck lids and doors) on a few select models. A characteristic molding time for SMC is of the order of two minutes, which is on the borderline of viability for automotive production rates. A recent breakthrough by one manufacturer has reduced this cycle time to one minute.

The next major step for composites in the automotive business is the extension of usage into truly structural applications such as primary body structure and chassis/suspension systems. These structures have to sustain the major road load inputs and crash loads and, in addition, must deliver an acceptable level of vehicle dynamics so that passengers enjoy a comfortable ride. These functional requirements must be totally satisfied for any new material to achieve extensive application in body structures and must do so in a cost-effective manner. Appropriate composite fabrication procedures must be applied or developed which satisfy high production rates but still maintain the critical control of fiber placement and distribution.

1. Composite Materials for Automotive Usage

By far the most comprehensive property data have been developed on aerospace composites, in particular carbon-fiber-reinforced epoxy designed for fabrication by hand layup from prepreg materials. Relatively extensive databases are available on these materials and it would be very convenient to be able to build from this database for less esoteric applications such as automotive structures. Carbon fibers are the preferred material in aerospace applications because of the superior combination of stiffness, strength and fatigue resistance exhibited by these fibers. Unfortunately for cost-sensitive mass-production industries, these properties are only attained at significant expense (typically carbon fibers cost US$25 or more per pound at 1988 prices). Intensive research efforts are being devoted to reducing these costs by utilizing a pitch-based precursor for production of these fibers, but the most optimistic cost predictions are around $10 per pound, which would severely limit the potential of these fibers for use in consumer-oriented industries. To illustrate the potential that carbon fiber could offer the automobile industry if a breakthrough ever occurred in the reduction of cost, it is interesting to summarize the data on the prototype carbon fiber LTD built by Ford to directly compare with a production steel vehicle (Beardmore et al. 1980). Kulkarni and Beardmore 1980). Although the vehicle was fabricated by hand-layup procedures several interesting features were evaluated. An exploded schematic showing the composite parts of the carbon fiber composite car, made of carbon-fiber-reinforced plastic (CFRP), is shown in Fig. 1. Virtually all structures were designed using aerospace techniques, utilizing 0°/90° and ±45° prepreg materials. The weight savings for the various structures are given in Table 1. While these weight savings (of the order of 55–65%) might be considered optimal because of the use of carbon fibers, other more cost-effective fibers can achieve a major proportion of these weight savings (see Sect. 2.3). Although the CFRP vehicle weighed 1138 kg compared to a similar production steel vehicle of 1705 kg, vehicle evaluation tests indicated no perceptible differences between the vehicles. Ride quality and vehicle dynamics were judged at least equal to top-quality production steel LTD cars. Thus, on a direct comparison basis, a vehicle with a structure made entirely of fiber-reinforced plastic (FRP) was proven at least equivalent to a steel vehicle from a vehicle dynamics viewpoint, at a weight of only 67% of that of the steel vehicle.

The CFRP car clearly showed that high-cost fibers (carbon) and high-cost fabrication techniques (hand layup) can yield a perfectly acceptable vehicle based on handling, performance and vehicle dynamics criteria. However, crash and durability performance were not demonstrated and these will need serious development work. An even bigger challenge is to translate that performance into realistic economics by the use of cost-effective fibers, resins and fabrication procedures.

The fiber with the greatest potential for automobile structural applications, based on optimal combination of cost and performance, is E-glass fiber (costing

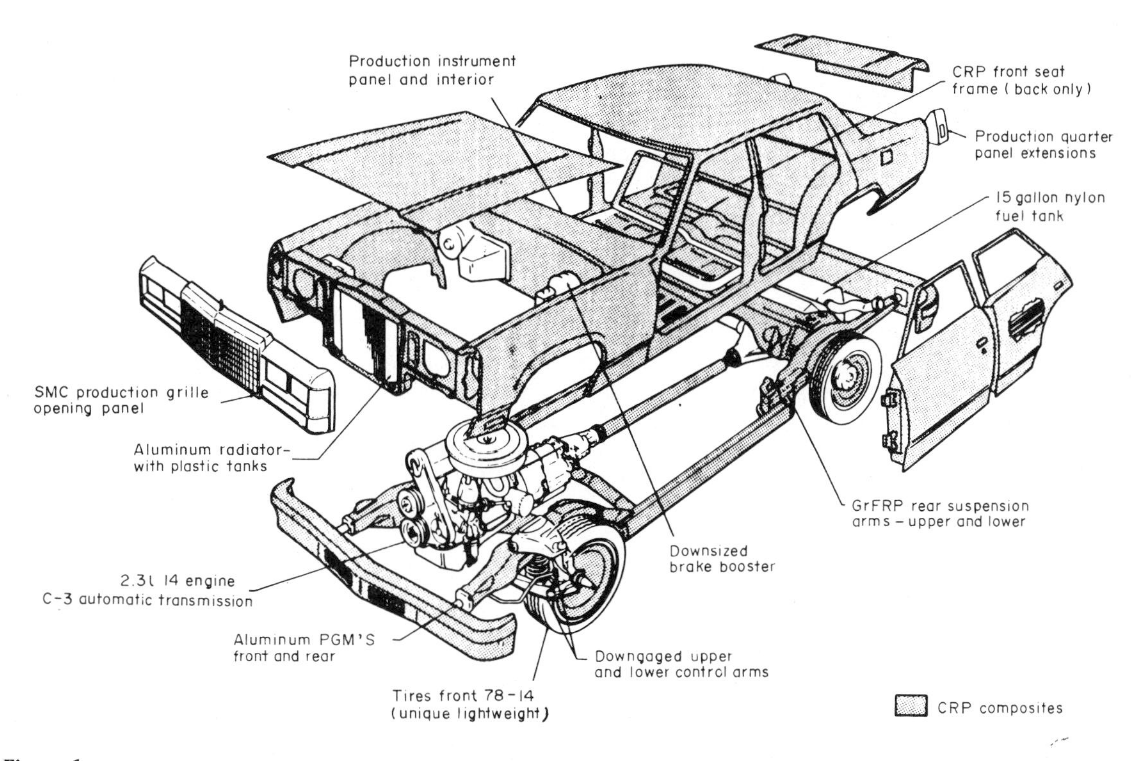

Figure 1
Exploded schematic view of a composite vehicle, with delineation of the composite components. GrFRP, graphite fiber reinforced plastic

Table 1
Major weight savings in a CFRP vehicle. Figures in brackets are weights in pounds

Component	Weight (kg)					
	Steel		CFRP		Reduction	
Body-in-white	192.3	(423.0)	72.7	(160.0)	115.0	(253.0)
Front end	43.2	(95.0)	13.6	(30.0)	29.5	(65.0)
Frame	128.6	(283.0)	93.6	(206.0)	35.0	(77.0)
Wheels (5)	41.7	(91.7)	22.3	(49.0)	19.4	(42.7)
Hood	22.3	(49.0)	7.8	(17.2)	14.7	(32.3)
Decklid	19.5	(42.8)	6.5	(14.3)	13.1	(28.9)
Doors (4)	64.1	(141.0)	25.2	(55.5)	38.9	(85.5)
Bumpers (2)	55.9	(123.0)	20.0	(44.0)	35.9	(79.0)
Driveshaft	9.6	(21.1)	6.8	(14.9)	2.8	(6.2)
Total vehicle	1705	(3750)	1138	(2504)	566	(1246)

approximately US$0.80 per pound at 1988 prices). Likewise, the resin systems likely to dominate, at least in the near term, are polyester and vinyl-ester resins based primarily on a cost–processibility trade-off. High-performance resins will find only specialized applications (in much the same way as carbon fibers) even though their ultimate properties may be somewhat superior.

The form of the glass fiber will be very application-specific and both chopped and continuous glass fibers will find extensive use. It is expected that most of the structural applications involving significant load inputs will utilize a combination of both chopped and continuous glass fiber with the particular proportions of each depending on the component or structure. Since all the fabrication processes anticipated to play a significant role in automotive production are capable of handling mixtures of continuous and chopped glass, this requirement should not present major restrictions. One potential development which is likely if glass fiber composites come to occupy a significant portion of the structural content of an automobile is the tailoring of glass fiber and corresponding specialty-resin development. Approximately 35 million vehicles per year are produced worldwide and consequently each pound of composite per vehicle implies the usage of 35 million pounds in the industry as a whole. This dictates that it should be economically feasible to have fiber and resin production tailored exclusively for the automobile industry. The advantage of such an approach is that these developments will lead to incremental improvements in specific composite materials, which in turn can promote usage.

Glass-fiber-reinforced composites must be capable of satisfying the functional requirements of the various elements of the structure in an automobile. The three primary criteria are fatigue (durability), energy absorption and ride quality (vehicle dynamics). Provided sufficient data are available on all three characteristics, it should then be possible to utilize these to satisfy the design requirements of the automobile. For example, unidirectional glass FRP materials typically have a well-defined fatigue limit of the order of 35–40% of the ultimate strength (Dharan 1975). By contrast the chopped-glass composite would have a fatigue limit closer to 25% of the ultimate strength (Smith and Owen 1969, Aotem and Hashin 1976) and would exhibit much greater scatter in properties. Clearly components and structures composed of a mixture of continuous and chopped fibers would have fatigue resistance specifically related to the proportions of each type of fiber. There is clear evidence that glass-fiber-reinforced composites can be designed to withstand the rigorous fatigue loads experienced under vehicle operating conditions. It must be emphasized, however, that fatigue design data are a critical function of the relevant manufacturing procedure.

In a similar manner, evidence is accumulating that fiber-reinforced-plastic composites can be efficient energy absorbing materials. Relative data from the collapse of tubes is given in Table 2 to illustrate this point and is based on the extensive work of Thornton (1979), Thornton and Edwards (1982) and Thornton et al. (1985). The fragmentation/fracture mechanism of energy absorption typical of glass-fiber-reinforced composites, as compared to the plastic deformation mechanism in materials, is actually very weight-effective. The real issue is the translation of this effective fracture mechanism into complex structures.

The third and less quantifiable requirement for composites in vehicle applications is vehicle dynamics. Glass-fiber-reinforced composites are inherently less stiff than steel, typically by factors of up to 10 (Table 3). However, there are two offsetting factors to compensate for these apparent deficiencies. First, an increase

Table 2
Energy absorption (typical properties)

Material	Relative energy absorption per unit weight
High-performance composites	100
Commercial composites	60–75
Mild steel	40

Table 3
Typical stiffness of composites. XMC is a sheet molding compound containing oriented continuous glass fiber; SMC-R50 is a sheet molding compound containing 50% chopped glass fiber

Material	Modulus (KPa)
Unidirectional CFRP	137.8
Unidirectional GFRP	41.3
Unidirectional Kevlar	75.8
XMC	31.0
SMC-R50	15.8
SMC-25	9.0

in section thickness can be used to partially offset the decreased material stiffness; also, the flexibility of composite fabrication processes allows the thickening of local areas as is required to optimize properties. Since the composite has a density approximately one-third that of steel, a significant increase in thickness can be achieved while maintaining an appreciable weight reduction. The second, and perhaps major, compensating factor is the additional stiffness attained in composite structures by virtue of part integration. This integration leads directly to the elimination of joints, which results in a significant increase in effective stiffness. It is becoming increasingly evident that this synergism is such that structures of acceptable stiffness and considerably reduced weight are feasible in glass-fiber-reinforced composites. As a rule of thumb, a glass FRP structure with significant part integration relative to the steel structure being replaced can be designed for a nominal stiffness of 50–60% of that of the steel structure. Such a design procedure should lead to adequate stiffness and typical weight reductions of 30–50%.

2. *Potential Applications*

The potential use of composites in structural applications in automobiles can be designated into two categories: the direct replacement of existing components, and the integration of multiple steel components into one composite component or structure. The second category, involving part integration, is by far the most cost-effective and ultimately will predominate. In the shorter term, however, much experience and confidence in the capabilities of composites is being generated by the singular component programs currently in production.

2.1 One-On-One Component Substitution

The significant differences in material costs between glass FRP and steel (approximately US$0.80 per pound versus US$0.25 per pound at 1988 prices) means that, even with significant weight saving, direct substitution in a singular component can rarely be cost effective. There are two examples currently in production which can be cited in the direct substitution category, namely driveshafts and leaf springs.

Composite driveshafts have been utilized in production on at least two vehicles in North America. One example is in the special drive line configuration of the Ford Econoline Van. These vehicles would normally utilize a two-piece steel driveshaft incorporating a connecting center bearing. The total length of the drive line dictates that a two-piece steel shaft must be used because unacceptable vibrations occur in a one-piece steel driveshaft. In contrast, a one-piece composite driveshaft will provide satisfactory vibratory characteristics since the lower weight combined with high stiffness satisfies the bending-frequency requirements. The driveshaft is fabricated by filament winding on a continuous machine operating at speeds up to 2 meters per minute. The longitudinal fibers (0°) are carbon (220 GPa modulus), to generate the required bending stiffness, and the $\pm 45°$ fibers are E-glass, to provide the torsional strength. The resin used is a vinyl ester which gives the appropriate combination of properties, processability and cost. This particular driveshaft is economically feasible only because it results in the elimination of the center bearing—such cost-effective usage is impossible in the case of single-piece steel (or aluminum) driveshafts.

The leaf spring is an example of a singular component utilizing glass-fiber-reinforced epoxy which has found a niche in the North American vehicle market. Current production in the USA is greater than 500 000 FRP springs per annum and their use to date has been a reliable demonstration of the feasibility and durability of composites. FRP leaf springs can be mass-produced by three basic techniques, although there may be several variants on each particular process. The basic materials are always the same, namely continuous E-glass fiber and epoxy resin. (Attempts to substitute more economical resins have all been unsuccessful because of creep requirements.) The three potential techniques are illustrated in Fig. 2. The compression-molding technique involves development of a preform (usually by a filament-winding

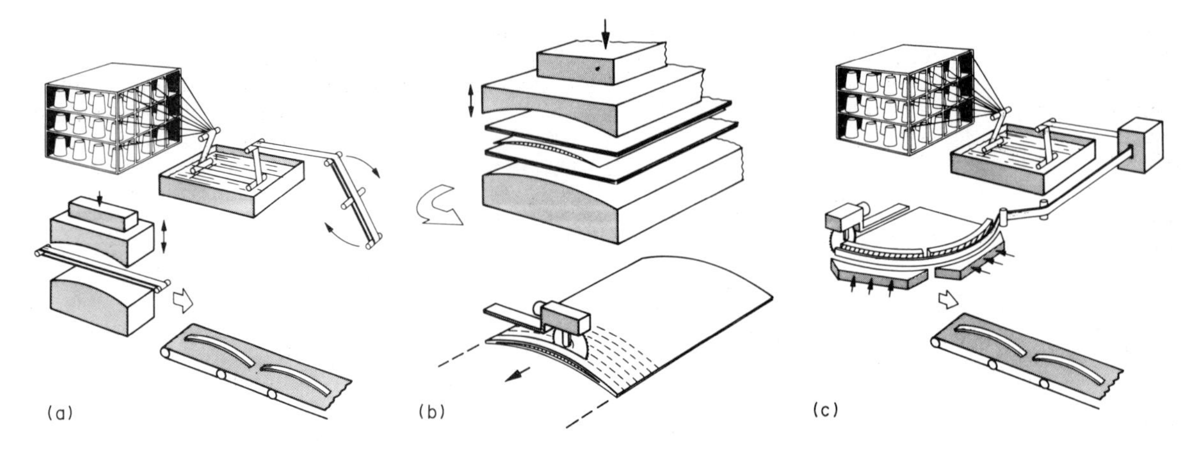

Figure 2
Different fabrication techniques for FRP leaf springs: (a) filament winding–compression molding, (b) compression molding and (c) pultrusion

procedure) followed by molding, and is the process currently used in the USA. An alternative procedure involves the automated layup of a wide slab of material, followed by molding under pressure and subsequent slitting of the slab to form springs. The third, least developed, technique is the pulforming procedure which has perhaps the greatest long-term potential for minimizing cost.

One-for-one replacements of steel components with FRP components will be exceedingly rare, primarily because of economic considerations but also because of complex attachment considerations. As a general principle, therefore, multiple-part integration to reduce assembly costs will be necessary to make FRP structures cost-effective relative to steel.

A simple example of this principle is demonstrated by the prototype composite integrated rear suspension for a Ford Escort (Morris 1986). The basic steel rear suspension and the replacement FRP integrated suspension are shown in Fig. 3. A simple extension of FRP leaf-spring design methodology allowed substitution for the complete steel rear suspension by allowing the FRP spring to perform the dual function of spring and suspension arms. The function and part integration resulted in a weight saving of approximately 4 kg (about 50% weight reduction). In addition to the weight saving and cost-effectiveness of this type of synergism, there are obvious improvements in package space, which is a major consideration in vehicle design.

2.2 Large Integrated Structures

The successful application of structural composites to large integrated automotive structures is more dependent on the ability to use rapid and economic fabrication processes than on any other single factor. The fabrication process must also be capable of close control of composite properties to achieve lightweight, efficient structures. Currently, the only commercial process which comes close to satisfying these requirements is compression molding of sheet molding components (SMC) or some variant of the process.

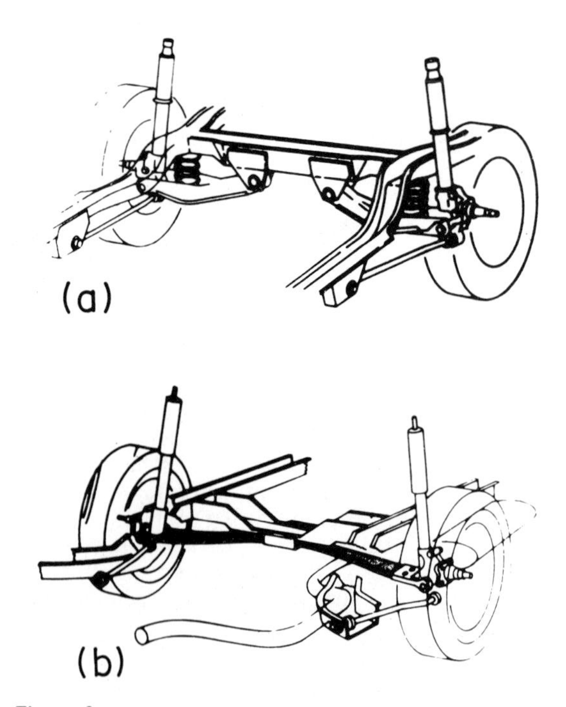

Figure 3
Comparison of (a) production Ford Escort steel rear suspension and (b) corresponding integrated composite suspension

There are, however, processes still at the development stage which hold distinct potential for the future in terms of combining high production rates, precise fiber control and high degrees of part integration. In particular, high speed resin-transfer molding (HSRTM) offers these potential benefits, provided technical developments can be achieved. The following examples demonstrate these fabrication approaches and part-integration concepts which can be utilized in automotive production and compare the utilization of both SMC and HSRTM procedures. It is not the intent here to describe these two processes in detail but for illustrative purposes the techniques are summarized in Fig. 4. Further details on these and other fabrication procedures may be obtained from the ASM Composite Materials Handbook (1988).

2.3 Primary Body Structure

The primary body structure of an automobile consists of approximately 250–350 major steel parts and utilizes approximately 300 assembly robots. For comparison of the two alternate composite construction techniques, consider the body side assembly. In Fig 5(a), a typical SMC approach is illustrated in which the complete body side consists of two moldings which would be bonded together. The HSRTM procedure for the same structure is shown in Fig. 5(b) and the major difference is the elimination of the adhesive bonding and the incorporation of a foam core. Utilizing either of these composite fabrication techniques for the other major segments of the body shell would lead to a typical body construction assembly as illus-

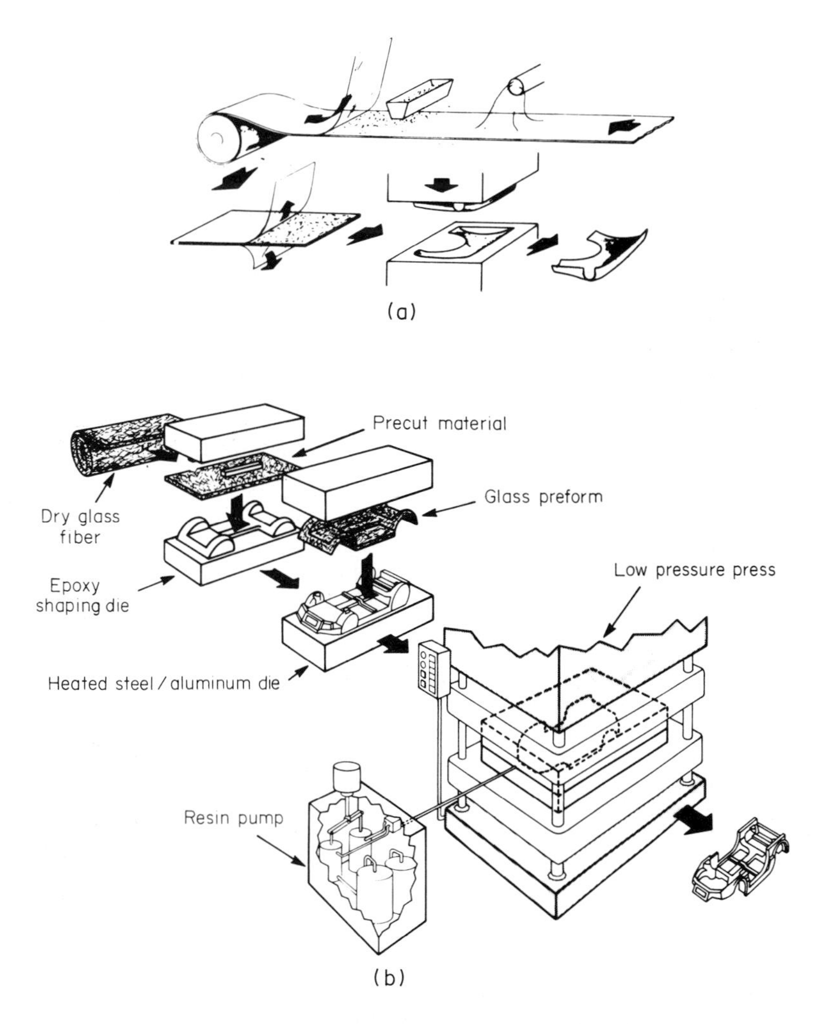

Figure 4
Depiction of (a) SMC material preparation and component fabrication and (b) high-speed resin-transfer molding process

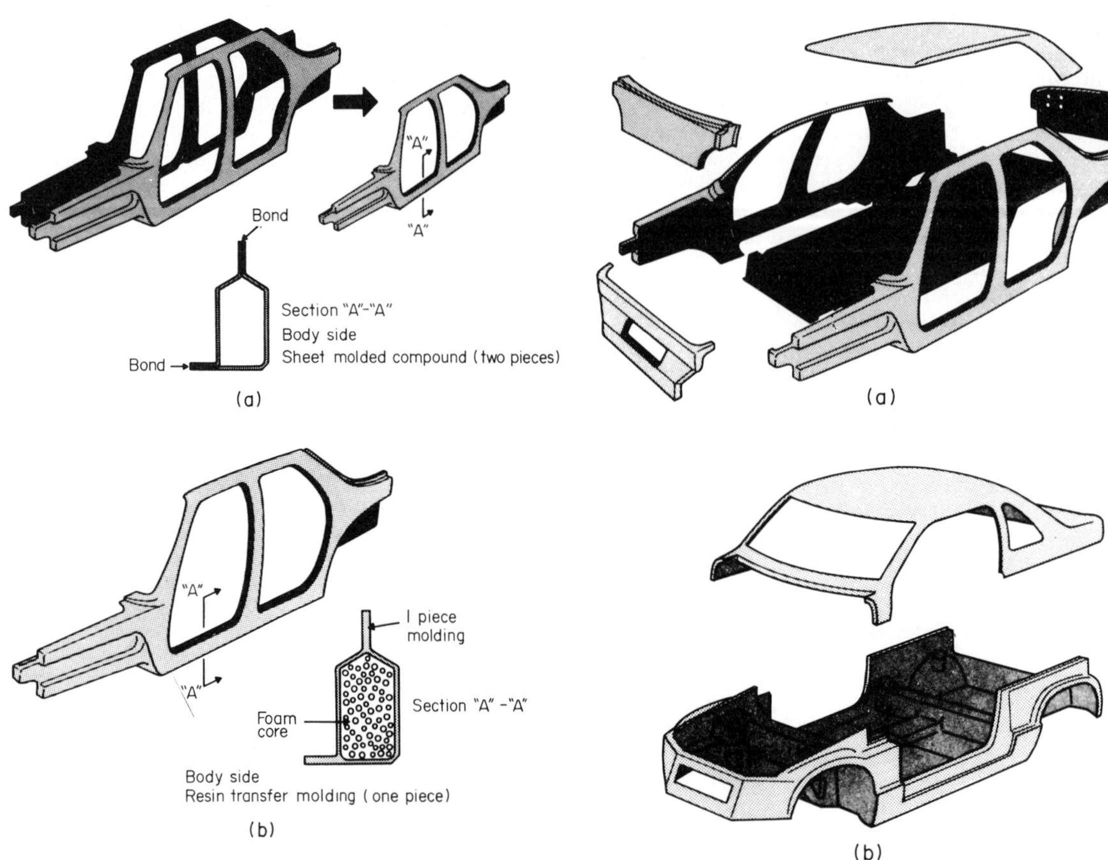

Figure 5
Typical assembly of composite body side panel by (a) adhesive bonding of inner and outer SMC molded panels and (b) HSRTM panels with foam core

Figure 6
Composite body shell. (a) Typical assembly, (b) assembly using two major moldings

trated in Fig. 6(a). In terms of reduction in parts, the molded SMC body structure could consist of somewhere between 10 and 20 major parts (compared to approximately 300 major steel parts) and the HSRTM structure could be composed of somewhere between 2 and 10 major parts. Note that the degree of integration is higher for HSRTM reflecting the greater versatility of this procedure. Clearly the lower limit of just two major parts for the HSRTM technique would require assembly as illustrated in Fig. 6(b).

While the attraction for the all-out application of composites to the total body shell, as discussed above, is appealing from the viewpoint of the materials connoisseur, reality dictates that the usage of composites in automobile structures will be evolutionary and progressive rather than revolutionary in nature. Therefore, it is to be expected that structural segments of vehicles will be first to utilize these procedures and there are several examples of prototype structures where these techniques have been used. Complex, high-load cross members have been developed for evaluation by both Chrysler (Farris 1987) and Ford (Johnson et al. 1987). The cross member is fabricated from glass-fiber-reinforced vinylester utilizing the resin-transfer molding (RTM) process. The unique cross member is projected to save approximately 30% in weight relative to the comparable steel member. The fabrication and testing of these high load, critical members is part of the gathering of detailed information on the realistic expectations that glass-fiber-reinforced plastic composite can satisfy the functionality required. Concurrently, the fabrication technique (RTM) has to be further developed to satisfy the economic constraints of the industry. The success of these prototypes is providing the impetus to drive the fabrication developments towards the required goal. A similar example concerning the fabrication of a highly integrated composite Escort front structure by RTM (Johnson et al. 1985) has been well documented and reported. In this prototype, 42 steel parts were replaced by one RTM molding resulting in a 30% reduction in weight, and structural stiffness in excess of that of the steel structure. This structure is currently being utilized to develop energy absorption capability in a typical vehicle design.

3. Summary

The extension of composites use to automotive integrated structures will require an expanded knowledge of the design parameters for glass-fiber-reinforced plastic materials together with major innovations in fabrication techniques. There is accumulating evidence, both in the laboratory and in prototype vehicles, which strongly indicates that glass-fiber-reinforced composites are capable of meeting the functional requirements of highly loaded automotive structures. The more imperative requirement is the cost-effective fabrication advancements that appear necessary to justify such increased use of composites. High-volume, less stringent performance components can be manufactured by variations on compression-molding techniques. However, the high volume, high-performance manufacturing techniques still need development and improved SMC materials and processes, and the HSRTM process holds promise in these areas.

Bibliography

Aotem A, Hashin Z 1976 Fatigue failure of angle ply laminates. *AIAA J.* 14: 868–72

Beardmore P, Harwood J J, Horton E J 1980 Design and manufacture of a GrFRP concept automobile. *Proc. Int. Conf. on Composite Materials, Paris,* August 1980

Dharan C K H 1975 Fatigue failure in graphite fiber and glass fiber–polymer composites. *J. Mater. Sci.* 10: 1665–70.

Farris R D 1987 Composite front crossmember for the Chrysler T-115 mini-van. *Proc. 3rd Annu. Conf. on Advanced Composites.* American Society for Metals, Metals park, Ohio, pp. 63–73

Johnson C F 1988 Compression molding; Resin transfer molding. In: *ASM Composite Engineered Materials Handbook,* Vol. 1. American Society for Metals, Metals Park, Ohio, pp. 559–68

Johnson C F, Chavka N G, Jeryan R J 1985 Resin transfer molding of complex automotive structures. *Proc. 41st Annu. Conf. SPI,* Paper 12a. Society of the Plastics Industry, New York

Kulkarni H T, Beardmore P 1980 Design methodology for automotive components using continuous fiber reinforced materials. *Composites* 12: 225–35

Morris C J 1986 Composite integrated rear suspension. *Compos. Struct.* 5: 233–42

Smith T R, Owen M J 1969 Fatigue properties of RP. *Mod. Plast.* 46(4): 124–29

Thornton P H 1979 Energy absorption in composite structures. *J. Compos. Mater.* 13: 262–74

Thornton P H, Edwards P J 1982 Energy absorption in composite tubes. *J. Compos. Mater.* 16: 521–45

Thornton P H, Harwood J J, Beardmore P 1985 Fiber reinforced plastic composites for energy absorption purposes. *Compos. Sci. Technol.* 24: 275–98

P. Beardmore
[Ford Motor Company, Dearborn, Michigan, USA]

B

Boron Fibers

Boron fibers were the first of the advanced fibers to be produced specifically for use in composite materials. Their development in the early 1960s marked an important improvement in specific properties of fibers when compared to those of glass fibers which at that time were the only other reinforcement available. Although the density of boron fibers is similar to that of glass fibers the former possess a Young's modulus nearly six times greater. These fibers were also found to possess exceptionally high compressive strengths. Boron fibers were produced initially for aerospace applications and seemed, at first, destined to find many uses in this area as reinforcements in resin and metal matrices. Although technically a success—their use in some aircraft structures, in the American space shuttle and in some sports goods—boron fibers have never realized the potential foreseen for them during the early stages of their development. The reason for this is the high cost of production when compared to carbon fibers, which can have similar properties, can be considerably cheaper and which appeared towards the end of the 1960s several years after boron fibers.

1. Production

It is not possible to draw boron by conventional wire-drawing techniques; rather, boron filaments are produced by chemical vapor deposition onto a tungsten wire substrate usually having a diameter of around 12 μm. Several chemical routes are possible but commercially, boron trichloride is mixed with hydrogen according to the following reaction:

$$2BCl_3(g) + 3H_2(g) \rightarrow 2B(s) + 6HCl(g)$$

Figure 1 shows schematically the production of boron fibers. The continuous tungsten wire passes into a deposition chamber which is usually vertical, where it is heated by dc electric current and inductive VHF techniques. The gas mixture is introduced at the top of the chamber and flows in the drawing direction. Temperature control is extremely important as it affects the rate of boron deposition. The temperature along the deposition chamber passes through a maximum at an early stage of deposition but as Wawner (1976) has shown this must not exceed 1350 °C as this leads to increased crystallite size and a drop in filament strength. Passage through the deposition chamber is of the order of one or two minutes and results in a fiber with a diameter of 140 μm.

Early production routes used multistage systems in which several reactors were arranged one after the other; however, problems associated with particles picked up at the mercury seals between chambers leading to a large nonuniformity in the properties of the fibers led to later industrial systems using single stages. Initially, also, several types of substrates were considered, including molybdenum, tantalum and carbon. Carbon was used in an attempt to reduce cost and density but the elongation of the boron fibers which occurs during the CVD process, attaining values of up to 5%, resulting in failure of the carbon fibers. This led to the development of "hot spots" which locally changed the rate of boron deposition and grain size and led to dramatic falls in strength. The failure of a carbon substrate could be avoided by a precoating of pyrolytic graphite. In this case, however, there was no bonding between the boron and the carbon fiber core. The 12 μm diameter tungsten wire was found to give best results. Complete boridization of the core occurs and an increase in diameter to 16 μm induces considerable residual stresses, putting the core into compression and the neighboring boron into tension as shown in Fig. 2. A final annealing process puts the fiber surface into compression.

Boron fibers can be successfully and relatively easily used to reinforce light alloys but in this case it is necessary to protect the fiber surface to avoid chemical reactions occuring, particularly during contact with the molten metal during composite manufacture. For this reason boron fibers have been commercially produced with both silicon carbide and boron carbide coatings. Such coatings led to the fibers being protected for significant lengths of time in molten alloys as well as an increase in strength due to a relaxation of internal stresses.

Early boron fibers coated with SiC had a diameter of 106 μm but were found to split longitudinally as shown by Kreider and Prewo (1972). This was avoided by increasing the diameter to 140 μm; however, splitting due to localized high stresses when fibers touched during composite manufacture was found to occur even with the larger diameter fiber, (Bunsell and Nguyen 1980).

2. Microstructure, Morphology and Defects

Boron deposition onto the tungsten substrate is affected by the surface roughness of the tungsten wire, resulting from the wire-drawing process, and leads to longitudinal ridges at the surface of the wire (Vega-Boggio and Vingsbo 1976a,b). The growth of the boron occurs as conical modules which nucleate preferentially on the raised parts of the tungsten substrate and give a modular surface oriented axially along the length of the fibers. Boron deposited on a smooth

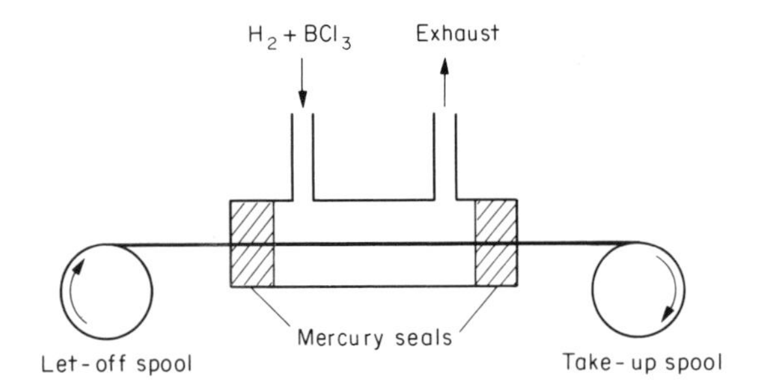

Figure 1
A schematic drawing of a decomposition chamber for continuous production of boron fibers

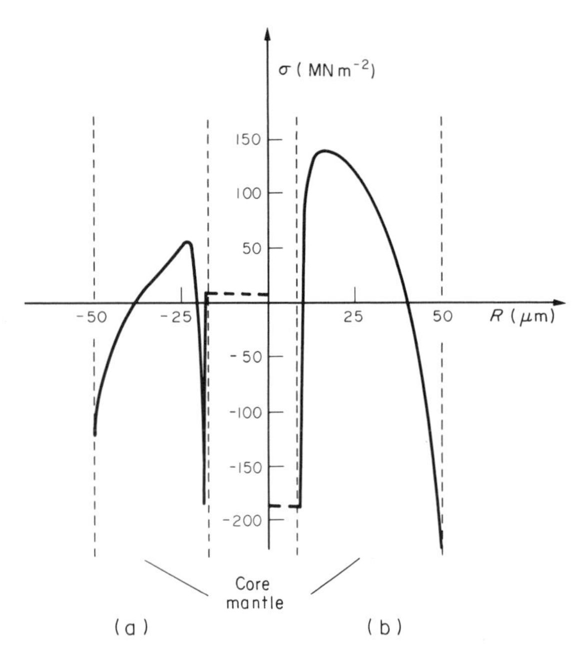

Figure 2
Longitudinal residual stress pattern for (a) carbon-substrate-based boron fibers, and (b) tungsten-substrate-based boron fibers

substrate such as a carbon filament shows no such surface features.

During deposition the boron atoms penetrate the tungsten core which is converted into WB_4 and W_2B_5 and which co-exist in the final fiber. Complete boridization of the core is important as the partial conversion of larger-diameter cores leads to reduced fiber strengths.

The original view that the deposited boron was in a glassy state has been replaced by the understanding that the structure is microcrystalline with a grain size of 2–3 nm consisting of α-rhombohedral and tetragonal polymorphs of boron (Lindquist et al. 1968, Bhardwaj and Krawitz 1983). Figure 3 shows the effects of different experimental conditions on the final appearance of boron fibers as well as some indications on the effects on fiber strength (Carlsson 1979). At low temperatures and high deposition rates a high-strength boron fiber with well-oriented boron nodules is formed. A lower temperature and a lower deposition rate yield a fiber of intermediate strength and with uneven boron nodules. This morphology is caused by a secondary nucleation process. Finally, low-strength boron fibers with large grain sizes are produced at high temperatures and low deposition rates. Figure 4 shows a typical histogram of the fracture stress of boron fibers.

Large surface nodules can be formed on the fibers by the presence of foreign particles but in the absence of such defects three main types of fault have been identified. The stress-concentrating effect of the geometry of the surface where two nodules join can be a

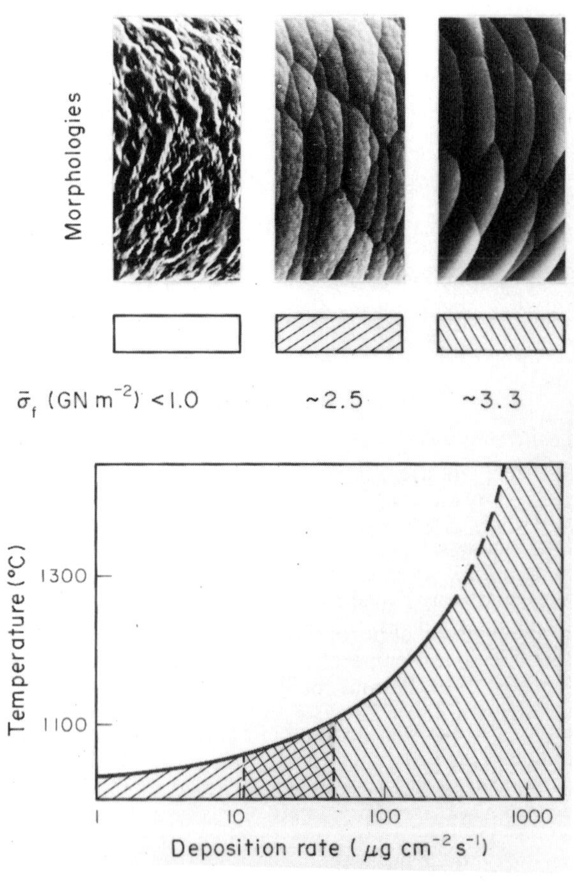

Figure 3
Main morphologies of boron fibers (magifications 500 ×), their average tensile fracture stress σ_f and the experimental conditions for obtaining them

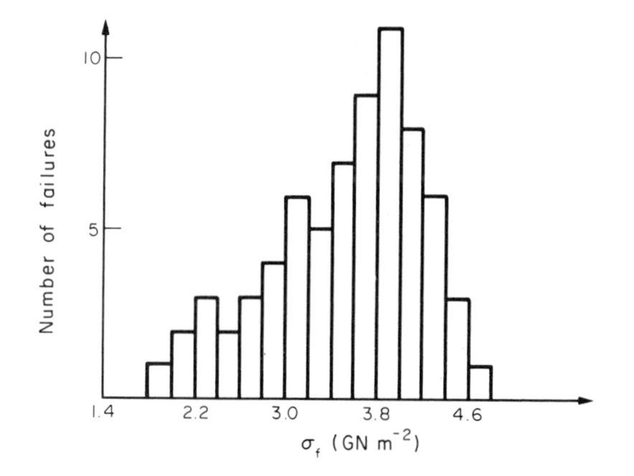

Figure 4
A typical histogram showing the fracture stress of boron fibers

deciding factor in limiting the strength of boron fibers, and surface etching to produce a smoother surface results in an increase in strength (Vega-Boggio and Vingsbo 1976b). Two types of defect are associated with the boron-core interface. Longitudinal voids formed by troughs of scratches on the tungsten core can occur as can proximity voids in the form of radial defects due to a lack of boron deposition between neighboring nucleation sites (Vega-Boggio and Vingsbo 1977). It has been shown that such defects limit the strengths of the fiber and obey the Griffith fracture criterion for brittle materials. The defects related to the core are considered to limit the ultimate strength of boron fibers.

3. Properties

The median strengths of boron fibers have been increased by the use of single-stage deposition chambers and this also led to less scatter. The average strength of boron fibers is greater than 3.45 GPa with a coefficient of variation of 15% or less, a Young's modulus of 478 GPa and a flexural modulus of 400 GPa. The specific gravity of boron fibers with a tungsten core is 2.6. The room-temperature shear modulus of boron fibers in 177 GPa and the average radial strength of the fiber is 480 GPa. The Poisson's ratio of boron fibers is 0.13. The thermal expansion coefficient of boron fibers is $4.9 \times 10^{-6}\ °C^{-1}$ up to 325 °C.

Boron fibers exhibit a linear axial stress–strain relationship at room temperature and up to 650 °C. At these temperatures reaction with the matrix or atmosphere limits the use of the fiber. At higher temperatures and stresses, however, boron fibers show nonelastic behavior which—up to at least 800 °C—can be explained in terms of an anelastic boron mantle and an elastic tungsten boride core (Erickensen 1974). A boron fiber formed into a loop and exposed briefly to an open flame which is then withdrawn will retain the imposed curvature. On reheating the fiber returns to its original shape, demonstrating that the deformation is anelastic and contains no plastic component. Boron fibers also show anelastic creep at room temperature, obeying the relationship:

$$t = \alpha \log t + c$$

where t is the strain and α and c are independent of time t (DiCarlo 1977).

Bibliography

Bhardwaj J, Krawitz A D 1983 The structure of boron in boron fibers. *J. Mater. Sci.* 18: 2639–49

Bunsell A R, Nguyen T T 1980 The radial strength of boron fibers and fiber splitting in boron-aluminium. *Compos. Fiber Sci. Technol.* 13: 363

Carlsson J-O 1979 Techniques for the preparation of boron fibers. *J. Mater. Sci.* 14: 225–64

DiCarlo J A 1977 Time-temperature-stress dependance of boron deformation. *Composite Materials: Testing and Design*, ASTM STP 617. American Society for Testing and Materials, Philadelphia, Pennsylvania, pp. 443–65.

Erickensen R 1974 Room temperature creep and failure of Borsic filaments. *Fiber Sci. Technol.* 7: 173

Kreider K G, Prewo K M 1972 The transverse strength of boron fibers. *Composite Materials: Testing and Design*, ASTM STP 497. American Society for Testing and Materials, Philadelphia, Pennsylvania, p. 539

Lindquist P, Hammond M, Bragg R 1968 Crystal structure of vapor deposited boron filaments. *J. Appl. Phys.* 39: 5152

Vega-Boggio J, Vingsbo O 1976a Application of the Griffith criterion to fracture of boron fibres. *J. Mater. Sci.* 11: 2242

Vega-Boggio J, Vingsbo O 1976b Tensile strength and crack nucleation in boron fibres. *J. Mater. Sci.* 11: 273

Vega-Boggio J, Vingsbo O 1977 Radial cracks in boron fibres. *J. Mater. Sci.* 12: 2519

Wawner F E 1988 Boron and silicon carbide/carbon fibres. In: Bunsell A R (ed.) *Fibre Reinforcement for Composite Materials*. Elsevier, Amsterdam, p. 372

A. R. Bunsell
[Ecole Nationale Supérieure des Mines de Paris, Evry, France]

C

Carbon–Carbon Composites

Graphite is an attractive material for elevated-temperature applications under inert, reducing or ablative environments. This is because of its high sublimation temperature (greater than 3500 °C) and improving strength with temperature up to 2500 °C. In its bulk forms (i.e., polycrystalline graphite or pyrolytic graphite) its utility for many applications is limited by low strain-to-failure, flaw sensitivity, anisotropy, variability in properties and difficulties of fabrication into large sizes and complex shapes.

The availability of carbon fibers in the late 1950s led to the development of improved graphite materials designated as carbon–carbon composites. These composites contain carbon or graphite fibers in a carbon or graphite matrix. The desirable properties of monolithic graphite were combined with the high strength and versatility of composites to produce this new class of materials.

Carbon–carbon composites range from simple unidirectional or random-fiber-reinforced constructions to woven multidirectional structures in block, hollow cylinder and other configurations. The variety of carbon fibers and multidirectional weaving techniques now available allow tailoring of carbon–carbon composites to meet complex design requirements.

1. Fabrication and Processing

The approach for producing structural carbon–carbon composites is to orient the required amounts of selected fiber forms to accommodate the design loads of the final component. Carbon–carbon composites can be prepared by a variety of fabrication approaches such as molding of random- or oriented-fiber composites with a char-yielding resin binder. Other types of multidirectional structures are produced by dry weaving, piercing of fabrics or by modified filament winding. Fully automated computer-controlled equipment for fabricating three-directional cylindrical, conical and contoured preforms has been developed both in the USA and in France.

All of the carbon–carbon preform structures must be densified by a process that fills the open volume of the preform with a dense, well-bonded carbon or graphite matrix. The actual densification process is dictated by the characteristics of the preform structure and the required properties of the final composite.

Methods of introducing the matrix into the composite include impregnation with char-yielding organic liquid followed by pyrolysis, and by chemical vapor deposition (CVD) of carbon from a hydrocarbon precursor gas. These processing methods are illustrated in Fig. 1.

Thermosetting resins are used for impregnating because they polymerize at low temperatures to form highly cross-linked nonmelting amorphous solids. As a result of pyrolysis, these resins form glassy carbon which does not completely graphitize at temperatures below 3000 °C. The thermosetting resins usually used are phenolic and furfuryl. Carbon yields at 1000 °C are about 50–60 wt %.

Another class of liquid impregants includes coal tar and petroleum pitches. Pitches have the advantage of low softening point, low melt viscosity and high coking value. They also tend to form graphitic coke structures.

The most widely used approach for introducing a carbon matrix into the fibrous preform is through impregnation with a liquid pitch or resin precursor. Impregnation is followed by carbonization in an inert atmosphere to convert the organic impregnant to coke. This conventional densification processing is conducted at atmospheric or reduced pressure and is usually repeated several times in order to reduce porosity.

There is also a variant known as high-pressure processing. The use of high isostatic pressure during the impregnation and coking stages of densification results in a more efficient process. Coke yields for pitch increase from 50 wt % at ambient pressure to 85 wt % at 70 MPa.

The chemical vapor deposition process for carbon–carbon densification involves gas-phase pyrolysis of a hydrocarbon such as methane or natural gas. In this process, the active carbon-bearing gases diffuse into the heated carbon substrate in such a manner as to achieve uniform carbon matrix deposition. The CVD process can be carried out under isothermal, thermal gradient or differential pressure conditions.

Selection of a particular densification process and specific process conditions will depend on the preform design, fiber type and matrix precursor. There are no standard carbon–carbon properties because the number of materials and process conditions are almost limitless. Some typical carbon–carbon properties compared to fine-grained graphite are given in Table 1.

2. Applications in Brakes

Carbon–carbon brake systems offer significant weight savings and life projections over steel/cermet brakes for both military and commercial aircraft. Carbon with a density 2.26 g cm^{-3} and of specific heat nearly two and a half times that of steel offers a 60% weight saving for a similar brake-temperature operating range. For Concorde, 600 kg is saved. The low wear rate of carbon–carbon has also increased the number of landings by two to four times as compared to steel/cermet brakes.

The properties sought in a brake-disk friction material are:

(a) high thermal capacity;

(b) good strength, impact resistance and strain-to-failure;

(c) adequate and consistent friction characteristics;

(d) high thermal conductivity; and

(e) low wear rate.

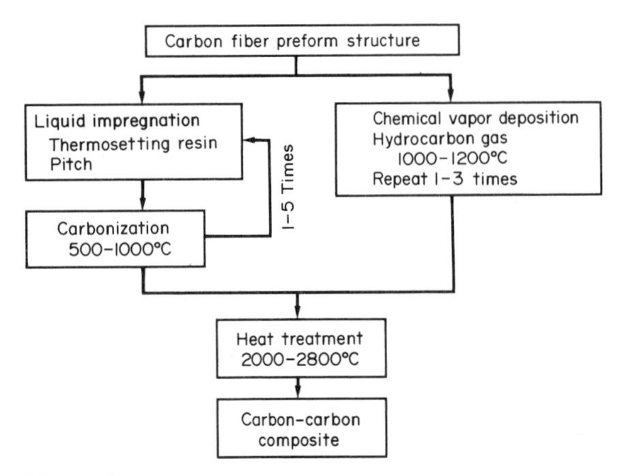

Figure 1
Carbon–carbon matrix processing approaches

Carbon–carbon composites are superior to high-strength graphite for brake applications because of their higher strength and toughness at about the same density, and (taking account of the anisotropy) about the same thermal conductivity.

3. Other Applications

Carbon–carbon composites offer great potential as high-performance engineering materials. However, high cost, a limited design database and underdeveloped design methodology have restricted their use.

Another limitation to the use of carbon–carbon composites has been their susceptibility to oxidation at high temperature. Recent developments have resulted in ceramic coatings that protect carbon–carbon composites from oxidation for hundreds of hours at

Table 1
Comparison of properties of fine-grained graphite with those of carbon–carbon composites

Property	Fine-grained graphite	Unidirectional fibers	Three-directional fibers
Elastic modulus (GPa)	10–15	120–150	40–100
Tensile strength (MPa)	40–60	600–700	200–350
Compressive strength (MPa)	110–200	500–800	150–200
Fracture energy (kJ m^{-2})	0.07–0.09	1.4–2.0	5–10
Oxidation resistance	very low	like glassy carbon	better than graphite

Table 2
Comparison of the suitability of tungsten, pyrolytic graphite and carbon–carbon composites for rocket nozzles

	Increased temperatures and pressures ⟶		
Requirements	Tungsten	Graphite pyrolytic	Carbon–carbon composites
Performance	temperature limited	acceptable	repeatable
Reliability	acceptable	questionable	high
Cost	high	high	moderate
Design	complex	complex	simple
Weight	heavy	medium	light
Growth potential	very limited	limited	high

Source: McAllister and Lachman (1983)

temperatures up to 1400 °C, and for short times up to 1750 °C.

The use of carbon–carbon composites for reentry-vehicle heat-shield applications was successfully demonstrated first in 1971, and the performance of carbon–carbon materials for missile nose-tip applications was reported at about the same time. The most publicized use of carbon–carbon composites as a reentry thermal protective material is its application to the leading edge surfaces of the Space Shutttle as a reusable radiative heat shield.

Representative applications in advanced rocket propulsion systems include nozzles, thrust chambers and ramjet combustion liners. The advantages of carbon–carbon composites lie in simpler designs and reduced nozzle weight and volume. A total weight saving of 30–50% has been predicted if a rocket nozzle is completely designed using carbon–carbon composites. A comparison of various rocket nozzle materials in Table 2 shows why carbon–carbon composites best meet the requirements for this application.

Other aerospace applications also utilize carbon–carbon composites because of their good thermal shock resistance and high sublimation temperature. For example, radioactive isotopes are used on many space vehicles to generate heat and electric power. Carbon–carbon capsules are used to protect these isotopes in the event that reentry into the earth's atmosphere is required.

Some commercial applications for carbon–carbon composites are beginning to develop. Various carbon–carbon composites have been evaluated for the internal fixation of bone fracture. Carbon is biocompatible and carbon–carbon composites can be tailored to be structurally compatible with bone because of similar Young's modulus and density.

Carbon–carbon hot pressing molds are now a commercial product; such molds can withstand higher pressure and offer longer life than polycrystalline bulk graphite.

Other applications that have been mentioned for carbon–carbon composites include high-temperature ducting systems, components for nuclear reactors, electrical contacts, seals and bearings, structural components for space transportation vehicles and components for advanced turbine engines.

As production techniques become refined and scaled up to higher volume capacities, carbon–carbon composites will become more economically acceptable for most of the applications mentioned above. The availability of low-cost carbon fiber will also have a major impact on the future of carbon–carbon products.

Bibliography

Batha H D, Rowe C R 1987 Structurally reinforced carbon–carbon composites. In: *Engineered Materials Handbook*, Vol. 1, Sect. 13. ASM International, Metals Park, Ohio

Fitzer E 1987 The future of carbon–carbon composites. *Carbon* 25(2): 163–90

McAllister L E 1987 Multidirectionally reinforced carbon/-graphite matrix composites. In: *Engineered Materials Handbook*, Vol. 1, Sect. 13. ASM International, Metals Park, Ohio

McAllister L E, Lachman W L 1983 Multidirectional carbon–carbon composites. In: Kelly A, Mileiko S T (eds.) 1983 *Fabrication of Composities*. North Holland, Amsterdam

Ruppe J P 1981 Today and the future in aircraft wheel and brake development. *Can. Aeronaut. Space J.* 27: 212–16

Stimson I L Fisher, R 1980 Design and engineering of carbon brakes. *Philos. Trans. R. Soc. London, Ser. A* 294: 583–90

L. E. McAllister
[Allied-Signal Inc., South Bend, Indiana, USA]

Carbon-Fiber-Reinforced Plastics

To utilize the high stiffness and strength that carbon fibers possess it is necessary to combine them with a matrix material that bonds well to the fiber surface and which transfers stress efficiently between the fibers. Fiber alignment, fiber content and the strength of the fiber–matrix interface all influence the performance of the composite. Furthermore, highly specialized processing techniques are necessary which take account of the handling characteristics of carbon fibers, which are different from those of other reinforcements. It is the processibility of polymers that makes them particularly attractive for use in a whole range of composite materials. For example, one form of carbon-fiber-reinforced plastic (CFRP) contains continuous carbon fibers, possibly in the form of a unidirectional or woven mat, which are impregnated by a thermosetting resin such as an epoxy. Prepreg material of this type can then be layed up to ensure the appropriate orientation of the fiber before curing of the resin is undertaken. Alternatively, if fiber orientation does not have to be so rigorously defined, then the processability of thermoplastic polymers means that, with the reinforcement in the form of short fiber (typically 0.2–10 μm), a molding compound of fiber and thermoplastic can be injection molded to produce directly finished components.

From the perspective of materials performance, the potential of CFRP is enormous: a civil aircraft constructed, where possible, entirely of CFRP instead of aluminum alloy would have a total weight reduction of 40%; a communications satellite built from CFRP can carry more telecommunications channels than permitted by previously specified materials; and a new generation of tactical aircraft can be planned, which will have a versatility regarded as impossible prior to the advent of CFRP, e.g., forward-swept wings for low stalling speeds, and very small turning-circle radii. However, the high cost of the fiber and the lack of cost-effective manufacturing processes are still inhibiting the more general use of CFRP, as is the newness of the art of composites design.

1. Matrix Materials

1.1 Thermosets

Examples of thermoset resins commonly used for carbon-fiber composites are listed in Table 1, together with their advantages and disadvantages. Long shelf life is an important advantage for prepregs, while low viscosity is vital to processing techniques such as pultrusion, filament winding and rapid resin injection molding (RRIM). Cure of the resin should not take long times or require extreme temperatures, while a high value of the glass transition temperature (T_g) (the temperature of the glassy–rubbery transition) is necessary if the composite is to be used at elevated temperatures. Carbon-fiber–epoxy composites represent about 90% of CFRP production. As can be seen in Table 1 the properties of an epoxy are dependent on the curing agent used and the functionality of the epoxy. The attractions of epoxy resins are that (a) they polymerize without the generation of condensation products which can cause porosity, (b) they exhibit little volumetric shrinkage during cure, which reduces internal stresses, and (c) they are resistant to most chemical environments.

The major disadvantages of epoxy resins are their flammability and their vulnerability to degradation in agressive media. Phenolic resins are, however, quite stable in both respects, although as structural composite materials, phenolic CFRPs are quite uninteresting, as their fatigue and impact properties are very poor. As indicated earlier, the prospect of using CFRPs as a substitute for light metal alloys is particularly attractive but there are problems to be overcome at high temperatures. Using polyimides, peak service temperatures of 400 °C can be reached. However, conventional polyimides present significant processing problems. Toxic solvents have to be evaporated and, during curing, condensation products have to be vented from the tooling. Cure cycles are also prohibitively long. PMR 15 is a polyimide developed to surmount the above problems. Three monomer reactants are mixed in an alcohol, the solution being used to impregnate the fiber. The alcohol is then evaporated, during which time a thermoplastic prepolymer is formed. This prepreg can then be cured using tooling identical to that used with epoxies, although much higher temperatures (~ 380 °C) are required.

Another compromise using polyimide chemistry to enhance CFRP performance is to form copolymers with epoxy resins. In this way flame retardancy and reduced moisture absorption are realized, but there is no substantial increase in high-temperature performance. Polybismaleimides represent a very good compromise in many respects and should command the applications arena between the epoxies and the polyimides.

1.2 Thermoplastics

A considerable number of thermoplastics are currently being used in CFRP, they may be either semicrystalline or amorphous. Semicrystalline thermoplastics offer greater solvent resistance, but are generally more difficult to process.

Table 1
Examples of thermosetting resins used for CFRP

Resin	Advantages	Disadvantages
Epoxides		
DGEBA[a] (aliphatic amine curing agent)	very low viscosity; room-temperature cure	very low T_g; short shelf life; flammability
DGEBA[a] (aromatic diamine curing agent)	medium viscosity; long shelf life	medium T_g; difficult cure
TGMDA[b] (aromatic diamine curing agent)	high T_g; long shelf life; toughness	high viscosity; difficult cure; moisture uptake; flammability
Novalac (phenolic)	nonflammability; high T_g	difficult cure; shrinkage; brittleness
PMR 15 (polyimide)	very high T_g; low moisture uptake; high strength	difficult cure; thermal fatigue

a diglycidylether of bisphenol A b tetraglycidylmethylenedianiline

In load-bearing applications, carbon fibers are most widely used in combination with nylon 66 which gives the highest strength and modulus over a fairly wide temperature range. A thermoplastic that is currently of considerable interest to fabricators looking to mass-produce articles is poly-1,4 butanediol terephthalate (PBT). It has a melting point of 224°C and, if a low molecular weight grade is used, it can be injection molded at 250°C. It crystallizes very rapidly so very short cycle times can be employed. CFRP based upon PBT exhibits a heat distortion temperature of 200 °C with adequate resistance to creep up to 120 °C.

The prospects of using high-temperature engineering thermoplastics, examples of which are shown in Table 2, with continuous-fiber reinforcements instead of thermosetting resins look particularly exciting. The advantages of thermoplastics are lower production costs, improved toughness and impact resistance and lower susceptibility to moisture uptake, which is found to reduce the value of T_g for epoxy matrices. Additional benefits are the ease of repair of the composites and the recovery of waste water or used material.

In Table 3 the properties of injection-molded nylon 66, polyethersulphone (PES) and polyether ether ketone (PEEK), each with 40 wt% of carbon fiber are compared. The nylon 66 compound exhibits superior properties at normal ambient temperature. It is with properties such as creep modulus (stress to induce 0.1% strain within 100 s) and tensile strength at elevated temperatures that the high-temperature thermoplastics reveal their merit.

Table 2
High-temperature engineering thermoplastics

Generic name (trade name; manufacturer)	Heat distortion temperature at 1.82 MPa (°C)
Polycarbonate (Lexan, Merlon; GE, Mobay)	133
Polyphenylene sulfide (Ryton; Phillips)	137
Polyether ether ketone (Victrex PEEK; ICI)	148
Polysulfone (Udel; Union Carbide)	174
Polyetherimide (Ultem; GE)	204
Polyethersulfone (Victrex PES; ICI)	204

2. *Fiber–Matrix Interphase*

The properties of the material between the fiber reinforcement and the plastic matrix have a strong influence on the mechanical properties of the CFRP. To produce a high-modulus composite requires maximizing adhesion between the fiber and matrix. However, if the polymer matrix is brittle (e.g., epoxies) there is a corresponding reduction in impact strength as mechanisms of energy dissipation such as debonding and fiber pullout become suppressed. If a tough polymer matrix is to be used (e.g., PEEK), where under

Table 3
Properties of three thermoplastics reinforced with carbon fiber (CF)[a]

Property	Material		
	40% CF–nylon 66	40% CF–PES	40% CF–PEEK
Density (gm cm^{-3})	1.34	1.52	1.45
Tensile strength (MPa)			
at −55°C			275
24 °C	246	176	227
50 °C	168	168	208
100 °C	108	140	165
140 °C	75	112	129
180 °C			72
Tensile elongation (%)	1.65	1.1	1.35
Tensile creep modulus (GPa)			
at 24 °C	26.2	22.1	26.5
120 °C	11.5	21.4	23.9
160 °C	9.0	21.1	10.5
200° C		17.5	6.6
Flexural strength (MPa)	413	244	338
Flexural modulus (GPa)	23.4	16.8	20.8
Compressive strength (MPa)	240	216	
Compressive modulus (GPa)	20.2	17.2	

a "Grafil" product, reference RG40

impact the principal mechanism of energy dissipation is yielding within the polymer, good adhesion between the fiber and matrix is required.

Unfortunately this region of the composite, sometimes known as the interphase, is extremely complex. Constituents of the interphase might include a modified fiber surface, the treatment being undertaken to promote better adhesion, or a coating of polymer present to ensure compatibility with the matrix polymer. In addition, the presence of the fiber surface may affect the cure of thermosets, resulting in chemically different polymer, or may affect the morphology of semicrystalline thermoplastics.

2.1 Surface Treatment

Surface-treatment processes are now established as an integral part of fiber manufacture. Most are either wet oxidative treatments using solutions of, for example, sodium hypochlorite, sodium bicarbonate or chromic acid, or dry treatments in which the oxidizing agent is ozone. The precise action of the treatment is undoubtedly complex, but is currently thought to include the introduction of chemically active sites onto which the polymer can graft, as well as chemical cleaning, an increase in rugosity of the fiber surface and the removal of any weak surface layer on the fiber.

A plethora of techniques has been proposed to measure fiber–polymer adhesion. Such techniques are experimentally tedious but, more significantly, the assumptions necessary to deduce adhesive strengths from the various types of measurement make them only comparative. The techniques include measurement of the composite short-beam interlaminar shear strength, the distribution of fiber lengths in a fully fragmented single fiber, the stress required to pull out a fiber embedded in the polymer and the stress required for a bead of polymer to become detached from a single fiber.

2.2 Surface Coating and Sizing

Polymer coatings and sizes are frequently applied to fibers immediately after manufacture. The light application of resin binds individual filaments together and maintains a compact bundle for processing into an intermediate product or final composite form. Size compositions applied to the surface can afford protection against filament damage and provide lubrication for a smooth delivery of fibers in continuous processing. The sizing or coating may be bulk polymer or resin specially functionalized to improve fiber wetting and adhesion while still retaining compatibility with the bulk polymer.

3. Properties of CFRP

The main points to appreciate when considering a carbon-fiber composite as compared to a conventional material such as a metal alloy are that the composite is heterogeneous and anisotropic. By heterogeneous it is meant that the properties vary from point to point within the material and by anisotropic it is meant that the properties depend on the direction in which they are measured. These two factors are important when considering not only mechanical properties (such as stiffness and strength) but also the physical, electrical and thermal characteristics of the composite.

The physical and mechanical properties of a CFRP are strongly dependent on the properties of the particular fiber used. There are a variety of fiber types available and the properties of fibers are being advanced continually by the development of new grades (see Table 4).

The section below outlines some of the main features of the mechanical, thermal and electrical properties of CFRP.

3.1 Mechanical Properties

CFRP containing continuous aligned fibers (which would typically have a volume fraction of 60%) are essentially elastic to failure (when loaded in tension parallel to the fibers) and exhibit no yield or plasticity region. There is, however, a slight nonlinearity in the stress–strain curve, with the stiffness increasing slightly with applied strain, as a result of improved orientation of the graphitic planes within the fibers. Although the failure strains in tension for CFRP may seem small when compared with those of other structural materials, the predictably elastic behavior under load allows a high proportion of the ultimate strength to be utilized in practice. Consequently, strain levels at useful working stresses are comparable with those of metals, alloys and glass-fiber composites. The properties of unidirectional CFRP are compared with similar performance characteristics of other commonly used structural materials in Table 5. A range of material properties typical of those required for design calculations is given in Table 6. These data confirm the anisotropy of the unidirectional material (which is much more marked in CFRP than in glass or aramid composite); the stiffness and strength of the materials

Table 4
Types of commercially available carbon fibers

Carbon fiber type	Tensile Strength (GPa)	Elastic modulus (GPa)	Breaking strain (%)
High strength (PAN)	3.5	240	1.5
High strain (PAN)	5.5	240	2.2
Intermediate modulus (PAN)	4.5	285	1.6
High modulus (PAN)	2.5	350	0.7
Ultrahigh modulus (PAN)	2.0	450	0.5
Isotropic (pitch)	1.0	35	2.5
Ultrahigh modulus (pitch)	2.0	550	0.4

Table 5
Some properties of fiber-reinforced composites (55–70 vol% fiber) and metals

Material	Density (gm cm^{-3})	Ultimate tensile strength (GPa)	Young's modulus (GPa)	Specific tensile strength (GPa)	Specific Young's modulus (GPa)
CF–Epoxy					
high strength	1.5	1.9	130	1.27	87
high modulus	1.6	1.2	210	0.94	119
E-glass–epoxy	2.0	1.0	42	0.50	21
Aramid–epoxy	1.4	1.8	77	1.30	56
Steel	7.8	1.0	210	0.13	27
Titanium	4.5	1.0	110	0.21	25
Aluminum L65	2.8	0.5	75	0.17	26

Table 6
Typical performance data for unidirectional laminates

Property	High strength	High modulus
0° Tensile strength (GPa)	1.90	1.15
0° Tensile modulus (GPa)	137	200
Elongation (%)	1.4	0.6
Interlaminar shear strength (MPa)	90	70
0° Flexural strength (GPa)	1.80	1.10
0° Flexural modulus (GPa)	130	190
0° Compression strength (GPa)	1.25	0.70
0° Compression modulus (GPa)	140	210
Density (g cm^{-3})	1.55	1.63
90° Tensile strength (MPa)	50	35
90° Tensile modulus (GPa)	10	7

are very much lower when they are loaded perpendicular to the fibers than when they are loaded parallel to the fibers.

It should also be noted that the unidirectional compression strength of the lamina is only about two-thirds of the tensile strength. In early carbon-fiber composite materials (from the early 1970s) the tensile strength and compression strength were approximately equal, with the failure mode in compression being shear failure. Improved fiber tensile properties have been achieved with a reduction in fiber diameter and, during this time, more ductile resins have come into favor. These factors have tended to promote fiber instability or microbuckling failure when unidirectional materials are loaded in compression. Hence the compression strength of the material is not a true measure of the fiber strength, but more of the ability of the matrix to support the fiber against buckling, and the integrity of the fiber–matrix interface. As a result of these developments, compressive performance of CFRP can be a limiting design condition.

There are very few practical applications for CFRP which require only unidirectional reinforcement. Laminates are assembled from unidirectional layers stacked at various orientations to one another. Layers are included at 90° to give adequate transverse properties and at 45° to give good shear properties. A laminate construction of 0°/90°/±45° plies has negligible anisotropy when loaded in-plane and is known as quasi-isotropic. Typical data for such a laminate are shown in Table 7. The mechanics of laminated composites are quite complicated and the interested reader is referred to the article *Laminates: Elastic Properties*.

As a result of the marked anisotropy, CFRP can experience large stress concentration factors at cutouts or mechanically fastened joints. Given that the material is predominantly elastic to failure this might be expected to present serious practical problems. Fortunately, certain damage mechanisms (such as matrix cracking parallel to the fibers and delamination cracking between layers) can occur in laminates so that the stress concentrations do not have the deleterious effects that might be expected. Even so, the designer must take care when dealing with cutouts or joints. Bonded joints must also be designed carefully to avoid failure of the laminate by interlaminar (through-thickness) failure.

The range of basic mechanical properties of CFRP is enhanced by excellent resistance to creep rupture and fatigue. Creep rupture times at equivalent stress levels are several orders of magnitude greater for CFRPs compared with those for glass-fiber-reinforced composites (GFRP), and also significantly better than for aluminum. CFRP is also capable of operating well under conditions of alternating stress. For high strength fibers, S–N fatigue curves are much flatter for CFRP compared with those for GFRP or aluminum, and a higher ratio of working stress to ultimate stress

Table 7
Performance data for multidirectional laminates with 60 vol% fiber and 0°/90°/±45° laminate construction

Laminate	Tensile strength (GPa)	Tensile modulus (GPa)
High strength	0.6	50
High modulus	0.35	75

can be utilized. However, the newer fiber-matrix systems are not yet characterized fully and there are signs that, with these materials, fatigue may be more of a consideration than in the past. In passing it should be noted that fatigue failure in composite materials is not caused by the initiation and growth of a dominant crack as it is in metallic materials; instead there is an accumulation of damage (matrix cracks, delamination, fiber breaks) which may result in a lowering of the stiffness and the residual strength. Much more work is needed to develop damage-tolerant design for CFRP composites.

The impact resistance of CFRP is generally considered low, and in fact the post-impact compression strength is considered to be a design-limiting factor. There is, however, scope for the modification of impact performance and fracture toughness by fiber hybridization with glass or aramid reinforcements leading to structures with much greater damage tolerance. Moreover, impact performance will improve as strain-to-failure of the fiber continues to increase.

3.2 Thermal Properties

CFRP is thermally conductive along the direction of the fibers. Conductivity perpendicular to the fibers is much less as it is dominated by the polymer matrix. The ability to dissipate heat is thought to contribute to the very good fatigue properties of CFRP. The conductivity of fibers increases with graphite content and highly graphitized fibers, such as high-modulus fibers, have thermal conductivity values of 700 $W m^{-1} K^{-1}$ in the longitudinal direction which surpass the value for steel (50 $W m^{-1} K^{-1}$).

Coefficients of thermal expansion for carbon-fiber composites are dependent on the fiber type and on temperature. The fibers themselves have small negative coefficients of thermal expansion and this leads to the possibility of designing composites with almost zero overall thermal expansion. This dimensional stability, which can be available over extremes of temperature, is very attractive in applications such as antennas for aerospace vehicles. As with the thermal conductivity, the coefficient of thermal expansion is anisotropic—the value measured perpendicular to the fibers is dominated by the polymer matrix and is usually much higher (see Table 8) than the value measured parallel to the fibers. This can have important practical consequences in that residual stresses are left trapped in a laminate after it has cooled from the processing temperature. These stresses can be large enough to promote matrix cracking, especially in systems such as PMR-15 which, as described previously, is processed at high temperatures. For such systems to achieve wider use this problem has to be overcome either by toughening the resin further, or by further development of a damage-tolerant approach to design.

3.3 Electrical Properties

CFRP is electrically conductive along the direction of the fiber. As with thermal conductivity, electrical conductivity increases with the level of graphitization of the fiber.

There are practical applications which make use of the electrical properties of carbon fiber, notably to provide resistance heating effects (in nonmetallic tooling, for example) or static charge dissipation in composite structures.

The electrochemical behavior of graphite can lead to problems when carbon-fiber composites are in contact with metal alloys, as they may be in an aircraft structure, for example. Graphite is cathodic to most metal alloys and this can lead to appreciable corrosion (especially in mechanically fastened assemblies) unless adequate precautions are taken.

4. Forms of Carbon Fibers

Continuous-filament fibers represent the primary production route for manufacture and other products are derived from this form by secondary processes. The conversion of fibers into some conveniently handleable intermediate product plays an important part in the effective utilization of the fibers as a reinforcement. Specific products are needed to meet the individual processing requirements of different fabrication techniques.

Table 8
Typical thermal expansion data for CFRP with 60 vol% fiber

CFRP	Coefficient of thermal expansion[a] ($10^{-6} K^{-1}$) Longitudinal	Transverse
High modulus–epoxy	−0.25 to −0.60	20–65
High strength–epoxy	0.30 to −0.30	20–65

a over the temperature range 100–400 K

4.1 Continuous-Filament Tow

Continuous fibers are grouped together in tow bundles, each containing a specified number of individual parallel filaments. The tows are produced in different bundle sizes, ranging in discrete steps from 400 to 320 000 filaments per tow. The main sizes used are in the 3000–12 000 range; finer tows are employed only for specialized applications. Heavier tows offer some advantages for rapid volume conversion to composite and can lead to economies of scale for both fiber production and composite manufacture by certain processes.

Twisted tows, containing typically 15 turns per meter, arise from processes employing twisted precursor fibers. The presence of twist reduces the efficiency of the fiber reinforcement but can enhance the handleability of the tow, aiding conversion to secondary products such as woven fabrics and braids.

4.2 Woven Fabrics

Carbon fibers can be woven into a variety of unidirectional and bidirectional fabrics capable of being processed efficiently into composite materials. The weave construction (e.g., plain, satin, twill) determines the handle of the fabric and controls its ability to conform to contoured shapes.

4.3 Preimpregnated Products

Forms of preimpregnated materials (prepregs) include continuous tows, unidirectional tapes and sheets, and woven fabrics, A resin is applied to the fibers using a solvent, melt or film impregnation technique. The resins used are specially formulated with a latent cure, so that at room temperature the cure reaction proceeds only very slowly and the material can be handled and stored at ambient temperature for a reasonable length of time. Curing temperatures are typically between 100 and 180 °C. A choice can be made from the wide range of commercial prepregs to suit the end use and fabrication technique employed.

4.4 Nonwoven Fabrics

Carbon-fiber mats, felts and papers manufactured by conventional nonwoven processes from randomly distributed short fibers (about 6 mm in length) provide multidirectional reinforcement but convert inefficiently into composite form. New products containing longer fibers (about 25 mm) are now being developed, exhibiting good handling characteristics and generating improved composite performance.

4.5 Molding Materials

Short-length fibers chopped from continuous tows are widely used for the manufacture of molding compounds in both thermoplastic and thermosetting matrices. The application of special resin sizes and coatings facilitates processing into molding materials and other products such as papers and felts.

Thermoplastic molding compounds are supplied as granules containing short-length (0.2 mm) fiber reinforcements for processing on conventional injection-molding equipment.

Sheet molding compounds (SMCs) and dough molding compounds (DMCs) are already well established with glass fibers. The use of a small specific quantity of carbon-fiber reinforcement is envisaged in SMC with careful placement of aligned continuous fibers for optimum reinforcing effect at minimum additional cost. Carbon-fiber-reinforced DMCs in epoxy and polyester systems achieve excellent stiffness but strength levels achieved so far are disappointing.

Aligned short fibers in the form of prepreg sheets (ASSM) are available from special processes developed in the UK and Germany. The products can be molded to generate high performance in composites with highly curved or contoured shapes by virtue of the aligned fiber reinforcement coupled with the ability to promote controlled fiber movement for mold conformity.

5. Fabrication Techniques

Fabrication techniques have an important bearing on the properties obtained from carbon-fiber composites, and efficient processes must be devised for handling the fibers to ensure the maximum contribution to mechanical performance. The manufacture of GRP provides some useful experience although two major points of difference arise. First, the handling characteristics of the two fibers are different; secondly, the aim in the manufacture of carbon-fiber composites is to obtain the best possible mechanical performance, requiring the fiber to be incorporated in the precise direction and in the exact amount necessary to meet the stresses applied.

The basic principles of the manufacture of carbon-fiber composites are similar to those used for GFRP (see *Glass-Reinforced Plastics: Thermosetting Resins*), although changes in the actual techniques are necessary in most cases. Processes unique to CFRP production are also being developed.

5.1 Compression Molding

Three suitable material forms containing carbon-fiber reinforcement are available: DMCs, SMCs, and preimpregnated tapes, sheets and fabrics.

Certain considerations which affect plant and mold tool design and process procedures are necessary for the manufacture of CFRPs. The thermal conductivity and low thermal-expansion characteristics of carbon fibers introduce the need for carefully controlled heating and cooling cycles and special attention to molding shrinkage (and expansion) to achieve dimensional accuracy in the final part. SMCs usually comprise hybrid reinforcements, mainly glass, with a small amount of carbon included for specific stiffening. The effectiveness of the carbon fiber is influenced by careful placement for maximum reinforcing effect and for minimum risk of misalignment caused by resin and glass-fiber flow during molding. Alignment accuracy, parallelism of mold-tool platens and additional considerations arise with the use of prepreg materials.

5.2 Autoclave and Vacuum-Bag Molding

The production of high-quality composites from prepreg materials (tapes and fabrics) is achieved by the consolidation on a single-sided tool of a laminated preform. The method makes use of a flexible bag or blanket through which pressure is applied to the prepreg layup on the mold. The mold is supported on a platen inside an autoclave which is pressurized with gas heated to the required cure temperature.

Unidirectional and woven materials can be molded equally successfully and the choice depends on the end use requirements and the layup techniques employed. Highly automated procedures for the precise placements of prepreg layers in laminate preforms are now well developed and are being introduced on a production basis in aircraft component manufacture. Computer-controlled tape-laying and ply-profiling machines are a major step towards a fully automated integrated composite-manufacturing scheme.

Current trends are towards prepregs with simplified cure cycles and zero resin-bleed characteristics. The careful use of viscosity modifiers in resin formulation offers greater flow control which is important for large or thick composite moldings in which significant temperature variations may exist.

Vacuum molding represents a simplified version of the autoclave technique where the pressure chamber is substituted by an oven and consolidation is achieved by virtue of the vacuum drawn inside the flexible membrane. The relatively low net pressure applied requires the resin to have easy flow characteristics.

5.3 Filament Winding

The filament-winding process is appropriate for the production of hollow carbon-fiber components which have an axis of rotation. Excellent mechanical properties are achieved by virtue of the accurate positioning of the fibers and the high volume fractions possible in the composite. The process is highly automated and can be programmed to lay down specific reinforcement patterns to comply with complex design requirements in multistressed components.

Carbon fibers can be processed successfully by both wet-winding and prepreg-winding techniques. Special attention to the gradual buildup of winding tensions and to the choice of smooth-surfaced materials for guide points is necessary in order to promote even, high-speed delivery of carbon fibers with minimum risk of abrasion and filamentation.

5.4 Mandrel Wrapping

Prepreg wrapping is used widely in the manufacture of sports goods, for the production of parallel and taped tubular shapes. Special purpose-built equipment is available for semiautomated production. Mechanical performance is determined by the layup pattern of the tube, although simple configurations are required for successful volume throughput. Unidirectional tapes and woven fabrics are equally applicable. Good material handleability is essential; unidirectional prepregs are often combined with a light scrim support to improve transverse properties.

5.5 Pultrusion

Pultrusion is a well-established technique for the production of profiles of glass-reinforced polyester composites and offers scope for CFRP manufacture. The process is simple, continuous, can be highly automated and does not require a high level of financial or technological investment. The use of carbon-fiber reinforcement with high-performance systems, such as epoxies, presents a completely new set of considerations for a continuous molding operation. The interacting effects of thermal expansion and resin shrinkage impose strict requirements for the tool design both in terms of dimensional accuracy and material of construction.

High-quality, high-performance pultruded stock is commercially available, Hybrid reinforcement configurations, new matrix materials (e.g., thermoplastics offering postformability) and methods for introducing off-axis reinforcement are being developed to extend the capabilities of the process.

5.6 Wet Layup

The contact molding method is simple and is a low cost method, but is prone to variability in quality and performance. The method is therefore not attractive for the widespread use of carbon fibers. However, the addition of a small proportion of carbon fibers as a specific reinforcement for GFRP structures can be both cost and performance effective. For example, unidirectional woven tapes strategically incorporated into GFRP laminates promote maximum stiffness for minimum added cost, and nonwoven fabrics applied as surfacing tissues for GFRP structures provide chemical resistance and electrical conductivity.

5.7 Other Methods

Thermoplastic molding compounds are processed by injection molding. Long-fiber reinforcements, combined with polyethersulfone, polyether ether ketone and other polymers can be converted to composites by thermoforming operations (see *Glass-Reinforced Plastics: Thermoplastic Resins*).

6. Applications

Applications for CFRPs fall generally into three broad categories: aerospace (aircraft, space and satellite structures), sports goods, and engineering and industrial equipment. The overriding application is that of metals substitution for improved mechanical performance and weight reduction. The stiffness-to-weight ratio for high-modulus CFRPs is the highest available for any structural material and hence carbon-fiber composites are applied where stiffness, strength and light weight are essential.

7. *Future Perspective*

Today, about 4000 t of carbon fiber per year are consumed worldwide. Compared to glass fiber used in composite materials, estimated at five to six million tonnes per year, this is indeed a modest market. The latter is now a mature product and advances at the approximate rate of world economic growth. Before carbon fiber arrives at this phase of its history, it must first establish a concrete role in the engineering consciousness and achieve an economic scale of production. Carbon fiber costs US$30 a kilogram at 1988 prices. Potentially, using polyacrylonitrile (PAN) produced on a commercial scale, it could reduce in price to US$10 a kilogram. On the basis of modulus, volume for volume, this would make CFRP less expensive than GFRP. If the early promise of using pitch as a precursor to carbon fiber should be fulfilled, the cost expectations would exceed this prediction by far.

The most important factor influencing the future of CFRP is the cost of energy. By the late 1980s, earlier anxieties about energy supplies had abated, but ultimately the driving force in the materials economy will be the energy cost of materials plus the energy cost of their use. CFRP offers considerable advantages in both respects compared to conventional materials.

Bibliography

Anon 1981 *Processing and Uses of Carbon Fiber Reinforced Plastics.* VDI, Dusseldorf

Anon 1983 Fiber composites; design, manufacture and performance. *Composites* 14(2): 87–139

Anon 1986 *Carbon Fibers, Uses and Prospects.* Noyes Data Corporation, Park Ridge, New Jersey

Clegg D W 1986 *Mechanical Properties of Reinforced Thermoplastics.* Elsevier, London

Goodman S 1986 *Handbook of Thermoset Plastics.* Noyes Data Corporation, Park Ridge, New Jersey

Kelly A, Mileiko S T 1983 *Fabrication of Composites.* North-Holland, Amsterdam

P. J. Mills
[ICI, Wilton, UK]

P. A. Smith
[University of Surrey, Guildford, UK]

Carbon Fibers

Carbon fibers of high axial strength and high Young's modulus have become important in modern technology for use in producing strong, stiff and light reinforced plastics. In this article, the potential for making carbon fibers of high axial Young's modulus is shown, and the production methods used for several types of precursors are described. Fiber structure and properties are also discussed.

1. *Potential for High-Modulus Carbon Fiber*

The elastic constants of the graphite lattice (a layer structure, Fig. 1) are 1060 $GN\,m^{-2}$ for the layer planes and 36.5 $GN\,m^{-2}$ perpendicular to these planes. From these values, a calculation of the variation of Young's modulus as a function of the angle between the layer planes and the direction of applied stress (Fig. 2) shows that a high degree of orientation of the layer planes to the direction of applied stress is necessary for a high modulus. The problem in making high-modulus carbon fibers is to produce very pronounced preferred orientation of the layer planes along the fiber axis.

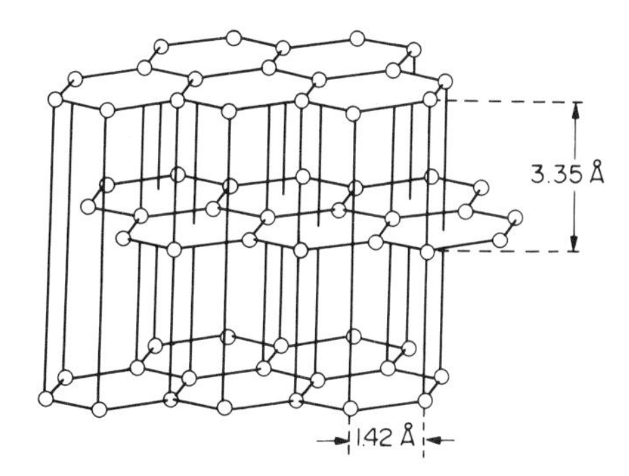

Figure 1
Graphite lattice structure

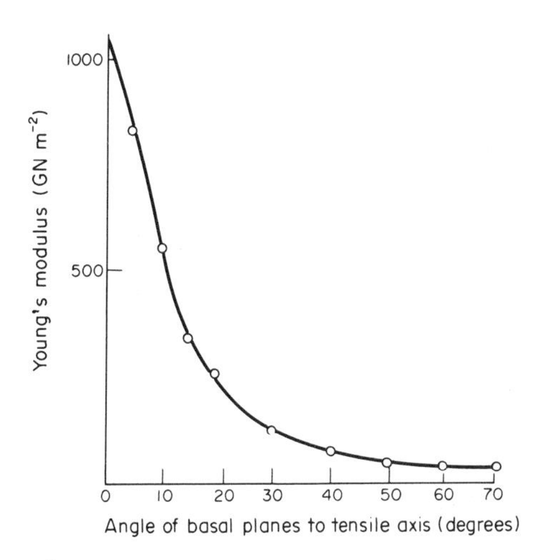

Figure 2
Young's modulus as a function of the angle of basal planes to tensile axis

2. Production of High-Modulus Carbon Fibers

2.1 Carbon Fibers from Cellulose

When a nonmelting textile fiber such as cellulosic rayon is pyrolyzed up to 1000 °C in an inert atmosphere, hydrogen and oxygen are lost as H_2O, CO_2 and CO, and a shrunken carbon fiber remains. Carbon fibers thus made from rayon have an axial modulus of only 35 $GN\,m^{-2}$ because the orientation of the layer planes is completely random. Tang and Bacon (1964) reasoned that cleavage of the glycosidic linkage occurs, resulting in complete loss of crystallinity. Hence, the subsequent formation in the latter stages of the pyrolysis is of an all-carbon structure of random orientation, despite an original orientation of the cellulose molecules. Later it was shown by Bacon and Schalamon (1969) that, if such carbon fibers are heated above 2500°C, sufficient plasticity is developed that on the application of tension, the fibers can be elongated; the layer planes are pulled into an axial preferred orientation, and fibers with Young's moduli exceeding 350 $GN\,m^{-2}$ can be produced.

This process was used commercially for a few years by the Union Carbide Corporation, but has now been abandoned for the simpler processes starting from polyacrylonitrile (PAN) fibers and mesophase pitch which do not use the troublesome very high temperature stretching.

2.2 Carbon Fibers from Polyacrylonitrile Fibers

PAN fibers ($\{CH_2CH(CN)\}_n$) are now preferred for producing most high-modulus carbon fibers. They are made in large quantities for normal textile use (e.g., Courtelle and Orlon) and are available in suitable small diameters. The yield of carbon is about 45% and the process requires no stretching at high temperature. PAN fibers are a good precursor for carbon fibers since PAN fibers have an all-carbon backbone and (of special importance) form a ladder polymer by intramolecular rearrangement when heated to 200–300°C (Scheme 1). A linear heterocyclic structure is formed which has an enhanced thermal stability. When the heating is carried out in an inert atmosphere the fibers develop a deep copper color, but heating in air gives black fibers because of an uptake of oxygen. This initial heating in air is termed the stabilization process. It has been shown (Watt and Johnson 1975) that in stabilization the ladder polymer is formed first and then undergoes oxidation and dehydrogenation to give an oxidized structure (Scheme 2). Other proposed mechanisms for the heat treatment and stabilization of PAN fibers have been reviewed by Donnet and Bansal

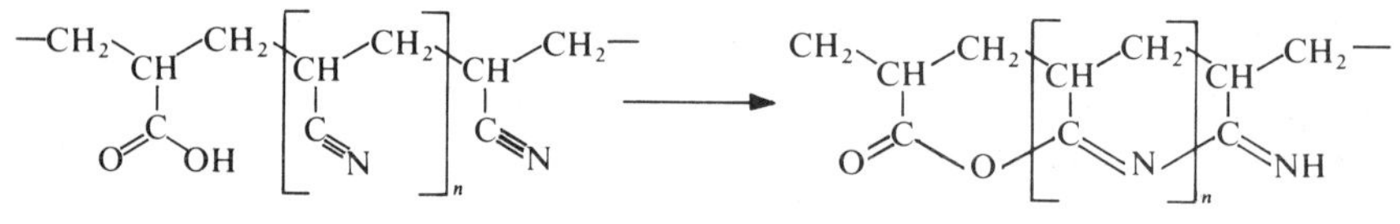

Scheme 1

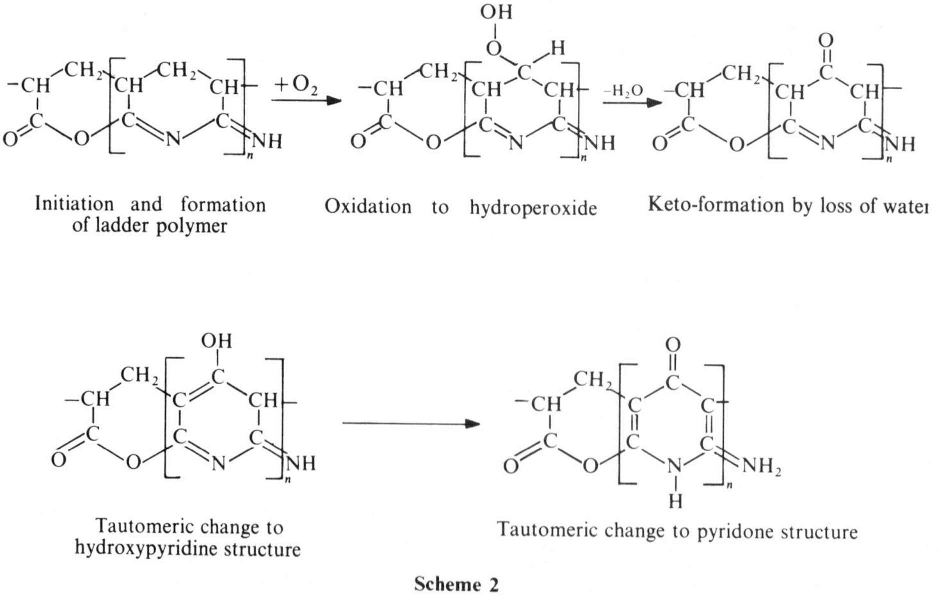

Initiation and formation of ladder polymer

Oxidation to hydroperoxide

Keto-formation by loss of water

Tautomeric change to hydroxypyridine structure

Tautomeric change to pyridone structure

Scheme 2

(1984). It is important that during the stabilization process the fibers are restrained from shrinking, or even extended, as shown by Watt and Johnson (1969). All synthetic textile fibers are stretched several times during manufacture to achieve an axial orientation of the linear molecules and thereby give the fibers the required properties for normal textile uses. PAN fibers are stretched about eight times at 100 °C after initial coagulation into a nonoriented fiber. It follows that on heating above 100 °C the fibers will shrink in length (entropic shrinkage) and lose some of their orientation. Since the ladder polymer is the template for the formation of the graphitic layer planes, it is necessary to maintain the preferred orientation of the molecules. The length changes of tows of a commercial PAN fiber after heating for 5 h in air under different loads has been measured (Watt and Johnson 1969), and Fig. 3 gives the modulus of carbon fibers as a function of these length changes during stabilization and subsequent heat treatment to 1000, 1500 and 2500 °C with no tension applied.

The linear shrinkage of stabilized PAN fibers on conversion to carbon fibers is only about 13%, but the diameter shrinkage is about 45%, indicative of the formation of a much more closely packed structure. It has been shown by x-ray diffraction that during stabilization the density increases and a diffuse oriented arc appears at a Bragg angle corresponding to $d = 0.35$ nm, the oriented PAN diffraction decreasing markedly in intensity. This shows the formation of a partially oriented sheet structure. The ladder-polymer formation and oxidation reactions are both exothermic; the former producing 33 kJ mol^{-1}, the latter about ten times this value. Hence, it is essential that the rate of heat release is controlled, otherwise the temperature rise of the fiber tows may be so great that the fibers will stick together. Thus in commercial practice continuous tows of PAN fibers are stabilized by heating in air at temperatures of 220–250 °C while moving through an oven with a very good air circulation, and subsequently pyrolyzed by passing through a series of furnaces at increasing temperatures with just sufficient tension to pull the tows through. The pyrolysis reactions in and between the oxidized ladder evolve H_2, NH_3, HCN and N_2, and at 1000 °C an oriented carbon fiber, albeit containing 5% substitutional nitrogen, results; the nitrogen is eliminated by 1500 °C. Changes in Young's modulus and tensile strength as a function of the final heat treatment temperature are shown in Fig. 4. Thus a spectrum of carbon fibers of different strengths and moduli is possible (Moreton et al. 1967).

Table 1 lists some properties of PAN-based carbon fibers which are marketed commercially. Space does not permit the inclusion of many other grades of fiber available from fiber producers. Furthermore, strength properties continue to improve while grade designations continually change. Carbon-fiber users are urged to contact fiber producers directly for the latest available fibers and their properties.

There are also a few ultrahigh modulus PAN-based fibers which are commercially available (which have Young's moduli of 500–600 GN m^{-2}), but these values are still considerably less than the value of 1060 GN m^{-2} for a single graphite crystal parallel to the layer planes. This is because the orientation of the planes in the carbon fibers approaches a maximum only in the surface layers. The enhanced orientation in the surface layers is almost certainly due to a similar effect in the precursor PAN fibers and also to the difficulty of achieving uniform stabilization across the PAN fiber; that is, there is a decreasing oxygen gradient from circumference to center in the cross section of the stabilized fiber.

2.3 Carbon Fibers from Pitch

At about the same time that high-strength carbon fibers were being developed from PAN, carbon fibers

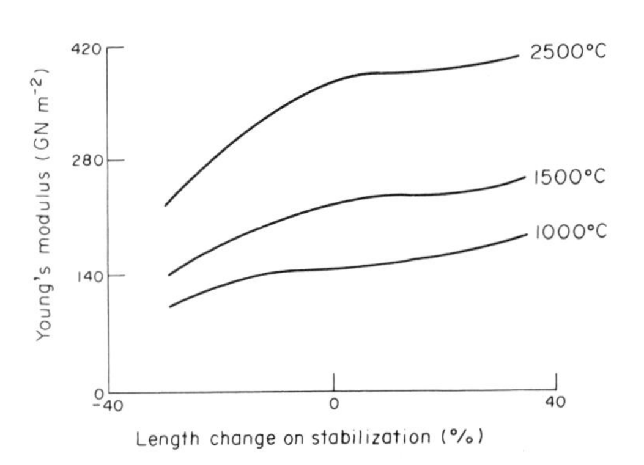

Figure 3
Young's modulus as a function of the length changes of PAN fibers on stabilization (after Watt and Johnson 1969, courtesy of the Royal Society of London)

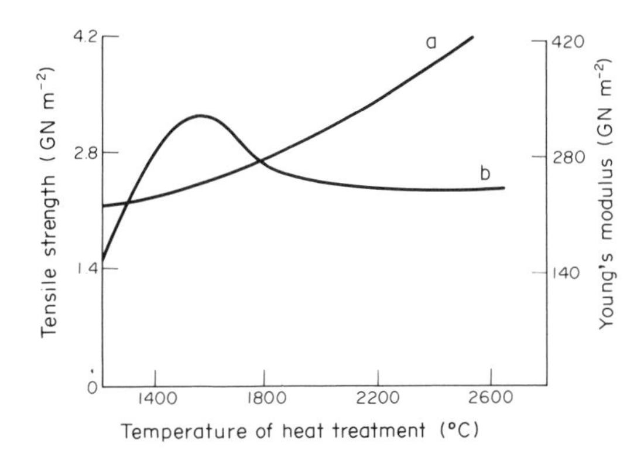

Figure 4
(a) Young's modulus and (b) tensile strength of carbon fibers from PAN fibers as a function of heat-treatment temperature (after Moreton et al. 1967, courtesy of the Royal Society of London)

Table 1
Properties of typical commercially available standard modulus (SM), intermediate modulus (IM) and high modulus (HM) grades of PAN-based fibers. (The fibers listed are available as continuous strands with 6000 or 12000 filaments per strand. Fiber producers should be contacted directly for the filament counts available for specific fiber grades)

Manufacturer and trade name	Type of fiber	Grade designation	Density ($g\,cm^{-3}$)	Young's modulus ($GN\,m^{-2}$)	Tensile strength ($GN\,m^{-2}$)	Strain at failure (%)
BASF	SM	G30	1.78	234	3.79	1.62
(Celion)	IM	G40	1.77	300	4.97	1.66
	HM	G50	1.78	358	2.48	0.7
Hercules	SM	AS4	1.80	235	3.80	1.53
(Magnamite)	IM	IM6	1.73	276	4.38	1.50
	HM	HMU	1.84	380	2.76	0.70
Hysol Grafil	SM	XA-S	1.83	227	4.48	1.95
(Grafil)	IM	IM-S	1.76	290	3.10	1.07
	HM	HM-S	1.86	379	3.00	0.74
Amoco	SM	T-650/35	1.77	241	4.55	1.75
(Thornel)	IM	T-650/42	1.78	290	5.03	1.7
	HM	T-50	1.81	390	2.42	0.70
Toray	SM	T-300	1.76	230	3.53	1.5
(Torayca)	IM	T-400H	1.80	250	4.50	1.8
	HM	M-40	1.81	392	2.65	0.6

were reported made from low-melting-point isotropic pitches (Otani 1965). Fibers were melt-spun, chemically cross-linked with oxygen and ozone to prevent them from remelting on further heat treatment, and finally carbonized. The fibers were isotropic, were not graphitizable and had poor tensile strengths and Young's moduli.

It was subsequently found by Hawthorne et al. (1970) that the crystallites in isotropic pitch-based carbon fibers could be oriented by hot-stretching them by as much as 200% at temperatures between 2200 and 2900 °C. Their tensile modulus could be increased to 620 $GN\,m^{-2}$ but the highly oriented carbon fibers were still not graphitizable, although the precursor pitches were. This hot-stretching step had the same practical difficulties that were encountered in the rayon-based process.

Shortly thereafter, Union Carbide (Barr et al. 1976) found that it was easier to orient the carbon structure at the mesophase or liquid crystal stage, and then to thermoset, carbonize and graphitize the fibers. Since the axial preferred orientation was maintained during all parts of the process, troublesome hot-stretching techniques were not required. By heat treating these fibers to graphitizing temperatures (>2800 °C), graphitic fibers were obtained with Young's moduli approaching the in-plane value for perfect graphite, ~1000 $GN\,m^{-2}$.

The formation of a liquid crystal (mesophase) as the first step in the ordering process during the thermal polymerization of pitch to coke and carbon was a remarkable discovery made by Brooks and Taylor (1965). It is at this stage that pitch, a complex mixture of thousands of different species of hydrocarbon and heterocyclic molecules, becomes an incipient graphitic structure with long-range order and orientation.

This transformation occurs as follows: as pitch is heated above 400 °C, either at constant temperature or with gradually increasing temperature, the molecules double or triple in size by dehydrogenative condensation reactions and become large enough (average molecular weights of approximately 1000) and flat enough to form layers of associated molecules which constitute a somewhat more dense, ordered nematic liquid-crystal phase or mesophase. At first the mesophase appears as small liquid spheres which collide and coalesce as shown in the schematic diagram in Fig. 5. Eventually the entire isotropic pitch can convert to this more ordered, more viscous phase which, with continued heating, would ultimately become a semisolid nematic glass, namely coke.

The aromatic molecules in the liquid carbonaceous mesophase can be depicted as a collection of slippery playing cards. The liquid mesophase is thus easily orientable above its melting point by shear or elongational forces, and can be extended and drawn into highly oriented filaments or fibers. The melt spinning–orienting process is illustrated in Fig. 6. Finally, the fibers are oxidized (to cross-link and thus prevent them from remelting), carbonized and, if de-

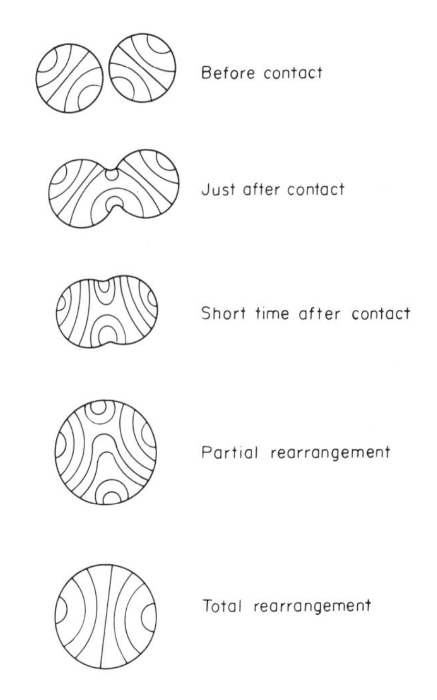

Figure 5
Schematic drawing of the collision and coalescence of mesophase spheres (after Singer 1981, courtesy of Butterworth Scientific Ltd., London)

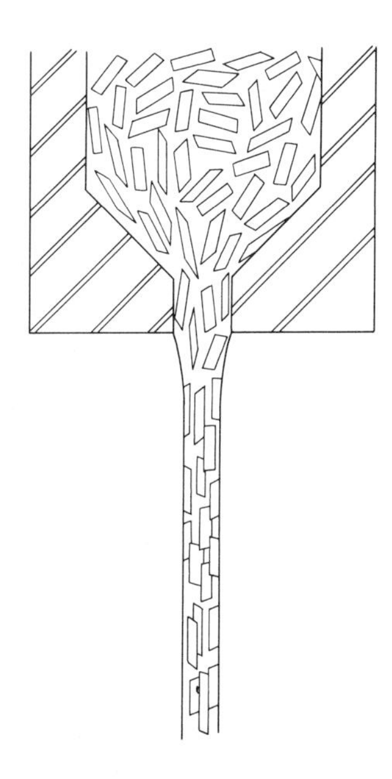

Figure 6
Schematic drawing of the orienting process during the spinning of mesophase pitch (after Singer 1981, courtesy of Butterworth Scientific Ltd., London)

sired, graphitized. The mesophase pitch as-spun fibers possess a built-in thern.odynamic driving force to convert thermally, without any tension or stretching, into a graphitic fiber possessing a high degree of axial preferred orientation and resultant high Young's modulus.

Other easily orientable pitches (Otani et al. 1972) and other processes for making mesophase pitch have been described (Diefendorf and Riggs 1980). The most comprehensive review of carbon fibers derived from mesophase pitches has been published by Rand (1985).

The structural characteristics of mesophase pitch-based fibers are unique compared to other carbon fibers derived from rayon, PAN or isotropic pitch. They are more graphitizable and possess larger crystallites and oriented domains than do fibers from other precursors. The filaments have a fibrillar-like longitudinal appearance and the large elongated domains are evident under polarized light. The transverse structure can be either circumferential, radial or random in character. Figure 7 shows SEM photomicrographs of the radial and circumferential (onion skin) types of carbon fiber.

Table 2 lists densities and tensile properties of mesophase pitch-based carbon fibers which are now available from Amoco Corporation. It is apparent from the high densities and high moduli that such fibers are capable of achieving a high degree of graphitic perfection (perfect graphite has a density of $2.25\ \mathrm{g\,cm^{-3}}$) and of axial orientation of carbon layer planes. The ease in attaining Young's moduli approaching $1000\ \mathrm{GN\,m^{-2}}$ has made these fibers particularly useful for stiffness-critical applications of composites.

Although the tensile strengths of these carbon fibers may not be as high as some of those reported in Table 1 for PAN-based fibers, their highly graphitic character is an advantage for those applications where high electrical and thermal conductivity, low axial thermal expansion, and high-temperature oxidation resistance are important. The negative room-temperature thermal expansion coefficient ($-1.4 \times 10^{-6}\ \mathrm{K^{-1}}$) of the ultrahigh-modulus fibers is an advantage for fabricating composites for which a net zero thermal expansion is desirable.

The evolution (without stretching) of the structural and electronic characteristics of mesophase pitch-based fibers with heat-treatment temperature has been described by Bright and Singer (1979). X-ray interlayer d-spacings decrease to below 0.337 nm and crystallite sizes (L_c) increase to ~ 20 nm at the highest heat-treatment temperatures, while the specific values depend upon fiber structure.

The electronic properties approach those of perfect graphite with increasing heat-treatment temperature, but there is still a significant gap between them. The transverse magnetoresistance exhibits large negative values for the pregraphitic materials and positive magnetoresistance for only the most highly graphitic

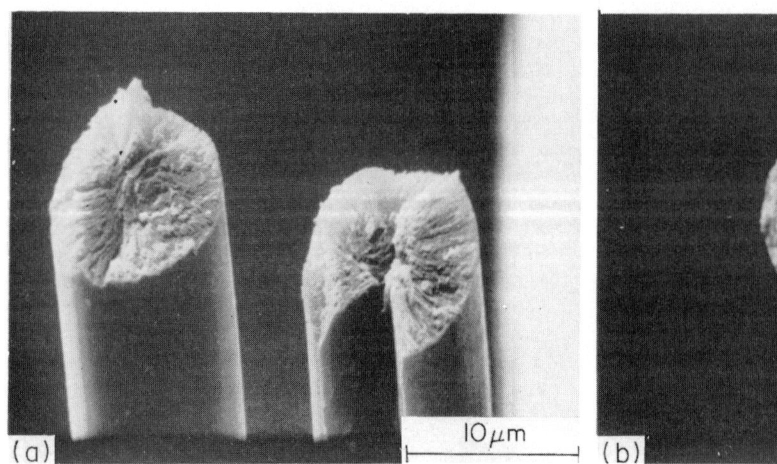

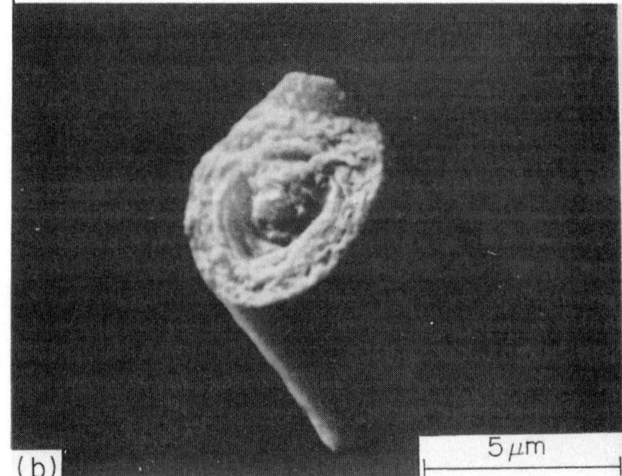

Figure 7
SEM photomicrographs of fracture surfaces of (a) radial structure and (b) onion-skin structure carbon fibers (after Singer 1981)

Table 2
Properties of mesophase pitch-based carbon fibers available from Amoco Corporation

Grade designation	Filaments per strand	Density ($g\,cm^{-3}$)	Young's modulus ($GN\,m^{-2}$)	Tensile strength ($GN\,m^{-2}$)
P-25W	4000	1.90	160	1.40
P-55S	2000	2.0	380	1.90
P-75S	2000	2.0	520	2.1
P-100S	2000	2.15	724	2.2
P-120S	2000	2.18	827	2.2

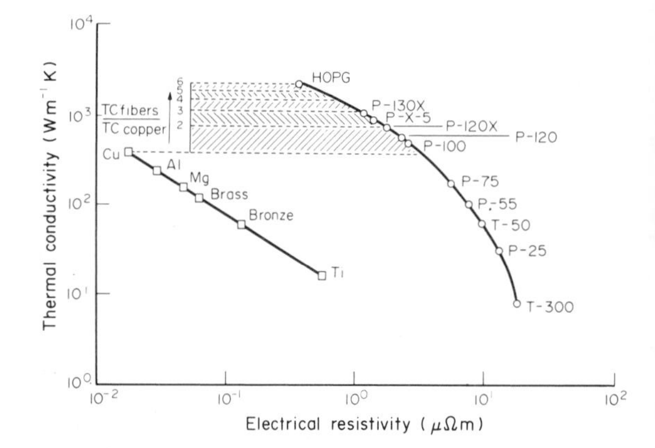

Figure 8
Comparison of thermal conductivity (TC) and electrical resistivity of carbon fibers and metals (after Kowalski 1987, courtesy of SAMPE)

fiber. Many of the electronic and structural properties correlate with a $R_{300K}:R_{4.2K}$ resistivity ratio, a quantity which has been previously observed to behave as an order parameter (Robson et al. 1973). Bright (1979) has developed a successful theory for the magnetoresistance of these pregraphitic fibers which begins with a two-dimensional band structure and takes into account the number of defects in the structure.

The thermal conductivities of ultrahigh-modulus mesophase pitch-based fibers are remarkably high and in some cases surpass that of copper at room temperature (Singer 1981). An interesting comparison (Fig. 8) of the relationship between thermal conductivity and electrical resistivity for carbon fibers and metals has been devised by Schulz and presented by Kowalski (1987). Note that the highest modulus P-130X fiber has almost three times the room-temperature thermal conductivity of copper. It will be of interest to see how closely the thermal conductivity of mesophase pitch-based fibers can be made to approach that of highly oriented pyrolytic graphite (HOPG), which has approximately six times the thermal conductivity of copper at room temperature.

Graphitic fibers such as those derived from mesophase pitch have also been excellent candidates for intercalation with various electron donors and acceptors (e.g., potassium and bromine) to enhance electrical conductivity (Murday et al. 1984). Enhancements by factors of ten to twenty have been reported. Applications of such highly conductive, lightweight carbon wire have been suggested for power transmission lines, aircraft wiring, and microelectronic circuitry. Thermal stability and cost are two considerations which may hinder large-scale applications.

2.4 Carbon Fibers by Pyrolytic Deposition

There now exists a substantial literature on the production and properties of carbon fibers prepared by pyrolytic deposition of hydrocarbons from the gas phase. Methane, benzene and naphthalene have been used at temperatures of ~1100 °C using hydrogen as a diluent.

Oberlin et al. (1976) describe the formation process in two parts: a primary process in which a thin tube of carbon 10–50 nm in diameter forms by catalytic growth on a submicroscopic iron particle and a secondary process of thickening of the filament by carbon chemical vapor deposition on the thin-tube substrate. They have developed a model of fiber growth from benzene which involves the surface diffusion of carbon species on the catalyst particles.

Their studies of structure by high-resolution electron microscopy indicate that the initial tubes consist of turbostratic stacks of carbon layers parallel to the fiber axis and arranged in concentric sheets like the annual ring structure of a tree. The secondary deposit of carbon layers by chemical vapor deposition is similar to ordinary pyrolytic carbon, that is, carbon layers approximately parallel to the substrate but made up of small turbostratic stacks which have many more defects than the catalytically formed carbon contained in the core. A description of the apparatus, general procedures, and the mechanism of producing carbon fibers from methane and iron catalyst particles has been given by Tibbetts et al. (1987). Filaments with diameters in the range of 5–30 μm and with lengths up to 5 cm were produced by these authors. No significant structural differences are observed between filaments prepared from methane and those prepared from benzene.

The unique structure of these carbon filaments has a number of property ramifications. Like other pyrolytic carbons, the filaments are highly graphitizable. For example, electron diffraction shows that the three-dimensionally ordered graphite structure partially develops at heat-treatment temperatures (HTT) as low as 2400°C (Endo et al. 1976). Further evidence for the onset of three-dimensional order comes from the observed positive transverse magnetoresistance for 2200 °C heat-treated filaments (Endo et al. 1982).

These authors have also shown that the degree of axial preferred orientation of the filaments at first increases with HTT and then begins to decrease above 2800 °C. This observation is explained by the polygonization of the filament cross section and concomitant bending of the layer planes along the polygonal surfaces. A schematic drawing of such a polygonized filament is shown in Fig. 9 (Endo et al. 1977).

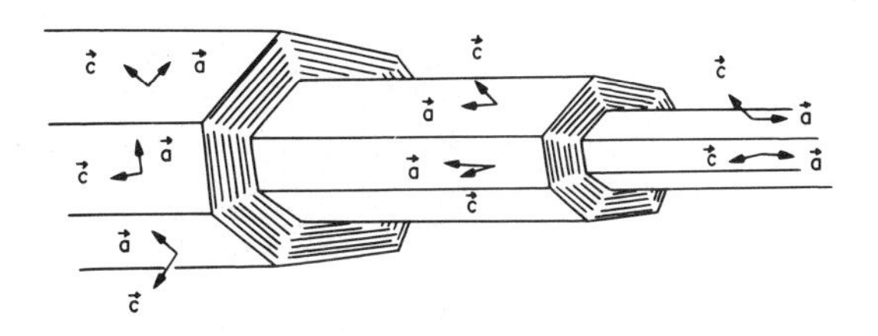

Figure 9
Schematic drawing of a pyrolytically deposited fiber heated to 2800 °C (adapted from Endo et al. 1977)

The mechanical properties of as-deposited benzene-derived fibers have been reported by Koyama et al. (1972). They report a density of 2.03 g cm^{-3} and tensile strengths and Young's moduli spanning the ranges 0.39–3.31 GN m^{-2} and 193–386 GN m^{-2}, respectively. The strength increases significantly with decreasing filament diameter.

The average strength and Young's modulus for methane-derived filaments have been determined to be 2.92 GN m^{-2} and 237 GN m^{-2}, respectively (Tibbetts and Beetz 1987). These properties are comparable to those of some commercially available PAN-based carbon fibers.

Although at present, pyrolytically deposited fibers are of short length and variable diameter, they are the most graphitizable carbon fibers known. When they are heat treated to temperatures of 3000 °C and above, their axial electrical and thermal properties approach those of HOPG. Such as-deposited or graphitized carbon fibers could be useful in certain composite applications.

3. Models of Carbon Fiber Structure

From studies of x-ray diffraction and electron microscopy of carbon fibers, various structural models have been proposed for the arrangement of carbon layers and pores in carbon fibers. Guigon et al. (1984) defined a basic structural unit (Fig. 10) which appears as a wrinkled aromatic carbon layer sheet and which is characterized by three parameters, a length, thickness, and a radius of curvature of wrinkles or folds.

These layer sheets are apparently crumpled parallel to the fiber axis and the folds are probably entangled as depicted by Bennett and Johnson (1978) in Fig. 11.

The representation in Fig. 11 is consistent with a number of aspects of carbon-fiber structure, the outside-to-inside inhomogeneity, the longitudinal slit-shaped pores, and the coexistence of small and large domains of carbon layers having greatly different degrees of orientational and graphitic perfection. In general, the domain size and overall graphitic perfection increase as one proceeds from rayon to PAN to mesophase pitch to benzene- or methane-derived fibers. Of course, increased heat-treatment tempera-

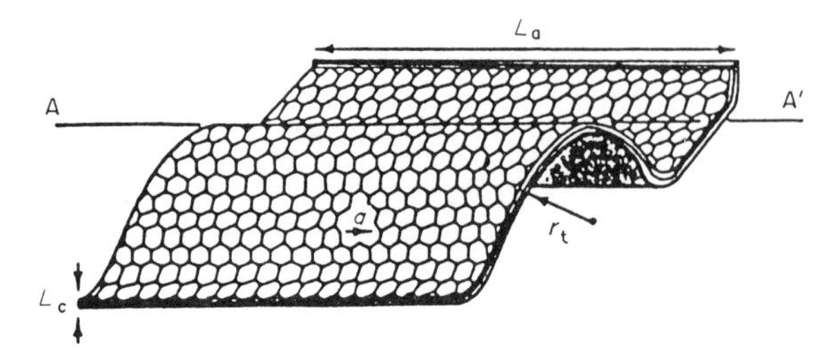

Figure 10
Schematic representation of a basic structural unit of a carbon fiber (after Guigon et al. 1984, courtesy of Elsevier Applied Science Publishers Ltd.)

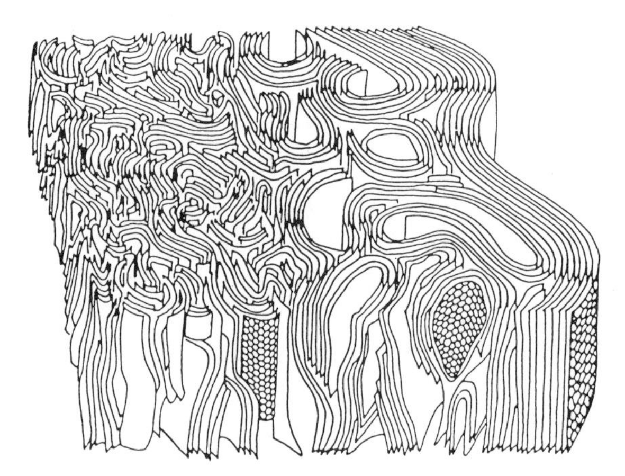

Figure 11
Schematic drawing of a three-dimensional model of a carbon fiber (after Bennett and Johnson 1978, courtesy of Society for Chemical Industry, London)

ture can improve structural order and axial preferred orientation in almost all carbon fibers.

4. Properties and Applications

In this section, the unique characteristics of carbon fibers are discussed with emphasis on which structures and precursors are most suitable for particular applications.

At present, the high-strength standard and intermediate modulus PAN-based continuous fibers listed in Table 1 have the best balance of tensile and compressive properties for structural applications. Whether the strength-limiting concentration of flaws due to pores and other intrinsic structural defects can be reduced even further remains to be seen. PAN-based carbon fibers are available with various surface treatments and sizings to improve their compatibility with many different thermosetting and thermoplastic resins in composites. Such composites, particularly with epoxy resins, have found applications in a variety of strong, lightweight aircraft and space structures, and in sports equipment as well.

Mesophase pitch-based fibers have advantages in those applications which require combinations of extremely high stiffness, high thermal and electrical conductivity, low thermal expansion, and superior oxidation resistance at high temperatures. These attributes are a result of the higher density and ease of graphitizability of these fibers. Mesophase pitch-based fibers have been found to be particularly suitable for use in carbon–carbon composites in which high modulus and high thermal conductivity are important.

New pitches, processes and fiber morphologies are under investigation in a number of laboratories (Yamada et al. 1981, Otani 1984, Hamada et al. 1987, Edie et al. 1986, Matsumoto 1985). Improvements in strength and other properties may be expected from these new pitch-based fiber technologies.

The applications of pyrolytically-deposited carbon fibers will depend upon future developments of longer lengths, better uniformity and more efficient mass-production processes. The as-deposited fibers ($\sim 1100\,°C$) could be utilized as an intermediate-modulus fiber, while the high-temperature heat-treated fibers would be used when the near-perfect graphite values of thermal and electric conductivity are important. Recent measurements (Nysten et al. 1985) have indicated that certain recently developed mesophase pitch-based fibers exhibit room-temperature electrical and thermal properties which are remarkably close to those for heat-treated pyrolytically-deposited carbon fibers.

Bibliography

Bacon R 1973 Carbon fibers from rayon precursors. In: Walker P L, Thrower P A (eds.) 1973 *Chemistry and Physics of Carbon*, Vol. 9. Dekker, New York, pp. 1–102

Bacon R, Schalamon W A 1969 Physical properties of high modulus graphite fibers made from a rayon precursor. *Appl. Polym. Symp.* 9: 285–92

Barr J B, Chwastiak S, Didchenko R, Lewis I C, Lewis R T, Singer L S 1976 High modulus carbon fibers from pitch precursor. *Appl. Polym. Symp.* 29: 161–73

Bennett S C, Johnson D J 1978 Structural heterogeneity in carbon fibers. *Proc. 5th London Carbon and Graphite Conf.*, Vol. 1. Society for Chemical Industry, London, pp. 377–86

Bright A A 1979 Negative magnetoresistance of pregraphitic carbons. *Phys. Rev. B* 20: 5142–49

Bright A A, Singer L S 1979 The electronic and structural characteristics of carbon fibers from mesophase pitch. *Carbon* 17: 59–69

Brooks J D, Taylor G H 1965 The formation of graphitizing carbons from the liquid phase. *Carbon* 3: 185–93

Del Monte J 1981 *Technology of Carbon and Graphite Fiber Composites*. Van Nostrand Reinhold, New York

Diefendorf R J, Riggs D M 1980 Forming optically anisotropic pitches. US Patent No. 4,208,267

Donnet J-P, Bansal R C 1984 *Carbon Fibers*. Dekker, New York

Edie D D, Fox N K, Barnett B C, Fain C C 1986 Melt spun non-circular carbon fibers. *Carbon* 24: 477–82

Endo M, Hishiyama Y, Koyama T 1982 Magnetoresistance effect in graphitising carbon fibers prepared by benzene decomposition. *J. Phys. D* 15: 353–63
Endo M, Koyama T, Hishiyama Y 1976 Structural improvement of carbon fibers prepared from benzene. *Jpn. J. Appl. Phys.* 15: 2073–76
Endo K, Oberlin A, Koyama T 1977 High resolution electron microscopy of graphitizable carbon fiber prepared by benzene decomposition. *Jpn. J. Appl. Phys.* 16: 1519–23
Guigon M, Oberlin A, Desarmot G 1984 Microtexture and structure of some high-modulus, PAN-base carbon fibers. *Fiber Sci. Technol.* 20: 177–98
Hamada T, Nishida T, Sajiki Y, Matsumoto M, Endo M 1987 Structures and physical properties of carbon fibers from coal tar mesophase pitch. *J. Mater Res.* 2: 850–57
Hawthorne H M, Baker C, Bentall R H, Linger K R 1970 High strength, high modulus graphite fibers from pitch. *Nature* 227: 946–47
Kowalski I M 1987 New high performance domestically produced carbon fibers. *SAMPE* 32: 953–63
Koyama T, Endo M, Onuma Y 1972 Carbon fibers obtained by thermal decomposition of vaporized hydrocarbon. *Jap. J. Appl. Phys.* 11: 445–49
Matsumoto T 1985 Mesophase pitch and its carbon fibers. *Pure Appl. Chem.* 57: 1537–41
Moreton R, Watt W, Johnson W 1967 Carbon fibres of high strength and high breaking strain. *Nature* (*London*) 213: 690–91
Murday J S, Dominguez D D, Moran J A Jr, Lee W D, Eaton R 1984 An assessment of graphitized carbon fiber use for electrical power transmission. *Synth. Met.* 9: 396–424
Nysten B, Piraux L, Issi J-P 1985 Use of thermal conductivity measurements as a method to characterize carbon fibers. *Proc. Thermal Conductivity Conf.* Plenum, New York
Oberlin A, Endo M, Koyama T 1976 Filamentous growth of carbon through benzene decomposition. *J. Cryst. Growth* 32: 335–49
Otani S 1965 On the carbon fiber from the molten pyrolysis product. *Carbon* 3: 31–38
Otani S 1984 Dormant mesophase pitch. US Patent No. 4,472,265
Otani S, Watanabe S, Ogino H, Iijima K, Koitabashi T 1972 High modulus carbon fibers from pitch materials. *Bull. Chem. Soc. Jpn.* 45: 3710–14
Pamington D (ed.) 1988 *Carbon and High Performance Fibers Directory*, 4th edn. (compiled by Lovell D R). Pammac Directories, High Wycombe, UK
Rand B 1985 Carbon fibers from mesophase pitch. In: Watt W, Perov B V 1985 *Handbook of Composites*, Vol. 1, *Strong Fibers*. Elsevier, Amsterdam, pp. 495–575
Reynolds W N 1973 Structure and physical properties of carbon fibers. In: Walker P L, Thrower P A (eds.) 1973 *Chemistry and Physics of Carbon*, Vol. 11. Dekker, New York, pp. 1–67
Riggs D M, Shuford R J, Lewis R W 1982 Graphite fibers and composites. In: Lubin G (ed.) 1982 *Handbook of Composites*. Van Nostrand Reinhold, New York, pp. 196–271
Robson D, Assabghy F Y I, Cooper E G, Ingram D J E 1973 Electronic properties of high-temperature carbon fibers and their correlations. *J. Phys. D* 6: 1822–34
Singer L S 1981 Carbon fibers from mesophase pitch. *Fuel* 60: 839–47
Tang M M, Bacon R 1964 Carbonization of cellulose fibers 1: Low temperature pyrolysis. *Carbon* 2: 211–20
Tibbetts G G, Beetz C P, Jr 1987 Mechanical properties of vapor-grown fibers. *J. Phys. D* 20: 292–7
Tibbetts G G, Devour M G, Rodda E G 1987 An adsorption–diffusion isotherm and its application to the growth of carbon filaments on iron catalyst particles. *Carbon* 25: 367–75
Watt W, Johnson W 1969 The effect of length changes during oxidation of polyacrylonitrile fibers on the Young's modulus of carbon fibers. *Appl. Polym. Symp.* 9: 215–27
Watt W, Johnson W 1975 Mechanism of oxidation of polyacrylonitrile fibers. *Nature* (*London*) 257: 210–12
Yamada Y, Matsumoto S, Fukuda K, Honda H 1981 Optically anisotropic texture in tetrahydroquinoline soluble matter of carbonaceous mesophase. *Tanso* 107: 144–46

L. S. Singer
[Consultant (formerly Union Carbide), Berea, Ohio, USA]

Composite Armor

A list of possible composite armor materials may include traditional materials such as aluminum and steel or less traditional ones such as silk and leather. In the past, many combinations have been applied and found satisfactory, usually over a narrow range of applicability. Recent efforts to increase ballistic protection and reduce the weight necessary for this protection have led to a new focus in the armor community. Two classes of materials have emerged as having great potential over a wide range: fibrous composites and ceramics. However, a vital aspect of armor design is an understanding of the properties which control ballistic performance. Unfortunately, for these new materials, the fundamental research necessary for this understanding has only recently received attention. This is mainly a consequence of the fact that the processes occurring during the projectile penetration event occur at extremely high strain rates. Few materials have been fully characterized under these conditions, least of all ceramics. In this article the material properties important in armor design are examined, briefly for fibrous composites, and in greater detail for ceramics. Ceramics will receive more focused attention because of the lack of published information on their behavior at high strain rates.

Before proceeding further, it is necessary to describe how armors are rated for performance. Two fundamental quantities always considered in describing the performance of any armor system are weight and ballistic protection. Ballistic protection is characterized in terms of the projectile used and the maximum projectile velocity withstood by the armor. A popular method is the v_{50} method, in which v_{50} is defined as the velocity at which 50% of the projectiles are defeated, and 50% penetrate completely through the armor. Many other methods are also used to determine when the armor has been defeated. Most often, a thin aluminum plate is placed behind the armor, and when either the projectile or debris from the armor penetrate the plate, the armor is considered to have been defeated. Weight efficiency is determined relative to an accepted standard material. The efficiency is expressed as the ratio of standard material weight necessary to defeat the projectile to the required weight of the new armor required to defeat the same projectile. The weight is usually computed per unit area or areal density of the armor. The standard material used by the US military is rolled homogeneous armor (RHA), a high strength steel defined by US specification MIL-A-12560.

The important point to be emphasized here is that many methods are used in armor performance characterization. One should be certain of the method used before making direct comparisons.

1. Fibrous Composite Armors

The use of fibers or fabrics as armor dates from the sixth century AD when the Chinese used silk as padding for heavy metal outer armors. It soon became apparent, however, that such metal armor was so encumbering that a lightly armored, more mobile adversary had a tactical advantage. An efficient armor composed entirely of lightweight materials was needed. The use of silk as protection against handguns was first introduced in Russia in around 1914. Unfortunately, widespread use of composites for body armor did not occur until the Korean War when the US Army introduced a protective vest composed of hardened nylon plates for use by ground troops. This vest was only effective against low-velocity fragmentation and could not stop projectiles fired from rifles. The Vietnam War brought about a new vest capable of defeating high-energy rifle bullets. This vest consisted of ceramic plates backed by a fiberglass laminate. Its 60 $kg\,m^{-2}$ weight was soon found to be an unacceptable burden for the average infantryman, so a lighter vest composed entirely of Kevlar was procured in the late 1970s and is still in use today. In addition to body armor for the military, many law enforcement and private security agencies use fibrous composites for protection. In this application, the threat is usually limited to low-energy handgun projectiles. The vest has the additional requirements of concealment and comfort. The role of fibrous composites as armor will be expanded in years to come.

As armored-vehicle designers are forced to reduce weight and increase protection, fibers will be combined with ceramics, metals and plastics to create more effective armors. These new armor systems will come in two forms, either as structural load-bearing components or as nonstructural parasitic appliqués.

The principal advantage of fibers is their high strength-to-weight ratio. Kevlar, for example, approaches a tensile strength of 2.75 GPa. Other properties such as tensile modulus, elongation, tenacity, energy absorption and sound speed have proven to be important in characterizing impact. Impact, however, is a complex problem and one should not rely on any single property as a predictor; often combinations of properties are the key. For example, glass fibers with relatively low impact strength, and polyester resins with poor tenacity, can be combined to produce a laminate of high impact strength. In addition, properties can be severely affected by strain rate. Most fibers show increasing strength and modulus with increasing strain rate, but elongation to break varies. Examination of fiber properties at the strain rate expected is important in ballistic design.

The application of fibers to ballistic protection generally takes two forms. Fabric or soft armor is constructed from multiple layers of woven fabric without a resin binder. The layers are sewn together using a common sewing thread in a straight or zigzag pattern. Composite laminate armors consist of multilayered fabrics combined with a resin binder. The application of the resin greatly affects the performance of the fabric and must be chosen to meet the given application. Depending on fiber and resin content, the laminate may be classified as either structural or nonstructural. Structural laminates usually contain a greater percentage of resin than of fiber.

Most ballistic threats which fiber composites are designed to resist are either soft lead projectiles or fragment-simulating projectiles (FSP). Ball ammunition common to handguns is typically fabricated from lead and easily "mushrooms" when striking a composite target. FSPs, on the other hand, are machined from hard steel and are used to simulate the effects of blunt-ended fragments propelled from larger exploding warheads. Fabric construction is used for the soft lead projectiles, whereas fragmentation is generally stopped using a laminate. Neither construction has proven particularly effective against the sharp point typical of armor-piercing projectiles. The sharp point can easily push aside the woven fibers and penetrate the composite.

1.1 Failure of Fibrous Composites

As a first step in the study of failure mechanisms in composites, it is instructive to examine the response of a single fiber to impact. In this discussion, it is assumed the fiber is oriented such that its longitudinal axis is perpendicular to the path of the projectile. At the moment of impact, a compressive wave is propagated outward along the longitudinal axis of the fiber. The speed of this wave, c, is defined by

$$c = \left(\frac{E}{\rho}\right)^{1/2} \tag{1}$$

where E is the fiber modulus and ρ is the density. The wave reaches the end of the fiber and is reflected as a tensile wave. Reflection of the wave as a tensile wave is necessary in order to satisfy the boundary condition of zero strain at the fiber ends. The resulting tensile strain is now experienced by the fiber in contact with the projectile, and the magnitude of strain is dictated by the impact velocity of the projectile. Continued wave reflections intensify the tensile strain, with energy absorbed by the fiber being proportional to this tensile strain. At the same time, a second wave propagates along the transverse axis (across the diameter) of the fiber moving parallel to the projectile and at the same velocity. Because of this transverse wave, the fiber deflects in the direction of the projectile path and as a result increases the energy absorbed by the fiber. The fiber continues to deflect and absorb energy until the projectile is decelerated to a stop or the fiber strains past its dynamic yield point and breaks. If the impact velocity is sufficiently high, the fiber cannot respond fast enough to exhibit a strain. The lowest velocity for which this occurs is the critical velocity and is defined as the velocity for which the fiber will break without straining. Critical velocity is a function of wave speed and in turn a function of fiber modulus and density. Increasing the wave speed of a fiber will allow quicker strain response and thus increase its critical velocity.

The mode by which the fiber separates is also important in determining its resistance to penetration. Three principal modes have been identified: melting, brittle fracture and plastic deformation with longitudinal splitting. Nylon fibers melt but exhibit large deformations before breaking. High values of elongation equate to large work-to-rupture values and thus more energy absorption by the fiber. Brittle fracture with low elongation-to-break is typical of glass fibers. Kevlar fibers do not show as high a value of elongation as nylon, but have proven to be ballistically superior. This may seem to be a contradiction until one examines the fiber breaking mode. Kevlar fibers show some degree of plastic deformation before rupturing. More important, however, is the fibers' ability to split along the longitudinal axis. Longitudinal splitting not only absorbs energy but acts as a crack arrester as well.

Failure in composite laminates is a far more complex process. The process is three-dimensional in nature, characterized by interactions occurring in the individual fabric planes and across those planes to adjacent fabric layers. Additional strain-wave reflections at fiber crossovers serve to distribute impact energy over a larger area, thus increasing ballistic resistance. Because of the numerous crossovers found in woven fabrics, the energy is quickly dissipated as the wave encounters more and more fiber crossovers and the impact damage is usually restricted to a small area adjacent to the point of impact. Energy transfer to adjacent layers (through the thickness) is facilitated by use of a resin binder. Strain waves are transmitted from the fabric layer to the resin matrix and then to adjacent fabric layers.

At the same time, waves are reflected back into the fabric or matrix at each interface. The amplitudes of these waves are determined by a quantity called mechanical impedance I, defined as:

$$I = \rho c \tag{2}$$

where ρ is the density and c is the wave speed defined in Eqn. (1). Waves may be reflected or transmitted according to the following relationships;

$$P_t = \left(\frac{2I_t}{I_t + I_o}\right) P_o \tag{3}$$

$$P_r = \left(\frac{I_t - I_o}{I_t + I_o}\right) P_o \tag{4}$$

where P_o, P_r and P_t are the pressure amplitudes of the incident wave, reflected wave and transmitted wave, respectively. I_o and I_t are the impedance of the incident and acceptor materials, respectively. As an example, suppose that Kevlar with an impedance of $1784\,\mathrm{g\,cm^{-3}\,s^{-1}}$ is used with an epoxy resin with an impedance of $320\,\mathrm{g\,cm^{-3}\,s^{-1}}$. Further suppose that the wave is travelling from the Kevlar to the epoxy. At the interface, a wave of amplitude $-0.696\,P_o$ is reflected back into the glass and a wave of amplitude $0.304\,P_0$ is transmitted to the epoxy matrix (a negative sign indicates a tensile wave).

Transmission of waves to the resin matrix increases the energy-absorption capability of the laminate. Cracks are formed in the matrix which propagate parallel to the fabric layers. This process is called delamination and is characterized by separation of the fabric layers. Impact energy is dissipated in the formation and advancement of these cracks and in the deflection caused by separation of the individual fabric layers.

1.2 Designing with Fibrous Composites

No armor design will be applicable to all situations. The designer is often faced with compromises and must tailor the armor to the specific application. The ballistic performance of the system often depends on the interaction of its various components. It is difficult to separate out the characteristics important in each case but some generalizations can be made. The following paragraphs will examine some of the properties important in selecting components in a fibrous armor design. The properties to be examined include fiber selection, yarn denier, weave, resin binder and environmental susceptibility.

As was seen earlier, ballistic efficiency resulting from individual fiber properties is difficult to characterize. Modulus, wave speed and elongation to break, however, tend to be the predominating factors involved. Increasing these properties will result in a greater energy absorption capability of the armor and thus an increase in protection.

Yarn denier is a measure of the diameter of the spun yarn used to produce the fabric. Denier numbers increase with increasing yarn diameters. Fabrics woven from high-denier yarns will have fewer fiber crossovers and less ability to distribute energy. Ballistic protection will usually decrease with increasing yarn denier with all other factors held constant.

Following the same logic, a close weave with many crossovers will be more efficient than a loose weave. In fabric armors, this seems to hold for handgun projectiles, but fragmentation-type projectiles show a lesser dependence, the difference being the area involved in the impact. Armors made with resin binders do not show a strong dependence on weave construction. The effect of weave is probably masked by the resin–fiber bonding which distributes the impact energy more efficiently.

Ductile resins such as vinyl esters perform better than more brittle resins like epoxy. Resins showing increased ductility will absorb more energy both in crack initiation and propagation. Evidence suggests that, for the best ballistic efficiency, the resin content should be 20–25 wt%. Laminates with less resin content show increased ballistic protection, but the deformations are severe and unacceptable for most applications. Some applications require the laminate to serve as a structural component as well as to give ballistic protection. The best mechanical properties occur at much higher resin contents, so a compromise must be reached between mechanical and ballistic performance.

Because of environmental factors, the fiber with the best ballistic properties may not make the best armor. Many fibers degrade when exposed to sunlight, moisture, seawater, temperature extremes and petroleum products. Military, and to a lesser extent civilian, armor designs, must take environmental factors into account. Fibers must be protected, or resistant fibers selected, when designing any such armor. In addition, cost, weight and space requirements may determine the design. No military commander will opt for an armor that is too heavy or costly when a lighter, less-expensive material will suffice. An armor whose cost does not justify the additional weight saving will not be implemented. For maximum ballistic efficiency, fibrous armors must be allowed to deflect or delaminate. For this reason, fibrous armors must never be combined with rigid components in a way which restricts deflection. When minimal space is allowed for the armor, fibrous armors are not the best choice.

2. Ceramic Composite Armors

If the definition of ceramics is limited to refractory materials, the history of their use as armor materials is a short one. A broader definition, however, extends their use to several thousand years. Stone, and more recently concrete, have long been recognized for their effectiveness as defensive materials against fragmentation, projectiles and blast. In the defensive posture, weight is not an issue. Offensive operations, on the other hand, require a high degree of mobility. Materials selected for this kind of mission must be light and effective. This requirement lead the US Navy to evaluate ceramics for use on warships during World War II. It was found that plate glass backed by Doron (a glass-reinforced polyester) was an efficient combination against fragmenting warheads and resulted in weight savings over the existing steel armor. The question of structural integrity, however, prevented its acceptance. During the Vietnam War, the search for lightweight armors again turned to ceramics. In the early 1960s, Goodyear Aerospace Corporation and the US Army Materials and Mechanics Research Center investigated the use of ceramics for aircraft and river patrol boats, and for protection of ground combat personnel and aircrews. In 1970, Goodyear was issued a patent for an armor design consisting of a ceramic front face backed by a resin-impregnated glass fabric. The bonding agent between the ceramic and glass fabric was identified as an elastomeric adhesive such as polysulfide.

The principal advantages of ceramics over conventional steel are their high compressive strength and hardness. Hardened and sharp-pointed armor-piercing projectiles are effective against softer materials. This effectiveness is lost in impacts with ceramics because the dynamic stress limits in ceramics are much greater than in the projectiles. The sharp point is quickly eroded away leaving a less-effective blunt cylinder. The loss of projectile mass reduces the energy of impact and increases the protection afforded by the ceramic system. The system aspect of ceramic armor design is important and must be emphasized. Because of the inherent brittleness of ceramics, they may never be suitable as a "stand alone" armor. Ceramics are often combined with other more ductile materials to achieve the required durability and structural integrity. The following section will examine the failure mechanisms in ceramics during impact from a theoretical point of view, followed by a discussion of important aspects of ceramic armor design. Design aspects are the result of the authors' observations and experience and are not founded on strict theoretical principles. Much fundamental research is needed before ceramic armor design is reduced to an exact science.

2.1 Failure Mechanisms in Ceramics

Failure of the ceramic during impact is a multistage process. These stages are: (a) erosion of the projectile tip and formation of surface waves; (b) failure of the ceramic in compression; (c) penetration into the ceramic and formation of the fracture conoid; (d) failure of the ceramic in tension; and (e) removal of ceramic debris. Discussion of each stage will be taken in turn. The discussion assumes the projectile velocity is sufficiently small to prevent hydrodynamic flow in the ceramic. Projectile material may, however, fail by flowing if the impact stress is sufficiently high.

When ceramics are confined, their compressive strength will increase, as outlined in the following section. The confinement does not have to be by outside forces but merely by the presence of ceramic material adjacent to the point of impact. The first few microseconds of the impact event are characterized by this increased compressive strength and the projectile is quickly overwhelmed. The projectile tip is eroded away and initially flows radially outward and then backward. Surface waves are propagated radially outward from the point of impact. When these waves reach the lateral boundaries of the ceramic, they are reflected back in the form of relief waves.

Relief waves reach the impact point, relieving the confinement of the ceramic. A decreased compressive yield strength results. The projectile is no longer severely overmatched and begins to penetrate into the ceramic. At the same time, combinations of compressive waves and relief waves begin to form the fracture conoid. In its final form, the fracture conoid will form a cone of fracture surfaces and debris with an apex angle of the order of 60° and a base on the rear surface of the ceramic opposite the point of impact.

The compressive wave mentioned earlier forms a spherical front which travels ahead of the projectile parallel to the direction of impact, again propagating at the wave speed. When this wave reaches the rear surface of the ceramic it is reflected as a tensile wave as described earlier in Eqn. (4). The ceramic, being very weak in tension, fails; this tensile failure propagates back toward the point of impact. When the tensile fracture zones meet with the fracture conoid propagating rearward, the conoid is complete. The final stage is characterized by the pushing of the remaining projectile and ceramic debris from the conoid out of the rear of the ceramic.

2.2 Ceramic Behavior Under Confining Pressure at High Loading Rates

Although the ceramics envisioned for armor use are typically extremely brittle, their mechanical behavior nevertheless is remarkably sensitive to two parameters related to such applications: confining pressure and strain rate. Relevant knowledge regarding these factors can be outlined as follows.

Uniaxial compressive stress–strain curves for ceramics such as alumina show that these materials exhibit no plasticity and fail by rubblization. As increasing confining pressure is applied, the modulus is unaffected, but the failure strength increases significantly; this increase is caused by the superposition of compressive stresses which reduce the opening of microcracks parallel to the nominal loading axis. If the failed (rubblized) ceramic is subsequently tested, increases in confinement raise both the slope of the stress curve (reflecting compaction of the fragments) and also the compressive strength. This behaviour is summarized for AD-85 alumina as a function of confining pressure in Fig. 1. The compressive strength of solid alumina contrasts markedly with that of the same material in powder form.

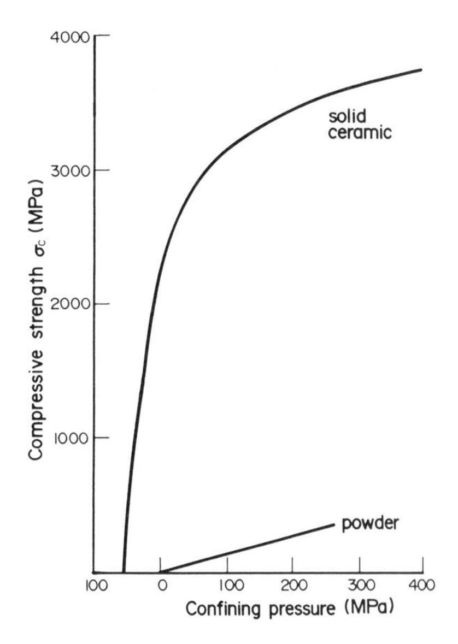

Figure 1
Compressive strength versus confining pressure for AD-85 alumina

Such pressure dependence is not accounted for by the familiar von Mises criterion which describes the flow of metals, since the criterion is pressure independent. Instead, the Coulomb–Mohr equation (or a higher order variant)

$$(J_2)^{1/2} = k - aP \tag{5}$$

is used, where J_2 is the second invariant of the deviatoric stress tensor, P is the hydrostatic pressure, and k and a are constants.

It should be noted that while confinement in the experiments described above is provided by a hydrostatic test chamber, the impact of a penetrator on a target provides confinement in and of itself because of the transient unixial strain conditions which are obtained. The high intrinsic compressive strength of the ceramic, strongly abetted by confinement, severely overmatches the strength of the penetrator, resulting in erosion of the latter as it is turned and flows in the opposite direction (failure). This is true even after the ceramic has failed, as long as the resulting powder is physically confined (Fig. 1).

Compressive strength (σ_c) versus strain-rate ($\dot{\varepsilon}$) data for several technological ceramics are shown in Fig. 2. It is evident that for $\dot{\varepsilon} \gtrsim 100\,s^{-1}$, some of the materials (SiC and hot-pressed Si_3N_4) exhibit a marked strain-rate strengthening, whereby

$$\sigma_c \propto \dot{\varepsilon}^n \tag{6}$$

where $n = 0.33$, while for others (Al_2O_3, reaction bonded Si_3N_4) no such strengthening is observed.

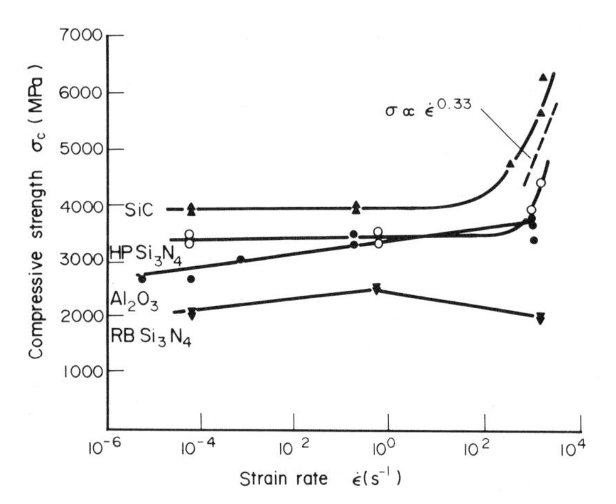

Figure 2
Compressive strength versus strain rate for several ceramics (HP, hot pressed, RB, reaction bonded)

These results suggest that there exists a threshold strain rate ($\dot{\varepsilon}^*$) for dynamic hardening which varies from one material to another (and which for Al_2O_3 and reaction-bonded Si_3N_4 obviously must exceed $2 \times 10^3\ s^{-1}$). The dynamic fracture analysis of Grady (1982) and Kipp et al. (1980) indicates that this should be so, and predicts Eqn. (6) above. Specifically, it is shown that the failure stress corresponding to the coalesence of a multitude of microcracks is given by

$$\sigma_c = (\rho_0 c_0 K_{IC}^2 \dot{\varepsilon})^{1/3} \tag{7}$$

where ρ_0 is the density, c_0 is the wave speed, and K_{IC} is the fracture toughness. Physically, Eqn. (7) reflects the fact that cracks possess "inertia," related to the finite time interval required for a stress wave to travel the length of a crack.

It should be noted that not all strain-rate-hardening brittle materials give evidence that $n = 0.33$. Recent work involving fiber-reinforced ceramic-matrix composites, in which failure occurs through more complex micromechanics (fiber kinking and shear-band propagation) than the simple coalescence of a relatively homogeneous array of microcracks, has shown that the strain rate exponent n can approach unity, depending upon the specific microstructure. This potential variability is important, since it will be shown that fragment size under impulsive loading is a sensitive function of n, and the higher the value of n, the lower is the strain rate (or ballistic velocity, for armor applications) at which the ultimate strength is achieved. It thus appears that appropriate reinforcement design would permit the fabrication of composite ceramics with dynamic strengths much higher than their monolithic matrices alone.

Theoretical estimates of the lower bound strain rate $\dot{\varepsilon}^*$ for inertial control of microfracture in monolithic materials yield

$$\dot{\varepsilon}^* = \frac{K_{IC}}{\rho_0 c_0 r_0^{3/2}} \tag{8}$$

which is a function of the initial crack size r_0. Substituting appropriate values of ρ_0, c_0, r_0 and K_{IC} for armor-type ceramics yield $10^3 \lesssim \dot{\varepsilon}^* \lesssim 10^4\ s^{-1}$, in basic agreement with Fig. 2.

It is possible to carry these concepts further, to arrive at the beginnings of an understanding of the factors which control dynamic fragmentation of brittle materials under ballistic loading conditions. In particular, it has been shown that the mean size L_M of such fragments can be described by:

$$L_M = A\dot{\varepsilon}^{(m/m+3)} \tag{9}$$

where A is a constant and m is the Weibull modulus for dynamic flaw activation. However, m turns out to be related to the dynamic compressive strength (Eqn. (6)) as well, according to:

$$n = 3/(m + 3) \tag{10}$$

Thus, if n should vary for different materials, so should m, hence the average fragment size. Fragment dimensions are potentially very important in penetrator erosion, since they determine the extent to which the fractured ceramic can behave like a fluid in the path of the penetrator as well as the density and surface area of eroding particles.

The fracture behavior of ceramic materials can be summarized as shown in Fig. 3. For $\dot{\varepsilon} \lesssim 10^2\ s^{-1}$,

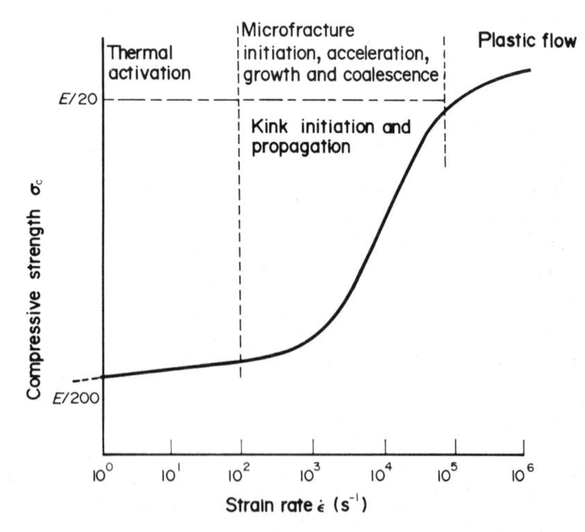

Figure 3
Schematic of strength–strain rate behavior of ceramic composites. E is Young's modulus. By adjusting the material microstructure, the steeply rising portion of the curve may be moved about within the range $10^3 \lesssim \dot{\varepsilon} \gtrsim 10^5\ s^{-1}$. The $E/20$ line is not necessarily the HEL; there is evidence that microfracture can occur below and above the HEL for brittle materials

fracture is controlled by thermal activation of microcracks, a process that is unimportant in ballistic penetration. At higher loading rates, inertial and flaw propagation (microcrack growth, fiber kinking) effects provide significant strain-rate hardening. For $\dot{\varepsilon} > 10^4\ s^{-1}$, microfracture is suppressed, and plastic flow begins to occur as the Hugoniot elastic limit (HEL) is exceeded. Recent work on Al_2O_3 suggests that the HEL represents true plastic flow, and that microfracture, although probably plasticity related, is not initiated until stresses exceed at least twice the value of the HEL.

2.3 Designing with Ceramics

A typical ceramic armor design will consist of a forward spall cover, ceramic, adhesive and backing plate. Many variations of this design are possible of course, but a description of this design will aid in illustrating the important aspects of ceramic armor design. Most industrial ceramics suitable for use in armors are available in a square or rectangular configuration. This shape will be assumed in the following discussion and will be referred to as ceramic tiles. Two important criteria not discussed previously apply to ceramic armors more than any other type of armor. These criteria are multihit capability and collateral damage. Multihit requirements are usually specified in terms of the maximum number of projectile caliber diameters allowed between impact locations without adversely affecting the protection afforded by the design. Collateral damage is a phenomenon often seen in improperly designed ceramic armor. Tiles adjacent to the impacted tile are cracked or damaged by the transmission of shock waves or debris ejected from the impact. This effect greatly reduces the multihit capability and overall desirability of the design. The armor designer should strive to reduce susceptibility to collateral damage.

Forward spall describes the ejection of projectile erosion products and ceramic debris out of the armor opposite to the direction of impact. Often this is of no concern and is beneficial in that it reduces the amount of energy the armor must absorb. If, however, the armor is to be used in close proximity to personnel, forward spall must be reduced. This can most easily be done by covering the ceramic with a flexible material such as a glass-reinforced plastic.

Ceramic material properties important in armor design include compressive strength, impedance, density and tensile strength. Ballistic efficiency can be correlated with the average of confined and unconfined compressive strengths

$$(Y + l_H)/2 \tag{11}$$

where Y is the unconfined compressive strength and l_H is the Hugoniot elastic limit, a measure of the confined strength. High tensile strength is desirable to delay the onset of tensile failure on the tile rear face described earlier. Delaying this failure only a few microseconds will significantly increase the perforation resistance of the ceramic. Low density allows the use of thicker tiles for the same areal density and reduces the stress on the back plate. Impedance should be high, as this is a measure of distortion produced in the ceramic. In short, the desired ceramic should have high compressive and tensile strength, high impedance and low density.

It has been observed that a single tile performs better than a stack of multiple tiles of equivalent thickness. The reason is that the single tile has fewer free boundaries where tensile cracks can form, particularly with respect to the time for wave travel to the rear surface of the tile. This time difference is large enough to greatly influence penetration resistance. The lateral dimensions of the tile should be as large as is consistent with acceptable multihit criteria. The larger the distance from the impact point to the sides of the tile, the longer it takes for relief waves to reach the point under the impacting projectile. The ceramic is then no longer confined and will fail in compression more readily. This represents a paradox for the ceramic-armor designer. A tile with large lateral dimensions is ballistically more efficient but has less capability for multihit. The thickness needed is dependent on the type of projectile and velocity. Generally, since lateral dimensions are larger than the thickness, thickness will be the predominant design factor. The final design thickness is often dictated by the corners and intersections of adjacent tiles. At corners where four tiles intersect, the ballistic efficiency may be reduced by as much as 30% compared to center-of-tile impacts. This reduction is a function of the gap size, so gap size should be reduced to the minimum value that does not cause collateral damage.

As mentioned earlier, collateral damage is a tremendous problem in ceramic armors. It has been observed that this problem can virtually be eliminated by placing an elastomer in the lateral gaps between tiles. The elastomer helps to dissipate the shock transmitted between tiles, but more importantly it prevents debris from impacting into adjacent tiles. A thin layer of elastomer between tiles and the back plate will also aid in mitigating transmitted shock waves and further reduce collateral damage. This aspect is more important if a metallic back plate is used, and less so if fibrous composites or other soft materials are used.

Backing plates are necessary in ceramic armors because of the inability of ceramics to provide structural integrity. In addition to a structural role, the backing plate catches the residual debris that is pushed out of the rear of the fracture conoid. In order to perform this task, the back plate must be a ductile material. Popular choices are aluminum and fibrous composites. Care must be taken not to allow too much deflection in the backing plate. Severe distortion can affect the armor's ability to sustain multiple impacts and remain structurally sound. The ratio of ceramic to back plate thickness should be kept between 50 and 60%.

Adhesives are often used to secure the ceramic to the backing plate. In general, the adhesive has little effect on the ballistic performance of thick tiles, but has a significant effect on thin tiles. In thin tiles, the adhesive thickness should be kept as thin as possible to still secure the tile to the backing plate. Typical thicknesses are on the order of 0.5 mm. Care should be taken to ensure the adhesive is not too spongy and that it adequately transfers stress from the ceramic to the back plate. In other words, the adhesive must not sacrifice the rigid support provided by the back plate.

Bibliography

Agarwal B D, Broutman L J, 1980 *Analysis and Performance of Fiber Composites.* Wiley, New York.
Grady D E 1982 Local inertial effects in dynamic fragmentation. *J. Appl. Phys.* 53: 322–25
Grady D E, Lipkin J 1980 Criteria for impulsive rock fracture. *Geophys. Res. Lett.* 7: 255–58
Kipp M E, Grady D E, Chen E P 1980 Strain-rate dependent fracture initiation. *Int. J. Fract.* 16: 471–78
Kolsky H 1963 *Stress Waves in Solids.* Dover, New York
Lankford J 1987 Temperature, strain-rate, and fiber orientation effects in the compressive fracture of SiC fiber-reinforced glass-matrix composites. *Composites* 18: 145–52
Liable R C (ed.) 1980 *Methods and Phenomena* 5: *Ballistic Materials and Penetration Mechanics.* Elsevier, New York
Lindholm U S, Yeakley L M, Nagy A 1974 The dynamic strength and fracture properties of Dresser basalt. *Int. J. Rock. Mech. Min. Sci. Geomech. Abstr.* 11: 181–91
Zukas J A, Nicholas T, Swift H F, Greszczk L B, Curran D R 1982 *Impact Dynamics.* Wiley, New York

J. Lankford and W. Gray
[Southwest Research Institute, San Antonio, Texas, USA]

Continuous-Filament Glass Fibers

Some of the earliest Egyptian glass vessels were built up from glass fibers wound by hand round a core of shaped clay. With the development of glass blowing, the use of fibres was essentially that of decoration of other products. Around the beginning of this century a commercial process for glass "silk" production was being developed in Germany in which fibers were drawn out of holes bored in refractory furnaces and wound onto wheels or drums.

Over the last 50 to 60 years the glass-fiber industry has grown out of all recognition and today the vast majority of glass fibers fall into one of two main categories:

(a) staple fiber, of discrete lengths and generally not straight, produced by some type of blowing or spinning process in the form of a loose mat or "wool"; and

(b) continuous-filament fiber, straight and of very great length, produced by a drawing and winding process.

Fibers in category (a) are used for thermal insulation and are not discussed further in this article; those in the latter category are used for load carrying reinforcement in composite materials such as fiber-reinforced plastics and cement, and are described below.

A further small but important range of products lies somewhere between these two main categories. Semi-continuous filaments are drawn from a bushing and twisted into yarn or made into a fine mat (tissue) by processes similar to papermaking. The fibers are heavily damaged and these staple fiber tissues, yarns or mats woven from yarns are not normally used as the main load-carrying fibers in composites. However, they do have some type of "reinforcing" role and are referred to in Sect. 4.

1. Fiber-Product Forms

In order to handle large volumes of fiber conveniently, and incorporate them in a composite material without balling, fluffing or excessive voidage, it is nearly always essential to have fibers which are straight and which are available in bundles rather than as individual filaments. Further, the strength of glass fibers can be easily and drastically reduced by abrasion. To obtain useful reinforcing fiber the surface must be protected from too much damage. These requirements have been met in the development of the continuous-filament process in which straight fibers are produced by drawing continuously from a bushing (see Fig. 1), protection is achieved by a size coating and fibers are grouped together in bundles or strands containing ~100 filaments before being wound onto a rotating mandrel.

Reinforcement fibers are made available to the end user in a variety of forms depending on the type of product to be made, the nature of the matrix material

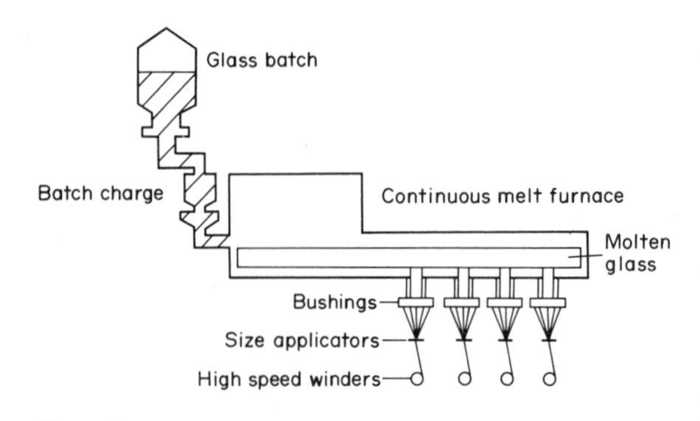

Figure 1
Direct melt process for glass fiber

and the manufacturing process used. The main varieties are as follows.

(a) Chopped strands—short lengths of fiber in bundles containing ~100 individual filaments held together by the size coating and used in automated mixing, molding and pressing processes with thermoplastic or thermosetting resin matrices or for cement reinforcement. Bundle or chopped-strand lengths may range from ~3 mm to 40 mm.

(b) Chopped strand mat—chopped strands, generally in the 30–40 mm length range, are distributed across a conveyor with a small amount of additional organic binder to form a loosely bound open mat which may be draped over a former and readily impregnated by resin in the hand lay-up process. Size and binder formulations are selected to be compatible with either epoxy or polyester resins.

(c) Rovings—a number of strands are grouped together and wound, without twisting, onto a cylindrical package to give a long continuous rope or tow which may be used for filament winding of pipes or cylinders for weaving, or for chopping or spraying, with resin or cement, onto a former or mold.

(d) Yarn—essentially twisted strands or rovings used in more complex weaving processes.

Within the category of continuous-filament reinforcement fibers there are two further main distinctions:

(a) glass fiber of a traditional range of compositions used for reinforcing plastics and rubbers, and

(b) a range of fibers based on extremely alkali-resistant glass compositions used for reinforcing cement and concrete.

The manufacture and application of all types of glass fiber are described in Mohr and Rowe (1978). A comprehensive account of continuous-filament manufacture and materials has been given by Lowenstein (1983), and useful surveys of materials, properties, uses and processes by Mettes (1969) and Gagin (1980).

2. *Fiber Production*

In a modern large-scale production unit the raw materials, or glass batch, are melted in an oil or gas-fired tank and fed along refractory canals to electrically heated platinum–rhodium alloy bushings. Several hundred individual filaments are then drawn simultaneously from specially shaped tips in the base of each bushing (Fig. 1). In an older process, still used in some small production units, premelted glass marbles were used to feed the bushing directly—called a marble melter.

A constant head of molten glass is maintained in the tank, refractory canals (known as the forehearth) and bushings, and the temperature of the glass in the bushings is controlled to very fine limits to provide a consistent viscosity. The diameter of the bushing tips is generally in the range 1–2 mm.

Molten glass flows through the tips under gravity, the rate of flow being governed by the viscosity and head of molten glass and the diameter of the tip. Glass beads flowing through the tips are gathered and the attached fibers are passed around a rapidly rotating collett. Fibers are then drawn away at linear speeds between 1000 and 2000 m min^{-1}, diameter being controlled by the drawing speed.

Just below the bushing and before they come into contact with each other or any other equipment, the fibers pass lightly over an applicator roller and are coated with an aqueous polymer emulsion or size, which protects them from abrasion. They are then gathered together in strands of 100–200 or more filaments before being wound onto a cardboard sleeve on the rotating collett to form a cake. Cakes are heated to dry and/or cure the polymer emulsion, this binds the individual filaments together in their strands which form the basic raw material for the production of the various continuous filament reinforcement products such as rovings, chopped strands, chopped strand mats and yarns, described above.

In commercial production a bushing may contain up to 2000 tips. Breakage of one single filament will cause a drip of glass across the remaining filaments, rapidly leading to breakdown of the drawing operation for the whole bushing, with severe effects on production cost. Fiber breakage can be due to mechanical causes, e.g., snagging or abrasion; to particles in the glass, e.g., undissolved batch, refractory dust or devitrification; or to hydrodynamic instabilities due to viscosity/winding speed variations, etc.

Stable fiberizing occurs within a narrow range of viscosity near to 100 Pa s, this governs the fiberizing temperature and hence bushing life. High-viscosity compositions require both higher fiberizing and melting temperatures. To avoid devitrification problems the fiberizing temperature also must exceed the liquidus temperature by an adequate margin. Manufacturing costs thus increase rapidly as the fiberizing temperature rises and these factors impose marked limitations on the range of glass compositions which can be successfully and economically produced as continuous-filament fiber.

3. *Size Coatings*

An essential component of continuous-filament glass-fiber products is the polymeric coating or size. This performs three main functions.

(a) It prevents the breakage of filaments and protects them from abrasive damage which would weaken them drastically. This is essential at all stages of fiber production, subsequent processing and composite manufacture.

(b) It binds the individual filaments into a convenient geometrical form for handling or for use as reinforcing elements. Fiber bundles may be retained in this form in the final composite, or may be required to disperse into single filaments in some matrices or processes.

(c) It improves the properties of the final composite in some way, either by increasing the strength of the fiber–matrix interfacial bond and improving its resistance to degradation in moist conditions or by affecting the fiber strength.

In order to fulfill these functions the size normally contains the following major components.

(*a*) *Film-forming polymer*. On drying/curing, this binds filaments together and provides a protective coating. Poly(vinyl acetate) emulsions have formed the basis for many sizes and are still widely used, different properties being obtained by varying the molecular weight, plasticizing, cross-linking or introducing comonomers. In aqueous environments glass surfaces become anionic and recent size developments have been based on water-soluble polymers with cationic centers which are attracted to the glass surface and remain strongly bonded when dried. Film formers based on alkanolamines reacted with epoxy functional groups are typical. Other types of film former include acrylic copolymers and polyesters in emulsified forms.

(*b*) *Lubricant*. This helps to protect both the glass surface and the polymer films from abrasion. A variety of proprietary chemicals are used, modern lubricants often being based on a cationic surface active agent (to bond to the anionic glass) with an aliphatic chain attached.

(*c*) *Active agent*. In fibers for reinforcement of plastics this is a coupling agent which provides a "chemical" bond or link between the glass-fiber surface and the matrix in reinforced plastic composites. Modern coupling agents are complex organosilanes of the form $X_3Si(CH_2)_nY$, where X is a hydrolyzable group which provides the bond to the glass surface and Y is an organofunctional group chosen to react into the matrix. Typically, vinyl or methacryloxy groups are used for polyester resins and amino groups for epoxy or nylon matrix composites. The choice of an incorrect coupling agent for a particular resin can lead to poor fiber-matrix bonding. The hydrolyzable groups on the silane molecule are converted in solution, prior to size application. The glass-fiber surface itself also hydrolyzes extremely rapidly and the OH groups of the hydrolyzed silane adsorb onto the hydroxy groups on the glass surface. On drying, it is assumed that at least some of these hydroxyls condense to form covalent siloxane bridges between the glass and the organofunctional end of the silane molecule. In the first generation of commercial fibers for cement reinforcement the size had virtually no effect on fiber–cement bonding or on fiber strength retention once the fibers had been incorporated in cement. More recently, however, second-generation alkali-resistant fibers have become available for which the size contains a chemical inhibitor which reduces the rate of chemical reaction between cement and glass and hence improves the strength retention of the fibers in a cement environment at modest cost penalty (Proctor 1985).

4. Glass Compositions

All commercial continuous-filament fibers are made from silicate glasses containing silica (SiO_2) as the major component in the composition. The mechanical properties of silicate glasses are not sensitively dependent on glass composition, but chemical activity—which controls such important parameters as durability, leaching and strength retention in corrosive environments—is closely related to composition. A number of glass compositions have been used in continuous-filament-fiber manufacture and are identified by various code letters or abbreviations which denote composition ranges rather than a precise formulation. The main compositions are given in Table 1.

4.1 Compositions for Plastic and Rubber Reinforcement

Nearly all continuous-filament fiber for the reinforcement of plastics and rubber is now manufactured from E-glass (electrical glass), which is a calcium alumino borosilicate type of composition, of low alkali content, selected from within the range given in Table 1. This type of glass was originally developed to impart good electrical properties to laminates, the very low alkali content ensuring little leaching of the glass in wet conditions and hence the maintenance of good electrical resistivity. E-glass compositions also possess good strength and fiberizing characteristics.

Although E-glass has good durability in wet conditions of near neutrality it is susceptible to degradation in highly alkaline and acid conditions. New alkali-resistant compositions for cement reinforcement are discussed below.

To meet acid corrosion a boron-free E-glass has been developed recently and is known as E-CR (E-corrosion resistant) and is now available commercially. The composition is essentially in the range given for E-glass in Table 1 but contains no B_2O_3, and with the SiO_2 content a little higher at about 60 wt%.

A-glass (alkali glass) corresponds to a typical window-glass composition. A-glass fibers were used quite

Table 1
Compositions of continuous-filament glass fibers

	Composition (wt%)											
	SiO_2	Al_2O_3	Fe_2O_3	B_2O_3	ZrO_2	MgO	CaO	Na_2O	K_2O	Li_2O	TiO_2	F_2
A-glass (typical)	73	1	0.1			4	8	13	0.5			
E-glass (range)	52–56	12–16	0–0.5	8–13		0–6	16–25		← <1 total →		0	0–1.5
AR glass (range)	60–70	0–5			15–20		0–10	10–15			0–5	
C-glass (range)	59–64	3.5–5.5	0.1–0.3	6.5–7		2.5–3.5	13.5–14.5	8.5–10.5	0.4–0.7			
S and R glasses (range)	50–85	10–35				← 4–25 total →		0				

extensively for reinforcement of plastics until the late 1970s but leaching of alkali from the glass in wet conditions led to poor electrical properties and degradation at the fiber–resin interface. A-glass fiber is not now manufactured in any quantity.

Prior to the development of E-CR glass, C-glass (chemical glass) was developed to give greater chemical resistance—particularly acid resistance—than E-glass. C-glass has been commercially available for many years, not normally as a continuous-filament reinforcement fiber, but as staple sliver and tissues, used in battery separators, corrosion protection coatings, gel coats, etc.

The calcium alumino silicate range of glass compositions has been explored to obtain glasses with rather higher strength and stiffness in fiber form (S and R glasses, Table 1). S-glass is available commercially in the USA at a considerable cost premium over E-glass.

In the 1950s and early 1960s attempts were made to develop glass fibers with significantly greater stiffness (Young's modulus) than conventional silicate compositions. Increases in modulus by up to a factor of 2 (i.e., to about 140 $GN\,m^{-2}$) were achieved in experimental fibers containing the highly toxic beryllia (BeO) but this work was superseded by the development of graphite, carbon, boron and Kevlar fibers.

4.2 Compositions for Cement Reinforcement

Conventional silicate and borosilicate glass fibers of the compositions just described are subject to severe corrosive attack in highly alkaline solutions (pH 12.5–13.5) such as those found in hydrating Portland cement. There is a rapid and drastic loss of fiber strength.

Zirconia additions were known to improve the durability of glasses generally, including conferring a degree of alkali resistance to sodium silicate glass compositions, and it has proved possible to develop a range of highly alkali-resistant glasses which can be fiberized within the constraints outlined in Section 3.2 (Proctor and Yale 1980, Majumdar 1982).

The composition range covering a number of commercial alkali-resistant (AR) glass fibers is given in Table 1. In these commercially available, high zirconia, alkali-resistant glasses, strength retention in cement is essentially controlled by zirconia content up to about 17 wt%. Above this level there is less improvement and the consequent increase in fiberizing temperature leads to significantly higher costs. Commercially available fibers containing some 10% rare-earth oxides, in addition to the components shown in Table 1, have recently been announced. Again, significant increase in cost accompanies the improved strength retention achieved by this approach.

5. *Structure and Properties*

The most widely held view of the structure of silicate glasses is based on the fundamental unit of a silicon–oxygen tetrahedron in which a central silicon ion is surrounded by four oxygen ions. A continuous, but irregular (random) spatial network is built up from these tetrahedra joined corner to corner via a common oxygen ion. Other cations fit into holes in the structure and cause a number of broken linkages in the network. There is little, if any, evidence for an oriented structure in glass fibers.

Glasses solidify gradually as they are cooled and the equilibrium structural arrangement, density and atomic separation is a function of rate of cooling. Glass fibers of the diameters used in continuous filaments cool and solidify extremely rapidly (of the order of milliseconds); their density and structurally sensitive properties such as Young's modulus are about 10% lower than corresponding bulk glass samples of the same composition. These effects have been reviewed by Otto (1961). Typical values are given in Table 2.

Macroscopically, glass fibers behave entirely elastically and at normal-use temperatures the stress–strain curve is linear to failure. Under prolonged load there is a small viscoelastic deformation due to the movement of cations into less energetic positions in the strained silica network. The magnitude of this effect is only ~5% of the elastic strain, it occurs over 2–3 days and is recoverable on removal of load. At higher tempera-

Table 2
Properties of continuous-filament glass fibers

	Liquidus temperature (°C)	Working temperature ($\eta = 100$ Pa s) (°C)	Density (g cm^{-3})	Coefficient of thermal expansion (°C^{-1})	Refractive index	Young's modulus (GN m^{-2})	Strength (MN m^{-2})	
							undamaged filament	strand from roving
A-glass	1140	1220	2.46	7.8×10^{-6}	1.52	72	3500	
E-glass	1400	1210	2.54	4.9×10^{-6}	1.55	72	3600	1700–2700
AR glass	1180–1200	1280–1320	2.7	7.5×10^{-6}	1.56	70–75	3600	1500–1900
C-glass			~2.5					
S and R glasses			~2.5			~85	~4500	2000–3000

tures, creep may occur as a true viscous flow phenomena, controlled by the viscosity of the glass.

Because of the continuous network structure and the absence of definable structural imperfections or deformation mechanisms, the strength of glass is very high. However, due to the continuity and elastic nature of the material, cracks propagate readily and the strength of the material is very sensitive to the presence of any stress concentrators such as solid particle inclusions or surface cracks (Proctor 1971). It is for this reason that the avoidance of devitrification during fiberizing and the protective role of the size coating are so important. Despite the protection of the size, the strength of undamaged single filaments is reduced considerably when they are brought together into a strand (Table 2), and this may be reduced still further by rough handling.

The strength of a given length of fiber depends on the probability of finding a stress-raising flaw of given severity in that length. Statistically, therefore, strengths of damaged, practical fibers decreases as the gauge length or stressed length increases. This is an important parameter in understanding composite behavior in relation to the strength of fibre–matrix bonding as discussed by Rosen (1970).

Any chemical attack which damages the surface can lead to significant strength loss. All glass compositions are attacked to a greater or less extent by acids, alkalis and even water, the degree of attack varying across the pH range and from one composition to another. E-glass is particularly susceptible to attack by acid solutions in the range 0.01–10 N, and by alkali as mentioned in Sect. 4.1. This has led to the development of acid-resistant C-glass, E-CR glass and the alkali-resistant high-zirconia glasses for cement reinforcement, which also turn out to be very acid resistant.

Finally, like many other brittle ceramic materials, glass is subject to a stress-enhanced corrosion reaction with water. Under a constant applied stress and in the natural moisture of the atmosphere, cracks may grow from flaws on the glass-fiber surface and lead to failure. Thus fibers may fail after some period of time under load and the time to failure is reduced as the stress level is raised—a phenomenon known as static fatigue. Published data on static fatigue is sparse, but for prolonged loading over a period of years stresses in fiber should be limited to about 25% of the short-term strength of the fiber.

Stress corrosion can also cause crack growth during a normal strength test in which the stress is being increased at a constant rate (Weiderhorn 1974). This is known as dynamic fatigue and results in the measured strength of the glass fiber being dependent on the loading rate, and being higher at faster loading rates. An approximate guide to the extent of this effect for common glass-fiber compositions is an increase in strength of about 10% for an increase in loading rate by a factor 10.

The strength of glass fibers is thus a complex parameter, being dependent on the initial degree of perfection of the glass surface, possible further damage due to chemical attack or mechanical abrasion, and finally upon the conditions and rate of stressing during the strength measurement.

Bibliography

Gagin L V 1980 The development of fibreglass—A history of compositions and materials. *Can. Clay Ceram.* 53(4): 10–14
Lowenstein K L 1983 *The Manufacturing Technology of Continuous Glass Fibres*, 2nd edn. Elsevier, Amsterdam
Majumdar A J 1982 Alkali resistant glass fibres. In: Watt W, Perov B (eds.) 1982 *Handbook of Fibrous Composites*, Vol. I: *Production and Properties of Strong Fibres*. North-Holland, Amsterdam
Mettes D G 1969 Glass fibers. In: Lubin G (ed.) 1969 *Handbook of Fiberglass and Advanced Composites*. Van Nostrand Reinhold, New York, pp. 143–81
Mohr J G, Rowe W P 1978 *Fiber Glass*. Van Nostrand Reinhold, New York
Otto W H 1961 Compaction effects in glass fibers. *J. Am. Ceram. Soc.* 44(2): 68–72
Proctor B A 1971 Sources of weakness in reinforcing fibres. *Composites* 2(2): 85–92
Proctor B A, Yale B 1980 Glass fibres for cement reinforcement. *Philos. Trans. R. Soc. London, Ser.* A 294: 427–36

Proctor B A 1985 Alkali resistant fibres for reinforcement of cement. In: *Proc. NATO Advanced Study Institute*. Plenum, New York, pp. 555–73

Rosen B W 1970 Strength of uniaxial fibrous composites. In: Wendt F W, Liebowitz H, Perrowe N (eds.) 1970 *Mechanics of Composite Materials*. Pergamon, Oxford, pp. 621–51

Weiderhorn S M 1974 Sub-critical crack growth in ceramics. In: Bradt R C, Hasselman D P H, Lange F F (eds.) 1974 *Fracture Mechanics of Ceramics*, Vol. 2. Plenum, New York, pp. 613–46

B. A. Proctor
[Pilkington, Ormskirk, UK]

Creep of Composites

The time-dependent deformation of a fiber-reinforced composite is a function of both the distribution of stress between the fiber and matrix, which may be dependent on strain, and of their individual modes of deformation. Although the creep behavior of the constituent phases may be described by relatively simple steady-state equations, the composite responds in a complex manner and often does not exhibit a steady-state creep rate. Among the most important factors governing composite creep deformation are (a) the fiber aspect ratio, (b) the fiber radius, and (c) the inclination of applied stress to the fiber axes.

1. Continuous Fibers

When the fiber aspect ratio is very large and the stress is applied parallel to the fibers, material continuity requires that both phases deform at the same rate; otherwise cavities will be formed. Consequently, the applied stress will be partitioned between the fibers and the matrix in the proportion required to equalize their rate of straining. There are two modes of composite creep, dependent on the different deformation mechanisms of the individual phases, that are of practical importance.

1.1 Creeping Matrix, Elastic Fibers

When the fibers deform elastically, constrained by a creeping matrix, then the proportion of stress carried by the fibers increases with strain and there is a complementary off-loading of stress from the matrix. In principle, equilibrium can be achieved at a limiting strain ε_c when all of the applied stress σ is carried by the fibers (i.e. $\varepsilon_c = \sigma/E_f V_f$), where E_f and V_f are the Young's modulus and the volume fraction of fibers, respectively). However, this is an asymptotic strain as $t \to \infty$, since matrix creep slows down as the component of stress on the matrix decreases. McLean (1983) has modelled this type of behavior for the case of power-law matrix creep, $\dot{\varepsilon} = A_m \sigma^{n_m}$, where A_m and n_m are constants, showing that for the composite:

$$\dot{\varepsilon} = \alpha A_m \sigma^{n_m} [1 - \varepsilon/\varepsilon_c]^{n_m} \quad (1)$$

where $\alpha^{-1} = [1 + (E_f V_f / E_m V_m)](1 - V_f)^{n_m}$ and the subscripts f and m refer to fiber and matrix, respectively.

The creep of polymers generally occurs by linear viscous flow where the strain rate is proportional to the applied stress, $\dot{\varepsilon} = A_m \sigma$. Indeed, the creep behavior of polymers is often characterized by a "creep modulus" $1/A_m$, which has the dimensions of MPa s^{-1}. When reinforced by elastically deforming fibers, the time-dependent deformation of polymeric composites is described by the expression

$$\varepsilon = \varepsilon_c \left[1 - \left(1 - \frac{\varepsilon_0}{\varepsilon_c} \right) \exp\left(- \frac{\alpha A_m \sigma_c t}{\varepsilon_c} \right) \right] \quad (2)$$

where ε_0 is the instantaneous strain on loading. Equation (2) shows that use of a creep-modulus description for polymeric composites would lead to a parameter that increases with increasing strain rather than the constant value typical of the polymer matrix.

For composites with metal matrices Eqn. (1) integrates to

$$\varepsilon = \varepsilon_c - (\varepsilon_c - \varepsilon_0) \left[1 - \left(1 - \frac{\varepsilon_0}{\varepsilon_c} \right)^{n_m - 1} \frac{(n_m - 1) \alpha A_m \sigma_c^{n_m} t}{\varepsilon_c} \right] \quad (3)$$

In both cases there is no steady-state creep rate; rather $\dot{\varepsilon}$ steadily decreases so that a simple power- or exponential-law representation is not possible. For some purposes the strain rate at or the time to achieve a specific strain, say 1%, before fracture intervenes is a useful measure. This is shown in Fig. 1 for parameters relevant to the $\gamma - \gamma'$ Cr_3C_2 in-situ composite. At stresses above about 250 MPa at 825 °C there is a linear relationship on a log–log plot compatible with power-law representation. However, at lower stresses the creep rate falls dramatically. Consequently power-law extrapolation of relatively short-term data seriously underestimates the creep performance of composites with elastic fibers.

The elastic strain in the fibers is reversible and can be recovered by heat treatment of the composite. This phenomenon, referred to as the length-memory effect (Khan et al. 1980), has been advocated as a method of extending the useful lifetime of high-temperature composites.

The elastic fiber, creeping matrix model is most appropriate to systems in which the matrix and reinforcement operate at widely different homologous temperatures. For example, the most advanced in-situ composites operate within 200 °C of the matrix melting temperature while the fibers are at less than half of their melting temperatures; similarly SiC in aluminum and carbon-fiber-reinforced resins are likely to satisfy the requirements of this model. When the constituent phases operate at similar homologous temperatures, both phases may deform by creep.

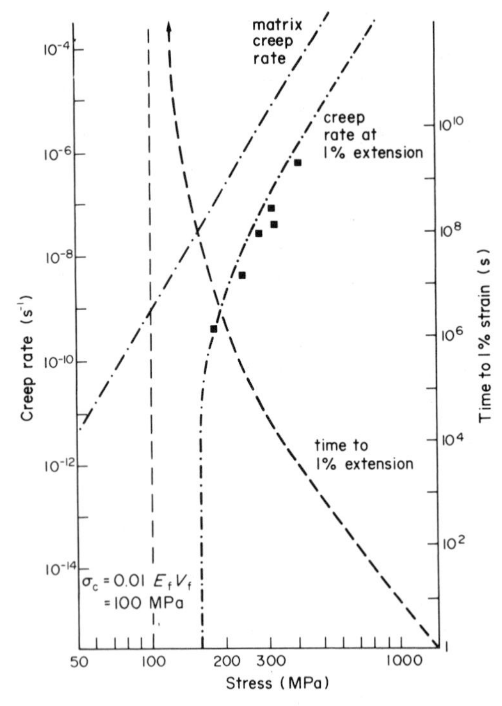

Figure 1
Calculated variations in creep rate at 1% strain and time to 1% strain as a function of stress for a composite of elastic fibers in a creeping matrix. Values of the parameters appropriate to the $\gamma-\gamma'$ Cr_3C_2 in situ composite at 825 °C (for which experimental data are shown) are: $V_f=0.11$, $n=6$, $E_f=10^5$ MPa, $E_m=2\times10^4$ MPa, $A_m=10^{-21}$ MPa^{-6} s^{-1}, $\varepsilon=0.01$

1.2 Creeping Matrix, Creeping Fibers

If a steady-state creep rate can be established in each phase individually, then the composite will also exhibit steady-state creep. However, the dependence of this creep rate on the principal parameters, stress and temperature, is more complex than for the single phases. Thus, if the steady-state creep rates of matrix and fibers can be represented by power laws:

$$\dot{\varepsilon}=B_i\sigma^{n_i}\exp(-Q_i/RT) \tag{4}$$

where the subscript i denotes either the matrix (m) or the fiber (f), B_i and n_i are constants, Q_i is an activation energy and RT is the thermal energy, then application of the rule of mixtures gives

$$\sigma=\left[\frac{\dot{\varepsilon}}{B_f\exp(-Q_f/RT)}\right]^{1/n_f}V_f+\left[\frac{\dot{\varepsilon}}{B_m\exp(-Q_m/RT)}\right]^{1/n_m}V_m \tag{5}$$

One important consequence of this result, which was first derived by McDanels et al. (1967), is that steady-state creep data of composites cannot be represented by a simple power law; if this is attempted then there will be apparent variations in n and Q with stress and temperature, as indicated in Fig. 2, between the limits (n_f, Q_f) and (n_m, Q_m). Thus, identification of a stress exponent or activation energy with creep of a composite, even when both phases are creeping, has little physical significance and may be misleading if used to extrapolate to different test conditions.

2. *Short Fibers*

When the fibers are of finite length, their aspect ratio becomes a critical parameter in the composite creep behavior. The most complete treatment available (Kelly and Street 1972) also distinguishes between composites with rigid (or elastic) and with creeping fibers. In both cases the distance from the end of the fibers over which matrix shear occurs to transmit load to the central part of the fibers is important. The stress distribution along the fibers, from 0 at the ends to a maximum in the central region, reduces the effective length of fiber that behaves as in a continuous composite either by elastic fiber deformation constrained by slow matrix flow or by coupled creep of fiber and matrix. However, matrix flow around the fiber ends provides a perturbation to the continuous composite behavior that can significantly increase the composite creep rate.

For rigid fibers of length l and diameter d in a matrix that deforms by power-law creep ($\dot{\varepsilon}=A_m\sigma^{n_m}$), the composite creep by flow around fiber ends is given by the expression

$$\dot{\varepsilon}=A_m\sigma^{n_m}[\Phi(l/d)^{(1+1/n_m)}V_f+V_m]^{-n_m} \tag{6}$$

where Φ is a load transfer parameter dependent on V_f and n_m such that

$$\Phi=(0.667)^{1/n_m}\left(\frac{n_m}{2n_m+1}\right)(0.95\,V_f^{-1/2}-1)^{-1/n_m} \tag{7}$$

This strain rate will be additional to that for continuous elastic fiber, creeping matrix composites normalized to account for the reduced length of active fibers.

In the case of creeping fibers the relative stress sensitivities for creep of matrix and fibers determines the length from the fiber ends over which the stress varies and how this alters with strain; for rigid fibers there is a linear increase in stress to the centers of the fibers. This leads to a more complex set of equations, the concept of which is similar to that for elastic fibers.

3. *Fiber Fracture and Internal Stresses*

In the above discussion, which assumes a constant microstructure, there is a progessive reduction in creep rate until a steady state is achieved. However, if the fiber aspect ratio decreases due to, for example, spheroidization by diffusional processes or to fracture of

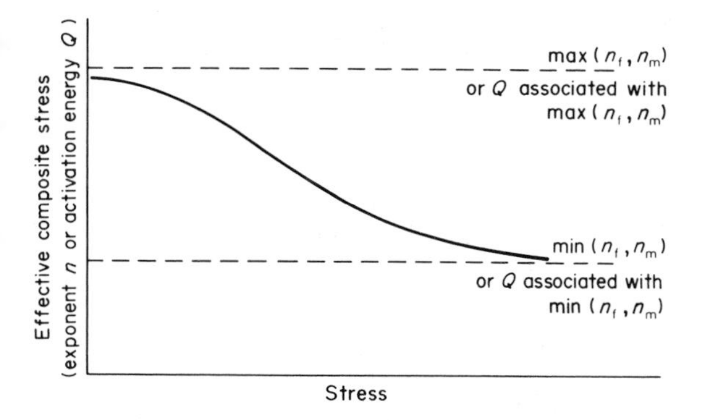

Figure 2
Schematic variation with stress of the apparent stress exponent n and activation energy Q for creep of a composite consisting of creeping fibers and creeping matrix

the reinforcing fibers, this will lead to an additional acceleration in creep rate that can give a type of tertiary creep. McLean (1986) has modelled the creep of aligned composites with brittle fibers that fracture at a distribution of local stresses that can be associated with the probability of defects occurring within the fibers. The creep behavior can be expressed as a series of coupled differential equations in terms of two state variables that have a clear physical significance: σ_f, the stress carried by the elastically deformed fibers and λ, the fiber aspect ratio.

$$\left.\begin{aligned}
\dot{\varepsilon} &= \alpha A_m \sigma^{n_m}\left[1 - V_f\frac{\sigma_f}{\sigma}\right]^{n_m} \\
\dot{\sigma}_f &= E_f\dot{\varepsilon} - R\sigma_f^{n_m}[1 + b\lambda^{(1+1/n_m)}]^{-n_m} \\
\dot{\lambda} &= -\frac{\lambda \ln[\lambda_0/\lambda_f]}{\varepsilon_f - \varepsilon_{th}}\dot{\varepsilon} \qquad \text{for } \varepsilon > \varepsilon_{th} \\
&= 0 \qquad \text{for } \varepsilon < \varepsilon_{th}
\end{aligned}\right\} \quad (8)$$

where $R = E_f A_m/(1 - V_f)^{n_m}$, $b = \Phi V_f/(1 - V_f)$, ε_{th} and ε_f are the strains when fiber cracking initiates and fracture occurs, respectively, and λ_0, λ_f are the initial and final fiber aspect ratios.

Equations (8) reduce to Eqn. 1 when $\lambda = \infty$ for all strains and to Eqns. (6) and (7) for a finite initial fiber aspect ratio and $\lambda = 0$ for all strains (i.e., $\varepsilon_{th} = \varepsilon_f$). Thus Eqns. 8 are quite general.

The differential formulation of Eqns. (8) is particularly suitable to be integrated numerically using different initial boundary conditions. In particular, the effects of differential thermal expansion of matrix and reinforcement during processing often leads to the development of large internal stresses σ_f within the unloaded composite and this can have an important effect on the creep performance.

4. Effect of Fiber Diameter

The above discussion assumes that composite creep behavior is a volume fraction weighted average of the properties of the constituent phases in their bulk forms; there is no parameter to account for microstructural dimensions. Many synthetic composites are adequately described by this rule of mixtures approach. However, in in-situ composites, which can contain submicrometer diameter reinforcement, there can be substantial creep strengthening through microstructural refinement. This is likely to be due to a dispersion strengthening of the matrix although two less direct factors may be contributing; namely (a) finer fibers having more whiskerlike characteristics and (b) the matrix being compressed due to differential thermal contraction during processing.

5. Off-Axis Loading

The models described above deal with axially loaded fiber-reinforced composites. When a stress σ is inclined at some angle ϕ to the fibers, then only the axial component of stress, $\sigma \cos^2 \phi$, contributes to these deformation modes. In addition there is a shear component, $\sigma \sin \phi \cos \phi$, parallel to the fibers which can lead to strain by matrix shear, and a tensile component normal to the fibers, $\sigma \sin^2 \phi$, which gives matrix flow around the fibers; in neither case is fiber deformation required. Simple summation of the strains resulting from each of the stress components gives an adequate description of creep in nonaxially loaded composites.

See also: Fatigue of Composites; Strength of Composites

Bibliography

Kelly A, Street K N 1972 Creep of discontinuous fibre composites. II: Theory for the steady-state. *Proc. R. Soc. London, Ser. A* 328: 283–93

Khan T, Stohr J F, Bibring H 1980 COTAC 744—An optimised ds composite for turbine blades. In: Tien J K, Wlodek S T, Morrow H, Gell M, Maurer G E (eds.) 1980 *Superalloys 1980*. American Society for Metals, Metals Park, Ohio

McDanels D L, Signorelli R A, Weeton J W 1967. *Analysis of Stress Rupture and Creep Properties of Tungsten-Fiber-Reinforced Copper Composites*, NASA TND-4173. National Aeronautics and Space Administration, Washington, DC

McLean M 1983 *Directionally Solidified Materials for High Temperature Service*, Book No. 296. Metals Society, London

McLean M 1986 Time dependent deformation of metal-matrix Composites. In: Montgomery F R (ed.) 1986 *Proc. 4th Conf. of the Irish Durability and Fracture Committee*.

M. McLean
[Imperial College of Science, Technology and Medicine, London, UK]

D

Dental Composites

Dental composites are classified by the position of the teeth in which they are used. Composites designed for use in the front teeth are called anterior composites and those designed for the back teeth are called posterior composites. The performance requirements of the two types are different and are reflected in their morphology, structure and physical properties.

Anterior composites need to have aesthetic qualities, such as variable shades and degrees of translucency. They also have to exhibit a high gloss after simple routine polishing procedures, combined with a resistance to wear by toothbrushing.

The need for aesthetics is of less significance in posterior composites. For these materials the dominant requirement is the ability to withstand high chewing forces over a long period in the hostile oral environment without fracturing and showing significant wear. To satisfy the specification set by the American Dental Association (ADA), wear has to be less than 25 µm per annum on the noncontact area of the chewing surface and less than 50 µm per annum on the contact area of the chewing surface over a period of 5 years. A further requirement of posterior composites is that they have to be x-ray opaque to allow a clear radiographic visualization of the material in the tooth for diagnostic purposes.

1. Structure, Morphology and Mechanical Properties

1.1 Composite Structure and Components

A dental composite is specifically a structure consisting of an inorganic particulate phase dispersed in an organic resin matrix. The resin matrix is made up of two components, prepolymer and comonomer. The prepolymer is usually a high molecular weight aromatic resin used to impart strength and stiffness and increase the refractive index of the resin. Because the viscosity of the prepolymer is high, a low-viscosity comonomer is used as a diluent to enable high volume fractions of the inorganic filler to be admixed. Typical examples of prepolymers are BIS-GMA and urethane dimethacrylate. Triethyleneglycoldimethacrylate is the most widely used comonomer.

Inorganic-phase particles range from fumed silica with primary dimensions of between 20 nm and 0.05 µm to radio-opaque glasses with sizes in the range 0.5–60 µm. Elements such as barium, strontium and lanthanum are incorporated into the network modifier position within the glass structure to make the glass x-ray opaque. These elements, because of their relatively high atomic number, increase the refractive index of the glass and, because of their relatively high atomic radius, disrupt the glass network and increase solubility.

To satisfy aesthetic requirements, dental composites need to have a range in optical translucency to match that of tooth structure. Clear composites, to match the appearance of the incisal edges of teeth, can be obtained by using either very fine inorganic particles or by having identical refractive index in filler and resin. Varying the mismatch in refractive index between filler and resin gives a control on optical translucency and opacity. This is done by varying the refractive index of the resin only, by changing the prepolymer-to-comonomer ratio in the resin until the required mismatch with the glass is obtained.

1.2 Morphology and Mechanical Properties

Dental composites are characterized by particle size and quantity. Three main groups, described below, exist.

(*a*) *Conventional.* These contain glass fillers in the range 1–60 µm in volumes of up to 60%. These composites have been superceded in recent years by the other two groups because they failed to fulfill the majority of the performance requirements, in particular aesthetics, surface gloss and wear resistance. They were the first dental composites and were developed in the early 1960s.

(*b*) *Microfilled.* These contain fumed silica particles only and, because of their high surface area, can only be incorporated in volumes of up to 40% in the resin. To produce a handleable paste some manufacturers produce prepolymerized blocks of composite containing agglomorated silica particles. These are fractured and ground down to 10 µm silica-impregnated polymer particles which are then redispersed into the silica filled resin paste. Because these composites contain very fine particles with dimensions below the wavelength of light they are relatively translucent and similar in appearance to tooth structure and can be easily polished to a high gloss. The mechanical properties of these composites are too low for use in back teeth but are acceptable as restorative materials in the front teeth. Typical values are within the ranges shown in Table 1.

(*c*) *Hybrid.* These contain up to three different size distributions of particles. The fine mode is made up from fumed silica with particle sizes ranging from 20 nm to 0.5–3 µm, and the large mode is also made up of glass particles but in the size range 5–15 µm. The total volume of the fillers in the composite paste is typically between 60% and 80%. The main benefit of

Table 1
Mechanical properties of microfilled anterior composites and hybrid posterior composites

Property	Value for microfilled anterior composites	Value for hybrid posterior composites
Flexural modulus	5–12 GPa	14–25 GPa
Flexural strength	40–80 MPa	100–170 MPa
Diametral strength	20–50 MPa	40–65 MPa
Surface hardness (Vickers)	30–50	60–120
Fracture toughness		1.5–2.3 $MN\,m^{-3/2}$

the three particle sizes is that a lower viscosity is obtained compared to that from a single particle size distribution (Farris 1968) and higher than expected filler loadings can be achieved, resulting in elastic moduli in the range 18–25 GPa. The other mechanical properties are also very high and are shown in Table 1. The three commercial composites specified by the ADA guidelines are from this category or group.

2. Methods of Activating Polymerization

Two methods of activation exist, resulting in the materials described below.

(*a*) *Chemically activated materials.* These are two-component systems: one composite paste contains a peroxide catalyst, the other paste an amine activator. On mixing the two components, setting or polymerization commences and proceeds throughout the bulk of the material.

(*b*) *Light-activated materials.* These are supplied as a single-component paste containing a light-sensitive catalyst. Activation is by exposure to light at the blue end of the visible spectrum (about 470 nm wavelength). Current catalyst systems are based on camphoroquinone and an amine (Dart and Nemcek 1978) and supercede the earlier uv-activated system (Waller 1972) which suffered from poor subsurface polymerization. The single-paste systems obviate the need for mixing and thus eliminate a major source of bubbles, voids and incomplete cure due to poor mixing. In the blue-light activated systems the degree of polymerization is proportional to the intensity of the light and time of exposure, and decreases with depth into the composite (Watts 1984) in a complex manner depending on the scattering and absorption characteristics of the composite. A major source of scattering is the degree of mismatch in refractive index between the radio-opaque glass and the resin.

3. Wear Resistance

Although industry in general uses composites for applications where resistance to abrasion is a major requirement, little has been published on how to systematically select material properties for optimum wear resistance. The major reason is that wear is not only dependent on the material properties but also on the environment in which the wearing surface functions.

3.1 The Oral Environment

The conditions under which dental composites have to perform can be subdivided as follows.

(*a*) *Biochemical.* The chemical and biological conditions within the mouth make it a particularly hostile environment by accelerating the hydrolytic degradation processes. These are mainly related to the water-absorption properties and hydrophilic nature of the composite and can result in the breakdown of the bond between the filler and the resin. The two main causes of this breakdown are the silane coupling-agent degradation and the filler surface dissolving in the aqueous environment, especially if it has become acidic due to diet or biological activity (Soderholm 1983, Charles 1958). Failure of the bond reduces the mechanical properties of the composite which in turn decreases the wear resistance.

(*b*) *Mechanical stresses.* Mastication or chewing produces stresses in the dental composite. The magnitudes of these stresses vary considerably, depending on diet and on the position of the tooth, the further back in the mouth the greater the load imposed on the tooth during chewing. A summary of the conditions under which teeth operate is given in Table 2 and is mainly based on a review by Harrison and Lewis (1975).

These data show that dental composites have to withstand a large number of cyclic stresses which would indicate that fatigue-related mechanisms, and in particular slow crack growth, must be involved in the wear processes.

3.2 In Vitro Wear Testing

The complex nature of both the environment and the imposed mechanical stresses in the mouth have made their simulation in an *in vitro* wear machine very difficult. Various researchers have attempted this, the most sophisticated method was developed by De Long

Table 2
The physical conditions in a masticatory cycle

Rate of chewing	60–80 cycles min^{-1}
Forces	
maximum biting force over all teeth	640 N
maximum biting force on a single tooth	265 N
normal forces on a single tooth	3–18 N
Contact time	
time maximum pressure sustained (20% of one masticatory cycle)	0.07 s
Total contact time	
normal	10 min
(over 24 h) Bruxist	30 min–> 3 hr
Mean sliding distance	1.0 mm
Contact area (1st molars)	15 mm
Maximum stress	20 MPa
Application of maximum stress	3000 times per day

and Douglas (1983) in which the complete masticatory cycle was reproduced mechanically and carried out in an environmental chamber under conditions which, according to the researchers, approached most closely an artificial oral environment.

Harrison and Draughn (1976) and Harrison (1977) used machines constructed to partially mimic the masticatory cycle. Their results showed that the wear measured for a conventional composite was six times less than that for mercury amalgam, clearly at odds with clinical findings.

Subsurface damage was first shown to exist in *in vivo* worn composites by a silver staining technique (Wu and Cobb 1981) and this type of damage was reproduced *in vitro* using a pin-and-disk wear machine (McKinney and Wu 1982). Bailey and Rice (1981), using a sliding wear mechanism, found that the wear rate for composites increased dramatically above a critical value of contact stress, whereas for amalgams they remained constant over a range of contact stresses.

3.3 Wear-Related Mechanical Properties

Hardness (H) and modulus (E) increase with volume fraction (V_f) of a particular filler, whilst fracture toughness (K_{Ic}), fracture energy (G_{Ic}), and strength (σ) may increase or decrease depending on a number of factors such as interfacial bond strength, particle shape and size, and the ductility of the resin and resin filler interface.

The friction properties of the composite surface may also play a major role in wear resistance, smooth surfaces are likely to have different wear characteristics from rough surfaces.

These complicated factors make the problem of relating mechanical properties a difficult one. Researchers in other areas of composite wear have used the empirically derived relationship shown below with some success (Friedrich 1983, Hornbogen 1975, Zum Gahr 1979):

$$w=\frac{1}{H}+\frac{\psi H^{1/2}}{G_{Ic}}$$

with $\psi=0$ for $p<P_{crit}$, and $\psi>0$ for $p>P_{crit}$. The quantity p is the contact stress on the wearing surface imposed by the counterpart, P_{crit} is the critical stress above which fracture takes place, and w is the wear rate.

There is also some evidence that fracture toughness may be an important parameter for consideration if abrasive wear is a dominant mechanism (Lancaster 1969). One of the three posterior composites currently satisfying the ADA requirement for *in vivo* wear resistance has the highest value of fracture toughness of all the posterior composites (Lloyd and Adamson 1987).

Bibliography

Bailey W F, Rice S L 1981 Comparative sliding–wear behaviour of a dental amalgam and a composite restorative as a function of contact stress. *J. Dent. Res.* 60: 731–32
Charles R J 1958 Static fatigue of glass. *J. Appl. Phys.* 29: 1549–53
Dart E C, Nemcek J 1978 Photopolymerizable composition consisting of ethylenically unsaturated monomer, photosensitizing compound and reducing agent. US Patent No. 4,071,424
De Long R, Douglas W H 1983 Development of an artificial oral environment for the testing of dental restoratives: Biaxial force and movement control. *J. Dent. Res.* 62: 32–36
Farris R J 1968 Prediction of the viscosity of multimodal suspensions from unimodal viscosity data. *Trans. Soc. Rheol.* 12: 281–301
Friedrich K 1983 Abrasiveverchleiss verstarkter Thermoplaste. *Plasverarbeiter* 34(1): 27–30
Harrison A 1977 Effect of packing pressure on abrasion resistance of dental amalgams. *J. Dent. Res.* 56: 613–15
Harrison A, Draughn R A 1976 Abrasive wear, tensile strength and hardness of composite resins—Is there a relationship? *J. Prosthet. Dent.* 36: 395–98
Harrison A, Lewis T T 1975 The development of an abrasion testing machine for dental materials. *J. Biomed. Mater. Res.* 9: 341–53
Hornbogen E 1975 Role of fracture toughness in wear of metals. *Wear* 33: 251–59
Jorgensen K D 1980 Restorative resins: Abrasion versus mechanical properties. *Scand. J. Dent. Res.* 88: 557–68
Lancaster J K 1969 Abrasive wear of polymers. *Wear* 14: 223–36
Lloyd C H, Adamson M 1987 The development of fracture toughness and fracture strength in posterior restorative materials. *Dent. Mater.* 3: 225–31

McCabe J F, Smith B H 1981 A method for measuring the wear of restorative materials in vitro. *Br. Dent. J.* 151: 123–26
McKinney J E, Wu W 1982 Relationship between subsurface damage and wear of dental restorative composites. *J. Dent. Res.* 61: 1083–88
Soderholm K J 1983 Leaking of fillers in dental composites. *J. Dent. Res.* 62: 126–30
Waller D C 1972 Photopolymerizable dental treatment composition based on aromatic dimethylacrylate for polymerization in situ. US Patent No. 3,709,866
Watts D C 1984 Characteristics of visible light activated composites. *Br. Dent. J.* 156: 203–15
Wu W, Cobb E N 1981 A silver staining technique for investigating wear of restorative dental composites. *J. Biomed. Mater. Res.* 15: 343–48
Zum Gahr K H 1979 How microstructure affects abrasive wear resistance. *Metal Progr.* 116(4): 46–52

T. A. Roberts
[Advanced Healthcare, Tonbridge, UK]

F

Failure of Composites: Stress Concentrations, Cracks and Notches

In designing structural components from brittle fibrous composites (e.g., glass fibers in epoxy), it is assumed that for an acceptable level of survival probability the operating stress does not exceed the strength of the composite. Unfortunately, the situation is more complex in practice. A fibrous composite containing a notch or hole in monotonic loading exhibits premature cracking at stresses significantly lower than the ultimate strength; this may take the form of interfacial shear cracking, delamination or splitting, fiber fracture, matrix cracking and so forth. The formation of a damaged region is due to the concentration of localized tensile and shear stresses close to the notch front. The precise mode of failure depends on the orientation of fibers and stacking geometry of the laminate; it is also sensitive to the properties of the matrix and fiber–matrix interface, and to the stress state and environment.

1. Failure Modes in Monotonic Loading

A typical unidirectional fibrous composite (glass fibers in epoxy) loaded in tension exhibits stable delamination or splitting at the root of a notch. As the split extends parallel to the fibers and direction of applied load, a matrix crack propagates from the notch tip, passing still-intact fibers. A second split nucleates at the crack tip and the crack becomes arrested. Further crack growth in the original notch orientation requires extension of the two splits, followed by fracture of the intact fibers somewhere along their debonded length. The extent of delamination, fiber debonding and fiber fracture in the localized damage zone is related to the applied load. The split that forms can be considered as rendering the sharpest inherent defect or crack effectively equal to a hole or notch (Mandell 1971).

In a monotonic tensile test on a crossply (0°/90°) laminate, the transverse (90°) layers hinder longitudinal splitting at the root of a notch. Those layers perpendicular to the direction of applied load, and adjacent to an interfacial shear crack, may also delaminate slightly during shearing between the two sliding surfaces of a longitudinal split. Premature fracture of the localized 0° load-bearing fibers may lead to subcritical crack growth in the original notch direction. Behind the crack front lies an array of longitudinal, parallel-shear cracks, which mark the position of successive crack arrest points. The length of the delaminations and spacing between them are related in some way to the laminate stacking geometry.

For a quasi-isotropic laminate (0°/±45°/90°), delamination in the 45° layers at the root of a notch reduces the concentration of stress on the 0° plies. An increase in thickness of the 45° layer decreases the constraining effect on the 0° plies, which permits further delamination (Bishop and MacLaughlin 1979). The formation of a damage zone at the notch tip having a lower modulus than the surrounding material can be thought of as decreasing the intensification of localized stress. In effect, the notched strength and fracture toughness of the laminate is increased. For some, perhaps all, laminates containing glass, Kevlar or carbon fibers, the damage zone can be considered as increasing the effective radius ρ of the notch (Potter 1978). The fracture stress of a notched composite σ_f can then be estimated using measurements of ultimate strength σ_u, together with a suitable stress concentration factor K_t:

$$K_f = \frac{\sigma_u}{\sigma_f} = 1 + (c/\rho)^{1/2}\left[2\left(\frac{E_{11}}{E_{22}}\right)\nu_{12} + \left(\frac{E_{11}}{G_{12}}\right)\right]^{1/2} \quad (1)$$

where c is the length of the notch, E_{11} and E_{22} are the longitudinal and transverse Young's moduli, ν_{12} is the Poisson ratio, and G_{12} is the shear modulus.

At the microscopic level, the breakage of a fiber at the tip of a notch may induce the sequential failure of longitudinal (0°) fibers by the transfer of load from the broken fiber to an adjacent intact fiber. The localized concentration of stress on the unbroken fiber is sensitive to the strength of the fiber–matrix interface, and the matrix. For example, a toughened matrix, epoxy dispersed with elastomeric spheres, would be expected to reduce the amount of delamination at the notch front, and this is observed. However, reducing splitting raises the localized stress at the notch tip, and the fracture toughness corresponding to crack propagation perpendicular to the 0° fibers is lowered. The initiation of cracking parallel to the original notch direction may be prevented if, as a result of the stress distribution due to the notch, the initial difference in stress carried by a fiber that has just broken and the adjacent intact fiber is sufficiently large. The interaction between the distribution of stress in the vicinity of a notch tip and the concentration of localized stress in an unbroken fiber adjacent to one that has just failed leads to an effect of notch size on strength and fracture toughness (Potter 1978). The smaller the notch, the more localized the perturbed stress field becomes and the greater is the applied stress required to initiate fiber fracture and transfibrillar crack propagation. It is this subtle balance between the transfer of load to an intact fiber next to a broken fiber, the distribution of flaws in the fiber and the variability of fiber strength, combined with the localized stress field ahead of a notch tip that determines the strength and notch sensitivity of the laminate.

The fracture toughness K_c can be related to the damage zone size C_0 and notch geometry as follows:

$$K_c = \sigma_u[\pi(C + C_0)(1 - \xi^2)]^{1/2} \quad (2)$$

where $\xi = C/(C + d_0)$ and d_0 is a critical distance ahead of the discontinuity (Nuismer and Whitney 1975).

The concept of a damage zone is invoked so that K_c reaches a maximum value as the notch length C becomes very small, that is, as $K_c \rightarrow \sigma_u(\pi C_0)^{1/2}$. Typical values of K_c, σ_u and C_0 for various laminates and stacking geometries are listed in Table 1.

2. Cyclic Failure and Residual Strength

Load cycling of fibrous composites containing brittle fibers, carbon fibers or glass fibers in epoxy, for example, brings about fiber–matrix decohesion and delamination at notches and inherent flaws. These damage zones are larger than those induced in monotonic loading. The nucleation and growth of damage zones in cyclic failure reduces the localized stress, together with a corresponding improvement in residual strength. The residual strength of a laminate depends on the stress level, and increases with time in cyclic loading, eventually reaching the unnotched or inherent strength in monotonic fracture. Fatigue lifetime and residual strength are therefore affected by interactions between microstructure, distribution of flaws, and the formation of damage zones, shear cracking and so forth. The micromechanisms of failure are sensitive to the chemistry of the resin and nature of the fiber–matrix bond. Extrinsic variables, such as temperature, humidity, and time in an aging experiment, are likely to have considerable influence on the properties of the fiber–matrix interface, and therefore on residual strength and fatigue lifetime of the laminate.

See also: Multiple Fracture; Strength of Composites;

Bibliography

Bishop S M, McLaughlin K S 1979 *Thickness Effects and Fracture Mechanisms in Notched Carbon Fibre Com-*

Table 1
Typical strength and fracture toughness data for various laminates

Laminate construction[a]	Tensile strength σ_u ($MN\,m^{-2}$)	Measured fracture toughness K_c ($MN\,m^{-3/2}$)	Estimated damage zone size C_0 (mm)
High tensile strength carbon fiber–epoxy			
$(0°)$	1045	90.0	1.86
$(0°/90°)_{2S}$	530	42.1	1.78
$(0°/\pm45°)_{2S}$	460	33.2	1.02
$(0°/\pm45°/90°)_{2S}$	270	32.8	1.49
E-glass-fiber–epoxy			
$(0°)$	1035		
$(0_2°/\pm45°)_{2S}$	600		
$(0°/90°)_{2S}$	540	30.7	
$(0°/\pm45°/90°)_{2S}$	350	24.3	2.54
Boron fiber–epoxy			
$(0°)$	1325		
$(0_2°/\pm45°)_{2S}$	700	67	2.54
$(0_2°/\pm45°/90°)_{2S}$	420	38.7	2.79
Kevlar 49 fiber–epoxy			
$(0°)$	1378		
$(0°/90°)$	578		
$(0°/\pm45°/90°)_{2S}$	393	~30	
$(\pm45°)$	119	~14	
Boron fiber–aluminum			
$(0°)$	~1000	90	
$(0°/45°)_{2S}$		50	

a The subscript S denotes a stacking sequence symmetrical about the midplane of the laminate: the subscript 2 indicates a laminate constructed from two sets of repeating sequences of laminae

posites, Royal Aircraft Establishment Technical Report 79051. HMSO, London
Mandell J F 1971 Fracture toughness of fiber reinforced plastics. Ph.D. thesis, Massachusetts Institute of Technology
Nuismer R J, Whitney J M 1975 Uniaxial failure of composite laminates containing stress concentrations. *Fracture Mechanics of Composites*, ASTM Special Technical Publication 593. American Society for Testing and Materials, Philadelphia, Pennsylvania, pp. 117–42
Potter R T 1978 On the mechanism of tensile fracture in notched fiber reinforced plastics. *Proc. R. Soc. London, Ser. A* 361: 325–41

P. W. R. Beaumont
[University of Cambridge, Cambridge, UK]

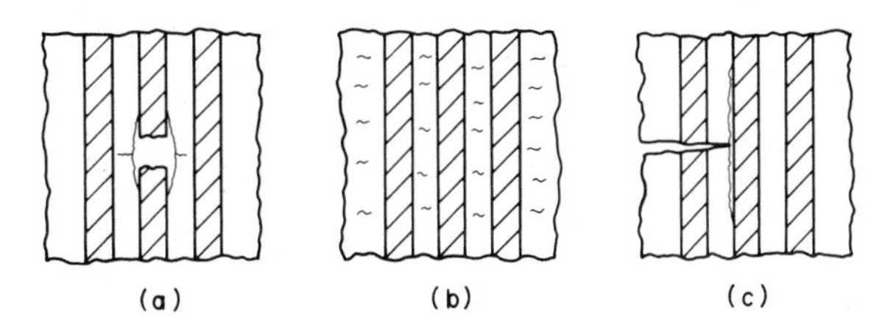

Figure 1
Fatigue damage mechanisms in unidirectional composites under loading parallel to fibers: (a) fiber breakage, interfacial debonding; (b) matrix cracking; (c) interfacial shear failure

Fatigue of Composites

Fatigue of metals has been studied for over a century and despite significant advances it remains a major cause of catastrophic failure of structures. Composites, on the other hand, have high potential for fatigue resistance and can, in certain cases, be designed to eliminate the fatigue problem. The fatigue properties of composites are anisotropic, i.e., directionally dependent, and can be dangerously low in some directions. This warrants careful use of composites based on proper understanding of the mechanisms that govern the fatigue behavior. The mechanisms are admittedly complex, but once analyzed and understood can provide a key to developing a new generation of engineered materials.

The composites considered here are continuous-fiber laminates having fibers of glass or carbon in a thermosetting polymeric matrix. The fiber orientations considered are the unidirectional, the bidirectional (angle-plied and cross-plied laminates) and some combinations of these.

1. Unidirectional Composites

Mechanisms of fatigue damage in unidirectional composites depend on the loading mode, for example tensile or compressive, and on whether the loading is parallel to or inclined to the fiber direction. For illustration, only tensile loads are considered here and the mechanisms for parallel loading and for inclined loading are described separately.

1.1 Loading Parallel to Fibers

The mechanisms may be divided into three types (see Fig. 1). Fiber breakage (Fig. 1(a)) occurs at a local stress exceeding the strength of the weakest fiber in the composite. An isolated fiber break causes shear-stress concentration at the fiber–matrix interface near the broken fiber tip. The interface may then fail, leading to debonding of the fiber from the surrounding matrix. The debond length depends on the shear strength of the interface and is usually small, of the order of a few fiber diameters. The debonded area acts as a stress concentration site for the longitudinal tensile stress. The magnified tensile stress may exceed the fracture stress of the matrix, leading to a transverse crack in the matrix.

The matrix undergoes a fatigue process of crack initiation and crack propagation and generates cracks normal to the longitudinal tensile stress. These cracks are randomly distributed and initially restricted by fibers (Fig. 1(b)). If the cyclic strain in the matrix is sufficiently low, the cracks would remain arrested by the fibers. When the local strains are higher than a certain threshold, the cracks break the fibers and propagate. In this progressive crack-growth mechanism the fiber–matrix interface will also fail due to severe shear stresses generated at the crack tip (Fig. 1(c)).

Final failure results when the progressive crack-growth mechanism has generated a sufficiently large crack (which may be only of the order of a millimeter or less for brittle composites). The fracture surface of a specimen looks messy or broom-like if the fiber–matrix interface is weak and is increasingly neat for stronger interfaces.

1.2 Fatigue-Life Diagram

The mechanisms of damage described above may operate simultaneously. However, observations indicate that the predominant mechanism leading to failure may be effective in a limited range of the applied cyclic strain. This is illustrated in a fatigue-life diagram shown schematically in Fig. 2. The horizontal axis is the number of load cycles to failure on a logarithmic scale and the vertical axis plots the maximum strain, i.e., the maximum stress divided by the modulus of elasticity of the composite, applied initially to a test specimen. Strain instead of stress is plotted since strain is roughly the same in fibers and in the matrix while stress differs in the two phases depending on the volume fraction of fibers and the elastic moduli of the two phases.

The lower limit to the diagram is given by the fatigue limit of the matrix ε_m, i.e., the threshold strain

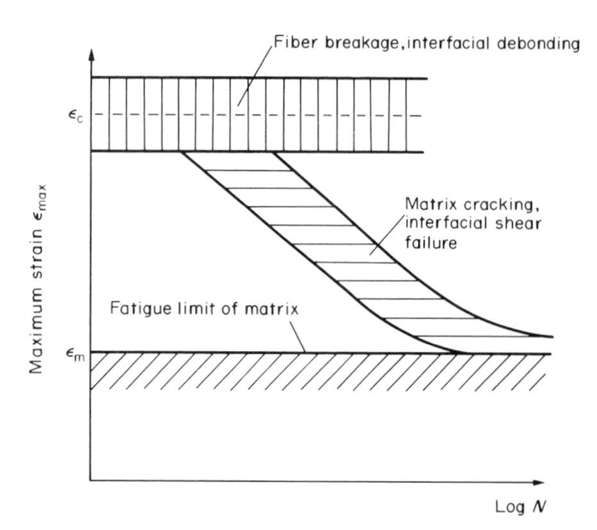

Figure 2
Fatigue-life diagram for unidirectional composites under loading parallel to fibers

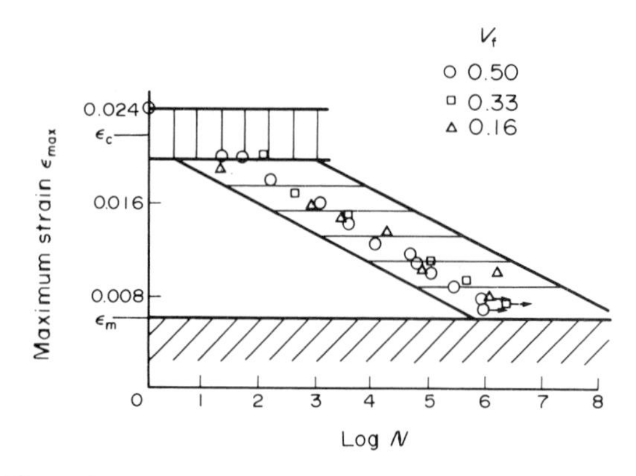

Figure 3
Fatigue-life diagram for a glass–epoxy under loading parallel to fibers. V_f is the volume fraction of fibers

below which the matrix cracks remain arrested by the fibers. This strain is observed to be approximately the fatigue strain limit of the unreinforced matrix material. The upper limit to the diagram is given by the strain-to-failure of the composite ε_c, which is also the strain-to-failure of fibers in a composite reinforced by stiff fibers. The diagram shows a scatter band on the failure strain since this quantity is usually subjected to significant scatter. The mechanism governing static failure, i.e., failure not preceded by significant cycle-dependent growth process, is fiber breakage with associated interfacial debonding. This mechanism is indicated in the scatter band on the failure strain.

The progressive damage mechanism is matrix cracking with associated interfacial shear failure, as described above, and this governs fatigue life. The sloping band of scatter on fatigue life in the diagram is due to this mechanism.

1.3 Effect of Fiber Stiffness

Consider two unidirectional composites with the same matrix and different fibers. The fatigue limit of the two composites will be the same and given by the fatigue limit of the matrix ε_m. The upper limit of the fatigue diagram, given by the composite failure strain, which is equal to the fiber failure strain, will be different for the two composites. Thus the range of strain in which progressive fatigue damage occurs will be different in the two composites. In a particular case where the composite-failure strain and the fatigue-limit strain are equal, the range of strain with progressive fatigue damage will be zero. In such a case fatigue damage will be absent and only static failure will be possible.

Figures 3 and 4 show data and fatigue life diagrams of unidirectionally reinforced epoxy with glass fibers

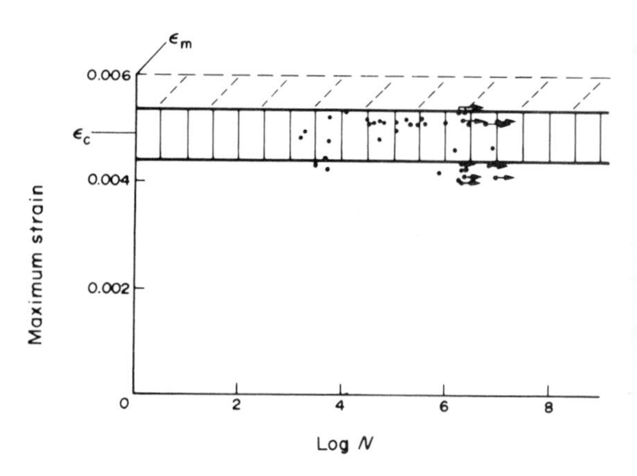

Figure 4
Fatigue-life diagram for a carbon–epoxy under loading parallel to fibers

and with carbon fibers, respectively. The fatigue-limit strain in both composites is 0.6% while the mean failure strains are 2.20% for glass–epoxy and 0.48% for carbon–epoxy. It is seen that the glass–epoxy composite has a wide range of strain with progressive fatigue damage while the carbon–epoxy composite of highly stiff fibers has its fatigue damage totally suppressed. Other carbon–epoxy composites with less stiff fibers and having failure strain of about 1% show some progressive fatigue damage.

1.4 Loading Inclined to Fibers

When the cyclic loading axis is inclined at angles of more than a few degress to the fiber axis, the predominant damage mechanism is matrix cracking along the fiber–matrix interface. The static failure band in the fatigue-life diagram is then lost and the fatigue limit drops with increasing off-axis angle. The lowest fatigue limit is given by the strain for transverse fiber debonding, i.e., failure of the fiber–matrix interface by growth of an interfacial crack in opening mode. This occurs at the off-axis angle of 90°, i.e., when loading is transverse to the fiber direction. Figure 5 shows a schematic fatigue-life diagram for off-axis fatigue of unidirectional composites.

The anisotropy of fatigue properties of composites is illustrated dramatically by Fig. 6 which shows the decrease of the fatigue limit strain with the off-axis angle. The strain below which a composite is safe against fatigue failure when loaded normal to fibers is only 0.1% for glass–epoxy composites, the data for which has been plotted in Fig. 6. This is one-sixth of the same strain for loading along the fibers. However, the ratio of the allowable stresses in the two directions is 1:24, when the elastic moduli in the two directions differ by a factor of 4, a typical value for glass–epoxy composites.

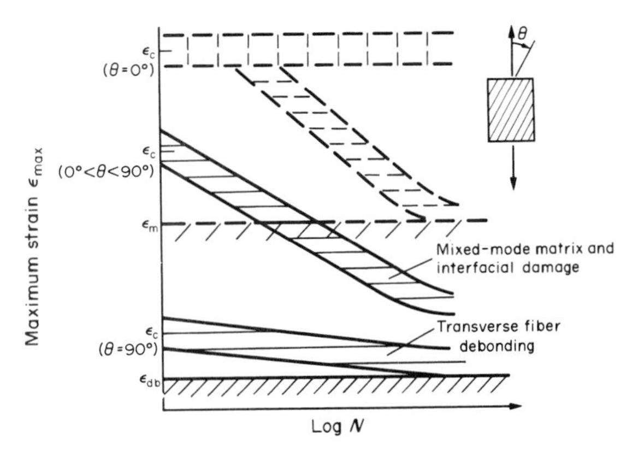

Figure 5
Fatigue-life diagram of unidirectional composites under loading inclined to fibers. ε_{db} is the strain to debonding of fibers from matrix

2. *Bidirectional Composites*

The inferior fatigue properties of unidirectional composites in the direction normal to fibers can be improved by building up laminates with plies of unidirectional composites stacked in two orientations. Such composites, called angle-plied composites, when loaded in direction bisecting the angle between fibers, suffer damage similar to that in a unidirectional composite loaded inclined to fibers. However, the rate of progression of damage is reduced due to the constraint provided by plies of one orientation to cracking in plies of the other orientation. The constraint is highly effective at low angles between fibers but loses effect increasingly with increasing angle. This is illustrated by Fig. 7 where the fatigue-limit strain of angle-plied laminates is plotted against the half-angle between fibers. For comparison the dotted line shows the fatigue limit of the unidirectional composite of the same material, glass–epoxy, against the off-axis angle. Significant improvement in the fatigue limit is seen for angles up to about 45°.

An important class of bidirectional composites is the cross-plied laminates where two orthogonal fiber directions are used. When loaded along one fiber direction a cross-plied laminate develops cracks along fibers that are loaded transversely. These transverse cracks are now constrained by plies with fibers normal to the crack planes, and the degree of the constraint depends on the thickness of the cracked ply (equal to the crack length) and the stiffness properties of the constraining plies. The load shed by a cracked ply is carried by the constraining plies over a distance determined by the constraint conditions. This distance determines the position of another transverse crack.

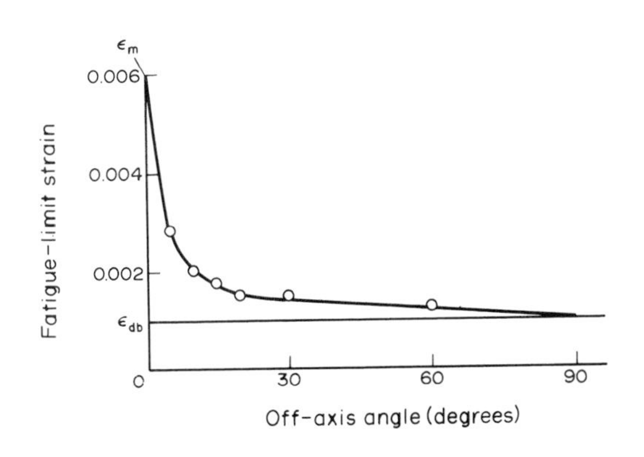

Figure 6
Variation of the fatigue limit with the off-axis angle. ε_{db} is the strain to debonding of fibers from matrix

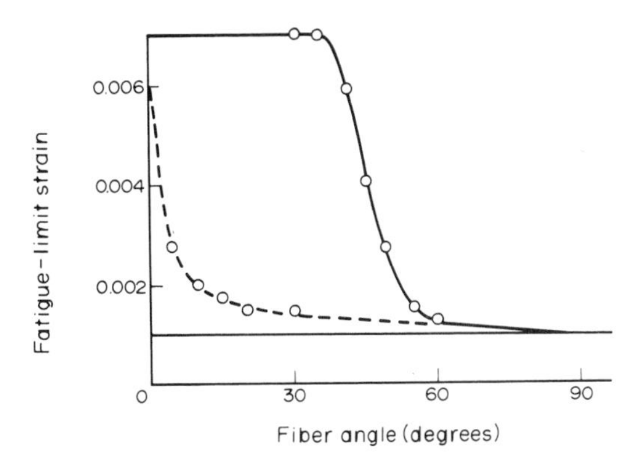

Figure 7
Variation of the fatigue limit with the fiber angle in an angle-plied laminate of glass–epoxy. The dashed line corresponds to the fatigue limit variation of Fig. 6

Thus a crack-density progression process occurs, leading to a saturation crack density.

Load cycling beyond attainment of the transverse crack saturation may lead to diversion of the transverse crack tips into the interfaces between plies. An interlaminar crack may thus form and grow causing an eventual delamination.

The fatigue-life diagram of a cross-plied laminate has the static failure band as the upper limit which is given by the failure strain of fibers. The lower bound, i.e., the fatigue limit, is determined by the strain to initiation of transverse cracking.

3. General Laminates

Combinations of the unidirectional, the angle-plied and the cross-plied orientations are used in various configurations to satisfy the performance requirements a structure may be subjected to. The fatigue properties have been studied primarily under loading along one of the in-plane symmetry directions of laminates. These studies carried out over a decade have formed the basis for an understanding of the development of damage pictured schematically in Fig. 8 (Reifsnider et al. 1983).

The early stage of damage development is dominated by cracking of the matrix along fibers in plies that are not aligned with the symmetry direction in which loading takes place. This cracking has been called primary matrix cracking and has been found to occur as an array of parallel cracks restricted to ply thickness and spanning the width of a test specimen. The number of cracks increase monotonically with the number of load cycles until a saturation density is reached. The saturation of cracks in all off-axis plies occurs unless the weakened laminate breaks at the maximum load. If crack saturation does occur in all off-axis plies, the resulting crack pattern has been found to be characteristic of the laminate configuration and the ply properties and independent of the load amplitude. The damage state associated with the characteristic crack pattern has been called the characteristic damage state (CDS) and signifies the termination of the first stage of matrix cracking damage. The following stage begins by initiation of cracks transverse to the primary cracks and lying in a ply adjacent to the ply with primary cracks. These cracks, called the secondary matrix cracks, extend short distances and appear to be initiators of the interlaminar cracks. The interlaminar cracks are initially distributed in the interlaminar plane and are confined to small areas. The resulting local delamination spreads with continued loading, merges with neighboring delamination and leads to large-scale delamination. The final stage is dominated by fiber breakage and ultimate failure occurs when the locally failed regions have sufficiently weakened the laminate to cause failure under the maximum load.

4. Properties Degradation

Mechanisms of fatigue damage in composites result in distribution of cracks of various orientation in the volume of the material. This leads to degradation of the overall material properties. An example is shown in Fig. 9, where the elastic modulus of a composite normalized by its initial value has been plotted against the fatigue cycles. The modulus degradation shows three stages corresponding to the damage development shown schematically in Fig. 8. In stage I extending until CDS the modulus degradation is abrupt. In stage II, which follows CDS and is characterized by a steady damage development, the modulus degradation is gradual. In stage III the damage development is unstable and gives rise to an erratic drop in the modulus.

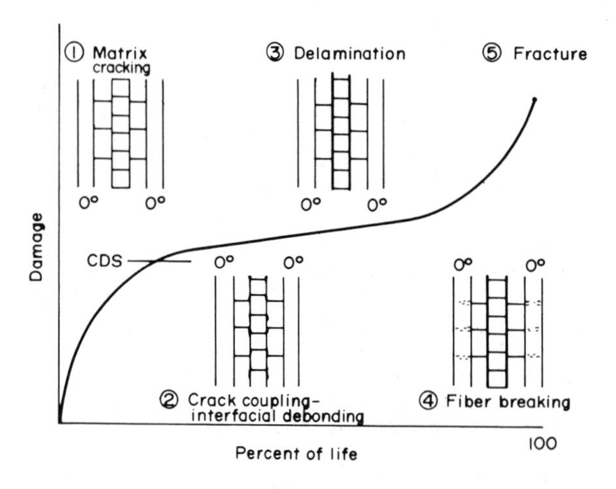

Figure 8
Development of damage in composite laminates

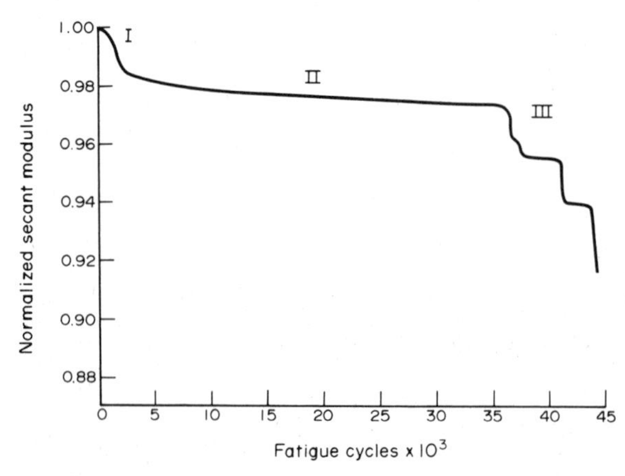

Figure 9
Degradation of the elastic modulus of a cross-plied laminate of carbon–epoxy under fatigue

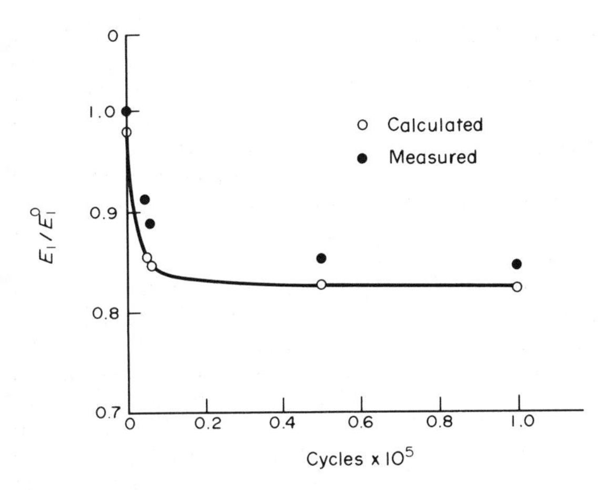

Figure 10
Predicted and measured degradation of the elastic modulus of a glass–epoxy laminate under fatigue

The overall properties degradation in stages I and II has been modelled by regarding the composite with damage as a continuum with changing microstructure (Talreja 1985). A phenomenological theory of constitutive behavior then provides relationships between the severity of damage and the overall stiffness properties of a composite. An example of such relationships for intralaminar cracking is

$$E_1 = E_1^0 - 2t_c^2/ts[a + b(\nu_{12}^0)^2 - c\nu_{12}^0]$$

where E_1 and E_1^0 are the current and initial values, respectively, of the elastic modulus along a symmetry axis labelled 1 in an orthotropic laminate, ν_{12}^0 is the initial Poisson's ratio for straining in the direction 1 with contraction in the orthogonal direction 2, t_c and t are thicknesses of the cracked ply and of the laminate, respectively, s is the spacing of the intralaminar cracks and a, b and c are the material constants. These constants have been determined for glass–epoxy and for carbon–epoxy laminates (Talreja 1987). Prediction of the modulus degradation in accordance with the equation is illustrated by Fig. 10.

See also: Failure of Composites; Multiple Fracture

Bibliography

Reifsnider K L, Henneke E G, Stinchcomb W W, Duke J C 1983 Damage mechanics and NDE of composite laminates. In: Hashin Z, Herakovich C T (eds.) 1983 *Mechanics of Composite Materials—Recent Advances*. Pergamon, New York, pp. 399–420.

Talreja R 1985 A continuum mechanics characterization of damage in composite materials. *Proc. R. Soc. London, Ser. A* 399: 195–216.

Talreja R 1986 Stiffness properties of composite laminates with matrix cracking and interior delamination. *Eng. Fract. Mech.* 25: 751–62

Talreja R 1987 *Fatigue of Composite Materials*. Technomic Lancaster

R. Talreja
[Georgia Institute of Technology, Atlanta, Georgia, USA]

Fiber-Reinforced Cements

Cement mortar and concrete differ from most other matrices used in fiber composites in that their essential characteristics are those of low tensile strength (generally less than 7 MPa) and very low strain to failure in tension (generally less than 0.05%) which result in materials which are rather brittle when subjected to impact loads.

The binder in the matrix is cement paste produced by mixing cement powder and water which harden after an exothermic reaction to form complex chemical compounds that are dimensionally unstable due to water movement at the molecular level within the crystal structure. Inert mineral fillers known as aggregates are added at up to 70% by volume to increase stability and reduce the cost, and therefore the space available for fiber inclusion may be severely limited.

Fiber volumes in excess of 10% of the composite volume are difficult to include in mortar and, in the case of concrete, 2% by volume of fiber is considered to be a high proportion which is generally insufficient to permit significant reinforcement of the matrix before cracking occurs, particularly where individual fibers may be spaced up to 40 mm apart by the aggregate particles.

The type of fiber composite described in this article is therefore one in which the fiber has little effect on the properties of the relatively stiff matrix until cracking has occurred at the microlevel. The fibers then become effective, resulting in a composite with reduced stiffness and increased strain to failure providing increased toughness compared with the unreinforced matrix.

Notable exceptions to this type of behavior are asbestos-cement and the cellulose fiber reinforced thin sheets introduced to the market in the mid-1980s. In these materials the fibers are used to increase the matrix cracking strain to more than 1000×10^{-6}.

1. Principles of Reinforcement in Tension

Owing to the relatively stiff matrix and its low strain to failure, fibers are not normally included in cement matrices to increase the matrix cracking stress. Therefore the merits of fiber inclusion lie in the load-carrying ability of the fibers after microcracking of the matrix has been initiated.

1.1 Critical Fiber Volume

The critical fiber volume $V_{f(crit)}$ is defined as the volume of fibers which, after matrix cracking, will carry the load which the composite sustained before cracking. For the simplest case of continuous aligned fibers with frictional bond, the critical fiber volume can be expressed in terms of the elastic modulus of the composite E_c, the ultimate strain of the matrix ε_{mu} and the fiber strength σ_{fu}:

$$V_{f(crit)} = \frac{E_c \varepsilon_{mu}}{\sigma_{fu}} \quad (1)$$

Figure 1 shows stress–strain curves for composites with less than the critical volume (OAB) and greater than the critical volume of fiber (OACD).

Efficiency factors for fiber orientation and fiber length increase $V_{f(crit)}$ considerably above that calculated from Eqn. (1). The result is that most concretes are unable to contain the critical volume of short chopped fibers and still maintain adequate workability for full compaction to be achieved. However, many techniques are available which enable the inclusion of more than the critical volume of fiber in fine grained cement mortars.

The dashed curve in Fig. 1 results when high volumes of very small diameter, stiff, well bonded fibers are included. The existing flaws are reinforced and may be restrained from propagating at the stress at which they would propagate in an unreinforced matrix, thus leading to an increased cracking strain in the matrix. Typical examples are asbestos-cement and cellulose fibers in autoclaved calcium silicate.

1.2 Stress–Strain Curve, Multiple Cracking and Ultimate Strength

Figure 1 shows the idealized stress–strain curve (OACD) for a fiber-reinforced brittle matrix composite with more than the critical volume of fiber. The assumptions used to calculate the curve are that the fibers are long, aligned with the stress, and that the bond τ is purely frictional. The matrix is assumed to crack at A at its normal failure strain ε_{mu} and the material breaks down into cracks in region A–C with a final average spacing S given by

$$S = 1.364 \frac{V_m \sigma_{mu} A_f}{V_f \tau P_f} \quad (2)$$

where V_m, V_f are the volume fraction of the matrix and fiber respectively; A_f, P_f are the cross-sectional area and perimeter of the fiber; and σ_{mu} is the cracking stress of the matrix. It will be noted that crack spacing is not dependent on the elastic modulus of the fiber. Region C–D represents extension of the fibers as they slip through the matrix without further matrix cracking.

The ultimate strength of the composite depends only on the fiber strength σ_{fu} and volume V_f whereas the ultimate composite strain is less than the fiber failure strain by a factor depending on the volume and elastic modulus of the matrix and of the fiber, and the matrix failure strain.

The multiple cracking represented in Fig. 1 is a desirable situation because it changes a basically brittle material with a single fracture surface and low energy requirement to fracture into a pseudoductile material which can absorb transient minor overloads and shocks with little visible damage.

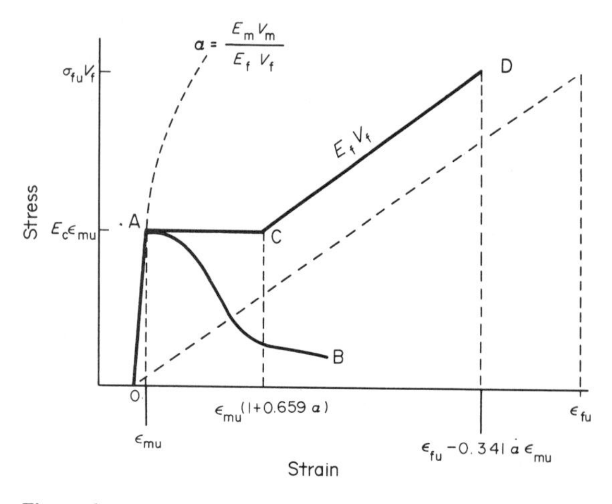

Figure 1
Idealized tensile stress–strain curves for a brittle cement matrix composite

1.3 Efficiency Factors for Fiber Length and Orientation

The efficiency of fibers in reinforcing the matrix in practice may be very much less than that assumed in the calculation of the idealized stress–strain curve. Efficiency depends on many factors including the orientation of the fibers relative to the direction of stress, whether the matrix is cracked or uncracked and the fiber length relative to the critical length (defined as twice the length of fiber embedment which would cause fiber failure in a pullout test).

For short fibers in a three-dimensional orientation, the combined efficiency could be less than one-fifth of the efficiency in the aligned fiber case (Hannant 1978).

2. Principles of Reinforcement in Flexure

The critical fiber volume for flexural strengthening can be less than half that required for tensile strengthening so that the load which a beam of the composite can carry in bending can be usefully increased, even though the tensile strength of the materials has not been altered by the addition of fibers. In fact, the flexural strength is commonly between two and three times that predicted from the tensile strength using an elastic analysis. The main reason for this is that the postcracking stress–strain curve in tension (AB or ACD in Fig. 1) is very different from that in compression (similar slope to OA in Fig. 1). As a result, the neutral axis moves towards the compression surface of the beam and conventional beam theory becomes inadequate because it does not allow for a plastic stress block in tension combined with an elastic stress block in compression (Hannant 1978).

The relationship between idealized uniaxial tensile stress–strain curves and flexural stress–strain curves is shown in Fig. 2.

The importance of achieving adequate post-cracking tensile strain capacity in maintaining a high flexural strength is demonstrated in Fig. 2. For instance, curve OAC is a tensile stress–strain curve and OAC′ is the associated bending curve. If the tensile strain reduces from C to A at a constant stress, then the bending strength will reduce from C′ to A with an increasing rate of reduction as A is approached and the material becomes essentially elastic. Thus, tensile strength on its own cannot be used to predict bending strength and composites which suffer a reduction in strain to failure as a result of natural weathering will also have a reduced bending strength.

3. Asbestos-Cement

Since 1900, the most important example of a fiber cement composite has been asbestos-cement. The proportion by weight of asbestos fiber is normally between 9 and 12% for flat or corrugated sheet, 11–14% for pressure pipes and 20–30% for fire-resistant boards and the binder is normally a Portland cement. Fillers such as finely ground silica at about 40% by weight may also be included in autoclaved processes where the temperature may reach 180 °C.

The most widely used method of manufacture for asbestos-cement was developed from papermaking principles around 1900 and is known as the Hatschek process. A slurry, or suspension of asbestos fiber and cement in water at about 6% by weight of solids, is continuously agitated and allowed to filter out on a fine screen cylinder. The filtration rate is critical and coarser cement than normal (typically 280 $m^2 kg^{-1}$) is used to minimize filtration losses. Also, although chrysotile fibers form the bulk of the fiber, most formulations have included a smaller proportion of the amphibole fiber wherever possible. This is because of the dewatering characteristics of crocidolite and amosite. Mixtures of cement and chrysotile drain very slowly and production rates are consequently slow. By using a blend of chrysotile and amphibole fibers (usually between 20% and 40%), considerable improvements in the drainage rate, plus the additional bonus of enhanced reinforcement, have been achieved.

Other types of fiber such as cellulose derived from wood pulp or newsprint have also been added to the slurry to produce different effects in the wet or hardened sheet.

There are several other manufacturing processes such as the Mazza process for pipes and the Magnani process for corrugated sheets. Also the material may be extruded and injection molded.

The tensile strength varies between about 10 MPa and 25 MPa depending on fiber content, density and direction of test whereas the flexural strengths may be between 30 MPa and 60 MPa.

Durability is exceptionally good, a lifetime in excess of 50 years under external weathering being possible. A disadvantage is the brittle nature of the material under impact situations due to its relatively low strain to failure.

Major applications have included corrugated sheeting and cladding for low-cost agricultural and industrial buildings, and flat sheeting for internal and external applications has also proved very successful.

Pressure pipes are another worldwide market and the ease with which the material can be formed into complex shapes has enabled a wide range of special purpose products to be supplied.

The future for asbestos-cement products depends largely on the attitudes taken by Government Health and Safety organizations to the health problems associated with asbestos fibers.

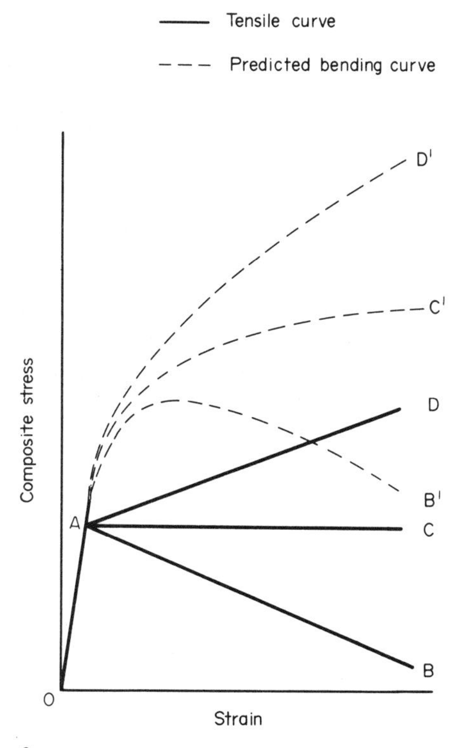

Figure 2
Apparent bending (flexure) curves predicted for assumed direct tensile curves (after Laws and Ali 1977)

4. Glass-Reinforced Cement

The early development of glass fibers in low-alkali cements took place in China and Russia in the 1960s but because of the susceptibility to attack of E-glass fibers by the higher-alkali Portland cements used in the West, considerable investment in the UK took place in the 1970s to produce alkali-resistant zirconia glass on a commercial scale.

The fibers are commonly included in the matrix in strands consisting of 204 filaments of 10 μm diameter fibers.

The three basic techniques are premixing which gives a random three-dimensional array unless the mix is subsequently extruded or pressed, the spray method which gives a random two-dimensional arrangement, and filament winding which gives a one-dimensional array or a multi-angled oriented array. The volume of fibers is generally greater than the critical volume for tensile strengthening.

The spray method is the most popular and consists of leading a continuous roving up to a compressed-air-operated gun which chops the roving into lengths of between 10 mm and 50 mm and blows the cut lengths at high speed simultaneously with a cement paste spray onto a forming surface which may contain a filter sheet and vacuum water extraction in certain applications.

The fiber-cement sheet can be built up to the required thickness and demolded immediately, an additional advantage being that it has sufficient wet strength to be bent round radiused corners to give a variety of product shapes such as corrugated or folded sheets, pipes, ducts or tubes. Colored fine aggregate can be trowelled into the surface to produce a decorative finish.

The process can be readily automated for standard sheets. The optimum fiber volume for flexural strength using the spray technique is about 6% because of increase in porosity of the composite at higher fiber volumes.

One of the problems with this type of glass-reinforced cement has been the reduction in strain to failure with time under water storage or natural weathering conditions. This effect results in the material becoming essentially brittle under these conditions but air storage in a dry atmosphere causes little change in the initial properties even after periods of ten years.

Continuing developments in glass-cement systems show promise of reducing rate of embrittlement under natural weathering conditions.

Typical properties are tensile strengths between 7 MPa and 15 MPa and bending strengths between 13 MPa and 35 MPa.

Applications include cladding panels, permanent formwork for concrete construction, pipes, flat sheeting, sewer linings and surface coatings for dry blockwork.

5. *Steel-Fiber Concrete*

Steel fibers have been used mainly in bulk concrete applications as opposed to cement mortars.

It has been found from experience that, for traditional mixing and compaction techniques to be applicable, the maximum aggregate size should not exceed 10 mm and there should be a high content of fine material. In these circumstances 1–1.5% by volume of fibers typically 50 mm long by 0.5 mm diameter can be mixed with little modification of existing plant.

The fibers are generally deformed to give better "keying" to the matrix but, as the fibers are arranged randomly in two and three dimensions, the critical fiber volume for tensile strengthening is rarely exceeded. However, the flexural strength can be increased up to about 15 MPa and toughness is greatly increased.

Higher volumes of longer fibers can be included by spray techniques known as "gunite" or "shotcrete" and these materials are often used as tunnel linings or for rock slope stabilization.

The increased toughness is utilized in highway and airfield pavement overlays which have reduced thickness and greater impact resistance than traditional construction. The use of stainless steel has proved especially advantageous in refractory applications.

Fiber production processes include spinning from the melt which has reduced the cost of the fiber thus increasing its economic viability.

6. *Organic Fibers in Cement*

6.1 Man-Made Fibers

Polymer fibers used in cement matrices cover the complete range of fiber elastic moduli. The moduli range from 2 GPa to 12 GPa for polypropylene to 200 GPa to 300 GPa for carbon fibers but the basic principles of reinforcement remain the same for all the fiber types.

Polypropylene, because of its high strength combined with excellent alkali resistance and low price, has had the greatest application in cement and concrete. It is generally used as fibrillated film between 50 μm and 100 μm thick made from isotactic polypropylene with draw ratios between 5 and 20 to produce a high degree of molecular alignment. The stretched fibrillated film with strengths between 300 MPa and 500 MPa may be used in concrete as chopped twine in volumes generally less than 1% or as layers of flat opened networks in fine grained mortar at volumes between 5% and 20%.

The chopped twine is mixed with the concrete in standard mixers but as the volume of fiber is well below the critical volume for tensile strengthening the major benefits are in handling the fresh material or in impact situations such as piling or pontoons where the fibers serve to hold the cracked material together when damaged.

The manufacture of thin cement sheets containing many layers of film requires special production machinery. The sheets, which are intended as alternatives to asbestos-cement, are characterized by high toughness and flexural strength (up to 40 MPa) resulting from the closely spaced multiple cracking.

The higher elastic modulus fibers such as carbon fibers and Kevlar show considerable promise as cement reinforcements but are too expensive for bulk applications. Cheaper, high-modulus polymers, such as high-modulus polyethylene (20 GPa to 100 GPa) will fibrillate in a similar manner to polypropylene and show potential should commercial production become available.

6.2 Natural Fibers

Cellulose fibers produced from wood pulp have been used for many years as additives in asbestos-cement products but their precise physical properties seem to vary widely depending on the authority quoted and this variability may also depend on the pulping techniques used. Fiber lengths between 1 mm and 4 mm have been quoted and the helical structure of the fibers gives a variable diameter between 10 μm and 50 μm. Tensile strengths for the fiber between 50 MPa and 600 MPa have been quoted with moduli between 10 GPa and 30 GPa.

Flat and corrugated sheets containing cellulose fibers in autoclaved calcium silicate have been marketed as alternatives to asbestos-cement. The strength of these products is lower in the wet than the dry state due to the sensitivity of cellulose fibers to water absorption.

Other natural fibers such as sisal, agave, piassava, coconut and bagasse fibers have been used in cement in an attempt to produce cheap cladding sheets for houses in developing countries but although some success has been achieved, the process is labor intensive and there are many problems still to be overcome.

Bibliography

Building Research Establishment 1979 *Properties of GRC: 10 Year Results*, Information Paper IP 36/79. Building Research Station, Watford

Cook D J 1980 Concrete and cement composites reinforced with natural fibres. *Fibrous Concrete—Concrete International 1980*. Construction Press, London, pp. 99–114.

Hannant D J 1978 *Fiber Cements and Fiber Concretes*. Wiley, Chichester

Laws V, Ali M A 1977 The tensile stress-strain curve of brittle matrices reinforced with glass-fiber. *Conf. Proc. Fiber Reinforced Materials*. Institute of Civil Engineers, London, pp. 101–9

D. J. Hannant,
[University of Surrey, Guildford, UK]

Fiber-Reinforced Ceramics

The development of fiber-reinforced ceramics has been prompted by the need for materials with the advantages of ceramics combined with increased toughness and strength, and a reduced variability of strength. For a variety of technological and commercial reasons, a very considerable increase in the amount of research and development carried out on ceramic-matrix fiber composites (CMC) began in the early 1980s. The increase in effort and rate of development made this one of the most rapidly advancing fields in composite technology by the mid-1980s, a field which is continuing to develop rapidly. By the late 1980s, three different generic types of CMC had been developed, characterized by the methods used to incorporate fibers or whiskers into the matrix. These are: continuous-fiber reinforced glass-ceramic and glass systems produced by a solid-state slurry impregnation route followed by hot-pressing; CMC produced from continuous woven-fiber preforms infiltrated with a ceramic matrix by gas or liquid-phase routes; and hot-pressed whisker toughened ceramics (Phillips 1987). In addition there are other promising fabrication routes which are at a less well developed stage. The mechanical properties of these different generic types of material, as well as the advantages and disadvantages of the different fabrication routes, differ, as to some extent, do the applications of the materials.

The useful properties of ceramics include retention of strength at high temperature, chemical inertness, low density, hardness and high electrical resistance. Their principal disadvantage is their brittleness: failure strain, fracture energy and fracture toughness (K_{IC}) being low compared with tough plastics and metals. This renders them susceptible to damage by thermal or mechanical shock, makes them easily weakened by damage introduced during service, and causes them to have a large variability in strength. Fiber reinforcement can increase their fracture energies by several orders of magnitude, to values approaching 10^5 J m^{-2}, through energy-absorbing mechanisms similar to those in reinforced polymers, with consequent increases in apparent values of K_{IC} to ~ 30 MPa m$^{1/2}$, as shown in Table 1. Reductions in variability of strength, as characterized by the Weibull parameter m, are typically from an m of less than 10 for unreinforced ceramics to 20–30 for the best CMC. This is equivalent, very approximately, to reducing the coefficient of variation (CV) in strength from more than 0.12 down to 0.04. This is important for the design of highly stressed engineering components, as an increase in m and a reduction in CV reduce the required safety factors and thus enable the component to be designed to operate at higher stresses.

1. Fabrication Routes

Currently, all successful techniques for the manufacture of CMC require processing at temperatures of the order of 1000 °C and upwards. Chemical and thermal expansion compatibility between fibers and matrix are therefore important. Since strains induced by thermal

Table 1
Representative strength and toughness data for monolithic and some fiber-reinforced ceramics

	Strength (MPa)	Fracture energy ($J\ m^{-2}$)	K_{IC} ($MPa\ m^{1/2}$)
Glasses	100	2–4	0.5
Engineering ceramics (untoughened)	500–1200	40–100	3–5
Zirconia toughened	600–2000		6–12
Whisker toughened	300–800		6–9
Short-fiber reinforced glass	50–150	600–800	7
Continuous-fiber reinforced glass and glass-ceramic	1600	10^4–10^5	20–30

expansion differences as the composite cools can cause the composite to crack and fragment, the fibers and matrices which can be combined successfully are limited. In general a fiber with the same or a higher thermal expansion coefficient than the matrix is preferred. High-temperature chemical reactions between the fiber and matrix during fabrication can have significant effects on the properties of the composite: the most severe are either the degradation of the fibers or production of too strong a fiber–matrix bond resulting in a brittle, low-strength composite. To control the bond strength and improve toughness an interlayer between the fiber and matrix may be necessary; for example, the fibers may be coated with graphite prior to composite manufacture.

Glass and glass-ceramic matrix composites are most successfully manufactured by a slurry impregnation route (Phillips 1983, 1985). Fibers, in the form of multifilament continuous tow, are passed through a slurry of finely powdered matrix material in a mixture of solvent and binder. As the tow passes out of the slurry, the solvent evaporates and the resulting tape can be wound onto a drum to produce handleable prepreg sheets consisting of intimately mixed fibers and powder held together by the binder. The prepreg sheets are then stacked as a series of plies, in a similar way to conventional polymer prepreg technology, and hot-pressed at temperatures of about 1000 °C or more and pressures of about 5–10 MPa, to produce a laminate with a low-porosity matrix. This process has produced the highest strength CMC and has enabled the manufacture of composites containing up to 60 vol% of unidirectional fibers with strengths equal to those predicted by the law of mixtures. The most successful composites produced in this way to date consist of multifilament carbon (graphite) or silicon carbide (e.g., Nicalon) fibers in borosilicate glass or lithium aluminosilicate (LAS) glass-ceramic matrices. The carbon-fiber composites oxidize in air above about 450 °C while the SiC-fiber-composites can be employed to around 1100 °C. Softening of borosilicate glass restricts the maximum temperature of this matrix to around 580 °C while LAS glass-ceramics can, in principle, operate to around 1000 °C–1200°C. The glass-ceramic composite can be produced either from a glass-ceramic powder or from a glass powder which is then heat treated to devitrify it. The latter process offers the prospect of lower-temperature processing than the former, but the fabrication process is more difficult to optimize because of the dimensional changes which occur when the glass crystallizes. The principal advantages of the slurry impregnation manufacturing process are the high strengths achievable and the relatively short manufacturing times; the main disadvantages are the need for high temperatures and pressures and the difficulty of manufacturing complicated shapes.

In the gas- and liquid-phase infiltration fabrication routes, a fiber preform is infiltrated with a medium which deposits the ceramic matrix. The preform may be made by stacking together sheets of woven fabric cut to the required shape, or may consist of a multidimensional woven or knitted structure. The most mature infiltration technique is through chemical vapor infiltration (CVI) (Naslain and Langlais 1985). The preform is heated in a chemical reactor vessel and a gas or mixture of gases is passed through the vessel to cause the deposition of the ceramic matrix within the fiber network, resulting in the slow growth of the matrix as an interpenetrating network. A typical example of such a reaction is:

$$CH_3SiCl_3\,(g) + H_2\,(g) \rightarrow SiC\,(s) + HCl\,(g)$$

Other ceramics which have been deposited in this way include B_4C, TiC, BN, Si_3N_4, Al_2O_3 and ZrO_2. Typical temperatures required for this process are around 1000 °C.

In order to achieve low matrix porosity it is necessary to maintain an open-pore network for as long as possible as the ceramic matrix grows. In practice this is difficult to achieve and it is necessary to carry out the impregnation several times, with the surface of the composite material having to be removed by machining between each stage, to permit gas to continue to enter the composite. In practice, the minimum porosities which are achieved are between 10 and 20 vol% of the matrix volume. Because of matrix porosity the strength of these composites tends to be around half of that expected from the law of mixtures. Figure 1 shows how the strength of a ceramic composite can be affected by porosity. The advantages of this fabrication route are: lower temperatures than those needed for powder hot-pressing routes, the potential for the easier production of more complex shapes by near-net shape processing, and materials which are more nearly isotropic because of their use of multidirectional preforms, and which are very tough. A major disadvan-

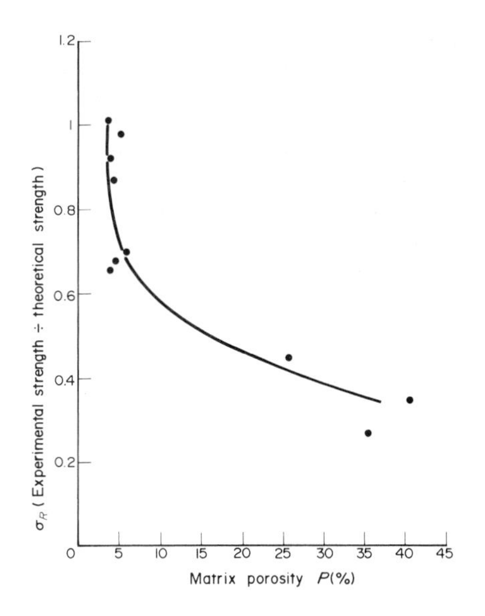

Figure 1
The variation of experimental strength, expressed as a function of theoretical strength, with matrix porosity for a ceramic-matrix composite (Phillips 1983)

tage is the long time required to manufacture a component and the necessity for intermediate stages of machining: it can take several weeks or even months to manufacture a component the size of a gas turbine blade. The long fabrication times can be mitigated against by using a batch production process in which a large number of components are manufactured simultaneously in one reaction vessel. Another disadvantage is the relatively poor utilization of fiber strength because of the high porosity of the final product. The most important material produced by this route is SiC-fiber-reinforced SiC. The procedure has also been used to modify carbon–carbon composites to produce a composite in which the outer surface is a hybrid matrix with improved oxidation resistance. Examples of such mixed matrices are SiC/C and B_4C/C.

Similar composites have been manufactured at the laboratory scale by liquid-phase infiltration employing sol-gel and polymer pyrolysis processes. Both of these require the initial synthesis of liquid organic compounds which can be converted to a ceramic after infiltrating a fiber preform. The sol-gel route requires the synthesis of an alkoxide—an organic salt of an alcohol and a metal. After hydrolysis and polycondensation, the resulting gel is dehydrated, leaving a porous ceramic of high chemical reactivity and sintering ability. An example of polymer pyrolysis is the conversion of polycarbosilane to SiC.

Whisker-toughened ceramics are the most successful category of discontinuous-fiber-reinforced ceramics. In principle discontinuous-fiber-reinforced ceramics are the simplest CMC to manufacture. Short fibers or whiskers are mixed intimately with powdered matrix material and then either hot-pressed uniaxially or hot-isostatically-pressed after isostatic cold-pressing. Temperatures in excess of 1500 °C and pressures in excess of 100 MPa are required for consolidation of ceramic matrices such as alumina. Following this process, the reinforcing phase in the resulting composite is randomly oriented in the pressing plane. Agglomeration of the fibers or whiskers makes it difficult to produce homogeneous materials with reinforcing volume fractions greater than about 30%.

Composites of this type containing short random fibers, as opposed to whiskers, have tended to be of lower strength than the unreinforced matrix but can have substantially greater toughness: techniques exist for aligning the fibers before consolidation and composites produced in this way can have higher strengths than the matrix. Whisker-toughened ceramics containing randomly oriented whiskers can be substantially stronger than the unreinforced matrix and have greater toughness. Successful systems of this type are SiC whiskers in Si_3N_4 and SiC whiskers in Al_2O_3. As an example of the improvement which can be obtained by whisker toughening, an unreinforced Al_2O_3 had a strength of 500 MPa and a K_{IC} of 4.5 MPa $m^{1/2}$ while 30 vol% of SiC whiskers increased these to 650 MPa and 9.0 MPa $m^{1/2}$, respectively (Wei and Becher 1985).

The development of whisker-toughened ceramics has occurred at the same time as the development and exploitation of a completely different but very important competing method of toughening ceramics, which employs the martensitic transformation of zirconia (Butler 1985). There is promising research aimed at combining whisker toughening and zirconia toughening and recent results appear to indicate synergistic effects, the combined toughening effect being greater than the sum of the individual effects. Advantages of whisker toughening over zirconia toughening include the lower densities of whisker-toughened materials and the retention of the toughening mechanisms to high temperatures—a major disadvantage of zirconia toughening is that the contribution to toughness from the martensitic transformation decreases as temperature increases. However, during the late 1980s a potential problem has arisen for whisker reinforced materials. Serious concerns have been expressed about the possible toxicity of whiskers with diameters less than 1 μm. It is unclear at present whether this will be confirmed as a serious problem for the whisker toughening of ceramics.

Other promising fabrication techniques, but at a less advanced stage, include the manufacture of fiber-reinforced silicon nitride by the nitridation of a silicon matrix to silicon nitride, and the oxidation of an aluminum matrix to alumina. Interesting composites have also been produced by the in-situ growth of a whisker-like phase, for example the growth of needle-like β-Si_3N_4 crystals in Si_3N_4 (Suzuki 1987).

2. *Applications*

The main demand for the development of CMC has been the requirements for higher-temperature materials, particularly in aerospace as components of gas turbine engines, for rocket motors and for hot spots on spacecraft reentering the atmosphere. There are, however, potentially important applications where extreme temperature capability is not required. For example, fiber-reinforced glasses and glass ceramics have mechanical properties comparable with high-performance fiber-reinforced polymers, and could be used at intermediate temperatures higher than attainable with polymeric systems. They are more stable to ionizing radiations than fiber-reinforced plastics and this could make them attractive for nuclear and space environments. They are less susceptible to hygrothermal effects and therefore could be attractive for aircraft or marine applications. Their hardness could make them more erosion-resistant than fiber-reinforced plastics and therefore attractive as radome materials, while hardness combined with a tailorable toughness could make them useful in armor. The chemical inertness of CMC could make them suitable in specialized chemical plants and they are already marketed as biomedical implant materials. Whisker-toughened ceramics are already marketed as cutting tool and die materials.

3. *Properties: Matrix Microcracking and Toughness*

The maximum strength, toughness and anisotropy of properties of the different classes of CMC differ widely and a detailed discussion of their properties is not appropriate here. However, a key question concerning the higher-strength systems is the importance of matrix microcracking.

The strain to failure of a ceramic matrix is usually less than that of the reinforcing fiber and, on loading, the matrix of the composite can crack at loads lower than the ultimate load. For a unidirectional composite stressed in the fiber direction this is manifested as an array of regularly spaced cracks. To a first approximation, the stress at which matrix cracking would be expected to occur on the basis of a simple isostrain model is

$$(\sigma_m)_u\left[1+V_f\left(\frac{E_f}{E_m}-1\right)\right]$$

where $(\sigma_m)_u$ is the strength of the unreinforced matrix, V_f is the fiber volume fraction, and E_f and E_m are the fiber and matrix Young's moduli, respectively. A composite with a multiply-cracked matrix can retain useful properties even under fatigue conditions, but the cracks lead to easier ingress of aggressive environments. However, suppression of matrix cracking to higher stresses can occur and the theories of Aveston et al. (1971), Marshall et al. (1985) and McCartney and Kelly (McCartney 1986) provide an explanation of crack inhibition. In practice the use of higher fiber volume fractions and, in the case of laminates, thin plies provide the best methods of minimizing multiple matrix cracking.

The importance of matrix microcracking in determining the life of a composite under service conditions has not yet been established. Matrix microcracking provides paths for the easier ingress of aggressive environments and hence more severe attack on fibers. By analogy with advanced polymer-composite systems it can also be expected to initiate a mechanism of fatigue failure, in which matrix cracks in one ply initiate orthogonal interlaminar cracks. Current data indicate weakening effects in SiC/LAS composites under both static and fatigue conditions at modest temperatures ($\sim 800\,^{\circ}C$) due to matrix microcracking (Prewo 1986, 1987), although strengths increase again at higher temperatures. Further research is needed to understand and resolve this problem since, if it becomes apparent that the SiC/glass-ceramic systems can be operated only at stresses below the matrix microcracking stress, it will imply a substantial decrease in their acceptable operating stresses.

Another issue concerns the toughness of fiber-reinforced ceramics. Increased toughness is one of the most important driving forces behind the development of CMC. The fracture energies listed in Table 1 are a measure of the amount of energy absorbed in creating unit area of macroscopic fracture surface. The energy absorption processes which occur when fiber-reinforced ceramics fracture are similar to those in polymer-matrix composites. In unidirectional continuous-fiber composites, and short-fiber composites, these are processes such as fiber pullout and debonding, and in laminated materials there are further contributions from multiple and delocalized interlaminar and intralaminar crack propagation. These can result in considerable energy absorption (see Fig. 2). On theoretical grounds, the energy absorption from mechanisms such as pullout and debonding decreases as the fiber diameter decreases. Consequently the fracture energies of whisker-toughened ceramics, containing a reinforcing phase of diameter $< 1\,\mu m$, are low compared with those of fiber-reinforced ceramics, containing a reinforcing phase of diameter $\sim 10\,\mu m$ and upwards. In whisker-toughened ceramics a substantial contribution to toughness derives from crack deflection processes and the conventional mechanisms of pullout and debonding may play a relatively minor role, although this still remains an issue requiring further research. In continuous-fiber systems the fracture process is complex and may be delocalized with extended damage (see Fig. 3). Under these circumstances, linear elastic fracture mechanics is inapplicable and the values of K_{IC} in Table 1 for continuous-fiber systems should be regarded, as for similar polymer-matrix composites, as a measure of damage tolerance rather than a design quantity.

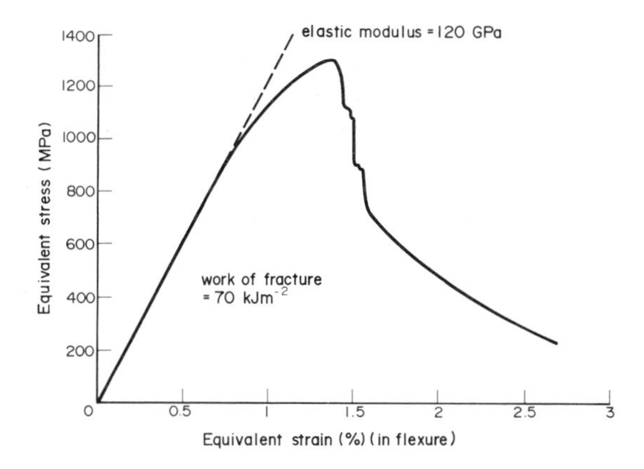

Figure 2
Load–extension behavior in flexure of a SiC fiber-glass matrix composite (courtesy D. M. Dawson and R.W. Davidge)

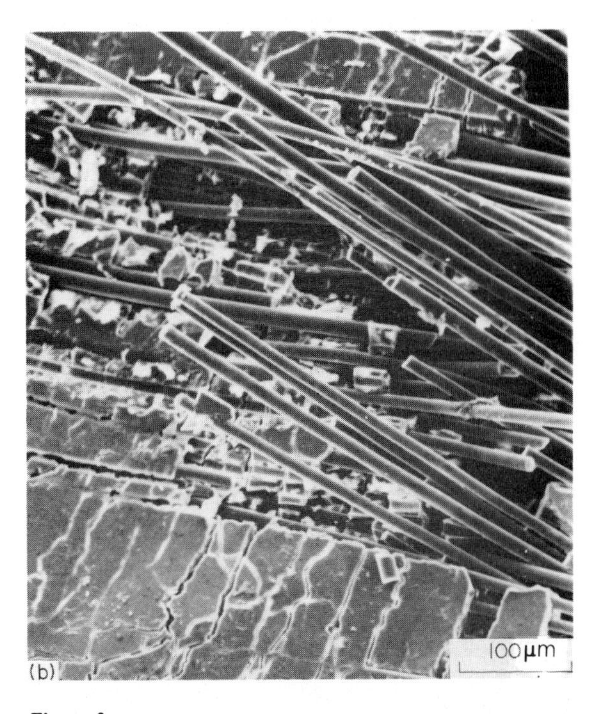

Figure 3
Examples of fracture morphology of a SiC fiber-glass matrix composite (courtesy of D.M. Dawson and R.W. Davidge)

4. Temperature Limitations

Currently the main practical temperature limitation on CMC is due to the lack of available ceramic fibers with good properties above 1000–1200 °C (Mah et al. 1987). Above this temperature range, degradation or creep of existing ceramic fibers becomes excessive. Carbon (graphite) fibers provide a higher temperature capability, maintaining their strengths in inert atmospheres to temperatures in excess of 2000 °C, but successful high-temperature ceramic-matrix composites utilizing carbon fibers and able to operate to high temperatures in oxidizing atmospheres have not yet been developed. At present carbon–carbon composites offer the best prospect of very high temperature capability, and much research is under way to enhance oxidation resistance. Promising approaches include the use of oxidation inhibitors such as borates incorporated in the carbon matrix for use at temperatures up to around 1000–1100 °C, and oxidation barriers such as borate glasses, Al_2O_3 and SiC coatings for higher temperatures.

The prospects for developing ultrahigh temperature composites, materials which can operate in air for long periods of time at temperatures above 1200 °C, have been assessed by Hillig (1985) who considered the necessary criteria which must be satisfied for a composite to survive and have useful properties under stress at high temperatures in air. These are, as a minimum: stability with respect to volatilization; low internal chemical reactivity; retention of stiffness; and a creep rate for the fiber of less than $10^{-7} s^{-1}$. He assumed that only an oxide matrix would have sufficient stability against oxidation; a nonoxide fiber would be necessary on grounds of stiffness and bonding; and that the matrix would provide a measure of oxidation protection for the fiber. He also pointed out that a nonoxide matrix might be suitable with an oxide coating to protect it; and that a further barrier between fibers and matrix may be necessary even for an oxide matrix. Although there is a dearth of good-quality thermochemical data he was able to compile some, and calculate or infer others, and concluded that there are a number of oxides, carbides, borides and nitrides, as well as carbon, which might function together to temperatures in the range 1700–2100 °C. Of the non-oxides only carbon, Si_3N_4 and SiC are currently available as fibers or whiskers. A conclusion which can be drawn from this survey is that SiC-whisker-reinforced Al_2O_3 is one of the most promising practical systems. Such a material already exists but displays a disappointing decrease in strength at 1200 °C. This highlights the uncertainty in predicting the high-temperature performance of ceramic composites. Much work is currently under way throughout the world to attempt to develop new fibers and extend the temperature range of ceramic-matrix–fiber composites.

See also: Manufacturing Methods for Composites: An Overview; Whiskers

Bibliography

Aveston J, Cooper G A, Kelly A 1971 Single and multiple fracture. *Proc. Conf. on Properties of Fiber Composites*. IPC, Guildford, UK, pp. 15–24

Butler E P 1985 Transformation-toughened zirconia ceramics. *Mater. Sci. Technol.* 1: 417–32

Ceramic Bulletin 1986 Issue devoted to papers on ceramic matrix composites. 65(2)

Hillig W B 1986 Prospects for ultra-high-temperature ceramic composites. In: Tessler R T, Messing G L, Pantano C G, Newnham R E (eds.) 1986 *Proc. Conf. on Tailoring Multiphase and Composite Ceramics*. Plenum, New York, pp. 697–712

Hillig W B 1987 Strength and toughness of ceramic matrix composites. *Annu. Rev. Mater. Sci.* 17: 341–83

McCartney L N 1987 Mechanics of matrix cracking in brittle-matrix fibre-reinforced composites. *Proc. R. Soc. Lon., Ser. A* 409: 329–50

Mah T, Mendiratta M G, Katz A P, Mazdiyasni K S 1987 Recent developments in fibre-reinforced high temperature ceramic composites. *Ceram. Bull.* 66: 304–08

Marshall D B, Cox B N, Evans A G 1985 The mechanics of matrix cracking in brittle-matrix fiber composites. *Acta. Metall.* 33: 2013–21

Naslain R, Lamon J, Doumeingts D (eds.) 1993 High temperature ceramic matrix composites. *Proc. 6th European Conf. Composite Materials*. Woodhead, Cambridge

Naslain R, Langlais F 1986 CVD processing of ceramic–ceramic composite materials. In: Tessler R T, Messing G L, Pantano C G, Newnham R E 1986 *Proc. Conf. on Tailoring Multiphase and Composite Ceramics*. Plenum, New York, pp. 145–64

Phillips D C 1983 Fibre reinforced ceramics. In: Kelly A, Mileiko S T (eds.) 1983 *Handbook of Composites*, Vol. 4: *Fabrication of Composites*. North Holland, Amsterdam, pp. 373–428

Phillips D C 1985 Fibre reinforced ceramics. In: Davidge R W 1985 *Survey of the Technological Requirements for High Temperature Materials Research and Development*, Sect. 3: *Ceramic Composites for High Temperature Engineering Applications*. Commission of the European Communities, EUR 9565, pp. 48–73

Phillips D C 1987 High temperature fibre composites. In: Matthews F L, Buskell N C R, Hodgkinson J M, Morton J 1987 *Proc. 6th Int. Conf. on Composite Materials/2nd European Conf. on Composite Materials*, Vol. 2 Elsevier, London, pp. 2.1–2.32

Prewo K M 1986 Tension and flexural strength of silicon carbide fiber reinforced glass ceramics. *J. Mater. Sci.* 21: 3590–600

Prewo K M 1987 Fatigue and stress rupture of silicon carbide fiber-reinforced glass ceramics. *J. Mater. Sci.* 22: 2695–2701

Suzuki H 1987 A perspective on new ceramics and ceramic composites. *Philos. Trans. R. Soc. Lon., Ser. A* 322: 465–68

Wei G C, Becher P F 1985 Development of SiC-whisker-reinforced ceramics. *Am. Ceram. Soc. Bull.* 64(2): 298–304

D. C. Phillips
[Kobe Steel, Guildford, UK]

Fiber-Reinforced Polymer Systems: Extrusion

The incorporation of discontinuous reinforcing fibers, such as chopped fiberglass or natural cellulose, into a polymer matrix produces a composite whose mechanical properties fall between the unreinforced polymer itself and high-performance composites reinforced with continuous fiber strands. An advantage of the short-fiber composite is that it can be fabricated into useful parts by the same high-output flow processes as for the unreinforced polymer (i.e., extrusion, molding and drawing). A longer fiber produces higher reinforcement. However, the fiber's effect on the melt rheology and conversely the influence of the flow on the composite structure, especially fiber directionality, are then more severe.

1. Flow Characteristics

Compared with particulate additives, the 500% increase in shear viscosity at low deformation rate ($< 0.01\ s^{-1}$), when 15 vol% fibers of 40:1 length/diameter (l/d) are added to a polymer melt, is dramatic. In an extrusion die, where the shear rate is higher ($\sim 100\ s^{-1}$), viscosity doubles. (The extruder head pressure usually rises further to produce an increased output.) One explanation for the reduced viscosity buildup in fast flows is a blunting of the velocity profile. At a high 50 vol % loading of 250 l/d fibers, the composite flows as a plug through channels of uniform cross section. Shear is then limited to a localized region near the walls of the channel.

When the channel cross section varies, the entire flowing mass must deform. The extensional flow field produced as the area reduces and the flow accelerates in the entrance to an extrusion die generates very large tensile stresses. Each fiber must be pulled past its neighbor; this induces high shear stresses in the intervening matrix. It has been shown that the rheology of fiber-reinforced polymers is dominated by these large hydrodynamic interactions between fibers (Czarnecki and White 1980).

Despite the high stresses, very low extrudate swell is characteristic of fiber-reinforced systems. However, the high shear stresses between fibers may cause rough, torn surfaces on the extrudate.

2. Fiber-Orientation Control

In shear fields, fibers rotate end over end with an unsteady motion that places them in a time-averaged position nearly parallel to the flow direction. Fiber orientation near the surface of an extrudate is of this type. The structure in the interior is dominated by extensional strains. In a converging die, the positive velocity gradient dv_x/dx stretches the material elements and aligns the fibers parallel to the x or flow direction. An increasing cross section would turn the velocity gradient negative and rotate the fibers perpendicular to the flow.

The equations for fiber rotation in two special extensional flows (Goldsmith and Mason 1967) reduce to

$$d\theta/dt = k\lambda(dv_x/dx)\sin 2\theta \qquad (1)$$

where θ is the mean fiber angle measured from the flow direction, λ is a coefficient measuring the orientability of the fibers, and k is a constant taking a value of $\frac{3}{4}$ in uniaxial extension ($dv_y/dy = dv_z/dz = -\frac{1}{2}(dv_x/dx)$) or $\frac{1}{2}$ in planar extension ($dv_y/dy = -dv_x/dx$; $v_z = 0$). Uniaxial extension applies to extrusion through a die orifice and planar extension to the special tube dies described below. Integrating along the flow the between channel positions 1 and 2 yields

$$\tan\theta_2/\tan\theta_1 = (A_2/A_1)^{2k\lambda} \qquad (2)$$

where A is the channel cross section and λ is unity for long, straight fibers rotating freely, but decreases to 0.4 as concentration is increased or fiber length is reduced.

The overall balance between surface and internal fiber orientation in the extruded part depends upon the channel geometry and the material composition, which affects both the value of λ and the extent of shear generation. Since fiber reinforcement varies strongly with directionality, these considerations are important to the design performance of the extruded part. A balanced orientation distribution is required to avoid warpage. A more exact approach to predicting flow-induced fiber orientation distributions is through numerical finite-element calculations. These also have the potential for handling complex fiber interaction effects that arise in the concentrated suspensions representative of commercial extrusion compounds (Givler et al. 1983, Folgar and Tucker 1984). Reviews of fiber orientation effects in plastic composites are available in the literature (McNally 1977, Folkes 1982, Goettler 1984, 1985).

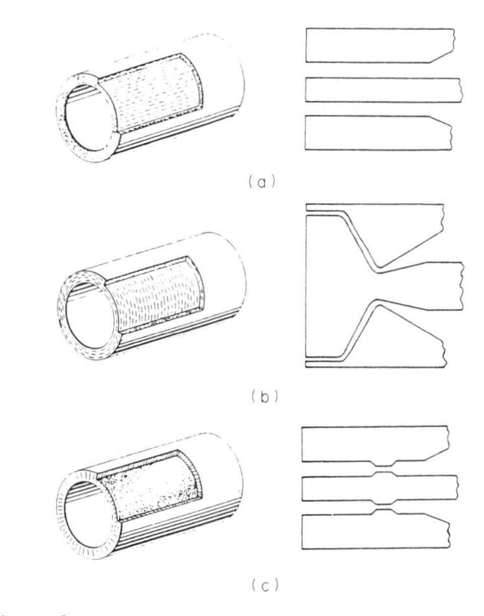

Figure 1
Tube extrusion dies for control of fiber-orientation patterns: (a) conventional converging, (b) expanding diameter and (c) expanding thickness.

3. Applications

In a profile or rod extruded through a converging die, the fiber orientations due to shear and extension are each directed parallel to the flow. The extrudate has both high axial strength and stiffness, typically 3–5 times the crosswise properties. Production of a smooth surface requires a long, heated die.

When the reinforcement is needed in a transverse direction, the fibers may be rotated by using a special die geometry to establish the proper extensional strains in the flowing material. Types of fiber orientation produced by different expanding tube dies are shown in Fig. 1 (Goettler et al. 1979). The fiber pattern generated in die (b) provides the needed hoop strength to double the pressure rating of a short-fiber-reinforced hose. At the same time, flexibility and surface appearance are improved by the absence of axial orientation. These principles have been further extended to coextrusion flows utilizing a lubricating nonreinforced melt layer at the wall (Doshi et al. 1986). A technique has been developed for offsetting the mandrel of an expanding extrusion die to bend a short-fiber reinforced rubber hose as it is being extruded, thus eliminating the costly postextrusion forming of curved hose shapes (Goettler et al. 1981).

See also: Glass-Reinforced Plastics: Thermoplastic Resins; Glass-Reinforced Plastics: Thermosetting Resins

Bibliography

Czarnecki L, White J L 1980 Shear flow rheological properties, fiber damage, and mastication characteristics of aramid-, glass- and cellulose-fiber-reinforced polystyrene melts. *J. Appl. Polym. Sci.* 25: 1217–44

Doshi S R, Charrier J-M, Dealy J M, Hamel F A 1986 Coextrusion of short fiber-reinforced plastic pipes. *Soc. Plastics Engineers 44th ANTEC*, p. 944

Folgar F, Tucker C L 1984 Orientation behavior of fibers in concentrated suspensions. *J. Reinf. Plast. Compos.* 3: 98

Folkes M J 1982 *Short Fibre Reinforced Thermoplastics.* Research Studies Press, Chichester, UK

Givler R C, Crochet M J, Pipes R B 1983 Numerical prediction of fiber orientation in dilute suspensions. *J. Compos. Mater.* 17: 330–43

Goettler L A 1984 Mechanical property enhancement in short-fiber composites through the control of fiber orientation during fabrication. *Polym. Compos.* 5: 60–71

Goettler L A 1986 The effects of processing variables on the mechanical properties of reinforced thermoplastics. In: Clegg D W, Collyer A A (eds.) 1986 *Mechanical Properties of Reinforced Thermoplastics.* Elsevier, London, pp. 151–204

Goettler L A, Lambright A J, Leib R I, DiMauro P J 1984 Extrusion-shaping of curved hose reinforced with short cellulose fibers. *Rubber Chem. Technol.* 54: 277–301

Goettler L A, Leib R I, Lambright A J 1979 Short fiber reinforced hose—a new concept in production and performance. *Rubber Chem. Technol.* 52: 838–63

Goldsmith H L, Mason S G 1967 The microrheology of dispersions. In: Eirich F (ed.) 1967 *Rheology: Theory and Applications*, Vol. 4. Academic Press, New York, pp. 86–250

McNally D 1977 Short fiber orientation and its effects on the properties of thermoplastic composite materials. *Polym.-Plast. Technol. Eng.* 8(2): 101–54

L. A. Goettler
[Monsanto Company, Akron, Ohio, USA]

Fibers and Textiles: An Overview

The first fibers used by man were natural fibers such as cotton, wool, silk, flax, hemp and sisal. The first man-made fiber was probably glass; the Egyptians made vessels out of coarse glass fibers. The first use of glass fibers in textiles took place in the eighteenth century. The idea that man could make an organic fiber is attributed to Hooke, who in 1664 suggested that a "glutinous substance" could be drawn into fibers. However, it was not until the latter part of the nineteenth century that Despassis was awarded a French patent for the production of rayon fibers by the cuprammonium process. Then, in 1893, a British patent was granted to Cross, Bevan and Beadle for making rayon fibers by the viscose process.

It was only at the beginning of the twentieth century, however, that man-made fibers started supplementing and replacing natural fibers. The first important man-made fibers were the cellulose-derived fibers, rayon and cellulose acetate. Then came nylon, the first truly synthetic fiber, followed by polyesters, polyacrylics and polyolefins. Also man-made elastomeric, glass and aramid (aromatic polyamide) fibers became important commercial products. Finally, the high-temperature, high-performance inorganic fibers including carbon, alumina, boron, silicon carbide and silicon nitride, some in the form of whiskers, have been produced and are used primarily as reinforcement fibers in composites.

Man-made fibers are now available, ranging in properties from the high-elongation and low-modulus elastomeric fibers, through the medium-elongation and medium-modulus fibers such as polyamides and polyesters, to the low-elongation, high-modulus carbon, aramid and inorganic fibers. With such a wide variety of man-made fibers available, the volume of synthetic fibers consumed in worldwide is now greater than that of natural fibers.

1. Classification

The major fibers used in textiles can be classified into the groups given in Fig. 1.

Although most man-made fibers are composed of one class of polymers only, some biconstituent or bicomponent fibers are manufactured; these fibers may be spun from different polymers side by side or in other configurations. One purpose of such fibers is to produce a crimp in the fibers by exploiting the differential thermal properties of the two polymers. Also, differential coloring effects may be produced in this manner.

In fabric manufacture, blends of fibers are often used to obtain the desired fabric properties. An example is a blend of polyester and cotton that is treated with a formaldehyde reactant to give a durable press fabric. The tensile strength and abrasion resistance of the polyester fibers counteract the loss of these properties in the treated cotton fibers. Another example is a blend of nylon and wool, which is often used for clothing fabrics where both warmth and wear resistance are important.

2. Physical Properties

Some of the more important physical properties of the most extensively used fibers are given in Table 1.

Fibers used in textile structures must have sufficient length to be made into twisted yarns, but can be as short as 1.3 cm in staple fibers. The diameter of fibers with the exception of some inorganic fibers generally

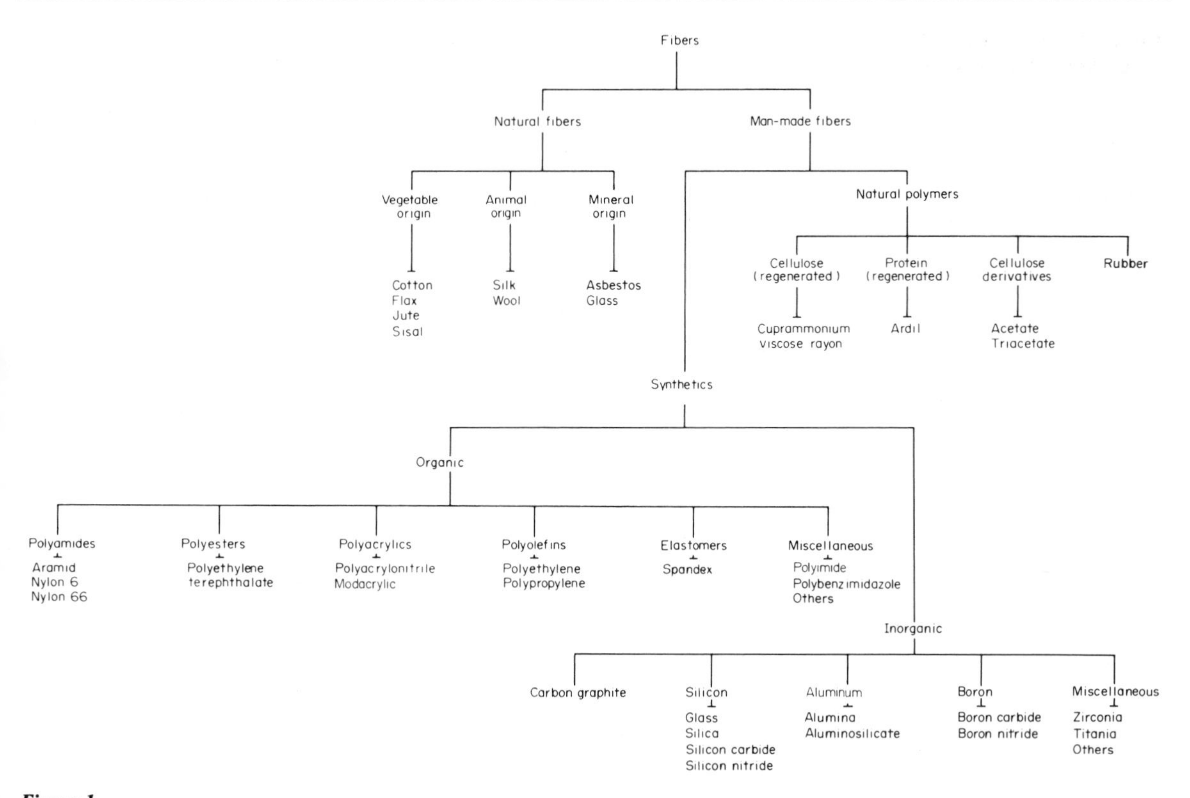

Figure 1
Classification of fibers used in textiles

varies from 5 to 45 μm. However, nylon, polyester and polyacrylic fibers are now being manufactured with diameters approaching 1 μm; these fibers give fabrics that are soft and have a good drape. Even smaller diameter fibers have been produced by solvent removal of one component of a bicomponent fiber.

The tensile strength of a fiber increases with decreasing fiber diameter as is illustrated in Table 2. Not only does the tensile strength increase with decreasing fiber diameter, but the modulus generally increases as well. The generally accepted explanation for these phenomena is that there are fewer imperfections and faults as the fiber diameter is reduced. Probably the ultimate in small fiber diameter and in perfection, with resulting high tensile and modulus values, are the inorganic whisker fibers (see *Whiskers*). These high-performance fibers with their high-temperature resistance are used as reinforcing fibers in metallic and ceramic composites used in space vehicles.

The cross-sectional shape of fibers varies considerably. The cross section of a cotton fiber is a collapsed circle, which is somewhat flat and irregular. Silk has a triangular cross section, which accounts for its sheen and feel. Wool is circular in shape with scales on the surface. Rayon can be circular or serrated-circular, and it often has a center or core which differs from the outer layer. Man-made fibers, particularly melt-spun fibers, can also be made oval, multilobal and even hollow-circular. The shape of the fiber cross section affects the optical properties of the fiber, including luster and translucency. It also affects the feel or hand of fabric made from the fiber.

The surface of fibers also varies considerably. Cotton has a surface that has many irregularities or rugosities. Wool has a scalar structure, which has a tendency to make a wool fabric felt or shrink during washing. Most man-made fibers have relatively smooth surfaces; however, rayon may be serrated.

Some fibers, such as wool, have a natural undulating physical structure or crimp, whereas man-made fibers are frequently subjected to various mechanical and heat-setting processes to provide crimp. These processes include false twist, stuffer box, gear crimping, knife-edge, antitwisting, knit-de-knit and air-jet crimping. Most of these processes are self-descriptive, and all involve heating the fiber or yarn while it is being held in a crimped configuration. Fabrics made from crimped fibers tend to have increased stretch, bulk and warmth.

The density of fibers (see Table 1) varies from about 3.9 g cm^{-3} for alumina to 0.9 g cm^{-3} for polypropylene, with polyester being about 1.4 g cm^{-3} and cotton

Table 1
Physical properties of textile fibers (after Goswami et al. 1977)

Fiber type	Diameter (μm)	Density (g cm^{-3})	Tenacity (gf tex^{-1})	Breaking extension (%)	Initial modulus (gf tex^{-1})	Work of rupture (g tex^{-1})	Moisture regain[a] (%)	Melting point (°C)
Natural								
cotton	11–22	1.52	35	7	500	1.3	7	
flax	5–40	1.52	55	3	1840	0.8	7	
jute	8–30	1.52	50	2	1750	0.5	12	decomposes
sisal	8–40	1.52	40	2	2500	0.5	8	
wool	18–44	1.31	12	40	250	3	14	
silk	10–15	1.34	40	23	750	6	10	
glass	≥5	2.54	76	2–5	3000	1	0	800
asbestos	0.01–0.30	2.5			1300		1	1500
Regenerated								
viscose rayon	≥12	1.45–1.54	20	20	500	3	13	decomposes
acetate	≥15	1.32	13	24	350	2	6	230
triacetate	≥15	1.32	12	30	300	2	4	230
Synthetic								
Nylon 6	≥14	1.14	32–65	30–55	250	6–7	2.8–5	225
Nylon 66	≥14	1.14	32–65	16–66	250	6–7	2.8–5	250
Qiana	≥10	1.03	25	26–36			2.5	274
Nomex	≥12	1.38	36–50	22–32		7.5	6.5	decomposes above 380 °C
Kevlar	≥12	1.44	200	2–4	4500–8000		4	decomposes above 500 °C
polyester	≥12	1.34–1.38	25–54	12–55	1000	2–9	0.4	250
acrylic	≥12	1.16–1.17	18–30	20–50	650	5–	1.5	sticks at 235 °C
polypropylene		0.91	60	20	800	8	0.1	165
polyethylene		0.95	30–60	10–45		3	0	115
spandex		1.21	6–8	444–555		18	1.3	230

a At 65% relative humidity

Table 2
Influence of diameter on tensile strength of glass fibers; fiber length held constant (after Anderegg 1939)

Diameter of test fibers (μm)	Ultimate tensile strength (GPa)
19.0	0.7
15.2	0.9
12.7	1.0
10.2	1.3
8.6	1.7
5.1	2.8
2.5	6.0

about 1.5 g cm^{-3}. The low density of polypropylene makes it desirable for marine ropes, since they will float in water.

Fibers must have sufficient strength to be useful in textile structures. The strength of textile fibers is usually referred to as tenacity, which is defined as the force required to rupture or break the fiber per unit linear density. Tenacity is generally measured in grams-force per denier (where denier is the weight in grams of 9 km of fiber or yarn), or as grams-force or newtons per tex (where tex is the weight in grams per kilometer of fiber or yarn). The tenacity of fibers varies from slightly less than 6–8 gf tex^{-1} for the elastomeric spandex fibers to 76 gf tex^{-1} and even higher for some inorganic fibers, with nylons and polyesters generally being 25–65 gf tex^{-1}.

In addition to having sufficient strength for its various end uses, a fiber must also have sufficient extensibility to withstand the stresses to which it will be subjected in service. The breaking extension or elongation-at-break varies from less than 1% for some inorganic fibers to over 500% for some spandex fibers, with nylons and polyesters generally varying from 12 to 66%. For many end uses, an elongation of less than 2% is not sufficient, so the use of fibers with elongations of this order is limited. However, for some composite applications, fiber elongations of less than 1% can be tolerated.

Another fiber property that is important is recovery from deformation. This property of the fiber is one that is also imparted to yarns and to fabric made from them; for example, the ability of a fabric to recover from wrinkling depends to a large extent on the recovery from deformation of the individual fibers. Recovery can be measured by imposing deformation in a cyclic manner and determining the elongation recovered. The initial modulus or slope of the first part of the stress–strain curve can also be used as a measure of this property. Generally, fibers with low elongation (e.g., glass fibers) have high modulus values, and those with high elongation such as spandex fibers have low moduli. Nylons and polyester fibers generally have intermediate elongation and modulus values. The area under a fiber stress–strain curve represent the toughness of the fiber and, if measured to rupture, is called the work of rupture. Values of work of rupture are generally high for spandex, medium to high for nylons and polyesters, and low for glass fibers.

Other important properties of fibers include abrasion resistance, water absorption, dimensional stability, thermal characteristics, flammability, density, and resistance to attack (e.g., from chemicals, sunlight and mildew). Generally, fibers with a low work-of-rupture value, such as glass fibers, have poor abrasive resistance, whereas fibers such as nylon and polyester have good abrasion resistance. Water absorption is an important factor for comfort in clothes. Fibers with high water absorption such as cotton and rayon are usually considered more comfortable than polyamide, polyester, or polyolefin fibers, which have low water absorption. Since it is important that fabrics neither shrink nor stretch excessively in an irreversible manner, man-made fibers are often heat-set. This gives the fibers dimensional stability provided that they are not exposed to a temperature in excess of that of the heat-setting. Some cellulosic fabrics shrink during washing if they have been stretched during manufacture. Finishing of cellulosic fabrics with a formaldehyde reactant may impart dimensional stability as well as recovery from wrinkling.

While some fibers such as the cellulosics do not melt but will burn, others (e.g., nylons, polyesters and polyolefins) melt at relatively low temperatures, but burn with difficulty because the molten polymer falls away from the flame.

3. Morphology

The morphology, or spatial arrangement of the individual polymer molecules and groups of molecules, has a bearing on fiber properties. The naturally occurring fibers already have a built-in morphology, which can sometimes be altered. For example, cotton can be mercerized by treatment with caustic to swell the fiber and give it higher luster, and it can also be cross-linked by various formaldehyde reactants to enhance its recovery from wrinkling. Similarly, cellulose can be chemically modified and spun into rayon or acetate fibers, and wool can be chemically treated to decrease its felting properties.

4. Production

A large number of man-made fibers with a variety of properties have been produced from polymers by various spinning techniques, including melt, dry, wet and emulsion spinning. Polyamide, polyester and polyolefin fibers are formed by forcing the melted polymer through an orifice into air. Polyacrylonitrile fibers are spun from a solvent solution of polyacrylonitrile by either dry spinning (into air), or wet spinning (into a coagulating bath). Spandex fibers can be formed from polyurethanes by either dry or wet spinning.

Generally, after being spun, the fibers are mechanically stretched or drawn. In this process, the molecules are oriented in a direction parallel to the axis of the fiber. The resulting aligned molecules may become closely packed, so that crystallization can occur in some areas. This crystallization stabilizes the fiber and imparts improved properties of strength and recovery from deformation. However, some areas are left relatively unaligned, and in these amorphous areas the molecules are loosely held together by secondary forces including hydrogen bonding, van der Waals forces and dipole–dipole interactions. These permit more flexibility of movement than in the crystalline regions. The amorphous regions also allow the penetration of dyes and finishes into the fiber.

Fibers are usually processed into yarns, and the yarns are woven or knitted into fabrics. Filament fibers can be made into yarns with little or no twisting, but staple fibers, being short, must be twisted together to form long continuous yarns. Often sizes, which are usually water-soluble polymers, are applied to the yarns to facilitate weaving. Fibers can also be randomly arranged in so-called nonwoven textile structures, without first being made into yarns. Nonwovens can be formed by spun bonding, needle punching or resin bonding. In spun bonding, bonds between the fibers are formed by heating the fibers to near their melting point. Fibers, especially the inorganic fibers, are used as reinforcing fibers in composites in which the matrix may be an organic resin, metal or ceramic. In order to obtain optimum reinforcement, the fibers are usually laid as nearly parallel to one another as possible.

5. Dyes and Dyeing

Fibers, yarns and fabrics may be colored with dyes (a) containing anionic groups (acid, direct and mordant dyes), (b) containing cationic groups (basic dyes), or (c) requiring chemical reaction during dyeing (azo, reactive, sulfur and vat dyes). Special colorants such as disperse dyes and pigments may also be used. All of the dye molecules have a chromophore group, which is responsible for the color, and usually a functional group which increases the affinity of the color for the fiber.

Acid dyes are water-soluble anionic dyes that are applied to textiles from acid solutions. Fibers which develop a positive charge in acid solutions (e.g., wool, silk and nylon) can be dyed with acid dyes. Direct dyes frequently contain azo groups connecting aromatic chromophores and are mostly used on cellulosic fabrics. Mordant dyes are acid dyes that are chelated with metal ions such as aluminum, chromium, cobalt and copper, and are used chiefly for wool and silk. These dyes are normally applied to a fiber and then mordanted.

Basic dyes are dyes which yield colored cations in aqueous solutions. They can be applied in neutral or mildly acidic dye baths to acrylic, cellulosic, nylon and protein fibers. These dyes may be treated with complexing agents such as tannin to improve their fastness to washing.

Azo dyes are formed in the fiber by interaction between two separate dye components. They are chiefly applied to cotton, but can be used with man-made fibers. Reactive dyes contain groups that can react covalently with a chemical group in the fiber to make the color durable to washing and dry cleaning. They are chiefly applied to cellulosic fibers, with the reaction occurring with the hydroxyl group of the cellulose, but are also used with nylon and protein fibers. Sulfur dyes are products formed by the reaction of aromatic compounds with sodium polysulfide. They are reduced under alkaline conditions before application to a fiber, and then oxidized after the fiber is dyed. They are chiefly applied to cellulosic fibers. Vat dyes are generally water-insoluble. They are reduced under alkaline conditions to form a water-soluble, colorless leuco form, which is applied to the fiber, and then oxidized to an insoluble colored state within the fiber. Vat dyes are chiefly used with cellulosic fibers, but can also be applied to acrylic and nylon fibers.

Disperse dyes are nonpolar, relatively insoluble molecules, usually containing azo or anthraquinone groups, and are applied to the fiber from an aqueous dispersion in water. They were developed originally to dye acetate fibers but are now also used in dyeing other man-made fibers. Pigments are insoluble colors that are normally added to the polymer from which the fiber is spun, or bonded to the fiber surface with a resin binder.

6. *Printing*

Fabrics may be printed with a design using any of the classes of colorants already described. The colors are dispersed in a thickened printing paste and applied to the fabric in a design. The printed fabric is then treated in some fashion such as by heating or an oxidation process to fix the colors in the fabric. With pigments, a resin binder is used to bind the pigment to the fabric.

7. *Finishing*

Fabrics are finished in a great variety of ways; the finishing may be to improve either the esthetic or performance properties, or both.

7.1 Removal of Impurities

In processing fibers to fabrics, sizes are often applied to the warp yarns. These sizes may be removed in a final washing step. With natural fibers, impurities such as waxes, grease and vegetable matter may remain in the fabric and are removed in a final scouring and bleaching process.

7.2 Changing the Hand or Feel

In the case of woollen fabrics, a felting or compacting operation may be the finishing step. Singeing or passing a fabric over a gas flame or a heated metal plate may be used to burn off the fiber ends of a fabric made from staple yarns. If a fabric is limp, it may be stiffened by adding a nonpermanent starch or other sizing, or a permanent size may be obtained by applying a resin finish.

7.3 Changing the Appearance or Luster

Fabrics may be whitened by bleaching and/or by the application of a fluorescent whitener. A glossy fabric surface can be produced by calendering or passing it through rollers. If a friction calender is used, a glazed finish is obtained. With a Schreiner calender, a soft luster and hand are produced. Embossed designs can be produced with an embossing calender. With the addition of resins to the fabric, a durable calendered finish is produced. Napped fabrics are made by brushing the fibers on the surface of the fabric, and napped or pile fabrics may be sheared to make the nap uniform in thickness.

7.4 Control of Biological Degradation

Cellulosic fabrics are especially susceptible to damage by mildew and bacteria. Various antimildew and antibacterial chemicals including chlorinated phenols and organometallic compounds are applied to these fabrics to retard biological degradation. In addition, treatment of cellulosic fabrics with formaldehyde reactants such as melamine–formaldehyde will also protect the fabrics from biological degradation. Wool is susceptible to damage from moth larvae unless properly treated.

7.5 Durable Press

Since cellulosic fibers do not recover well from deformation, formaldehyde reactants are used to treat these fabrics to improve this property and consequently the recovery of the fabric from wrinkling. If sufficient formaldehyde reactant is applied, the resulting fabric will have both wet and dry recovery from wrinkling, and consequently will have durable press properties.

The formaldehyde reactants used to produce a durable press have at least two functional groups capable of reacting with the hydroxyl groups of cellulose. Such reactants are normally applied in an aqueous solution with a metal salt catalyst by immersion of the fabric in a pad bath, followed by passage through a pad roll. The treated fabric is dried and then cured in an oven. The most frequently used formaldehyde reactant is dimethyloldihydroxyethyleneurea (DMDHEU). This agent can react with the hydroxyl groups of cellulose and cross-link adjacent cellulose chains. Other formaldehyde reactants that are used include the dimethylocarbamates and dimethylolethyleneurea. The methyl, ethylene glycol or diethylene glycol ethers of DMDHEU are often used because the release of formaldehyde from the etherified products is less than from DMDHEU.

Since the formaldehyde reactant tends to lower the tensile strength and abrasion resistance of cellulosic fibers, blends with polyester are commonly used to improve both of these properties. Also, ammonia treatment of the cotton in durable press fabric, either before or after treatment, is now being practised. With the ammonia treatment, less formaldehyde reactant is required to obtain durable press properties.

In the manufacture of garments from durable press fabrics, the garment is often fabricated before the curing operation is carried out, and then cured with the creases in place. This operation is referred to as postcuring.

7.6 Shrinkage Control

The tendency of some fabrics, especially wool and cellulosic, to shrink during washing can be minimized by finishing treatments. The shrinkage of wool fabrics caused by felting can be reduced by treatment with chlorine, or by resin treatment, or by a combination of the two. The shrinkage of cellulosic fabrics during washing can be reduced by compressive shrinkage of the fabric during finishing; an example of this kind of treatment is Sanforizing . Another method of controlling shrinkage is with formaldehyde reactants such as those used for durable press. In fact, durable press treatments will reduce the shrinkage of cellulosic fabrics during washing.

7.7 Water and/or Oil Repellency

There are a number of chemical compounds that will impart water and/or oil repellency to fabrics. Durable water repellency can be obtained with silicone polymers and with long hydrocarbon chain pyridinium chloride derivatives. Both water and oil repellency can be obtained if a fabric is treated with certain polymeric fluorocarbon containing materials. With these compounds, it is important that the fluorine groups be present at the end of the polymer side chains to impart a low-energy surface to the fabric. The most effective compounds have a terminal trifluoromethyl group in the polymer.

7.8 Soil Release

Fabrics with a certain degree of soil release can be obtained by the application of hydrophilic polymers (e.g., polymers or copolymers of acrylic or methacrylic acid, or hybrid fluorocarbon polymers). These finishes can be applied along with the formaldehyde reactants to give durable press fabrics with soil-release properties.

7.9 Fire Retardancy

Most textile fibers with the exception of the inorganic fibers are flammable to some degree. Some of the aramids are thermally stable. Some fibers such as wool and some modacrylics are inherently flame retardant. Others such as nylon, polyester and polyolefins will melt and the molten beads will drop away from the flame. Cellulosic fibers will burn readily unless chemically treated (e.g., with organic compounds containing halogen and/or phosphorus such as tetrakis (hydroxymethyl) phosphonium chloride). The flammability of man-made fibers can be reduced by the addition of flame retardants to the polymer before spinning, although only limited commercial success has been obtained with these fibers. The flame retardant can also be applied to the fabric, as was done in the unfortunate treatment of children's sleepwear with tris(2,3-dibromopropyl) phosphate, which was found to be a potential carcinogen.

7.10 Antistatic Properties

The build-up of static electricity on fibers, with resultant clinging and/or shock, is mainly a problem with hydrophobic man-made fibers. This problem can be minimized by the addition of antistatic agents, which discharge the electrostatic charges that tend to build up on fiber surfaces. These agents are surface active materials, and include ethylene oxide derivatives and polyamines.

See also: Woven Fabrics: Properties; Three-Dimensional Fabrics for Composites

Bibliography

Anderegg F O 1939 Strength of Glass Fibers. *Ind. Eng. Chem.* 31: 290

Goswami B C, Martindale J G, Scardino F L 1977 *Textile Yarns: Technology, Structure and Applications.* Wiley-Interscience, New York

Hearle J W S, Grosberg P, Backer S 1969 *Structural Mechanics of Fibers, Yarns and Fabrics.* Wiley-Interscience, New York

Lewin M, Sello S B 1983 *Chemical Processing of Fibers and Fabrics*, Vols. 1 and 2. Dekker, New York

Mark H F, Atlas S M, Cernia E 1967 *Man-Made Fibers: Science and Technology.* Wiley-Interscience, New York

Mark H F, Wooding N S, Atlas S H 1971 *Chemical Aftertreatment of Textiles.* Wiley-Interscience, New York

Needles H L 1981 *Handbook of Textile Fibers, Dyes and Finishes.* Garland, New York

Tortora P G 1978 *Understanding Textiles.* Macmillan, New York

T. F. Cooke
[Textile Research Institute, Princeton, New Jersey, USA]

Fibrous Composites: Thermomechanical Properties

The development of fibrous composites is one of the great advances in materials. This article is concerned with the thermomechanical properties of one of the most important types, unidirectional composites, in which the reinforcing phase is an array of straight, parallel, continuous fibers.

Although all materials are heterogeneous when examined on a sufficiently small scale, they are invariably treated as homogeneous continua when used a practical engineering materials. This is true of unidirectional composites, which are anisotropic on the macroscopic level. A body of analyses, often called "micromechanics," has been developed relating effective properties of composites to those of their constituents, fibers and matrices. The analyses for elastic, viscoelastic, plastic and thermal properties of unidirectional composites are considered here, along with macroscopic strength theories.

1. Effective Properties

The practice is to treat composites as if they were homogeneous, anisotropic materials. The properties used to represent the materials are commonly referred to as effective properties. The use of effective properties implies that fibers are distributed throughout the material either in a regular array or in a statistically random manner, producing statistical "uniformity." In the latter case, the material is said to be statistically homogeneous.

The concept of effective properties involves an averaging of properties over some region of material. To justify this averaging approach, the dimensions of the region, usually called a representative volume or representative volume element, must be large with respect to the characteristic dimensions of material heterogeneity, such as fiber diameter and fiber spacing. It is further assumed that when effective properties are used the dimensions of the body are large with respect to those of the representative volume element.

2. Elastic Properties

The body of analyses dealing with the elastic properties of composites is far more developed than any other aspect of micromechanics. Numerous approaches have been proposed, and several are considered in this section. Because of the complexity of the subject, discussion focuses on composites with isotropic fibers and matrices. However, several important types of fibers, such as carbon and aramid, are strongly anisotropic and are usually considered to be transversely isotropic. The expressions for elastic properties of composites having isotropic fibers can, with care, easily be converted to include these important reinforcements (Hashin 1979).

This article considers unidirectional composites in which each of the phases in the material is bounded by parallel cylindrical surfaces, so that each cross section perpendicular to the fiber direction is identical. Figure 1 illustrates this geometry. The 1 axis is taken to be parallel to the cylindrical surfaces of the phases that make up the material. This is commonly called the fiber, axial or longitudinal direction of the material.

Figure 2 shows some cross-sectional geometries which have been examined. In the most general case, the shapes of the boundaries separating the two constituent materials or phases of the composite are completely arbitrary; this is commonly referred to as arbitrary phase geometry (Fig. 2(a)). Figure 2(b) represents a special case of arbitrary phase geometry in which fibers of equal diameter are randomly distributed throughout the composite. Figures 2(c) and 2(d) show circular fibers arranged in regular and hexagonal arrays, respectively. The square array is a special case of the rectangular array. Although no real process produces truly random fiber dispersion, case 2(b) is generally considered to be a reasonable representation of many practical materials of interest.

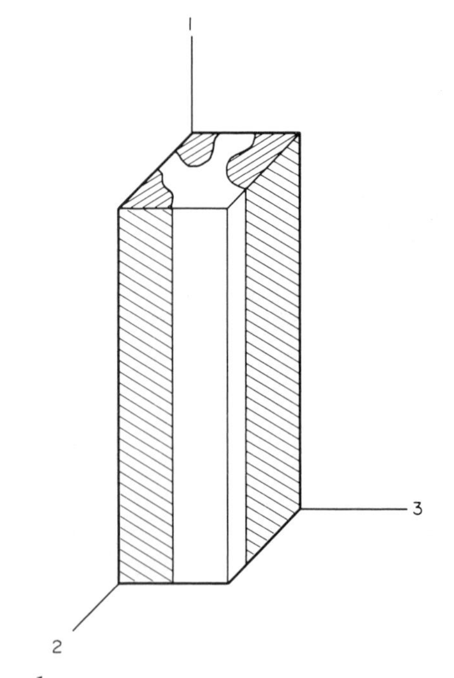

Figure 1
Geometry of a unidirectional composite

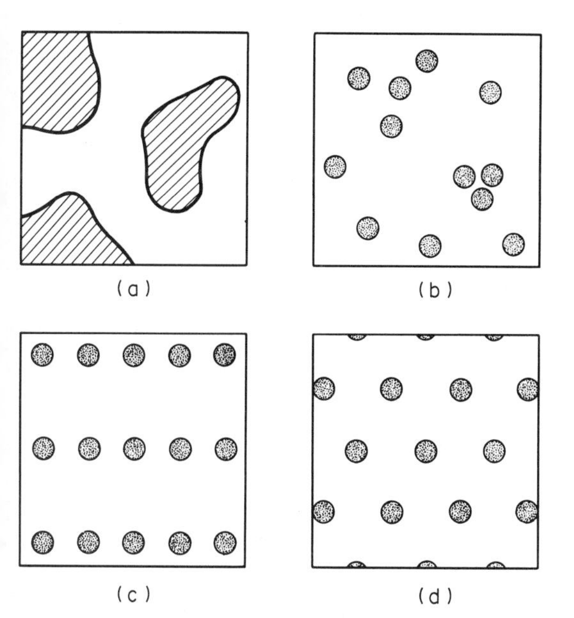

Figure 2
Cross-sectional phase geometries of composites: (a) arbitrary phase geometry; (b) randomly distributed fibers; (c) rectangular fiber array; (d) hexagonal fiber array

The rectangular fiber array produces a macroscopically orthotropic material which has nine independent elastic constants. When the material geometry is random, or when fibers of equal diameter are arranged in a hexagonal array, the composite is transversely isotropic and has five independent elastic constants. The latter class of materials is the focus of this article.

2.1 *Effective Elastic Constants*

The effective elastic constants of a composite material are commonly defined as the quantities that relate the average stress $\bar{\sigma}_{ij}$ to the average strain $\bar{\varepsilon}_{ij}$ in a representative volume element V.

Mathematically, average stress and strain tensor components are defined as

$$\bar{\sigma}_{ij} = (1/V)\int_V \sigma_{ij}\, dV \tag{1}$$

$$\bar{\varepsilon}_{ij} = (1/V)\int_V \varepsilon_{ij}\, dV \tag{2}$$

The effective elastic moduli, or effective stiffnesses, are defined by the relation

$$\bar{\sigma}_{ij} = \hat{C}_{ijkl}\bar{\varepsilon}_{kl} \tag{3}$$

where repeated indices indicate summation.

For simplicity, compact notation is used, in which 11 is replaced by 1, 22 by 2, 33 by 3, 23 by 4, 31 by 5 and 12 by 6.

For a transversely isotropic body, which has five independent elastic constants, Eqn. (3) can be expressed in the following simplified form:

$$\left.\begin{aligned}
\bar{\sigma}_1 &= \hat{C}_{11}\bar{\varepsilon}_1 + \hat{C}_{12}\bar{\varepsilon}_2 + \hat{C}_{12}\bar{\varepsilon}_3\\
\bar{\sigma}_2 &= \hat{C}_{12}\bar{\varepsilon}_1 + \hat{C}_{22}\bar{\varepsilon}_2 + \hat{C}_{23}\bar{\varepsilon}_3\\
\bar{\sigma}_3 &= \hat{C}_{12}\bar{\varepsilon}_1 + \hat{C}_{23}\bar{\varepsilon}_2 + \hat{C}_{22}\bar{\varepsilon}_3\\
\bar{\sigma}_4 &= (\hat{C}_{22} - \hat{C}_{23})\bar{\varepsilon}_4\\
\bar{\sigma}_5 &= 2\hat{C}_{66}\bar{\varepsilon}_5\\
\bar{\sigma}_6 &= 2\hat{C}_{66}\bar{\varepsilon}_6
\end{aligned}\right\} \tag{4}$$

The effective elastic constants $\hat{C}_{11}$, $\hat{C}_{12}$, $\hat{C}_{22}$, $\hat{C}_{23}$ and $\hat{C}_{66}$ are related to the more familiar engineering elastic moduli by the following expressions:

$$\left.\begin{aligned}
\hat{E}_a &= \hat{E}_{11} = \hat{C}_{11} - (2\hat{C}_{12}^2/\hat{C}_{22} + \hat{C}_{23})\\
\hat{E}_t &= \hat{E}_{22} = \hat{E}_{33}\\
&= \hat{C}_{22} + \{[\hat{C}_{12}^2(\hat{C}_{23} - \hat{C}_{22})\\
&\quad + \hat{C}_{23}(\hat{C}_{12}^2 - \hat{C}_{11} - \hat{C}_{23})]/(\hat{C}_{11}\hat{C}_{22} - \hat{C}_{12}^2)\}\\
\hat{G}_a &= \hat{G}_{12} = \hat{G}_{13} = \hat{C}_{66}\\
\hat{G}_t &= \hat{G}_{23} = \tfrac{1}{2}(\hat{C}_{22} - \hat{C}_{23})\\
\hat{\nu}_a &= \hat{\nu}_{12} = \hat{\nu}_{13} = \hat{C}_{12}/(\hat{C}_{22} + \hat{C}_{23})\\
\hat{\nu}_t &= \hat{\nu}_{23} = (\hat{C}_{11}\hat{C}_{23} - \hat{C}_{12}^2)/(\hat{C}_{11}\hat{C}_{22} - \hat{C}_{12}^2)
\end{aligned}\right\} \tag{5}$$

$\hat{E}_a$ and $\hat{E}_t$ are the effective extensional (Young's) moduli in the axial and transverse directions, respectively; $\hat{G}_a$ is the effective shear modulus in planes parallel to the fiber axis; $\hat{G}_t$ is the effective shear modulus in the plane transverse to the fiber direction; and $\hat{\nu}_a$ and $\hat{\nu}_t$ are the Poisson's ratios in plane parallel to and perpendicular to the fiber direction, respectively. The convention adopted here is that $\hat{\nu}_{12}$ is equal to the magnitude of the transverse strain divided by the axial strain when a stress is applied in the axial direction. $\hat{\nu}_{23}$ is defined in a similar manner.

Another important elastic constant is the plane-strain bulk modulus $\hat{k}$, which is related to the $\hat{C}_{ij}$ by the expression

$$\hat{k} = \tfrac{1}{2}(\hat{C}_{22} + \hat{C}_{23}) \tag{6}$$

The modulus $\hat{k}$ is one of the basic elastic constants evaluated using micromechanical analyses. The following expressions are useful for evaluating other engineering constants:

$$\left.\begin{aligned}
\hat{E}_t &= 2(1 + \hat{\nu}_t)\hat{G}_t\\
\hat{\nu}_t &= (\hat{k} - q\hat{G}_t)/(\hat{k} + q\hat{G}_t)
\end{aligned}\right\} \tag{7}$$

where

$$q = 1 + (4\hat{k}\hat{\nu}_a^2/\hat{E}_a) \tag{8}$$

The basic difficulties in developing analytical expressions for elastic constants are the complex and, in practice, undefined composite internal geometry. The problem has been approached in a number of ways, as discussed by Ashton et al. (1969), Hashin (1972), Sendeckyj (1974) and Christensen (1979). These include: finite-difference, finite-element and series solutions for regular arrays; "self-consistent" models consisting of a fiber (or a fiber surrounded by a concentric cylinder of matrix material) in a homogeneous material with the effective properties of the composite; semiempirical equations; rigorous bounding methods; and special models. Three widely recognized approaches are considered here: bounding methods, a special model called the composite cylinder assemblage, and the semiempirical Halpin–Tsai equations.

2.2 Bounds for Effective Elastic Constants for Arbitrary Phase Geometry

Upper and lower bounds on the elastic constants of transversely isotropic unidirectional composites with arbitrary internal geometry have been determined (Hill 1964a, Hashin 1965). These expressions involve only the elastic constants of the two phases and the fiber volume fraction c. (The matrix volume fraction is $1-c$.)

The following symbols and conventions are used in expressions for mechanical properties: plus and minus signs denote upper and lower bounds, respectively; subscripts f and m indicate fiber and matrix, respectively.

Upper and lower bounds on composite axial Young's modulus, axial Poisson's ratio, axial shear modulus, transverse plane strain bulk modulus and transverse shear modulus are given by the following expressions:

$$\hat{E}_a(+) = cE_f + (1-c)E_m + \frac{4c(1-c)(\nu_f-\nu_m)^2}{c/k_m + (1-c)/k_f + 1/G_f} \tag{9a}$$

$$\hat{E}_a(-) = cE_f + (1-c)E_m + \frac{4c(1-c)(\nu_f-\nu_m)^2}{c/k_m + (1-c)/k_f + 1/G_m} \tag{9b}$$

$$\hat{\nu}_a(+) = c\nu_f + (1-c)\nu_m + \frac{c(1-c)(\nu_f-\nu_m)(1/k_m - 1/k_f)}{c/k_m + (1-c)/k_f + 1/G_f} \tag{10a}$$

$$\hat{\nu}_a(-) = c\nu_f + (1-c)\nu_m + \frac{c(1-c)(\nu_f-\nu_m)(1/k_m - 1/k_f)}{c/k_m + (1-c)/k_f + 1/G_m} \tag{10b}$$

$$\hat{G}_a(+) = G_f + \frac{1-c}{1/(G_m-G_f) + c/2G_f} \tag{11a}$$

$$\hat{G}_a(-) = G_m + \frac{c}{1/(G_f-G_m) + (1-c)/2G_m} \tag{11b}$$

$$\hat{k}(+) = k_f + \frac{1-c}{1/(k_m-k_f) + c/(k_f+G_f)} \tag{12a}$$

$$\hat{k}(-) = k_m + \frac{c}{1/(k_f-k_m) + (1-c)/(k_m+G_m)} \tag{12b}$$

$$\hat{G}_t(+) = G_f + \frac{1-c}{1/(G_m-G_f) + c(k_f+2G_f)/2G_f(k_f+G_f)} \tag{13a}$$

$$\hat{G}_t(-) = G_m + \frac{c}{1/(G_f-G_m) + (1-c)(k_m+2G_m)/2G_m(k_m+G_m)} \tag{13b}$$

These relations hold when $k_f > k_m$ and $G_f > G_m$.

It has been shown that, with the exception of the expressions for transverse shear modulus, these bounds are the best possible when only phase elastic constants and volume fractions are specified (Hashin 1972). To improve on these bounds, more information on internal geometry must be specified. It is not known whether the bounds on $\hat{G}_t$ are the best possible.

For most materials, the bounds on axial Young's modulus tend to be reasonably close together, and the rule-of-mixtures predictions are generally accurate enough for practical purposes; that is,

$$\hat{E}_a \cong cE_f + (1-c)E_m \tag{14}$$

This is also true, to a lesser extent, for the axial Poisson's ratio $\hat{\nu}_a$; that is,

$$\hat{\nu}_a \cong c\nu_f + (1-c)\nu_m \tag{15}$$

The rule of mixtures generally provides a poor estimate of $\hat{k}_t$, $\hat{G}_a$ and $\hat{G}_t$.

2.3 Composite Cylinder Assemblage

If all of the bounds on elastic moduli discussed above were reasonably close for all material combinations, the subject of elastic properties would be closed. However, this is not the case. To improve on the bounds, it is necessary to specify internal geometry. A variety of models have been proposed (Christensen 1979). One particular model provides closed-form solutions for four of the five elastic constants of a macroscopically transversely isotropic material. This is the composite cylinder assemblage (CCA) (Hashin 1972).

The CCA model consists of an infinite number of concentric circular cylinders of different diameters that are packed together so that they completely fill the volume of the composite (Fig. 3). The inner material of each cylinder has the elastic properties of the fiber, and the outer material those of the matrix. The inner and outer diameters of composite cylinder n are denoted a_n and b_n, respectively. The ratio a_n/b_n is the same for each cylinder and is equal to $c^{1/2}$, where c is the fiber volume fraction; that is, the fiber volume fraction of

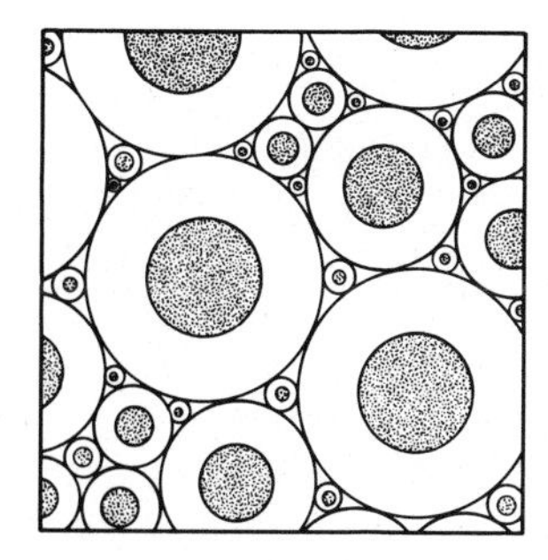

Figure 3
The composite-cylinder-assemblage model

each cylinder, $(a_n/b_n)^2$, is equal to the overall average fiber volume fraction of the composite.

The range of diameters of most real fibers is relatively small, so that the CCA model does not reflect the actual internal geometry found in practice. However, it is argued that it does reflect a certain randomness which is characteristic of composites.

The model provides closed-form solutions for $\hat{E}_a$, $\hat{\nu}_a$, $\hat{G}_a$ and $\hat{k}$, and bounds on the transverse shear modulus $\hat{G}_t$. However, the lower bound on $\hat{G}_t$ obtained for arbitrary phase geometry, Eqn. (13b), is better than that resulting from the CCA model. Since the arbitrary-phase-geometry model is perfectly general, and includes the CCA as a special case, its bounds also apply to the CCA. Therefore, Eqn. (13b) should be used for the lower bound on $\hat{G}_t$ and only the upper CCA bound $\hat{G}_t(+)$ is presented.

The expressions for elastic constants obtained from the composite-cylinder-assemblage model are:

$$\hat{E}_a = cE_f + (1-c)E_m + \frac{4c(1-c)(\nu_f - \nu_m)}{c/k_m + (1-c)/k_f + 1/G_m} \quad (16)$$

$$\hat{\nu}_a = c\nu_f + (1-c)\nu_m + \frac{c(1-c)(\nu_f - \nu_m)(1/k_m - 1/k_f)}{c/k_m + (1-c)/k_f + 1/G_m} \quad (17)$$

$$\hat{G}_a = G_m\left[\frac{G_m(1-c) + G_f(1+c)}{G_m(1+c) + G_f(1-c)}\right] \quad (18)$$

$$\hat{k} = \frac{k_m(k_f + G_m)(1-c) + ck_f(k_m + G_m)}{(1-c)(k_f + G_m) + c(k_m + G_m)} \quad (19)$$

$$\hat{G}_t(+) = G_m\left[\frac{(1+c^3\alpha_1)(\rho_1 + c\beta_1) - 3c(1-c)^2\beta_1^2}{(1+c^3\alpha_1)(\rho_1 - c) - 3c(1-c)^2\beta_1^2}\right] \quad (20)$$

where $\alpha_1 = (\beta_1 - \gamma_1\beta_2)/(1 + \gamma_1\beta_2)$, $\rho_1 = (\gamma_1 + \beta_1)/(\gamma_1 - 1)$, $\gamma_1 = G_f/G_m$, $\beta_1 = 1/(3 - 4\nu_m)$, and $\beta_2 = 1/(3 - 4\nu_f)$.

2.4 Halpin–Tsai Equations

The Halpin–Tsai equations represent a semiempirical approach to the problem of the significant separation between upper and lower bounds observed for some properties when the fiber and matrix elastic constants differ significantly (Ashton et al. 1969). The equations employ the rule-of-mixtures approximation for axial extensional modulus and Poisson's ratio (Eqns. (14) and 15)). The expressions for the three other elastic constants are assumed to be of the form

$$\hat{p} = p_m(1 + c\xi\eta)/(1 - c\eta) \quad (21)$$

where

$$\eta = \left(\frac{p_f - 1}{p_m}\right) \Big/ \left(\frac{p_f}{p_m} + \xi\right) \quad (22)$$

Here, $\hat{p}$ stands for the composite moduli $\hat{E}_t$, $\hat{G}_a$ or $\hat{G}_t$, and p_f and p_m are the corresponding fiber and matrix moduli, respectively. The reinforcing factors ξ are arbitrary constants which are, in general, different for the three elastic constants. For example, the Halpin–Tsai expression for the transverse modulus is

$$\hat{E}_t = E_m(1 + c\xi_E\eta_E)/(1 - c\eta_E) \quad (23)$$

where

$$\eta_E = \left(\frac{E_f}{E_m} - 1\right) \Big/ \left(\frac{E_f}{E_m} + \xi_t\right) \quad (24)$$

and ξ_E is the transverse-modulus reinforcing factor.

3. *Viscoelastic Properties*

Polymers are widely used as matrix materials and as reinforcements. As these materials have viscoelastic characteristics, their influence on the time-dependent properties of composites is an important subject.

The analysis of viscoelastic behaviour is more complex than that of elastic behavior. Fortunately, the elastic–viscoelastic correspondence principle permits the use of analytical elasticity solutions in many cases. A thorough discussion of the subject is beyond the scope of this article. Generally speaking, however, the correspondence principle states that the solution of static elastic problems can be converted to Laplace-transformed solutions of viscoelastic solutions by replacing elastic moduli with transformed viscoelastic moduli multiplied by the transform parameter.

For example, expressions for the transformed composite effective relaxation moduli $\bar{C}_{ij}(s)$ can be obtained from expressions for composite effective elastic moduli $\hat{C}_{ij}$ by replacing the phase elastic moduli by phase relaxation moduli multiplied by the transform parameter s. A similar relationship exists between the transformed composite effective creep compliance $\bar{S}_{ij}(s)$ and the composite effective elastic compliance $\hat{S}_{ij}$.

Use of the correspondence principle requires that there be an "exact" or approximate analytical solution for the appropriate effective elastic property. The existence of upper and lower bounds on viscoelastic properties has been proven only for a very few special cases (Christensen 1979).

As an example, the Halpin–Tsai approximate expression for the transverse extensional modulus given by Eqns. (23) and (24) can be used to obtain an approximate expression for the transformed transverse modulus $\tilde{E}_t(s)$:

$$\tilde{E}_t(s) = \tilde{E}_m(s)[1 + \xi_E \tilde{\eta}_E(s)c]/[1 - c\tilde{\eta}_E(s)] \quad (25)$$

where

$$\tilde{\eta}_E(s) = \left[\frac{\tilde{E}_f(s)}{\tilde{E}_m(s)} - 1\right] \Big/ \left[\frac{\tilde{E}_f(s)}{\tilde{E}_m(s)} + \xi_E\right] \quad (26)$$

and $\tilde{E}_f(s)$ and $\tilde{E}_m(s)$ are the Laplace transforms of the fiber and matrix viscoelastic extensional moduli, respectively.

The correspondence princple can also be used to obtain the composite effective complex moduli $\hat{C}^*(\omega)$ for steady-state harmonic oscillation, where ω is the frequency. For example, the Halpin–Tsai relations, Eqns. (23) and (24), can also be used to obtain an approximate expression for the composite effective complex transverse modulus, $\hat{E}_t^*(\omega)$:

$$\hat{E}_t^*(\omega) = E_m^*(\omega)[1 + c\xi_E \eta_E^*(\omega)]/[1 - \eta_E^*(\omega)c] \quad (27)$$

where

$$\eta_E^* = [E_f^*(\omega)/E_m^*(\omega) - 1]/[E_f^*(\omega)/E_m^*(\omega) + \xi_E] \quad (28)$$

4. Plastic Behavior

The increasing interest in metal-matrix composites is focusing attention on the subject of plasticity, which has not received as much consideration as other aspects of mechanical behavior. The contribution of metal matrices to composite stiffness, even in the axial direction, can be significant and matrix plasticity effects can have an appreciable influence on composite behavior.

Composite plastic load–deformation behavior is far more difficult to describe than for the elastic or even viscoelastic cases. In the latter two cases, it is possible to use effective composite properties to obtain relations between average stresses and strains and their time derivatives. However, to describe plastic behavior, it is necessary to define the complete internal state of stress for every value of applied load. This requires specification of a geometric model and is extremely complex, even for simple loading conditions. The problem has been approached both analytically and by the use of finite-element models (Hill 1964b, Mulhern et al. 1967, Dvorak and Rao 1976a, Adams 1970, Foye 1973). Although definition of complete load–deformation behavior is difficult, it has been possible to establish analytical bounds on limit loads, and this is discussed in Sect. 4.1.

The complexity of the subject precludes a detailed discussion of plastic load–deformation behavior. However, micromechanical analyses have provided important insights into the general characteristics of composites with plastic constituents and these will be examined in this section. The discussion is based primarily on the work of Drucker (1975) and Dvorak and Rao (1976a, b).

The elastic constants of fiber and matrix usually differ, so the internal stress distribution is not homogeneous, even under hydrostatic loading. Unexpected results follow from this; for example, a composite with elastic fibers in an elastic, perfectly plastic matrix shows an initial stress–strain curve with work hardening caused by the formation of plastic zones which spread as the load is increased. The application of hydrostatic stress produces irreversible volume changes due to yielding of the matrix, although the matrix is plastically incompressible.

Dvorak and Rao (1976b) showed that plasticity effects are often important when composites whose constituents have different coefficients of thermal expansion undergo significant temperature change.

A simple model for the behavior under axial load of a composite with elastic fibers and an elastic, perfectly plastic matrix proposed by Spencer (1972) illustrates some of the effects of plasticity on composite behavior. Figure 4 shows the stress–strain behavior of the composite; the fibers, which have an elastic moduli E_f; and the matrix, which has an elastic modulus E_m, a tensile yield stress Y_m and a compressive yield stress $-Y_m$. Assuming that the composite is originally stress-free, the initial modulus is approximately $cE_f + (1-c)E_m$. The matrix yields when the applied stress is $[cE_f + (1-c)E_m]\ Y_m/E_m$ (point A), and the effective composite modulus drops to cE_f. The slope of the unloading curve BC equals the initial elastic slope. At point C, the matrix stress reaches $-Y_m$ and it yields in compression, whereupon the effective composite modulus again drops to cE_f. At point D, where the applied stress is zero, the matrix is under a state of residual compressive stress $-Y_m$, the fibers are in tension and there is a macroscopic residual deformation. Subsequent application of tensile stress unloads the matrix and it behaves elastically, so that the slope of the composite stress–strain curve is, again, $cE_f + (1-c)E_m$. At point E, the matrix yields in tension. The cycle EBCD can then be repeated indefinitely.

This simple model does not display the apparent work hardening predicted by more detailed analyses (Dvorak and Rao 1976a, b). Composite behavior under shear and transverse extensional loadings is more complex and is not easily represented by simple models like the one presented here (Adams 1970, Foye 1973).

4.1 Composite Limit Loads

Limit analysis permits the definition of bounds on composite failure loads without consideration of the

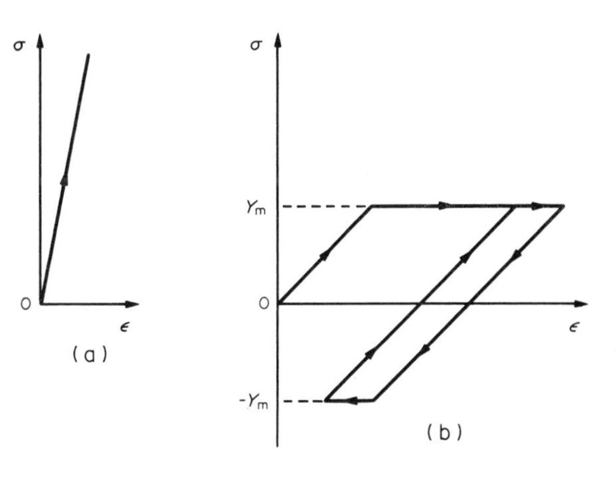

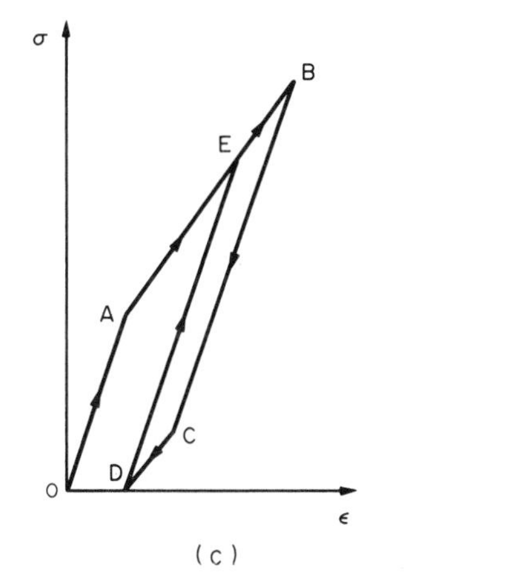

Figure 4
Stress–strain curves for (a) elastic fibers, (b) an elastic, perfectly plastic matrix and (c) the resulting composite

details of the internal state of stress. In this respect, it is somewhat analogous to the bounding approach to composite elastic properties discussed in Sect 2.2. Lower bounds on the surface tractions producing unrestricted plastic flow are obtained by defining statically admissible stress fields that do not exceed the constituent yield stresses. Upper bounds are established by assuming kinematically admissible velocity fields and equating work done by surface tractions to the internal plastic dissipation (Hashin 1972).

For example, Shu and Rosen (1967) constructed a composite-cylinder-assemblage model consisting of rigid circular fibers in an elastic, perfectly plastic matrix that obeys the von Mises yield criterion. They found upper and lower bounds on limit loads for various surface tractions. In particular, they found that a lower bound (LB) on the limit load associated with an axial-shear surface traction. $(\sigma_{12}^{\mathrm{L}})_{\mathrm{LB}}$ is simply the matrix yield stress in shear σ_{y}. They also obtained an upper bound (UB), subsequently improved by Majumdar and McLaughlin (1973). The upper bound obtained by the latter is

$$(\sigma_{12}^{\mathrm{L}})_{\mathrm{UB}} = \sigma_{\mathrm{y}}\left[1 + c\left(\frac{4}{\pi} - 1\right)\right] \tag{29}$$

This expression predicts that the composite in-plane shear strength can never be more than 27% greater than the matrix shear strength when it is reinforced with circular cylindrical fibers, no matter how strong they are.

In general, surface tractions that produce failure in the unreinforced matrix are lower bounds on composite strengths (Hashin 1972). Other upper bounds are more complex than that of Eqn. (29) and are not presented in this article. One other result should be noted: Hashin (1972) showed that the shear strength of a longitudinal plane not passing through any fibers is simply the matrix shear strength.

5. *Coefficients of Thermal Expansion*

The relations between average stress and average strain in a statistically homogeneous composite material undergoing a uniform temperature change θ are

$$\hat{\varepsilon}_{ij} = \hat{S}_{ijkl}\bar{\sigma}_{kl} + \hat{\alpha}_{ij}\theta \tag{30}$$

and

$$\bar{\sigma}_{ij} = \hat{C}_{ijkl}\bar{\varepsilon}_{kl} - \hat{C}_{ijkl}\hat{\alpha}_{kl}\theta \tag{31}$$

where the $\hat{\alpha}_{kl}$ are the effective thermal-expansion coefficients. For a transversely isotropic or square symmetric material there are two independent effective thermal-expansion coefficients, the axial $\hat{\alpha}_{\mathrm{a}}$ and the transverse $\hat{\alpha}_{\mathrm{t}}$.

Methods for establishing relations between composite effective expansion coefficients and constituent properties are similar to those used for elastic constants. For example, Schapery (1968) and Rosen and Hashin (1970) used thermoelastic extremum principles to derive upper and lower bounds. A direct method for two-phase composites, proposed by Levin, is reported in Rosen and Hashin (1970).

For a transversely isotropic composite having two phases which are elastically and thermally isotropic, the coefficients of thermal expansion are

$$\hat{\alpha}_{\mathrm{a}} = \bar{\alpha} + \left(\frac{\alpha_{\mathrm{f}} - \alpha_{\mathrm{m}}}{1/K_{\mathrm{f}} - 1/K_{\mathrm{m}}}\right)\left[\frac{3(1 - 2\hat{\nu}_{\mathrm{a}})}{\hat{E}_{\mathrm{a}}} - \left(\frac{1}{K}\right)\right] \tag{32}$$

$$\hat{\alpha}_{\mathrm{t}} = \bar{\alpha} + \left(\frac{\alpha_{\mathrm{f}} - \alpha_{\mathrm{m}}}{1/K_{\mathrm{f}} - 1/K_{\mathrm{m}}}\right) \times \left[\frac{3}{2\hat{K}} - \frac{3\hat{\nu}_{\mathrm{a}}(1 - 2\hat{\nu}_{\mathrm{a}})}{E_{\mathrm{a}}} - \left(\frac{1}{K}\right)\right] \tag{33}$$

where

$$\bar{\alpha} = c\alpha_{\mathrm{f}} + (1 - c)\alpha_{\mathrm{m}} \tag{34}$$

$$\left(\frac{1}{K}\right) = \frac{c}{K_{\mathrm{f}}} + \frac{(1 - c)}{K_{\mathrm{m}}} \tag{35}$$

and α_f is the fiber coefficient of thermal expansion, α_m is the matrix coefficient of thermal expansion, K_f is the fiber bulk modulus and K_m is the matrix bulk modulus.

These expressions for thermal expansion coefficients require knowledge of the composite effective axial modulus $\hat{E}_a$, axial Poisson's ratio $\hat{\nu}_a$ and transverse bulk modulus $\hat{k}$, as well as constituent thermal expansion constants and bulk moduli. The composite elastic properties can be measured experimentally or evaluated from analytical expressions. Use of bounds for composite effective moduli results in bounds on effective thermal-expansion coefficients.

6. Thermal Conductivity

Interest in composite thermal conductivity has been stimulated by the development of pitch-based carbon fibers having axial conductivities several times greater than that of pure copper (Nysten et al. 1985).

The effective axial thermal conductivity of a transversely isotropic composite is given by Hashin (1972)

$$\hat{Q}_a = cQ_f + (1-c)Q_m \tag{36}$$

where Q_f and Q_m are the fiber and matrix thermal conductivities. Eqn. (36) is, of course, simply the rule of mixtures.

The problem of transverse thermal conductivity is mathematically analogous to that of axial shear. For the case of arbitrary phase geometry, the bounds are (Hashin 1972):

$$\hat{Q}_t(+) = Q_t + \frac{1-c}{1/(Q_m - Q_f) + c/2Q_f} \tag{37}$$

$$\hat{Q}_t(-) = Q_m + \frac{c}{1/(Q_f - Q_m) + (1-c)/2Q_m} \tag{38}$$

For the CCA model, the solution is exact, and is given by Hashin (1972):

$$\hat{Q}_t = Q_m\left[\frac{Q_m(1-c) + Q_f(1+c)}{Q_m(1+c) + Q_f(1-c)}\right] \tag{39}$$

7. Macroscopic Failure Criteria

Failure of composites results from complex internal processes that depend on the kind of loading applied and numerous constituent material parameters including stress–strain and statistical strength characteristics, interfacial bond strength and internal geometry. Consequently, it is not reasonable to expect that a precise, simple expression should exist for the strength of composites under arbitrary loading. For convenience, a number of empirical failure criteria for unidirectional composites under arbitrary load have been proposed. The most widely used—maximum stress, maximum strain and interaction—are examined here. Stresses are referred to the principal axes of the material.

The primary advantage of the maximum stress and strain failure criteria is their great simplicity. Required strength parameters can be determined from tests involving single-stress components. They also provide information on the mode of failure

7.1 Maximum-Stress and Maximum-Strain Failure Criteria

The maximum-stress failure criterion assumes that failure occurs when any one of the applied stresses equals the ultimate stress obtained when that stress alone is applied to the composite. That is, it completely ignores any interaction effects between the stresses. For a state-of-plane stress, for which $\bar{\sigma}_{13} = \bar{\sigma}_{23} = \bar{\sigma}_{33} = 0$, the maximum-stress failure criterion is

$$\left.\begin{array}{l} \bar{\sigma}^c_{11} \leqslant \bar{\sigma}_{11} \leqslant \bar{\sigma}^t_{11} \\ \bar{\sigma}^c_{22} \leqslant \bar{\sigma}_{22} \leqslant \bar{\sigma}^t_{22} \\ |\bar{\sigma}_{12}| \leqslant \sigma^u_{12} \end{array}\right\} \tag{40}$$

where the superscripts c and t indicate compression and tension failure strengths, respectively, and $\bar{\sigma}^u_{12}$ is the axial shear strength. Extension of this criterion to a more general state-of-stress is obvious.

The maximum-strain criterion assumes that failure occurs when any macroscopic strain value reaches the ultimate strain achieved in a corresponding single-component tension, compression or shear test. For the case of plane stress ($\bar{\sigma}_{13} = \bar{\sigma}_{23} = \bar{\sigma}_{33} = 0$) the criterion is

$$\left.\begin{array}{l} \bar{\varepsilon}^c_{11} \leqslant \bar{\varepsilon}_{11} \leqslant \bar{\varepsilon}^t_{11} \\ \bar{\varepsilon}^c_{22} \leqslant \bar{\varepsilon}_{22} \leqslant \bar{\varepsilon}^t_{22} \\ |\bar{\varepsilon}_{12}| \leqslant \varepsilon^u_{12} \end{array}\right\} \tag{41}$$

where $\bar{\varepsilon}^c_{11}$ is the failure strain under a compressive axial stress, $\bar{\varepsilon}^t_{11}$ is the failure strain under a tensile axial stress, $\bar{\varepsilon}^c_{22}$ and $\bar{\varepsilon}^t_{22}$ are similarly defined for transverse extensional loading, and $\bar{\varepsilon}^u_{12}$ is the ultimate strain under pure shear loading.

7.2 Interaction Failure Criteria

Numerous interaction failure criteria have been proposed. For example, Sendeckyj (1972) reports eighteen. A description of features of these criteria is beyond the scope of this article. Instead, one of the more general and widely used formulations is considered—that of Tsai and Wu (1971). For convenience, contracted notation will be used.

The most general form of the Tsai–Wu criterion is

$$F_i\bar{\sigma}_j + F_{ij}\bar{\sigma}_i\bar{\sigma}_j = 1 \tag{42}$$

where a repeated index indicates summation. This defines a failure surface in stress space. To ensure that the criterion predicts finite failure stresses, the surface must be closed. (Infinite failure stresses, even for

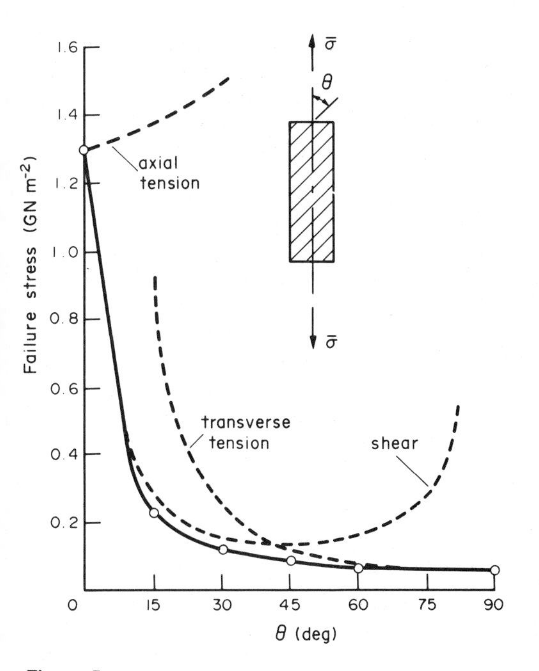

Figure 5
Comparison of the Tsai–Wu and maximum-stress failure criteria with off-axis strength of unidirectional boron–epoxy composites:——, Tsai–Wu prediction; ---, maximum stress; o, experiment (after Pipes and Cole 1973)

hydrostatic loading, are generally rejected on physical grounds.) This requires that $F_{ii}F_{jj} - F_{ij}^2 \geqslant 0$ for $i, j = 1, 2, \ldots 6$, where repeated indices are not summed.

For thin layers in a state-of-plane stress, with $\bar{\sigma}_3 = \bar{\sigma}_4 = \bar{\sigma}_5 = 0$, Eqn. (42) reduces to

$$F_1\bar{\sigma}_1 + F_2\bar{\sigma}_2 + F_{11}\bar{\sigma}_1^2 + F_{22}\bar{\sigma}_2^2 + 2F_{12}\bar{\sigma}_1\bar{\sigma}_2 + F_{66}\bar{\sigma}_6^2 = 1 \quad (43)$$

which has six independent constants. The four parameters F_1, F_{11}, F_2 and F_{22} can be determined from axial and transverse tension and compression tests, and an axial shear test is used to evaluate F_{66}. This leaves F_{12}, which can only be determined from a multiaxial strength test in which $\bar{\sigma}_1$ and $\bar{\sigma}_2$ are both present. To avoid performing multiaxial tests, it is often assumed that $F_{12} = 0$.

Figure 5 compares predictions of the Tsai–Wu and maximum-stress failure criteria for a series of off-axis tensile tests performed on unidirectional boron–epoxy composites.

See also: Creep of Composites; Fatigue of Composites; Strength of Composites; Multiaxial Stress Failure

Bibliography

Adams D F 1970 Inelastic analysis of a unidirectional composite subjected to transverse normal loading. *J. Compos. Mater.* 4: 310–28

Ashton J E, Halpin J C, Petit P H 1969 *Primer on Composite Materials: Analysis.* Technomic, Westport, Connecticut

Christensen R M 1979 *Mechanics of Composite Materials.* Wiley, New York

Drucker D C 1975 Yielding, flow and fracture. In: Herakovich C T (ed.) 1975 *Inelastic Behavior of Composite Materials.* American Society of Mechanical Engineers, New York, pp. 1–15

Dvorak G J, Rao M S M 1976a Axisymmetric plasticity theory of fibrous composites. *Int. J. Eng. Sci.* 14: 361–73

Dvorak G J, Rao M S M 1976b Thermal stresses in heat-treated fibrous composites. *J. Appl. Mech.* 43: 619–24

Foye R L 1973 Theoretical post-yielding behavior of composite laminates: I—Inelastic micromechanics. *J. Compos. Mater.* 7: 178–93

Hashin Z 1965 On elastic behaviour of fibre reinforced materials of arbitrary transverse phase geometry. *J. Mech. Phys. Solids* 13: 119–34

Hashin Z 1972 *Theory of Fiber Reinforced Materials.* National Aeronautics and Space Administration, Hampton, Virginia

Hashin Z 1979 Analysis of properties of fiber composites with anisotropic constituents. *J. Appl. Mech.* 46: 543–50

Hill R 1964a Theory of mechanical properties of fibre-strengthened materials: I. Elastic behaviour. *J. Mech. Phys. Solids.* 12: 199–212

Hill R 1964b Theory of mechanical properties of fibre-strengthened materials: II. Inelastic behaviour. *J. Mech. Phys. Solids* 12: 213–18

Majumdar S, McLaughlin P V 1973 Upper bounds to inplane shear strength of unidirectional fiber-reinforced composites. *J. Appl. Mech.* 40: 824–25

Majumdar S, McLaughlin P V 1975 Effects of phase geometry and volume fraction on the plane stress limit analysis of a unidirectional fiber-reinforced composite. *Int. J. Solids Struct.* 11: 777–91

Mulhern J F, Rogers T G, Spencer A J M 1967 Cyclic extension of an elastic fibre with an elastic–plastic coating. *J. Inst. Maths. Appl.* 3: 21–40

Nysten B, Piraux L, Issi J-P 1985 Thermal conductivity of pitch-derived fibers. *J. Phys. D* 18: 1307–10

Pipes R B, Cole B W 1973 On the off-axis strength test for anisotropic materials. *J. Compos. Mater.* 7: 246–56

Rosen B W, Hashin Z 1970 Effective thermal expansion coefficients and specific heats of composite materials. *Int. J. Eng. Sci.* 8: 157–73

Schapery R A 1968 Thermal expansion coefficients of composite materials based on energy principles. *J. Compos. Mater.* 2: 380–404

Schapery R A 1974 Viscoelastic behavior and analysis of composite materials. In: Sendeckyj G P (ed.) 1974 *Composite Materials,* Vol. 2, *Mechanics of Composite Materials.* Academic Press, New York, pp. 86–168

Sendeckyj G P 1972 A brief survey of empirical multiaxial strength criteria for composites. *Composite Materials: Testing and Design.* American Socieity for Testing and Materials, Philadelphia, Pennsylvania, pp. 41–51

Sendeckyj G P 1974 Elastic behavior of composites. In: Sendeckyj G P (ed.) 1974 *Composite Materials.* Vol. 2, *Mechanics of Composite Materials.* Academic Press, New York, pp. 45–83

Shu L S, Rosen B W 1967 Strength of fiber-reinforced composites by limit analysis methods. *J. Compos. Mater.* 4: 366–81

Spencer A J M 1972 *Deformations of Fibre-reinforced Materials.* Clarendon Press, Oxford
Tsai S W, Wu E M 1971 A general theory of strength for anisotropic materials. *J. Compos. Mater.* 5: 58–80

C. Zweben
[Martin Marietta Astro Space Division,
Philadelphia, Pennsylvania, USA]

Friction and Wear Applications of Composites

When two sliding surfaces come into contact, there is a progressive loss of material from one or both. This is known as wear. Such interaction between surfaces also gives rise to friction. The science of friction, wear and lubrication is called tribology (Furey 1986). Tribology always deals with pairs of surfaces. Thus friction and wear are not just physical properties of individual materials: they are system properties involving the interactions within pairs of sliding surfaces and between them and the environment. Contrary to general belief, friction and wear are not always undesirable. For instance, high friction is essential in brake linings; and a lead pencil would be useless without some wear.

In addition to suitable friction and wear characteristics, a material for a tribocomponent (whether a clutch plate or a bearing) must have a precise balance of physical and mechanical properties: thermal expansion, damping capacity, conformability, strength, stiffness and fatigue life. The required combination of properties is rarely found in single-phase materials. Most materials used in tribological applications are composites containing two or more phases.

Depending on the aspect ratio (ratio of length to diameter) of the second phase, a composite may be particle-dispersed or fiber-reinforced. In the former case the aspect ratio of the second phase is near unity. In general, such composites show no increase in strength: their value lies in substantially improved tribological properties. Increased modulus and tensile strength result when the second phase has a high aspect ratio (typically over 100) and a modulus higher than that of the matrix. In such fiber-reinforced composites, both mechanical and tribological properties depend upon the orientation of the fibers, whereas in particulate composites they are more or less the same in all directions.

1. Antifriction Materials

Lamellar solids such as graphite and molybdenum disulfide (MoS_2) are widely used as lubricants in composites for antifriction applications. Their lubricating property derives from their anisotropic crystal structure. Graphite, for example, has a hexagonal structure with a high c/a ratio ($c = 3.40$ Å, $a = 1.42$ Å). The carbon atoms in its basal planes are held together with strong covalent bonds while the basal planes themselves are held together by weak Van der Waals forces, resulting in interplanar mechanical weakness (Wood 1986). The presence of water vapor and crystal defects influences the interlamellar shearing of graphite, which provides lubrication. The crystal structure of MoS_2 is also hexagonal ($c = 12.29$ Å, $a = 3.16$ Å) and each molybdenum atom is surrounded by a triangular prism of sulfur atoms (Braithwaite 1964).

Among polymeric materials, polytetrafluoroethylene (PTFE) is well known for its antifriction property. This follows from its smooth molecular profile and low intermolecular cohesion (Tabor 1978). PTFE has no unsaturated bonds and is not easily polarized. During sliding contact it forms a thin transfer film of itself on the counterface; but this film does not adhere well to the counterface. Unfortunately the low intermolecular cohesion responsible for easy drawing of molecular chains out of the crystalline portions of the polymer, which gives rise to low friction (~ 0.1 against a steel counterface), results also in unacceptable amounts of wear.

It is impossible to achieve the desired combination of low friction and low wear using PTFE alone. Without sacrificing the characteristic low friction of PTFE, its wear resistance can be improved (by up to a factor of 1000) by adding fillers to the PTFE matrix. The improved wear-performance of filled PTFE may be due to the formation of a continuous and strongly adhering layer of transfer film on the counterface. The type of debris produced during sliding contact also determines the nature of the transfer film. Unfilled PTFE gives wear debris in the form of fragmented sheets which appear as loose films on the wear track. By contrast, the debris of filled PTFE is generally in the form of loose particles which are easily locked into the crevices of the rough counterface to produce a continuous film (Bahadur and Tabor 1983). A variety of fillers have been used in PTFE, with varying degrees of success; but the choice of a filler and of the proportion in which it is used appears to be somewhat empirical. PTFE is not used only as the matrix: fibers and powder are also used as solid lubricant fillers in composites for antifriction applications.

No material can act as a solid lubricant under all operating conditions. For instance, graphite loses its lubricating property in vacuum, as the complete absence of adsorbed vapors makes it difficult to shear its layers. Similarly, PTFE loses its antifriction property if the roughness of the counterface exceeds 1 μm, if the sliding speed is too high or if the bulk temperature is too low. Presumably these factors interfere either with the easy drawing of molecular chains out of the material or with the formation of transfer films (Pooley and Tabor 1982). The choice of a solid lubricant depends on the environmental and other conditions in which a tribocomponent has to function.

1.1 Plain Bearings

The most widespread application of antifriction materials is in plain bearings. Sometimes a bearing is expected to perform without an external fluid lubricant; that is, with a 'built-in solid lubricant. Even when a fluid lubricant film is provided, at some stage it will break, permitting the opposing faces to come into contact. The material for a plain bearing must therefore have a low coefficient of friction against a counterface which is usually of steel, together with good fatigue life under high load and velocity. Further, it should have adequate mechanical strength; and to allow asperities to be smeared out and debris produced during wear to become embedded, softness and/or low melting temperature (or recrystallization temperature) are necessary.

These diverse requirements have been met by multiphase materials with at least one soft phase: examples are white-metal alloys (Babbitt metals) based on needle-shaped intermetallic Cu_6Sn_5 in a low-melting-point tin-rich matrix; leaded bronzes containing islands of lead in a bronze matrix; Al–Sn alloys containing 6–20% tin; sintered porous metals impregnated with oil; and polymer composites. With proper control over the size and morphology of phases it is even possible to dispense with the soft phase: aluminum alloy with 11% Si and 1% Cu is a case in point. In composition this is similar to aluminum piston-alloy, but its microstructure consists of finely divided near-spherical (as against needle-shaped) silicon particles (Pratt 1973).

Recent reports indicate the potential of graphite-dispersed aluminum-alloy composites as bearing materials. In the absence of complete fluid-film lubrication, such composites show much better seizure and gall resistance than conventional aluminum bearing alloys. In applications where weight is an important consideration (e.g., aircraft bearings), the advantage of using aluminum-alloy–graphite composites in place of bronze can be overwhelming. One way to produce these composites is by the casting route, which involves mixing the molten alloy with graphite particles to make a uniform suspension. The problem of graphite rejection by liquid aluminum is always faced here, caused by density differences (Al: 2.7×10^3, graphite: $2.3 \times 10^3\ kg\,m^{-3}$) and poor wettability between the two. This can be overcome by the use of metal coatings (e.g., Ni and Cu) on the particles of graphite and by the addition of reactive elements (e.g., Mg and Ti) to the melt. Commercially viable technologies are yet to be developed for the production of sound castings with uniformly dispersed graphite.

The measure of performance of a bearing is load P (more accurately, stress) multiplied by linear velocity v; This is designated Pv. The Pv limits of PTFE bearings can be considerably enhanced by the use of fillers (see Fig. 1; Neale 1973). The reduction in strength associated with the dispersion of particulate fillers (e.g., lead oxide) can be offset by the use of fiber reinforcements

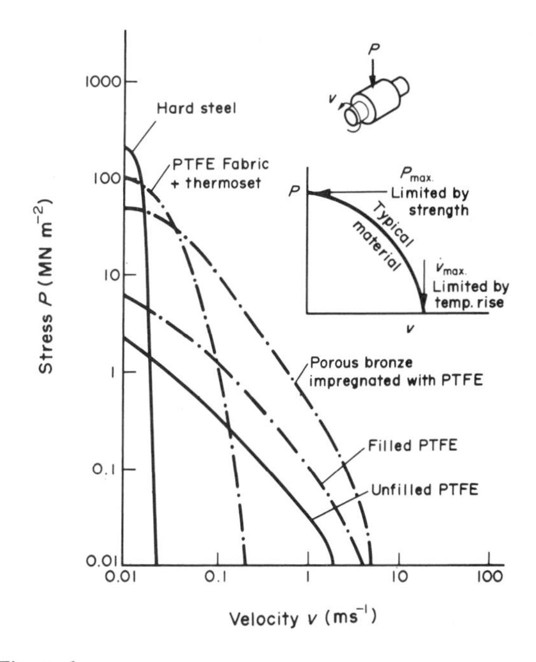

Figure 1
Pv limits for PTFE-based composite bearings (after Lancaster 1973)

(e.g., glass) Graphite fibers can provide, although expensively, lubrication as well as strength and stiffness. Bronze powder is sometimes added to PTFE to improve thermal conductivity. PTFE can also serve as filler, for example in a matrix of porous bronze. Polyimide, unlike other polymers, is capable of being used in composites meant for good heat stability upto 300 °C.

2. Friction Materials

A material with a reasonably high coefficient of friction is needed when a torque has to be transmitted from a power source to a driven component, as in a clutch, or when a moving body has to be decelerated or brought to a halt, as in a brake. This is usually accomplished by providing a pad or liner of friction material which is pressed against a counterface. Apart from having a high coefficient of friction, a friction material should be resistant to wear, should provide stable friction over a wide range of operating conditions and should not cause excessive wear of the counterface. In brake linings the frictional work, which causes the desired retardation, is dissipated as heat, raising the temperature of both the material and the counterface. This heat buildup at some stage results in the loss of frictional resistance (brake fade). It is thus essential for the friction material to resist fade and to return to its prefade condition upon cooling (recovery).

Attempts to improve the friction characteristics of a material usually result in a deterioration of its other properties. Friction materials are complex composites in which a large number of phases are used to arrive at an optimal combination of interdependent properties (see Table 1). In many formulations, asbestos in fibrous or woven form is the major constituent; however, health regulations governing the use of asbestos fibers have led to the development of formulations based on glass or steel fibers.

The performance of an organic-based friction material depends on the formation of a continuous film, 1–7 μm thick, on its own surface and on that of the counterface, which is usually cast iron. This film comprises elements of both the friction material and the counterface (Liu et al. 1979). When the temperature of the sliding interface increases because of higher speed or higher braking force, thermally unstable organic ingredients begin to decompose. The friction film is ultimately destroyed, excessive wear results, and frictional resistance is lost. The wear rate of a friction material is controlled by abrasive, adhesive and fatigue mechanisms; as higher temperatures are reached, thermal decomposition may also set in (Jacko et al. 1984).

Aircraft brakes are an interesting application. Weight is a crucial consideration, and the temperatures generated are significantly higher than those reached in automobile brakes. Friction materials must therefore combine low density with high thermal stability and thermal conductivity. Graphite is an excellent choice, as its low impact strength and tensile strength can be overcome by using it in carbon–carbon composites. Brakes made of such composites weigh only one-third the weight of equivalent steel brakes: this results in a saving, in the Concorde aircraft, of some 600 kg. Carbon–carbon composites are in fact unique among composite materials. They are high-performance structural materials and were first developed for that role.

3. Composites for Antiabrasion Applications

Abrasion is the removal of material from a relatively soft surface by the ploughing or cutting action of hard grit particles. It is encountered by mining and mineral dressing equipment, agricultural implements, materials used in coal conversion, and so on. The microstructures of antiabrasion materials consist of hard second-phase particles in a relatively ductile matrix. Such materials include high-chromium white irons containing large volume fractions of M_7C_3-type carbides in an austenite or martensite matrix and cobalt-based powder metallurgy alloys processed to produce carbides (stellites). Entire components or hard facings may be made of such materials.

The ductile matrix is usually worn away by the cutting or ploughing action of the abrasive, while the hard phase, which is generally brittle, wears out by fracture. It is the mode of this fracture that determines the resistance of a material to abrasion. The mode of fracture is in turn determined by such factors as the shape of the abrasive particle, the fracture toughness and relative hardness of the hard phase, local stresses (which can be much higher than nominal stresses), and the nature of the environment.

Quartz, the most commonly encountered natural abrasive, is not as hard as, for example, the carbide phase mentioned above. Like most natural abrasives, it has semirounded particles. The carbide phase is much more resistant to such abrasives than is the austenetic matrix. The matrix will therefore be worn away first, leaving protrusions of carbides on the

Table 1
Typical formulations of friction materials

Type	Formulation	Applications
Organic[a]		
Asbestos-based	Woven asbestos mat impregnated with 30–40% resin, usually phenolic rubber, and/or organic friction particles; minor additions of Zn or Cu to improve thermal conductivity, and of graphite and ceramic particles to balance friction characteristics	Cranes, lifts, excavators; molded variety used in automobiles
Asbestos-free	Steel fibers (up to 70%) or glass fibers (up to 40%) in place of asbestos	As above
Sintered metal[b]	Tin 8–10%, lead 5–12%, iron 2–8%, graphite up to 8%; asbestos, silica or carbides up to 10%; balance Fe or Cu	Heavy-duty brakes and clutches; earth moving equipment; aircraft brake pads
Carbon–carbon composites[c]	Woven graphite cloth impregnated with resins, cured and carbonised or matrix built by vaporphase deposition of carbon on carbon fibers	Brake pads in supersonic aircraft

a Jacko et al 1984, b Hausner et al. 1970, c Weaver 1972

surface. The concentrated stresses at the sharp edges of the carbides introduced by the action of abrasive will result in microfracture at the carbide edges, eventually leading to the rounding of such edges (see Fig. 2). These two processes, namely preferential matrix removal and rounding of carbide edges, will probably occur simultaneously during the course of wear. Beyond a certain volume fraction of the carbide phase—specifically, the intercarbide spacing in relation to the size of the abrasive particles—the hard-phase protrusions will completely protect the matrix from further abrasion (see Fig. 3).

Fracture events will, however, progressively remove the carbide (Prasad and Kosel 1983, 1984) until once again the matrix is exposed to the abrasive. The rate of carbide removal can be lowered if its fracture toughness is high. The role of the matrix should not be underestimated. It provides support to the hard phase and imparts ductility to the composite. If it provides insufficient support, the unsupported hard-phase edges become susceptible to fragmentation.

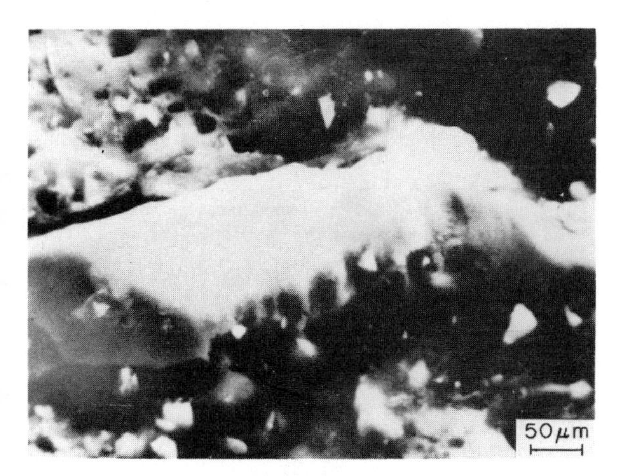

Figure 2
SEM micrograph of high-Cr white-iron surface showing microfracture of carbide edges. Surface is deeply etched, with carbide protrusions, following an in-situ SEM scratch test with quartz abrasive (after Prasad and Kosel 1983)

4. Metal-Cutting Tools

Machining costs, which are usually an important part of the total cost of manufacture, can be significantly lowered by raising the cutting speed or by increasing the depth of cut. At higher cutting speeds, however, tool wear is greater. The limits of cutting speed, or of tool life at a given cutting speed, are therefore governed by tool wear. A tool material must possess several characteristics in addition to high wear resistance: hot hardness, toughness, low friction and favorable cost. These requirements have been met to a large extent by cemented carbides. Tungstens carbide (WC), the most important constituent of a cemented carbide, was discovered in 1890. The methods of fabricating tool bits from this material (by sintering the compacts of WC–Co mixtures) were developed in the 1920s. Other important classes of tool material include high-speed steel (HSS), ceramics and diamonds (Komanduri 1986)

When a tool has been in use for some time, wear becomes evident in two regions (see Fig. 4). It appears first on the clearance face of the tool, extending from the cutting edge: this is known as flank wear. Wear may also be found on the rake face of the tool, over which the chip of material moves. Here the zone of wear begins not at the cutting edge but some distance (0.2–0.5 mm) above it: such wear is called cratering. Both kinds of wear arise mainly from the diffusion of elements from the tool to the workpiece.

When steels are cut at high speed, temperatures of the order of 1000 °C occur at the position of the crater on the rake face. At these temperatures the WC in a straight WC–Co composite diffuses rapidly into the workpiece, cratering increases, and the cutting ability of the tool is drastically reduced. This problem was solved in the late 1930s by incorporating one or more of the cubic carbides (TiC, TaC and NbC) in WC–Co composites. In tool alloys these carbides are present in the form of solid solutions of TiC and TaC in WC.

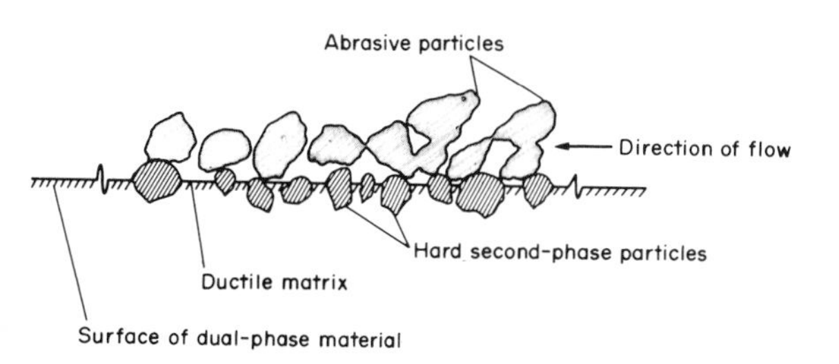

Figure 3
Schematic illustration of hard second-phase particles protecting the ductile matrix from abrasion

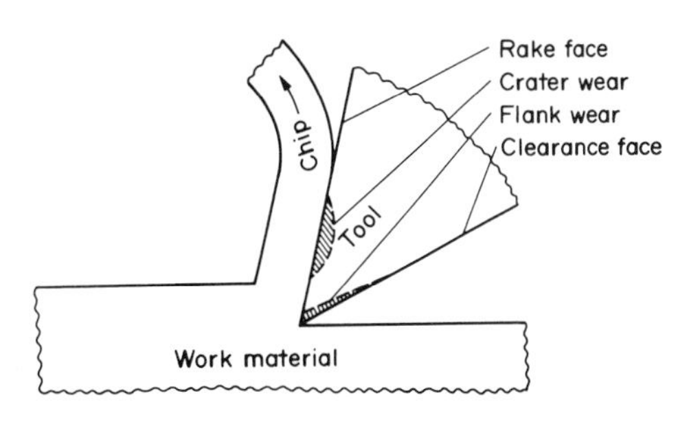

Figure 4
Schematic illustration of tool wear during machining

Typical steel-cutting-grade cemented-carbide tool bits contain, in addition to WC and Co, 10–25% TiC and 5–10% TaC; while the nonsteel grades are straight WC–Co composites containing up to 95% WC. The permissible speeds of cutting steel can be raised by increasing the TiC or TaC content. Tools based entirely on TiC–Co are available commercially, but their low toughness has prevented them from replacing multicarbide–Co tools on a large scale. Hard refractory coatings (of TiC, for example) can be deposited on WC–Co composites by chemical vapor deposition (CVD). These coated tool bits, while retaining the essential toughness of the WC-based substrate, permit higher cutting speeds or give longer tool life at the same cutting speed.

5. Biomedical Materials

Materials for artificial joints should have low friction, high resistance to fatigue, wear and creep, and biocompatibility. The human hip, for example, is a ball-and-socket joint whose prosthetic replacement is a stainless steel or Co–Cr–Mo alloy ball moving in a polymer socket. PTFE was used to form the socket, but it produced large amounts of stringy debris which caused extreme reactions in the tissue surrounding the implant. Trials with filled PTFE were also unsuccessful: they either abraded the stainless steel head or triggered severe foreign-body reactions in patients (Dumbleton 1981).

Ultra high molecular-weight polyethylene (UHMWPE) has become the accepted polymer for the socket. This material has low friction, low wear and good biocompatibility. The useful life of UHMWPE joints is expected to be around ten years; and to increase this, a carbon-fiber-reinforced UHMWPE composite has been developed (Farling and Greer 1978). Components are compression-molded directly into the shapes required. Laboratory simulations indicate that the wear rate of carbon-fiber-reinforced UHMWPE is lower than that of unfilled UHMWPE by a factor of between four and ten. The fiber–matrix interface here is crucial: its breaking down may result in fibers being ejected from the matrix, possibly accelerating wear and causing foreign-body reactions.

5.1 Dental Restorative Materials

Human teeth are constantly subjected to various kinds of wear. Mild abrasion by soft calcite particles occurs when teeth are brushed; and impact and fatigue occurs during mastication. Wear may be accelerated by pathological conditions. Materials used to fill cavities in teeth must have, among other properties, high resistance to wear and an aesthetically acceptable appearance. Dental amalgam is widely used but has the disadvantage of being unaesthetic in front teeth. Polymeric fillings proved to be unsatisfactory because of rapid wear and excessive polymerization shrinkage; but particle-filled polymer composites with much improved wear properties were developed later. Current dental composites are mostly based on BIS–GMA, a condensate of bisphenol A and glycidyl methacrylate with quartz or glass filler. Typically, the inorganic phase is 75% by weight or 50% by volume.

The performance of a dental composite depends on the type and severity of the wear that it must undergo. With soft abrasives such as calcite, the polymeric matrix wears out preferentially, leaving protrusions of hard filler particles. At high magnifications the surface looks somewhat like the schematic illustration in Fig. 3. Filler removal is dependent on the strength of the particle-matrix bond. Silane coupling (Hair and Filbert 1986) substantially improves this bond and provides better support to the filler. In the molar and premolar regions, hard grit particles chewed involuntarily, or impact and fatigue experienced during mastication, can cause extensive fracture of the brittle phase. Here composites are slightly inferior to amalgam: but as frontal fillings they offer exceptional resistance to wear as well as a pleasing appearance.

Composites are finding use in a growing range of applications in which their tribological properties, among other things, are of significance. Electrical contact materials, pneumatic tyres, and pistons and cylinder liners in internal combustion engines are some of these.

See also: Carbon–Carbon Composites; Dental Composites; Solid Fiber Composites as Biomedical Materials

Bibliography

Bahadur S, Tabor D 1983 The wear of filled polytetrafluoroethylene. In: Ludema K C (ed.) 1983 *Proc. Int. Conf. on Wear of Materials.* American Society of Mechanical Engineers, New York, pp. 564–70

Bever M B (ed.) 1986 *Encyclopedia of Materials Science and Engineering.* Pergamon, Oxford

Bowden F P, Tabor, D 1986 *The Friction and Lubrication of solids.* Clarendon, Oxford

Braithwaite E R 1964 *Solid Lubricants and Surfaces.* Clarendon, Oxford

Dumbleton J H 1981 *Tribology of Natural and Artificial Joints.* Elsevier, Amsterdam

Farling G M, Greer K 1978 An improved bearing material for joint replacement prostheses: Carbon fibre-reinforced ultra high molecular weight polyethylene. In: Hastings G W, Williams D F (eds.) 1980 *Mechanical Properties of Biomaterials.* Wiley, Chichester, pp. 53–64

Furey M J 1986 Tribology. In: Bever 1986, Vol. 7, pp. 5145–57

Hair M L, Filbert A M 1986 Glass surfaces. In: Bever 1986, Vol. 3, pp. 2004–8

Hausner H H, Roll K H, Johnson P K (eds.) 1970 *Perspectives in Powder Metallurgy Fundamentals, Methods and Applications,* Vol. 4: *Friction and Antifriction Materials.* Plenum, New York

Jacko M G, Tsang P H S, Rhee S K 1984 automotive friction materials: Evolution during the past decade. *Wear* 100: 503–15

Komanduri R 1986 Cutting tool materials. In: Bever 1986, Vol. 2, pp. 1003–12

Lancaster J K 1983 Composites for increased wear resistance: Current achievements and future prospects. *Tribology in the 80's*, National Aeronautics and Space Administration conference publication 2300. NASA, Cleveland, Ohio, Vol. 1, pp. 333–35
Lawn B R, Wilshaw T R 1975 Identation fracture: Principles and applications. *J. Mater. Sci.* 10: 1049–81
Liu T, Rhee S K, Lawson K E 1979 A study of wear rates and transfer films on friction materials. In: Ludema K C, Glaeser W A, Rhee S K (eds.) 1979 *Proc. Int. Conf. on Wear of Materials.* American Society of Mechanical Engineers, New York, pp. 595–600
Neale M J (ed.) 1973 *Tribology Handbook.* Butterworth, London
Pooley C M, Tabor D 1982 Friction and molecular structure: the behaviour of some thermoplastics. *Proc. R. Soc. London Ser. A* 392: 251–74
Prasad S V, Calvert P D 1980 Abrasive wear of particle-filled polymers. *J. Mater. Sci.* 15: 1746–54
Prasad S V, Kosel T H 1983 A study of carbide removal mechanisms during quartz abrasion, I: In situ scratch test studies. *Wear* 92: 253–68
Prasad S V, Kosel T H 1984 A study of carbide removal mechanisms during quartz abrasion, II: Effect of abrasive particle shape. *Wear* 95: 87–102
Prasad S V, Rohatgi P K 1987 Tribological properties of Al alloy particle composites. *J. Met.* 39: 22–26
Pratt G C 1973 Materials for plain bearings. *Int. Metall. Rev.* 18: 174–201
Pratt G C 1977 The wear properties of polymer composites. In: Richardson M O (ed.) 1977 *Polymer Engineering Composites.* Elsevier, London, pp. 237–61
Sarkar A D 1980 *Friction and Wear.* Academic Press, London
Tabor D 1978 The wear of non-metallic materials: A brief review. In: Dowson D, Godet M, Taylor C M (eds.) 1978 *Proc. 3rd Leeds-Lyon Symp. on Tribology: The Wear of Non-metallic Materials.* Mechanical Engineering Publications, Surrey, pp, 3–8
Trent E M 1979 Wear of metal cutting steels. In: Scott D (ed.) 1979 *Treatise on Materials Science and Technology*, Vol. 13: *Wear.* Academic Press, New York, pp. 443–89
Wakelin R J 1974 Tribology: The friction, lubrication and wear of moving parts. In: Huggins R A, Bube R H, Roberts R W (eds.) *Annu. Rev. Mater. Sci.* Annual Reviews Inc., Palo Alto, California, 4: 221–53
Weaver J V 1972 Advanced materials for aircraft brakes. *Aeronaut. J.* 76: 695–98
Wood A A R 1986 Carbon and graphite. In: Bever 1986, Vol. 1, pp. 495–500

S. V. Prasad
[Council of Scientific and Industrial Research, Regional Research Laboratory, Bhopal, India]

G

Glass-Reinforced Plastics: Thermoplastic Resins

Thermoplastic resins are one of several classes of polymeric materials which may be effectively reinforced with glass fiber. The principal distinction of the thermoplastics is that they may be repeatably softened or melted by heating, and that they resolidify when cooled. They are thus amenable to processing by methods which involve shaping in the liquid or semiliquid state. No chemical changes need be involved during fabrication operations. This is considered attractive in that processing conditions are often less critical than for thermosetting-resin-based systems.

Glass-fiber-reinforced thermoplastics (GFRTP) fall into two distinct categories: molding compounds which contain relatively short fibers and are most usually shaped by injection molding, and thermoformable sheet materials which contain longer chopped strand or even continuous-fiber reinforcement (Bader 1983, Folkes 1982). The former are well-established engineering materials which may be fabricated into complex load-bearing components, typically of less than 1 kg mass, although much larger moldings may be made. Their attraction is a simple "one-shot" processing operation to form a single component which may replace a complicated assembly of metal parts. Economies result from reduction in the number of parts in the system, reduced mass, less corrosion and better wear resistance. The thermoformable sheet materials are shaped by stamping the preheated sheet between cold metal dies. The process is generally similar to metal stamping and it is considered that this process should therefore be easier to integrate into automobile and other mass-production operations than the much slower processes using thermosetting materials. The range of possible shapes is more limited than with the molding compound but parts may be at least as complex as those produced by deep-drawing of metals.

The glass fibers used for reinforcement are almost invariably of E glass and are typically 10–20 μm in diameter. They are drawn in bundles of several hundred filaments (strands) which are bound together with a protective size or finish. This finish usually incorporates a silane coupling agent which controls the strength of the bond formed at the fiber–matrix interface. Heavier bundles of glass fiber are made up by combining several strands; these are termed rovings.

1. GFRTP Injection Molding Compounds

A molding compound, in which the glass fibers and the polymer resin are intimately blended together, is first prepared. The second and final operation is to injection-mold the component from the granular molding compound. The volume fraction and aspect ratio (length : diameter) of the fiber, together with the distribution of any other added phases, is controlled by the compounding and subsequent molding operations. There are two distinct processes, one of which produces compounds containing much longer fibers.

Most compounds are produced by extrusion blending: a dry mixture of polymer granules (typically about 3 mm in diameter and length) and chopped glass strand of similar length is charged into the hopper of a single- or double-screw extruder. The charge is passed through the machine and extruded through a "spaghetti" die and chopped into granules of similar dimensions to the original charge. During the extrusion process the charge is melted and subjected to considerable shear and mixing. This produces a homogenous compound but causes the glass strand to break up into individual filaments, typically less than 500 μm in length. A variation of this process is to feed continuous-fiber rovings directly into the molten polymer in a vented-barrel extruder. This causes less fiber breakage but average fiber length seldom exceeds 1 mm. The injection-molding machine also incorporates a screw-plasticizing section which causes further fiber breakage. The significance of this is discussed below.

In the alternative compounding process a continuous-fiber roving is fed, either into a cross-head die on an extruder, or drawn through a tapered heated die fed with solid (powder) or liquid polymer. The product is a continuous length of fiber roving completely impregnated with the polymer. This may then be cut into lengths suitable for charging into the injection-molding machine, typically 5–10 mm. The distinction between the two processes is that in extrusion compounding the fibers are typically much less than 1 mm long whereas in the drawn product they are as long as the chopped pellet: any length may be selected, limited only by the feed characteristics of the molding machine to be used. The characteristics of the two types of pellet are shown in Fig. 1. These "long-fiber" compounds will suffer some fiber breakup during injection molding, but this is controllable and the final fiber aspect ratio in the molding is much greater than in the extrusion-compounded material.

In the injection-molding operation, the charge must be melted and then injected at high pressure (50–200 MPa) into the mold-cavity, where the charge freezes and is ultimately ejected as the finished part. Critical parameters are the screw back pressure, the injection pressure, injection rate, gate dimensions, charge and mold temperatures and the time interval over which the full pressure is applied. High injection

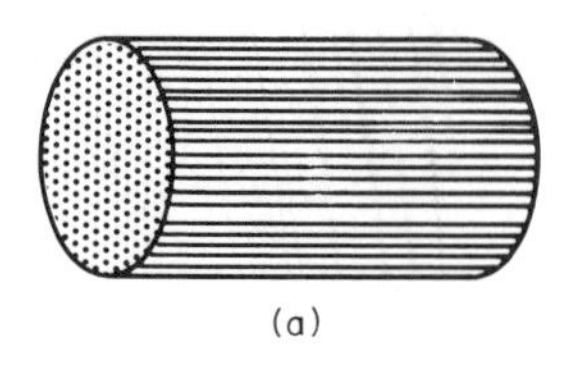

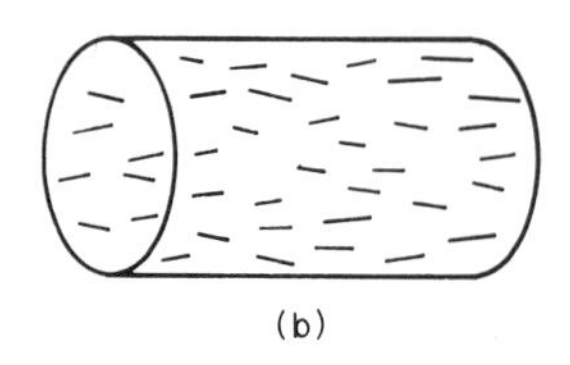

Figure 1
Injection-molded pellets: (a) continuous fiber; (b) short fiber

pressures ensure that the mold is filled, especially to the extremities of intricate passages. Thermoplastics are compressed at high pressures and this helps to compensate for the shrinkage which occurs when they subsequently freeze. This ensures that the part conforms dimensionally to the mold cavity. High flow rates exploit the pseudoplasticity of thermoplastics; the apparent viscosity is lowered under high shear-rates and less pressure is therefore needed to fill the mold. The pressure must be maintained until the material in the gate has frozen, otherwise part of the charge might flow back out of the mold and lead to shrinkage and the formation of sink marks. Surface finish is strongly influenced by the flow parameters and by the charge and mold temperatures. Considerable fiber breakage can occur if the shear rates are very high during molding, high back pressure is especially bad and over-small gates and high flow rates should generally be avoided. A further factor affecting the performance of the molding is the fiber orientation distribution, which is influenced by the flow pattern of the molten charge into the mold cavity. As a rule, convergent flow tends to align both fibers and molecules in the flow direction, whereas divergent flow aligns them normal to the flow. This effect is strongly influenced by surface effects, so that often three or five regions of differing orientation are observed through thin sections. Nonuniform fiber orientation may result in distortion due to differential contraction on cooling from the molding temperature. Where flow paths intersect, for example downstream from cored holes, weld or knit surfaces will be formed. These are not bridged by fibers and constitute planes of weakness. Mold design must ensure that these features occur in noncritically-stressed parts of the molding.

In discontinuous-fiber-reinforced composites the reinforcement is effected by the transfer of stress, through shear at the fiber–matrix interface, to the fiber. The tensile stress in the fiber builds up from zero at the fiber ends at a rate dependent on the shear stiffness of the matrix and the strength of the interface (Cox 1952, Kelly and Tyson 1965) until, ideally, the strain in both fiber and matrix are equal (Fig. 2.). The length of fiber over which this stress buildup occurs is termed the transfer length L_t. Clearly a fiber for which $L<2L_t$ will be unable to build up the full stress in its middle region. The transfer length increases with applied strain, as does the plateau stress, the limitation being the strain to failure of the fiber when $2L_t$ is called the critical length L_c. If all the fibers were perfectly aligned in the direction of the principal tensile stress and were all of exactly the critical length, then their stiffness-reinforcing efficiency would be exactly half that of continuous aligned fibers. The reinforcing efficiency increases as the fiber length is increased, so that when $L=5L_c$ the efficiency is 90% and at $10L_c$ it is 95%. For a typical polyamide thermoplastic reinforced with 12 μm diameter glass. L_t has been estimated (Bowyer and Bader 1972) at 60 μm at a strain of 1% and 120 μm at 2%, so that for 95% efficiency the fibers should exceed 2.4 mm in length. This may be the case in the long-fiber materials but in the normal molding compounds the fiber length seldom exceeds 250 μm. This is clearly inadequate and is reflected in a comparison of the properties of the two types of material (Table 1). In addition to stiffness and strength enhancement the longer fibers confer much better impact resistance and fracture toughness.

The overall effect of discontinuous-fiber reinforcement is that the stiffness is increased in proportion to

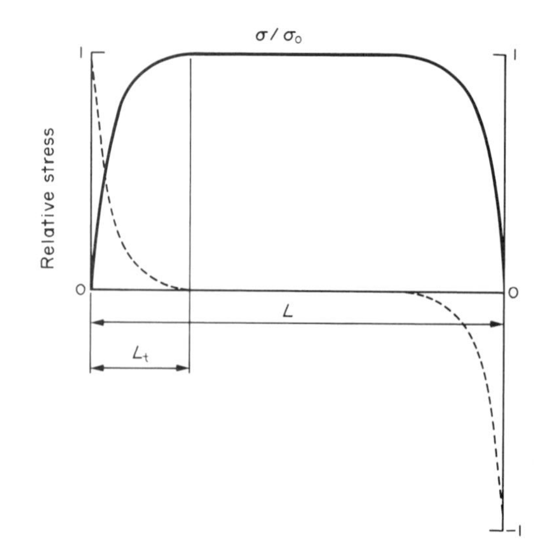

Figure 2
Distribution of tensile stress (solid line) and interface shear stress (dashed line) in a short fiber in an elastic matrix under uniform tension. The plateau stress σ_0 is equal to $E_f\varepsilon_c$, where ε_c is the uniform strain in the composite

Table 1
Typical properties of GFRTP materials in the form of injection-molded test bars

Material and reinforcement[a]	Relative density	Young's modulus (GPa)	Tensile strength MPa	Heat distortion temperature (°C)	In-mold shrinkage (%)
Polypropylene					
none	0.91	1.35	34	60	1.8
20% SGF	1.14	5.7	45	130	0.35
20% SGF/ chemically coupled	1.14	5.7	103	155	0.35
Polyamide 6.6					
none	1.14	3.2	80	75	2.0
20% SGF	1.46	10	210	260	0.50
20% LGF	1.46	10	220	260	—
Polycarbonate					
none	1.20	2.3	60	140	0.70
20% SGF	1.45	9.0	135	160	0.20

a Reinforcement level is vol % SGF, short glass fiber (~250 μm); LGF, long glass fiber (~ mm)

the volume fraction of fiber, but to a lesser extent than for continuous fibers. The tensile strength is increased but in a nonlinear fashion due to a reduction in the overall strain to failure. The low-strain stiffness of these materials may be estimated using a modified rule-of-mixtures equation of the form:

$$E_c = KE_f V_f + E_m(1 - V_f)$$

Where E_f and E_m are the Young's moduli of fiber and matrix respectively, V_f is the fiber volume fraction and K is a constant which is less than unity and depends on the interface strength and the fiber length and orientation distribution. K is typically of the order of 0.6 for molded test bars. Impact resistance tends to be lower than for unfilled polymer but fiber reinforcement enhances the notched toughness in many cases. The coefficient of thermal expansion is greatly reduced, making the moldings more compatible with metallic parts and the heat-distortion temperature is raised. Shrinkage in the mold is much less so that dimensionally accurate moldings may be more easily produced. A new generation of GFRTP materials incorporating longer fibers has recently become available (Gore et al. 1986). These materials offer modest strength improvement and much better impact performance than the conventional extrusion-compounded materials.

The balance of properties in GFRTP is very much influenced by the strength of the fiber–matrix bond. Strongly polar polymers, especially those which can form hydrogen bonds, tend to form relatively high-strength interfaces, whereas nonpolar materials such as polypropylene bond much less readily. In the case of these nonpolar materials it is often necessary to modify the molecular structure, e.g., by copolymerization, to achieve a satisfactory performance. The bond is also strongly affected by water, which is absorbed to some extent by all polymers, especially polar materials such as the polyamides, and is also adsorbed onto the surface of the glass fibers. Control of the bond may be effected by the choice of coupling agent incorporated into the finish applied to the glass fiber. For optimum performance it is necessary to use a finish designed specifically for the polymer matrix in question. A further aspect of the glass finish is that the extent to which the fibers are bound into the strand may be varied. A strongly bound strand will tend to remain as a bundle during processing, whereas a lightly bound strand will separate into individual filaments.

Current trends are towards tailor-made compounds of greater complexity, incorporating further additions such as particulate reinforcement, fire retarders and rubber toughening agents. A number of commercial materials incorporate both fiber to enhance stiffness and strength and dispersions of fine (~1 μm) elastomeric particles which actually reduce the stiffness but enhance toughness. These materials appear to have a more useful spectrum of properties than the simple two-component composites.

The mechanical properties of some typical GFRTP materials including both short- and long-fiber types are given in Table 1.

2. *Thermoforming Sheet Materials*

The concept of these materials is to produce a semi-finished product in the form of sheet, which may be heated and then pressed, stamped or otherwise formed into its final shape. The process is designed to resemble that of cold pressing of metals and is considered to be more compatible with mass-production fabrication, in for example the automobile industry, than the slower injection-molding or press-molding methods used with thermosetting-resin-based materials such as sheet-molding compound.

A well-established form of the material is as sheet, typically ~5 mm thick, made by laminating together alternate layers of thermoplastic film and glass-fiber reinforcement between heated rolls. The thermoplastics are usually based on either polypropylene or polyamides, and the reinforcement in the form of chopped-strand mat, continuous random mat (swirl mat) or woven cloth. One face of the sheet is usually a heavier layer of glass mat, which helps support the material when it has been heated to above the melting point of the polymer. To finally shape the component a suitable blank (or blanks) is cut from the sheet, heated in an oven to melt the polymer and then quickly transfered to a press, where it is stamped between cold dies. The material flows between the dies to fill the cavity and then freezes within a few seconds. Quite deep shapes may be formed by this process, e.g., valve covers and sumps for automobile engines, but the extent of possible flow is dependent on the form of the reinforcing web: chopped strand flows most readily, but can lead to uneven fiber distribution, whereas swirl mat generally gives better distribution but more limited flow. Woven cloth is suitable only for forming gentle curvatures with little change in section thickness.

Although these materials have been available for more than a decade, their use is still very limited. The reasons for this are that rather less fiber may be incorporated in them (usually about 30%) than is the case for the thermoset-based sheet-molding compound and bulk-molding compound. Stiffness and strength are therefore inferior while costs are relatively high.

Recent developments in high-performance thermoplastics reinforced with carbon fibers (e.g., ICI's APC-2, carbon-fiber/polyether ether ketone material) are leading to parallel developments in glass-reinforced materials. The proprietary process leads to the production of very thin sheets of collimated fibers impregnated with the polymer. The sheet is typically only 0.1–0.25 mm thick and contains 40–60% of aligned fibers. Final fabrication of components is achieved by laminating a number of sheets together in a manner similar to that used with thermoset-based prepreg materials to form multiaxial laminates. However, the economic constraints of using glass-based materials for general engineering applications are quite different from the application of the high-performance carbon-fiber materials in aerospace, and their adoption will depend on the development of rapid automated laminating processes. Autoclave and hot-press molding may not be suitable, but a number of techniques based on continuous welding of strips of material together, possibly using filament winding or pultrusion techniques, are being developed. The candidate polymers include the polyamides and polyolefins. A further development (Beever et al. 1988) has been the production of simple sections, e.g., flat strips 5–10 mm thick and more than 100 mm wide, which may be postformed by heat and pressure into more complex shapes such as angle, hat and beam sections. These developments are still in their infancy but offer future possibilities for precisely engineered laminates with high fiber content and excellent mechanical properties.

See also: Glass-Reinforced Plastics: Thermosetting Resins; High-Performance Composites with Thermoplastic Matrices

Bibliography

Bader M G 1983 Reinforced thermoplastics. In: Kelly A, Mileiko S T (eds.) 1983 *Handbook of Composites*, Vol. IV, *Fabrication of Composites*. North-Holland, Amsterdam, p. 177
Beever W H, Rhodes V H, Wareham J R 1988 *SAMPE J.* 24(1): 8
Bowyer W H, Bader M G 1972 *J. Mater. Sci.* 7: 1315
Cox H L. 1952 *Br. J. Appl. Phys.* 3: 72
Folkes M J 1982 *Short Fibre Reinforced Thermoplastics*. Research Studies Press, London
Gore C R, Cuff G, Cianelli D A 1986 *Mat. Eng.* 103 (3): 107
Kelly A, Tyson W R 1965 *J. Mech. Phys. Solids* 13(6): 329

M. G. Bader,
[University of Surrey, Guildford, UK]

Glass-Reinforced Plastics: Thermosetting Resins

Glass-reinforced plastics (GRP) consist of fibers of low-alkali E-glass embedded in a thermosetting plastics matrix, such as polyester or epoxide resin. During fabrication, glass fibers are incorporated into the liquid thermosetting resin to which catalyst and hardener have been added. These cause the resin to polymerize and after curing give a solid plastic matrix reinforced with glass fibers. The fibers enhance the low stiffness and strength of the resin, whose main purpose is to transmit the load into the stiffer, brittle fibers and to protect them from damage.

Single E-glass fibers, typically 8–15 μm in diameter, are easily damaged and difficult to handle. For protection, they are coated with a silane or poly(vinylacetate) size, which binds the fibers into strands containing about 200 filaments and acts as a coupling agent to provide a good fiber resin bond. To facilitate fabrication of GRP components, glass strands are incorporated into rovings, mats and fabrics. Glass fiber rovings consist of up to 120 untwisted strands, usually supplied wound together on a spool and suitable for the unidirectional (UD) fiber reinforcement of resins. Woven rovings (WR) are glass fiber rovings woven into a coarse fabric, usually with a balanced square weave. Glass fabrics are woven on textile machinery from twisted glass fibers (glass yarn) and are available in several weaves such as plain, square, twill and satin. Chopped strand mat (CSM) consists of chopped glass strands about 30 mm long held randomly orientated in a mat by a small amount of resin binder.

1. Glass-Fiber-Reinforced Resins

The principal resins used in GRP materials are unsaturated polyester resins, with epoxides being used in more limited quantities for high-technology applications, and phenolics, furanes and vinyl esters finding use for specialist applications. Polyesters are less dangerous to handle, easier to catalyze and much cheaper than the other resins. They may be cold cured and are suitable for low-pressure, ambient-temperature fabrication techniques. Epoxide resins have better mechanical properties than polyesters, with higher stiffness and strength, particularly at high temperatures, and lower shrinkage on curing. Phenolics and vinyl esters are used in high-temperature applications, with continuous temperatures up to 260°C, and furanes are noted for their chemical resistance at high temperatures. These specialist resins must be cured at high temperature and GRP components require hot-press molding. The higher performance resins are generally used with UD-glass fibers and glass fabric reinforcement, while CSM and WR reinforcements are used almost exclusively with polyester resins. Recent surveys show that polyesters accounted for 92% of resin consumption in the USA and 99% in the UK, with the remainder being mainly epoxide. Thus the important GRP materials in high-volume applications are CSM–polyester, WR–polyester and, to a lesser extent, UD-glass–polyester. Glass fabric–epoxide and UD-glass–epoxide composites are used in low-volume, high-technology applications.

Glass fiber content is related to reinforcement type and is shown, by weight, in Table 1 for the main GRP materials. A high fraction of aligned fibers (60–80% by weight) can be packed into a UD composite, while more resin is needed to impregnate all the fibers in CSM and WR laminates, giving glass contents as low as 25–35% by weight in CSM–polyester materials. In addition to the fibers, resin, catalyst and hardener, glass–resin composites may contain fire-retardant additives to reduce surface flame spread, uv absorbers to improve outdoor durability, pigments, dyestuffs and thixotropic additives which thicken resins to aid fabrication on inclined surfaces.

2. Glass-Reinforced-Plastic Molding Compounds

Several GRP molding compounds are now available in which the resin and glass fibers are premixed before molding. Component fabrication is achieved by placing the uncured compound in a closed mold and hot-press molding. The three main types of molding compound are sheet molding compound (SMC), dough molding compound (DMC), sometimes referred to as bulk molding compound (BMC), and preimpregnated glass fiber sheet (prepreg).

Table 1
Typical short-term mechanical property data for GRP materials

		DMC	SMC	CSM/ polyester	WR/ polyester	UD-glass/epoxide longitudinal	UD-glass/epoxide transverse
Glass content (wt %)							
	average	20	30	30	50	70	
	range	15–25	20–40	25–35	45–60	60–80	
Modulus (GPa)							
tensile	average	9	13	7.7	16	42	12
	range	8–11	9–16	6–9	12–22	30–55	8–20
flexural	average	8	11	6.3	13		
	range	7–9	7–14	5–8	10–18		
shear		3	3	3	4	5	
Strength (MPa)							
tensile	average	45	85	95	250	750	50
	range	35–60	50–120	60–150	195–350	600–1000	25–75
flexural	average	100	180	170	290	1200	
	range	85–120	90–240	100–300	160–500	1000–1500	
in-plane shear	average			80	95	65	
	range			60–100	80–120	50–80	
interlaminar shear	average		15	25	20	40	
	range		12–20	20–30	10–30	30–50	
Impact strength[a] ($kJ\,m^{-2}$)		20–40	50–75	40–80	100–200		

a Measured by Izod-type test, unnotched specimen, according to British Standard BS2782. Sect. 306A

Sheet molding compound consists of E-glass fibers reinforcing a mixture of catalyzed polyester resin and a mineral filler, such as particulate, chalk, limestone or clay. The filler thickens the resin sufficiently for uncured sheets of SMC to be stored, handled and cut to the correct size for molding. Sheet molding compound usually contains chopped glass strands 20–50 mm in length randomly orientated in the plane of the sheet and is supplied in several nominal glass weights, containing 20, 25, 30 or 35% glass fiber, with filler contents of 30–50% by weight. Fillers reduce material costs and modify the handling characteristics of the SMC sheet and the flow properties in the mold. High-performance SMC materials may also contain aligned chopped glass fibers or continuous fibers, at glass contents up to 70% by weight. These high-performance materials, sometimes termed HMC or XMC, are being used in automotive and other demanding high-volume applications. Their further development and a rapid growth in their usage are expected.

Dough molding compound contains catalyzed polyester resin, up to 50% mineral filler and 15–25% short glass fibers, 3–12 mm in length. The usual filler is chalk though some resins are chemically thickened with alkali metal oxides. Shorter fibers and less glass than SMC means that DMC flows more easily and is thus used to make small articles of intricate shape by hot-press or injection molding.

Prepregs consist of sheets of glass-fiber reinforcement about 1 mm thick preimpregnated with catalyzed resin and ready for hot-press molding. The reinforcement is usually UD fibers or glass fabric, with epoxide rather than polyester resin and component thickness is achieved by laminating together several sheets in the mold. The glass-fiber content is high, typically 50–80% by weight. Accurate alignment of fibers is possible with prepregs, because there is minimal flow during molding and hence they are used mainly in high-technology applications. Prepreg tapes are also available for selective stiffening of complex moldings and for the fabrication of GRP tubes and vessels by a dry filament winding process.

3. *Sandwich Materials and Laminates*

Different types of GRP material are frequently combined together in laminates or with metals, wood or foamed plastics as sandwich materials. Glass fibers may also be combined with carbon or polyamide fibers in a hybrid composite (see *Hybrid Fiber–Resin Composites*). A composite structure may thus be built up with material properties tailored to design loads. A widely used GRP laminate consists of CSM combined with WR or UD fibers. The CSM–polyester plies improve the interlaminar shear strength and provide a corrosion barrier to the WR or UD fibers which carry the main loads. To simplify fabrication, glass reinforcement may be supplied in a combination mat, consisting of CSM and WR or UD fibers stitched together. With UD reinforcement such as prepreg, or in filament wound structures, fibers may be orientated in different directions through the thickness. A common construction is a cross-ply laminate, with alternate plies orientated at 0° and 90° to a fixed axis, or an angle-ply laminate with plies at $\pm\alpha°$ for some fixed angle α.

Glass-reinforced-plastic sandwich materials are used widely in panel applications where the main loading is flexural. An ideal sandwich panel should consist of thin stiff skins with a thick low-density or low-cost core. For low and medium technology applications, GRP skins of CSM–polyester or WR–polyester are combined with cores of end-grain balsa wood or foamed plastic such as PVC or polyurethane. A typical construction for a meter-square cladding panel might be CSM–polyester skins 3 mm thick with a balsa core 25 mm thick. For high-technology applications where weight saving is a priority, special low-density honeycomb core materials have been developed. These consist of a hexagonal cell structure arranged in a honeycomb pattern as shown in Fig. 1.

The main honeycomb core materials are aluminum foil and a thin resin-impregnated paper. High-performance GRP skin materials are used with honeycomb cores, such as glass fabric–epoxide and cross-ply or angle-ply UD-glass–epoxide laminates. A typical construction for a meter-square aircraft floor panel might be cross-plied skins 1 mm thick, with an aluminum honeycomb core 10 mm thick.

4. *Applications of Glass-Reinforced Plastics*

Glass-reinforced-plastic materials are used in low-, medium- and high-stress applications in industries as diverse as boat building, chemical plant, motor vehicles and aerospace. Important properties of these materials which have led to their wide usage are ease of fabrication into complex shapes, high strength-to-

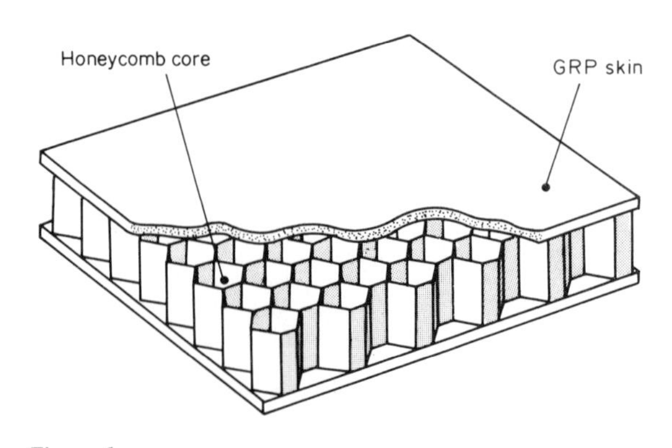

Figure 1
Schematic diagram of GRP/honeycomb sandwich material

weight ratio, excellent corrosion resistance, good weathering properties, low taint and toxicity for foodstuffs and good thermal and electrical insulation. The selection of a particular material is determined mainly by material properties, costs and the suitability of fabrication and quality-control procedures for use in a particular industry. For large-scale production automated fabrication methods based on filament winding, pultrusion and SMC press molding are used, with hand lay-up and semiautomated methods such as resin injection with preforms retained for smaller numbers of large moldings.

In low-stress applications such as cladding panels, shower cabinets, small boats and meter boxes, GRP materials act as space-filling panels, supporting their own weight but not subjected to any significant external loads. The main design requirements are a good surface finish, good weathering properties and the ability to withstand accidental damage. The most widely used material is CSM–polyester with fabrication by hand lay-up, spray-up or vacuum-bag molding. Sheet molding compound materials are used for longer production runs of larger moldings such as business-machine cases and cladding panels, and DMC for smaller components such as electrical switchgear, lampholders and meter boxes. Glass-reinforced-plastic sandwich materials containing balsa or foamed-plastic core with CSM–polyester skins are used for boats and for some of the larger panel applications.

In medium-stress applications, GRP materials may be subjected to significant loads of short duration or low sustained loads. Stress analysis is usually based on the short-term behavior of the material and product design may be governed by codes of practice or standards. Large design factors are imposed so that stresses in the material are often less than 10% of the short-term strength. Applications in this category include: pressure vessels and storage tanks, with associated pipework, for use in chemical plant and the food processing industry; larger marine applications such as workboats, small hovercraft, minesweepers and small submersibles; and land-transport applications such as rail coachwork and larger automotive moldings for car bumpers, body panels and truck cabs. Resins are usually polyester, reinforced by WR or UD fibers with CSM–polyester surfaces for improved weather and chemical resistance. Most of the vessels, tanks and the marine applications are fabricated by hand lay-up, with filament winding becoming increasingly important for chemical plant pipework, sewage pipes and cylindrical tanks and pressure vessels. Sheet molding compound is used for the volume production of such items as car bumpers, lorry cab panels and rail seat shells. Structural GRP sections are becoming increasingly important; for example, box beams, channels and tubes fabricated by pultrusion for use in frame structures, ladders, walkways, ducting, etc., and hollow rotationally-molded tapered sections for the volume production of GRP street lighting columns. Where low weight is required, as in storage tank covers, hovercraft body panels and rail carriage doors, GRP–foam sandwich materials are used with fabrication by hand lay-up or vacuum-bag methods.

The high-stress applications of GRP materials are in the aerospace and defense industries. Here, low weight is important and hence materials are highly stressed and used efficiently. Design factors are kept as low as 1.5 in some applications and detailed stress analysis is thus required. Typical applications include power boats, gliders, which are fabricated by hand lay-up using WR–polyester or UD-glass–polyester, and helicopter rotor blades and rocket motor casings, which are filament wound usually with epoxide resin. Sandwich materials consisting of angle-ply prepreg GRP skins with foam or honeycomb cores find use in lightweight structures such as aircraft flooring, radomes and radar antennae.

5. *Fabrication by Hand Laminating*

Hand laminating or contact molding is the traditional fabrication technique for GRP materials. It is the method used for about 50% of the UK and 30% of the US consumption of GRP materials. Glass-fiber reinforcement, in the form of CSM, WR, fabric or combination mat, is placed on a mold, impregnated with catalyzed resin and consolidated by hand rolling or brushing. The required thickness is built up by adding further layers of reinforcement and resin. Alternatively, chopped glass rovings are sprayed with a controlled quantity of resin from a special gun onto the mold and then consolidated by hand. The resulting "spray-up" material has a random distribution of fibers in the plane of the molding, giving similar mechanical properties to CSM–polyester. After an initial gelation stage, the GRP molding may be removed from the mold and left to cure, during which time the material attains its full mechanical properties. Curing takes about a week at 20°C, but is accelerated by postcuring at higher temperatures. Typical postcure times vary from 3 h at 80 °C to 30 h at 40 °C, and depend on the resin used.

Contact-molded laminates have a single molded surface usually protected by a 0.5 mm thick resin layer termed the gel coat, which is brushed or sprayed onto the mold before the main laminate. For additional protection, the gel coat may be backed by a thin glass surfacing tissue. The main advantage of hand laminating is its versatility. There is no size limitation and metal inserts or additional glass reinforcement can be added where they are needed. Because it is a low-pressure, ambient-temperature process, inexpensive molds of wood, plaster or GRP may be used, which makes it suitable for small production runs. The main disadvantages are that moldings usually only have one good surface and the process is labor intensive with quality dependent on the laminator. Control of glass content, thickness and the cure schedule is poor, which can lead to variability in mechanical properties in a single molding or between moldings in a batch.

6. *Automated Fabrication Techniques*

To eliminate the variability in hand laminating and to permit longer production runs, several automated and semiautomated fabrication techniques have been developed. The semiautomated methods are based on matched die molding which permits critical control of thickness and glass content, and produces a molding with two good surfaces. Cold-press molding makes use of relatively inexpensive tools in a hydraulic press with low pressures up to about 200 kPa. Glass-fiber reinforcement such as CSM is cut or preformed to shape and placed by hand in the mold with the required quantity of catalyzed polyester resin. The mold is closed and the resin allowed to cure for about 5 min. Two variations on cold-press molding are vacuum-bag molding and resin injection. In the vacuum-bag method, the matched molds, or an open mold, are sealed inside a rubber bag, which is evacuated so that atmospheric pressure is exerted over the surface of the molding. Resin injection or transfer molding involves the injection of catalyzed resin into the closed mold containing a preformed glass fiber mat. Injection pressures are up to 200 kPa and may be combined with vacuum assistance to improve resin flow. This fabrication method is becoming increasingly important, since it allows accurate placement of reinforcing fibers in the molding and is also suitable for the production of large moldings such as car body shells.

In hot-press molding, matched metal dies are used with a hot curing polyester resin. Pressures are 1 MPa and temperatures in the range 100–130°C, with a mold cycle time of 2–5 min. Because of the higher pressures, entrapped air is forced out of the mold giving lower void content than with cold-press molding. In conventional hot-press molding, the operator must still handle the glass fibers and the liquid resin, but the process can be made cleaner by using molding compounds. These have a high viscosity and thus require higher molding pressures, typically 1–7 MPa, with temperatures in the range 120–170 °C. Prepregs are cut accurately and each ply is positioned in the mold to have the required orientation. Because of the short fibers SMC, and more particularly DMC, are able to flow into the mold. With these materials, a weighed charge is positioned by hand centrally in the mold and on closing flows into the mold causing significant fiber reorientation. Sheet molding compound has the better mechanical properties and is used in larger panel applications, with DMC retained for smaller, bulkier components. The use of in-mold coatings to improve surface finish and fast curing resins giving cycle times as low as $1\frac{1}{2}$ min, has meant that SMC is now widely used for the volume production of automotive components such as body panels and bumpers. Because of the good flow characteristics. DMC components may also be fabricated by injection molding.

Fully automated methods have been developed for the large volume production of GRP pipe, sheet and rod or tube stock. In the filament winding process, glass-fiber rovings are fed under tension through a resin bath onto a revolving mandrel. Several layers of reinforcement, with the same or different winding patterns, may be placed on the mandrel in a controlled way. The mandrel is removed after curing, which usually takes place at elevated temperatures. A variation, which eliminates the resin bath, is dry winding using UD prepreg tape. Filament winding is used for GRP pipes and cylindrical tanks and enables high glass contents to be achieved with close control of fiber orientation. In order to improve chemical resistance, the glass fiber reinforcement may be wound over a chemically resistant liner of CSM–polyester or a thermoplastic such as PVC or polypropylene.

Continuous laminating processes have been developed for the manufacture of GRP sheet for applications such as electrical insulating panels, corrugated panels and translucent roof sheets. A moving sheet of CSM or WR reinforcement is impregnated with catalyzed polyester resin and sandwiched between layers of plastic film. This is then passed continuously through a die, to impart the required shape, and a curing oven before being cut into lengths. Pultrusion is a similar process suitable for the production of long lengths of GRP in such forms as rods, tubes or I beams. In pultrusion, glass rovings are impregnated with resin from a bath or by injection and pulled through a shaped, heated die. Other forms of glass reinforcement such as WR or continuous-filament mat may also be pulled through the die with the UD rovings to improve the transverse mechanical properties of the pultruded section. Another recently developed automated process is rotational molding, which is used for hollow lighting and telegraph poles. Here the laminating pressure is derived from centrifugal force in the rotating mold. Methods are also being developed to automate hand lamination, for example, by machine-controlled spray-up.

Automated fabrication techniques are suitable for volume production of good quality components through careful control of glass and resin contents, fiber orientation and the cure cycle. Many of these fabrication methods are suitable for control by microprocessor. which can lead to further cost savings and help overcome the styrene emission problem by eliminating manual operations. The main disadvantage is the high capital cost of the processing equipment. Some of the versatility of GRP materials is also lost, since these methods produce GRP stock in the form of pipes or sheets which require bonding or jointing to make into a component, whereas contact or press molding are often integral GRP structures.

7. *Glass-Reinforced-Plastic Material Properties*

On loading a GRP material such as CSM–polyester in a short-term tensile test, it is found that the behavior is approximately linearly elastic to failure, which occurs by a brittle fracture at a strain of 1.5–2.5%. The short-term mechanical properties may thus be characterized by an elastic modulus and a failure stress or strength. A more detailed study of failure (see *Strength of Composites*) shows that there is progressive damage to the material at stresses below the ultimate. The first sign of damage under tensile loading is transverse fiber debonding, that is, separation takes place between the resin and those fibers perpendicular to the load direction. In CSM–polyester and WR–polyester, this occurs typically at 0.3% strain and at a stress level of about 30% of the tensile strength. Under increased loading, the debonds initiate resin cracks and as these cracks spread, more of the load is transferred to the fibers which eventually fracture or pull out at ultimate failure. Resin cracking occurs at approximately 50–70% of the tensile strength and is observed as crazing and by a change of slope in the stress–strain curve. In this article, ultimate failure stress data are given although it should be noted that, for many applications, failure data based on alternative criteria such as resin cracking or fiber debonding may be necessary. Limited data of this type are given, for example by Johnson (1978), or alternatively a suitable design factor may be applied to the ultimate strength data.

With a knowledge of the appropriate moduli and strengths, elastic design formulae may be used for stress and deformation analysis of GRP materials under short-term loads. For CSM–polyester and SMC materials, conventional isotropic design formulae may be used, while for WR, fabric or UD reinforcement, account must be taken of the anisotropy in stiffness and strength properties arising from fiber orientation. For GRP materials, it is not usually possible to quote a single value of modulus and strength because of the inhomogeneous structure of the material. Thus, in addition to the effects of anisotropy, mechanical properties also depend on the quantity and type of glass reinforcement and on the fabrication method used.

Comparative short-term mechanical property data are summarized in Table 1 for the principal GRP materials and typical physical properties are given in Table 2. The tables are abstracted from Johnson, who has collated material property data on several types of GRP material from UK resin and glass fiber suppliers, GRP fabricators and research laboratories. Table 1 gives mean values of modulus and strength for DMC, SMC, CSM–polyester, WR–polyester and UD-glass–epoxide at a typical glass content for each type of material. Below each mean value approximate bounds are given for the quantities. These bounds are usually wide because they refer to a full range of glass weight fractions, to different resins and glass fibers, fabrication techniques and test methods.

Reference to Table 1 shows that DMC, SMC and CSM–polyester, which all contain chopped fibers at low glass contents, have corresponding low moduli and strength values. The higher moduli and lower strength values for the molding compounds, compared with CSM–polyester at similar glass contents, are explained by the mineral filler which enhances the modulus but lowers the strength of the molding compound. Higher glass contents are found in WR–polyester laminates with average tensile modulus and strength about twice as high as the corresponding values for CSM–polyester. The data given in Table 1 refer to a fiber direction in a plain weave WR, with an equal number of rovings in the warp and weft directions. The highest glass contents are found in UD-glass–epoxide composites and these materials show significant anisotropy in properties. Average tensile moduli range from 42 GPa along the fibers to 12 GPa transverse to them, with corresponding tensile strengths of 750 MPa and 50 MPa. Data on other GRP materials lie within the spectrum of properties shown in the table. Glass fabric reinforced resins and

Table 2
Typical physical property data for GRP materials

	SMC and CSM polyster	Fabric and WR/ polyester	UD-glass/ epoxide
Density (kg m^{-3})	1300–1600 (SMC, 1600–1900)	1500–1900	1800–2000
Linear thermal expansion coefficient (10^{-6} K^{-1})	18–35	10–16	5–15
Thermal conductivity (W m^{-1} K^{-1})	0.16–0.26	0.2–0.3	0.28–0.35
Specific heat (J kg^{-1} K^{-1})	1200–1400		950
Maximum heat distortion temperature (BS2782, 102 G) (°C)	175	250	300
Dielectric strength (BS2782, 201 A, C) (kV mm^{-1})	9–12	13–16	
Permittivity at 1 MHz (BS2782, 207 B)	4.3–4.7	4.1–5.2	
Power factor at 1 MHz (BS2782, 205 B)	0.015	0.016	
Dry insulation resistance (MΩ) (BS2782, 204 A)	10^6	10^6	

UD-glass–polyester, fabricated by pultrusion or filament winding, usually have glass contents and mechanical properties intermediate between the WR–polyester and UD-glass–epoxide values shown. Other points to note from Table 1 are the low interlaminar shear strength of all GRP materials and the differences observed between tensile and flexural properties, with flexural modulus usually below, and flexural strength above, the tensile values, as a result of the inhomogeneous structure of the material.

Under long-term loads, fatigue loads and when subjected to high temperature or a wet environment, GRP stiffness and strength properties are below the short-term values given in Table 1 (see *Fatigue of Composites*; *Nonmechanical Properties of Composites*)

8. Calculation of GRP and Sandwich Properties

A wide range of theoretical formulae are available for predicting the mechanical properties of composite materials in terms of microstructure and fiber and matrix properties (see, for example, Hull 1981). Although these methods give satisfactory predictions of GRP properties, particularly moduli, they are usually only applicable to UD fiber reinforcement. The more rigorous results are also difficult to apply and often require correlation coefficients to be determined. It is therefore of interest to consider a simpler "rule-of-mixtures" approach which accounts for the main trends in the GRP properties reported here. Let E denote the Young's modulus of the composite, and E_g and E_r the Young's moduli of the glass fibers and resin matrix. If v_g, v_r are the volume fractions of glass fiber and resin in the composite then the Halpin–Tsai equation may be written in the modified form

$$E=\alpha E_g v_g + E_r v_r \tag{1}$$

where α is a parameter which depends on the efficiency of the reinforcement. Suitable values are $\alpha=1$ for UD reinforcement, $\alpha=0.5$ for WR, with E then referring to the modulus along the fiber directions, and $\alpha=0.3$ for CSM, which has a random fiber distribution. Taking $E_g=75$ GPa, $E_r=3$ GPa as typical modulus values for glass fiber and polyester resin and assuming fiber volume fractions v_g are 0.2 for CSM laminates, 0.35 for WR laminates and 0.5 for UD fibers, the following estimates for the GRP modulus are obtained from Eqn. (1): $E=6.9$ GPa for CSM–polyester, $E=15.1$ GPa for WR–polyester and $E=39$ GPa along the fibers in UD-glass–polyester. Reference to Table 1 shows that these predicted values agree well with typical GRP data.

An analogous expression to Eqn. (1) may be used for estimating GRP strength properties, in the form

$$\sigma=\alpha\sigma_g v_g + \sigma_r v_r \tag{2}$$

Here σ is the tensile strength of the composite, σ_g that of the glass fiber and σ_r the tensile stress in the resin at the fiber failure strain. Because of damage, the strength of a glass fiber in a composite may be only half the fresh drawn fiber strength (Parkyn 1970, Chap. 10). Taking as a nominal fiber strength $\sigma_g=1500$ MPa, with $\sigma_r=30$ MPa and the v_g, v_r values used above, Eqn. (2) gives the following estimates of GRP tensile strength properties: $\sigma=114$ MPa for CSM–polyester, $\sigma=278$ MPa for WR–polyester and $\sigma=765$ MPa along the fiber direction in UD-glass–polyester. Again, these values are seen to be in reasonable agreement with measured data.

The mechanical property data and the prediction methods described above refer to GRP materials containing a single type of reinforcement. In practice, GRP structures often contain several types of reinforcement laminated together or consist of angle-ply or sandwich materials, and it is not feasible to provide detailed mechanical property data on each material combination. However, theoretical formulae are available which enable the designer to calculate the stiffness and strength properties of combined materials, and these have been found to be in good agreement with measured properties. Laminated plate theory permits the calculation of the full anisotropic stiffness and strength properties of laminated composite materials, consisting of plies of different materials or the same material oriented in different directions (see *Elastic Properties of Laminates*). The analysis method is rigorous and requires a complete knowledge of the anisotropic stiffness and strength properties of each lamina. The calculations are lengthy and best carried out by a small computer program. A number of inexpensive microcomputer programs for laminate analysis are now commerically available (see, for example, ESDU (1985, 1987), Tsai (1987)). A simpler, strength of materials method for estimating combined laminate tensile and flexural properties is described by Johnson (1978) and by Allen (1969) for calculating the properties of sandwich structures. When the flexural rigidity of the core material is negligible compared with that of the stiffer skin materials, a simple design formula may be derived for the flexural rigidity per unit width, D, of a sandwich panel, which is an important quantity for the design of sandwich structures. For a symmetric sandwich with thin skins

$$D=\tfrac{1}{2}Eh(h+c)^2 \tag{3}$$

where E is Young's modulus of the skin material in the load direction, h is the skin thickness and c the core thickness. Similar design formulae are available for calculating strength properties and the minimum weight design of sandwich panels (see, for example, Johnson and Sims 1986).

See also: Glass-Reinforced Plastics: Thermoplastic Resins

Bibliography

Allen H G 1969 *Analysis and Design of Structural Sandwich Panels*. Pergamon, Oxford

Anon 1981 Materials survey. *Mod. Plast. Int.* 11(1): 34–35
ESDU 1985 *Software for Engineers 2022, Stiffness and Properties of Laminated Plates.* ESDU International, London
ESDU 1987 *Software for Engineers 2033, Failure of Composite Laminates.* ESDU International, London
Holmes M, Just D J 1983 *GRP in Structural Engineering.* Elsevier, London
Hull D 1981 *An Introduction to Composite Materials.* Cambridge University Press, Cambridge
Johnson A F 1978 *Engineering Design Properties of GRP.* British Plastics Federation, London
Johnson A F, Sims G D 1986 Mechanical properties and design of sandwich materials. *Composites* 17(4): 321–28
Parkyn B (ed.) 1970 *Glass Reinforced Plastics.* Iliffe, London
Rubber and Plastics Research Association 1977 *Survey of Fibre-Reinforced Plastics Markets and Needs.* RAPRA, Shawbury, UK
Tsai S W 1987 *Think Composites Software: User Manual "Composites Design,"* 3rd edn. Think Composites, Dayton, Ohio
Whitney J M 1987 *Structural Analysis of Laminated Anisotropic Plates.* Technomic, Basel

A. F. Johnson
[DFVLR, Stuttgart, Germany]

H

Halpin–Tsai Equations

The Halpin–Tsai equations constitute a method of estimating the elastic moduli of an aligned linearly elastic orthotropic composite where it is assumed (a) that the fibers are parallel to axis 1 in a matrix; (b) that the Young's modulus in this direction is given by

$$E_{11} = E_f V_f + E_m V_m$$

(where E_f, E_m are the respective Young's moduli of the fibers and matrix, and V_f, V_m are the respective volume fractions); and (c) that the principal Poisson's ratio is

$$\nu_{12} = \nu_f V_f + \nu_m V_m$$

(where ν_f, ν_m are the respective Poisson's ratios of the fibers and matrix). Then the composite moduli E_{22}, G_{12} or G_{23} can be estimated from the Halpin–Tsai equations,

$$\frac{\bar{p}}{p_m} = \frac{1 + \xi \eta V_f}{1 - \eta V_f}$$

where

$$\eta = \left(\frac{p_f}{p_m} - 1\right) \bigg/ \left(\frac{p_f}{p_m} + \xi\right)$$

Here $\bar{p}$ is the composite modulus E_{22}, G_{12} or G_{13}; p_f, p_m are the corresponding fiber and matrix moduli E or G; and ξ depends on the geometry of the fiber arrangement.

See also: Fibrous Composites: Thermomechanical Properties

Bibliography

Ashton J E, Halpin J C. Petit P H 1969 *Primer on Composite Materials*. Technomic Publishing, Westport, Connecticut

A. Kelly
[University of Surrey, Guildford, UK]

Helicopter Applications of Composites

Modern helicopter design depends on the successful combination of materials to fulfill a very wide range of critical functions. In the past few years, demands from the aerospace industry have resulted in considerable enhancement of the favorable properties of the fiber-reinforced polymer-based materials generally known as composites. These materials exhibit properties of high strength, toughness and low weight, and maintain these properties despite the application of extremes of climatic conditions, vibration and loading. The primary objectives for pursuing the development and application of composite materials to helicopters include improved airworthiness and safety, reduced empty weight fraction, reduced purchase and operating costs and improved operational performance capability.

1. Composite Materials

The most commonly used fiber types used in composite materials are glass, Kevlar and carbon, although boron fibers have been used successfully in rotor blades. The fibers are embedded in a matrix material which is generally an epoxy resin, although for special requirements polyimides and thermoplastics may be used.

Epoxy resin offers good interlaminar shear strength, and, given suitable surface preparation of the fibers, effectively wets the fiber tows resulting in high fiber-to-resin ratios. The choice of fiber results from their main characteristics: glass is low in cost, Kevlar is tough and damage tolerant, and carbon has very high strength and stiffness. It is not uncommon to have combinations of fibers in a structure either by use of hybrid materials or by combining material types at the manufacturing stage.

The aerospace industry uses material in the form of prepreg, in which the fiber in unidirectional or woven form is embedded in the uncured resin and hardener mix. The thin prepreg material sandwiched between a backing sheet and protective film is available in reels of several widths. It must be stored at low temperatures, typically −20 °C, and has a useful shelf life of about 6 months. In use, the prepreg is unreeled and cut to shape, the protective films are removed, and the prepreg parts are stacked in a mold to the desired thickness. Care is taken to avoid contamination, trapped air pockets and wrinkles in the fibers.

The component is then cured under carefully controlled conditions of temperature and pressure. The advantage of a prepreg system is that the material properties are reproducible. The requirements of the designer in placing load-carrying material only where it is needed can be realized in manufacture by orienting the prepreg in the mold tool, thereby achieving a high-performance lightweight structure.

Composite materials can offer a better solution to the strength, stiffness and fatigue requirements of helicopter structures, but there are a number of disadvantages which require consideration by the designer. Composite materials are generally electrical insulators, and this is a problem in the case of lightning strike. Special prepregs incorporating metal-coated fibers, for example aluminum-coated glass, nickel-coated carbon or stainless steel fibers, are used in areas at risk.

The low electrical conductivity also poses problems in electromagnetic shielding. Modern helicopters, particularly military types, have sophisticated electronic systems on board which are susceptible to interference from electromagnetic radiation. The conventional aluminum structure provides a natural shielding from interference effects. Special measures taken for composite structures include the bonding of metal foils on the interior surfaces of composite structures to provide some shielding. The same technique can provide a ground plane for aerials associated with the electronic systems.

Composites have considerable corrosion resistance, but carbon and aluminum are widely spaced on the electrolytic table. Aluminum fasteners or components must be replaced by stainless steel or titanium varieties if corrosion problems are to be avoided.

2. Applications

The helicopter industry was quick to react to the possibility of using composite materials for rotor blades—examples were flying in the 1960s. The attractiveness of composite materials stemmed initially from the enhanced strain and fatigue performance available, but it was quickly recognised that the versatility of the manufacturing processes for composite components allowed substantially improved aerodynamic designs to be introduced. Subsequently the application of composites spread to all the main load-carrying structures of a modern helicopter. Figure 1 is representative of a medium transport helicopter of 9000–13000 kg all-up-weight. Composite materials have been applied successfully in three main areas: main and tail rotor blades, rotor hubs and airframe.

By the late 1980s, composite materials accounted for 25–30% of the total helicopter structure, although aircraft were in development with significantly higher proportions of composite materials mainly due to all-composite airframe structures.

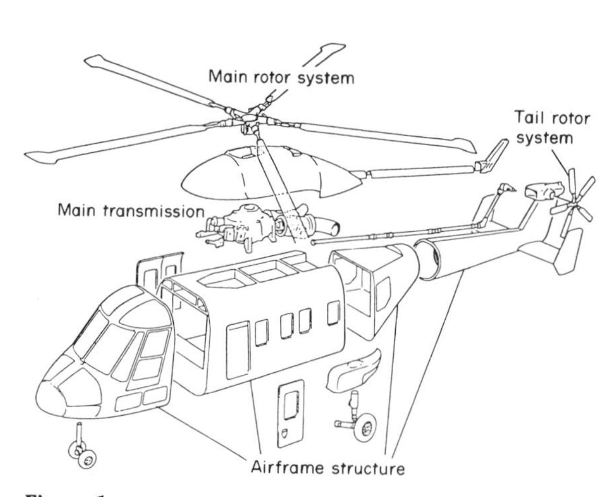

Figure 1
Helicopter load-carrying structures

2.1 Rotors

The main rotor blade is a component of the helicopter that has benefited considerably from the application of new materials. Most people have an appreciation of the functions that a helicopter can perform but few have an understanding of the principles of its operation, which relies on the rotor which lifts, propels and controls the helicopter.

The main rotor is driven at a constant speed of rotation, with a tip speed of approximately 200 m s^{-1}. To hover, the same pitch is applied to each of the blades such that lift is generated to balance the downward loads. This is known as collective pitch. When flying forward the blade on the advancing side experiences a high relative airspeed and hence increased lift, while the blade on the retreating side experiences low airspeed and thus reduced lift (Fig. 2). To balance the lift forces, the pitch of the blade must be varied during each revolution as the blade moves around the rotor azimuth from advancing to retreating conditions. Further pitch changes are introduced effectively to tilt the rotor disk to generate propulsion and control forces. These pitch changes are known as cyclical-pitch variations.

These cyclical-pitch and velocity variations result in changes in the aerodynamic conditions and therefore blade loadings to which the blade is subjected. Compressibility problems on the advancing side of the blade and stall problems on the retreating side may also be encountered. The cyclical forces are extremely demanding from the viewpoint of fatigue, particularly considering the need to cater for near infinite life (in excess of 40 000 h for a civil helicopter). The blade material must have good fatigue performance; the

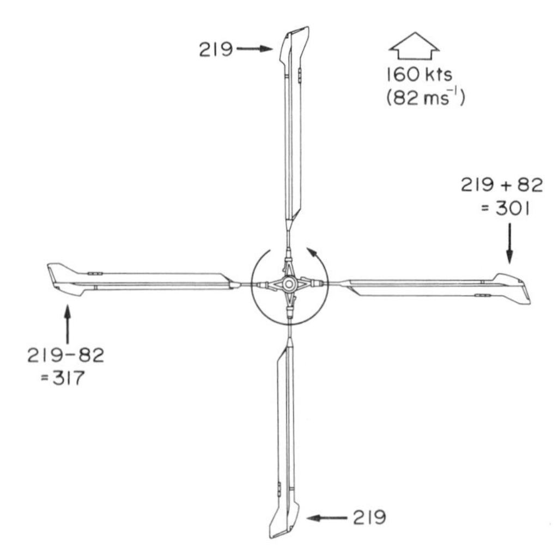

Figure 2
Relative airspeed (m s^{-1}) normal to blade tips at a forward speed of 160 knot

structural requirements of a rotor blade are best met by glass-fiber and/or carbon-fiber composites.

A typical blade construction, Fig. 3, has three major components: the spar, which carries the great majority of the loads; the erosion shield; and trailing edge, carrying some load but primarily completing the aerofoil shapes. In this example the trailing edge construction consists of green Nomex honeycomb to which precured glass (one layer of 45° woven material) and carbon (four layers of 45° unidirectional material) skins are bonded in an assembly process.

The spar is of the modular D type with walls constructed principally of a preimpregnated glass and carbon unidirectional fiber hybrid sandwich with inner and outer wraps of 45° unidirectional carbon and one outer layer of glass fiber; it contains a foam core which is used in the molding process to provide consolidating pressure.

Torsional stiffness is derived from the 45° carbon in the spar (and in the trailing edge), flap stiffness from the unidirectional glass and carbon in the spar, and lag stiffness also from the unidirectional material in the spar.

Future main rotor-blade developments for application into the 1990s and beyond include the concept of aeroelastically tailored blade designs and "one-shot" fully automated manufacturing processes.

The aeroelastic concept is one which utilizes blade structural and material properties to passively control blade motion. Motion is controlled, principally in torsion, as the blade passes around the rotor azimuth such that, for example, aerodynamic performance is maximized, rotor blade stresses are reduced or vibration is reduced. Passively controlled behavior is likely to require radially distributed aerofoil sections/advanced tips in combination with highly asymmetric blade cross sections and complex hybrid fiber-reinforced composite-material lay-ups. One-shot automated manufacturing processes are required to further reduce manufacturing costs, reduce component lead times and provide precise control and consistency of quality.

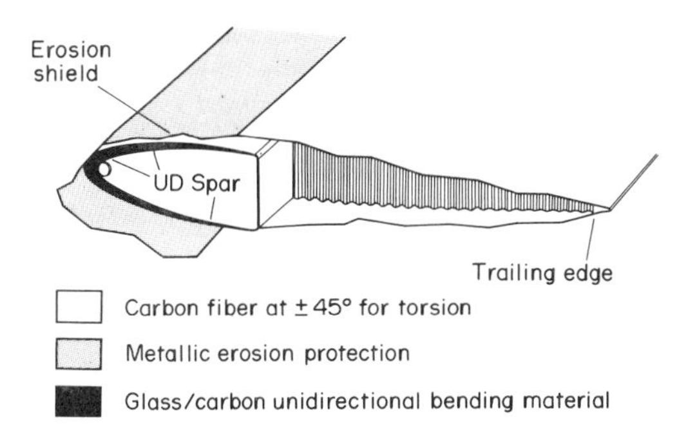

Figure 3
Composite main rotor blade: general section

The new material options potentially available for the next generation rotors are glass, carbon and Kevlar fiber composites employing either toughened epoxy or thermoplastic matrix systems.

2.2 Airframe Structures

The helicopter airframe structure comprises a forward lower fuselage and cockpit canopy, a center structure/main cabin, a rear fuselage and tailboom, tail pylon and tailplane.

There are a myriad of design issues that impact the airframe structure and the relative importance of these issues varies across the main structural areas. Amongst the design issues are the need for minimizing weight cost and drag, robustness, crashworthiness, electrical integrity (ground plane, antenna acceptance and lightning conductivity) EMC (electromagnetic compatibility) integrity and acceptable acoustic and structural dynamic properties. Military helicopter structures also need to exhibit low radar and visual signatures, ballistic damage tolerance and nuclear-, biological- and chemical-weapon damage tolerance.

The majority of helicopter manufacturers predominantly have used aluminum alloys for the primary load-carrying airframe structural items, and glass-reinforced epoxy composites for secondary and tertiary structural applications (e.g., fairings and access panels).

More recent designs, for example the Westland–Agusta EH101, use composites extensively for both primary and secondary structures. A particularly interesting application is the cockpit glazing structure of the EH101. The geometry of the glazing structure is essentially double curvature, and this requirement, together with that for maximum visibility, suggested that a composite construction would offer considerable benefits. It was rapidly established that substantial savings of up to 15% in weight, material costs, production and tooling costs would result from a composite design, and additionally a smoother aerodynamic profile would reduce drag.

The aerodynamic loadings on the structure were known to be insignificant when compared to the birdstrike requirements. The glazing structure was required to stop a 2 kg bird travelling at 80 m s^{-1} from penetrating the cockpit and disabling the aircraft.

The poor impact properties of carbon-fiber-reinforced epoxy precluded the use of carbon-fiber composites. When subjected to the energy levels of a birdstrike, a carbon-fiber-composite glazing bar shatters into fragments at the point of impact. Kevlar composite has good impact properties resulting from the toughness of the Kevlar fiber. The resulting design used hybrid materials employing carbon fiber for its stiffness properties and Kevlar fibers for impact resistance. The glazing bars are a hollow trapezoidal cross section constructed from woven Kevlar prepreg, with carbon prepreg tows concentrated in the four corners. Kevlar-composite-faced honeycomb sandwich panels are bonded to the ribs to complete the structure, and the glass/polycarbonate glazing panels are bonded to the glazing bars.

2.3 Rotor Hubs

Rotor hubs fabricated from metals utilize complex mechanisms and bearing systems to accommodate the main motions of the blade root-end, i.e., flap, lead-lag and pitch changes. Such mechanisms are necessary because of the essentially isotropic nature of the materials commonly used. Composite materials, on the other hand, may be designed with significant anisotropy because the type, amount and orientation of the fibers in the composite material can be selected at will. The designer has control of properties such as the ratio of bending to torsional stiffness and this allows designs for rotor hubs in which elastic deformation can accommodate the necessary blade motions.

The concept of the bearingless main rotor hub represents a major step in the evolution of rotor systems. The complete absence of hinges offers a number of benefits: reduced maintenance, reduced aerodynamic drag, reduced weight, reduced parts count, improved damage tolerance and increased overhaul/replacement lives. Reduction of both manufacturing and operating costs is achieved. Bearingless rotor research has identified a number of potential designs, bounded by two fundamentally different approaches. First, the design approach employing established materials (i.e., glass-fiber epoxy composites); this approach necessitates flexures of complex cross section. Second, the design approach employing new materials enabling a more simple or elegant cross section to the flexure.

Bearingless hub systems for either tail or main rotors have been developed by most of the major helicopter manufacturers including Boeing Vertol, Sikorsky, MBB and Aerospatiale, whose Starflex rotor head eliminates 97 bearings, seals and lubricators. The main rotor hub for the Agusta–Westland EH101 (Fig. 4) represents a state-of-the-art articulated system employing composite materials (i.e., glass- fiber-reinforced epoxy) and elastomeric bearings. The elastomeric bearings carry blade tension loads while allowing flap, lag and torsional motion. The centrifugal load from the elastomeric bearings is carried by the hub plate, which is a composite structural plate fitted to the mast by a splined metallic core.

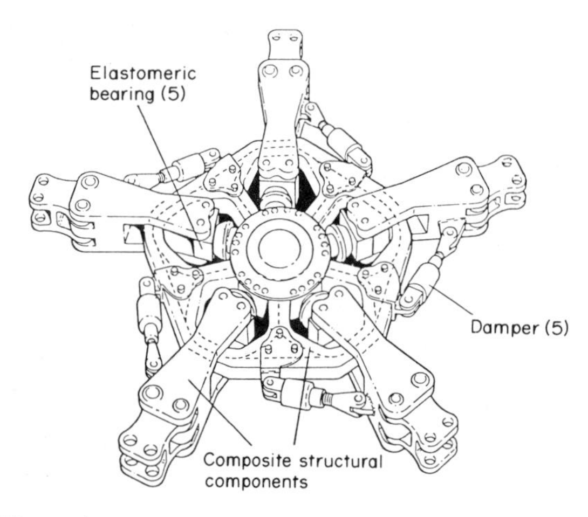

Figure 4
Agusta–Westland EH101 main rotor hub

A tension link incorporating the blade folding system attaches the blades to the hub and comprises top and bottom plates bonded to and sandwiching, a metallic frame. Maximization of safety has been a basic design aim with the hub; in particular, the composite hub has load path redundancy and the composite materials provide low rates of crack propagation.

3. Operational Considerations

Before a new material can find widespread application, its properties need to be fully determined; that is, it must be characterized. This requirement contributes very significantly to the apparently long gestation period of perhaps 10–20 years between the invention of a material and its general use. Some of the factors of vital importance which need to be fully understood include the variability in basic material properties, the variability which may result from associated manufacturing processes and the behavior of a material when subject to long-term usage throughout its required operational environment.

The increasing number of new materials available place much pressure on the procedures for substantiation and these pressures are compounded by the commercial need for multisource supplies.

Selection of unique materials is prohibitively expensive and involves high commercial risk. Second sourcing is strategically essential and not just good commercial practice.

An integrated multidisciplinary management and engineering approach is necessary to develop and apply composite materials if the commercial potential is to be fully exploited. Straightforward material substitution is not usually the best policy, since the use of composites often dictates a molding process with a material which is anisotropic. The material properties differ substantially in different axes, and indeed the relationship between the axes can be chosen by the designer. Thus the designer has additional design freedoms. There may, however, be additional constraints imposed by difficulties in manufacturing components economically. The low material cost and high conversion cost of metal components contrast with the high-material-cost composites which demand a low conversion cost.

4. Design

A particular benefit of composite materials is the greater freedom of design allowed by the molding process. To capitalize on the structural benefits available from this freedom a significant investment in sophisticated computer-based design tools is necessary. Extensive mechanisation of the design process is a definite advantage enabling rapid design iterations, and design integration with manufacturing.

The complicated profiles of "molded" designs stimulate the use of powerful computer-aided design techniques. Using the very complicated profile of a rotor blade as an example, the first step is to define the blade sections at a number of spanwise locations. The sections are then patched radially using a variety of curve types which serve to define the blade geometry by a grid of some 2000 data points. Given the blade geometry it has proved to be a very simple and rapid process to transmit the information to CAD centers. At this stage the detailed material lay-up information is specified and the various metal fittings can be drawn and dimensioned.

The blade has to be aerodynamically defined, drawn, stressed (with the input of substantial aeroelastic information) and, if necessary, redrawn, restressed, etc. Substantial strides have been taken by the industry to develop interlinking routines in which data can be freely transmitted from one specialized area to another, using common formats and coordinate systems.

5. Manufacturing Engineering

High-performance composite structures must be manufactured using metal mold tools. For a component such as a main rotor blade, the mold tool can be 15 m long! This demands a substantial tool-making capability. Since the blade geometry is defined, there has been a natural merging of the traditional lofting and jig and tool functions. This integration of design and tooling functions is extended by the direct computer linking of the tooling planner to the CNC machine.

Thus a multidisciplinary approach to design and manufacture is essential. Mechanizing the interface between design and manufacture for tool and process development and production is a definite advantage.

6. Manufacture

Manufacturing components from composite materials is labor intensive, difficult from a quality point of view and employs high-value plant (presses, tools, autoclaves, cleanrooms) on low-cycle usage. Automation is attractive for even low-volume component production. Manufacturing costs and lead times are reduced and product quality and consistency are improved by automation. The high temperature and pressure processes applicable to composites require particular attention to tool design to ensure that component geometry and internal stresses are properly controlled.

Design-to-cost techniques are particularly appropriate to composite production. A cost-effective product in composite materials can only result from a program which interactively considers the performance criteria, design, technology, materials and manufacturing techniques which are available.

Bibliography

Agarwal B, Broutman L 1980 *Analysis and Performance of Fiber Composites*. Wiley, New York

Cassier A, Siefler J C 1978 Trends in helicopter head design. In: *12th Int. Helicopter Forum*. SNI Aerospatiale, Paris

Weatherhead R G 1980 *FRP Technology*. Burgess–International Ideas, Englewood, New Jersey

White R W 1981 Composite technology in the UK helicopter industry. *J. Am. Helicopt. Soc.* 26(4): 34

D. Holt,
[Westland Aerostructures, East Cowes,
Isle of Wight, UK]

High-Modulus High-Strength Organic Fibers

High-modulus high-strength organic fibers are polymeric fibers characterized by high specific tensile modulus and tensile strength when compared to metals and ceramics, as illustrated graphically in Fig. 1. These fibers, especially the du Pont aramid fiber, Kevlar, have found utility in a broad range of composite and specialty applications from secondary aircraft structure to antiballistic vests. All share a serious deficiency in low compressive strength, which precludes their use in composites for most primary-structure applications and requires well-designed composite assemblies for many others. In applications which require minimum weight and low fiber thermal and electrical conductivity, however, high-modulus organic fibers are often the material of choice.

1. Historical Background

From the late 1960s through the mid-1970s a series of announcements from US industrial laboratories (du Pont, Monsanto, Eastman, Celanese) made it apparent that the level of achievable fiber tensile properties had taken a major jump—tensile modulus up by a factor of about 5, tensile strength up by a factor of 2–3—when compared to the commodity industrial fibers such as PET or nylon. All of these commodity fibers were composed of stiff, highly aromatic molecules and were characterized by very high axial molecular orientation in the solid state (orientation function >0.95). These discoveries constituted a major milestone in fiber and polymer technology.

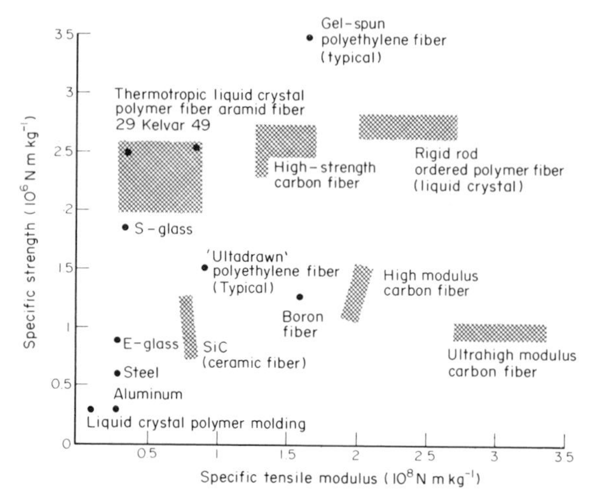

Figure 1
Specific tensile modulus and specific strength for various fibers

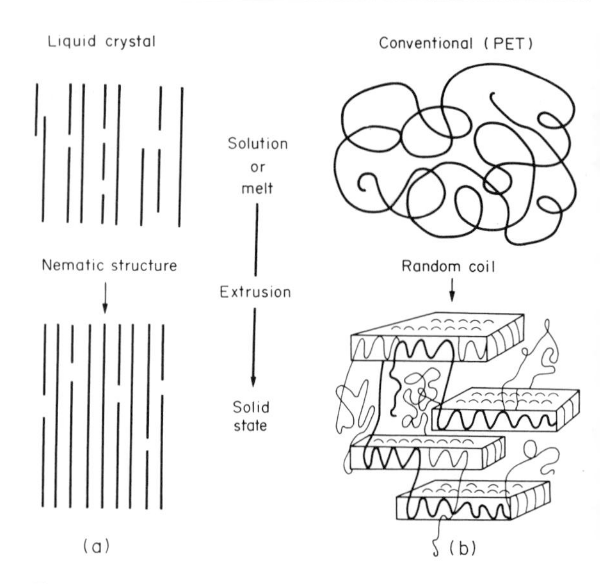

Figure 2
Formation of: (a) liquid-crystal fibers with extended chain structure, high chain continuity and high mechanical properties, (b) conventional (PET) fibers with lamellar structure, low chain continuity and low mechanical properties

As exciting as the discovery that fibers with tensile properties approaching their theoretical values could be spun was the recognition that many of the stiff molecules exhibited nematic liquid-crystalline behavior in solution or in the melt. The nematic behavior of these stiff polymers leads to low elongational viscosity (ease of orientation) and a solid-state structure similar to a highly oriented assembly of closely packed rods (extended-chain structure). Figure 2 illustrates the processing of these polymers in contrast to a conventional polymer such as PET.

In the mid-1970s it was shown that a variety of essentially paralinked aromatic polyester copolymers exhibit nematic behavior in the melt. These thermotropic copolyesters could be easily melt-spun into highly oriented fibers of high tensile modulus and high tensile strength.

The potential for both high modulus and liquid crystallinity in linear synthetic polymers had been predicted well ahead of discovery, the former by Mark in 1936 and the latter by Flory in 1956. Mark calulated an upper modulus limit of about 350 GPa. Although it has taken forty years to reach them, values as high as 300 GPa have now been realized in continuous fiber.

Flory, using lattice-model-generated phase diagrams for monodispersed rigid rods in solution, accurately showed the type of phase relationships observed for lyotopic nematic polymers such as poly (p-phenylene terephthalamide) (PPT) or poly (p-phenylene benzo-bis-thiazole) (PBZT). He also showed that liquid-crystalline behavior was possible for the melt of a stiff-chain polymer.

Polyethylene fibers, carefully processed to minimize internal defects and drawn slowly to very high molecular orientations, are being commercially produced by the Allied Signal Corporation, and others, to mechanical property levels fully competitive with the fibers based on liquid-crystalline polymers. Attempts to extend this technology to other conventional polymers has not proven successful.

It is now evident that two generic routes to high-modulus high-strength fibers exist, namely (a) the spinning of stiff, nematogenic polymers which easily form highly oriented structures in the solid state, and (b) the morphological manipulation of flexible, conventional polymers into highly oriented extended-chain fibers through complex processing.

Two useful subclassifications are the "almost mesogens" such as the aramid/hydrazid copolymers developed by Monsanto, and the true molecular rods such as PBZT championed by the USAF. The former polymers are "conventional" in quiescent solution but exhibit nematic-like behavior under elongational flow. The latter are characterized by a persistence length in solution which is equivalent to the fully extended length of the molecular chain.

2. Chemical Structures and High-Modulus Fiber Formation

The unifying characteristics of stiff polymers yielding high-modulus fibers are: a high degree of aromaticity, high planarity, and essential linearity in the chain backbone. Most common are paraphenylene or 2,6-naphthylene moieties linked by ester or amide bonds. Even stiffer structures can be produced by incorporating heterocyclic rings in the chain backbone such as bibenzooxazoles or bibenzothiazoles. Examples of such structures are shown in Fig. 3, grouped by mesogenic behavior. The heterocyclic polymers suffer in general from intractibility: they are infusible and exceedingly difficult to dissolve, making fabrication an arduous task. Alternatively, only a minor portion of backbone aromatic moieties can have other than linear alignment if high modulus is to be obtained.

The rod-like nature of para-oriented aromatic polymers, in addition to making the polymer difficult to dissolve or melt, dramatically increases the viscosity of a solution or melt over that of nonrigid structurally isomeric polymers such as poly (m-phenylene isophthalamide), the meta-analogue of poly (p-phenylene terephthalamide) (Kevlar). The solution to this problem is either to spin from very dilute dopes (>6%) or to process from the nematogenic state if possible.

Several important benefits are derived from polymer liquid-crystalline behavior, involving the process and the product. For the lyotropic polymers (those which are liquid crystalline in solution), as the concentration is raised above the critical value the viscosity falls rapidly until a minimum is reached at a substantially higher concentration than the critical value. Thus it is possible to obtain a much higher concentration of polymer in the spinning solution, thereby decreasing the quantity of spinning solvent to be recycled or disposed of. For polymers of useful molecular weight, the concentration of isotropic solutions is limited to 4–6 wt%, whereas concentrations of 15–25 wt% are common for liquid-crystalline solutions. Similar benefits of low viscosity for a given molecular weight are also realized in thermotropic systems.

A second consequence of liquid crystallinity in stiff-chain polymers is the ease of attaining very high mechanical properties directly in spinning. This is a result of the low elongational viscosity associated with nematogens, which leads to the easy formation of highly axially oriented extended-chain structures during the spinning operation. While very useful property sets are obtained as-spun, mechanical properties can often be improved through annealing. In the aramid case, annealing leads to a doubling of modulus through an orientation-perfection mechanism. In the case of the thermotropic copolyesters (those which are liquid crystalline when molten), strength improves during annealing through a solid-state polymerization mechanism. Only in the case of the "almost mesogens," such as the hydrazide–amide copolymers, is the processing similar to conventional fiber production, i.e., spinning followed by drawing to improve axial orientation. This latter process is also utilized in the production of gel-spun polyethylene where the production of a gel fiber precursor allows drawing to very high draw ratios (20–40), and very high mechanical properties.

3. *Major High-Modulus High-Strength Organic Fibers*

The structural formulae shown in Fig. 3 include the compositions which have received the most developmental attention as high-modulus fibers. Table 1

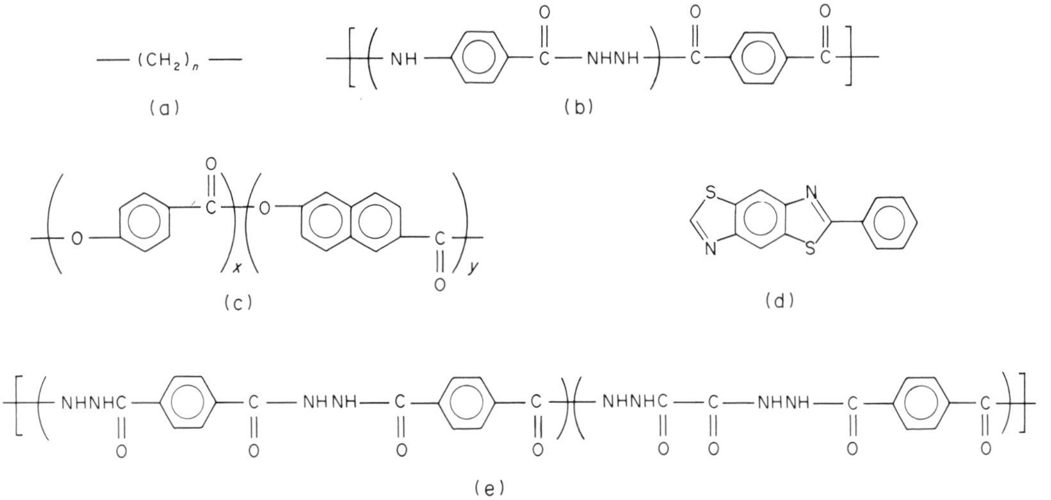

$-(CH_2)_n-$

(a)

NH —C(=O)— NHNH —C(=O)— —C(=O)—

(b)

O —C(=O)— O —C(=O)— x y

(c)

S N N S

(d)

NHNHC(=O) —C(=O)— NHNH —C(=O)— —C(=O)— NHNHC(=O) —C(=O)— NHNH —C(=O)— —C(=O)—

(e)

Figure 3
Examples of typical polymers suitable for high-modulus high-strength fibers: (a) polyethylene, (b) aramid: poly(p-phenylene terephthalamide), (c) thermotropic copolyester, (d) ordered polymer: poly(p-phenylene benzobisthiazole), (e) copolyhydrazide

Table 1
Properties of high-modulus high-strength organic fibers

Fiber	Typical elastic modulus (GPa)	Typical tensile strength (GPa)	Lateral strength (GPa)	Density ($g\,cm^{-3}$)	T_g (°C)
Gel PE	140	3	0.1	0.97	−85
Kevlar 49	120	3	0.6	1.4	360
Kevlar 29	60	3	0.6	1.4	360
Thermotropic copolyester	60	3	0.1	1.4	150
Ordered polymer	265	3	0.3	1.5	
X-500	100	2.5	N/A	1.5	N/A

N/A, not available

summarizes and compares important mechanical and other end-use critical properties of these fibers. Note that for all the fibers listed, lateral mechanical properties, as exemplified by the compressive strength, are several orders of magnitude lower than axial tensile properties. This manifests itself in poor compressive performance for this class of fibers. All of these fibers share excellent environmental stability, low coefficient of thermal expansion ($\sim -4\times10^{-6}\,K^{-1}$) and excellent dimensional stability. Not listed in Table 1 is fiber price, which in all cases is over US$10 per pound and in most cases is over US$20 per pound, a factor which has limited the acceptance of these fibers in many markets. To some extent, the high price reflects the absence of a volume-generating large market such as tire cord, although the combination of specialized monomers and/or process complexity which characterizes the production of high-modulus organic fibers probably precludes low cost variants. As such, these fibers are finding acceptance in specialty markets such as ballistic protection fabrics, lightweight rigid composites for aerospace applications, and ropes and cables for the anchoring of large structures such as oil-drilling platforms.

3.1 Aramid Fibers

Aramid fibers are commercially produced by du Pont (Kevlar), AKZO (Twarlon) and Teijin (Technora). Within the aramid class there are two basic variants, a low-modulus type (~60 GPa) and a high-modulus type (~120 GPa). Examples of these are Kevlar 29 and Kevlar 49, respectively. All of the aramids possess similar tensile strengths of about 3 GPa. Recently, du Pont has introduced Kevlar 149, a somewhat stronger and stiffer variant. Kevlar 29 is used primarily for such products as tensile members (ropes, cables and webbings) and ballistic cloth, while Kevlar 49 and 149 are reinforcing fibers for high-performance fiber-reinforced composites. To overcome the poor compressive performance of these fibers while utilizing their high specific tensile properties and toughness, they are often used in hybrid constructions with glass or carbon.

The aramid fibers are spun from liquid-crystal dopes through a dry-jet wet-spinning process which yields high tensile property fiber directly. The process is fundamentally a melt-spinning of the dope (~20% solids in 100% sulfuric acid) which is crystalline at room temperature and is extruded at elevated temperature into a coagulation bath (water) to remove the residual solvent. The difference between the high and low modulus variants of the aramids is purely structural. Kevlar 29 may be converted to Kevlar 49 through an annealing process which improves overall chain orientation and structural perfection. Advantages of the aramid fibers include high use temperatures (stable up to about 300°C in the absence of hydrolysis agents) and very low creep. Du Pont scientists have estimated that Kevlar fibers will sustain over half their one-second breaking load for longer than 100 years based on the absolute-rate theory of creep failure for a fixed load.

3.2 Polyethylene Fibers

A number of commercial high-modulus polyethylene fibers have entered the market place during 1986–88, including Spectra (Allied Signal), Dyneema (DSM/Toyobo) and Tekilon (Mitsui Toatsu). Because of the low density of polyethylene ($0.95\,g\,cm^{-3}$ versus $1.4\,g\,cm^{-3}$ for the other fibers) the specific tensile properties of polyethylene are the highest achieved with any organic material (see Fig. 1). The lateral properties of high-modulus polyethylene are the same order of magnitude as those found for the fibers spun from liquid-crystalline polymers, leading to poor compressive performance. Polyethylene fibers also exhibit high levels of creep and are limited, by the low polyethylene melting point, to a use temperature of around 100°C. Chemically, polyethylene is highly inert and is the best performer of the high-modulus fiber candidates in alkaline environments.

High-modulus polyethylene fibers are produced via a gel-spinning process in which a low concentration solution of ultrahigh molecular weight ($M > 2 \times 10^6$) polyethylene is extruded to form a gel precursor fiber. This precursor fiber is then hot-drawn to draw ratios of 20 or more to produce a very highly oriented fiber with an extended-chain fibrillar microstructure. The high-molecular-weight polymer is necessary to achieve the desired high strength but cannot be conventionally melt-spun because of its high melt viscosity. Lower-molecular-weight-polyethylenes are melt spinnable but do not give the high values of strength after drawing. It is thought that the gel-state precursor fiber has a minimum of internal flaws which allows drawing to the very high draw ratios necessary to obtain the desired properties.

3.3 Thermotropic Copolyester Fibers

The thermotropic copolyesters can be easily melt-spun to moduli similar to Kevlar 29 (see Fig. 1), although a range of moduli spanning the aramid axial tensile-property range can be made. Strength can be enhanced through annealing to the 3 GPa range representative of all the high-modulus high-strength organic fibers. The product closest to commercial reality is the fiber Vectran, being evaluated in a joint feasibility study by Hoechst Celanese and the Kuraray company. Vectran is based on the commercial Hoechst Celanese Vectra product line of thermotropic copolyesters for plastics applications. Advantages of this product include very high cut resistance (up to 8 times that of the aramids) and excellent hydrolytic stability. Use temperatures are intermediate between Kevlar and polyethylene, in the range 150–200°C depending on composition.

3.4 Other High-Modulus Organic Fibers

Two other approaches to high-strength high-modulus organic fibers worthy of note are the aromatic heterocyclic ordered polymers developed under the auspices of the USAF Materials Laboratory and the "almost mesogens" represented by the hydrazid copolymers represented by the X-500 products developed by Monsanto in the early 1970s. The former are true molecular rods and have exhibited the highest axial tensile moduli achieved in an organic fiber, values in the region of 300 GPa. Low compressive properties and high inherent cost, however, have slowed commercial development of these fibers. The X-500 fibers were introduced by Monsanto as a competitor to Kevlar. These fibers are conventionally wet-spun and then drawn to high tensile properties. Compared to the aramids, the X-500 fibers offered no advantage and were withdrawn from the market by the mid-1970s.

4. Structure–Property Relationships

A distinctive feature of high-modulus organic fibers is the very high axial molecular orientation, which leads to extreme anisotropy of mechanical properties. In the transverse direction, the strength is only about 20% of the strength along the fiber axis and the modulus is less than 10% of the fiber-axis tensile modulus. In essence, the filaments may be described as a hierarchy composed of fibrillar structures covering a range of several orders of magnitude in diameter. Figure 4 shows a structural model developed for the thermotropic copolyester fibers but which is a reasonable representation of the microstructure of all the fibers discussed. The fibrils are highly aligned along the fiber axis with relatively few connections between the fibrillar entities. In tension, it has been shown that both strength and modulus can be attributed to molecular parameters (chain modulus, chain interactions, molecular weight) and that the microstructure plays a minor role. This suggests that the fibrils are very long compared to the effective testing gauge lengths.

In compression, the situation is much the same as in the transverse direction. The compressive yield as determined in 60 vol% Kevlar–epoxy composites was

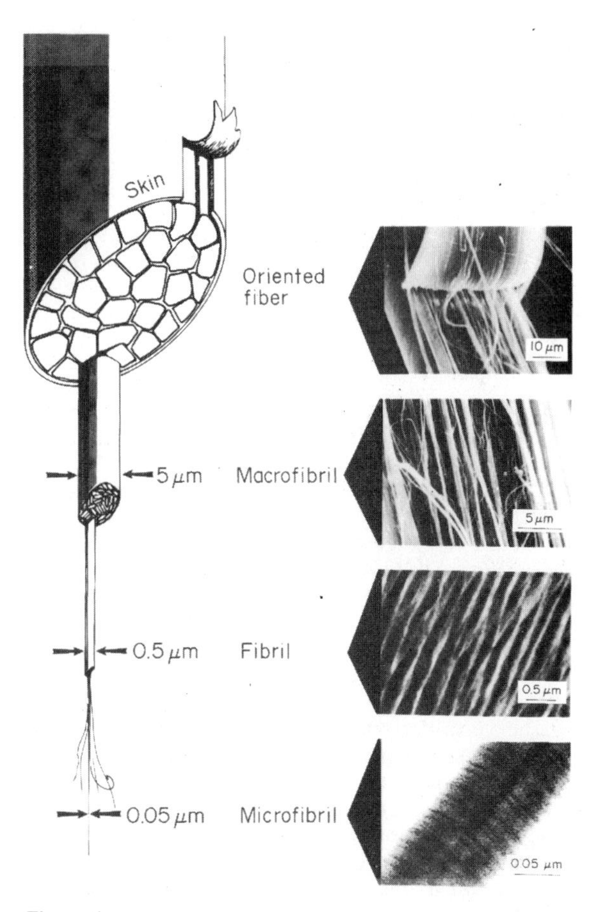

Figure 4
Structural model of a typical thermotropic copolyester fiber

less than 20% of the tensile stress for failure. In compression, the analogy to pushing on a steel cable is not a gross simplification. Also related to the low lateral properties is the relatively poor flex fatigue and abrasion resistance exhibited by all these fibers.

5. Applications

High-modulus high-strength organic fibers have found markets in areas where very high performance is required and high costs can be tolerated. Examples of such markets include primary and secondary structures for aerospace applications, ballistics protection for the military, prosthetic devices, sporting goods and ropes and cables for anchoring very large structures. The effectiveness of high-modulus organic fibers in these applications is a function of their unique balance of properties, including low density, high tensile properties, chemical inertness and good thermal stability. This combination allows the production of lightweight, stiff and strong structures whose performance cannot be matched by metals, ceramics or other fiber-reinforced composites.

The increased usage of high-modulus organic fibers is limited by their poor compressive performance as a class and their very high cost. While thermal stability limits the use of polyethylene fibers to a degree, the predominance of applications require thermal stability from about −20°C to perhaps 100°C, which all of these fibers possess. To some extent, the compressive problems can be overcome by hybridization with other reinforcing fibers or through composite design (cross-lamination, three-dimensional woven structures, etc.). The largest factor restricting the market for high-modulus high-strength organic fibers is cost.

6. Conclusions

The technologies leading to high-modulus high-strength organic fibers are relatively well understood and fibers with tensile moduli of greater than 70 GPa at a strength level of about 3 GPa can be produced from literally hundreds of different polymers. In the limit of the very high molecular orientation inherent in these fibers, all possess similar microstructure and high property anisotropy. Thermal stability varies over a broad range as expected from the range of molecular chemistries being employed, but all these fibers are stable in the neighborhood of room temperature. The limitations to the commercial acceptance, and hence to the replacement, of other materials by high-modulus high-strength organic fibers or their composites are: (a) the low compressive strength which limits the utility of the fibers to applications where the use is in tension, and (b) the high costs which make less sophisticated approaches to the solution of materials problems attractive on a cost–performance basis.

By 1988, Kevlar usage had grown to greater than 2.5×10^7 kg annually, all in the specialty markets described above. For this market size to increase substantially, the cost–performance of high-modulus organic fibers must be improved. Routes to such an improvement include the increase of lateral strength of these highly oriented fibers, and reduction of monomer costs through new or improved chemical syntheses. In the absence of success in these areas, it is likely that usage of high-modulus high-strength organic fibers will be relatively limited and that few of the fiber chemistries known will survive outside the laboratory.

Bibliography

Black W B 1979 Stiff-chain aromatic polymer solutions, melts and fibers. In: Miller R L (ed.) 1979 *Flow-Induced Crystallization*. Gordon and Breach, New York, pp. 245–306

Black W B 1980 High Modulus/high strength organic fibers. *Annu. Rev. Mater. Sci.* 10: 311–62

Calundann G W, Jaffe M 1982 Anisotropic polymers, their synthesis and properties. *Proc. Robert A. Welch Conf. on Chemical Research* 26: 247

Ciferri A, Ward I M (eds.) 1979 *Ultrahigh Modulus Polymers*. Applied Science, London

du Pont 1978 Kevlar 49 Aramid. du Pont Technical Bulletin K-2. E I du Pont de Nemours, Wilmington, Delaware

Jaffe M 1987 High modulus polymers. In: *The Encyclopedia of Polymer Science and Engineering*. Wiley, New Yok, pp. 699–722

Jones R S, Jaffe M 1985 High performance aramid fibers. In: Lewin M, Preston J (eds.) 1985 *High Technology Fibers*. Dekker, New York, pp. 349–392

Kavesh S, Prevorsek D C 1983 High tenacity, high modulus polethylene and polypropylene fibers and intermediates therefore. US Patent No. 4,413,110

Morgan P W 1977 Synthesis and Properties of aromatic and extended chain polyamides. *Macromolecules* 10: 1381–90

Sawyer L C, Jaffe M 1986 The Structure of Thermotropic Copolyesters. *J. Mater. Sci.* 21: 1897–1913

Smith P, Lemstra P J, Kalb B, Pennings A J 1979 Ultrahigh strength polyethylene filaments by solution spinning and hot drawing. *Polym. Bull.* 1: 733–36

M. Jaffe

[Hoechst Celanese Research Division, Summit, New Jersey, USA]

High-Performance Composites with Thermoplastic Matrices

Throughout the 1970s and 1980s, thermosetting resins have been the principal type of matrix used for high-performance fiber composites. The main disadvantages of these resins are: a low strain to failure and brittleness, lengthy cure cycles, moisture absorption and its deleterious effects on mechanical properties, and the preclusion of forming operations and certain kinds of repair once the matrix has been fully cured.

Some of these problems can be overcome by using a thermoplastic polymer for the matrix. Compared with thermosets these materials usually have a higher strain to failure and a lower moisture absorption, they can be shaped when heated and repaired, do not require a lengthy cure cycle and have a very long storage life at ambient temperature. Their principal disadvantage has been the difficulty in fabricating fiber-reinforced composites. This arises because, at a given temperature, the viscosity of a thermoplastic is much greater than that of an uncured thermoset. It may not be possible to reduce this sufficiently for fiber impregnation, by raising the temperature, before the polymer degrades. There are also problems of bonding to the fibrous reinforcement. Early composites with thermoplastic matrices contained an excessive void volume, and damaged or misaligned fibers, and were inferior in performance to thermoset-based specimens. The approach to the fabrication of thermoplastic-matrix composites that has evolved has been to concentrate on producing fiber-reinforced sheets and tapes by, for instance, solvent and hot-melt impregnation methods, aided by modification of the basic polymer. These feedstocks can then be turned into artifacts by hot pressing, film stacking and hot pressing, tape winding and other techniques. Short-fiber-reinforced molding granules have been developed, but, because of the reduced fiber orientation in the final product, composites based on these materials, although having a better degree of isotropy, have poorer properties in a specific direction than unidirectionally reinforced composites.

Because of the difficulties initially encountered in fabricating thermoplastic-matrix composites the tendency has been, for high-performance continuous-fiber composites, to concentrate on thermoplastics which give maximum resistance to temperature rather than on the use of the commoner, commodity thermoplastics.

1. Types of Thermoplastic Used

Ideally the thermoplastic polymer matrix should have a high softening point and resistance against pyrolysis and chemical degradation. These are properties that may be shown by crystalline polymers with a rigid backbone chain, regularly spaced substituents and strong interchain forces due to Van der Waals or polar interactions. Polymer properties usually show a marked reduction at the glass transition temperature, T_g, but crystalline materials have useful property retention until close to T_m, the crystalline melting point, as the crystalline phase holds the molecular chains together. Practically, it is necessary to compromise between the desired properties and the ability to process the polymer without causing degradation. Aromatic segments in the main polymer chain can be used to increase chain stiffness and reduce susceptibility to pyrolysis while ether linkages increase chain flexibility. Further details are given by Marks and Atlas (1965), Woodhams (1985) and Rose (1984).

Among the polymers used as matrices are polysulfones, aromatic polyketones, modified polyimides, polysulfides and liquid crystal systems. To obtain a processible polysulfone, polyether sulfone (PES) or polyaryl sulfone, the backbone polymer chain of the simplest aromatic polysulfone, is modified by the incorporation of ether linkages. Commercial polymers of this type include Udel and Radel (Amoco Chemicals), Ultrason S (BASF) and the polyether sulfones Victrex and Ultrason E, produced by ICI and BASF, respectively. Despite their regular structure, polysulfones are essentially amorphous. These materials are tough, resistant to nuclear but not ultraviolet radiation and are subject to environmental stress cracking. The resistance to most acids, alkalis and aliphatic hydrocarbons is good but the polymers are attacked by concentrated sulfuric acid and aircraft hydraulic fluids. Polysulfones are soluble in dimethyl formamide, dimethyl acetamide, dichloromethane and N-methyl pyrrolidone. Temperatures in excess of 300°C are required for their processing.

Polyketones are represented by ICI's polyether ether ketone, Victrex PEEK, and their polyether ketone, Victrex PEK. BASF and Hoechst also manufacture polyether ketones known as Ultrapek and Hostatec, respectively. Unlike the polysulfones, polyketones are crystalline. Both PEEK and PEK have excellent resistance to nuclear radiation and a wide range of chemicals and solvents, though they are attacked by concentrated sulfuric and hydrofluoric acids. The materials are tough and not susceptible to environmental stress cracking. Processing is more difficult than for polysulfones and a temperature of approximately 390°C is required. In addition it may be necessary to control the degree of crystallinity in the final product, to give the desired properties, by cooling at a specified correct rate. There is evidence that reprocessed PEEK may undergo some thermal cross-linking.

Modified polyimides were developed in attempts to retain the heat resistance of polyimides while rendering their processing more tractable. Torlon, produced by Amoco Chemicals, is a polyamide imide (PAI), which is not attacked by aliphatic, aromatic chlorinated or fluorinated hydrocarbons, dilute acids, aldehydes, ketones, esters or ethers, and is very resistant to nuclear and ultraviolet radiation. It is soluble in N-methyl pyrrolidone, is attacked by alkalis and is notch sensitive. Its melt temperature is 355°C and the processed polymer needs a prolonged period of heat treatment to promote chain extension. The time involved can be up to 100 hours or more. Polyether imide (PEI) (General Electric Plastics), in which imide groups are interspersed with ether linkages, is sold under the name of Ultem. It has a good resistance to nuclear and ultraviolet radiation and to mineral acids, dilute bases and aliphatic hydrocarbons, but is soluble in halogenated hydrocarbons. It is notch sensitive and subject to environmental stress cracking. When PEI is burnt, the smoke generation is low and the gases produced nontoxic. The processing temperature is in the range 350–425°C.

Probably the most widely produced high-performance thermoplastic is polyphenylene sulfide (PPS). Phillips Petroleum market the material under the name of Ryton. Bayer also provide the polymer, Ciba Geigy have introduced a variant called Craston, while Solvay in conjunction with Tohpren, and Celanese together with Kureha Chemicals, are entering the market. The Celanese product is more linear and purer than Ryton. Idemitsu have recently developed a very high molecular weight, ultrapure product. The linear version of the polymer is highly crystalline and may cross-link oxidatively. The resistance to acids, aqueous bases, halogenated hydrocarbons and alcohols is good but the polymer is attacked by concentrated sulfuric acid and some amines. It is soluble in chloronaphthalene at elevated temperatures. Though brittle, its environmental stress resistance is good, as is its resistance to nuclear, but not ultraviolet, radiation. Smoke generation is low and nontoxic gases are produced on burning. Polyphenylene sulfide is the simplest member of the polyarylene family. Recently two new members known as polyarylene sulfide polymers 1 and 2 (PAS 1 and PAS 2) have been produced by Phillips Petroleum. The first is a semicrystalline material and the second is amorphous in nature. Mechanical properties appear similar to those of polyphenylene sulfide though the strain to failure of PAS 2 is higher, at 8%. The glass transition temperatures of PAS 1 and PAS 2 are 145 and 215 °C, respectively and the crystalline melting point of PAS 1 is 340 °C.

Liquid-crystal polymers are based on aromatic thermoplastic polyesters. Varieties available include Xydar, produced by Dartco, Vectra made by Celanese and Ultrax (BASF). The polymers are self-reinforcing structures in which molecular alignment is frozen-in on cooling, giving a significantly higher modulus and strength than those of the other polymers considered here. To ease processing, the molecular backbone chain is kinked by inserting hydroxy naphthoic acid so that the molecules do not fit together too well. The strain to failure is low, presumably because of the fibrous nature of the materials, and behavior is anisotropic. The heat and chemical resistance are excellent and in organic and chlorinated solvents the behavior of Vectra and Xydar is superior to that of the amorphous polymers such as polysulfone, polyether sulfone and polyether imide. In addition the materials are not attacked by acids, bases, alcohols or esters. The resistance to environmental stress cracking and nuclear radiation is good. The ability of the highly ordered molecular chains to slide over one another in the molten state while retaining their relative orientation enables these materials to be processed relatively easily. The melt temperature of a typical injection or extrusion grade of Vectra is 275–330 °C. For Xydar the figure is about 400 °C and the melt viscosity is approximately two orders of magnitude greater than that of Vectra under similar conditions.

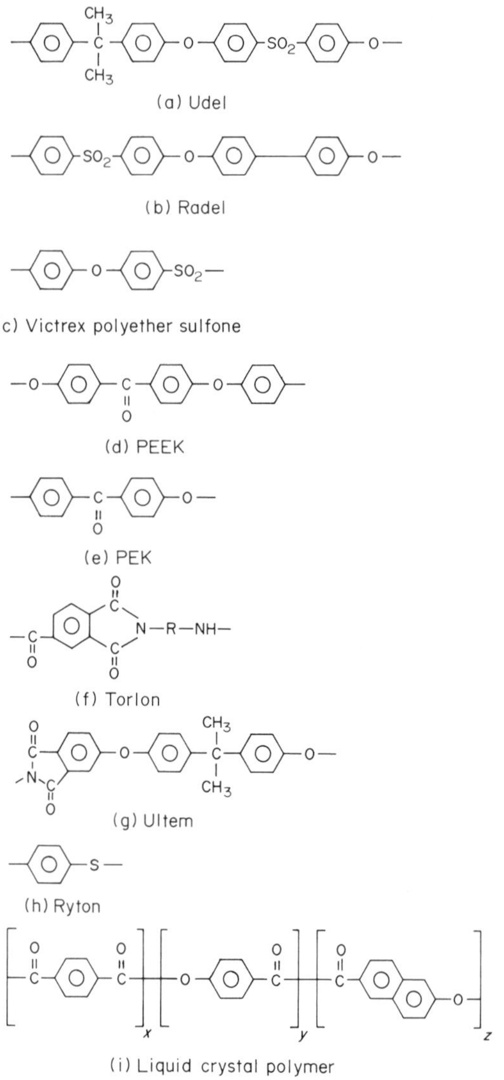

(a) Udel

(b) Radel

(c) Victrex polyether sulfone

(d) PEEK

(e) PEK

(f) Torlon

(g) Ultem

(h) Ryton

(i) Liquid crystal polymer

Figure 1
Basic structural repeat units of thermoplastic polymers

Further details of the various polymers discussed here, except the liquid-crystal systems, are available in Brydson (1982). Liquid-crystal systems are discussed by Collyer and Clegg (1986a). The basic repeat units of many of the systems are illustrated in Fig. 1.

2. *Properties of High-Performance Thermoplastics*

Some typical room-temperature properties of unreinforced grades of various polymers are listed in Table 1. The information has been taken from trade data and

Table 1
Properties of unreinforced polymers: * indicates notched specimens, + indicates specimens exposed for 100 000 h

	Polysulfone	Polyether sulfone	Polyether ether ketone	Polyether ketone	Polyether imide	Polyamide imide	Polyphenylene sulfide	Vectra (LCP)	Xydar (LCP)
Crystalline	no	no	semi-		no	no	semi-	liquid crystal	liquid crystal
Density (Mg m^{-3})	1.25[a]	1.37	1.26–1.32		1.27	1.4[d]	1.36[i]	1.37–1.4	1.35–1.4[j]
σ_f(MPa)	106	129	170		145	197	154[i]	169–245	110–134[j]
E_f(GPa)	2.7	2.6	3.6		3.3	4.8[d]	3.5[i]	9–15.2	11–13[j]
σ_t(MPa)	60–75[a]	84	93	110	105	93[d]	84[i]	165–188	80–123[j]
E_t(MPa)	2.5[a]	3.2[e]	3.2[e]		3.0		3.3[e]	9.7–19.3	
ε_f(%)	50–100	40–80	50		60	17[f]	4[i]	1.3–3.0	3.3–4.9[j]
Izod impact (J m^{-1})	70	76–84*	83*		50*	53[f]	21*[i]	45–530	75–210[j]
Fracture toughness (MN m$^{-3/2}$)	2.2[b]		7.5[b]						
G_{Ic}(KJ m^{-2})			6.6[l]			1.1	1.4		
LOI	32	36	35		47	43	44[i]	35–50	42–47[j]
HDT (1.8 MPa) (°C)	174[a]	200	150	165	200	273[f]	136[e]	222	316–355[j]
T_g(°C)	190[c]	230[c]	143	165	217		93[c]		
T_m(°C)			334	365			285	280	423[j]
Continuous service temp. (°C)	150	180^{+}	250^{+}		170	230–260[g]	200–240[h]		240[k]
α (°C^{-1} × 10^{-6})	55.8	55	47		62	63[d]	54	−5–+75	
Water absorption in 24 h at RT (%)	0.2[d]	0.43	0.1[d]		0.25[d]	0.3[d]	0.2[d]	0.02–0.04	

HDT: heat distortion temperature a Pye 1982 b Gotham and Hough 1984 c Muzzy and Kays 1984 d English 1985 e Specmat (UK) Ltd trade data 1987 f Collyer and Clegg 1984 g Collyer and Clegg 1985 h Woodhams 1985 i O'Connor 1987 j Collyer and Clegg 1986a k Collyer and Clegg 1986b l Leach and Moore 1985
All other entries are taken from trade data

the other sources indicated. The values quoted depend on the grade of polymer (e.g., molecular weight, additives to control crystallinity, etc.), the nature of the test specimen (e.g., degree of crystallinity, preferred orientation) and the technique used to determine the property. The properties of liquid-crystal polymers are anisotropic and here the data refer to the direction of molecular alignment, except for the coefficient of thermal expansion where the larger value is for the transverse direction. The limiting oxygen index (LOI) indicates the percentage of oxygen in an oxygen–nitrogen mixture necessary to sustain combustion. Since air contains 22% oxygen all the polymers detailed would be self-extinguishing in air. The continuous service temperature is based on the Underwriters' Laboratories (USA) test method. This gives the temperature at which 50% of the measured property (e.g., tensile strength) will be retained after a period of up to 100 000 hours exposure under static conditions. Stress, with or without an aggressive chemical environment, could reduce this figure. It is not possible to condense details of resistance to specific chemicals or, where it is available, fatigue and creep data. The manufacturers' data sheets should be consulted in these cases.

In 1984 the estimated Western-world production figures for some of the high-performance polymers listed here were as follows: polyphenylene sulfide 8200 t, polysulfone 6500 t, polyether sulfone 700 t, polyether imide 300 t and polyether ether ketone 100 t. The total market is projected to grow by 460% by 1995. For the year 1984 8% of all high-performance composites were based on thermoplastic matrices. This represents 400–600 t of thermoplastic. The cost of the unfilled grades (early 1987) ranged from £4000 t^{-1} for polysulfone, through £15 000–£30 000 t^{-1} for liquid-crystal polymers to £30 000 t^{-1} plus for PEEK.

3. Fabrication of Thermoplastic Preforms

Mineral-filled, short-glass and carbon-fiber-reinforced thermoplastic materials suitable for injection molding, or bulk materials from which components can be machined, have been available for some years. Though the strength and modulus are increased, and thermal expansion and shrinkage reduced, these types of material are not high-performance composites. To qualify for inclusion in this category reinforcement with continuous, aligned or woven glass, carbon or aramid fibers, or possibly a fiber mat, is required. Usually the fabricator buys reinforced thermoplastic preform (analogous to a thermoset prepreg) in the form of a sheet, tape or pultruded section and uses this to produce the final artifact, though some manufacturers make the final, shaped product from separate fibers and polymer. Developments involving liquid-crystal polymers and long, but not continuous, fiber reinforcement are discussed by Cogswell (1992).

In making a preform the aim is to produce a sheet or tape of constant thickness in which the fibers are aligned and uniformly distributed, and individually wetted by the matrix. Since they are amorphous and hence more readily soluble, solvent impregnation was first used with polysulfones, polyether sulfone and later polyamide imide. The sheet material produced must be thin so that the solvent can be removed before any further fabrication operations. Crystalline or semicrystalline polymers tend to have better solvent resistance and melt impregnation is used to fabricate feedstock. Sometimes, to improve alignment, unidirectional fibers are stitched together with thermoplastic fibers based on, for instance, PEEK. Intimate blends of carbon fibers and PEEK, PEK or polyether sulfone fibers can be knitted, woven or combined together as filaments. This leads to better fiber wet-out and an absence of polymer-rich areas. Another method of preparing a fabric or sheet is to impregnate with a suspension of fine (<5 μm) polymer particles suspended in water or another inert liquid. This method is applicable to polysulfones and polyether ketones as well as polyether imide and polyphenylene sulfide, but not to liquid-crystal polymers since these cannot be cryogenically ground to produce fine particles, because of their inherent toughness and fibrous nature. Pultruded sections or tape may be prepared using a similar approach. Fibers are pulled through powdered resin (30–250 μm) in a coating unit and then shaped and heated in a die. The production rate can be as high as 1 m s^{-1} with the fiber volume loading not deviating by more than ± 1 vol%. The method has been used with polysulfones, polyphenylene sulfide, PEEK, polyether imide and polyamide imide. Finally the fibers may be coated with a prepolymer which is then reacted to give a linear thermoplastic. Commercial manufacturers frequently refer to their process as proprietary.

It has been suggested that reinforced thermoplastic sheet and pultruded sections have a lower fiber loading (~ 55 vol%) and poorer fiber alignment than their thermoset counterparts. The preform products should be stored in clean, dry conditions, since the matrices will pick up small amounts of water (see Table 1). Unlike thermoset prepregs, which become tacky at room temperature, thermoplastic-based materials are stiff and do not become drapable until above their T_g, or, for crystalline polymers, above T_m. If the material is unidirectionally reinforced it will easily split unless handled carefully. A soldering iron is recommended for joining pieces of preform before solidification.

Starting materials based on all the polymers listed in Table 1 except PEK, which is a very recent product, and liquid-crystal polymers, are in principle available, though most commercial emphasis appears to be on reinforced PEEK, polyphenylene sulfide and polyether imide products. Phillips Petroleum produce pultruded structural sections and an aramid-fiber tape, both with a polyphenylene sulfide matrix, but the most widely used system appears to be ICI's APC 2 based on carbon fibers in a PEEK matrix. At least one

UK manufacturer produces finished articles based on liquid-crystal polymers.

4. Fabrication of Thermoplastic Composite Artifacts

Given a supply of preform the final step is to convert the material into the finished article. One method of doing this involves stacking layers of fiber-reinforced preform with the desired orientation, possibly interspersed with very fine layers of polymer film of either the same type as used in the preform or another kind. The stack is processed by pressing in a matched mold at a pressure of 1–10 MPa and at a temperature in excess of T_g, or T_m if appropriate. Processing times vary from 30–300 s while it has been suggested that 30 min may be required. The exact time will depend on the thickness of material used, and laying-up and cooling may increase the process time very considerably. Care must be taken in molding large, shaped components with shallow curvature because of the long slip path between adjacent plies which can result in fiber wrinkling when a compressive force is applied. In this case, shaping individual preform sheets before fusing them is recommended. As with thermoset fabrication, it is necessary to remove air and any solvents that have been used in preparing the preform before final consolidation. To do this, a properly designed mold with adequate venting or the application of a vacuum is required. Another method is to vacuum-form using an autoclave in a manner similar to that used for a thermoset prepreg. A disadvantage of this method is the high temperature that is required.

Certain sheet-metal-forming techniques can be used to form thermoplastic-composite artifacts. In hydroforming a preheated, consolidated blank is formed to the shape of the mold by using a hydraulically pressurized flexible diaphragm or block of rubber. Because of the hydraulic nature of the loading, an even pressure is applied over the whole surface of the charge, or blank, minimizing fiber damage. A typical forming time of 15 s is claimed. Another method which can be used to produce complex shapes is diaphragm forming. This is analagous to the superplastic forming of metal alloys. A charge of, for instance, APC 2 is laid up in the requisite way and encased between two sheets of Supral aluminum. Air is evacuated from the enclosed charge and the temperature raised to 380 °C when the aluminum sheets become superplastic. Differential pressure and a shaping tool can be used to

Table 2
Room-temperature properties of thermoplastic composites based on polysulfone, polyether sulfone and PEEK: * indicates notched specimens

	Polysulfone 40 vol% glass cloth[a]	Polysulfone 55 vol% TOHO HTA7[b]	Polyether sulfone 40 vol% glass fiber[a]	Polyether sulfone 60 vol% carbon fiber[d]	PEEK APC 2 61 vol% AS-4[e]	PEEK APC 2 61 vol% 1M-6[e]	PEEK 61 vol% Celion G30-50[f]	PEEK T 300[g]
Density (Mg m^{-3})					1.657			
σ_f(MPa)	172	1560	214	1551	1880	2170	2160	1796
E_f(GPa)	83	112	11	94	121	151	131	139
ILSS (MPa)		75		79			90	107
σ_t(MPa)	131		159	1232	2130	2700	1840	
E_t(GPa)	11.7		13.8	153	134	176	145	
ε(%)		1.45			1.45	1.48		1.29
σ_c(MPa)	166		152		1100	1100	1075	
σ_t transverse (MPa)					80			
E_t transverse (GPa)					8.9			
σ_f transverse (MPa)					137	160		171
E_f transverse (GPa)					8.9	9.3		
Izod impact (J m^{-1})	84.6*		79*					
Instrumented impact (KJ m^{-2})		60[c], 62[c]		78[c], 114[c]				75[c] (melt)
G_{Ic}(KJ m^{-2})					2.4	2.5		
HDT (°C)	185		216					
α (°C^{-1} × 10^{-6})	2.34		2.52					

ILSS: interlaminar shear strength; G_{Ic}: critical work of fracture a English 1987 b Weiss and Huttner 1987 c Stori and Magnus 1983 (first reading is for a solution-impregnated specimen, second for a melt impregnated one, both of which have 50 vol% fiber loading) d McMahon 1984 e Anon 1986 f Clemans *et al.* 1987 g Owens and Lind 1984 shape the sandwich structure. The method can presumably be used with other types of thermoplastic, provided that the polymer can withstand the temperature at which the aluminum becomes superplastic. This process can be used to produce complex shapes and is very competitive for short runs. The cycle time can be as low as 20–40 min. One other technique, borrowed from the metal-forming industry, is roll forming. Continuous lengths of preform are heated, consolidated and formed, and further consolidated using shaped rollers. In this way corrugated sheet and "top hat" sections, among others, can be made. The throughput speed is up to $0.17\ m\,s^{-1}$.

Other means of fabrication available are filament winding and tape laying. Filament winding is widely used in thermoset fabrication. The main differences when it is applied to thermoplastics are that a filament already impregnated with the matrix is used and additional heating, possibly of both the mandrel and filament, is required to fuse the filaments together. Methods of heating include infrared, lasers, microwaves, induction, contact, ultrasonic welding, flame and nonoxidizing hot gas. Advantages of thermoplastic filament-winding include the ability to deviate markedly from geodesic paths and to fabricate structures with reentrant surfaces. Tape laying is used to fabricate large, planar, or low curvature components. Because of the geometry, it is not possible to consolidate by applying tension to the tape. Instead, an external head must be used. In another method,

Table 3
Room-temperature properties of thermoplastic composites based on polyether imide and polyamide imide

	Polyether imide 55 vol% Carbon fiber[a]	Polyether imide 54 vol% glass fiber[a]	Polyether imide 52 vol% Kevlar fiber[a]	Polyamide imide 0/90 T300[b]	Polyamide imide 0/90 Celion 3000[b]
σ_f(MPa)	711	643	253	1000	1010
E_f(GPa)	51	26	54	63	59
ILSS (MPa)				75	100
σ_t(MPa)	527	355	396		680
E_t(GPa)	57	24	41		62
ε(%)	1.6				
σ_c(MPa)					560
Inst. impact ($KJ\,m^{-2}$)	28[c] (sol.)				

a Specmat (UK) Ltd. trade data 1987 b McMahon 1984 c Stori and Magnus 1983

Table 4
Room-temperature properties of thermoplastic composites based on polyphenylene sulfide

	70 vol% glass fiber[d]	66 vol% carbon fiber[a]	55 vol% pultruded[b] carbon fiber[b]	56 vol% carbon fiber[c]	61 vol% Celion G30–50[d]
Density ($Mg\,m^{-3}$)	2	1.6	1.52		
σ_f(MPa)	1159	1310	1173	1366	1770
E_f(GPa)	44.2	124	96.6	124	131
ILSS (MPa)		70			62
σ_t(MPa)	911	1656	1380	1172	1820
E_t(MPa)	49.7	135	117.3	130	117
ε(%)			1.1	0.8	
σ_c(MPa)	759	655	586		634
Izod impact ($J\,m^{-1}$)		1586	2078		

a Phillips Petroleum trade data b O'Connor 1987 c Specmat (UK) Ltd. trade data 1987 d McMahon 1984

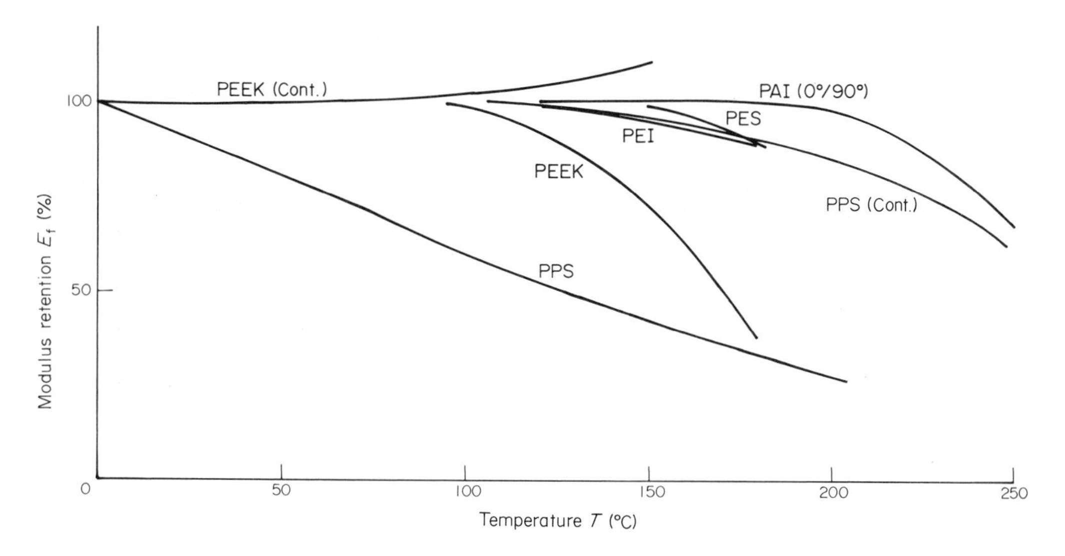

Figure 2
Percentage modulus retention of thermoplastic-matrix composites plotted against temperature. Continuous and 0°/90° reinforcement is carbon fiber; in other cases the reinforcement is 30% short glass fiber

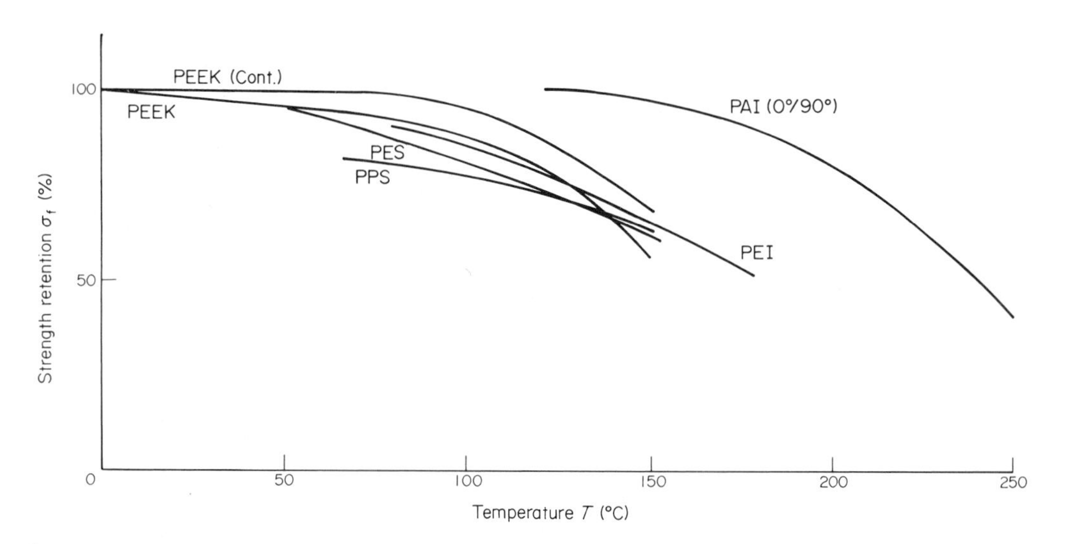

Figure 3
Percentage strength retention of thermoplastic-matrix composites plotted against temperature. Continuous and 0°/90° reinforcement is carbon fiber; in other cases the reinforcement is 30% short glass fiber

preform tape is spirally wound between two parallel mandrels with continuous butt welding to create a single-ply tube with any fiber orientation between 0° and 90°. The welding speed for PEEK is 0.03 $m\,s^{-1}$.

Stamping has been suggested as a means of forming components. The tooling costs are moderate and the process suitable for long runs and large, simply shaped components. Phillips Petroleum have developed a polyphenylene sulfide reinforced with chopped or swirl mat glass or woven and nonwoven glass fiber. A temperature of 315–343 °C and a pressure of 14–42 MPa are required for stamping. The cycle time is 20–120 s.

Further details and references on all methods of fabrication, especially when applied to reinforced PEEK (APC 2), are given by Cattanach and Cogswell (1986).

5. Composite Properties

Some room-temperature thermomechanical properties of carbon-, glass- and aramid-fiber composites with thermoplastic matrices are given in Tables 2–4. Unless otherwise stated, the specimens are unidirectionally reinforced. Unfortunately no data covering one type of fiber with all the different matrices is available and thus it is not possible to compare the effectiveness of the various thermoplastic matrices. Information on high-temperature properties and the performance of laminates is available for PEEK and polyphenylene sulfide materials in the trade and open literature. The percentage retention of modulus and strength with temperature for some systems is shown in Figs. 2 and 3.

Bibliography

Anon 1986 Development of thermoplastic composites for airframe construction. *Aerospace Design and Components* Oct–Nov 1986, 16–21

Brydson J A 1982 *Plastics Materials*, 4th edn. Butterworth, London

Cattanach J B, Cogswell F N 1986 Processing with aromatic polymer composites. In: Pritchard G (ed.) 1986 *Developments in Reinforced Plastics*, 5. Elsevier, London

Clemans S R, Western E D, Handermann A C 1987 *Hybrid Yarns for High Performance Thermoplastic Composites*, Materials Science Monographs 41. Society for the Advancement of Materials and Process Engineering, Azusa, California

Cogswell F N 1987 The next generation of injection moulding materials. *Plast. Rubber Int.* 12: 36–39

Cogswell F N 1992 *Thermoplastic Aromatic Polymer Composites*. Butterworth-Heinemann, Oxford

Collyer A A, Clegg D W 1985a Polyimides—mechanical properties versus processability. *High Performance Plastics* 2(7): 1–5

Collyer A A, Clegg D W 1985b Radiation resistant polymers. *High Performance Plastics* 2(2): 2

Collyer A A, Clegg D W 1986a Self reinforcing polymers. *High Performance Plastics* 4(1): 2–7

Collyer A A, Clegg D W 1986b Xydar electrical component. *High Performance Plastics* 3(5): 8

English L K 1985 Aerospace composites. *Mater. Eng.* 102(4): 32–36

English L K 1987 Fabricating the future with composite materials, part 3: Matrix resins. *Mater. Eng.* 104(2): 33–37

Gotham K V, Hough M C 1984 *Durability of High Temperature Thermoplastics*. RAPRA, Shrewsbury, UK

Leach D C, Moore D R 1985 Toughness of aromatic polymer composites reinforced with carbon fibers. *Compos. Sci. Technol.* 23: 131–61

McMahon P E 1984 Thermoplastic fiber composites. In: Pritchard G (ed.) 1984 *Developments in Reinforced Plastics*, 4. Elsevier, London, Chap. 1

Marks H F, Atlas S H 1965 Principles of polymer stability. *Polym. Eng. Sci.* 5: 204–07

Muzzy J D, Kays A O 1984 Thermoplastic versus thermosetting structural composites. *Polym. Comp.* 5: 169 72

O'Connor J E 1987 Polyphenylene sulfide pultruded type composite structures. *SAMPE Q.* 18: 32–38

Owens G A, Lind D J 1984 *Fibre Reinforced Composites*, Conf. Proc. Paper 12. Plastics and Rubber Institute, London

Pye A M 1982 High performance engineering plastics. *Mater. Eng.* 3: 407–09

Rose J B 1984 High temperature engineering thermoplastics from aromatic polymers. *Plast. Rubber Int.* 10: 11–15

Stori A, Magnus E 1983 An evaluation of the impact properties of carbon fibre reinforced composites with various matrix materials. In: Marshall I H (ed.) 1983 *Composite Structures*. Elsevier, London, paper 24

Weiss R, Huttner W 1987 *High Performance Carbon Fiber Reinforced Polysulfone*, Materials Science Monographs 41. Society for the Advancement of Materials and Process Engineering, AZUSA, California

Woodhams R T 1985 History and development of engineering resins. *Polym. Eng. Sci.* 25: 446–52

N. L. Hancox
[AEA Technology, Didcot, UK]

Hybrid Fiber–Resin Composites

Reference to hybrid composites most frequently relates to the kinds of fiber-reinforced materials, usually resin-based, in which two types of fibers are incorporated into a single matrix. The concept is a simple extension of the composites principle of combining two or more materials so as to optimize their value to the engineer, permitting the exploitation of their better qualities while simultaneously mitigating the effects of their less desirable properties. As such, the definition is much more restrictive than the reality. Any combination of dissimilar materials could in fact be thought of as a hybrid. A classic example is the type of structural material in which a metal or paper honeycomb or a rigid plastic foam is bonded to thin skins of some high-performance, fiber-reinforced plastic (FRP), the skins carrying the high surface tensile and compressive loads and the core providing lightweight (and cheap) structural stability. The combination of sheets of aluminum alloy with laminates of fiber-reinforced resin, as in the commercial product ARALL (aramid-reinforced aluminium, Davis, 1985) is a related variety of hybrid, and the mixing of fibrous and particulate fillers in a single resin or metal matrix is to form another species of hybrid composite. In this article, however, the emphasis is placed on mixed-fiber, resin-based hybrid composites, since most work has been done on such materials.

Some hybrids of current interest represent attempts to reduce the cost of expensive composites with reinforcements like carbon fibers (CFRP) by incorporating a proportion of cheaper, lower-quality fibers such as glass without too seriously reducing the mechanical properties of the original composite. Of equal importance is the reverse principle, that of stiffening a glass-reinforced plastic (GRP) structure with a small quantity of judiciously placed carbon or aromatic polyamide fiber, without inflicting too great a cost penalty. In high technology fields the question of cost may be insignificant by comparison with the advantages of optimizing properties. In aerospace applications, a familiar purpose of using hybrids is to utilize the natural toughness of Kevlar- or glass-fiber-reinforced plastics to offset a perceived brittleness of typical CFRP. From the designer's point of view the important aspect of using hybrids is that provided there is adequate understanding of the underlying mechanisms of stiffening, strengthening and toughening, they allow even closer tailoring of composite properties to suit specific requirements than can be achieved with single-fiber types of composites.

In addition to the choice of individual fiber species for hybridization, there is also the important question of arrangement or dispersion of the separate species within the composite. The coarsest structural level (Fig. 1) is the skin-core structure, used where the outer plies must sustain tensile and compression loads in bending applications or are intended to protect a load-bearing core against impact or abrasion damage. In substantially tensile or tension–compression applications, the plies may be arranged in some balanced arrangement throughout the sheet thickness so as to achieve a predictable elastic response. Again, however, it may be advantageous to place the plies of the higher strain-to-failure component on the outside to improve resistance to external damage or crack initiation. Technically this type of ply-by-ply hybridization is perhaps the easiest (and cheapest) level of hybridization to achieve, since each species of ply can be cut from a single batch of prepreg or tape. It also allows more flexibility in tailoring structure and composition than the tow-by-tow level of hybridization which calls for the use of prefabricated hybrid reinforcement, either woven or nonwoven, of given composition, and provides a more homogeneous dispersion of reinforcement species. The finest level of hybridization is that at which the two species of fibers are intimately mixed, ideally in random fashion. This degree of intimacy is difficult to achieve with continuous fibers, but the short fiber fabrication process described by Parratt and Potter (1980) and developed for practical applications by Messerschmitt–Bölkow–Blohm Gmb–(Richter 1980) readily permits the preparation of random, intimate mixtures of two (or more) species of fiber in prepreg sheets. A less obvious form of hybridization is based on the local effect of selectively introducing strips of an extensible composite like GRP to act as crack arresters at appropriate locations in a less extensible laminate like CFRP (Bunsell and Harris 1976). Such insertions, which are currently referred to as "softening strips" (Sun and Luo 1985) act as efficient crack arresters provided their widths are sufficient to dissipate the energy of a crack moving rapidly in the CFRP by localized debonding and splitting.

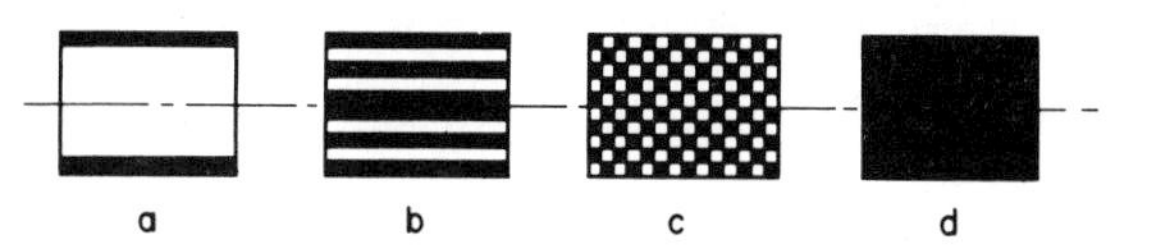

Figure 1
Schematic illustrations of different levels of hybridization: a, skin-core structure; b, ply-by-ply lamination; c, tow-by-tow hybridization (may be ordered or random); d, intimate mixture at fiber scale (random)

1. Elastic Reponse of Hybrids

The modelling and calculation of properties of hybrids will depend strongly on the geometry of hybridization and the stress distribution throughout the composite. These two features alone permit the determination of elastic properties, to within limits acceptable for most design purposes, by application of the principles of composite mechanics. For the prediction of elastic properties of hybrids, the methods of applied mechanics should, in principle, give satisfactory results, since the laminate theory (Tsai et al. 1987) is generally applicable and takes into account the anisotropic properties of each ply, whatever its orientation or composition. In composites with an intimately-mixed fiber distribution the longitudinal modulus E_{11} of a unidirectional sample should be satisfactorily approximated by a rule of mixtures sum, whether the modulus is measured in tension or in bending. The problem in such composites is to describe the composition accurately since it may happen that, by virtue of their different processing characteristics, the end-point compositions, i.e. "pure" GRP and "pure" CFRP in a GRP–CFRP hybrid, will not necessarily have the same resin content, so that a ternary composition diagram, as used by Harris and Bunsell (1975), is needed to represent the results. This problem has been discussed in detail by Phillips (1981).

When a mixed-ply hybrid laminate is tested in bending, geometrical effects may complicate matters, since the measured flexural modulus is highly sensitive to the number and dispersion of the plies. This has been studied theoretically and experimentally by Wagner et al. (1982a, b) who observed that apparent deviations from the rule of mixtures could be found in the flexural modulus of CFRP–GRP hybrid laminates, and that the deviation increased as the segregation of glass and carbon layers increased. These deviations, which were positive if the outer plies were of carbon and negative if they were of glass, were, however, predictable by their analysis, a straightforward model based on the formulae of Stavsky and Hoff (1969) for the bending stiffness of a composite beam:

$$EI_{\text{hybrid}} = \sum_{i=1}^{n} E_i I_i \tag{1}$$

This being so, there is therefore no logic in describing these effects as "synergistic." The observed deviations disappeared as the number of plies increased beyond about 15, or at very high or low fiber volume fractions, when the observed and predicted moduli tended to a rule of mixtures value. Chamis and Lark (1977) have shown that fair agreement can be obtained between measured values of the flexural modulus and those calculated by the formula

$$E_{\text{hybrid}} = \frac{1}{3t} \sum_{i=1}^{k} [(z_{i+1})^3 - (z_i)^3] E_i \qquad (2)$$

where t is the laminate thickness, z_i is the distance to the bottom of the ith ply and z_{i+1} is the distance to the top of the ith ply. For more thorough analysis of general laminates, the classical thin laminate theory provides adequate predictions of both elastic and thermal strains.

2. Other Mechanical Properties

By contrast with elasticity calculations, predictions of strength, failure strain, toughness and fatigue properties are intrinsically more complex because they depend on micromechanisms of damage accumulation which are in turn influenced by the construction of the laminate and the scale of dispersion of the mixed fibers. A common type of hybrid is made by laminating some balanced sequence of plies within each of which there is only a single species of fiber. The composition-dependence of the strength of a unidirectional composite of this kind is obtained by analogy with the response of a single fiber composite by considering the failure strains of the two separate constituents (Aveston and Kelly 1979). When a small amount of fiber with low failure strain ε_1 is added to a composite of otherwise high-failure-strain fibers ε_h, these low-elongation (LE) fibers will experience multiple failure before the failure strain of the high-elongation (HE) fibers is reached. By analogy with the case of, say, a flexible resin containing only one species of brittle fibers, the strength of the hybrid composite σ_H will be given at first (Fig. 2) by the residual load-bearing capacity of the HE composite, assuming the LE component makes no contribution at all, viz $\sigma_H = \sigma_{hu} V_h$, where V_h is the volume fraction of the HE component ($V_l + V_h = 1$) and σ_{hu} is the strength of the pure HE fibers (likewise, σ_{lu} is the strength of the pure LE fibers). At the other extreme, if a small amount of HE fiber is added to an LE composite, the strength is again reduced because the HE fibers cannot carry the load when the LE component fails. Failure of the LE fibers therefore leads to single, catastrophic fracture of the composite. The hybrid strength in this mode is

$$\sigma_H = \sigma'_h V_h + \sigma_{lu} V_l \qquad (3)$$

and the intercept of this line on the HE axis ($V_h = 1$) is the stress σ'_h in the HE component at the failure strain of the LE component, or $\varepsilon_{lu} E_h$ (ε_{lu} and ε_{hu} are the failure strains of pure LE fibers and pure HE fibers, respectively). The critical composition for the tran-

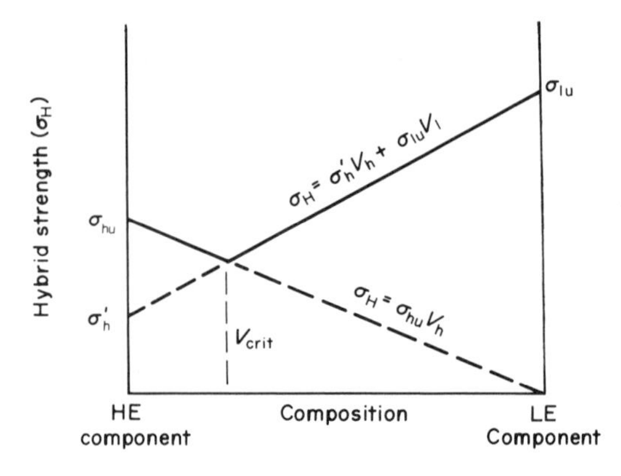

Figure 2
Schematic illustration of variation of hybrid strength σ_H with composition for mixtures of two components, one of high failure strain (HE) and one of low failure strain (LE)

sition from single to multiple fracture is thus

$$V_{\text{crit}} = \frac{\sigma_{lu}}{\sigma_{lu} + \sigma_{hu} - \varepsilon_{lu} E_h} \qquad (4)$$

and this transition point also defines the hybrid composition of minimum strength. Figure 3 illustrates that this simple model is adequate for such hybrid combinations as CFRP–GRP and CFRP–KFRP. It also shows that in the case of the GRP–CFRP combination, for which the individual failure strains (ε_{lu} and ε_{hu}) are markedly different, the hybrid composite failure strain varies nonlinearly with composition and falls below the straight line joining ε_{lu} and ε_{hu}. This type of variation is characteristic (see Kretsis 1987 for a survey of experimental results) and has given rise to the peculiar notion of "failure strain enhancement," also known as the "hybrid effect." This idea stems from the fact that since there is a continuous variation of composite failure strain between the two end points, some hybrid compositions must inevitably fail at strains greater than the failure strain of the lower extensibility component ε_{lu}. There have been several successful attempts to model this effect on the basis of energy arguments (Aveston and Silwood 1976), statistical models (Manders and Bader 1981a, b, Zweben 1977), and in terms of the dynamics of load transfer (Xing et al. 1982), but proliferation of the idea of a so-called synergistic or hybrid effect is unnecessary and misleading. There is of course no genuine enhancement of failure strain—the effect is simply a consequence of the progressive multiple fracture of a group of reinforcing filaments for which the failure strains are statistically distributed, as demonstrated with the aid of acoustic emission monitoring for the simpler case of brittle carbon fibers in a ductile resin matrix by Fuwa et al. (1975). As such, the statistical treatments of Zweben (1977) and Manders and Bader

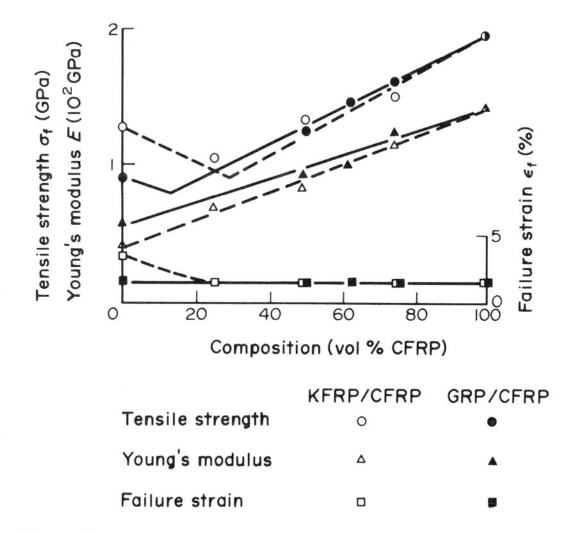

Figure 3
Variation of experimental values of tensile strength σ_f, Young's modulus E and failure strain ε_f for two families of hybrids, KFRP–CFRP and GRP–CFRP (Fernando et al. 1988, Dickson et al. 1988). The lines associated with the strength points are those given by the model illustrated in Fig. 2

(1981a, b) appear to be the most suitable approaches to modelling the variation of failure strain with composition. Zweben's analysis, for example, shows that the failure process is affected by the statistical spread of failure strains of the two fiber species and predicts that the addition of HE fiber to an LE composite raises the strain level needed to propagate fiber breaks because the HE fibers behave like crack arresters at a micromechanical level. The situation becomes still more complicated when the fibers are intimately mixed because micromechanisms of failure change as the pattern of nearest neighbors around a given fiber alters and the strength is governed either by failure of single fibers or of small bundles of fibers, as modelled by Parratt and Potter (1980).

Work by Piggott and Harris (1981) on the compression behavior of hybrids containing various carbon/glass and carbon/Kevlar mixtures indicates no simple pattern that would permit prediction of compression strength and modulus although the strengths of most of the hybrids studied fell reasonably close to rule of mixtures values. This work emphasizes that changes in local failure modes resulting from hybridization are probably responsible for this uncertainty. Results relating to interlaminar shear strength (ILSS) are not clear cut, but it has been shown that the addition of high-modulus carbon fiber to a high-strength carbon–epoxy composite does not significantly alter the ILSS (Arrington and Harris 1978).

As far as toughness and related characteristics are concerned, a good deal again depends on the type of test used and the specific failure mechanism. It is often suggested that the combination of a "tough" composite such as GRP or KFRP with a more "brittle" composite like CFRP results in amelioration of the fracture behavior of the CFRP without seriously affecting other mechanical properties. There seems some evidence, however, that when conventional fracture toughness measurements are made, the change in the critical stress intensity factor K_{Ic} (or the candidate fracture toughness K_Q) for a given change in hybrid composition is simply proportional to the concomitant change in tensile strength (Harris et al. 1988). When the chosen toughness parameter is work of fracture or fracture energy γ_F, there is some ambiguity. Energy arguments would suggest that the fracture energy of hybrids should be satisfactorily predicted by a mixtures rule, provided no dramatic change in failure mechanism occurs as the composition changes. This was in fact observed for intimately-mixed CFRP–GRP hybrids (Bunsell and Harris 1974, Harris and Bunsell 1975). Some positive and negative deviations from the rule of mixtures can apparently be explained in terms of unaccounted for compositional variations. But Stefanidis et al. (1985) observed a distinct negative deviation from the rule of mixtures in CFRP–KFRP laminates, not attributable to errors in estimation of composition, which they explained in terms of ply failure strains. Anstice and Beaumont (1983) have also been able to interpret deviations from linearity in terms of localized micromechanisms of energy absorption during failure.

The simplest expectation for the fatigue stress/life (S/log N) curve of a hybrid laminate would be that it would fall between those of the two single-fiber components, but published results do not offer an obvious picture on which to base predictive models. Early work by Phillips (1976) on a woven cloth CFRP–GRP composite tested in flexure showed that the fatigue stress for a life of 10^5 cycles was a linear function of the carbon/glass ratio, i.e., it follows the mixtures rule. Hofer et al. (1978) also found that the fatigue stress of unidirectional HTS-carbon–S-glass hybrids obeyed the rule of mixtures when in the as-manufactured state, but showed a positive deviation from linear when the composites were hygrothermally aged. Fernando et al. (1988) found that the failure stresses for lives of 10^5 and 10^6 cycles for unidirectional carbon–Kevlar-49 hybrids were linear functions of composition for both repeated tension and tension/compression loading. However, since the tensile strengths of the same series of hybrids were given by the failure strain model (as in Fig. 3), the fatigue ratio (fatigue stress for a given life divided by tensile failure stress) showed a marked positive deviation from the linear weighting rule. The same workers (Dickson et al. 1988) found that unidirectional CFRP–GRP fatigue strengths fell above a linear relationship, and the fatigue ratio therefore showed an even more marked positive synergistic effect. This suggests that factors controlling monotonic tensile (and compression) failure do not necessarily continue to determine failure under cyclic loading conditions, and that for fatigue applications, if for no other conditions, there appear to be positive benefits in using hybrids in place of single-fiber composites.

Bibliography

Anstice P D, Beaumont P W R 1983 Micromechanisms of fracture of fibrous composites in monotonic loading. *J. Mater. Sci. Lett.* 2: 617

Arrington M, Harris B 1978 Some properties of mixed fibre CFRP. *Composites* 9: 149–52

Aveston J, Sillwood J M S 1976 Synergistic fibre strengthening in hybrid composites. *J. Mater. Sci.* 11: 1877–83

Aveston J, Kelly A 1979 Tensile first-cracking strain and strength of hybrid composites and laminates. *Proc. R. Soc. London, Ser. A* 366: 599–623

Bunsell A R, Harris B 1974 Hybrid carbon and glass fibre composites. *Composites* 5: 157–64

Bunsell A R, Harris B 1976 Hybrid carbon/glass fibre composites. In: Scala E, Anderson E, Toth I, Noton B R (eds.) 1976 *Proc. 1st Int. Conf. Composite Materials*, 2. American Institute of Mining, Metallurgical and Petroleum Engineers, New York, pp. 174–90

Chamis C C, Lark R F 1977 Non-metallic hybrid composites: analysis design, application and fabrication. In: Renton W J (ed.) 1977 *Hybrid and Select Metal-Matrix Composites*. American Institute of Aeronautics and Astronautics, New York, pp. 13–51

Davis J W 1985 ARALL—from a development to a commercial material. In: Bartelds G, Schlickelman R J (eds.) 1985 *Progress in Advanced Materials and Processing*. Elsevier, Amsterdam, pp. 41–49

Dickson R F, Fernando G, Adam T, Reiter H, Harris B 1989 Fatigue behaviour of hybrid composites: II, Carbon/glass hybrids. *J. Mater. Sci.* 4, 227–33

Fernando G, Dickson R F, Adam T, Reiter H, Harris B 1988 Fatigue behaviour of hybrid composites: I Carbon/Kevlar hybrids. *J. Mater. Sci.* 23, 3732–43

Fuwa M, Bunsell A R, Harris B 1975 Tensile failure mechanisms in carbon fibre reinforced plastics. *J. Mater. Sci.* 10: 2062–70

Harris B, Bunsell A R 1975 Impact properties of glass fibre/carbon hybrid composites. *Composites* 6: 197–201.

Harris B, Dorey S E, Cooke R G 1988 Strength and toughness of fibre composites. *Compos. Sci. Technol.* 31, 121–41

Hofer K E, Stander M, Bennett L C 1978 Degradation and enhancement of fatigue behaviour of glass-graphite-epoxy hybrid composites after accelerated ageing. *Polym. Eng. Sci.* 18: 120–27

Kretsis G 1987 A review of the tensile compressive flexural and shear properties of hybrid fibre reinforced plastics. *Composites* 18: 13–23

Manders P W, Bader M G 1981a Strength of carbon/glass hybrids, I-Failure strain enhancement *J. Mater. Sci.* 16: 2246–56

Manders P W, Bader M G 1981b Strength of carbon/glass hybrids II-Statistical model. *J. Mater. Sci.* 16: 2246–56

Parratt N J, Potter K D 1980 Mechanical behaviour of intimately mixed hybrid composites. In: Bunsell A R, Bathias C, Martrenchar A, Menkes D, Verchery G (eds.) 1980 *Advances in Composite Materials* 1. Pergamon, Oxford, pp. 313–26

Phillips L N 1976 On the usefulness of glass/carbon hybrids. *Proc. Reinforced Plastics Congress, 1976, Brighton.* British Plastics Federation, London, pp. 207–11

Phillips M G 1981 Composition parameters for hybrid composite materials. *Composites* 12: 113–16

Piggott M R, Harris B 1981 Compression strength of hybrid fibre-reinforced plastics. *J. Mater. Sci.* 16: 687–93

Richter H 1980 Single fibre and hybrid composites with aligned discontinuous fibres in a polymer matrix. In: Bunsell A R, Bathias C, Martrenchar A, Menkes D, Verchery G (eds.) *Advances in Composite Materials*, Vol. 1. Pergamon, Oxford, pp. 387–98

Stavsky Y, Hoff N J 1969 Mechanics of composite structures. In: Dietz A G N (ed.) 1969 Engineering Laminates. MIT Press, Cambridge, Massachusetts, pp. 5–59

Stephanidis S, Mai Y W, Cottrell B 1985 Specific work of fracture of carbon/Kevlar hybrid fibre composites. *J. Mater. Sci. Lett.* 4: 1033–35

Sun C T, Luo J. 1985 Failure loads for notched graphite epoxy laminates with a softening strip. *Compos Sci. Technol.* 22: 121–34

Tsai S W, Massard T N, Susuki I 1987 *Composites Design*, 3rd edn. Think Composites, Dayton, USA

Wagner H D, Roman I, Marom G 1982a Analysis of elastic properties of symmetrically-laminated beams in bending. *Fibre Sci. Technol.* 16: 295–308

Wagner H D, Roman I, Marom G 1982b Hybrid effects in the bending stiffness of graphite/glass reinforced composites. *J. Mater. Sci.* 17: 1359–63

Xing J, Hsiao G C, Chou T W 1982 A dynamic explanation of the hybrid effect. *J. Compos. Mater.* 15: 443–60

Zweben C 1977 Tensile strength of hybrid composites. *J. Mater. Sci.* 12: 1325–37

B. Harris
[University of Bath, Bath, UK]

I

In Situ Composites: Fabrication

The term in situ composites has been applied to aligned two-phase materials in which the reinforcing phase is produced simultaneously with the matrix (e.g. directionally solidified eutectic, monotectic, and cellular–dendritic polyphase alloys). The use of the term in situ was intended to contrast these melt-grown materials with those in which the reinforcement is formed by one process and the matrix added in another. However, this simple distinction has become more diffuse. Solid phases can also be used as starting materials, providing composite microstructures by control of disproportionation or deformation. Additionally, aligned composite growth can be induced by isothermal reaction of a crystalline-solid solution with a reactive component in the liquid, solid or gaseous state. These in situ composites generally involve oxygen in the gaseous state, which preferentially oxidizes one of the components of a solid solution or compound phase (Cense et al. 1977).

In situ pathways or compositing techniques are each confined to a thermodynamic or kinetic space, within which control of composition and process variables achieves the desired microduplex structure. There are thousands of fabricated in situ composites, consisting of metals, intermetallics, ceramics, semiconductors and organic compounds. It is the control over the duplex microstructure which endows these materials with interesting new properties—magnetic, electrical, optical, thermal, chemical and mechanical. By combining the properties of two or more anisotropic phases in one body, the pathway has been opened for the development of new classes of materials with a wide range of uses. Potential applications range from aerospace structures and turbine components (including fibrous electrostatic atomizers) to aligned superconductors and electronic submicrometer conductor arrays.

1. Eutectic, Monotectic and Off-Eutectic Solidification

The schematic phase diagram in Fig. 1 shows the eutectic invariant point at temperature T_e and composition C_e. At this point, solid phases α and β simultaneously solidify from the liquid L. For many eutectic or monotectic systems, if this solidification is done directionally, an aligned two-phase solid can be produced (Lemkey 1984). At least three conditions must be simultaneously satisfied, however, to produce the desired parallel duplex microstructure: first, the heat must be removed from the melt in a unidirectional manner; second, a sufficiently positive temperature gradient must be maintained ahead of the solidifying interface to prevent nucleation and maintain interface planarity; and third, cooperative nucleation and growth processes must occur between the phases. The α and β phases usually grow normal to this interface, with certain crystallographic directions in one or both of the phases tending to line up parallel to the solidification direction. On examining this process, it becomes clear that the major solidification parameters the thermal gradient G at the liquid–solid interface and the growth rate R or the velocity at which the liquid–solid interface advances—are keys to the control of structure and hence the chemical, mechanical and physical properties.

With the normal temperature gradients ($G \simeq 10$–$100\ \mathrm{K\,cm^{-1}}$) used in unidirectional solidification, aligned composites with inter-rod or lamellar spacings λ typically in the range 0.1–10 μm are produced. This comes about because the most stable value of λ is determined during the solidification process by the balance between that component of the total driving force needed to provide the excess free energy for forming interphase surface area and that required to drive the mass-transport process in the liquid by diffusion. This free-energy balance leads to the relationship

$$\lambda^2 R = \text{constant} \quad (1)$$

where λ is the spacing of the microstructure (Fig. 1), R is the solidification velocity and the constant is dependent on the material system and on such parameters as the diffusion coefficient and the interphase surface energy.

If the melt composition is far removed from the eutectic point, solidification of one phase before the other occurs in the form of coarse primary dendrites (Fig. 2). In most cases these dendrites will produce undesirable cast properties; however, under appropriate growth conditions, compositions deviating in both directions from the binary eutectic composition C_e can be directionally solidified to produce aligned eutectic-like composites with varying volume fractions of the phases.

Another approach to vary the volume fraction of the phases as well as their composition lies in the biphase solidification of monovariant eutectic alloys. These compositions lie on or near the eutectic troughs in the liquidus surface of ternary alloys. The solidification considerations to promote planar coupled growth are identical to those for off-eutectic composite growth (Flemings et al. 1972) and are given by the simplified constitutional supercooling equation

$$G/R \geqslant \Delta T/D \quad (2)$$

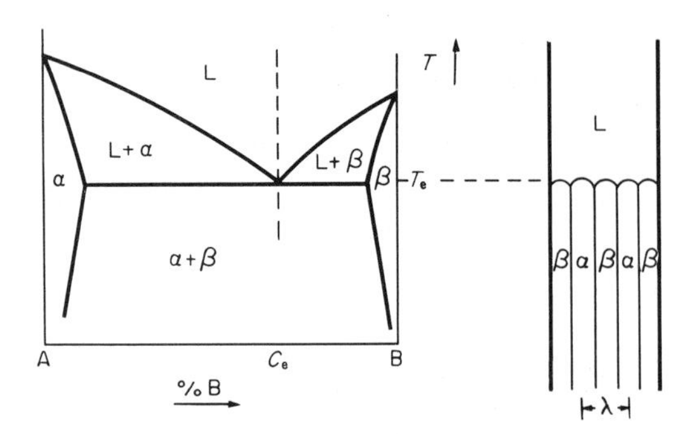

Figure 1
Schematic phase diagram characteristic of a eutectic system

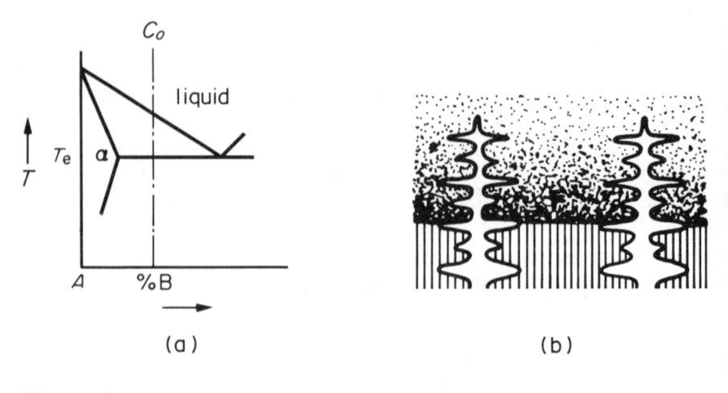

Figure 2
Dendritic solidification of Cu–Nb: (a) phase diagram for the alloy, (b) growth of aligned dendrites

where G is the temperature gradient in the liquid, D is the diffusion coefficient and ΔT is the temperature range between start and completion of solidification. (Note that in the simple binary system of Fig. 1, ΔT is zero and increases monotonically as the composition departs from C_e.)

The ability to depart from the precise eutectic composition is extremely important from the technological point of view, as it permits the addition of various alloying elements to modify both the chemistry of the phases and their volume fractions. This 'alloyability' in multicomponent systems played an important role in the development of high-temperature in situ composites with complex nickel-base superalloy matrices.

2. Co-deformation

There are alternative approaches which may be used to fabricate composite materials by directional solidification in alloys which do or do not possess a eutectic-like phase transformation. Consider the phase diagram of Fig. 2(a) for an alloy C_o, rich in A, which freezes above the eutectic temperature T_e. When the liquid of this alloy begins to solidify, it will form dendrites of α growing into the liquid as long branched structures. By maintaining unidirectional heat flow, it should be possible to force the dendrites to grow as an aligned array in the heat flow direction as shown in Fig. 2(b). If the alloy consists of two ductile phases and is subsequently deformed by mechanical reduction (e.g., swaging, rolling or drawing) then another class of in-situ-formed filamentary composite is created (Bevk et al. 1982, Verhoeven et al. 1986). It can be characterized by an extremely dense dispersion of very small filaments (about 10 nm to 1 μm in diameter) embedded in the matrix. A Cu–Nb in-situ-formed composite with the copper matrix etched away, showing loose ribbon-like niobium filaments, is shown in Fig. 3. The sample shown was cold-drawn without any intermediate annealing to 99.5% reduction. The shape of the filaments reflects the fact that body-centered cubic (bcc) metals

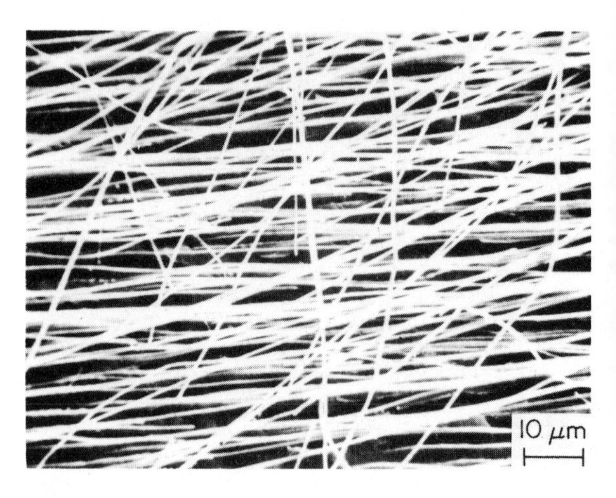

Figure 3
Dendritic structure of Cu–Nb in-situ-formed composite after 99.5% mechanical deformation

develop ⟨110⟩ textures during mechanical elongation and consequently deform in plane strain rather than in an axially symmetric mode. For any bcc fiber, slip will only be able to occur in two of the four ⟨111⟩ slip-directions so that the plane defined by these two slip-directions will become the longer direction of the fiber cross section. Besides the directional dendritic structures, the starting two-phase alloys may be produced either by means of powder metallurgy or by rapid cooling.

One of the most striking characteristics of these ultrafine filamentary structures prepared by deformation is their exceptional mechanical strength as illustrated in Fig. 4. This remarkable behavior is linked to the presence of densely spaced interfaces which inhibit dynamic recovery in both the matrix and filaments.

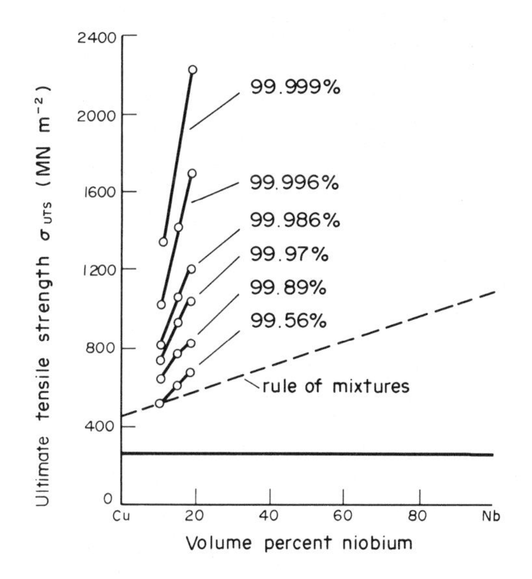

Figure 4
Ultimate tensile strength of in situ formed Cu–Nb composites. Percentages = percentage reduction in area; the horizontal line represents the strength of "pure" metals in the annealed state

3. Liquid-Phase Reactions

Composites with aluminum or copper matrices reinforced with fine carbide particles such as titanium carbide, tantalum carbide or niobium carbide can be formed by liquid-phase reactions in aluminum–titanium, aluminum–tantalum, aluminum–niobium or copper–titanium matrices (Fishman 1987). Stoichiometric refractory carbides 0.1–3 μm in size have been produced. The carbide precipitates form in situ and remain dispersed in the melt, which can be roller quenched, formed into powder or conventionally cast. A basis for understanding the reactions in the in situ processing technique lies in the method for producing bulk ceramic materials developed in the USSR, known as self-propagating high-temperature synthesis (SPHTS). It utilizes the exothermic heat of reaction, evolved when the elements making up the compound combine to spontaneously sustain the reaction. This process has been extensively exploited in the USSR since the 1960s, and most known refractory compounds have been produced in this way. In order to utilize these reactions to produce composites, a gas containing the reactant element is introduced into the molten alloy; a controlled exothermic reaction is thus initiated. If gas molecules consisting of unlike atoms, such as NH_3 or hydrocarbons C_mH_n, are decomposed at high temperatures on the surface of a metal M, frequently atoms of only one element (e.g., N or C) are absorbed and atoms of the other element are desorbed. Reactions such as this can be characterized by the following equations:

$$\text{Al–M(liquid)} + C_mH_n\text{(gas)} = \tfrac{n}{2}H_2\text{(gas)} + \text{MC(solid in liquid metal)}$$

or

$$\text{Al–M(liquid)} + NH_3\text{(gas)} = \tfrac{3}{2}H_2\text{(gas)} + \text{MN(solid in liquid metal)}$$

Additionally, ternary or quaternary metal carbides can form, such as (Ti, Ta)C. In comparison with SPHTS, the in situ reaction occurs on a less exothermic and more controlled scale. The exothermic heat of reaction is dissipated by the bulk of the liquid alloy, thereby reducing the efficiency of the reaction and the rate of formation of the intermetallics. In addition to the formation of very fine, stable precipitates, the processing advances include the ability to continuously cast sheets directly. Such materials have been found to have low cost, but offer strength and wear advantages over nonreinforced alloys or artificially produced metal-matrix composites.

4. Melt Oxidation

One of the new and exciting processes to be developed in the 1980s is the Lanxide process, which involves formation of ceramic–metal composites by the oxidation of bulk molten metal (Newkirk et al. 1986). The Lanxide process is thought to be generic in nature, although to date only alumina–aluminum and aluminum-nitride–aluminum composites have been reported. The process involves one or more metals giving up electrons to, or sharing electrons with, another element or combinations of elements to form a compound. The solid ceramic–metal body is produced by a directed growth process that results from the unusual oxidation behavior. The solid reaction product forms initially on the exposed surface of the molten metal and grows outward fed by transport of additional metal by capillary action through narrow channels in the ceramic oxidation product. One or more dopants, such as magnesium or silicon, are usually alloyed with the parent metal to stimulate the rapid oxidation procedure. Typically, dopant concentrations range from less than one percent to a few percent by weight. Temperature is an important parameter, as this oxidation phenomenon occurs within a limited temperature interval, located between 1370 K and 1670 K for the formation of alumina, for example.

Once started under appropriate conditions, the growth produces a composite of uniform microstructure, as long as the molten metal and oxidation are available and the proper temperature is maintained. Growth kinetics are not parabolic but appear to be linear with time. Growth rates depend upon the metal, oxidation and process temperature employed. Composites up to 10 cm thick with weights of 180 N have been grown (in times of about 24 hours) with no

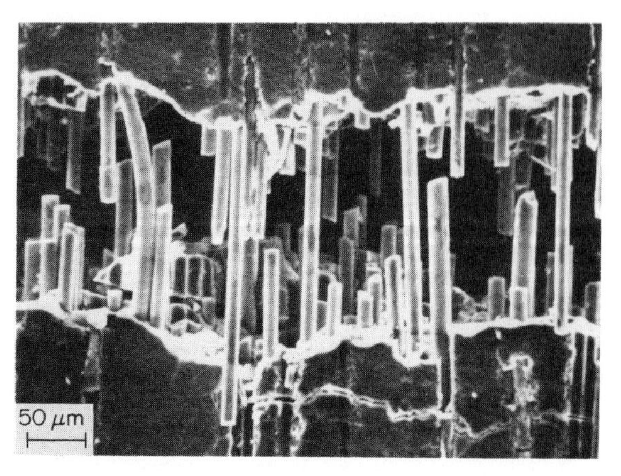

Figure 5
Aluminum alloy–Al_2O_3-matrix composite reinforced with Nicalon fibers produced by the Lanxide process, showing the extensive fiber pullout that contributes to toughening (after Mehrabian 1988)

indication of decrease in growth rate. The composite matrix can grow around a bed or preform of fibers and/or filler material, as shown in Fig. 5, without disturbing its arrangement. Thus, in situ composites of near-net shape can be produced. Alumina–aluminum–silicon-carbide-fiber composites have been produced with fracture toughness values of 18 MPa m$^{1/2}$ (Newkirk et al. 1987). The volume fractions of metal and ceramic, the scale of the micro-structure, the porosity and the degree of inter-connectivity of the metallic phase in the end product can be controlled by varying the processing parameters (temperature, time, dopant).

The mechanism of the unusual oxidative growth in the Lanxide process has not been completely defined. Central to any proposed mechanism is an explanation for the continued wicking of molten aluminum through the oxide that is required to sustain the process, and the nonprotective nature of the oxide formed. A tentative explanation has been proposed, based upon transmission electron microscopic observations that the majority of grain boundaries between alumina crystallites in the composite are low-angle. High-angle ceramic–ceramic grain intersections apparently contain thin channels of aluminum metal. It is thought that the surface energy for the alumina–aluminum is lower than that for either alumina–air or the grain boundary energy for high-angle alumina–alumina boundaries, and that such boundaries are therefore filled by the molten aluminum forming the channels necessary to sustain the growth process.

5. *Precipitation*

A new class of castable metal-matrix composites, XD composites, are produced by in situ precipitation of various reinforcements, such as borides and carbides of elements such as titanium and zirconium, directly in the molten matrix of interest. Such matrices as aluminum, nickel and copper, as well as various intermetallic aluminides, have been demonstrated. The reinforcements, which are uniformly distributed in the matrix, can be produced either in the form of particles or whiskers, depending upon the reinforcement chemistry or processing conditions. It is possible to simultaneously introduce a number of different reinforcements in a given matrix, thus allowing for the design of the microstructure to meet a specific material need (fine particles for strength and whiskers for creep rupture, for example).

The XD process has several attractive features. As a result of its compatibility with conventional ingot-processing techniques, it is relatively inexpensive. Because the reinforcements are formed in situ, they are not only thermodynamically stable but have non-contaminated interfaces with the matrix metal. Since precipitate size, chemistry and shape can be controlled, microstructures can be designed or engineered to provide optimum mechanical performance. For example, in alloys which are difficult to fabricate, such as the titanium–aluminides, the fine microstructures resulting from XD processing not only aid fabricability through toughness and ductility enhancement, but are also effective in increasing high-temperature strength and tensile modulus.

Bibliography

Bevk J, Sunder W A, Dublon G, Cohen D E 1982 Mechanical properties of Cu-based composites with in situ formed ultrafine filaments. In: Lemkey F D, Cline H E, McLean M (eds.) 1982 *In Situ Composites IV*. Elsevier, New York

Cense W A, Klerk M, Albers W 1977 Aligned in situ growth of the fibrous Ni–Nb_2O_5 composite by preferential internal oxidation. *J. Mater. Sci.* 12: 2184–88

Fishman S G 1987 In situ and near net shape processing of composites. *J. Met.* 39: 26–27

Flemings M C, Sharp R M, Rinaldi M D 1972 Growth of ternary composites from the melt: part I. *Metall. Trans., A* 3: 3133–38

Lemkey F D, 1984 Advanced in situ composites. In: Murr L E (ed.) 1984 *Industrial Materials Science and Engineering*. Dekker, New York, pp. 441–69

Mehrabian R 1988 New pathways to processing composites. In: Lemkey F D, Evans A G, Fishman S G, Strife J R (eds.) 1988 *High Temperature/High Performance Composites*. Materials Research Society, Pittsburgh, Pennsylvania

Newkirk M S, Urquhart A W, Zwicker H R, Brevel E 1986 Formation of Lanxide ceramic composite materials. *J. Mater. Res.* 1: 81–89

Newkirk M S, Lesher H D, White D R, Kennedy C R, Urquhart A W, Claar T D 1987 Preparation of Lanxide ceramic matrix composites: matrix formation by the directed oxidation of molten metals. *Ceram. Eng. Sci. Proc.* 8: 879–85

Verhoeven J D, Schmidt F A, Gibson E D, Spitzig W A 1986 Copper-refractory metal alloys. *J. Met.* 9: 20–24

F. D. Lemkey and S. G. Fishman
[United Technologies, East Hartford, Connecticut, USA]

In Situ Polymer Composites

Synthetic composite materials are formed by the introduction of fibers or particles into a matrix of resin, metal or ceramic. Long-fiber reinforcement of resins leads to increased strength and toughness over the properties that would be obtained from a monolithic piece of the (usually brittle) reinforcing material. Particles or short fibers can also be used to increase the modulus and yield point of polymers or other ductile materials. In this context, filled polymers can be compared to metal alloys where a second phase may toughen or harden a matrix. In the study of polymers the term alloy is normally used for blends where a dispersed rubbery phase toughens a glassy matrix or a hard phase acts as a cross-linker in a thermoplastic rubber. As well as composites where a solid phase is mixed into a liquid from a different class of material, and alloys where similar materials phase-separate during processing, there is a third class, called in situ composites, where dissimilar materials phase-separate during processing. Many biological materials are in this last category but the field is largely unexplored for synthetic materials. This article describes some of the relevant biological examples, reviews current synthetic efforts and discusses the potential for application of such systems. Since the field has not been reviewed in the past, the Bibliography contains a number of typical papers which give readers access to the wider literature.

1. General

A number of books and reviews have discussed the materials science of biological materials. Biological composite materials such as bones, teeth and shells have a polymeric matrix (protein and polysaccharide) reinforced by a mineral, usually hydroxyapatite or calcium carbonate, which forms within the matrix. Most biological materials are composites at all levels from the organization of individual macromolecules to the whole organism. Those materials which play a predominantly structural role, such as bone or tendon, comprise a swollen polymeric matrix reinforced with polymer fibers and/or a mineral filler. In bone, for example, a glycosaminoglycan matrix is reinforced with collagen fibers and with plate-like crystallites of hydroxyapatite, a calcium phosphate. The extent of mineralization varies depending on the exact function of the bone and on the species. A typical level is about 38 vol% mineral with a crystallite thickness of 4 nm and a width of 35 nm. The structure of mammalian tooth is similar but the mineral level is higher, about 86 vol%, and the individual crystallites are larger. Shells of invertebrates have a comparable range of structures but the mineral is normally calcium carbonate and the matrix is reinforced with chitin. In comparison with equivalent filled-polymer composites, biological materials have two striking characteristics: the shape, size, orientation and organization of the mineral in the matrix show a high degree of sophistication when compared to the random dispersion of particles in a filled-polymer composite; and the structures form by growth of the mineral phase in the polymer rather than by the synthetic route of the dispersion of particles into a liquid resin. Mann (1983) has reviewed the chemistry and physics of biological mineralization.

The simplest way to prepare a precipitate within a polymer is to dissolve the additive in the polymer at high temperature and induce precipitation by cooling. This method will be limited to organophilic solutes, which have a high degree of solubility in the polymer. A more versatile method is to codissolve the additive and polymer in a mutual solvent which is then evaporated. A much wider range of composites can be prepared if precipitation is induced by a chemical reaction within the polymer. Bone mineralization could be viewed in this way in that calcium and phosphate are introduced into the matrix in a soluble form and then released to combine and precipitate.

Precipitation within a polymer will differ from precipitation from solution in a number of ways. Firstly, there is no convection or flow within a polymer so particle collisions and agglomeration will be largely eliminated. Particle growth will be controlled to a great extent by diffusive processes within the matrix. Also the matrix can be oriented and so can transmit a perferred orientation to the precipitate. In addition the matrix may be semicrystalline or otherwise structured such that there may be an epitaxial relationship between the polymer and the precipitate. In crystallization from solution the solvent can influence the particle morphology through the dependence of growth kinetics on the surface energy of different crystal faces. This effect would also be expected in polymeric media and may be the explanation for preferred precipitation sites in a number of precipitation-induced diseases such as gout. Finally, it is known that small concentrations of polymers in solution can modify or "poison" crystal growth by adsorption to surface step sites. Such selective poisoning may be important in the control of bone growth.

2. Crystallization from Solution in Glassy Polymers

A number of cases of the growth of organic crystals from amorphous polymers have been studied by Narkis et al. (1979) and by Joseph et al. (1968). The glass transition temperature of the polymer is reduced by the organic solute. The solute shows a temperature-dependent solubility in the polymer, and will crystallize within the polymer at temperatures above the glass transition of the blend. The crystal morphology depends on the growth conditions. Acetanilide crystallizing from polystyrene forms rods which decrease in diameter from 10 μm to 0.1 μm as the crystallization temperature is reduced from 90 °C to 45 °C.

Moyle et al. (1987) have worked on such systems, focusing on organic crystals with interesting nonlinear optical properties, in conditions where composites may prove more tractable materials than single crystals. Several groups have prepared conducting composites by the precipitation of conducting charge-transfer complexes in polymers.

3. Crystallization from Solution in Crystalline Polymers

Crystalline polymers generally show little solubility for small molecules within the crystal structure and normally form eutectics with organic compounds if the two liquids are miscible. For example, the system benzoic acid–polyethylene has a eutectic point at 25% benzoic acid and 388 K, while polypropylene is immiscible with benzoic acid in the liquid state. The situation is somewhat more complex than for simple binary systems because polymers always contain some amorphous material which will solubilize a fraction of the organic solute. Smith and Pennings (1976) have studied mixtures of polyethylene and polypropylene with some aromatic halocarbons. The morphology of the solid is highly dependent on composition and crystallization rate. On the solute-rich side of the eutectic point, large crystals form, surrounded by clumps of polymer lamellae. Polymer-rich mixtures form the spherulitic structures characteristic of polymers with embedded small solute crystals. There are a few systems such as resorcinol–poly(ethylene oxide) where the solute and polymer form a stable mixed crystal (Myasnikova et al. 1980).

4. Precipitation by Reaction

Dispersed precipitate phases in polymers can be formed by dissolving an organophilic metal compound into the polymer and then subsequently treating the film with a reagent to precipitate a metal, oxide or other compound.

Mark et al. (1985) have blended silicone rubbers with liquid silicon tetraethoxide. The alkoxide was allowed to hydrolyze within the solid rubber by exposure to moist air such that it converted to finely divided silica which simultaneously acts as a filler and a cross-linker. Electron microscopy shows particles of 10–20 nm. Up to 17 wt% filler has been incorporated. Up to 1.5% titania has also been incorporated into polydimethylsiloxane (PDMS) by in situ hydrolysis of titanium tetra-*n*-propoxide.

A number of groups have explored the production of particles of magnetic metals in polymers by decomposition of metal carbonyls. Hess and Parker (1966) prepared cobalt particles of 10–100 nm by thermal decomposition of dicobalt octacarbonyl in solutions of various polymers. The polymer promoted the formation of single-domain particles rather than large multidomain particles. In some cases the particles formed chains which had high coercive forces and remanence ratios. Smith and Wychick (1980) made similar dispersions of up to 8% of 10 nm iron particles by thermolysis of iron pentacarbonyl in polymer solutions. This approach has been extended to precipitation into solid polymers by Reich and Goldberg (1983) who prepared iron and magnetic γ-Fe_2O_3 in a variety of polymers by carbonyl decomposition. Up to 20 wt% iron oxide was incorporated into polyvinylidene fluoride. Sobon et al. (1987) have studied the precipitation of iron oxides and iron into polymers by the hydrolysis or reduction of $FeCl_3$ or iron(III) acetylacetonate.

There have been a number of reports of conducting composites made by the decomposition of metal salts in polymers. Polyimides have been doped with silver, gold or palladium salts which decompose to metal during curing of the resin at 200–300 °C. In some cases the metal forms as a surface mirror which makes the film highly conducting at low metal contents (Auerbach 1984). Silver has also been formed in fine conducting lines in polyimide by laser decomposition of dissolved silver nitrate (Auerbach 1985). Copper sulfide and cadmium sulfide dispersions in polymers have been prepared and become highly conducting at 40 wt% sulfide, and electrically conducting copper-sulfide-treated polyacrylonitrile fiber is reported to be commercially available (Yamamoto et al. 1986). Copper has been precipitated into cellulose to 18 wt% by first dissolving the cellulose into a basic copper solution and then reducing it (Fitzgerald et al. 1986). A number of recent papers have described a counter-current diffusion method for the production of metal films embedded in polymers (Manring 1987).

5. Evaporative Precipitation

Kovacs and Vincett (1985) have published a number of papers on the evaporation of metals (especially selenium) or salts onto soft polymer surfaces. Inorganic vapors tend to form subsurface particles while organic materials remain partly embedded in the surface. This effect is determined by the various surface tensions. What is more surprising is that the inorganic particles form as a monolayer of uniformly sized particles a few nanometers below the surface. As evaporation proceeds the spherical particles increase in diameter but maintain a constant area coverage and do not form continuous films until, at about 0.4 μm, particle coalescence becomes too slow to balance the influx of metal.

6. *Glass and Metal Systems*

A typical glass–ceramic system is lithia–silica–alumina where a fine-grained precipitate of β-spodumene or β-quartz is formed by reheating the glass to a crystallization temperature of around 1090 °C. A nucleating addition of titania, or a metal such as gold, is frequently used. The titania phase-separates and then crystallizes as $Al_2Ti_2O_7$ during a lower-temperature (900 °C) nucleation treatment. The nucleation treatment is carried out just above the glass transition temperature to obtain abundant nuclei. The resultant glass-ceramic is from 50–100 vol% crystalline with a very fine crystal size, around 1 μm. Glass ceramics can be oriented by crystallization in a temperature gradient. This method has also been applied to a number of eutectics including glass-forming oxides and salts (Carpay and Cense 1974). Highly aligned lamellar or fibrous structures have been produced on a scale that decreased from 10 μm to 1 μm as the crystallization rate increased. Similar structures have been formed in metal alloys and are also termed in situ composites (Albers 1974).

7. *Potential Applications*

The most obvious applications for in situ polymer composites are as reinforced materials. The major drawback is that the processing is quite slow for bulk materials but not for thin films or fibers. In many cases it will also be easier to get property improvements by using a different polymer rather than trying to attain a sophisticated morphology. There would seem to be more promise in the combination of specific electrical, magnetic or optical properties of the precipitate with the toughness of polymers. This would be particularly true as we learn to control the particle size, volume fraction, distribution and orientation.

Bibliography

Albers W 1974 The growth and degree of dispersion of in situ composites. *Acta Electron.* 17: 75–86

Auerbach A 1984 A method for increasing the conductivity of silver-doped polyimide. *J. Electrochem. Soc.* 131: 937–38

Auerbach A 1985 A method for reducing metal salts complexed in a polymer host with a laser. *J. Electrochem. Soc.* 133: 1437–40

Carpay F M A, Cense W A 1974 In situ growth of composites from the vitreous state. *J. Cryst. Growth* 24/25: 551–54

Fitzgerald E, Gadd K F, Mortimore S, Murray W 1986 Small copper particles in a cellulose matrix. *J. Chem. Soc., Chem. Commun.* 21: 1588–89

Hess P H, Parker P H 1966 Polymers for stabilization of colloidal cobalt particles. *J. Appl. Polym. Sci.* 10: 1915–27

Joseph J R, Kardos J L, Nielsen L E 1968 Growth, morphology and reinforcement potential of low molecular weight crystals in amorphous polymer matrices. *J. Appl. Polym. Sci.* 12: 1151–65

Kovacs G J, Vincett P S 1985 Particles in polymers: Surface chemistry of their nucleation, growth, configuration, and interactions with the matrix. *Can. J. Chem.* 63: 196–203

Mann S 1983 Biomineralization in biological systems. *Struct. Bonding (Berlin)* 54: 125–74

Manring L E 1987 Electroless deposition of silver as an interlayer within polymer films. *Polym. Commun.* 28: 68–71

Mark J E, Ning Y-P, Jiang C-Y, Tang M-Y, Roth W C 1985 Electron microscopy of elastomers containing in-situ precipitated silica. *Polymer* 26: 2069–72.

Moyle B D, Ellul R E, Calvert P D 1987 Second harmonic generation by composite materials. *J. Mater. Sci. Lett.* 6: 167–70

Myasnikova R M, Titova E F, Obolonkova E S 1980 Study of 2:1 poly(ethylene oxide)–resorcinol molecular complex. *Polymer* 21: 403–07

Narkis M, Siegmann A, Puterman M, DiBenedetto A T 1979 Glassy polymer solutions: Morphology of in-situ crystallized additives. *J. Polym. Sci., Polym. Phys. Ed.* 17: 225–34

Reich S, Goldberg E P 1983 Poly(vinylidene fluoride)–γFe_2O_3 magnetic composites. *J. Polym. Sci., Polym. Phys. Ed.* 21: 869–79

Smith P, Pennings A J, 1976 Eutectic solidification of the pseudo binary system of polyethylene and 1,2,4,5-tetrachlorobenzene. *J. Mater. Sci.* 11: 1450–58

Smith T W, Wychick D 1980 Colloidal iron dispersions prepared via the polymer-catalyzed decomposition of iron pentacarbonyl. *J. Phys. Chem.* 84: 1621–29

Sobon C A, Bowen H K, Broad A, Calvert P D 1987 Precipitation of magnetic oxides in polymers. *J. Mater. Sci. Lett.* 6: 901–4

Strnad Z 1986 *Glass–Ceramic Materials*, Glass Science and Technology Vol. 8. Elsevier, Amsterdam

Vincent J F V 1982 *Structural Biomaterials*. Macmillan, London

Wainwright S A, Biggs W D, Currey J D, Gosline J M 1976 *Mechanical Design in Organisms*. Arnold, London

Yamamoto T, Kubota E, Taniguchi A, Kubota K, Tominaga Y 1986 Electrical conduction properties of CuS- and CdS-polymer composites prepared by using new organosols of CuS and CdS. *J. Mater. Sci. Lett.* 5: 132–34

P. Calvert
[Arizona Materials Laboratory, Tucson, Arizona, USA]

J, K

Joining of Composites

As fiber-reinforced materials, particularly continuous-fiber-reinforced plastics (FRP), become more widely used an important requirement for their full exploitation is the development of suitable attachment methods and general joint design philosophies.

Unlike isotropic materials, problems of joining FRP arise mainly due to the unique characteristics of the material, that is, the inherent weakness of the material in interlaminar shear, transverse tension and transverse compression. Also, since the material remains elastic to failure (in unidirectional material where fibers are oriented in one direction), there is no plastic stress–strain behavior, and therefore the mechanisms of stress relief around geometric discontinuities, such as holes, require special attention. Clearly all of these problems require full consideration before engineers can embark upon the design of an efficient joint, efficiency being measured in terms of high strength for low weight and/or cost.

The epoxy-resin-matrix FRP (which make up the bulk of the current high-technology matrices for FRP) cannot be welded, soldered or brazed. With the exception of components made using filament winding techniques, where there is usually no significant end attachment problem, most joints are effected by means of adhesive bonding, mechanical fasteners or a combination of the two.

With the recent introduction of thermoplastics into the field of structural plastics, especially for the aerospace industry, other methods of joining are being explored, one of particular interest that promises future potential is that of welding.

1. Adhesive Joints

The form of adhesive-bonded joints that has probably received most attention is a simple lap joint. This is attractive for two main reasons: it is simple to fabricate, and many of the problems of load introduction such as are found in mechanical joints are largely eliminated. However, even with the use of advanced adhesive technology only relatively low load-transfer rates can be realized because of the low shear strength of both the adhesive and the composite adherend, and the high thermal strains that can exist within the bonded regions due to the high cure temperature of adhesives. Also degradation of the adhesive, and the interface between adhesive and composite, can occur after exposure to certain environments. This is discussed in greater depth later in this section.

Failure of FRP adhesive joints can occur in any of four modes: adhesive failure at the interface between the adhesive and adherend, cohesive failure through shearing of the adhesive, peel failure due to out-of-plane loads tending to pull a joint or adherend apart in transverse tension, and adherend failure in axial tension or compression. Axial failure of the adherend occurs at or close to the end of a lap joint with failure initiating in axial tension or compression of the outer plies. Analysis shows that due to a shear lag phenomenon caused by shear deformation in the composite matrix, axial load distribution through a laminate's thickness is a function of axial distance from the joint. Consequently the most highly loaded plies are those on the face that is bonded, even outside the joint. A solution to the problem is to ensure the outer plies are oriented to give maximum strength, i.e., laid parallel to the load direction. Because of this mode of failure, high joint efficiencies are possible only in very thin laminates; for thick laminates high joint efficiencies are achieved by the use of multistepped or scarf (acute-angled-butt) joints, both of which are more difficult and costly to produce. The configurations of bonded joints available to a designer are numerous; the most common, together with their design limitations, are given in Fig. 1.

One of the problems associated with FRP is that of environmental degradation. This arises because of the readiness of both FRP adherends and adhesive to absorb moisture (water) when exposed to a humid atmosphere. This causes problems during the bonding operation, and also after bonding has been successfully achieved. During the bonding operation, which is usually carried out at an elevated temperature to ensure high-temperature structural properties, moisture diffuses out of the adherend into the adhesive producing high voiding in the adhesive and bond line. As a consequence the adhesive and cohesive properties of the adhesive are significantly reduced. Clearly, therefore, the bonding of "wet" adherends must be avoided, either by ensuring the bond surfaces do not absorb moisture or by drying the surfaces prior to bonding. The consequence of environmental exposure of assembled joints is discussed in the following paragraphs.

Central to any successful adhesive joint is joint preparation; this ensures that a maximum strength bond will exist between the adherend and adhesive layer interface. Since both FRP-to-FRP and FRP-to-metal joints will need to be used it is worth commenting on the essential differences between them.

The joint between an FRP adherend and an adhesive layer is, for epoxy-matrix FRP, fairly simple since they are essentially similar materials and usually the interface is not a plane of weakness since joint

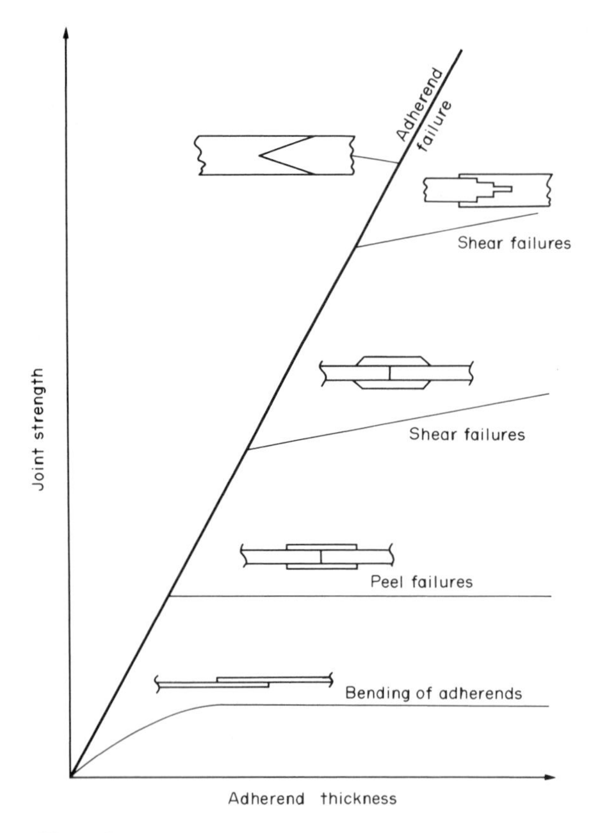

Figure 1
Bonded joint types and their design limitations.

failure more often occurs in the FRP adherend. Of major concern in the surface preparation is that of contamination prior to bonding. Provided this is avoided a good joint should be obtained. During the exposure of the joint to a humid environment moisture is absorbed by the FRP adherend, and a plasticization of the matrix and a lowering of the material glass-transition temperature occurs; this directly influences the matrix-dependent strengths properties. A similar effect is observed in most adhesives, causing some reduction in compliance and modulus. Moisture absorption also produces swelling of both adherend and adhesive which can result in differential swelling stresses and possible damage of the bond.

Metallic adherends are the most complex to prepare for bonding since it is necessary to etch and/or anodize the bonding surface to produce an oxide layer. Priming of the surface is often necessary to improve the bonding further and to provide a surface protection for production purposes. One of the main difficulties associated with metallic-to-FRP bonding is that of subsequent environmental exposure. Due to moisture ingress through the FRP and through the adhesive edges, hydrolizing of the metallic bond surface can occur. This is usually brought about when substandard surface treatment has occurred. Unlike the FRP-to-FRP bond the metallic-to-FRP bond can have a major plane of weakness at the metal–adhesive interface.

The adhesive joining of thermoplastics using standard epoxy adhesives is not as simple as joining thermosets, owing to the surface chemistry of the two being so different. The surface pretreatment of a thermoplastic is, therefore, more critical and needs special attention if good adhesion is to be achieved. One method that is currently showing great promise is that of the corona discharge method. This improves the "wettability" of the surface which gives a correspondingly improved bond over other abrasive type preparations.

2. *Mechanical Joints*

With the increasing interest in the use of mechanical fasteners for joining FRP components a variety of fastening techniques that are used for metals have been successfully applied to many forms of FRP. These range from self-tapping screws which are used for lightly loaded connections to quality engineering bolted connections used in heavily loaded structures. As with metals, bolted or demountable joints are needed in FRP structures where there is a call for inspection and servicing, modification, repair or replacement of damaged or worn components, or in situations where off-axis loads have to be accounted for and where bonding is uneconomic or impractical. Hence holes or cutouts in FRP laminated composites have to be catered for in design. Like metallic structures, FRP exhibit failure modes in tension, shear and bearing but, because of the complex failure mechanisms of FRP, two further modes are possible, namely cleavage and pullout. Figure 2 shows the location of each of the modes.

It is well known that holes can seriously weaken a laminate, particularly if the laminate is highly anisotropic. This is due in part to the high stress concentrations that occur in the region around such discontinuities, which in tension can be as large as 8 for highly anisotropic carbon-fiber-reinforced plastics (CFRP), compared with a much lower value of 3 normally associated with isotropic materials. Furthermore, isotropic materials yield before failure and the effect of stress concentrations on the net failing stress is small. Since unidirectional FRP do not respond in this way it is not surprising that mechanical fastening of unidirectional-fiber materials is inefficient. However, the problem can be over-come by reducing the degree of anisotropy in the region of a hole and imparting a degree of "plasticity" by the incorporation of fibers oriented in a number of different directions. In fact the failing stress of all three principal modes of failure, that is, tension, bearing and shear pullout, are increased by the inclusion of fibers oriented at $\pm 45^\circ$. Figure 3 shows the change in failure mode of a $0^\circ \pm 45^\circ$ laminate

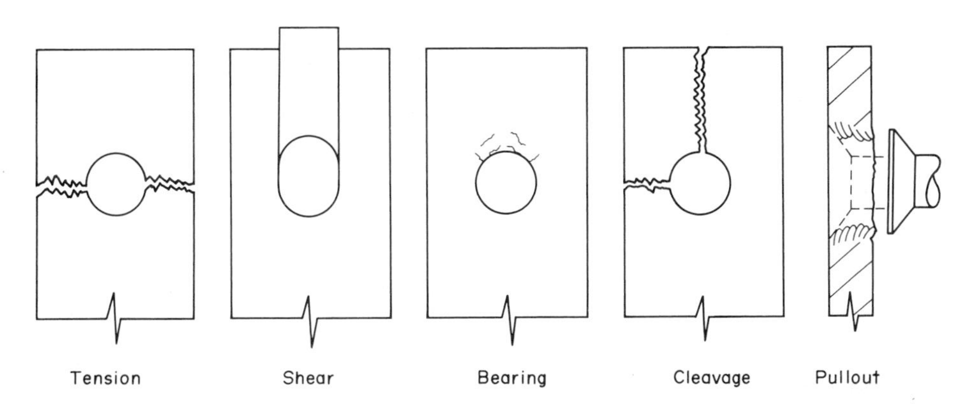

Figure 2
Modes of failure for mechanical joints in FRP

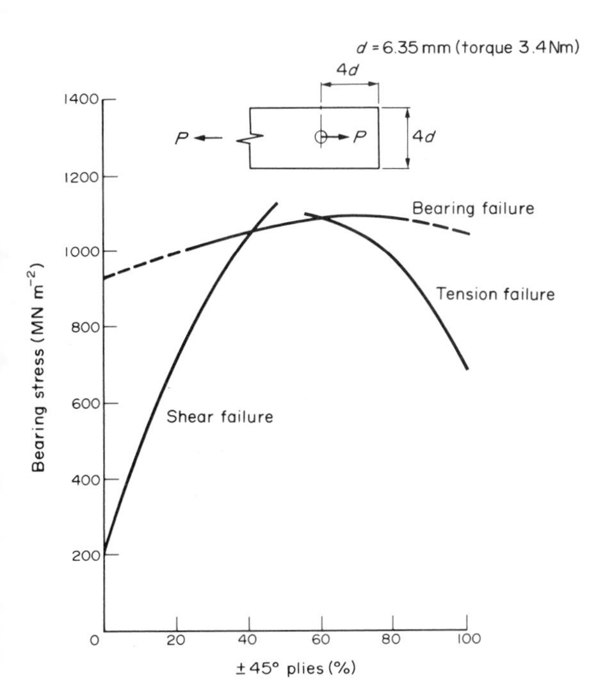

Figure 3
Influence of fiber orientation on failure mode ($0° \pm 45°$ CFRP)

as the contribution of $\pm 45°$ fibers is varied. If we express the efficiency of a joint in terms of its specific strength, then the efficiency of a carbon-fiber-reinforced epoxy joint can be greater than that of an L71 aluminum alloy (Cu $4\frac{3}{4}$%–Si $\frac{3}{4}$%–Mn $\frac{3}{4}$%–Mg $\frac{3}{4}$%) by a factor of 1.32 in tension, 3.7 in bearing and 1.1 in shear pullout.

Environmental degradation of a bolted joint, after exposure to a hot, wet environment, is most likely to occur in the shear and bearing strength properties. At present evidence exists for the bearing property only. The evidence shows for fiber-reinforced epoxies that temperature has a more significant effect than moisture but in the presence of both (at 127 °C) a strength loss of 40% is possible.

The information available for the design of mechanical joints for fiber-reinforced thermoplastics is at present incomplete. The key ingredient still required for establishing a design capability is the characterization of variables such as fiber lay-up and stacking sequence in terms of joint parameters. Despite the incompleteness of this work sufficient evidence exists to suggest the failure behavior of thermoplastics is much the same as for thermoset FRP and that similar high joint efficiencies can be obtained.

3. *Welding*

The special properties of the currently available thermoplastics, in particular damage tolerance and high resistance to hot wet environments, make them suitable matrices for use with fiber reinforcement and, therefore, a strong contender to thermosets in the engineering and structural plastics field. Thermoplastics are, compared with thermosets, still in their infancy and need further development of fabrication techniques and joining methods before their full potential can be realized.

One of the major features of thermoplastics is their ability to be reformed by heating, and for this reason they lend themselves to joining techniques involving remelting that could never be expected of thermosets. Hence particular attention is now being given to developing welding procedures that are acceptable as efficient load-transfer methods. The experimental work has, so far, examined one particular thermoplastic, a polyether ether ketone (PEEK) manufactured by ICI (UK). Currently several techniques are under review, including the following.

(a) Hot-plate welding—this method gives sound

welds and can achieve good lap shear and fracture properties.

(b) Resistance-heating welding—with this method a resistively heated implant technique is needed to produce good properties without disturbance of the fibers. Unfortunately it is limited, subject to further development, to small joint areas.

(c) Induction welding—this technique is capable of continuous welding and produces good mechanical properties. The method shows great promise for the production of large assemblies and potentially of varying joint areas.

All of the three processes show a strong potential for use in production with induction welding showing the most promise. Further investigation into these techniques is required, together with the development of nondestructive testing techniques to qualify them.

4. Shim Joints

Shim joints use thin metallic interleaved layers to reinforce the composite material in the region of the joint. The action of the shims is to transfer load from the composite material through interlaminar shear between the shim and the composite and then through pin bearing by means of conventional bolts. Failure of this type of joint is much the same as for metals but includes a failure mode due to excessive shear between the composite and metal shims. The multishim joint is a compromise between bonded joints and mechanical joints, and it attempts to make use of the advantages found in each. Unfortunately, because of fabrication difficulties and emphasis on overall cost and weight, shim joints are not always an attractive method of joining.

All of the joining techniques discussed can be used without undue difficulty for FRP, but the behavior of the joints must be properly understood and the limitations realized if structural loads are to be carried safely.

Bibliography

Collings T A 1977 The strength of bolted joints in multi-directional CFRP laminates. *Composites* 7: 43–55

Collings T A 1987 Experimentally determined strength of mechanically fastened joints. In: Matthews F L (ed.) 1987 *Joining Fibre-Reinforced Plastics*, Elsevier Applied Science, London, pp. 9–63

Erdogan F, Ratwani M 1971 Stress distribution in bonded joints *J. Compos. Mater* 5: 378–93

Hart-Smith L J 1987 Design of adhesively bonded joints. In: Matthews F L (ed.) 1987 *Joining Fibre-Reinforced Plastics*. Elsevier Applied Science, London, pp. 271–311

Parker B M 1978 The effect of hot-humid conditions on adhesive-bonded CFRP-CFRP joints. In: *Symp. on Jointing in Fibre Reinforced Plastics*. IPCS Science and Technology Press, pp. 95–103

Wong J P, Cole B W, Courtney A L 1969 Development of the shim-joint concept for composite structural members. *J. Aircr.* 6 (1): 18–24.

T. A. Collings,
[Defence Research Agency, Farnborough, UK]

Kirchhoff Assumption (Kirchhoff Hypothesis)

In the elementary theory of beams and plates, certain assumptions are made regarding the stress distribution. If the axis of bending is in the 1 2 plane, the stress σ_3 is assumed zero. It is also assumed that any line perpendicular to the beam or plate midplane before deformation remains perpendicular to the midplane after deformation and is unaltered in length. It follows that the strains ε_{13}, ε_{23} and ε_3 are zero, hence σ_{13} and σ_{23} can be neglected. These assumptions constitute the Kirchhoff hypothesis. For thin plates, such as the individual lamellae in a laminate, they correspond to a state of plane stress.

See also: Laminates: Elastic Properties

A. Kelly
[University of Surrey, Guildford, UK]

L

Laminates: Elastic Properties

A composite lamina can be regarded as a planar arrangement of unidirectional fibers or woven fibers in a matrix. A lamina based upon a polymer matrix material can be produced by a filament-winding technique. Laminates based upon metal and ceramic matrices can be fabricated by various solid routes (diffusion bonding, hot rolling, extrusion and drawing, hot isostatic pressing, electrochemical plating and slurry infiltration), liquid routes (liquid infiltration, squeeze casting, compocasting, sol-gel, and eutectic solidification) and gaseous routes (chemical vapor deposition (CVD) and chemical vapor infiltration (CVI)). A composite laminate is constructed by stacking and bonding together individual laminae with various orientations of principal material directions. Basic arrangements of fibers in a laminate may include unidirectional (0°), cross-ply (0°/90°), angle-ply ($+\theta/-\theta$) and combinations thereof.

Lamination theory is a relatively mature subject; its treatment can be found in the work of, Ashton et al. (1969), Vinson and Chou (1975), Jones (1975), Tsai and Hahn (1980), and Carlsson and Pipes (1987), and in the review article of Chou (1988).

1. Elastic Behavior

1.1 Stress–Strain Relations for Unidirectional Laminae

A unidirectional lamina can be treated as an orthotropic material. Four independent constants are needed to specify its elastic behavior. These constants, referring to the fiber (x_1) and transverse (x_2) directions are denoted by E_1 (longitudinal Young's modulus), E_2 (transverse Young's modulus), ν_{12} (Poisson's ratio due to loading in the x_1 direction and contraction in the x_2 direction), and G_{12} (in-plane shear modulus). The relationship between these orthotropic elastic properties and the elastic properties of the fiber and matrix materials and the fiber volume fraction can be found using techniques developed in detail in the composites literature. Let the isotropic properties of the fiber and matrix be denoted by E (Young's modulus), ν (Poisson's ratio), G (shear modulus) and K (bulk modulus). Volume fractions are denoted by V, and the subscripts f and m indicate fiber and matrix, respectively. The relations (1a), (1b) from Hashin and Rosen are quoted for their concise forms and hence ease of application.

It should be noted that the expressions in Eqns. (1) are true for a unidirectional composite treated as a three-dimensional, transversely isotropic material. Five independent elastic constants are needed to describe its properties. K_t^* and G_t^* in Eqns. (1) are, respectively, the transverse plane-strain bulk modulus and transverse shear modulus of the unidirectional composite. For completeness, the transverse Poisson's ratio is given by

$$\left.\begin{aligned}
E_1 &= E_f V_f + E_m V_m + \frac{4V_f V_m(\nu_f-\nu_m)^2}{\dfrac{V_m}{K_f}+\dfrac{V_f}{K_m}+\dfrac{1}{G_m}}\\
E_2 &= \frac{4K_t^* G_t^*}{K_t^* + G_t^*\left(1+\dfrac{4K_t^*\nu_{12}^2}{E_1}\right)}\\
\nu_{12} &= \nu_f V_f + \nu_m V_m + \frac{V_f V_m + (\nu_f-\nu_m)\left(\dfrac{1}{K_m}-\dfrac{1}{K_f}\right)}{\dfrac{V_m}{K_f}+\dfrac{V_f}{K_m}+\dfrac{1}{G_m}}\\
G_{12} &= G_m \frac{V_m G_m + (1+V_f)G_f}{(1+V_f)G_m + V_m G_f}
\end{aligned}\right\} \quad (1a)$$

where

$$\left.\begin{aligned}
K_t^* &= \frac{K_m K_f + (V_f K_f + V_m K_m)G_m}{V_m K_f + V_f K_m + G_m}\\
G_t^* &= G_m \frac{(\alpha+\beta_m V_f)(1+\rho V_f^3) - 3V_f V_m^2 \beta_m^2}{(\alpha - V_f)(1+\rho V_f^3) - 3V_f V_m^2 \beta_m^2}\\
\alpha &= (\gamma+\beta_m)/(\gamma-1), \qquad \beta = 1/(3-4\gamma)\\
\rho &= (\beta_m - \gamma\beta_f)/(1+\gamma\beta_f), \qquad \gamma = G_f/G_m
\end{aligned}\right\} \quad (1b)$$

$$\nu_t^* = \tfrac{1}{2}\left(\frac{E_2}{G_t^*}\right) - 1 \quad (1c)$$

The strain–stress relations for the unidirectional lamina in a plane-stress state, referring to the $x_1 - x_2$ plane, are given by

$$\begin{bmatrix} \varepsilon_1 \\ \varepsilon_2 \\ \gamma_{12} \end{bmatrix} = \begin{bmatrix} S_{11} & S_{12} & 0 \\ S_{12} & S_{22} & 0 \\ 0 & 0 & S_{66} \end{bmatrix} \begin{bmatrix} \sigma_1 \\ \sigma_2 \\ \tau_{12} \end{bmatrix} \quad (2)$$

Here, ε and σ denote the normal strain and stress components, respectively; γ and τ are the shear strain and stress, respectively. Again, x_1 and x_2 refer to the fiber and transverse directions, respectively. The elastic compliance constants S_{ij} (often referred to as the engineering constants) can be expressed in terms of the unidirectional lamina properties.

$$S_{11} = \frac{1}{E_1},\ S_{22} = \frac{1}{E_2},\ S_{12} = -\frac{\nu_{12}}{E_1} = -\frac{\nu_{21}}{E_2},\ S_{66} = \frac{1}{G_{12}} \quad (3)$$

Obviously the reciprocity of S_{12} holds. Four independent constants appear in Eqn. (2) and the lamina is termed specially orthotropic.

Eqn. (2) can be inverted to obtain the stress–strain relations:

$$\begin{bmatrix}\sigma_1\\ \sigma_2\\ \tau_{12}\end{bmatrix}=\begin{bmatrix}Q_{11} & Q_{12} & 0\\ Q_{12} & Q_{22} & 0\\ 0 & 0 & Q_{66}\end{bmatrix}\begin{bmatrix}\varepsilon_1\\ \varepsilon_2\\ \gamma_{12}\end{bmatrix} \tag{4}$$

where the Q_{ij} terms are known as the reduced stiffnesses and are related to the engineering constants of the unidirectional lamina as follows:

$$\left.\begin{aligned} Q_{11}&=\frac{E_1}{1-\nu_{12}\nu_{21}}\\ Q_{12}&=\frac{\nu_{12}E_2}{1-\nu_{12}\nu_{21}}=\frac{\nu_{21}E_1}{1-\nu_{12}\nu_{21}}\\ Q_{22}&=\frac{E_2}{1-\nu_{12}\nu_{21}}\\ Q_{66}&=G_{12}\end{aligned}\right\} \tag{5a}$$

It is worth noting that the engineering constants can also be expressed in terms of the reduced stiffnesses as

$$\left.\begin{aligned} E_1&=Q_{11}-\frac{Q_{12}^2}{Q_{22}}, \quad E_2=Q_{22}-\frac{Q_{12}^2}{Q_{11}}\\ \nu_{12}&=\frac{Q_{12}}{Q_{22}}, \quad G_{12}=Q_{66}\end{aligned}\right\} \tag{5b}$$

The interrelations among the different forms of the elastic constants as exemplified in Eqns. (3) and (5) are summarized in Table 1.

For a unidirectional lamina at an angle θ with respect to the reference axes x and y (Fig. 1), the stress–strain relations in the x and y coordinates are

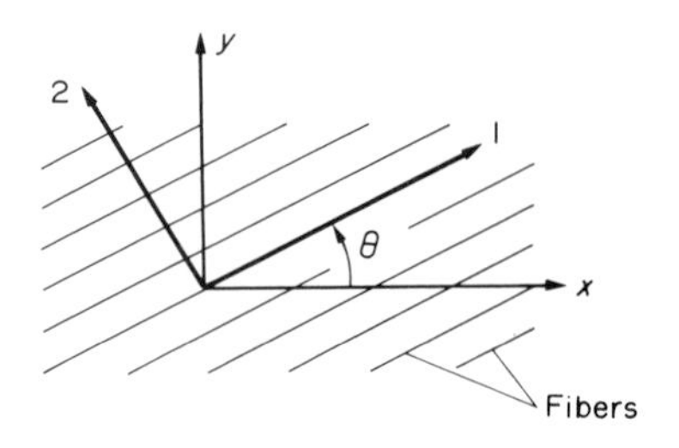

Figure 1
Fiber axis at an angle θ from the lamina reference axis

$$\begin{bmatrix}\sigma_x\\ \sigma_y\\ \tau_{xy}\end{bmatrix}=\begin{bmatrix}\bar{Q}_{11} & \bar{Q}_{12} & \bar{Q}_{16}\\ \bar{Q}_{12} & \bar{Q}_{22} & \bar{Q}_{26}\\ \bar{Q}_{16} & \bar{Q}_{26} & \bar{Q}_{66}\end{bmatrix}\begin{bmatrix}\varepsilon_x\\ \varepsilon_y\\ \gamma_{xy}\end{bmatrix} \tag{6}$$

where $\bar{Q}_{ij}$, the transformed reduced stiffness terms, are given by

$$\left.\begin{aligned} \bar{Q}_{11}&=Q_{11}\cos^4\theta+2(Q_{12}+2Q_{66})\sin^2\theta\cos^2\theta\\ &\quad+Q_{22}\sin^4\theta\\ \bar{Q}_{12}&=(Q_{11}+Q_{22}-4Q_{66})\sin^2\theta\cos^2\theta\\ &\quad+Q_{12}(\sin^4\theta+\cos^4\theta)\\ \bar{Q}_{22}&=Q_{11}\sin^4\theta+2(Q_{12}+2Q_{66})\sin^2\theta\cos^2\theta\\ &\quad+Q_{22}\cos^4\theta\\ \bar{Q}_{16}&=(Q_{11}-Q_{12}-2Q_{66})\sin\theta\cos^3\theta\\ &\quad+(Q_{12}-Q_{22}+2Q_{66})\sin^3\theta\cos\theta\\ \bar{Q}_{26}&=(Q_{11}-Q_{12}-2Q_{66})\sin^3\theta\cos\theta\\ &\quad+(Q_{12}-Q_{22}+2Q_{66})\sin\theta\cos^3\theta\\ \bar{Q}_{66}&=(Q_{11}+Q_{22}-2Q_{12}-2Q_{66})\sin^2\theta\cos^2\theta\\ &\quad+Q_{66}(\sin^4\theta+\cos^4\theta)\end{aligned}\right\} \tag{7}$$

The unidirectional lamina is termed generally orthotropic along the x and y axes.

Table 1
Interrelations among the different forms of the elastic constants

	Fiber direction	Transverse direction	In-plane		
Engineering constant	E_1	E_2	ν_{12}	ν_{21}	G_{12}
corresponding compliance	$1/S_{11}$	$1/S_{22}$	$-S_{12}/S_{11}$	$-S_{12}/S_{22}$	$1/S_{66}$
corresponding stiffness	$(Q_{11}Q_{22}-Q_{12}^2)/Q_{22}$	$(Q_{11}Q_{22}-Q_{12}^2)/Q_{11}$	Q_{12}/Q_{22}	Q_{22}/Q_{11}	Q_{66}
Compliance	S_{11}	S_{22}	S_{12}		S_{66}
corresponding stiffness	$Q_{22}/(Q_{11}Q_{22}-Q_{12}^2)$	$Q_{11}/(Q_{11}Q_{22}-Q_{12}^2)$	$Q_{12}/(Q_{11}Q_{22}-Q_{12}^2)$		$1/Q_{66}$
corresponding engineering constant	$1/E_1$	$1/E_2$	ν_{12}/E_1		$1/G_{12}$
Stiffness	Q_{11}	Q_{22}	Q_{12}		Q_{66}
corresponding engineering constant	$E_1/(1-\nu_{12}\nu_{21})$	$E/(1-\nu_{12}\nu_{21})$	$\nu_{12}E_2/(1-\nu_{12}\nu_{21})$		G_{12}
corresponding compliance	$S_{22}/(S_{11}S_{22}-S_{12}^2)$	$S_{11}/(S_{11}S_{22}-S_{12}^2)$	$-S_{12}/(S_{11}S_{22}-S_{12}^2)$		$1/S_{66}$

Equation (6) can be inverted to obtain the strain–stress relations in the following general form:

$$\begin{bmatrix} \varepsilon_x \\ \varepsilon_y \\ \gamma_{xy} \end{bmatrix} = \begin{bmatrix} \bar{S}_{11} & \bar{S}_{12} & \bar{S}_{16} \\ \bar{S}_{12} & \bar{S}_{22} & \bar{S}_{26} \\ \bar{S}_{16} & \bar{S}_{26} & \bar{S}_{66} \end{bmatrix} \begin{bmatrix} \sigma_x \\ \sigma_y \\ \tau_{xy} \end{bmatrix} \quad (8)$$

in which the $\bar{S}_{ij}$ terms are the transformed compliance constants and their relations to S_{ij} and θ are

$$\left.\begin{aligned}
\bar{S}_{11} &= S_{11}\cos^4\theta + (2S_{12} + S_{66})\sin^2\theta\cos^2\theta \\
&\quad + S_{22}\sin^4\theta \\
\bar{S}_{12} &= S_{12}(\sin^4\theta + \cos^4\theta) \\
&\quad + S_{11} + S_{22} - S_{66})\sin^2\theta\cos^2\theta \\
\bar{S}_{22} &= S_{11}\sin^4\theta + (2S_{12} + S_{66})\sin^2\theta\cos^2\theta \\
&\quad + S_{22}\cos^4\theta \\
\bar{S}_{16} &= (2S_{11} - 2S_{12} - S_{66})\sin\theta\cos^3\theta \\
&\quad - (2S_{22} - 2S_{12} - S_{66})\sin^3\theta\cos\theta \\
\bar{S}_{26} &= (2S_{11} - 2S_{12} - S_{66})\sin^3\theta\cos\theta \\
&\quad - (2S_{22} - 2S_{12} - S_{66})\sin\theta\cos^3\theta \\
\bar{S}_{66} &= 2(2S_{11} + 2S_{22} - 4S_{12} - S_{66})\sin^2\theta\cos^2\theta \\
&\quad + S_{66}(\sin^4\theta + \cos^4\theta)
\end{aligned}\right\} \quad (9)$$

Along the x and y axes, which are not aligned with the material principal directions, the engineering constants of the unidirectional lamina can be expressed as functions of the off-axis angle θ by using Eqns. (3) and (9):

$$\left.\begin{aligned}
\frac{1}{E_x} &= \frac{1}{E_1}\cos^4\theta + \left(\frac{1}{G_{12}} - \frac{2\nu_{12}}{E_1}\right)\sin^2\theta\cos^2\theta \\
&\quad + \frac{1}{E_2}\sin^4\theta \\
\nu_{xy} &= E_x\left[\frac{\nu_{12}}{E_1}(\sin^4\theta + \cos^4\theta)\right. \\
&\quad \left. - \left(\frac{1}{E_1} + \frac{1}{E_2} - \frac{1}{G_{12}}\right)\sin^2\theta\cos^2\theta\right] \\
\frac{1}{E_y} &= \frac{1}{E_1}\sin^4\theta + \left(\frac{1}{G_{12}} - \frac{2\nu_{12}}{E_1}\right)\sin^2\theta\cos^2\theta \\
&\quad + \frac{1}{E_2}\cos^4\theta \\
\frac{1}{G_{xy}} &= 2\left(\frac{2}{E_1} + \frac{2}{E_2} + \frac{4\nu_{12}}{E_1} - \frac{1}{G_{12}}\right)\sin^2\theta\cos^2\theta \\
&\quad + \frac{1}{G_{12}}(\sin^4\theta + \cos^4\theta)
\end{aligned}\right\} \quad (10)$$

1.2 Classical Lamination Theory

Based upon the constitutive relations for a lamina composed of a generally orthotropic material, Eqn. (6), the constitutive relations for a laminate formed by bonding several laminae together is presented in this section. The orientation and material system of each lamina are general. Figure 2 depicts the geometry of an n-layered laminate of thickness h; the x-y plane coincides with the laminate geometric middle-plane. Following the approach of the classical linear thin-plate theory, the following assumptions are made (Vinson and Chou 1975).

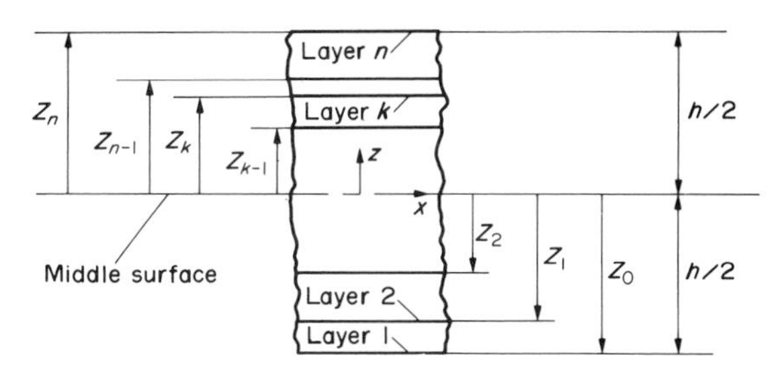

Figure 2
An n-layered laminate

Upon the application of load, a lineal element of the plate extending through the plate thickness normal to the middle surface (x-y plane) in the unstressed state undergoes, at most, a translation and a rotation with respect to the original coordinate system, and remains normal to the deformed middle surface. This assumption implies that the lineal element does not elongate or contract and remains straight when a load is applied.

The plate resists lateral and in-plane loads by bending, transverse-shear stress and in-plane action, not through block-like compression or tension in the plate in the thickness direction.

From the foregoing assumptions, also known as the Kirchhoff hypothesis for plates, the strain components can be derived:

$$\begin{bmatrix} \varepsilon_x \\ \varepsilon_y \\ \gamma_{xy} \end{bmatrix} = \begin{bmatrix} \varepsilon_x^0 \\ \varepsilon_y^0 \\ \gamma_{xy}^0 \end{bmatrix} + z \begin{bmatrix} \kappa_x \\ \kappa_y \\ \kappa_{xy} \end{bmatrix} \quad (11)$$

Here, $\varepsilon_x^0 \varepsilon_y^0$ and γ_{xy}^0 are the laminate middle-plane strains, which are expressed in terms of the middle plane displacements u_0 and v_0:

$$\varepsilon_x^0 = \frac{\partial u_0}{\partial x}, \qquad \varepsilon_y^0 = \frac{\partial v_0}{\partial y}, \qquad \gamma_{xy} = \frac{\partial u_0}{\partial y} + \frac{\partial v_0}{\partial x} \quad (12)$$

The middle-plane curvatures are related to the z-direction middle-plane displacement w_0:

$$\kappa_x = \frac{\partial^2 w_0}{\partial x^2}, \qquad \kappa_y = \frac{\partial^2 w_0}{\partial y^2}, \qquad \kappa_{xy} = 2\frac{\partial^2 w_0}{\partial x \partial y} \quad (13)$$

Note that κ_{xy} represents the twist curvature of the middle plane. Figure 3 depicts the deformation associated with a typical cross-sectional element in a thin plate.

Also, following the approach of the classical plate theory, the resultant forces and moments, instead of

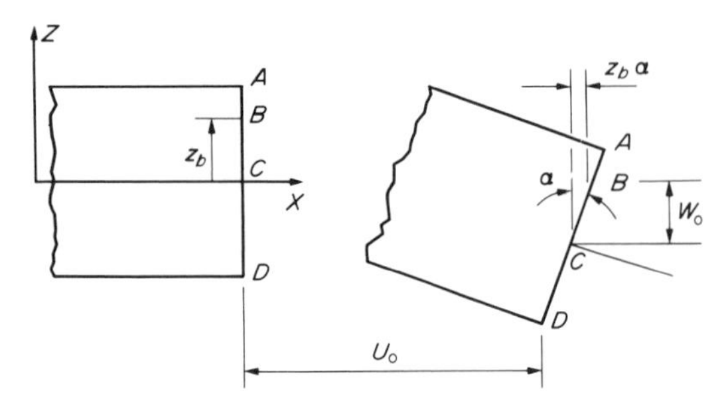

Figure 3
Deformation of a typical cross-sectional element in a thin laminated plate

the stresses, are utilized in the constitutive relations. Referring to Fig. 4(a), (b), the force and moment resultants of the laminate are obtained by integrating the stresses of each lamina through the laminate thickness h:

$$(N_x, N_y, N_{xy}) = \int_{-h/2}^{h/2} (\sigma_x, \sigma_y, \tau_{xy})\, dz \qquad (14)$$

$$(M_x, M_y, M_{xy}) = \int_{-h/2}^{h/2} (\sigma_x, \sigma_y, \tau_{xy})\, z\, dz \qquad (15)$$

Substitution of Eqns. (6) and (7) into Eqns. (11) and (12) results in the following:

$$\begin{bmatrix} N_x \\ N_y \\ N_{xy} \end{bmatrix} = \begin{bmatrix} A_{11} & A_{12} & A_{16} \\ A_{12} & A_{22} & A_{26} \\ A_{16} & A_{26} & A_{66} \end{bmatrix} \begin{bmatrix} \varepsilon_x^0 \\ \varepsilon_y^0 \\ \gamma_{xy}^0 \end{bmatrix} + \begin{bmatrix} B_{11} & B_{12} & B_{16} \\ B_{12} & B_{22} & B_{26} \\ B_{16} & B_{26} & B_{66} \end{bmatrix} \begin{bmatrix} \kappa_x \\ \kappa_y \\ \kappa_{xy} \end{bmatrix} \qquad (16)$$

$$\begin{bmatrix} M_x \\ M_y \\ M_{xy} \end{bmatrix} = \begin{bmatrix} B_{11} & B_{12} & B_{16} \\ B_{12} & B_{22} & B_{26} \\ B_{16} & B_{26} & B_{66} \end{bmatrix} \begin{bmatrix} \varepsilon_x^0 \\ \varepsilon_y^0 \\ \gamma_{xy}^0 \end{bmatrix} + \begin{bmatrix} D_{11} & D_{12} & D_{16} \\ D_{12} & D_{22} & D_{26} \\ D_{16} & D_{26} & D_{66} \end{bmatrix} \begin{bmatrix} \kappa_x \\ \kappa_y \\ \kappa_{xy} \end{bmatrix} \qquad (17)$$

where

$$\left.\begin{aligned} A_{ij} &= \sum_{k=1}^{n} (\bar{\mathbf{Q}}_{ij})_k (z_k - z_{k-1}) \\ B_{ij} &= \tfrac{1}{2} \sum_{k=1}^{n} (\bar{\mathbf{Q}}_{ij})_k (z_k^2 - z_{k-1}^2) \\ D_{ij} &= \tfrac{1}{3} \sum_{k=1}^{n} (\bar{\mathbf{Q}}_{ij})_k (z_k^3 - z_{k-1}^3) \end{aligned}\right\} \qquad (18)$$

and z_k and z_{k-1} are as shown in Fig. 2.

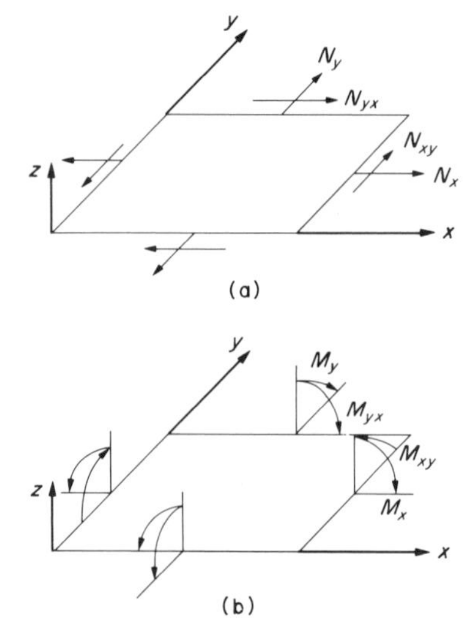

Figure 4
(a) In plane force resultants, (b) in-plane moment resultants

In Eqns. (16)–(18), A_{ij}, B_{ij} and D_{ij} are called extensional stiffness, extension–bending coupling stiffness and bending stiffness, respectively. The summation in Eqn. (18) is carried out over all the laminae; $(\bar{Q}_{ij})_k$ refers to the reduced stiffness of the kth layer.

The physical significance of the components of the (A), (B), and (D) matrices (those containing A, B and D terms, respectively) deserves careful examination. The (A) matrix relates the stress resultants to the midplane strains. The couplings between normal stress resultants and midplane shear strains, as well as shear-stress resultants and midplane normal strains, are due to the components A_{16} and A_{26}. There is also coupling, through the (B) matrix, between midplane-stress resultants and the bending and twisting of the laminate. In particular, the components B_{16} and B_{26} relate normal-stress resultants with the twisting of the laminate. The (B) matrix also plays a role in the coupling between the moment resultants and in-plane strains. Finally, the D_{16} and D_{26} terms are responsible for the interaction between bending moment and twisting.

The various coupling effects in laminated composites can be minimized or eliminated through suitable choices of the laminae stacking sequence. Table 2 shows the effect of stacking sequence on the (A), (B) and (D) matrices.

The constitutive relations of Eqns. (16) and (17) can be rearranged to other useful forms by partially or totally inverting them. The totally inverted form of Eqns. (16) and (17) is given in the following concise matrix expressions:

$$\begin{aligned} [\varepsilon^0] &= [A'][N] + [B'][M] \\ [\kappa] &= [C'][N] + [D'][M] \end{aligned} \qquad (19)$$

Table 2
Effect of stacking sequence on (A), (B), and (D) matrices

	Value of matrix element			
	A_{16}, A_{26}	B_{11}, B_{22}, B_{12}	B_{16}, B_{26}	D_{16}, D_{26}
Stacking sequence:				
0° or 90°	zero	zero	zero	zero
0°/90°	zero	—	zero	zero
$\ldots {}^{+}\theta/{}^{-}\theta/{}^{+}\theta/{}^{-}\theta \ldots$ (antisymmetrical)	zero	zero	—	zero
$\ldots {}^{+}\theta/{}^{-}\theta/{}^{-}\theta/{}^{+}\theta \ldots$ (symmetrical)	zero	zero	zero	—
same number of ${}^{+}\theta$ and ${}^{-}\theta$ layers	zero	—	—	—

where

$$\left.\begin{aligned} &[A']=[A^*]-[B^*][D^{*-1}][C^*] \\ &[B']=[B^*][D^{*-1}] \\ &[C']=-[D^{*-1}][C^*] \\ &[D']=[D^{*-1}] \\ \text{and} \\ &[A^*]=[A^{-1}] \\ &[B^*]=-[A^{-1}][B] \\ &[C^*]=[B][A^{-1}] \\ &[D^*]=[D]-[B][A^{-1}][B] \end{aligned}\right\} \tag{20}$$

An application of Eqn. (19) can be found, for instance, when the stress and moment results acting on a laminated plate are specified. Then, with the knowledge of the elastic constants, the middle-plane strain and curvature of the laminate can be found. The strain components of a specific lamina in terms of the plate reference axes can be derived from Eqn. (11), and the corresponding stresses from Eqn. (6). The existing criteria for laminar failure, due to combined in-plane stresses or strains, require the knowledge of stresses and strains along the fiber as well as the transverse directions. This information can be readily obtained by transformation of the stress and strain components onto the principal material directions. Thus, the correlation between external loading on the laminated plate and the failure of an individual lamina can be established.

2. *Hygrothermal Effect*

Deformation in laminated composites can also occur because of changes in temperature and the absorption of moisture. This is known as the hygrothermal effect. As polymers undergo both dimensional and property changes in a hygrothermal environment, so do composites utilizing polymers as the matrix. Since fibers are fairly insensitive to environmental changes, the environmental susceptibility of composites is mainly due to the matrix. Consequently, in a unidirectional composite the temperature–moisture environment has a much greater effect on the transverse and shear properties than on the longitudinal properties.

Consider the heat conduction and moisture diffusion in a composite along the z direction; the respective governing equations are

$$\frac{K_z^T}{\rho c}\frac{\partial^2 T}{\partial z^2}=\frac{\partial T}{\partial t}, \qquad K_z^H\frac{\partial^2 H}{\partial z^2}=\frac{\partial H}{\partial t} \tag{21}$$

Here, t is the time, $T(=T(z,t))$ the temperature, $H(=H(z,t))$ the moisture concentration, K_z^T the thermal conductivity, K_z^H the moisture diffusion coefficient, c the specific heat and ρ the mass density. The fact that Eqns. (21) are identical in form to each other indicates the similarity of the underlying processes.

The thermal diffusivity $K_z^T/(\rho c)$ and moisture diffusion coefficient K_z^H are used as measures of the rates at which temperature and moisture concentrations change within the material: in general, these parameters depend on temperature and moisture concentration. However, over the range of temperature and moisture concentration that prevails in typical applications of composites, the thermal diffusivity is about 10^6 times greater than the moisture diffusion coefficient. Consequently, thermal diffusion takes place at a rate much faster than moisture diffusion, and the temperature will reach equilibrium long before the moisture concentration does. This allows the heat-conduction and moisture-diffusion equations to be solved separately.

Deformations resulting from hygrothermal effects can be described by a modified set of linear equations: the total strain minus the nonmechanical strain is linearly related to the stress. This is done by replacing the terms ε_j in Eqn. (6) by $\varepsilon_j-\alpha_j T-\beta_j H$, where α_j is the coefficient of thermal expansion, and β_j the coefficient of hygroscopic expansion. It is understood that $j=1$, 2 and 6 along the three plate reference axes. This approach is based upon the assumption of elastic behavior. The nonmechanical strain is measured from a stress-free reference state, and the elastic moduli used in the calculation are taken at the final environmental conditions. For example, in the fabrication of polymer-matrix composite laminates, the curing of an individual ply results in different deformations along the fiber and transverse directions. The constraint of deformation of a single ply due to the presence of other plies in a multidirectional laminate gives rise to residual stresses. Since most of the cross-linking in the polymer occurs at the highest curing temperature, the polymer matrix can be considered as being still viscous enough to allow complete relaxation of residual stress. Thus the highest curing temperature can be regarded as the stress-free temperature.

By taking into account the nonmechanical strain in Eqn. (6) for hygrothermally induced deformation, the laminated-plate analysis developed in Sect. 12 can be modified to determine the overall elastic response. The stresses due to moisture absorption and temperature change are analogous to each other in that they are dilatational and self-equilibriating when the whole laminate is considered. In general, the longitudinal properties of polymer-matrix composites are far less sensitive to temperature and moisture than the transverse and shear properties of unidirectional composites, because of the excellent retention of mechanical properties by the fibers. The greatest reduction in properties occurs when temperature and moisture are combined, such as in hot and humid environments. However, the combination of temperature and moisture could render a laminate free of residual stresses. This can be understood by considering, for example, a 0°/90° cross-ply. The thermal stress induced by fabrication is tensile in the transverse direction of a ply, while the residual stresses induced by moisture absorption are compressive.

3. Strength

The strength of laminated composites is best understood for unidirectional systems. The tensile strength of unidirectional composites can be estimated by rule-of-mixtures-type theories. The behavior of unidirectional composites under compressive loading is considerably more difficult to understand. Experimental work on compression of fiber composites has revealed that fiber buckling, shear failure and interfacial failure are the major failure modes. Figure 5 shows a typical tensile stress–strain relation for unidirectional glass-fiber-reinforced plastics when they are deformed along the fiber (0°) and transverse (90°) directions. The fiber volume content in this example is 56%. The low tensile failure-stress in the 90° direction can be attributed to the strain concentrations at the filament–resin interface.

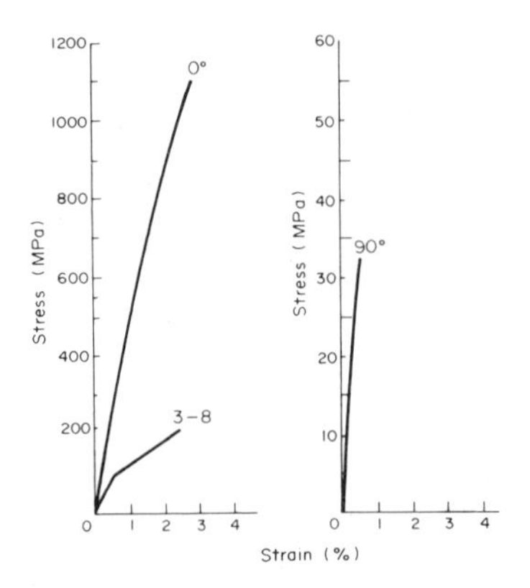

Figure 5
Stress–strain curves for unidirectional and cross-ply glass-fiber-reinforced plastics: 3–8 refers to 3 plies in the load direction and 8 normal to the load

The tensile stress–strain curve of a cross-ply with three plies in the 0° orientation and eight in the 90° orientation is also shown in Fig. 5. The unique feature of this curve is the change in slope or the knee behavior, which does not exist in unidirectional composites. The knee in the stress–strain curve represents fracturing of plies in the 90° orientation. It occurs at approximately 0.4–0.5% strain. It can be observed that the fracture strain for the unidirectional composite stressed at 90° to the filament direction is indeed 0.4%. Further verification of this behavior can be obtained by calculating the change in slope of the stress–strain curve assuming all the 90° layers have failed and can no longer contribute to the modulus when stressed beyond the knee.

4. Thick Laminates

The elastic behavior of thin laminate is adequately described by classical lamination theory. Exact elasticity solutions of thick laminates—those possessing in-plane dimensions to thickness ratios less than ten—have demonstrated the inadequacies of thin-plate theories to accurately describe their elastic behavior. One reason for the departure of thick-plate behavior from classical thin-plate theory prediction is the presence of transverse-shear deformation in thick laminates. The effect of transverse-shear deformation is pronounced in anisotropic materials with high ratios of in-plane Young's moduli to interlaminar-shear moduli; this is typical in laminated composites. Other assumptions of the classical theory, such as negligible transverse normal strains ($\varepsilon_z=0$), and the linear in-plane strain variation with the z-coordinate, all contribute to the limitations of the theory (Sect. 2.2).

The inclusion of transverse-shear deformation into the classical thin-plate theory is achieved by allowing the transverse-shear strains ε_{xz} and ε_{yz} to be nonzero.

This gives rise to definitions of the shear-force resultants:

$$(Q_x, Q_y) = \int (\tau_{xz}, \tau_{yz})\, dz \qquad (22)$$

These shear resultants are then related to the transverse-shear strains through appropriate constitutive relations. Often, however, shear correction factors are derived and inserted into the relations to satisfy appropriately the requirement of vanishing transverse-shear stress on the top and bottom of the thick plate.

A plate theory is developed from a starting point that assumes a certain form of the displacement field within the plate. Appropriately assumed displacement fields are simply power-series expansions of the displacements in the coordinate normal to the midplane of the plate. In classical plate theory, the assumed displacement field is quite simple (Fig. 3):

$$\left.\begin{aligned} u &= u^0(x, y) + z\alpha(x, y) \\ v &= v^0(x, y) + z\beta(x, y) \end{aligned}\right\} \qquad (23)$$

where in-plane strains vary only linearly with the z-coordinate. The definitions of α and β can be found in Fig. 3. Higher-order theories are simply theories which retain more terms in the assumed displacement field expansions. The accuracy of these theories is generally greater for a greater number of terms retained in the assumed field, but the complexity of the equations involved to obtain solutions places severe limits on the number of terms for which solutions are realistically attainable.

Numerous higher-order theories of plate deformation have been developed, differing primarily in the functional form of the displacement field assumed. Consider the following assumed field where u and v are given by Eqn. (23) and

$$w = w^0(x, y) + z\gamma(x, y) + z^2\xi(x, y) \qquad (24)$$

This constitutes a first-order theory for the in-plane displacements, whereas the effects of transverse normal strains are now accounted for by the extra terms of $\gamma(x, y)$ and $\xi(x, y)$ retained in w.

The format of the solution to higher-order systems generally involves the application of the principle of potential energy to derive the pertinent governing equations of equilibrium. Using the typical strain–displacement relations and the assumed displacement field, in conjunction with the equations of equilibrium, a set of partial differential equations are derived. The solution of these equations describes the elastic behavior of the plate. The number of equations will correspond to the number of terms retained in the assumed displacement form.

Although accounting for shear deformation and higher-order plate deformation in thick laminates involves a great deal more complexity than the classical approach, it is evident that the extra effort needed to describe accurately their fundamentally different elastic behavior is necessary for further understanding of these materials.

See also: Strength of Composites; Woven-Fabric Composites: Properties

Bibliography

Ashton J E, Halpin J C, Petit P H 1969 *Primer on Composite Materials: Analysis.* Technomic, Westport, Connecticut

Bailey J E, Curtis P T, Parvizi A 1979 On the transverse cracking and longitudinal splitting behavior of glass and carbon fiber reinforced epoxy cross-ply laminates and the effect of Poisson and thermally generated strain. *Proc. R. Soc. London, Ser. A* 366: 599–623

Bogetti T A, Gillespie J W Jr., Pipes R B 1987 A literature review in thick section composites. Technical Report 87–55, Center for Composite Materials, University of Delaware

Carlsson L A, Pipes R B 1987 *Experimental Characterization of Advanced Composite Materials.* Prentice–Hall, Englewood Cliffs, New Jersey

Chou T-W 1989 Flexible composites. *J. Mater. Sci.* 24: 761–83

Chou T-W, Kelly A 1976 Fiber composites. *Mater. Sci. Eng.* 25: 35–40

Chou T-W, Kelly A 1980a The effect of transverse shear on the longitudinal compressive strength of fiber composites. *J. Mater. Sci.* 15: 327–31

Chou T-W, Kelly A 1980b Mechanical properties of composites. *Annu. Rev. Mater. Sci.* 10: 229–59

Jamison R D 1986 Survey of recent research in thick composite laminates. US Naval Academy Report EW-21–86

Jones R M 1975 *Mechanics of Composite Materials.* McGraw–Hill, New York

Lo K H, Christensen R M, Wu E M 1977 A higher-order theory of plate deformation Part 1: Homogeneous plates. *J. Appl. Mech.* 44: 663–68

Lo K H, Christensen R M, Wu E M 1977 A higher-order theory of plate deformation Part 2: Laminated plates. *J. Appl. Mech.* 44: 669–76

Tsai S W, Hahn H T 1980 *Introduction to Composite Materials.* Technomic, Westport, Connecticut

Vinson J R, Chou T-W 1975 *Composite Materials and Their Use in Structures.* Wiley, New York

Walter J D 1978 Cord–rubber tire composites: theory and application. *Rubber Chem. Technol.* 51: 524–76

Whitney J M 1987 *Structural Analysis of Laminated Anisotropic Plates.* Technomic, Lancaster, Pennsylvania

Whitney J M, Sun C T 1973 A higher order theory for extensional motion of laminated composites. *J. Sound Vib.* 30: 85–97

T.-W. Chou
[University of Delaware, Newark, Delaware, USA]

Long-Term Degradation of Polymer-Matrix Composites

Most materials evolve with time and by interaction with their surroundings. In the case of steel, oxidation leads to rusting and this is accelerated in the presence of water; nylon fibers turn yellow under the effect of sunlight; rubber ages faster under the effect of ozone which makes it more brittle; wood splits because of the loss of water; and cement becomes increasingly brittle with age. These changes may be irritating but they can be accommodated if the rate of change and its effect on behavior is known, and any structure made from these materials designed appropriately.

The mechanical properties of a composite depend on the properties of its constituents and on their interaction, producing load transfer between fibers by shear of the surrounding matrix. In most environments organic-matrix composites will absorb water, and the presence of the water may cause changes in the constituents and in their interaction. The effect most commonly feared is the breaking of the interfacial, generally secondary, bonds between fiber and matrix, leading to a reduction in load transfer and a fall in mechanical properties of the composite.

The diffusion laws have generally been applied to the modelling of simple water uptake, and it has been shown that the simple Fick's law can be applied to carbon-fiber-reinforced epoxy resin subjected to humid environments (Shen and Springer 1976, Loos and Springer 1979). In this model the water is considered to remain in a single free phase driven to penetrate the resin by the water concentration gradient. Other studies have indicated, however, that non-Fickian processes do occur which can complicate the understanding of the role of the water and which may lead to irreversible changes in mechanical properties (Springer and Tsai 1967, Dewimille and Bunsell 1982). Water penetration can cause swelling and plastification of the resin, and ingress can occur by capillary action along the fiber–matrix interface (Hahn and Kim 1978, Deiasi and Whiteside 1978). Composite structures consisting of several layers of differing types of fiber lay-up, often with a thick resin coat or gel coat, can suffer from water being trapped in a particular layer or at the interface between layers. This effect may be irreversible because of osmosis, and blistering may result (Marine et al. 1988). This is particularly a problem with glass-fiber-reinforced boats for which considerable effort is made, by a judicious choice of resin materials and manufacturing techniques, to avoid this type of damage. In addition to changes in mechanical properties, modifications to other physical characteristics may be observed, such as changes in dielectric properties (Cotinaud et al. 1982).

1. *Diffusion Models*

1. *Fick's Law and Single-Phase Diffusion*

The most common approach to modelling water diffusion is to consider Fick's law applied to simple single free-phase diffusion. Fick's law assumes that the water is driven to penetrate the material by the water concentration gradient dc/dx (Crank 1956) so that the water uptake as a function of time is given by

$$\frac{\partial c}{\partial t}=\frac{d}{dx}\left(D\frac{dc}{dx}\right) \tag{1}$$

where D is a diffusion coefficient. D is considered to be constant in the direction of diffusion x so that Eqn. (1) can be rewritten as

$$\frac{\partial c}{\partial t}=D_x\frac{d^2c}{dx^2} \tag{2}$$

It has been shown (Crank 1956) that for an infinitely large plate of thickness h the mass M_t of water absorbed in time t across a unit surface area as a percentage of the mass of water absorbed at saturation M_∞ is given by

$$\frac{M_t}{M_\infty}=1-\frac{8}{\pi^2}\sum_{n=0}^{\infty}\frac{1}{(2n+1)^2}\exp[-D(2n+1)^2\pi^2t/h^2] \tag{3}$$

This equation has been simplified by Shen and Springer (1976), to show that the initial absorption is given by

$$\frac{M_t}{M_\infty}=\frac{4}{h}\left(\frac{Dt}{\pi}\right)^{1/2} \tag{4}$$

The initial linear relationship of M_t/M_∞ as a function of the square root of time is to be noted, and so water uptake is usually plotted as M_t/M_∞ versus the square root of time. Figure 1 shows classical Fickian behavior of a glass-fiber-reinforced epoxy resin.

Under most conditions water can penetrate into the material in several directions and so it is necessary to correct for edge effects.

Figure 2 shows that, in a unidirectional specimen, two diffusion coefficients must be considered accounting for penetration normal ($D_\perp$) and parallel ($D_{||}$) to the fibers, and a third coefficient D_r, for the unreinforced resin, also exists.

It has proved possible to derive the relationships between the diffusion coefficients of the composite and that of the resin by analogy to thermal and electrical diffusion, taking into account the fiber volume fraction V_f and assuming a square or hexagonal fiber-packing arrangement. In the case of a square array of fibers having a diameter d and distance between centers a, the volume fraction is given by

$$V_f=\frac{\pi d^2}{4a^2} \tag{5}$$

The relationships are therefore

$$D_{||}=(1-V_f)D_r \tag{6}$$

$$D_\perp=[1-2(V_f/\pi)^{1/2}]D_r \tag{7}$$

The analogy with electrical conductivity was first solved by Lord Rayleigh and described by Augl (1976) and the solution results in values of $D_{||}/D_r$ and $D_\perp/D_r$ which are slightly greater than those obtained by Springer and Tsai (1967).

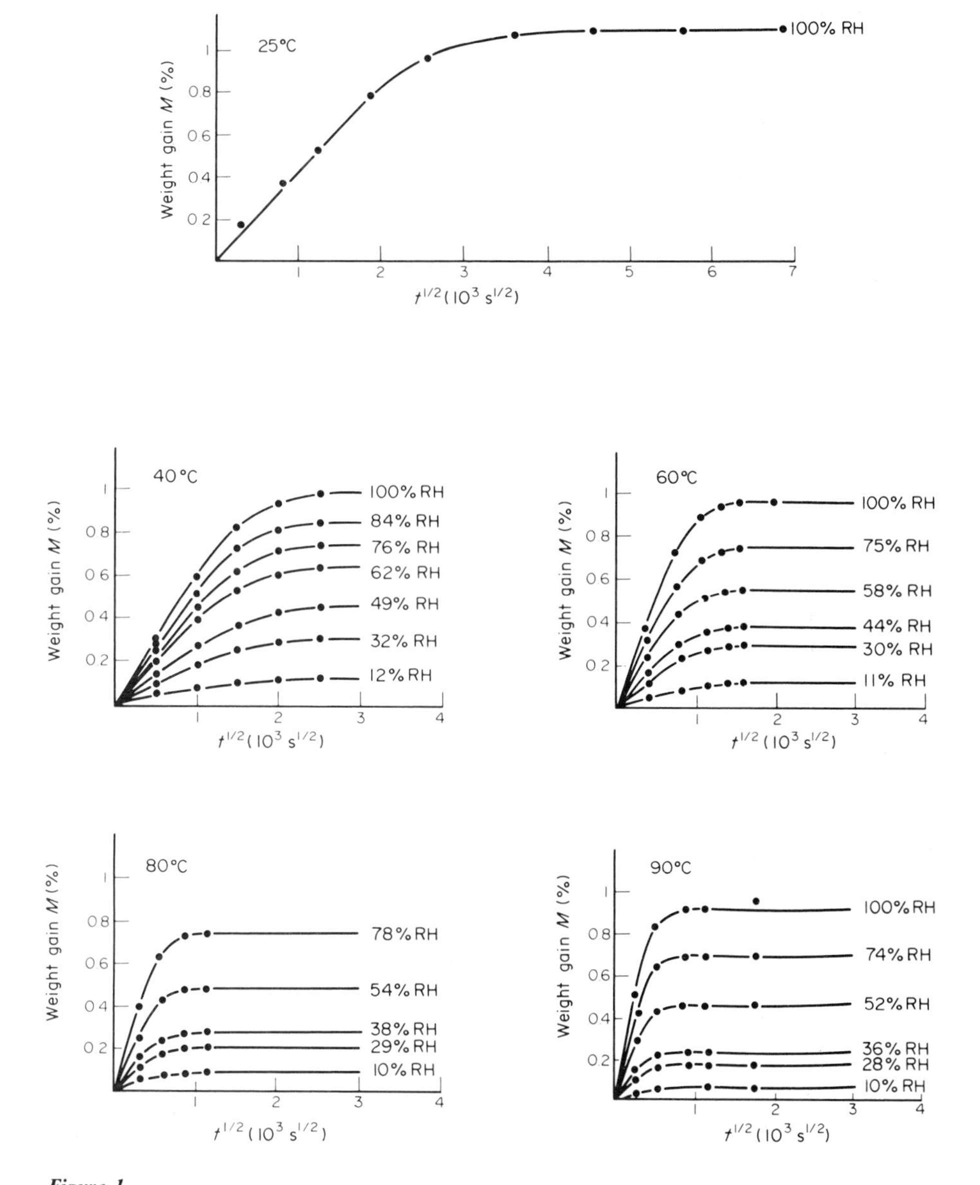

Figure 1
Absorption curves obtained under different conditions of temperature and relative humidity with glass-fiber-reinforced DGEBA epoxy resin, showing classical Fickian absorption.

1.2 The Langmuir Model of Diffusion

In cases where the simple Fickian model proves inadequate, such as when saturation is not achieved or two plateaux in the absorption curve are observed or where it is found that the absorption depends on the specimen thickness, it is sometimes useful to consider a more complex model of the Langmuir type. Carter and Kibler (1978) and Gurtin and Yatomi (1979) have applied the Langmuir model to the problem of water absorption and assumed that the diffusion coefficient remained independent of water concentration as in the Fickian model. However, the water was considered to be in two phases, one free to diffuse and the other trapped and so not free to move in the absorbing medium.

The analysis of diffusion involving chemical reaction, reversible or otherwise, has been treated by Crank (1956), and similar reasoning has led to use of the same solutions by Caskey and Pillinger (1975) for the case of hydrogen diffusion into metals.

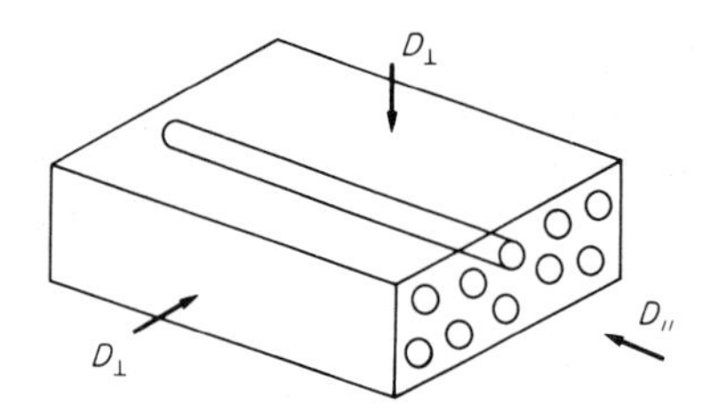

Figure 2
Coefficients of diffusivity necessary to describe water penetration into a composite

In addition to the two parameters D and M_∞ used in the simple free-phase diffusion model, the two-phase model introduces two other parameters: α, which is the probability of a trapped water molecule being released and γ, the probability of a free water molecule being trapped.

The analysis of this situation results in the following simplified equation which describes the rate of water uptake in a plate of thickness $2l$.

$$\frac{M_t}{M_\infty}=1-\frac{\gamma}{\gamma+\alpha}\mathrm{e}^{-\alpha t}-\frac{8}{\pi^2}\frac{\alpha}{\gamma+\alpha}\mathrm{e}^{-\gamma t}\sum_{n=1}^{\infty}\frac{1}{(2n-1)}\mathrm{e}^{-(\pi/2)^2Dt} \quad (8)$$

Figure 3 shows how Eqn. (8) describes the uptake of water into plates of differing thicknesses h. It should be noted that the Langmuir model reduces to the simple Fickian model where $\gamma=0$.

In passing to a two-phase model it is assumed that the absorbed water is in two states: the free phase having a water concentration $n(t)$ water molecules per unit volume can diffuse in the resin, but can be linked

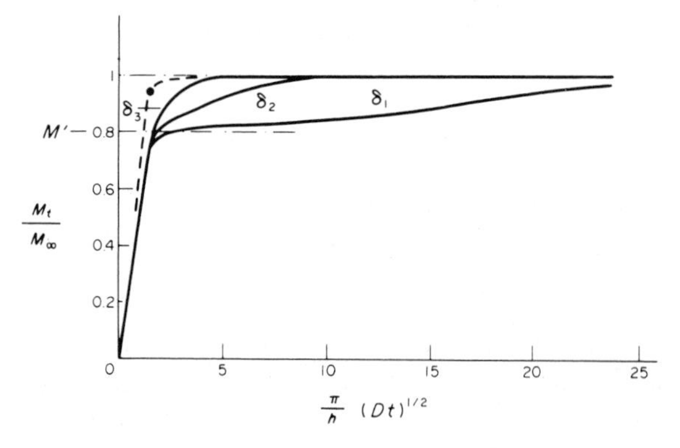

Figure 3
Two-phase diffusion as described by the Langmuir model, showing the influence of specimen thickness. The dielectric loss tangent $\delta=(h/\pi)(\alpha/D)^{1/2}$. $\delta_1=0.05$, $\delta_2=0.2$ and $\delta_3=0.5$. The broken curve follows Fick's law. $(\alpha/(\alpha+\gamma)=M'/M_\infty=0.8\Rightarrow\alpha/\gamma=4)$

or trapped in the resin with a probability γ, and the combined phase with concentration $N(t)$ able to break free with a probability α. At saturation:

$$n_\infty\gamma=N_\infty\alpha$$

that is to say, there is an equilibrium between those water molecules being trapped and those being freed.

As the parameters D and M_∞ are still invoked, the model involves four variables and, given the heterogeneity of a composite material, there exist numerous possible interpretations for the physical mechanisms involved in absorption. One is that water can diffuse freely and can interact chemically with the molecular structure of the resin or coupling agent; alternatively, water can be mechanically trapped, absorbed in the matrix or onto the fiber surface. As the interfacial zone is probably very different from the unreinforced resin, different diffusion behavior could also be expected, even without considering capillary infiltration. The physical processes involved in water absorption by the composite are certainly more complex than the processes considered in the Langmuir model but because of its additional flexibility the model is applicable to a wider range of cases.

2. *Physicochemical Aspects of Absorption*

Most polymers absorb water to a greater or lesser degree so that, even in the absence of defects or preferential routes for water uptake such as are provided by poor fiber–matrix adhesion, resin-matrix composites will absorb water in humid environments. Water diffusion occurs in most resins by hydrogen bonds being formed between the water and the molecular structure of the resin (Browning 1976).

Epoxide resins, which are the resins most widely used in high-performance composite structures, are based on the group

$$\begin{array}{c}\diagdown\quad\ \diagup\\ \mathrm{C{-}C}\\ \diagup\ \ \diagdown\diagup\ \ \diagdown\\ \mathrm{O}\end{array}$$

and are obtained by polycondensation of glycol epichlorohydrin

$$\mathrm{CH_2{-}CH{-}CH_2Cl}\quad(\text{epoxide ring via O between } \mathrm{CH_2} \text{ and } \mathrm{CH})$$

with, most commonly, propane diphenylol, also known as bisphenol-A, to give an epoxy resin usually known as DGEBA. The epoxide resins are mixed with a hardener, which can be of several types, to produce a cross-linked structure which provides many sites for hydrogen bonding for the water molecules such as hydroxyl groups (O–H), phenol groups (O–C), amine groups (N–O) and sulfone groups (O–S). Figure 4 shows the sites available on a commonly used epoxy consisting of a bisphenol-A epoxy resin cross-linked with a dicyandiamide hardener. The hydrolysis of epoxy resins is in itself a cause of degradation and is

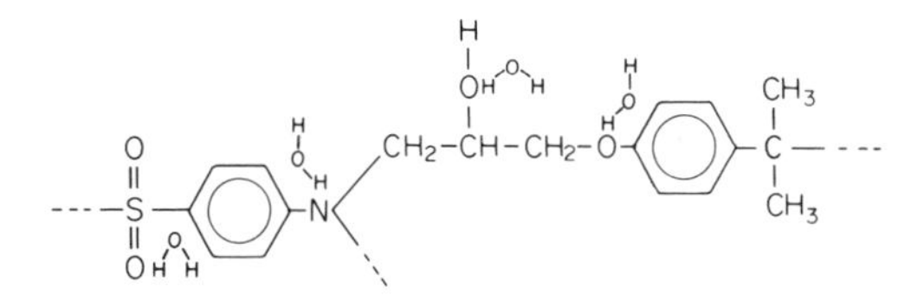

Figure 4
Sites available for water molecules to be combined by hydrogen bonding onto the molecular structure of a DGEBA epoxy resin cross-linked with a dicyandiamide hardener

not by any means limited to this class of matrix materials.

Certain fibers, such as those of glass, can be damaged by prolonged exposure to water which arrives at the interface. Glasses are made up of silica in which are dispersed metallic oxides including those of the alkali metals. These latter nonsilicate constituents represent microheterogeneities which are hydroscopic and hydrolyzable. The absorption of water by glass is therefore characterized by the hydration of these oxides. The most common form of glass fiber is made of E-glass, which contains only small amounts of alkaline-metal oxides and so is resistant to damage by water. However, the presence of water at the glass-fiber surface may lower its surface energy and promote crack growth. Water trapped at the interface may also allow components of the resin to go into solution and if an acidic environment is formed the fiber can be degraded.

Organic fibers such as the aramid family can absorb considerable quantities of water and although this does not lead to a deterioration of fiber properties the resulting swelling of the fiber may be a cause of composite degradation.

The size, or coating, put onto many fibers to protect them from abrasion and to ensure bonding with the matrix also serves to protect them and the composite from damage from water absorption by eliminating sites at which the water can accumulate. However, this is not always adequate.

2.1 Influence of the Matrix on Composite Degradation

One of the most widely employed composite materials used for high-performance structures is composed of glass fibers embedded in an epoxy matrix. Degradation of this type of composite depends greatly on the type of hardener used to cross-link the resin. Figure 1 shows the classical Fickian behavior of glass-fiber-reinforced bisphenol-A epoxy cross-linked with a diamine hardener. Several important observations may be made by reference to Fig. 1. Initial weight gain as a percentage of the dry-composite weight is a linear function of the square root of time. The composite reaches an equilibrium with its environment and saturates. The saturation level is a function of relative

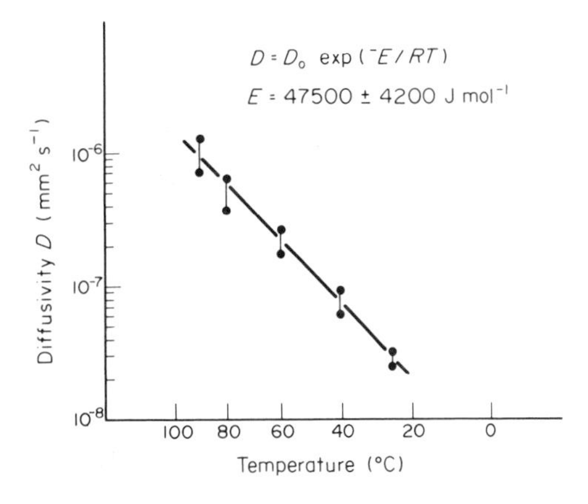

Figure 5
Thermoactivated diffusion into a glass-fiber-reinforced DGEBA epoxy resin, obeying an Arrhenius relationship with temperature. The quantity E is the activation energy

humidity and is independent of temperature. Figure 5 shows that the diffusivity D obeys a thermoactivated Arrhenius-type relationship and is independent of relative humidity. This behavior indicates that the water is absorbed by the resin and is linked to it by hydrogen bonds. The saturated composite reveals a somewhat increased tenacity due to a softening of the matrix which is accompanied by a fall in glass transition temperature (Bonniau and Bunsell 1981, Springer 1982). Damage does occur under severe conditions (90 °C–100 °C RH for two weeks) and appears as a progressive whitening of the material with no detectable weight gain. This is probably due to microcracking which begins at the surface.

A change to a dicyandiamide hardener produces, however, a significant change in water-uptake kinetics, and this behavior can only be explained by the two-phase model of absorption. Saturation is attained after considerably longer exposure than with the diamine-hardened resin but after equilibrium is reached both composites behave in a similar manner.

The use of an anhydride as the cross-linking agent produces a composite which, in water which is warmer than about 40 °C, degrades quite rapidly. Saturation is not achieved and severe weight loss after drying is observed, due to leaching of the resin into the water. An excess of the anhydride not used in the cross-linking process combines with the water to produce an acid which degrades the composite. The correct choice of hardener in these epoxy composites is therefore extremely important if they are to be used in warm humid conditions.

As mentioned above, the glass transition temperature T_g of organic-matrix composites is lowered with water uptake. In the case of thermosetting resins such as the epoxides, this decrease can be as much as 30 °C

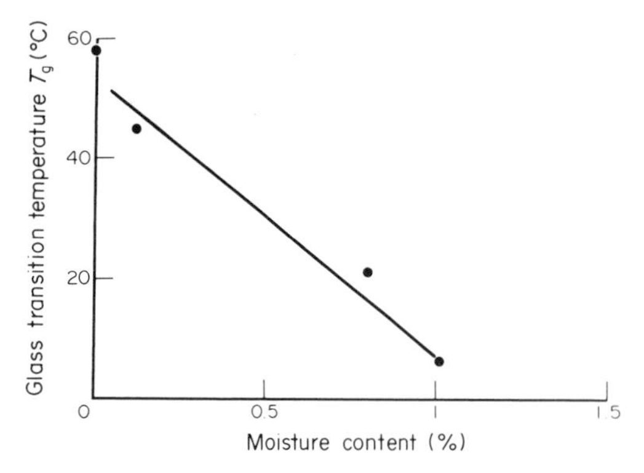

Figure 6
The glass-transition temperature of glass-fiber-reinforced nylon falls dramatically with water uptake. In the above curve the composite had been immersed in water at 90 °C and T_g was measured using a differential scanning calorimeter

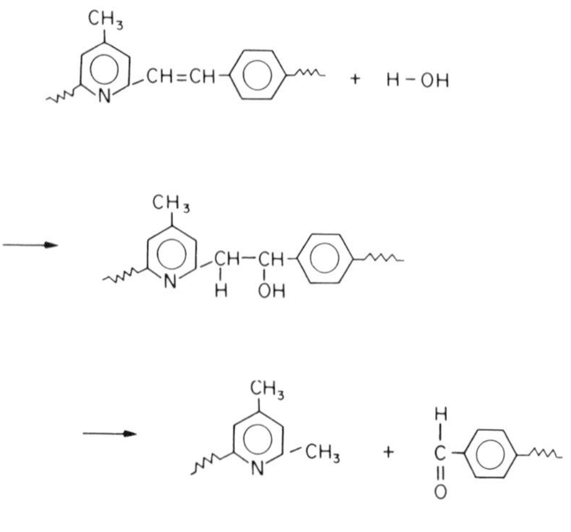

Figure 7
Hydrolysis reaction occurring in carbon-fiber-reinforced PSP resin

and is even more startling in the case of glass-fiber-reinforced polyamide, as is revealed by Fig. 6.

Families of resins able to withstand higher temperatures than the epoxy resins have been developed and in general are less susceptible to water uptake. Prolonged exposure of these resins to hot water can, however, lead to irreversible changes. Such a resin is polystyrylpyridine (PSP) which is usually reinforced with carbon fibers and destined for aerospace applications (Guetta 1987). This resin can operate at 250 °C for 1000 hours and can even withstand several hours at 400 °C. However, prolonged exposure to humid conditions, for example at 70 °C and 100% RH, reveals complex behavior. Initial water uptake appears to be Fickian. Infrared spectroscopy shows that the water is linked to the polar PSP functions by hydrogen bonding, with apparent saturation being reached at this stage. Longer exposures lead, however, to further water uptake and the creation of microcracks as well as the hydrolysis of ethylene double bonds left over after cross-linking of the polymer. This is illustrated in Fig. 7.

Degradation of this type of resin matrix can also occur during thermal cycling, particularly if water is present in microcracks. The carbon-fiber-reinforced PSP composites show increased water absorption after cycling from 70 °C to 150 °C due to the water being vaporized and forcing the microcracks further open. Cycling to 250 °C leads to a more marked fall in properties due to greater damage exacerbated by oxidation of the resin.

2.2 Mechanical Degradation due to Water Absorption

As has been mentioned above, water penetration into resin-matrix composites can lead to changes in mechanical properties. A softening of the matrix is commonly observed which in some cases can be beneficial, inducing a greater toughness. However, changes are usually associated with a fall in properties such as strength or modulus. These changes can appear to be reversible as, upon drying, the composite may regain most of its lost properties. Figure 8 illustrates this for unidirectional glass-reinforced epoxy subjected to three-point bending. It can be seen that drying the specimen almost reversed the effects seen at saturation. A second cycle in humid conditions usually reveals that water penetration occurs much more quickly and property loss is much more rapid than in the first cycle. This behavior is due to the destruction of the chemical

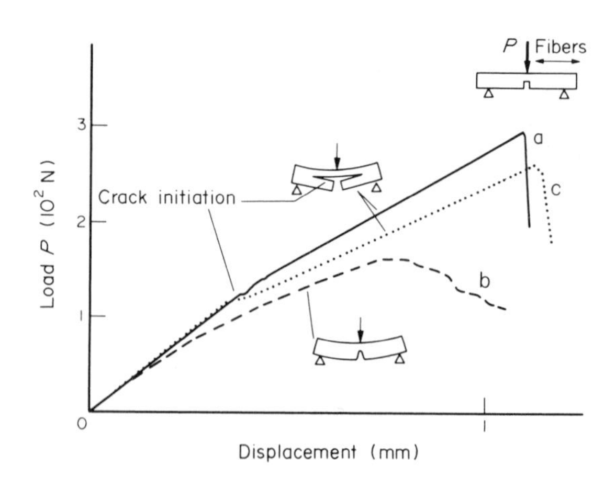

Figure 8
Load–deflection curves obtained with notched three-point bending tests on unidirectional glass-fiber-reinforced epoxy resin: (a) initial state, (b) after 96 hours in boiling water, and (c) after 96 hours in boiling water followed by drying

bonding at the interface by the arrival of water. Upon drying, load transfer between fiber and matrix is assured by frictional forces which may largely compensate for the loss of chemical bonding. A second exposure to humidity results in a rapid water uptake as the water can quickly penetrate the composite along the fiber–matrix interfaces and, acting as a lubricant, rapidly results in loss of mechanical properties.

Figures 9(a), (b) show the fracture surfaces of glass-fiber-reinforced bisphenol-A resin cross-linked with an amine hardener. The specimens were the Tattersall–Tappin type in which the square cross section is cut so as to leave an intact triangular section which breaks progressively from the apex. It can be seen from Fig. 9(a) that little pull-out occurred in the original specimen but after immersion in boiling water followed by drying, pull-out was very extensive. Figures 10(a), (b) reveal that although the resin was well adhered to the fibers in the original state there was no fiber–matrix adhesion after immersion and drying (Dewimille and Bunsell 1982).

Figure 9
(a) Fracture surface of "as received" glass-fiber-reinforced epoxy resin showing little fiber pull-out; (b) extensive fiber pull-out observed after immersion for 96 h at 100 °C, followed by drying, then fracturing

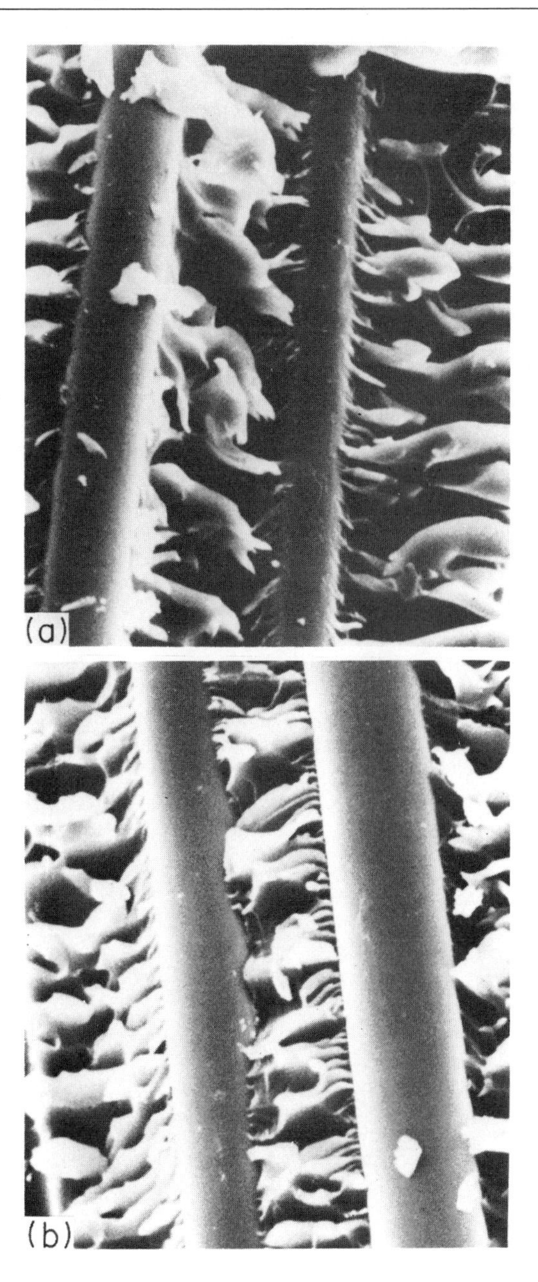

Figure 10
(a) Fracture surface of "as received" glass-fiber-reinforced epoxy resin reveals good bonding of the matrix to the fiber. (b) The fiber–matrix bond has been completely destroyed after immersion in boiling water for 96 hours followed by drying

2.3 Dielectric Changes

In addition to mechanical deterioration due to water uptake, other physical properties, most notably dielectric properties, may be affected. Glass-fiber-reinforced epoxy is an extremely good electrical insulator with low values of dielectric loss factor when dry. The absorption of the polar water molecules leads initially to a slow increase in dielectric loss tangent for fiber-reinforced-epoxy composites as a function of water uptake. However, as Fig. 11 shows, this can suddenly increase because of the development of microcracks (Cotinaud et al. 1982). Simultaneously a dramatic fall in electrical resistivity occurs and if a sufficient voltage is applied to the composite a continuous current can be detected. This behavior suggests that beyond a certain water concentration irreversible damage occurs, leading to microcracking or interfacial damage and the creation of a network of free water molecules. Changes in dielectric properties are, however, generally reversible as they are a function of water content. Little has been published on the effect of water absorption on aramid-fiber composites but as these fibers absorb water to a greater degree than the matrix resins it must be expected that such composites will show considerable changes to their electrical properties.

2.4. *Effect of Hydrostatic Pressure*

Composites are increasingly being used for underwater structures, sometimes at considerable depths. It might be expected that the high hydrostatic pressures experienced on the sea bed would influence water uptake by composites and hence their degradation. This does not seem to be the case, at least for glass-fiber-reinforced epoxy resin (Avena and Bunsell 1988). The effect of the pressure on water uptake is to reduce it slightly because of the consequent closing of microvoids and interfacial defects. As changes in mechanical properties are usually a function of the amount of absorbed water, a descent into deep water would, if anything, have a beneficial effect on degradation processes. Water absorption at pressures below one atmosphere is greatly influenced by the type and efficiency of the size on the fibers, and these effects are maintained irrespective of hydrostatic pressure.

The use of appropriate coupling agents is most important in other types of structure used at depth. Syntactic foams composed of hollow glass microspheres embedded in a resin to form an extremely light composite material can be easily damaged by hydrostatic pressure. The microspheres implode under pressure and, in the absence of an appropriate coupling agent which effectively isolates the broken sphere, such an implosion triggers microcrack growth and a chain reaction of implosions. A good bond between the glass microspheres and the resin resolves this problem and allows these materials to be used at very great depths (Avena 1987).

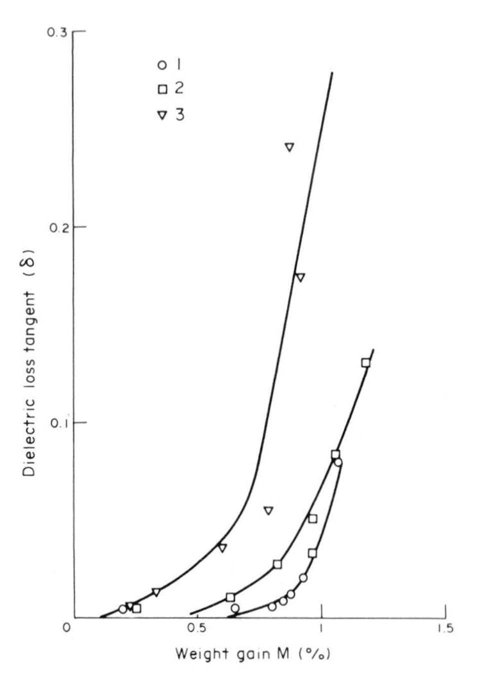

Figure 11
Variation of dielectric loss tangent δ as a function of water uptake (and thus weight gain) for E-glass-fiber-reinforced epoxy resin specimens cured with diamine (1), dicyandiamine (2) and anhydride hardeners (3)

2.5. *Accelerated Aging*

The exposure of resin-matrix composites to humid conditions can produce a variety of damage and absorption mechanisms, which are often thermally activated and can be reversible or irreversible. The distribution of the water, the quantity and the rate at which it is absorbed depend greatly on the temperature and on the damage processes which are induced in the composite. As the temperature increases, the amount and rate of water absorption increases greatly, and models assuming reversible simple diffusion are not always applicable. It seems, therefore, highly questionable to expect immersion in hot or boiling water to simply accelerate those processes which occur at lower temperatures at a slower rate. It is more likely that other damage mechanisms will be provoked which will dominate and lead to overly pessimistic interpretations (Dewimille and Bunsell 1983). To illustrate this point consider Fig. 12 which shows the time necessary for observable damage to occur in a glass-fiber-reinforced bisphenol-A epoxy resin cross-linked with an anhydride hardener. Immersion for only a few hours in boiling water produced debonding and cracking whereas at 50 °C such damage was observed only after 200 days. Specimens immersed for nearly three years in distilled water at room temperature retained their initial appearance and showed no indications of damage.

An interesting technique for accelerating the effects of water uptake involves placing the composite in a succession of humid environments. If, for example, the effects of saturation at 50% RH are required, this state

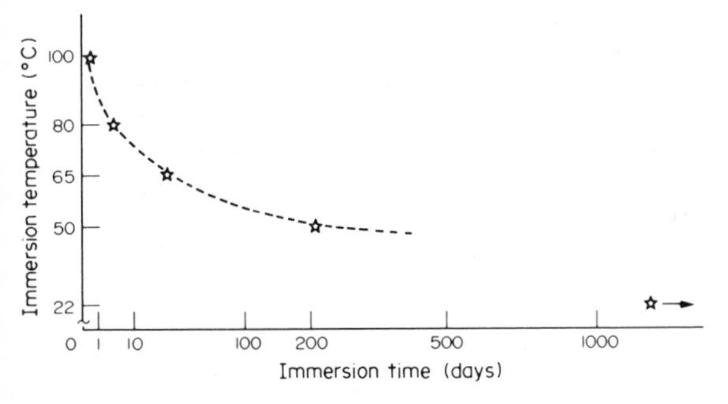

Figure 12
Time to first observable cracking during immersion of glass-fiber-reinforced epoxy resin cross-linked with an anhydride hardener. At 22 °C, results obtained after three years still showed no effects of damage

can be reached much more quickly than by simply leaving the composite in a static environment by first exposing it to 100% RH. The surface of the composite immediately becomes saturated and water begins to penetrate the specimen. After some time water reaches the center of the composite which, however, is not saturated. The composite is placed some time later in an environment at 30% RH, so that the outer surface begins to dry. Shortly afterwards, the composite is placed at 50% RH and left to saturate completely. A considerable saving in time can be gained in this way. However, a word of warning is necessary: if damage can occur in the composite under high water concentrations it is possible that the surface of the composite will be damaged during the first conditioning cycle. This can be misinterpreted later as damage that occurred under the final conditions of relative humidity.

Bibliography

Augl J M 1976 The effect of moisture on carbon fiber reinforced epoxy composites, Silver Spring report NSWC/WOL/TR 76–7

Avena A 1987 Comportement à long terme de matériaux composites en immersion à grande profondeur. Doctoral thesis, Ecole des Mines de Paris, Evry, France

Avena A, Bunsell A R 1988 Effect of hydrostatic pressure on the water absorption of glass fibre reinforced epoxy resin. *Composites* 19:355

Bonniau P, Bunsell A R 1981 A comparative study of water absorption theories applied to glass epoxy composites. *J. Compos. Mater.* 15:272

Browning C E 1976 The mechanisms of elevated temperature property losses in high performance structural epoxy resin matrix materials after exposure to high humidity environments, Technical Report AFML-TR-76-153. US Air Force Materials Laboratory

Carter F G, Kibler K G 1978 Entropy model for glass transition in wet resins and composites. *J. Compos. Mater.* 12:265

Caskey G R, Pillinger W L 1975 Effect of trapping on hydrogen permeation. *Metall. Trans. A* 6:467

Cotinaud M, Bonniau P, Bunsell A R 1982 The effect of water absorption on the electrical properties of glass fiber reinforced epoxy composites. *J. Mater. Sci.* 17:867

Crank J 1956 *The Mechanics of Diffusion.* Clarendon, Oxford

Deiasi R, Whiteside J B 1978 Effect of moisture on epoxy resins and composites. In: Vinson J R (ed.) 1978 *Advanced Composite Materials—Environmental Effects,* ASTM-STP 658. American Society for Testing and Materials, Philadelphia, Pennsylvania, pp. 2–20

Dewimille B, Bunsell A R 1982 The modelling of hydrothermal ageing in glass fibre-reinforced epoxy resins. *J. Phys. D* 15:2079

Dewimille B, Bunsell A R 1982 Vieilissement hygrothermique d'un composite verre–résine époxyde. *Ann. Compos.* 1:1

Dewimille B, Bunsell A R 1983 Accelerated ageing of a glass fibre-reinforced epoxy resin in water. *Composites* 14:35

Guetta B 1987 Vieilissement hygrothermique des composites à matrix PSP: étude cinétique, méchanique et spectroscopique. Doctoral thesis, Ecole des Mines de Paris, Evry, France

Gurtin M E, Yatomi C 1979 On a model for two phase diffusion in composite materials. *J. Compos. Mater.* 13:126

Hahn H T, Kim R Y 1978 Swelling of composite laminates. In: Vinson J R (ed.) 1978 *Advanced Composite Materials—Environmental Effects,* ASTM-STP 658. American Society for Testing and Materials, Philadelphia, Pennsylvania, pp. 98–120

Loos A D, Springer G S 1979 Moisture absorption of graphite–epoxy composites. *J. Compos. Mater.* 13:131

Marine R, Rockett T, Rose V 1988 Blistering in glass reinforced plastic marine materials: A review. University of Rhode Island Marine Technical Report, University of Rhode Island, Kingston, Rhode Island

Shen C H, Springer G S 1976 Moisture absorption and desorption of composite materials. *J. Compos. Mater.* 10:2

Springer G S 1982 Moisture absorption in fiber–resin composites. In: Pritchard G (ed.) 1982 *Development in Reinforced Plastics—2.* Applied Science, London, p. 43

Springer G S, Tsai S W 1967 Thermal conductivities of unidirectional materials. *J. Compos. Mater.* 1:166

A. R. Bunsell
[Ecole Nationale Supérieure des Mines de Paris, Evry, France]

M

Manufacturing Methods for Composites: An Overview

Examples of composites involving natural fibers, such as clay bricks reinforced with straw, and plaster reinforced with animal hair, can be traced back over many centuries. For significant engineering purposes, however, fiber composites can be said to originate with the development during the 1940s of processes for spinning glass fibers in commercial quantities. Subsequent development during the 1960s and 1970s of methods of manufacturing strong stiff fibers from carbon, aramid polymers and ceramics such as silicon carbide and aluminum oxide has now made feasible the production of a variety of composite materials in which the fibers are used to enhance the properties of metals and ceramics as well as polymers.

To illustrate the range of possibilities, Fig. 1 plots the strength and modulus of materials of construction in bulk form, and as fibers and whiskers. It can be seen that conventional materials occupy the lower two decades of strength whereas most fibers display strengths greater by one or two orders of magnitude. Whiskers, as a result of their high degree of crystal perfection, show even higher strengths which in some cases approach theoretical values as calculated from interatomic binding energies.

The physical and mechanical properties of a composite can, to a first approximation, be represented by a simple mixtures law:

$$P_c = P_f V_f + P_m V_m$$

where P_c represents the composite property (for example tensile strength, Young's modulus or density) and P_f and P_m are the values of the corresponding property for the fibers and matrix, respectively. V_f and V_m are the fractions of the total volume occupied by the fibers and the matrix, respectively. The value of V_f typically lies in the range 0.25–0.65, depending on whether the fibers are aligned and closely packed or randomly oriented with respect to each other. Since most synthetic fibers are spun continuously as multifilament bundles or rovings it is relatively easy to achieve high packing fractions in components. Figure 1 indicates that there is then considerable potential to increase the strength, and to a lesser extent the modulus, of many bulk engineering and constructional materials provided suitable manufacturing methods are available to combine the fibers and the matrix. Unlike spun fibers, whiskers tend to be only a few

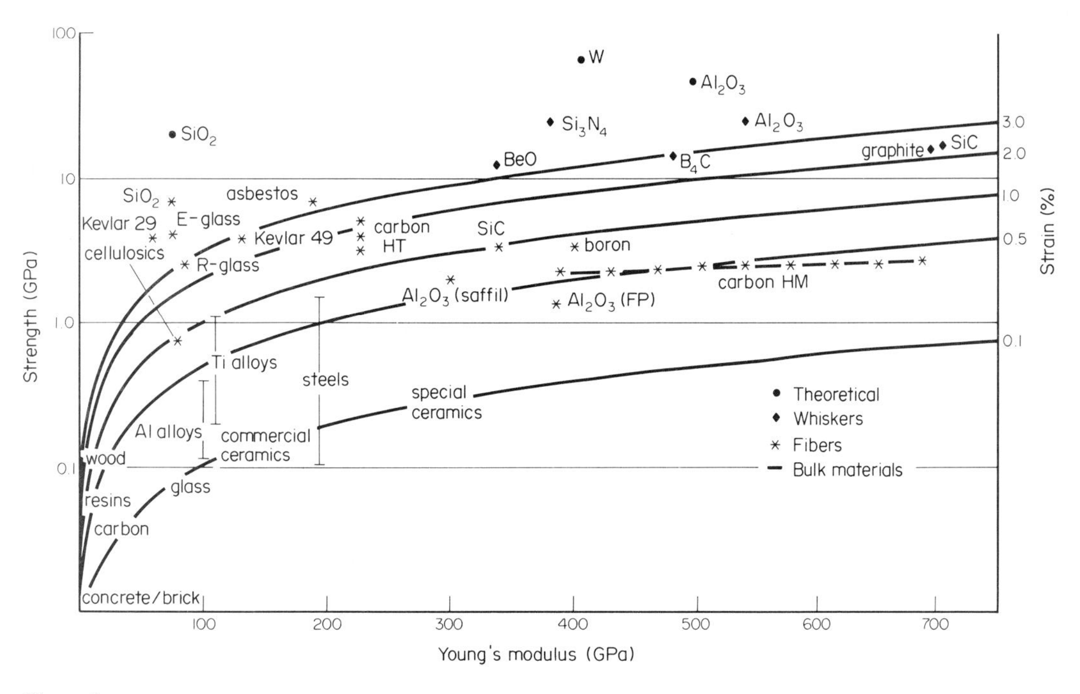

Figure 1
Mechanical properties of various material forms.

millimetres in length and their organization into parallel arrays is a slow and expensive process. Consequently, they are more frequently used as randomly oriented mats with low packing fractions and composite properties are correspondingly low despite the high intrinsic properties of the whiskers themselves.

At the present time only polymer-matrix composites have achieved significant market penetration and this is due partly to the relative ease with which fibers and liquid-thermosetting polymers can be combined and processed and partly due to the fact that suitable reinforcements for metal and ceramic matrices have become available only comparatively recently. Consequently, most of the manufacturing techniques described in this article for metal- and ceramic-matrix composites are still at the research and development stage whereas most of the methods outlined for polymer-based composites have become established for industrial production.

It is important to recognize that the manufacture of composite components differs in several respects from, say, the manufacture of metal parts, where material of a specified or controlled composition is supplied by a primary producer and is then shaped, joined and assembled by a fabricator. During this process, the basic materials properties either remain unaltered or, in the case of heat treatment, are changed in a predictable way. In composites manufacture, although some feedstocks of fiber and resin may be preassembled by a primary producer, the composite material itself is generally produced by the fabricator in the process of forming the component. As a result, the quality of the material may be a function of the type of fabrication process employed and the degree of quality control exercised by the fabricator.

Although the maximum reinforcing effect is achieved with arrays of continuous parallel fibers, the properties in the transverse direction are governed primarily by those of the matrix material and by the strength of the bond between the fiber surface and the matrix. Thus unidirectionally-reinforced materials are highly anisotropic in their properties and the anisotropy must be taken into account in the design of a component and may have an influence on its manufacturing conditions. Anisotropy may be reduced, of course, by randomizing fiber directions or by laminating sheets of parallel fibers with the fibers in each sheet oriented in different directions.

1. Reinforced Plastics

1.1 Feedstock Materials

Manufacturing processes for composites, particularly those with polymer matrices, are intimately bound up with the forms in which the fibers and matrices are available to the fabricator. The range of materials is very wide. For instance, there are at least three types of reinforcing fiber of commercial importance—glass, carbon and aramid—each available in different grades, and there are many types and formulations of resin, both thermosetting and thermoplastic. The fiber length, depending on the mechanical duty required of the component, may fall within the range 1–50 mm or may be effectively continuous; the fibers may all be aligned parallel or laminated in sheets at angles to each other or lie in a plane in random orientations. Thus an enormous number of combinations can be made, depending on the design and cost of the end result. The fabricator has the choice either to combine fibers and resin himself when making the component or to use various types of preassembled materials produced by specialist suppliers. A description of manufacturing processes can therefore usefully be prefaced by an account of the principal forms of feedstock materials. For more detailed descriptions of materials and manufacturing processes see Lubin (1982) and Mohr (1973).

(*a*) *Resins.* Apart from thermoplastics used for injection molding, which are often filled or reinforced with short (less than 1 mm) glass fibers, most reinforced plastics are based on thermosetting resins. A full description is beyond the scope of this article but polyesters (Dudgeon 1987), epoxies (Bauer 1980), phenolics (Knight 1980) and bismaleimides (Stenzenberger 1986) are among the principal types used for composites manufacture. From a manufacturing standpoint, resins of these types can be formulated to cure catalytically or thermocatalytically, and some polyester systems can be cured photolytically: curing can therefore be achieved over a range of temperatures from ambient up to about 200 °C. Choice of resin type depends upon a variety of design and manufacturing considerations, not least of which are the viscosity and flow characteristics of the resin as a function of time and temperature. An important step in the production of some types of composite feedstocks is the ability to partially cure (B-stage) a thermosetting resin to convert it from a liquid to a tacky solid, and to then inhibit further curing, usually by refrigeration, until the component has been assembled, when completion of the cure can be achieved by raising the temperature.

Although in terms of volume, thermosetting resins are used predominantly for making composites, interest has developed in recent years in continuous-fiber-reinforced thermoplastics. Candidate materials range from commodity thermoplastics such as polypropylene and nylon to high-performance engineering materials including polysulfone (PS), polyether sulfone (PES), polyetherimide (PEI) and polyether ether ketone (PEEK) (McMahon 1984, Cogswell 1986). Thermoplastic-matrix composites possess certain potential manufacturing advantages over thermosets. For example, feedstocks possess unlimited shelf-life at ambient temperature; the fabricator does not have to be concerned with proportioning and mixing resins, hardeners and accelerators as with thermosets; and the reversible thermal behavior of thermoplastics means that components can be fabricated more quickly because the lengthy cure schedules for thermosets, sometimes extending over several hours, are eliminated.

(*b*) *Fibers.* Rovings are the basic forms in which fibers are supplied, a roving being a number of strands or bundles of filaments wound into a package or creel, the length of the roving being up to several kilometres, depending on the package size. The term rovings is usually applied to glass fibers, whereas bundles of continuous carbon or aramid fibers are often referred to as tows, this name reflecting the textile fiber technology involved in their manufacture. A strand of glass fibers contains typically some 200 filaments, each approximately 10 μm in diameter, and the number of strands in the roving will depend upon the type of application for which the glass is employed (Waring 1970). Carbon fibers, with similar diameters to glass fibers, are produced in a variety of tow sizes ranging from a few hundred filaments per tow to 320 000 per tow. The commonest sizes in most manufacturers ranges are 3000, 6000, 10 000 and 12 000 ends (Lovell 1988). Fibers are usually sized during production, partly to protect the surface against mechanical damage during subsequent processing or handling, partly to promote wetting by matrix resins and partly to increase adhesion between the fiber and the matrix material. Glass fibers are usually sized with a complex organosilicon compound with other additives, carbon fibers are often treated with a dilute resin compatible with the matrix resin.

Rovings or tows can be woven into fabrics, and a range of fabric constructions are available commercially, such as plain weave, twills and various satin weave styles, woven with a choice of roving or tow size depending on the weight or areal density of fabric required (see *Woven-Fabric Composites: Properties*). Fabrics can be woven with different kinds of fiber, for example, carbon in the weft and glass in the warp direction, and this increases the range of properties available to the designer. One advantage of fabrics for reinforcing purposes is their ability to drape, or conform to curved surfaces without wrinkling. Special forms of woven rovings include braids and knitted constructions which can be used to reinforce tubular components and will readily accommodate bends or curves. Indeed it is now possible, with certain types of knitting machine, to produce fiber preforms tailored to the shape of the eventual component. Generally speaking, however, the more highly convoluted each filament becomes, as at crossover points in woven fabrics, or as loops in knitted fabrics, the lower its reinforcing efficiency.

Chopped fibers are used in lengths ranging from less than 1 mm as a reinforcement or filler in thermoplastics for injection-molding purposes, to approximately 50 mm randomly oriented in sheets or mats for the production of laminates with isotropic properties. For handling purposes the fibers in chopped strand mats are held together with a polymeric binder designed to be soluble in the matrix resin. Chopped strand mats composed of glass fibers are one of the most commonly used feedstocks for hand-placement manufacturing processes, being relatively cheap and easy to cut and drape over curved mold surfaces. Thin mats or tissues are often used in the surface layer of moldings to impart a smooth finish.

An alternative use for chopped fibers is in sprayed constructions. Guns are available that simultaneously spray liquid resin and fibers, the latter being derived from a continuous roving that is reduced to appropriate lengths by a chopper within the gun. Spray-up processes are used to build large shell structures on a suitable core and can be controlled robotically to eliminate exposure of operatives to the styrene vapor associated with the polyester resins widely used in glass-reinforced-plastics technology. Spraying of chopped fibers and resin is also sometimes used in conjunction with other manufacturing processes, such as the filament winding of pipes, to introduce a degree of isotropy in the properties of products.

To simplify the handling of fibers and resins for the fabricator, preimpregnated fibers (prepregs) are available from specialist suppliers. The simplest form, continuous warp sheet, consists of an array of parallel rovings or tows in which the filaments have been spread out laterally to produce a uniform distribution within the thickness of the sheet. The sheet is then impregnated with liquid resin to produce a known fiber/resin volume fraction. Prepregs, developed originally with thermosetting resins, are now also available with certain high melting-point thermoplastics. In the case of thermosets, the resin is partly cured (B-staged) to a slightly tacky condition after impregnation and the sheet surfaces are protected with siliconized paper during storage. Fabrics can also be obtained in the preimpregnated condition and such materials are used typically to build up shell structures by stacking as a preliminary to press or autoclave molding operations. Unidirectional warp sheet is frequently stacked with adjacent layers oriented with respect to each other in such a way as to reduce anisotropy in the final product, rather like the orthogonal grain directions in adjacent veneers in plywood. Warp sheet is also produced in tape of various widths for use in the automated assembly of thin shell structures such as aircraft wing skins and helicopter blades.

Sheet and dough molding compounds are materials which generally comprise polyester resin, an inert particulate filler and chopped fibers, typically in roughly equal proportions by volume. Sheet molding compounds (SMC) are produced by sandwiching a layer of chopped-strand mat between layers of resin mixed with a filler such as talc or chalk, and rolling to ensure penetration of the mixture into the bed of fibers. A material of paste-like consistency results, additives being employed to obtain the appropriate rheological characteristics to ensure the flow of the mixture into mold cavities. Sheets are usually supplied in thicknesses between 3 and 10 mm and are protected on each surface until use by a polyethylene film to prevent evaporation of volatile constituents of the resin.

In standard sheet-molding compounds containing 25–35% fiber, the fibers are generally short and are oriented randomly to give isotropic properties and good flow characteristics into deeply drawn sections. Mechanical properties are correspondingly poor but can be improved by increasing the fiber content at the expense of the filler. Fiber contents up to 65% can be achieved, in which case the materials are usually referred to as HMCs. Further improvements in mechanical strength can be attained with XMCs, which embody continuous fibers or rovings in two directions at the surface of the sheet. Clearly the presence of continuous fibers modifies the ability of such compounds to flow readily and to conform to complex surfaces.

Dough molding compounds (DMC) or bulk mold ing compounds (BMC) differ from SMCs principally in employing a shorter fiber length. The compounds are made by combining the constituents in a high-shear type mechanical mixer and the fibers, although short, may provide a greater degree of three-dimensional reinforcement than in sheet molding compounds.

It may sometimes be desirable to prefabricate the mass of reinforcing fibers to fit a shape to be molded, rather than to rely on flow processes to carry fibers to all parts of the mold. For example, in resin-transfer molding processes, where the mold is first filled with dry reinforcing fibers and the resin is then injected, ideally to flow around the fibers to fill the mold. However, some fibers may be displaced by the flow of the resin. In these circumstances, the fiber mass may first be shaped to fit the mold cavity, for example by spraying on to a porous screen along with a dilute binder, whose function is simply to hold the fibers to the required shape while impregnation with the matrix resin and molding take place. Such a fiber mass is called a preform.

1.2. Manufacturing Methods

A basic aim in almost all reinforced-plastic manufacturing processes is to avoid the entrapment of air or vapor in the form of bubbles or voids. These represent flaws, often located at the fiber–matrix interface or between plies or sheets of material, and generally reduce the strength of the cured composite. A rule of thumb for glass-fiber-reinforced plastics (GRP) is that the interlaminar-shear strength decreases by approximately 7% for every 1% of porosity. The use of low viscosity resins, the vacuum outgassing of resins prior to application and pressure consolidation of the composite are among methods adopted to reduce porosity in the final product.

(*a*) *Hand and spray placement*. In the simplest form, chopped-strand mat or woven rovings are laid over a polished mold surface previously treated with a release or nonstick agent, and liquid thermosetting resin is worked into the reinforcement by hand with the aid of a brush or roller. Polyester resins are most commonly used with glass fibers because of their low cost and good chemical resistance. Resin and curing agent are mixed prior to application and most systems are formulated to cure at ambient temperature. The principal advantages of hand-lay processes are versatility—there is little limitation on the size and complexity of the molding—and low capital cost, the major investment usually being the cost of the tooling. The major disadvantages are low fiber volume fraction and hence low mechanical properties, partly due to the absence of positive consolidation measures. Spray-up methods offer greater rates of material lay-down in addition to the advantages previously outlined. Both methods are well suited to one-offs or short production runs and an outstanding example of the importance of this type of technology is the boatbuilding industry where GRP has virtually superceded traditional construction materials. Hand-lay methods have been used for the construction of quite large craft such as minesweepers up to approximately 60 m in length (Dixon et al 1973).

(*b*) *Press molding*. Molding under pressure in heated matched male/female tools is the most widely used process for the volume production of reinforced plastic components. Most of the feedstocks described in Sect. 1.1 can be molded in this way, but the process is particularly appropriate for use with sheet and dough molding compounds. These materials are formulated first to undergo a rapid reduction in viscosity from their paste-like consistency as the charge of compound heats up by contact with the mold surfaces. This enables the compound to flow smoothly, under molding pressures of the order 3–7 MPa, into complex and deeply re-entrant parts of the mold. A rapid gelation and cure then ensues, total cycle times ranging typically from seconds to minutes, depending on the size and thickness of the molding. The process is extensively employed for the production of domestic articles, cabinets and containers for electronic equipment and office machinery, and for automotive parts in sizes up to those of truck doors and cab panels. Figure 2 illustrates schematically a simple press-molding operation.

Press molding is also used for components reinforced with continuous fibers, but the fibers inhibit flow in the mold and such materials are used primarily for the production of parts of constant wall-thickness and of limited curvature.

In production runs involving many thousands of components, steel or plated-steel tools are employed and for large complex parts can represent a substantial capital investment. For short production runs or the development of prototypes, molds can be made by

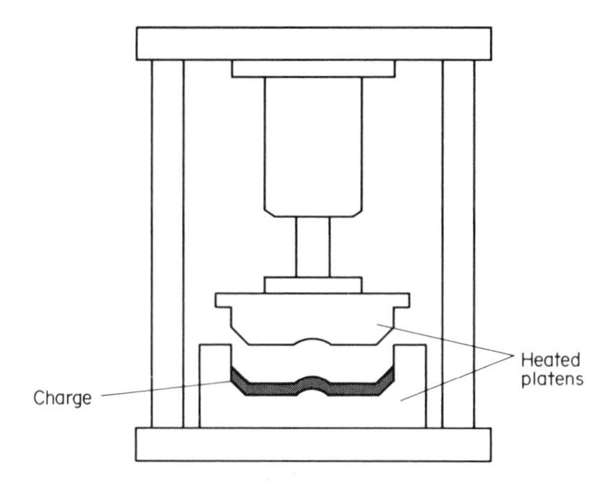

Figure 2
Simple press molding arrangement

casting a high-strength epoxide resin around a pattern at a relatively modest cost.

(*c*) *Vacuum molding.* This process makes use of atmospheric pressure to consolidate the material while curing, thereby obviating the need for a hydraulic press. The laminate, in the form of preimpregnated fibers or fabric, or as chopped-strand mat or woven rovings impregnated with liquid resin, is placed on a single mold surface and is overlaid by a flexible membrane which is sealed around the edges of the mold by a suitable clamping arrangement. The space between the mold and the membrane is then evacuated and the vacuum maintained until the resin has cured. Quite large, thin shell moldings can be made in this way at low cost.

(*d*) *Autoclave molding.* If a higher density and lower void content are required, a consolidating pressure greater than atmospheric pressure may be necessary. The component, laid up on a mold, is enclosed in a flexible bag tailored approximately to the desired shape and the assembly is enclosed in an autoclave—a pressure vessel designed to contain a gas at pressures generally up to approximately 1.5 MPa and fitted with a means of raising the internal temperature to that required to cure the resin. The flexible bag is first evacuated, thereby removing trapped air and organic vapors from the composite, after which the chamber is pressurized to provide additional consolidation during cure. The process is used extensively in the aircraft industry for the production of thin shell structures of high mechanical integrity, and large autoclaves capable of housing complete wing or tail sections have been installed.

(*e*) *Resin-transfer molding.* The molding processes thus far described start with an open mold or tool which is subsequently closed to surround and compress the charge of fibers and resin. In resin-transfer molding, the fibers only, sometimes as a preform, are enclosed in the mold and the resin is subsequently injected to infiltrate the fibers and fill the cavity. Low-viscosity resins are used to ensure low voidage and good penetration of resin to all parts of the mold, and in vacuum-assisted resin injection, a variant of the process, air in the mold is pumped out prior to injecting the resin. Resin-transfer molding can be used to produce quite large moldings, such as car body-shells, in small-scale production of a few hundred moldings per year. Compared to press-molding of similar-sized parts, the capital cost is low because the low pressures involved enable the molds themselves to be made from GRP.

(*f*) *Reaction injection molding.* Reaction injection molding (RIM) and reinforced reaction injection molding (RRIM) also use relatively low pressures (approximately 0.5–1 MPa) to produce large moldings, principally in polyurethane elastomers (Becker 1979). The principal difference from resin-transfer molding is that, instead of using a precatalysed resin with a relatively slow cure, the RIM process brings two fast-reacting components together and mixes them just prior to injection into the mold. The mixed system can be tailored to cure in the mold within 30–60 s, thus giving rise to component cycle times of the order 1–2 min. The basis of the process is illustrated in Fig. 3. The equipment comprises two reservoirs storing, typically, a diisocyanate and a polyol, which polymerize when intimately mixed. Figure 3(a) shows the mixing/injection stage during which the two reactants are pumped through an impingement chamber in the mixing head, and thence to the mold cavity. The impingement chamber produces highly turbulent flow conditions and hence rapid mixing of the two constituents, and this is followed by rapid gelation and cure in the mold. In the second stage, represented by Fig. 3(b), all the mixed polymer is ejected from the mixing head and the two reactants are separated and continuously recirculated ready for the next molding

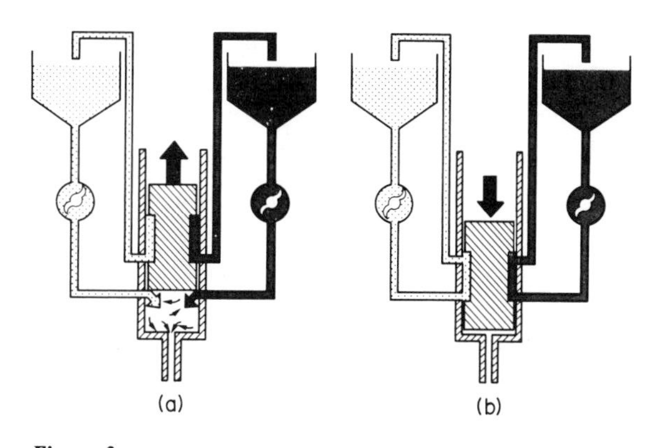

Figure 3
Reaction injection molding process: (a) mixing/injection stage, (b) mixed polymer ejection stage

cycle. The mixing head is thus self-cleaning and any cured polymer in the connecting pipe from the mixing head to the mold cavity is arranged to be removed as a piece of flash attached to the molding.

A variety of urethane formulations is available to give polymers with a range of hardness from soft and rubbery to hard and brittle. The moldings can be reinforced to some degree with short glass fibers by including them in one or both of the liquid reactants to form a slurry. The limitation on fiber length and hence reinforcing efficiency, and the quantity of fiber that can be incorporated, is set by the design of the mixing head and its ability to handle the fibers without risk of blockage or of undue agglomeration of the fibers resulting in a nonuniform distribution in the molding. RIM and RRIM processes compete in many ways with press-molded sheet and dough-molding compounds for markets in office and domestic equipment and for automotive products such as bumpers, body and trim panels, and both materials and equipment are the subject of intensive development at the present time.

(*g*) *Pultrusion.* The processes thus far described produce discrete moldings. Pultrusion enables profiles of constant cross-section to be manufactured continuously. The process is analogous to the extrusion of plastics and nonferrous metals in that the profile is shaped by continuous passage of the feedstock through a forming die, but in the case of pultrusion the reinforcing fibers are used to pull the material through the die. In Fig. 4, continuous rovings are passed through a bath of resin followed by a series of carding plates to encourage complete penetration of resin into the fiber bundles and to remove excess resin. The impregnated fibers are then drawn through a heated die in which the resin first gels and then cures. The finished profile finally passes through a traction mechanism before being cut to length.

Glass fibers and polyester resins are the most commonly employed materials because of their relatively low cost and good corrosion resistance. Preheating by radiofrequency methods is often used to increase the throughput, and production rates of several meters per minute are achieved depending on section thickness. The process is ideally suited to the production of structural elements of high strength and stiffness owing to the presence of the high proportion of axial fibers necessary to sustain large tractive forces, particularly when high volume fractions of fiber (up to approximately 0.7) are required in the cross section. It is, however, possible to incorporate varying cross-sectional geometry although of constant cross-sectional area.

(*h*) *Filament winding.* In this process, strong, stiff shell-structures can be made by winding bands of continuous fibers impregnated with resin on to a former or mandrel which can be withdrawn after the resin has been cured. The method is analogous to the technique of wire wrapping for the reinforcement of steel pressure-vessels and gun barrels and is extensively used for the manufacture of reinforced-plastic pipe, and for cylindrical tanks and vessels in GRP for the storage of aggressive chemicals and a wide variety of agricultural products. The size of component that can be manufactured is limited only by the capacity of the winding machine, and methods also exist for the building of large vertical silos in situ, involving the construction of a circular railway track upon which a carriage, dispensing fiber and resin, travels continuously around a vertical former.

Modern filament-winding machinery is capable of producing shapes that are much more complex than the simple axisymmetric structures indicated above, largely as a consequence of adopting robotics technology. Figure 5 shows schematically a gantry-type filament-winding machine with five axes or machine motions that are controlled independently by a microprocessor. Two such motions, rotation (θ_1) of a mandrel between head and tailstock, and translation (x) of a carriage backwards and forwards parallel to the axis of mandrel rotation, suffice for the winding of cylindrical shells at helix angles dictated by the relative rates of rotation and translation. The additional facility of movement of the feed eye along the y and z directions, however, enables a band of rovings to be laid in a

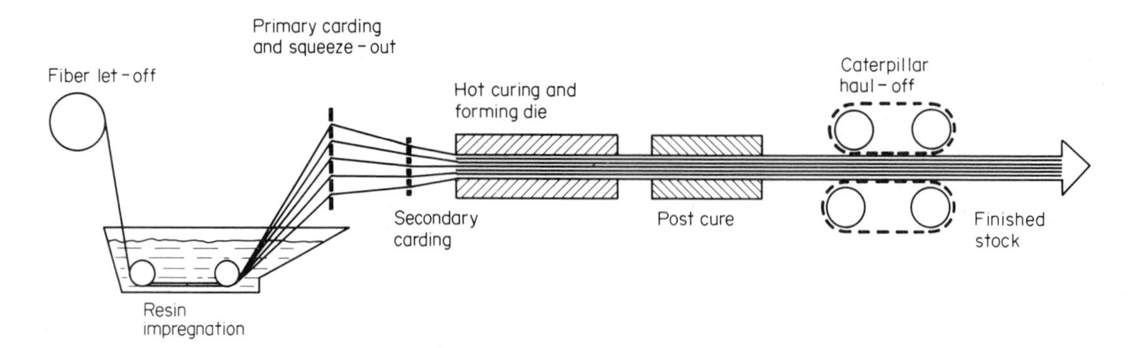

Figure 4
Schematic of pultrusion process

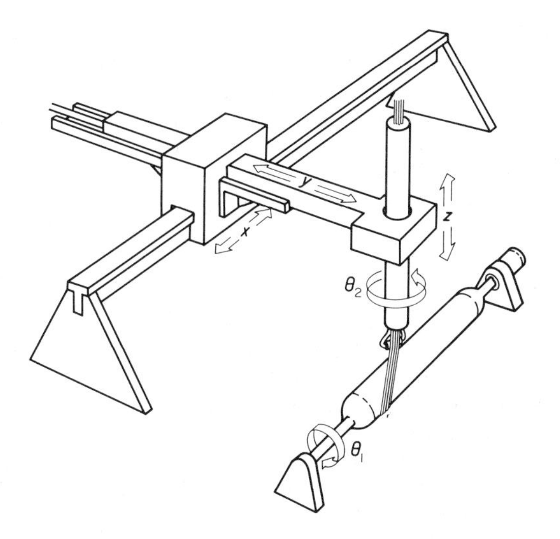

Figure 5
Basic motions of a 5-axis filament winding machine

controlled way on convex surfaces of any given shape, and rotation of the feed eye (θ_2) enables the plane of the band to be matched to the inclination of the mandrel surface at any given point.

For a stable winding it is necessary to identify geodesic paths, defined as the shortest distance between two points on a surface, to avoid the possibility of the band of fibers slipping from the desired trajectory. This condition, coupled with the sheer magnitude of the task of programming a multiaxis robot to traverse a constantly changing and nonrepetitive path, has led to the development of computer based methods aimed at integrating design and manufacturing procedures for filament-wound components by analogy with CADCAM processes for metal machining (Wells and McAnulty 1987). This type of technology opens up the possibility of winding components as complex as aircraft fuselages, steering wheels and bicycle frames.

2. *Reinforced Metals*

Reference to Fig. 1 shows that the mechanical properties of the nonferrous metals and alloys commonly used for structural purposes are considerably better than those of unreinforced polymers, and the improvements to be obtained by combining them with fibers or whiskers will therefore be less dramatic than with polymers. However, other worthwile improvements can be achieved by incorporating fibers or even particulates into a metal matrix. For instance, creep resistance at a particular temperature may be improved or the resulting material may be more wear-and-abrasion resistant, or less susceptible to fatigue failure. Such improvements, even if modest, may be of great technological importance if, for example, the temperature limit for continuous operation of an engine component or an airframe structural part can be increased. Considerable research is now being devoted to both the development of reinforcements suitable for metal matrices, and of manufacturing methods. Much of the drive behind the upsurge in metal-matrix composites development derives from the recent development of processes for producing refractory fibers and whiskers in commercial quantities (Bunsell 1987).

2.1 *Feedstock Materials*

(*a*) *Matrix materials.* The most common matrices are the low density metals, such as aluminum and aluminum alloys, and magnesium and its alloys. Some work has been carried out on Pb–Sn alloys, mainly for bearing applications, and there is interest in the reinforcement, for example, of titanium-, nickel- and iron-base alloys for higher-temperature performance. However, the problems encountered in achieving the thermodynamic stability of fibers in intimate contact with metals become more severe as the potential service temperature is raised, and the bulk of development work at present rests with the light alloys.

(*b*) *Reinforcements.* The principal reinforcements for metal matrices include continuous fibers of carbon, boron, aluminum oxide, silica, aluminosilicate compositions and silicon carbide. Some ceramic fibers are also available in short staple form, and whiskers of carbon, silicon carbide and silicon nitride can be obtained commercially in limited quantities. There is also interest in the use of refractory particles to modify alloy properties such as wear and abrasion resistance. In this case particle sizes and volume fractions are greater than those developed metallurgically in conventional alloys, and incorporation of the particles into the metal is achieved mechanically rather than by precipitation as a consequence of heat treatment.

Most metal-matrix composites consist of a dispersed reinforcing phase of fibers, whiskers or particles, with each reinforcing element ideally separated from the next by a region of metal. An alternative approach is to reinforce the metal, generally in sheet form, with a fiber-reinforced polymer. An example is ARALL (Vogelesang and Gunnink 1983), a laminate consisting of alternate plies of aluminum alloy bonded to sheets of aramid-fiber-reinforced resin. This has the advantage of a simple, low-temperature production process in which the continuous parallel fibers can display a high reinforcing efficiency. Applications are, however, restricted to thin shell-type structures.

Apart from sheet materials similar to ARALL, the production of feedstocks in which reinforcement and matrix can be purchased preassembled for the convenience of the fabricator is less well advanced than the production of polymer-matrix composites. Wires or tapes of aluminum containing up to 500 continuous silicon carbide filaments are available in development quantities and in some cases reinforcements can be obtained woven into fabrics or agglomerated into randomly oriented mats.

2.2 Manufacturing Methods

Many of the processes used for the production of unreinforced metal parts, or for the deposition of metal layers, have been explored or adapted with the aim of producing a continuous, void-free metal matrix with a controlled distribution of reinforcements. Fabrication methods can involve processing the metal in either the molten or the solid state, and components can be formed either by direct combination of matrix and reinforcement or by the production of a precursor composite which, in the form of composite wires, sheets or plies, is used to build up a component. In the latter case, the assemblage of plies must be consolidated and bonded in a subsequent process.

In liquid-metal techniques, composites can be prepared by infiltrating mats or fiber preforms with liquid metals or, under carefully controlled conditions, by physically mixing the reinforcement and the liquid metal together. A pseudoliquid route is offered by plasma or flame spraying in which metal-powder particles are heated above their melting point and are sprayed onto an array of fibers on a thin sheet of the same matrix metal. The resulting sheet of fiber-reinforced metal can then be stacked with other sheets and consolidated in a subsequent operation. An alternative approach is to build up a solid body by codepositing metal and particulate reinforcements by spraying techniques.

The simplest solid-state preparation route is to mix short fibers or particulates with metal powders. The mixture can then be processed by standard powder-metallurgical procedures. Alternatively, the metal may be coated onto the reinforcement by electrochemical or chemical-vapor deposition methods.

Many of these processes are still under development or evaluation and, as with polymer-matrix composites, the methods of manufacture that will eventually be adopted commercially will depend on production costs as well as technical suitability for particular components. For this reason, considerable attention is being focused on infiltration and casting processes because of their relatively low cost. For detailed reviews of manufacturing techniques see Mileiko (1983) and Chou et al. (1985).

(*a*) *Liquid-metal infiltration.* Direct impregnation of bundles of continuous filaments has been used to produce precursor wires or tapes that can be subsequently consolidated into component forms. A typical scheme involves drawing a bundle of filaments through a bath of molten metal and then through an orifice which shapes the impregnated bundle to a circular or rectangular cross section. For complete impregnation to take place it is important that the metal should wet the fiber surface, and considerable research has been devoted to coatings for fibers or additives to the metal that will reduce the contact angle of the molten metal at the fiber surface and promote wetting. Penetration of fiber bundles may be enhanced by infiltrating under vacuum or using inert-gas pressure over the molten metal (Masur et al. 1987).

(*b*) *Squeeze casting.* In this process a mass or preform of fibers or whiskers is infiltrated with molten metal under a high pressure derived from a hydraulic press. The method is illustrated schematically in Fig. 6 and can result in a fully dense composite with good dispersion of the reinforcement. Whisker concentrations of up to 30% by volume have been achieved, resulting typically in increases in strength and Young's modulus of a factor of two over unreinforced aluminum alloys. The method can be used to produce billets that can then be shaped by conventional metal hot-working techniques such as extrusion or forging, or it can be used directly for near net-shaping of components such as automotive connecting rods. With some materials it may be desirable for the infiltration to be carried out under vacuum, to avoid either the risk of oxidation of the reinforcement or the possibility of entrainment of air bubbles in the matrix.

(*c*) *Stir casting and compo-casting.* In some cases it is possible to introduce particulate or short-fiber reinforcements simply by feeding into the vortex region produced by vigorously stirring the molten metal. Long stirring times may be necessary to develop a good bond between the two components, and the volume fraction of the reinforcement may be limited to about 10% by the rapid increase in the viscosity of the mixture with the further addition of reinforcement.

Differences in density between molten metals and reinforcing fibers or whiskers may mean that separation of the two components will occur due to gravitational force if simple mixing of the fibers in molten metal is attempted. The tendency to separate can be reduced or eliminated in compo-casting whereby stirring takes place as the metal undergoes solidification. Under these conditions, mixed phases of solid

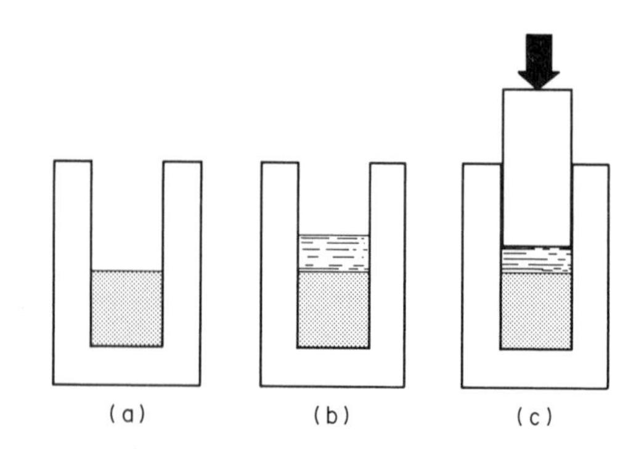

Figure 6
Schematic of squeeze casting process: (a) fiber preform placed in heated mold, (b) liquid metal poured into mold, (c) infiltration under pressure

and liquid metal can coexist in the form of a slurry in which reinforcement particles can be mixed and become trapped between the solid phase regions with uniform dispersion.

(*d*) *Precursor production.* Preimpregnated sheet that can be used to build up more complicated structures can be made in a variety of ways. The most common method using continuous monofilament or bundles of filaments is to assemble them with the required spacing by winding on to a drum or around pegs to produce a parallel array on a flat surface. The metal is then coated onto the filaments by plasma spraying, electroplating or by chemical or chemical vapor deposition tehchniques. Alternatively, bundles or tows of filaments can be coated by electrochemical or plasma-activated deposition processes. These various precursor forms must then be shaped, stacked, consolidated and bonded to form a continuous matrix phase.

(*e*) *Consolidation and bonding methods.* Densification and bonding of sheets of precursor materials or of mixtures of metal powder and reinforcement are usually carried out in the solid state under conditions of temperature and pressure such that consolidation results from a combination of plastic deformation and diffusion in the metal phase. Methods of placing the material under the appropriate conditions of temperature and pressure include conventional hot-pressing, hot and cold isostatic pressing, hot rolling and explosive welding. As an example see Hack and Amateau (1982) who describe diffusion bonding of carbon-fiber-reinforced aluminum alloy wires to aluminum alloy foils in a press at a temperature between 554 and 562 °C and a pressure of 24 MPa for a dwell time of 20 minutes. An account of the technology of boron-reinforced aluminum composites, including details of diffusion bonding, has been given by Miller and Robertson (1977).

Many of the composites incorporating short-fiber or particulate reinforcements are produced in the form of billets by the methods described above and may require subsequent processing by standard metallurgical techniques such as extrusion, forging or drawing to achieve their final form.

3. *Reinforced Ceramics*

Research into the properties of fiber-reinforced ceramics has been in progress since the 1950s, but until the advent of refractory, oxidation-resistant fibers such as alumina and silicon carbide in commercial quantities, much of the work was confined to developing an understanding of the principles of reinforcement and mechanisms of failure (Phillips 1983). As with metal-matrix composites, there is currently an increase in research and development activity, although little commercial application of materials thus far. This description of manufacturing methods consequently relates to laboratory-type processes.

3.1. Feedstock Materials

(*a*) *Matrix materials.* The choice of matrix depends primarily on the mechanical and thermal duty required of the composite. For potential applications in aerospace and in heat engines, oxidation resistance is of prime importance and most refractory oxides, carbides and borides have been considered as matrices. The most extensively employed, however, have been aluminum oxide, silicon carbide and silicon nitride, mainly because the technology of producing well-consolidated artifacts in the unreinforced state is better established for these materials. Glasses and silica have also been explored because of the relative ease with which matrix consolidation can be accomplished by processing in the liquid state. Glass–ceramic materials that can be combined with reinforcements while in a liquid state and then devitrified by heat treatment have also been the subject of considerable investigation.

Although not a ceramic, carbon is one of the most refractory materials in nonoxidizing atmospheres, and Fig. 1 indicates that substantial improvement to the mechanical properties of bulk carbon is feasible by fiber reinforcement. In samples unidirectionally reinforced with carbon fibers, tensile strengths of up to 1 GPa have been reported (Hill et al. 1974). For some applications, reinforcement in three dimensions is required and weaving or knitting processes have been developed for producing continuous-carbon-fiber preforms that can subsequently be invested with a carbon matrix.

(*b*) *Reinforcements.* Only a few of the fibers available commercially are suitable for ceramic-matrix composites by virtue of their compatibility and stability with the matrix during fabrication and at intended service temperatures. Continuous fibers include silicon carbide, boron, aluminum oxide, aluminosilicates and carbon (Bunsell 1987). In oxidizing atmospheres, only silicon carbide and the oxide fibers offer the prospect of high-temperature composites and, depending on the stress to which they are subjected, the upper limit of ceramic-matrix composites is currently in the region of 1200 °C.

(*c*) *Manufacturing methods.* Hot-pressing of sheet-material preforms is the route most widely adopted for ceramic-matrix composites, and hot-pressing conditions generally follow established procedures for consolidation and sintering of monolithic ceramics. Early work has been reviewed by Sambell (1970) and a method of producing sheet preforms is described by Sambell et al. (1974). More recently, chemical vapor deposition (CVD) methods have been adopted to infiltrate a fiber preform with, for example, silicon carbide derived from the gaseous-phase cracking of methyl trichlorosilane in hydrogen. Alternatively, the preform may be impregnated with a liquid organometallic compound which is then pyrolyzed to yield a refractory ceramic residue (Bernhart et al. 1985).

Similar methods are used for the manufacture of carbon-fiber-reinforced carbon materials (Fitzer 1987). Preforms are either coated with carbon produced in the vapor phase by reacting, say, methane and hydrogen, or are subjected to impregnation of a polymer possessing a high carbon yield when pyrolyzed. One of the major disadvantages of both CVD and pyrolysis routes is that successive depositions are necessary to produce a fully dense matrix. The processes are therefore slow and correspondingly expensive. Nevertheless, such processes are operated commercially, notably for the manufacture of carbon–carbon disk brakes for aircraft and for racing cars.

4. Conclusion

Within the scope of this article it has been possible to describe only briefly some of the methods used for manufacturing composite materials and components. In practically every case, however, the aim is to control the dispersion of the reinforcement in the matrix in such a way as to maximize the properties of the composite while minimizing the manufacturing cost. The great variety of materials and material forms involved in composites technology provides considerable promise for the evolution of materials with improved performance, but the availability of cost-effective manufacturing methods capable of providing products with a high degree of consistent and reproducible performance is a key factor in determining the rate at which they can compete with more traditional materials of construction.

Bibliography

Bauer R S 1980 The versatile epoxies. *Chem. Technol.* 9: 692–700

Becker W E 1979 *Reaction Injection Moulding.* Van Nostrand Reinhold, New York

Bernhart G, Chateigner S, Heraud L 1985 Les composites ceramique-ceramique, du concept a la piece. In: Bunsell A R, Lamicq P, Massiah A (eds.) 1985 *Developments in the Science and Technology of Composite Materials.* European Association for Composite Materials, Bordeaux, pp. 475–81

Bunsell A R 1987 Fibre reinforcements—past, present and future. In: Matthews F L, Baskell C R, Hodgkinson J M, Morton J (eds.) 1987 *Proc. 6th Int. Conf. on Composite Materials,* Vol. 5. Elsevier, London, pp. 1–13

Chou T W, Kelly A, Okura A 1985 Fibre-reinforced metal-matrix composites. *Composites* 16 (3): 187–206

Cogswell F N 1986 Continuous fibre reinforced thermoplastics. In: Collyer A A, Clegg D W (eds.) 1986 *Mechanical Properties of Reinforced Thermoplastics.* Elsevier, London, pp. 83–119

Dixon R H, Ramsay B W, Usher P J 1973 Design and build of the GRP hull of HMS Wilton. *Proc. Symp. on GRP Ship Construction.* Royal Institution of Naval Architects, London, p. 1

Dudgeon C A 1987 Polyester resins. In: *Materials Engineering Handbook,* Vol. 1. American Society for Metals, Metals Park, Ohio, pp. 90–96

Ewald G 1981 Curved pulforming—a new manufacturing process for composite automobile springs. *Proc. 36th Ann. Conf. Reinf. Plastic/Composites Inst.* Paper 16C. Society for Plastics Industry, New York

Fitzer E 1987 The future of carbon-carbon composites. *Carbon* 25(2): 163–90

Hack J E, Amateau M F 1977 Mechanical behaviour of metal matrix composites. *Proc. AIME Conf. Dallas, Texas, 16-18 Feb. 1982.* American Institute of Mining Engineers, New York

Hill J, Thomas C R, Walker E J 1974 Advanced carbon-carbon composites for structural applications. In: *Carbon Fibers, their Place in Modern Technology,* paper 19. The Plastics Institute, London. pp. 122–30

Knight G J 1980 High temperature properties of thermally stable resins. In: Pritchard G (ed.) 1980 *Developments in Reinforced Plastics,* Vol. 1. Applied Science, London, pp. 145–210

Lovell D R 1988 *Carbon and High Performance Fibres Directory,* 4th edn. Pammac Directories Ltd., High Wycombe UK, pp. 45–74

Lubin G (ed.) 1982 *Handbook of Composites.* Van Nostrand Reinhold, New York

McAdams L V, Gannon J A 1986 Epoxy resins. In: *Encyclopedia of Polymer Science and Engineering.* Vol. 6, 2nd edn. Wiley, New York, pp. 322–82

McMahon P E 1984 Thermoplastic carbon fibre composites. In: Pritchard G (ed.) 1984 *Developments in Reinforced Plastics,* 4. Elsevier, London, pp. 1–30

Masur L J, Mortensen A, Cornie J A, Flemings M C 1987 Pressure casting of fiber-reinforced metals. In: Matthews F L, Buskell N C R, Hodgkinson J M, Morton J (eds.) 1987 *Proc. 6th Int. Conf. on Composite Materials,* Vol. 2 Elsevier, London, pp. 320–29

Mileiko S T 1983 Fabrication of metal-matrix composites. In: Kelly A, Mileiko S T (eds.) 1983 *Handbook of Composites,* Vol. 4. North-Holland, Amsterdam, pp. 221–94

Miller M, Robertson A 1977 Boron/aluminium, and borsic/aluminium analysis, design, application and fabrication. In: Renton W J (ed.) 1977 *Hybrid and Select Metal Matrix Composites: A State of the Art Review.* American Institute of Aeronautics and Astronautics, New York, pp. 99–157

Mohr J G (ed.) 1973 *SPI Handbook of Technology and Engineering of Reinforced Plastics/Composites,* 2nd edn. Van Nostrand Reinhold, New York

Phillips D C 1983 Fibre reinforced ceramics. In: Kelly A, Mileiko S T (eds.) 1983 *Fabrication of Composites,* Vol. 4. Elsevier, Amsterdam, pp. 373–428

Sambell R A J 1970 The technology of ceramic-fibre ceramic-matrix composites. *Composites* 1(5): 276–85

Sambell R A J, Phillips D C, Bowen D H 1974 The technology of carbon fibre reinforced glasses and ceramics. In: *Carbon Fibres, their Place in Modern Technology,* paper 16. The Plastics Institute, London, pp. 105–13

Stenzenberger H 1986 Bismaleimide resins. In: Kinloch A J (ed.) 1986 *Structural Adhesives—Developments in Resins and Primers.* Elsevier, London, pp. 77–126

Vogelesang L B, Gunnink J W 1983 ARALL, a material for the next generation of aircraft—a state of the art. In: Jube G, Massiah A, Naslain R, Popot M (eds.) 1983 *High Performance Composites Materials: Proc. 4th Int. Conf. Society for the Advancement of Materials and Process Engineering, European Chapter.* SAMPE, Bordeaux, pp. 81–92

Waring L A R 1970 Reinforcement. In: Parkyn B (ed.) 1970 *Glass Reinforced Plastics*. Butterworth, London, pp. 121–45
Wells G M, McAnulty K F 1987 Computer aided filament winding using non-geodesic trajectories. In: Matthews F L, Buskell N C R, Hodgkinson J M, Morton J (eds.) 1987 *Proc. Sixth Int. Conf. on Composite Materials*, Vol. 1. Elsevier, London, pp. 161–73

D. H. Bowen
[AEA Technology, Didcot, UK]

Metal-Matrix Composites

Metal-matrix composites offer a number of property advantages over conventional materials. They share with other composites the flexibility to combine the properties of both fiber and matrix to gain some of the advantages of each. Also, the combination of properties can be tailored to match composite properties to component requirements by varying the constituents and their volume fraction content. The properties achieved can surpass those possible with conventional materials.

1. Metal Matrix Versus Polymer Matrix

Metal-matrix composites offer a combination of properties that makes them superior to competing materials for some applications. A comparison of the similarities and differences helps to illustrate these advantages. The matrix serves as a glue to bond fiber arrays into useful structural shapes for both polymer- and metal-matrix composites. The greater strength and stiffness of fibers are used to increase properties of the composite over those of the unreinforced matrix. The inorganic fibers used to reinforce polymers can also be used to reinforce metals. The uniaxial property contribution of the fiber is similar with both matrix materials, but the matrix contribution varies considerably. Metals have significantly greater strength and stiffness compared with polymers, and higher potential service temperatures. While density values for metals are typically higher than those of polymers, the modulus-to-density ratios of metals are over five times those of polymers used in composites. The higher property values of metallic matrices permit a greater matrix contribution to transverse properties which can reduce the need to orient fibers in cross-ply arrays: cross-ply orientation increases transverse properties but results in a decrease in longitudinal properties. A further advantage of metal as a matrix is the crack- and flaw-tolerance afforded by a plastically deformable matrix. Relaxation of stress concentrations by plastic deformation of metal matrices is an important advantage of a composite matrix. Excellent fatigue properties and improved toughness are benefits derived from the yielding of ductile matrices.

The family of metal matrices covers a wide spectrum of use temperatures, from lead alloys with peak use temperatures below 200 °C to refractory alloys with use capability at temperatures over 1200 °C. Despite their numerous advantages, metal-matrix composites have not achieved significant commercial application. This is the result of a lower level of technical development which is largely related to the fabrication-processing technology of metal-matrix composites, which is more complex than that of polymer-matrix composites.

Fabrication of polymer composites can be accomplished by liquid-phase infiltration of the matrix at room temperature into multistrand fiber yarn or tow. Consolidation and polymerization into a composite panel or member is accomplished at a few hundred degrees Celsius. Metal-matrix composites, with some exceptions, require higher temperature and pressure fabrication processing. Liquid-phase infiltration of metal matrices into a fiber array has been achieved with relatively few fiber–matrix combinations. The fibers produced in commerical quantities—carbon, boron, silicon carbide and aluminum oxide—are difficult to fabricate because they are poorly wetted by molten matrices; thus either infiltration is made difficult, or else it occurs readily but a fiber–matrix reaction occurs which significantly degrades fiber properties. Matrix-alloy compositions and coatings to improve wetting and to control reactions are under development, and preliminary results are encouraging. Solid-state fabrication methods have been used to circumvent the problems associated with liquid-phase processing. Diffusion bonding of matrix-alloy foil and fiber arrays has been the common fabrication method for metal-matrix composites (see Sect.3). The need for solid-state fabrication has limited fiber content to below 60 vol%. The interface spacing necessary to permit matrix flow has limited fiber content and has fostered the use of larger diameter fibers. Boron and silicon carbide monofilaments with diameters from 100 to 200 μm are typical for solid-state fabrication. Excellent properties have been achieved with composites produced with long-length fibers and solid-state bonding, but the necessary process times of several hours at high temperatures and high pressures are expensive and limit the size produced.

The excellent property potential of metal-matrix composites combined with service temperatures that extend well above those of polymer-matrix composites provides the incentive for continued technological development. However, an alternative approach has been the subject of intensive research effort more recently. The need for fiber-matrix composites that permit fabrication which is more rapid, with lower costs, has been emphasized.

Liquid-phase infiltration processing was one of the first approaches used to reduce the cost of metal-matrix composites. Such processing required fiber coatings and wetting aids to promote matrix penetration. This approach has produced composites with 40–70 vol% using smaller (6–20 μm) carbon and aluminum oxide yarns or tow filaments, and aluminum or magnesium alloys. However, processing parameters for reproducible properties have been elusive, particularly for larger-sized components. Even then, processing costs have not been competitive with polymer composites except for specialized requirements.

The following alternative technology approach is being developed for service temperatures up to about 250 °C. Composites with less than optimum properties are adequate for many applications: the very high strength and modulus values obtained with composites containing continuous fibers in oriented fiber arrays often are not needed. Low-cost discontinuous-length reinforcements and low-cost fabrication processing offer the potential for cost-effective composites. Efforts center on aluminum-alloy-matrix composites using powder metallurgy or casting to produce composite billets. The billets have random arrays of short-length chopped filaments, whiskers or particulate reinforcement contents, typically up to 20 vol%. Extrusion or rolling of composite billets reduces the cross section to more useful dimensions. Low-cost composites with useful potential properties have been produced and are now being applied.

Polymer-matrix composites—primarily epoxy matrix for applications below 150 °C—are gaining expanding commercial acceptance. Higher service temperatures require different matrices such as polyimides. The lower-cost metal-matrix composites are competitors for more recent applications and for specialized needs for aerospace components. The similarity of aluminum alloy and discontinuous-length reinforced aluminum components can ease the introduction of composites into manufacturing and assembly operations with existing equipment and personnel. Acceptance of low-cost metal composites is eased for applications from 150 to 300 °C because metal-matrix composite costs can be comparable or lower than those of competing materials. Metal-matrix composites are also expected to play a leading role in applications at temperatures above 300 °C.

2. *Toughness*

The plastic deformation of metals can provide excellent toughness. Further, large-diameter fibers with very strong bonding can use more of the fiber-strength properties than those with weak fiber bonding. While matrix toughness is beneficial, the option of fiber-disbond energy absorption is open for metal composites. More recent research efforts have used less ductile matrix compositions. Intermetallic matrix composites offer another range of design freedom with a combination of the properties of both metallic and ceramic matrices.

A common problem with all high-temperature composites is how to fabricate them. Penetration of the matrix into the fiber array is one difficulty; achieving the proper bond between fiber and matrix without degrading the properties of the composite is also fraught with difficulty. The potential property advantages possible have fostered continued effort to achieve success. Fabrication processing continues to be an area of intense technological development for metal- and ceramic-matrix composites.

3. *Solid-State Fabrication Methods*

Diffusion bonding, the most frequently used fabrication technique for metal-matrix composites with continuous-length fibers of diameter 100 μm, is shown schematically in Fig. 1. A monolayer array of fiber is prepared by winding on a drum with controlled spacing. The filament is held in place by a codeposited polymer binder such as polystyrene. The fiber mat, cut and removed from the drum, is vacuum hot-pressed between layers of matrix alloy to form a composite monotape. An alternative method of forming monotape is to plasma-spray matrix alloy onto the drum-wound fiber array instead of using a polymer binder. Where used, the polymer binder is removed during an early part of the diffusion-bonding cycle with low-pressure binder pyrolysis. Monolayer composite laminate is cut into the appropriate angle-ply-laminate shapes, stacked in proper sequence in a die and diffusion bonded to form a panel or component shape.

Monotape fabrication is bypassed in a single-pressing diffusion-bonding process called foil–filament array, wherein unconsolidated fiber-array plies and matrix-alloy foil or plasma-sprayed matrix–fiber laminates are cut, stacked in a die and the process completed in one hot-press operation. Diffusion bonding is an effective method of producing high-specific-property composites. However, the process is energy intensive and limited to small sizes of composite.

Powder-metallurgy methods have also been used to fabricate metal-matrix composites. Continuous-filament tows have been infiltrated by dry matrix-powder, followed by hot isostatic pressing. Alternatively, matrix-alloy powder has been dry-blended with short whiskers or chopped filament and hot-pressed to produce a random fiber orientation which can be partially oriented by hot mechanical deformation, usually extrusion or hot forming. This method has been successful in producing low fiber contents, typically 15–40 vol%. Another powder-metallurgy method consists of slip casting a slurry of matrix-alloy powder into a uniaxial bundle of fiber in a metal tube, sintering to partially densify and consolidating by hot isostatic pressing. Powder methods have been used effectively for the fabrication of large diameter (200–500 μm) laboratory test specimens of wire-reinforced composites with fiber contents up to 65 vol%.

Electrodeposition of matrix alloy onto fibers, followed by hot pressing, has produced laboratory test specimens of carbon-filament composites. Individual strands of filament in the tow or yarn must be separ-

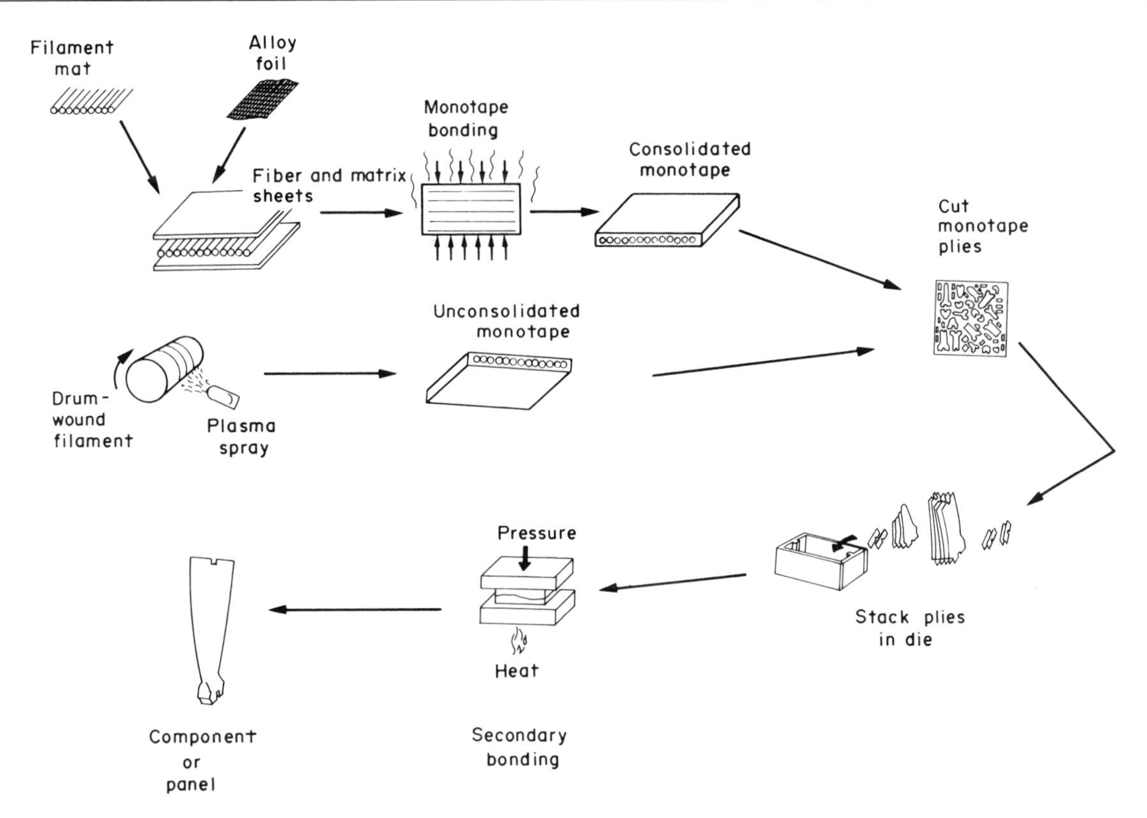

Figure 1
Diffusion bonding of metal-matrix composites

ated to coat each one uniformly. The matrix-alloy electrodeposition on a 6 μm diameter filament would be slightly over 1 μm thick to achieve a 50 vol% composite. Properties of hot-pressed bundles of matrix-coated filaments have been reasonably close to rule-of-mixtures predictions despite some filament breakage during hot pressing with the necessarily thin matrix coatings.

Coreduction by mechanical deformation of ductile–metal combinations is a method used for a limited and specialized group of metal composites: copper–niobium and titanium–beryllium are two examples. Copper–niobium composites are mechanically reduced in size by extrusion and drawing, plated with tin and reacted to form Nb–Sn for superconductivity. The Be–Ti combination is coreduced as an expedient means of producing beryllium filament within the titanium.

4. Liquid-Phase Fabrication Methods

Casting is one of the very-low-cost methods used to produce composites. Aluminum and magnesium alloys can be fabricated using this method because of limited reaction at the fiber–matrix interface. Discontinuous fibers, short whiskers or particulate reinforcement are being fabricated into billets using casting. Fabrication techniques have been evolved to distribute the reinforcement throughout the matrix. One of the methods blends the reinforcement into the matrix by stirring. The process is aided by carrying out the blending when the matrix has cooled to a viscous slushy state.

Liquid-phase fabrication has also been used for aluminum and magnesium alloys with continuous-length filaments. Silicon carbide or polycrystalline aluminum oxide filaments arranged into arrays or as mat have been infiltrated in gravity- or vacuum-assisted conditions. Another method combines liquid infiltration with solid-state diffusion bonding. Liquid-matrix infiltration into carbon-filament tow is aided by a vapor-deposited coating on the filament to increase wetting by the coating. The coating is deposited onto each filament in the tow as it passes through a reaction chamber. The coated-filament tow is immediately drawn through a molten-matrix-alloy bath to form a composite wire about 0.3 cm in diameter. Composite wires are formed into rod or sheet by solid-state diffusion bonding with filament-free matrix cladding.

Hot molding is the term used to describe low-pressure diffusion bonding of monolayer composites in a temperature range where the matrix alloy is partially molten, between the solidus and liquidus temperature of the alloy. This fabrication is possible where matrix–fiber reaction at the interface is not a problem. Such a condition can be achieved by coating otherwise vulnerable fibers such as boron with a diffusion barrier. Several barrier coatings, including boron nitride, silicon carbide and boron carbide, have demonstrated some success. Hot molding offers the potential to fabricate large components with low pressures while retaining the high specific properties associated with high-pressure diffusion bonding.

5. Vapor-Phase Fabrication Methods

A small effort has been undertaken to develop vapor-phase fabrication. The primary motivation is to produce higher fiber content composites. Where filament diameters are small, such as with carbon fibers, composites produced using liquid-phase fabrication methods are typically limited to 30 vol% or less. A matrix thickness of only 3 μm on each carbon fiber (6 μm diameter) would limit composite fiber content to less than 30 vol%.

Another incentive to develop vapor-phase fabrication processing is the need to develop strong matrix–fiber bonding without property-degrading reactions. Secondary processing to bond the matrix-coated filaments may be achieved at a lower temperature, thereby minimizing the potential reaction degradation.

A number of vapor-phase techniques have been tried with some success. These include activated sputtering, ion plating and chemical vapor deposition. The potential advantages are significant, but much work remains to be done before this type of processing can be considered ready for widespread use.

6. Technology Development

Representative composite systems and their properties are shown in Table 1. Aluminum-matrix composites have been studied more extensively than any other metal-matrix composites. Aluminum-matrix composites can be divided into two general types: the first is reinforced with monofilaments such as boron or silicon carbide and is fabricated by solid-state diffusion bonding; the second type contains multifilament yarn or tow, or discontinuous whisker or particulate reinforcement, and is fabricated by liquid-phase or powder-metallurgy methods. The monofilament composites have excellent specific strength and stiffness properties and have been investigated for high-performance aerospace components. The technology for monofilament composites is relatively static. The limitation is not property deficiency but high cost.

The current emphasis in metal-matrix composites is predominantly on low-cost using small-diameter multifilament or discontinuous-reinforced composites. The relative property advantage of this type of composite is illustrated in Fig. 2. The room-temperature strength is moderately higher than that of unreinforced aluminum. The strength, however, is retained to higher temperatures. The tensile strength at 250 °C is markedly superior to that of aluminum. Further, the room-temperature stiffness of the composite is about 50% higher than that of aluminum alloys. The modulus of elasticity approaches that of titanium. When the lower density of the aluminum composite is considered, the modulus–density ratio is increased to almost twice that of titanium. The silicon-carbide whisker length-to-diameter ratio is 20:1 or less, and similar properties can be obtained with particulate reinforcement with even lower aspect ratios. The

Table 1
Representative metal-matrix composites

Matrix	Fiber	Fiber diameter (μm)	Fiber content (vol%)	Fabrication method[a]	Use temperature (°C)	Density ($g\,cm^{-3}$)	Ultimate tensile strength (MPa)	Tensile modulus (GPa)
Aluminum	boron	200	50	DB	350	2.62	1482	221
	B_4 coated boron	140	50	HM	350	2.62	1517	221
	SiC (CVD)	140	50	DB, HM	350	2.96	1724	214
	SiC whiskers	2	20	PM	350	2.80	483	110
	Al_2O_3 tow	20	60	liquid	350	3.45	586	262
	carbon	6	40	liquid + DB	350	2.37	455	131
Magnesium	boron	140	50	liquid	300	218	1310	193
	carbon	6	40	liquid + DB	300	1.82	565	110
	Al_2O_3	20	50	liquid	300	2.80	517	200
Titanium	SiC-coated boron	145	35	DB	650	3.76	758	207
	SiC (CVD)	140	35	DB	650	4.00	862	193
Iron-, nickel base alloys	SiC	140	50	DB	800	5.41	1655	311
	tungsten wire	200	35	DB	1150	11.7	1793	304

a DB, diffusion bonding; HM, hot molding; PM, powder metallurgy

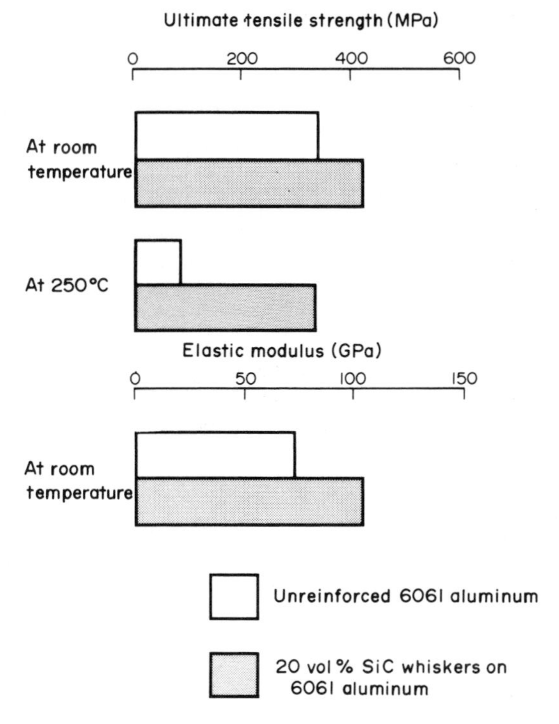

Figure 2
Variation with temperature of tensile strength and elastic modulus of low-cost discontinuous-reinforced aluminum composite

properties are relatively isotropic. Processing technology is under extensive study because the strain-to-failure values vary with processing and the lower values are a limitation to application. Consistently higher strain-to-failure values are the goal of research programs.

Magnesium composites, overlooked for some years, have recently attracted significant development activity. The relative ease of fabrication by liquid-matrix infiltration with the potential for low cost has overcome the fear of corrosion-induced property loss. Magnesium composites with small-diameter filaments, such as aluminum oxide and carbon filaments, are promising for stiffness improvement, in such areas as helicopter transmissions where vibration and noise is reduced.

Lead-matrix composites are candidates for some specialized needs. Pure lead electrodes for acid batteries have a longer electrochemical service life compared with lead-alloy electrodes because of their superior corrosion resistance. However, lead alloys are normally used to satisfy tensile- and creep-strength requirements. Exploratory developments with nonmetallic fibers such as carbon or Al_2O_3 attempt to combine the electrochemical benefits of pure lead with the necessary mechanical properties. Another specialized application of lead-matrix composite is for sound absorption panels in the walls of some commercial buildings. Increased creep strength is sought to resist sag over decades in moderately heated wall cavities.

Titanium is a basic structural material for aircraft turbine engines and the potential increase in stiffness and strength has fostered developmental studies. The reaction between reinforcement fibers and the matrix during processing has been a severe problem. Diffusion bonding at about 900 °C causes matrix–fiber reaction, limiting the tensile strength obtained, and the brittle reaction products can degrade fatigue strength. The latter is of concern with a notch-sensitive material such as titanium. Barrier coatings to control reactions have shown promise and work is continuing on them. A further incentive to use titanium composites is the potential for superplastic deformation combined with diffusion bonding. Demonstration titanium structural panels with selective reinforcement have been fabricated. Excellent modulus and strength values are possible, both parallel and perpendicular to the fiber, if the process reaction can be controlled. Successful resolution of the processing difficulties would make titanium composites the reasonable choice for applications between 350 °C and 600 °C.

Copper composites, used for early model-system studies that were a major factor in composite technology growth, also have application potential. The mutual insolubility of copper–tungsten, combined with good wetting, permits liquid-phase infiltration. The composite displays high thermal and electrical conductivity with creep strength at temperatures above 300 °C. The conductivity and creep strength may be useful for magnetohydrodynamics, while the thermal conductivity of tungsten (three times that of iron or nickel alloys) combined with creep strength may be useful in propulsion systems. Superconductivity of Nb-Sn in copper is another specialized potential application.

Iron, nickel or cobalt alloys reinforced with refractory filaments or wire offer the potential to combine the ductility and oxidation resistance of superalloys with the strength and creep resistance of refractory materials. For example, tungsten-alloy-wire-reinforced iron or nickel alloys have demonstrated as much as an eightfold increase in creep-rupture life at 1100 °C compared with the best nonreinforced superalloys. A further benefit of tungsten wire for composites is its thermal conductivity which is three times higher than that of nickel or iron alloys. Higher thermal conductivity can increase cooling effectiveness for high-temperature components such as turbine blades. Superalloys reinforced by silicon carbide or aluminum oxide monofilaments offer the potential for lower density and higher creep resistance than that possible with refractory-alloy wire. However, they are more vulnerable to surface-flaw-induced fracture. Furthermore, the large thermal-expansion mismatch between these brittle fibers and superalloys, combined with even minor surface defects, has caused fracture in a few temperature cycles. Tungsten-wire composites offer the opportunity to develop a high-temperature composite for service up to 1150 °C, which can serve as a prototype for other composites with greater potential properties but greater experimental difficulties.

Niobium alloys reinforced with tungsten wire are of interest for service temperatures up to 1250 °C. Niobium alloys are more compatible in thermal expansion and interfacial reaction with tungsten wire than are iron and nickel alloys. Oxidation resistance of niobium alloys is inadequate; however, surface protection could be afforded by coatings.

7. *Applications*

A number of applications have been achieved for metal-matrix composites and a much larger number of application demonstrations have been successful. The total number of applications, however, is very small compared to polymer composites and conventional materials. The areas of application include automotive parts, sports equipment and aerospace components. The largest areas of application are aerospace and military hardware. Boron–aluminum structural members have been used in the US Space Shuttle orbiter vehicles. Graphite–aluminum composites have demonstrated a combination of high stiffness and near-zero thermal expansion coefficient values. These are necessary to maintain geometry in space for space-telescope and antennas applications. High thermal conductivity is necessary to resist the distortion of components caused by exposure to sunlight and shade in space. Low-cost, discontinuous-reinforcement aluminum-matrix composites have demonstrated a number of military applications, including mobile bridges, torpedo frame members and missile structural members.

Very large portions of civil and military aircraft structures are candidates for application of low-cost discontinuous-reinforced aluminum. Demonstrations of merit have been completed and potential advantages are large. Newly developed aluminum alloys, including Al–Li and dispersion-strengthened aluminum, are being applied to aircraft structures because of their improved properties. Discontinuous silicon-carbide-reinforced aluminum composites have similar yield and tensile strength and significantly higher modulus and modulus–density ratio values compared with Al–Li alloys. The superior properties of the composite offer the potential to save as much as twice the weight projected for the use of Al–Li alloys for aircraft structures.

Another application is for internal combustion engine pistons for automotive use. An aluminum composite containing less than 10 vol% industrial-grade aluminum oxide fiber mat was substituted for an iron-base alloy insert portion of aluminum pistons. The iron-alloy ring insert, containing the upper piston-ring groove, was added to the cast aluminum piston. With the composite approach, selective reinforcement of the aluminum piston in the annulus area containing the upper piston-ring groove not only improved performance and wear but also eliminated the additional insert production steps. The piston's performance was improved while reducing its cost. This example best illustrates the status of applications of metal-matrix composites.

The acceptance of composite materials over conventional materials requires that superior properties be combined with equivalent or lower cost where conventional materials perform adequately. Of course, a significant portion of the technology of composites is aimed at satisfying severe and extremely high-performance requirements. For some applications in aerospace power and propulsion, composites can make it possible to achieve performance impossible with conventional materials. The use of composites for specialized needs expands the technology base and should broaden their acceptance. The increased emphasis on reduced-cost composites should further this expansion.

See also: Fibrous Composites: Thermomechanical Properties; In Situ Composites: Fabrication; Manufacturing Methods for Composites: An Overview; Particulate Composites; Whiskers

Bibliography

Clyne T W, Withers P J 1993 *Introduction to Metal Matrix Composites*. Cambridge University Press, Cambridge

Harrigan W, Strife J, Dhingra A (eds.) 1985 *Proc. 5th Int. Conf. on Composites*. The Metallurgical Society, Warrendale, Pennsylvania

Hayashi T, Kawata K, Umekawa S (eds.) 1982 *Progress in Science and Engineering of Composites*. Japanese Society for Composite Materials, Tokyo

Kelly A, Tyson W R 1965 Fiber strengthened materials. In: Zackary V F (ed.) 1965 *High Strength Materials*. Welby, New York, pp. 578–602

Kreider K G (ed.) 1974 *Composite Materials*, Vol. 4, *Metallic Matrix Composites*. Academic Press, New York

Lilholt H, Talreja N (eds.) 1982 Fatigue and creep of composite materials. *3rd Risø Int. Symp. on Metallurgy and Materials Science*. Risø National Laboratory, Roakelds, Denmark

Matthews F L, Buskell N C R, Hodgkinson J M, Morton J (eds.) 1987 *Ceramic Matrix Composites: Metal Matrix Composites*, Vol. 2, ICCM VI and ECCM 2. Elsevier, London

Mileiko S T 1983 Fabrication of metal matrix composites. In: Kelly A, Mileiko S T (eds.) 1983 *Fabrication of Composites*, North Holland–Elsevier, London

Noton B, Signorelli R, Street K, Phillips L (eds.) 1978 *Proc. 1978 Int. Conf. on Composite Materials*, ICCM II. The Metallurgical Society, Warrendale, Pennsylvania

Schoulens J F 1982 *Introduction to Metal Matrix Composite Materials*. MMCIAC, Santa Barbara, California

Weeton J W, Peters D, Thomas K 1987 *Engineers Guide to Composite Materials*. American Society for Metals, Metals Park, Ohio

R. A. Signorelli
[NASA Lewis Research Center, Cleveland, Ohio, USA]

Molded Fiber Products

Molded fiber is the product of a manufacturing technique in which fibers are dispersed in water and then accreted by filtration on a porous felting die which is in the shape of the desired article. The method has been called "pulp molding," and the product "molded fiber" and the misnamed "molded paper." The US Bureau of the Census uses the term "pressed and molded pulp goods." The products of the industry which are most visible and familiar are flower pots, food trays, meat packaging trays and egg containers. More sophisticated products such as radio speaker diaphragms, automobile components, luggage shells, tropical helmets and combustible cartridge cases are often not recognized as being molded fiber. Williams (1970) has described the process and its many unique advantages.

Tooling for molded fiber involves both expense and expertise, but once the equipment is at hand, the method provides economical mass production. Fibrous raw materials are readily available: strong synthetic fibers are employed when strength is required, refined wood fiber serves for the majority of goods, and for the less costly items, recycled fibers are entirely adequate. Desirable properties are designed into the products. Food trays are porous, will absorb a spill and after use are readily disposable in landfills. Rocket launcher fairings for aircraft are light, rigid and made to break into small pieces as the rockets are launched. Combustible ammunition components, made with a substantial portion of nitrocellulose fibers, are consumed completely on firing, contribute to the propellant charge and leave no unburned residue in the gun tube. As filters, molded fibers form an open structure of graded density which retains the contaminant load while protecting the dense filter bottom from blinding. Shapes which are stable at high temperature, of use in casting metals, can be made by combining ceramic fibers with inorganic binders. Metal fibers can be felted into useful shapes and the fibers bonded by sintering.

The *Annual Survey of Manufacturers* issued by the US Bureau of the Census gives the following values for pressed and molded pulp goods (SIC Code 26460), including bituminous impregnated pipe, in millions of dollars:

1964	94	1980	200
1965	117	1981	240
1966	122	1982	228
1977	226	1983	217
1978	209	1984	236
1979	208	1985	260

1. Manufacture

1.1 Preparation of the Fiber Slurry

Various fiber mixes such as kraft, ground wood, cotton linters, sulfite, rayon, glass fiber, polyester and nitrocellulose are utilized. Thick mats are ordinarily formed, so the stock must be kept open and free-draining. The fibers may be cut to a desired length, but heavy beating and "hydration" are almost never required. Additions of size, wet strength resin and resin emulsions are made at the beater. Fibers such as Hercules Pulpex, which cofelt and function as binders, are available. The finished stock is pumped from the beater to the storage chest and diluted to the desired consistency to supply the felting tank.

1.2 The Felting Vacuum and the Felting Die

Vacuum is ordinarily provided by a liquid piston pump which exhausts a tank preferably placed below the felting tank. The felting die is immersed in the slurry in the felting tank and the fibers drawn on by vacuum-induced flow to the desired weight. With the felting die removed from the slurry, the water collected in the vacuum tank is sent to white water storage, or, in some cases, returned to the felting tank. The felting die is made in the shape of the product and must be strong enough to stand up to the considerable force exerted on it when under vacuum. It may be a casting, or, for some cylindrical shapes, a metal or plastic pipe. The necessary water channels are provided by drilling $\frac{1}{8}$ in. (3.2 mm) holes on $\frac{1}{2}$ in. (13 mm) centers. These are interconnected at the outside surface by shallow grooves. The die is then smoothly covered with 40 mesh screen. A male felting die tends to make a thin felt at a sharp radius, while a female felting die thickens the felt at the radius. The latter effect is of value in strengthening corners and radii in luggage shells. The preform can be made thin where desired, as in the hinge of a radio speaker diaphragm, by restricting the drainage at that point in the felting die.

A problem encountered in preforming items of long configuration is unequal distribution of fibers from top to bottom. To correct this, the following procedure is recommended. The inside bottom of the preforming mandrel is closed with a metal disk carrying a metal tube through the center. The inside diameter of the metal tube should carry no more than half the water flow induced by the vacuum system. A second disk is placed near the top of the pipe which divides the mandrel into two chambers, 60% of the volume below and 40% of the volume above. The bottom plate is drilled with a series of holes with an aggregate diameter somewhat less than the diameter of the pipe. These holes are increased or decreased until the felted wall thickness becomes uniform.

1.3 The Felting Tank

The fiber concentration of the slurry, the freeness of

the stock, the vacuum applied to the felting tool and the time of immersion govern the weight of preform obtained. To maintain consistency at the proper level, stock and water are added in each cycle to supply the slurry which has been removed in the felting operation.

Another method of maintaining felting-tank consistency, one which keeps fibers well distributed, is to dilute the stock in the supply tank to the desired concentration, perhaps 0.2%, and pump this continuously at the base of the felting tank, allowing it to overflow symmetrically at the top circumference of the tank and from there return to the supply tank.

The operator periodically weighs the preforms and adjusts the felting time. To assist in holding the assigned weight, the vacuum tank may be equipped with a "sight" gauge in which the water level is marked with a float. The float, at the cutoff volume, optically activates equipment which brings the felting platform out of the slurry.

The felting die is brought out of the slurry with the vacuum still applied to remove water in the preform. A thin rubber blanket may be used at this point, to assist in dewatering. The vacuum is then cut off and air pressure is applied to free the felt. The preform is removed by hand or taken by a vacuum transfer.

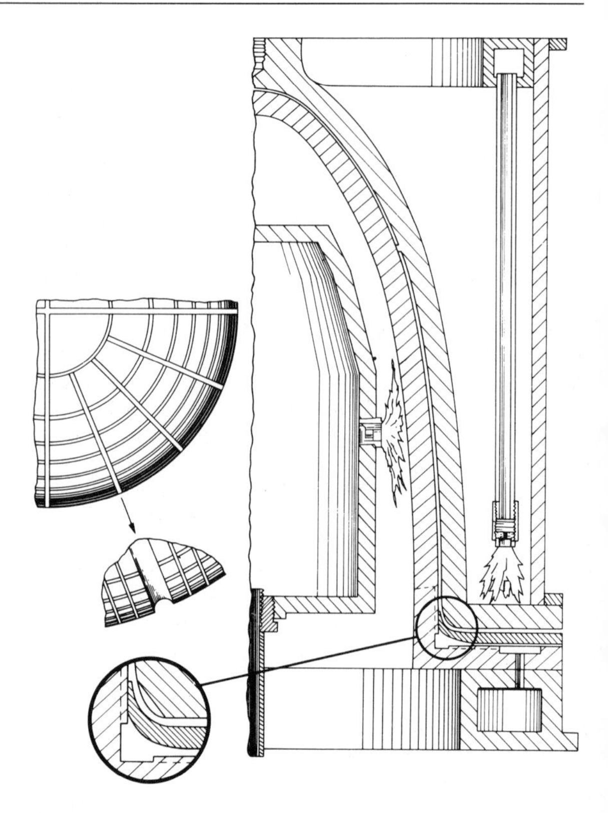

Figure 1
Matched metal drying dies to form break pattern in aircraft rocket launcher fairing (after De Luca 1966)

1.4 Drying the Preform

The transfer may place the preform on a belt which carries it into an oven for drying. The belt may carry drying forms to receive the felt and to maintain its form during drying. Very open stocks, as used in filtration products, and thin sections, as felted in speaker diaphragms, are dried by pulling heated air through the part. The wet felt may also be dried and pressed in matched metal dies. In this system it is usually the male die which is grooved, and possibly screened, with a vacuum connection to carry away water and steam. There is an unfortunate but natural tendency for the preform to elongate and even tear as the dies are closed. The intrinsic "green strength" or initial wet strength of the stock is important at this point. The use of the rubber blanket to reduce water content strengthens the preform. Shaped inflatable rubber bags and press-mounted drained dies are used to give firmer, drier felts which do not tear. The female die is given a flared "ironing edge" to reduce the drag on the part as the dies close.

The preform does not flow in the die dry; this is a pressing rather than a molding operation. As a consequence, the weight of the felt must be accurately governed. Too light a felt puts the press pressure on the top of the part, while too heavy a felt puts wedge pressure on the sides. Elongation of the preform as the dies close makes forming the bottom edge difficult, though not impossible. Post-trimming by machine is often required. Drying dies are heated using premixed gas, as shown in Fig. 1, or, where more exact temperature control is required, as in the drying of nitrocellulose, by steam or heated oil.

1.5 Special Forming and Drying Techniques

The fiber slurry may be injected under pressure into split screened and drained molds. After the part is formed, some drying may be carried out by following the slurry with heated air. The dies are then separated, the part is removed and drying is completed in the oven. Duck decoys have been made by this process, with subsequent hardening and water proofing by a varnish impregnation.

In the Lass process, large split screened felting dies, which when joined have an aperture at the lower end, are dipped in the fiber slurry and the slurry is drawn in to accrete the part. The dies are then withdrawn and heated air is pulled through the aperture to dry the molding, after which the dies are separated to obtain the part. Large-dimensional advertising signs have been made by this method (Lass 1943).

Preforms which carry from 8 to 20% thermoplastic resin from a beater treatment may be oven dried and then pressed to a dense exact shape in a heated die.

2. Specific Products

The radio speaker diaphragm is made in the form of a shallow cone. An outer ring is made thin to act as a hinge, while the center portion is made thicker to carry the voice coil and to reproduce high notes. Felting is carried out in large tanks with the fiber at low consistency to give optimum formation; diaphragms are held at the drying station to close weight and resonance tolerances. The apex of the cone which receives the voice coil must be molded to an exact diameter. A guided ring is slipped over this section as the drying oven is brought down over the screen-supported wet felt. The dried trimmed cones are dipped in lacquer and may be redipped at the apex to meet waterproofing and sound reproduction specifications.

In a process for felting annular or cylindrical depth-type filter cartridges, the dried cartridges were impregnated with phenolic resin, cured and trimmed (Anderson 1951). Such a cartridge is most open at the outside (upstream) end of the cylinder and most dense at the center-aperture base of the filter. As contaminated liquid flows through the cartridge, particles are removed and stored in the wall according to their size. Thus flow is maintained over a long period of use. The cartridge holds its shape under high pressure drop without a central support. In a modified process the cartridge felt was formed in a small tank fitted close to the felting form (Williams 1970). Slurry was brought into the tank during the felting operation. The composition of the slurry could be changed at will, allowing the filter bottom to be placed at any point in the wall; the rate at which the graded density opened was also brought under control.

Fiber molded shells for luggage are light and strong and have been the basis of a very successful line of carrying cases. If necessary, the shell is strengthened by the incorporation of glass fiber. Such fiber is available as a cut strand coated with an agent which causes it to bond to polyester resin during molding. When cut strand is added to a molding slurry, it separates, under agitation, into a multitude of small filaments. These felt to make a bulky preform and a great deal of the reinforcing action is lost. However, if catalyzed polyester resin is added to the slurry along with the cut strand, the resin associates with the glass as a result of the coating applied. Mild agitation is required to bring this about. Cellulose fibers do not take the resin and act as spacers to prevent the glass fibers from sticking together. As the batch is heated, the resin cures and the strands are stabilized. Excellent improvement in impact strength is obtained. If the glass content of the preform is raised to 90% in this method, open preforms are produced which, when dry, can be molded in matched metal trimming dies using the resin pour-on technique.

Dress tropical helmets have been made in large quantities using varnish-impregnated, cloth-covered, molded shells. The edge of the helmet was rolled and adhered, which gave considerable strength to a lightweight article. Combat helmet liners have been made by the same method.

Molded preforms filled with phenolic resin have been dried at low temperatures and then remolded at higher temperatures in polished dies. For a restaurant tray, three preforms have been used; the two outside at 60% resin to give a food and detergent resistant finish and the center preform at 15% resin to give the tray toughness.

Molded fiber frangible nose and tail fairings have been produced in large quantities for airborne rocket launcher systems. The matched metal drying dies press a grenade break pattern into fairings as shown in Fig. 1. The fairings withstand the aerodynamic loading created by high-speed combat aircraft. On firing the first rocket, the fairings disintegrate. The break pattern combined with the phenolic impregnation gives strength, rigidity and small pieces which do not damage the aircraft on break-up (De Luca 1966).

The combustible cartridge case has proved to have advantages. The case is molded of wood and nitrocellulose fibers and contains the propellant: it is itself part of the charge. Use of such a case permits more rapid firing of ammunition rounds. After firing there is no spent case to dispose of, which is a great advantage in the close quarters of a tank turret. Slurry molding and drying of the preform in matched metal dies give the case accurate dimensions. Since nitrocellulose fibers are used, die drying must be done with good control of temperature and time in the die. Dies are shielded to protect operators and safety equipment comes into action in case of a power failure. Preforms containing nitrocellulose fiber have been dried using steam heat at 121 °C without problems (De Luca 1967). The cartridge case may be postimpregnated with resin, or the binder may be introduced at the beater. Incorporation of a portion of fibrillated polyacrylic fiber improves the stability of the case (De Luca and Williams 1984).

See also: Paper and Paperboard

Bibliography

Anderson L E 1951 Method of making a filter element. US Patent Nos. 2,539,767 and 2,539,768 (30 January 1951)

De Luca P L 1966 Process for making fibrous articles. US Patent No. 3,250,839 (10 May 1966)

De Luca P L 1967 Cartridge case and method for the manufacture thereof. US Patent No. 3,320,886 (23 May 1967)

De Luca P L, Williams J C 1984 Fibrillated polyacrylic fiber in combustible cartridge cases. *I & EC Prod. Res. Dev.* 23(3): 438–40

Lass W P 1943 Lass dehydroform. *Pac. Pulp Pap.* July: 13–16

Williams J C 1970 Pulp molding. Three-dimensional paper products. In: Mosher R H (ed.) 1970 *Industrial and Specialty Papers*, Vol. 4. Chemical Publishing, New York, pp. 198–233

Williams J C 1971 Accretion apparatus. US Patent No. 3,585,107 (15 June 1971)

P. L. De Luca
[Armtec Defense Products, Coachella, California, USA]

J. C. Williams
[Technical Consultant, Washington DC, USA]

Multiaxial Stress Failure

There is a need for soundly based criteria for failure under multiaxial stress. Physically satisfying theories are of two types. The first is a maximum stress criterion where it is envisaged that there is a certain upper limit to the force (identified as producing either a shear or a tension) that a solid can withstand. The second is based on the idea of a maximum failure strain whereby it is envisaged that the relative displacement of the component parts of a solid cannot exceed a certain amount if cohesion is to remain (see *Fibrous Composites: Thermomechanical Properties*).

It is appropriate to review the classic case of *yield* criteria (in metals). The *Tresca* yield criterion states that yield (not fracture) will occur when the magnitude of the maximum shearing stress has a value $\frac{1}{2}\sigma_0$ which is a constant of the material. Denoting the principal stresses σ_1, σ_2, σ_3, in decreasing order of value, the maximum shearing stress is $\frac{1}{2}(\sigma_1 - \sigma_3)$ so Tresca's condition is

$$\sigma_1 - \sigma_3 = \sigma_0 \tag{1}$$

This gives equal stresses in tension and compression, which is approximately true in metals for yield though far from true for fracture. It is also assumed that the mean normal stress does not affect the process, which is approximately true in metals but far from true in plastics. In order to use Eqn. (1) one must find the principal stresses and subtract the smallest of these from the largest. This can be done but makes the use of Eqn. (1) unwieldy. The choice of a criterion should be invariant with respect to change of axes, which Tresca's criterion is. The invariance demanded by physical reasoning *suggests* that one should formulate a yield criterion in terms of the invariants of the deviatoric stress tensor $S_{ij} = \sigma_{ij} - \frac{1}{3}\sigma_{ii}$, with principal values S_1, S_2, S_3. The second simplest invariant is $(S_1^2 + S_2^2 + S_3^2)$. The von Mises yield criterion takes this invariant to be equal to a constant $\frac{2}{3}\sigma_0$, which on substituting for S_1, S_2, S_3 and on multiplying out gives

$$(\sigma_2 - \sigma_3)^2 + (\sigma_3 - \sigma_1)^2 + (\sigma_1 - \sigma_2)^2 = 2\sigma_0^2 \tag{2}$$

It is conventional to think of yield under combined stresses in terms of a yield surface expressed as

$$F(\sigma_1, \sigma_2, \sigma_3, \sigma_0) = 0 \tag{3}$$

This conventional yield surface for metals is illustrated in Fig. 1, where the modes of failure are indicated for the case $\sigma_3 = 0$. The straight lines such as AB and DE indicate the intersection of the yield surface with the plane $\sigma_3 = 0$ according to the Tresca criterion. This criterion does not give a smoothly varying representation as the criterion for yield and so von Mises is preferred, which gives, with $\sigma_3 = 0$,

$$\sigma_1^2 - \sigma_1\sigma_2 + \sigma_2^2 = \sigma_0^2 \tag{4}$$

and in section the ellipse in Fig. 1.

In three dimensions all points on the line $\sigma_1 = \sigma_2 = \sigma_3$ have zero stress deviation. Since the yield criterion involves only the principal stress deviations, the yield surface must be a cylinder in stress space with its axis along the line $\sigma_1 = \sigma_2 = \sigma_3$ equally inclined to the principal directions. The cross-section of the cylinder where it intersects the plane $\sigma_1 + \sigma_2 + \sigma_3 = 0$, which is perpendicular to its axis, is known as the *yield locus* and is a circle for von Mises criterion and a regular hexagon for Tresca. The cylindrical form of the yield surface is caused by the assumption that the mean *normal stress has no effect on the process of yield*. In the more general case when this is abandoned, as it must be for polymers, the surface given by Eqn. (2) is not necessarily a cylinder.

Hill (1948) applied these ideas to an anisotropic substance, again for the simple case of yield, and again assuming this to be unaffected by the mean normal stress. He considered an orthotropic material, that is, one showing three mutually perpendicular planes of symmetry, the intersection of these planes being the principal axes of anisotropy. Choosing these axes as reference axes, Hill proposed the simplest yield criterion that would reduce

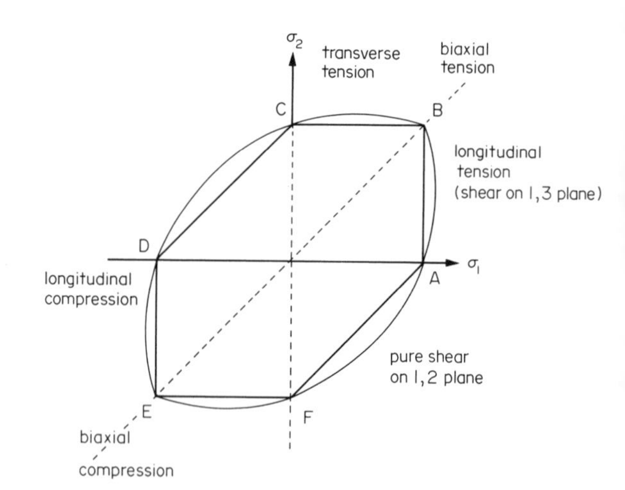

Figure 1
Section of yield surface in 1, 2 plane with $\sigma_3 = 0$

to von Mises criterion when the anisotropy is vanishingly small. The assumption of a homogeneous quadratic in the stresses is taken over which implies that the yield in tension and in compression in the same direction occurs at numerically equal stresses. In view of the assumed symmetry, terms in which any one shear stress occurs linearly must be rejected and a third assumption is that a pure hydrostatic pressure does not affect yielding. The yield criterion then becomes

$$F(\sigma_2-\sigma_3)^2+G(\sigma_3-\sigma_1)^2+H(\sigma_1-\sigma_2)^2 + 2L\tau_{23}^2+2M\tau_{31}^2+2N\tau_{12}^2=1 \quad (5)$$

with reference axes being the principal axes of anisotropy and F, G, H, L, M, N constants with τ_{ij} representing shear stresses. This expression transforms in the same way as the stress components when the axes are changed. If σ_{01}, σ_{02}, σ_{03} are the tensile yield stresses in the principal anisotropic directions, then by considering the effect of each stress component acting alone it is seen that $(G+H)=1/\sigma_{01}^2$; $(H+F)=1/\sigma_{02}^2$; $(F+G)=1/\sigma_{03}^2$; and if τ_{023}, τ_{031} and τ_{012} are the yield stresses in pure shear with respect to the principal axes of anisotropy we have $2L=1/\tau_{023}^2$; $2M=1/\tau_{031}^2$; $2N=1/\tau_{012}^2$.

It can be seen that to describe anisotropic yielding, *six* independent yield stresses must be measured. If there is rotational symmetry about an axis, say 3, then obviously $F=G=H$ and $L=M=N=3F$. The expression Eqn. (5) is then identical with Eqn. (2), with $2F=1/\sigma_0^2$.

Hill was well aware of the assumptions made in this anisotropic theory of yielding, namely: that yield stress in tension equals that in compression; that hydrostatic pressure does not affect yielding; and that six independent yield stresses need to be measured for an orthotropic material.

None of these theories has been applied to the *failure* of metals. However, a modified Hill-like criterion has been proposed for the failure of composite materials and is widely quoted. It is associated with the names of Tsai and Wu (1971). The derivation of Hill's approach detailed above points out that even in a material with orthotropic symmetry at least six independent failure stresses need to be determined. This has certainly not been fully appreciated or acted upon according to the composite literature.

What experiments there are follow better a maximum strain theory in this author's opinion for simple laminates. A careful review of (strength) failure theories up to the early 1980s is given by Rowlands (1985).

A theory using a formulation mathematically similar to Hill's Eqn. (5) has been widely used in producing computer codes for predicting failure. This has made it difficult to introduce ideas based on the physics of failure and of observed failure modes (Hart-Smith 1991a). Nevertheless, such must be the only sound approach, and however complicated it turns out to be it must be followed through. Useful numerical approximation can then be made with confidence.

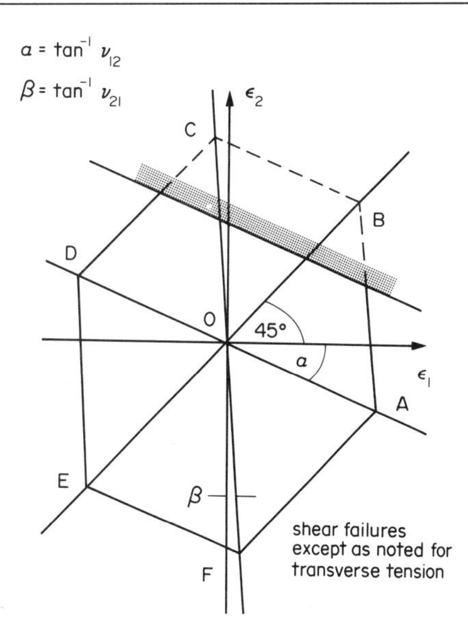

Figure 2
Shear failure envelope in strain coordinates. Transverse tension cutoff illustrated by line dotted on one side

By a series of careful experiments involving a critique of the test methods employed, Hart-Smith in a series of papers succeeded in producing a good approximation to a physically satisfying failure model. Carefully prepared and *carefully tested* unidirectional carbon-fiber composites fail both in tension and in compression by shear (Ewins and Potter 1980). Hart-Smith is able to substantiate in many cases a maximum shear stress condition for failure of (carbon fiber) fibrous composites. He first expresses the (Tresca type) shear failure criterion in the strain plane for laminated structures under in-plane loads. The failure envelope is then generalized to transversely isotropic materials.

If we proceed directly from Fig. 1 and draw the shear failure envelope for a transversely isotropic material with fibers say parallel to axis 1 but having equal tensile and compressive strengths, it will appear in strain space as in Fig. 2. Ignore, for the moment, the distinction between dashed and full lines of the irregular hexagon. It is unsymmetric because two Poisson ratios arise. If fibers are aligned along axis 1, and axis 3 is normal to the plane of the laminates, we have in the usual notation $\varepsilon_1=\sigma_1/E_{11}-\nu_{21}\sigma_2/E_{22}$; $\varepsilon_2=-\nu_{12}\sigma_1/E_{11}+\sigma_2/E_{22}$; and $\gamma_{12}=\tau_{12}/G_{12}$ with $\nu_{12}E_{22}=\nu_{21}E_{11}$, so ν_{12} is the "large" Poisson ratio. The

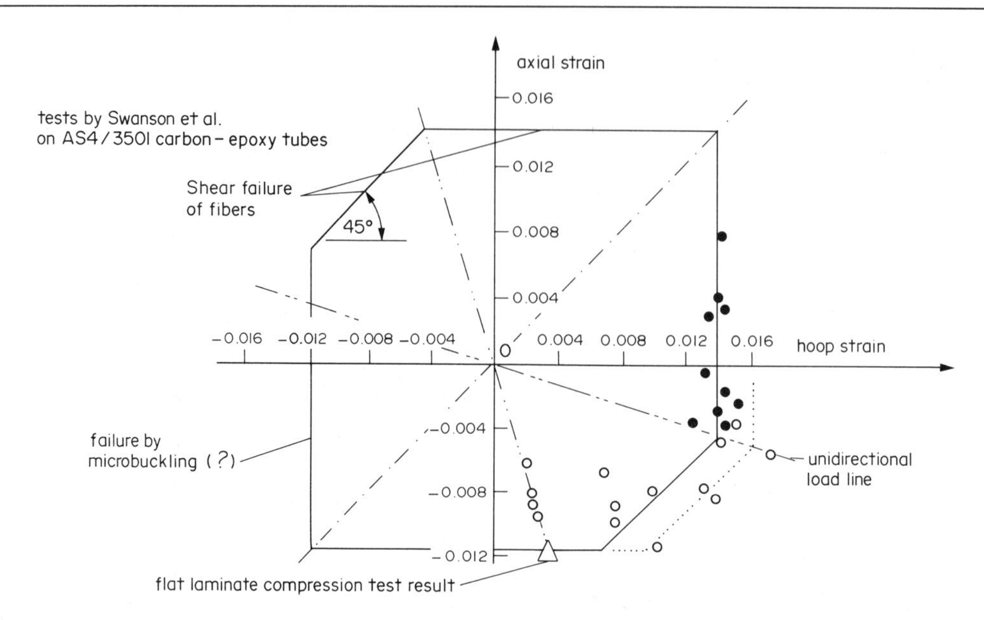

Figure 3
Comparison of test results with theory for the truncated maximum strain failure model for carbon–epoxy composites (reprinted with permission from Hart-Smith 1991a). ●, First set of tests; ○, second set of tests

lettering of the corners of the hexagon is similar in Figs. 1 and 2 so that point B corresponds to failure in biaxial tension and C to tensile failure transverse to the fibers. Clearly along FA failure occurs by in-plane shear (cf. Fig. 1). In Fig. 2 the line BC corresponds to failure in transverse tension and in Fig. 1 such a failure corresponds on the Tresca criterion to shear occurring on a plane parallel to axis 1 and inclined to axes 2 and 3. In a fiber composite consisting of a single lamina, transverse tensile failure may occur within the matrix. Hart-Smith points out that this should be recognized by inserting a cutoff at a particular strain, as indicated in Fig. 2. Other failure modes can be dealt with in an analogous fashion; for example, high-strain carbon-fiber laminae fail by instability under compressive loads at a lower strain than that which they can withstand in tension. This would be taken account of, for instance, by inserting a cutoff on Fig. 2 that is parallel to DE but closer to the origin.

Most composite structures contain patterns of fibers based on 0°, 90° and ±45° directions with perhaps 10% of the fibers in each of these directions, in order to prevent matrix-dominated failures from occurring; the percentage of +45° and −45° are usually the same. Under these conditions the failure envelope of the structure is best represented by removing the asymmetry in the pattern in Fig. 2 by reflecting it in the line EB and, since ν_{21} is usually very small, by making the lines parallel to the axes of strain. A further truncation is introduced in the in-plane shear quadrants to produce the failure envelope of the shape shown, together with some experimental points in Fig. 3.

In summary, Hart-Smith's approach is to work in terms of strain and to produce a failure envelope for a composite structure which takes into account realistically the possibilities of a variety of failure modes (matrix, interface) by introducing cutoffs of appropriate form for each different failure mode superimposed on what is essentially a Tresca-type failure criterion for the fibers in the composite. The results of this method agree quite well with experimental results.

Bibliography

Ewins P D, Potter R T 1980 Some observations on the nature of fibre-reinforced plastics and the implications for structural design. *Philos. Trans. R. Soc. London, Ser. A* 294: 507

Hart-Smith L J 1991a Predicting the strength of fibrous composites by an orthotropic generalization of the maximum shear-stress (Tresca) criterion. *Failure of Polymeric Composite Structures—Mechanisms and Criteria for the Prediction of Performance*. SERC/IMechE Annual Expert Meeting and references therein. Institution of Mechanical Engineers, London

Hart-Smith L J 1991b The role of biaxial stress in discriminating between meaningful and illusory composite failure theories. Presented to 9th DOD/NASA/FAAS Conference on Fibrous Composites in Structural Design, Lake Tahoe, Nevada, November 4–7 1991. McDonnell Douglas Corporation

Hart-Smith L J 1993 Should fibrous composite failure modes be interacted or superimposed? *Composites* 24: 53–5
Hill R 1948 A theory of the yielding and plastic flow of anisotropic metals. *Proc. R. Soc. London, Ser. A* 193: 281
Rowlands R E 1985 Strength (failure) theories and their experimental correlation. In: Sih G C, Skudra A M (eds.) 1985 *Failure Mechanics of Composites*, Handbook of Composites, Vol. 3. Elsevier, Amsterdam, p. 71
Tsai S W, Wu E M 1971 A general theory of strength for anisotropic materials. *J. Comp. Mater.* 5: 58

A. Kelly
[University of Surrey, Guildford, UK]

Multilayers, Metallic

Advanced thin-film deposition techniques enable the production of multilayered materials, consisting of alternating layers of two types, in which the layer thicknesses are both very small and very precisely controlled. These materials with synthetic periodicity (in the range <1 nm to ~ 100 nm) are variously termed "multilayers," "compositionally modulated materials" or "superlattices." Many aspects of these are treated by Barbee et al. 1988. They are a form of two-dimensional composite.

Metallic multilayers are distinguished from laminated composites by their very small layer thicknesses. They are normally in thin-film form, and may consist of a few hundred layers. Great progress has been made in developing new deposition techniques with better control of deposition rate and of deposited structure (Barbee 1985). Metallic multilayers have been made not only by evaporation, but also by sputtering, molecular beam epitaxy (MBE), chemical (and pulsed-laser photochemical) vapor deposition and electrodeposition. The most accurate techniques can produce layers only one atom thick, and can control deposition rates to better than 0.1%. This level of precision in the dimension perpendicular to the substrate greatly exceeds the precision possible in lateral patterning of thin films by lithography. Coarser less-precisely controlled multilayers can be produced by mechanical reduction of macroscopic laminates (Atzmon et al. 1985).

1. Mechanical Properties

One of the most remarkable properties associated with metallic multilayers is the enhanced elastic modulus for (among other modes) biaxial stretching of the thin films (Cammarata 1988). This "supermodulus effect" is found in compositionally modulated multilayers of cubic-close-packed (ccp) metals (e.g., copper–palladium) in (111) orientation. The biaxial modulus is normal, except near a particular modulation wavelength of 2–3 nm for which it shows a maximum, being up to four times greater than the modulus expected for the composite or for the homogeneous solid solution that could be formed by the elements. The effect remains controversial as the evidence is disputed by some and the origins are not agreed upon. Proposed origins are: electronic effects based on an interaction between the Fermi surface and a new Brillouin zone due to the layering; compressive and tensile strains in alternating layers due to coherency; and overall compressive strain in the multilayer due to interfacial stresses (Cammarata and Sieradzki 1989). The peak in modulus has been correlated with peaks in effective interdiffusivity and in thermopower.

Epitaxial multilayers of ccp and bcc metals (e.g., nickel–molybdenum) have shown modulus dehancements, associated with expansion at the interfaces (perhaps due to disordering) perpendicular to the film plane. The effect occurs at the same repeat distance as the change from positive to negative coefficient of resistivity (Falco and Schuller 1985).

Higher yield stress (measured by indentation) is found associated with the enhanced modulus in ccp multilayers. In addition, any multilayer of materials of different stiffness will impede dislocation motion (Koehler 1970). In, for example, aluminum–copper multilayers, tensile strength varies as $d^{-1/2}$ (where d is the layer thickness) up to a limit where d (for each metal) becomes less than 70 nm; that is, less than the critical thickness for dislocation generation (Tsakalakos and Jankowski 1986). Such effects may become technologically significant: for example, 1 mm thick sheet of aluminum–transition-metal multilayer has been produced by high-rate evaporation followed by rolling. Such sheet, with aluminum-layer thickness 20–1600 nm and transition-metal thickness 0.1–21 nm, shows high strength and high temperature stability (Bickerdike et al. 1985).

Bibliography

Atzmon M, Unruh K M, Johnson W L 1985 Formation and characterization of amorphous erbium-based alloys prepared by near-isothermal rolling of elemental composites. *J. Appl. Phys.* 58: 3865–70
Barbee T W 1985 Synthesis of multilayer structures by physical vapor deposition techniques. In: Chang and Giessen 1985, pp. 313–37
Barbee T W, Spaepen F, Greer A L (eds.) 1988 *Multilayers: Synthesis, Properties and Non-Electronic Applications*. Materials Research Society, Pittsburgh, PA
Bickerdike R L, Clark D, Easterbrook J N, Hughes G, Mair W N, Partridge P G, Ranson H C 1985 Micro-

structures and tensile properties of vapour deposited aluminium alloys. Part 1: Layered microstructures. *Int. J. Rapid Solidification* 1: 305–25

Cammarata R C 1988 Elastic properties of artificially layered thin films. In: Barbee et al. 1988, pp. 315–25

Cammarata R C, Sieradzki K 1989 Effects of surface stress on the elastic moduli of thin films and superlattices. *Phys. Rev. Lett.* 62: 2005–8

Chang L L, Giessen B C (eds.) 1985 *Synthetic Modulated Structures*. Academic Press, Orlando, FL

Falco C M, Schuller I K 1985 Electronic and magnetic properties of metallic superlattices. In: Chang and Giessen 1985, pp. 339–64

Koehler J S 1970 Attempt to design a strong solid. *Phys. Rev. B* 2: 547–51

Tsakalakos T, Jankowski A F 1986 Mechanical properties of composition-modulated metallic foils. *Ann. Rev. Mater. Sci.* 16: 293–313

A. L. Greer
[University of Cambridge, Cambridge, UK]

Multiple Fracture

Multiple fracture is of very widespread occurrence, and occurs in situations as diverse as the cracking of a brittle lacquer on a ductile metal when the latter is extended; and the cracking of carbide whiskers in a eutectic with a nickel-base alloy when a specimen is extended in tension. As the specimen is extended the surface of the specimen becomes covered with an array of parallel cracks in the case of the brittle lacquer and, in the case of the carbide whiskers, each whisker is seen to be broken into a set of short lengths. It is sometimes called *transverse cracking*, particularly when it occurs in laminates; and is called *matrix microcracking* when it occurs in fiber-reinforced ceramics. When it occurs in fibrous systems, the fibers do not necessarily have to be aligned.

The phenomenon arises whenever one component in a multicomponent system breaks at a smaller strain than the other *and* there is sufficient of the high-elongation component so that it is able to bear the total load which is then thrown upon it, say in a tensile test.

So, for instance, if 0°/90° balanced laminate is strained in tension and there are sufficient of the 0° plies, as is most often the case, then after quite a small strain, of the same order as the transverse failure strain in a single ply, the specimen is found to contain a set of parallel and closely spaced cracks within the 90° ply. Bailey et al. (1979) give a beautiful set of examples.

The phenomenon was first named multiple fracture by Cooper (Cooper and Sillwood 1972) and was at first extensively investigated in cement matrix composites (Aveston et al. 1971). Aveston et al. described the essentials of the process and also dealt with the effect of the cracking on the stress–strain curve. A review of multiple fracture in laminated composites has been given by Abrate (1991).

The understanding and control of the effect is of great importance for the use of laminated composites since the array of cracks lowers the longitudinal modulus, and alters the Poisson ratio and the thermal expansion coefficient. It has likewise a very important effect on fatigue properties since fatigue loading also produces multiple fracture (see *Fatigue of Composites*).

Cracking can be induced by changes of temperature. Measurements of the Poisson ratio can be used to estimate the density of cracking. In order to do this well an exact theory of the elastic properties of a laminate containing parallel arrays of cracks is necessary and has been given by McCartney (1992).

The effect is also of very great importance in composites with a ceramic matrix, which are designed to be used at elevated temperature since cracking of the matrix may allow the ingress of gases which leads to oxidation of the fibers or other injurious reactions. Matrix cracking of much the same type, though on a vastly different scale of size, is also important in normal reinforced concrete.

Cracking may be reduced by making the 0° plies thinner in a laminate or by decreasing the diameter of the fibers in a unidirectional specimen; decreasing the diameter of the fibers is equivalent to reducing the interfiber spacing for a given volume fraction of fibers.

The effect has been explained either by using an approach based on the energetics of the process of cracking (Aveston et al. 1971; cf. McCartney 1987) or as a result of statistical variation of the strength whereby thinner things are stronger (Manders et al. 1983).

See also: Fatigue of Composites; Fiber-Reinforced Cements; Hybrid Fiber–Resin Composites; Strength of Composites

Bibliography

Abrate S 1991 Matrix cracking in laminated composites: A review. *Compos. Eng.* 1: 337–53

Aveston J A, Cooper G A, Kelly A 1971 Single and multiple fracture. In: *Conf. Proc. National Physical Laboratory: The Properties of Fibre Composites*. IPC Science and Technology, Guildford, UK, pp. 15–26

Bailey J E, Curtis P T, Parvizi A 1979 On transverse cracking and longitudinal splitting behaviour of glass and carbon reinforced epoxy cross ply laminates and the effect of Poisson and thermally generated strain. *Proc. R. Soc. London, Ser. A* 366: 599

Cooper G A, Sillwood J M 1972 Multiple fracture in a steel reinforced epoxy resin composite. *J. Mater. Sci.* 7: 325–33

McCartney L N 1987 Mechanics of matrix cracking in brittle matrix fibre reinforced composites. *Proc. R. Soc. London, Ser. A* 409: 329–50
McCartney L N 1992 Theory of stress transfer in a 0°–90°–0° cross-ply laminate containing a parallel array of transverse cracks. *J. Mech. Phys. Solids* 40: 27–68
Manders P W, Chou T W, Jones F R, Rock J W 1983 Statistical analyses of multiple fracture in 0/90/0 glass fibre/epoxy resin laminates. *J. Mater. Sci.* 18: 2876–89

A. Kelly
[University of Surrey, Guildford, UK]

N

Nanocomposites

In materials science, a "composite" implies that the material has a structure comprising two or more different (usually solid) phases arranged so as to bring benefits they do not possess individually. Thus, the term "nanocomposite" implies that the physical arrangement of the different phases is on a scale of less than 100 nm (Roy et al. 1986). Its widest use is in the development of novel ceramic and cermet materials using nanoscale powders and, less widely, for materials made by mechanical alloying of metallic materials (Shingu 1991).

The description can be applied principally to a range of geometries of mixing phase A with phase B, including (Niihara 1991)

(a) nanoscale particles, platelets or fibers of A in a microscale matrix of B; and

(b) nanoscale grain mixture of A and B (so-called nano–nano composite);

each with various positional arrangements of A relative to the grains of B (Clarke 1992); and finally the special case of

(c) alternating nanoscale thickness layers of A and B.

Nanocomposites are for the most part still technical curiosities, but offer real possibilities for exploitation once cost-effectiveness has been established.

1. Benefits of Nanoscale Composite Microstructures

What advantages can be gained by reducing the scale of composite material microstructures to less than 100 nm? The remarkable properties of wood, bone and shells are examples of structures which have compositelike components at the nanometer scale (Birchall 1984, Jeronimidis 1986, Yasrebi et al. 1990). If the structures of these materials could be fabricated synthetically on an engineering scale, there could be significant benefits (e.g., in stiffness–toughness combinations). Based on surface free energy calculations, Kamigaito (1991) has suggested critical upper size limits below which significant changes in properties can be achieved, and gives examples of <5 nm for catalytic activity, <20 nm for making a hard magnetic material soft, <50 nm for producing refractive index changes or <100 nm for achieving superparamagnetism and providing strengthening or restricting matrix dislocation movement. Birringer and Gleiter (1989) have reviewed the subject in general terms.

2. Types of Nanocomposite

2.1 Nanoscale Mechanically Reinforced Ceramics

To widen the range of successful applications for ceramic materials, improvements are required in strength and toughness, typically through either fibrous microscale reinforcement or the development of finer grain size materials with reduced defect content. The latter has limitations because even if all strength-limiting features in the microstructure are reduced in scale, there is a micrometer-scale limit posed by surface damage. The poor properties of nanoscale grain size glass-ceramics (e.g., lithium aluminosilicates) are an example. However, if toughness can also be improved, restricting propagation of microcracks, there are prospects of higher strength with resistance to surface damage. The principal development is nanoscale particulate reinforcement of micrometer- or submicrometer-scale grain size ceramic matrices.

In a successful example first demonstrated by Niihara and co-workers (Niihara 1991), a fine-grained polycrystalline alumina has a distribution of nanoscale silicon carbide particles. Having a lower expansion coefficient than Al_2O_3, SiC shrinks less on cooling from the densification temperature, and is therefore in hydrostatic compression. The matrix is in tangential tension around each particle, but in radial compression, encouraging microcracks to deflect towards and become pinned by the particles. To avoid spontaneous internal fracture under the applied stress fields, the stored elastic energy associated with each particle must be insufficient to supply the fracture surface energy. High strength and toughness in the Al_2O_3–SiC system is achieved when the particles are <50 nm.

A key factor is obtaining and retaining the dispersion of particles, which naturally tend to demix and agglomerate during processing. Further, if they are swept to matrix grain boundaries by grain growth during consolidation, they leave grain centers unreinforced, and conversely, if the grain boundaries are not reinforced, the commonplace intergranular fracture mode will not be impeded. Careful design of processing is required. A variety of fabrication routes has been used to attack this problem, including intensive milling, pyrolysis of organometallic compounds, sol–gel and chemical vapor deposition (CVD) methods (e.g., Haarland et al. 1987, Roy 1987, Niihara 1991).

Other types of nanocomposite ceramic reported in the literature include Si_3N_4–SiC, AlN–Si_3N_4–SiC, β-sialon–β-SiC, Al_2O_3–Si_3N_4, mullite–ZrO_2 and mullite–TiO_2, mullite–SiC, MgO–SiC, SiC–diamond, Al_2O_3–diamond, Al_2O_3–TiC, Al_2O_3–SiC_w–TiC and YBCO–CuO superconductors (in each case the nanophase is listed last). CVD methods have been used to prepare thin-layer nanocomposites, including C–SiC for thermal barrier purposes, and Si_3N_4–z-BN, TiN–z-BN and BN–Si_3N_4. Sol–gel methods have been used to prepare nanoscale ceramic films on porous ceramic substrates for ultra or hyperfiltration membranes (Burggraaf et al. 1989).

2.2 Nanoscale Functional Electrical Composite Ceramics

Electrical functions have attracted less activity than improving strength and toughness, but some examples include soft ferrite transformers, $BaTiO_3$ multilayer capacitors, lead zirconium titanate transducers, and doped $BaTiO_3$ PTC (positive temperature coefficient) thermistors (Newnham and Trolier-McKinstry 1990). Modified electrical characteristics arise because of the dominance of grain boundary effects; for example, ferroelectric glass-ceramics in which individual grains, each smaller than the domain size, are developed in a glassy matrix, giving up to an order of magnitude lower dielectric loss since domain wall movement does not occur. Piezoelectric Al_2O_3/PZT–PMN–MnO_2–NiO nanocomposites have also been reported showing parallel increases in strength, mechanical Q-factor and electromechanical coupling coefficient. The term nanocomposite has also been applied to special low-permittivity dielectrics produced by sol–gel methods in which void space is used to reduce permittivity to well below that for the normal fully dense solid. These materials have been proposed for use in interchip packaging in integrated circuits. Thin-film nanocomposite YBCO superconductors have been prepared by sol–gel and OMCVD (organometallic chemical vapor deposition) processes.

2.3 Modified Nanocomposite Layer Minerals

Many silicate minerals occur in sheetlike form, and some exist with especially weak interlayer bonding, allowing the sheet structure to be split into individual thin layers by other species. The process is known as intercalation (Giannelis et al. 1990, Yamanaka 1991). The clay montmorillonite contains hydrated exchangeable cations in the interlayer space. Neutral, polar "guest" molecules can exchange for these cations, expanding the interlayer space. Various guest precursors can be used to form oxides on subsequent heating, stabilizing the intercalation and essentially "pillaring" the layer structure apart to leave a controlled pore structure. A potential application is as functional molecular sieves. Tubular silicate-layered silicates have also been developed, and have potential as supports for metal or metal oxide catalysts. Similar processes have been used to make polymer–ceramic composites (see also Sect. 2.7).

2.4 Ceramic–Metal Nanocomposites

A variety of processes has been used to make ceramic or glass matrices containing nano-dispersed metallic particles. Intensive milling has been used to make Al_2O_3–Me (where Me = V, Cr, Mn, Co, Ni, Cu, Zn, Nb, W, Si, Fe or Fe alloys). Pyrolysis methods have been used to make Me–BN nanocomposites. The sol–gel method has been used to make Al_2O_3–iron which becomes superparamagnetic below a temperature of 10 K, and SiO_2 glass–iron, –copper or –nickel for modification of electrical conduction or optical properties. Film-sputtering methods have been used to make Al_2O_3–Mo. In contrast, glass containing nanometer-sized copper particles deposited by ion bombardment is not normally termed a nanocomposite.

2.5 Metal–Metal and Cermet Nanocomposites

The phase development in cobalt-hardened gold electrodeposited alloys has been described as a nanocomposite by Kahn (1992) and examined for magnetic properties. Schlump et al. (1989) report on the development of a wide range of nanocrystalline multicomponent materials including Ti–Ni, Ti–Co, Ti–Cr, W–Fe, W–Co, Cu–Ta, Fe–Ta–W, high-speed steel–NbC, TiC–Co, WC–Ni–Co, all prepared by intensive milling of powder mixtures, followed by pressing and sintering or hot isostatic pressing. They demonstrate significant increases in hardness of WC–Ni–Co materials made by this route compared with normal powder routes. Intermetallic Al_3Nb–NbC and Ag–MgO have also been produced. Thin-film nanocomposites have been prepared by various vapor depositions methods, using codeposition of two or more different phases on a nanoscale; for example, Mo–Al_2O_3 and Al–Mo produced by sputtering, Ag–Fe_3O_4 by laser evaporation. Chow et al. (1989) demonstrate the codeposition of silica and metal alloys to give a nanoscale composite coating of silica fibers in a metal matrix—one of the few examples of nanoscale fibers being produced.

2.6 Multilayer Composites

By using pulsing techniques, nano-thickness alternating layers of dissimilar materials are readily produced by sputtering, CVD or evaporation techniques, and such materials have value as mechanical and microelectronic structures. Examples include molybdenum–aluminum with greater hardness than aluminum alone. However, these are time-consuming fabrication processes for other than thin layers, and as yet there are few commercial applications.

2.7 Polymer–Ceramic Composites

Polycerams or ormocers (*or*ganically *mo*dified *cer*amics) can be formed, combining the advantages of both ceramics and polymers (e.g., both stiffness and toughness). The inclusion of polymers means that the processing has to occur at low temperatures, and thus sol–gel techniques become essential. Examples include poly(*p*-phenylene benzobisthiazole) intimately mixed with sol–gel derived glass (Kovar and Lusignea 1988), and alumina-, silica- and zirconia-containing polymers (Schmidt 1990). Possible uses include structured insulating layers with low permittivity and high resistivity for microelectronics (Popall et al. 1990), and scratch-resistant polytetrafluoroethylene (PTFE) coatings (Doyle et al. 1988). Poly(vinyl alcohol) (PVA)–hydrated calcium aluminate, and poly(vinyl pyrrolidine)–clay nanocomposites have also been reported using intercalation techniques and self-assembling characteristics (Giannelis 1992).

3. Thermodynamic Considerations

The development of nanoscale microstructures is more difficult than coarser scale microstructures. The fine grains have high surface energy, and the natural tendency in microstructure development during processing is the elimination of surface or interface energy by grain growth. Nanocomposite materials therefore tend to be less stable structurally than their microscale equivalents, require more care in processing, and are more easily destroyed by heating. In some cases, a nanoscale precipitate can be advantageously exsolved within a material by heat treatment (e.g., the tetragonal phase in some partially stabilized zirconias), but the majority of useful nanocomposite systems require the assembly of chemically very different species (e.g., Al_2O_3 and SiC). Vapor-phase or liquid-phase assembly techniques are therefore advantageous for intimate mixing on the nanoscale, but clearly have shape limitations if the material is required in bulk. For the latter, intensive milling of mixtures of fine powders followed by conventional processing techniques, aided perhaps by hot pressing or hot isostatic pressing seems to be the most popular route at present, especially for the more refractory types of ceramic or cermet. It is probably true that effective stable nanocomposite materials in general require that the components are essentially immiscible and do not mutually dissolve during processing or use.

Bibliography

Birchall J D 1984 Shells, cements and ceramics. *Br. Ceram. Trans. J.* 83: 158–65

Birringer R, Gleiter H 1989 Nanocrystalline materials. In: Cahn R W (ed.) 1989 *Encyclopedia of Materials Science and Engineering*, Suppl. Vol. 1. Pergamon, Oxford, pp. 339–49

Burggraaf A J, Theunissen C S A M, Winnubst A J A 1989 Synthesis and properties of nanophase ceramics and composites. In: de With G, Terpstra R A, Metselaar R (eds.) 1989 *Proc. Euroceramics Conf.* Elsevier, London pp. 1.8–1.12

Chow G-M, Klemens P G, Strutt P R 1989 Nanometer-size fibre composite synthesis by laser-induced reactions. *J. Appl. Phys.* 66: 3304–8

Clarke D R 1992 Interpenetrating phase composites. *J. Am. Ceram. Soc.* 75: 739–59

Doyle W F, Fabes B D, Root J C, Simmons K D, Chiang Y M, Uhlmann D R 1988 PTFE-silicate composites via sol–gel processes. In: Mackenzie J D, Ulrich D R (eds.) 1988 *Ultrastructure Processing of Advanced Ceramics*. Wiley, New York, pp. 953–62

Giannelis E P 1992 A new strategy for synthesizing polymer–ceramic nanocomposites. *J. Opt. Mater.* 44: 28–30

Giannelis E P, Mehrotra V, Russel M W 1990 Intercalation chemistry: A novel approach to materials design. *Mater. Res. Soc. Symp.* 180: 685–96

Haarland R S, Lee B L, Park S-Y 1987 SiC/Al_2O_3 gel-derived monolithic nanocomposites. *Ceram. Eng. Sci. Proc.* 8: 872–8

Jeronimidis G 1986 Natural composite materials. In: Bever M B (ed.) 1986 *Encyclopedia of Materials Science and Engineering*, Vol. 4. Pergamon, Oxford, pp. 3128–31

Kahn D 1992 Magnetic properties and structure of cobalt-hardened gold. *Proc. SUR/FIN '92 Conf.*, Atlanta, GA, Vol. 1. American Electroplaters and Surface Finishers Society, Orlando, FL, pp. 305–41

Kamigaito O 1991 What can be improved by nanometre composites? *J. Jpn. Soc. Powder Powder Metall.* 38: 315–21

Kovar R F, Lusignea R W 1988 Ordered polymer/sol–gel glass microcomposites. In: Mackenzie J D, Ulrich D R (eds.) 1988 *Ultrastructure Processing of Advanced Ceramics*. Wiley, New York, pp. 715–24

Newnham R E, Trolier-McKinstry S E 1990 Structure–property relationships in ferroic nanocomposites. *Ceram. Trans.* 8: 235–52

Niihara K 1991 New design concept of structural ceramics: Ceramic nanocomposites. *Nippon Seramikkusu Kyokai Gakujutsu Ronbunshi* 99: 974–82

Popall M, Meyer H, Schmidt H K 1990 Inorganic–organic composites (ORMOCERs) as structured layers for microelectronics. *Mater. Res. Soc. Symp.* 180: 995–1001

Roy R 1987 Ceramics by the solution sol-gel route. *Science* 238(4834): 1664–9

Roy R, Roy R A, Roy D M 1986 Alternative perspectives on "quasicrystallinity"—Nonuniformity and nanocomposites. *Mater. Lett.* 4: 323–8

Schlump W, Deuerler F, Wagner N, Grewe H 1989 Manufacture of nanocrystalline structures in multi-component systems. In: Bildstein H, Ortner H M (eds.) 1989 *Proc. 12th Int. Plansee Seminar '89*, Vol. 2. Metallwerk Plansee, Reutte, Austria, pp. 117–49

Schmidt H K 1990 Aspects of chemistry and chemical processing of organically modified ceramics. *Mater. Res. Soc. Symp.* 180: 961–73

Shingu P H 1991 Mechanical alloying opens new era in alloy development R/D of powder metallurgy of light metals. *Proc. Conf. Science and Engineering of Light Metals, RASELM '91*. Japanese Institute for Light Metals, Tokyo, pp. 677–84

Yamanaka S 1991 Design and synthesis of functional layered nanocomposites. *Bull. Am. Ceram. Soc.* 70(6): 1056–8

Yasrebi M, Kim G. H, Gunnison K E, Milius D L, Sarikaya M, Aksay I A 1990 Biomimetic processing of ceramics and ceramic–metal composites. *Mater. Res. Soc. Symp.* 180: 625–35

R. Morrell
[National Physical Laboratory, Teddington, UK]

Natural Composites

Interest in composite materials in general and fiber-reinforced systems in particular has been growing steadily since the Second World War. Their potential in terms of specific strength and stiffness makes them competitive substitutes for metals and wood in many traditional applications.

The ability to exploit a combination of materials to obtain mechanical properties that neither individual component has or to compensate for shortcomings is not new. Nature has been designing for hundreds of millions of years biological structures based on composite materials. Although many animals and plants are often taken for granted, the way in which the properties of materials and structural design interact is not trivial and there are many principles and specific solutions which are worthy of attention.

1. Biological Materials

Biomaterials are generally very complex and we still know very little about them. In order to discuss their properties in the appropriate context, we have to assume that the biological structures which survive are those where the combination of materials and structural requirements has reached a successful level of optimization. Other factors are obviously important but the ability to carry loads safely is fundamental.

Biological tissues are based on relatively few chemical substances. Plants can stand up because of cellulose and lignin. Structural materials in animals are mainly made of proteins (e.g., collagen, elastin and keratin) in combination with various polysaccharides, calcium minerals (in bone and teeth) or complex phenolic compounds (in hard insect cuticles). Water is also very important in controlling stiffness, strength and toughness. To obtain the wide range of mechanical properties observed in practice, composite materials become the rule rather than the exception.

Long-chain molecules of proteins and sugars are synthesized inside living cells and then transferred to the outside where, under various control mechanisms, they are assembled into microfibrils which are the next elementary units above the macromolecules. Microfibrils are present for example in cellulose, chitin, keratin and collagen. Their diameters are typically in the range 5–20 nm and their length is unknown.

In most cases, a great number of intermediate levels of organization have been found between the microfibrils and the macroscopic elements to the extent that the concept of fibrous composite is applicable at many levels within the same structure. In wood and bone, for example, the wood cells and the osteons could be considered as the reinforcing "fibers" of these materials; however, they are not simple fibers but rather complicated structures in their own right. Similar arguments apply to collagen, teeth and many other substances.

This hierarchy of structural elements is quite common in natural composite materials and, although the reasons for it are not clear (synthesis and deposition being some of the factors involved), this type of subdivision has important consequences. Firstly the high inherent stiffness of the smallest organized units such as crystallites cannot be fully utilized because they have to be cemented together to produce the next structural element of the hierarchy. This degradation of Young's modulus (the whole cannot be stiffer than its parts) is partly due to a decrease in the volume fraction of the high-modulus component and, for discontinuous fibrous systems, to transfer length effects. In cellulose, for example, the Young's modulus of the crystallites has been estimated at 250 GPa but the experimental value in well aligned parallel systems of microfibrils is only of the order of 80 GPa. Collagen tendon has, when wet, a Young's modulus of 2–3 GPa but even in the dried condition, the stiffness is much lower than one might expect from the structure. This may seem a rather inefficient way of using materials which are metabolically expensive but "stiffness at all costs" is not a structural strategy that is often found in nature.

Secondly, the complex design of biological fibers introduces an element of redundancy which is probably desirable from a safety point of view. It is true that the absolute strength of the composite fibers is lower than that of the microfibrils they are made of, but such fibers will be far less sensitive to defects which may lead to fracture; their practical strengths are very high in relation to their stiffness, most of them having breaking strains of 3–5% or more. They rarely fail in a brittle manner because the interaction between the subelements in each structural level is such as to allow nonelastic deformations before fracture. Biological reinforcing fibers can combine resilience (energy storage) and toughness (energy absorption).

Some man-made fibers such as carbon can also be obtained in a high-modulus variety, with low strength, and a low-modulus type with higher breaking strain (see *Carbon Fibers*). The concept of interaction between fibers of different strengths and stiffnesses within the same composite is beginning to be applied to artificial materials (see *Hybrid Fiber–Resin Composites*) but hybrid systems in nature are almost universal.

2. *Fiber Arrangements in Biological Composites*

Depending on the specific requirements for stiffness, strength and toughness dictated by the function of living structures, almost any conceivable arrangement of fibers can be found. The mechanical properties determined by the various geometrical patterns can be altered further by the choice of fibers (e.g., straight or crimped), the type of matrix (rigid or compliant), the aggregation with a mineral phase and the overall water content.

Tissues which are carrying tensile stresses in one direction only have parallel systems of fibers and the degree of their coupling depends to a great extent on how brittle the structure can afford to be. The stronger the bonding between elements at any particular level, the more brittle the material will be at that same level. Since the most likely mode of failure of a material in tension is by crack propagation, it is generally safer to limit the bonding between fibers, except in small structures where the critical crack lengths will, in any case, be of the order of the dimensions of the structure (e.g., teeth or hard insect shells).

However, unidirectional tension members are comparatively few. A great many biological materials are designed to carry loads in more than one direction. Random distributions of collagen fibers are found in skin (although partially orientated regions may be present when unidirectional stresses dominate) and in the walls of unstretched blood vessels. These are examples of extensible composite materials where straining induces reorientation of fibers changing the mechanical properties of the tissues. Most biological membranes have nonlinear stress–strain curves characterized by an increase in Young's modulus with stress.

Some worms function as cylindrical tubes under fluid pressure and one finds that the collagen fibers form a cross-helical pattern with an angle of about 55° to the longitudinal axis. This represents the optimum fiber angle for carrying simultaneously the hoop and axial stresses. Furthermore, this allows a certain degree of change of shape at constant volume. Similar considerations apply to other "hydrostatic structures" like sharks which also have the same fiber geometry in their skin.

Plant cell walls are a good example of laminated composite material; the shape of the cells is roughly tubular with various laminae of cellulose microfibrils glued together to form the wall. Each lamina has a characteristic fiber orientation which can be random, cross-helical or single-helical. The same kind of plywood construction is found in the hard cuticle of arthropods where the stacking sequence of the chitin laminae produces a pseudo-isotropic material.

Tissues which require hardness (such as those in mollusc shells or teeth) or which carry compressive loads (such as those in bones or teeth) are reinforced further with minerals in the form of platelets or crystallites of different aspect ratios.

3. *Strength and Fracture*

Strength and fracture of materials have often been considered separately but bitter experience with engineering structures has shown that these two factors are inseparable. What controls the safe use of materials in real life is their strength in the presence of defects and stress concentrations.

Fracture toughness, i.e., the resistance to crack propagation, is particularly important when tensile stresses are present. Most biological materials used in tension are part of flexible structures designed for strength rather than stiffness (e.g., tendons, skin and blood vessels). The degree of damage tolerance that they have is remarkable and, on the whole, these tissues can be considered as tough materials where the resistance to tearing and fracture is very much dependent on the fact that they are composites.

Toughness in fibrous materials can be obtained in two ways, either by having mechanisms which are capable of absorbing energy irreversibly during crack propagation (see *Toughness of Fibrous Composites*) or by eliminating the stress concentration effect of a notch or defect. The buildup of stress near a crack tip depends on the shear stiffness of the substance in which the fibers are embedded. If the shear stiffness is high, there will be a stress concentration and the material will require mechanisms such as fiber pull-out to achieve adequate toughness. On the other hand, many flexible biological tissues seem to rely more on the fact that a system of loosely connected fibers, continuous or of very high aspect ratios, can prevent dangerous stress concentrations even in the presence of sharp notches. The lack of shear stiffness in the matrix prevents load transfer from one broken fiber to the next. However, negligible shear stiffness is acceptable only in materials designed to carry tensile loads; in bending or compression there is a need for rigidity. Tendon, for example, works very much like a flexible rope where one of the main requirements is safety against crack propagation (achieved through virtually independent parallel strands) rather than maximum flexural rigidity. When unwanted stiffening creeps in, as in pathological calcification of artery walls, for example, these materials become more brittle.

Bending structures, on the other hand, need to be rigid so that in this case low shear stiffness is not the answer to toughness. Wood, bone, teeth and shells are much more like artificial composites in this respect and their energy-absorbing mechanisms are similar. Fiber pull-out is very common although in certain cases more complicated than in man-made materials because of the structure of the fibers themselves.

The strength requirements for compression structures are mainly determined by their ability to resist buckling. One can show that, in this case, the materials efficiency criteria are controlled by $E^{1/2}/\rho$ and $E^{1/3}/\rho$ where E is the Young's modulus and ρ the density. Increases in stiffness can be obtained through mineralization as in bone and teeth or by aligning fibers as closely as possible to the load axis. The first method increases the weight of material and is often combined with cellular construction which brings substantial reductions in density (cancellous bone); the second method is dangerous because of the poor compressive properties of fibrous materials. Higher plants could increase their stiffness by aligning the microfibrils in the cell walls as closely as possible but the effect of decreasing density by having lumens full of air is much more effective and also cheaper.

4. Interactions Between Materials and Structures

Natural structures are safe structures which are expected to last for the lifespan of the animal or plant. At the same time, nature has to optimize its designs because stiff and strong materials are expensive. Overdesigning is not popular. The use of composite materials allows flexibility and full advantage of this comes from the fact that biomaterials are grown under stress; this means that the loading conditions of the structure as a whole can be used effectively as blueprints for the most efficient use of fiber reinforcement. This is particularly noticeable in the design of joints and other unavoidable stress raisers.

As mentioned before, by their very nature, fibrous composites are better materials in tension than in compression and their use in many applications is often limited by this fact. Biological composites are not different in this respect but nature uses various solutions, such as cellular materials and mineralization or prestressing, to solve this problem. The excess of tensile strength available can be profitably used to prestress in tension the regions of the structure (or material) which are more vulnerable to compressive loads. Also, the use of fluids as compression members (as in turgid plants, worms and sharks) will result in lighter structures.

All this greatly depends on the interaction between structural function and design of materials. It is an integrated approach which has proved successful and which modern technology may well use for inspiration.

See also: Natural-Fiber-Based Composites

Bibliography

Ashby M F 1983 The mechanical properties of cellular solids. *Metall. Trans. A.* 14A: 1755–69

Aspen R M 1986 Relation between structure and mechanical behaviour of fibre-reinforced composite materials at large strains. *Proc. R. Soc. London, Ser. A.* 406: 287–98

Baer E, Gathercole L J, Keller A 1975 Structural hierarchies in tendon collagen: An interim summary. In: Atkins E D T, Keller A (eds.) 1975 *Structure of Fibrous Biopolymers.* Butterworth, London, pp. 189–95

Glimcher M J, Krane S M 1968 The organization and structure of bone and the mechanism of calcification. In: Ramachandran G N, Gould B S (eds.) 1968 *Treatise on Collagen*, Vol. 2B. Academic Press, London, pp. 68–251

Gordon J E 1978 *Structures.* Penguin, Harmondsworth

Helmcke J G 1967 Ultrastructure of enamel. In: Miles A E W (ed.) 1967 *Structure and Chemical Organization of Teeth*, Vol. 2. Academic Press, New York, pp. 135–63

Hukins D W L (ed.) 1984 *Connective Tissue Matrix.* Macmillan, London

Kelly A, Macmillan N H 1986 *Strong Solids*, 3rd edn. Clarendon Press, Oxford

Mark R E 1967 *Cell Wall Mechanics of Wood Tracheids.* Yale University Press, New Haven, Connecticut

Vincent J F V, Currey J D (eds.) 1980 *The Mechanical Properties of Biological Materials.* Cambridge University Press, London

Wainwright S A, Biggs W D, Currey J D, Gosline J M 1976 *Mechanical Design in Organisms.* Edward Arnold, London

G. Jeronimidis
[University of Reading, Reading, UK]

Natural-Fiber-Based Composites

Natural fibers such as jute, sisal, sunhemp, banana and coir are grown in many parts of the world. Some of them have aspect ratios (ratio of length to diameter) greater than 1000 and can be easily woven. These fibers are extensively used for cordage and twine, sacks, fishnets, matting and rope, and as filling for mattresses and cushions (rubberized coir is an example). Recent reports indicate that plant-based natural fibers may be used as reinforcements in polymer composites, replacing to some extent more expensive and nonrenewable synthetic fibers such as glass.

1. Structure and Properties of Natural Fibers

Plant-based natural fibers are lignocellulosic, consisting of cellulose microfibrils in an amorphous matrix of lignin and hemicellulose. They consist of several hollow fibrils which run all along their length. Each fibril exhibits a complex layered structure, with a thin primary wall encircling a thicker secondary layer, and is similar to the structure of a single wood-pulp fiber (see Coté 1986).

The secondary layer is made up of three distinct layers, the middle one being by far the thickest and the most important in determining mechanical properties. In this layer parallel cellulose microfibrils are wound helically around the fibrils; the angle between the fiber axis and the microfibrils is termed the microfibril angle. Natural fibers are themselves cellulose-fiber-reinforced materials in which the microfibril angle and cellulose content determine the mechanical behavior of the fiber. Stress–strain curves of these fibers have an initial portion which is more-or-less linear, followed by a curvilinear portion to the point of maximum stress. Fibers with high microfibril angle such as coir ($\sim 45°$) generally exhibit high elongation to break ($\sim 35\%$), whereas fibers such as sunhemp, with high cellulose content ($\sim 80\%$) and low microfibril angle ($\sim 10°$), possess high tensile strength (~ 400 MPa) and low elongation ($\sim 1\%$).

2. *Interfacial Bonding with Polymers*

The overall performance of any fiber-reinforced polymer composite depends to a large extent upon the fiber–matrix interface, which in turn is governed by the surface topography of the fiber and by the chemical compatibility of fiber surface and resin matrix. Natural-fiber surfaces are fairly irregular, which should in principle enhance the fiber–matrix interfacial bond. However, in many cases this advantage is likely to be offset by chemical incompatibility between the fiber surface and the resin. Some natural fibers have an outer waxy layer of fatty acids and their condensation products; coir fiber, for example, has a 3–5 μm thick waxy layer. Fatty acids, being long-chain aliphatic compounds, are not compatible with common resins such as polyester. It is thus difficult to incorporate coir fibers in polyester without introducing significant amounts of porosity in the composite. This problem may be solved by removing the outer waxy layer of the coir fiber by a simple alkali pretreatment similar to the mercerization of cotton (Prasad et al. 1982). Such treatment was not found to be a prerequisite for polyester composites reinforced with sunhemp or jute fibers.

3. *Behavior of Composites under Impact and Tension*

The principal advantage of natural-fiber-reinforced polymer composites stems from their ability to absorb tremendous amounts of energy during impact fracture. When a fiber-reinforced composite fractures, three types of surface are created: (a) fiber cross-sectional surfaces, (b) matrix surfaces, and (c) surfaces between matrix and fiber. The toughness contribution of each of these "new" surfaces will depend on the work done to create a unit area of the surface. Assuming that the fibers break randomly, some energy is absorbed while pulling the broken fibers out of the matrix after fracture. In addition, another energy-absorbing mechanism has been proposed to account for the redistribution of strain energy from fiber to matrix after a fiber breaks. The toughness of a composite is the summation of the contributions of these three different sources.

In polyester-matrix composites reinforced with unidirectionally aligned sunhemp or jute fibers, the work of fracture shows a linear increase with fiber volume fraction V_f (Sanadi et al. 1986a, Roe and Ansell 1985). The work of fracture (Izod) for a 0.24 V_f sunhemp–polyester composite was reported to be 21 kJ m^{-2}; this is 15 times higher than the work of fracture for polyester resin alone. Although the high toughness of sunhemp–polyester composites can be due to the fiber pullout work and the work done in creating new surfaces at the fiber–matrix interface (Sanadi et al. 1986a), one has also to consider the complex fracture mode of natural fibers as compared to the planar failure obtained in polymers reinforced with glass and carbon fibers. Since natural fibers are themselves cellulose fibril-reinforced composite materials, their fracture modes include uncoiling of fibrils, fibril pullout, plastic deformation of fibrils, fibril splitting and diversion of the crack at the fibril–fibril interface. These fracture mechanisms, which are not seen in glass-fiber-reinforced plastics (GFRP), contribute to the high toughness of natural-fiber-reinforced composites (Sanadi et al. 1986b).

Natural fibers possessing a range of tensile properties (strength, modulus and elongation) are available. High-modulus (E) and high-strength (σ) fibers such as jute, sunhemp and sisal can therefore act as reinforcements in polyester (sunhemp: $E \sim 40$ GPa, $\sigma \sim 400$ MPa; polyester: $E = 3.5$ GPa, $\sigma = 48$ MPa). For polyester reinforced with jute or sunhemp fiber, both modulus and tensile strength were reported to increase linearly with V_f according to the rule of mixtures.

4. *Problems and Prospects*

Plant-based materials have always been important in the plastics industry. The early phenolics were modified with wood flour to lower their cost and improve processibility, and plywood is a composite of thin layers of wood joined by adhesive (see *Plywood*). However, there is considerable reluctance to use natural-fiber-reinforced composites for structural applications because it is very difficult to obtain reproducible mechanical properties with natural fibres. The tensile strengths of fiber samples obtained from different sources can sometimes vary by a factor of three.

Natural fibers are extracted from the plant by a

process known as retting (see Ross 1986). As this process is not well standardized, fibers with varying degrees of flaws are generally produced. Mechanical properties are also dependent on age and moisture content. Unless proper standards are developed and adopted in the extraction of natural fibers from plants, their potential in designing tough composites will be limited.

Natural-fiber-reinforced polymers absorb more moisture than do glass-fiber-reinforced plastics. This could be a serious drawback if these composites are to be used in outdoor applications where they are frequently exposed to rain. This difficulty may be minimized if the outer surface of the composite is protected by applying a gel coat or by hybridizing the outer layers with glass fibers. The natural fibers will, however, remain exposed to the moisture that diffuses through polyester resin.

On technoeconomic grounds, considering particularly specific stiffness per unit cost, composites like sunhemp–polyester (Sanadi et al. 1985) and jute–polyester (Roe and Ansell 1985) are far superior to GFRP. In applications where stiffness is important but strength is not a priority, such as a suitcase or a tabletop, the advantages of natural-fiber-reinforced polymers can be overwhelming.

See also: Natural Composites

Bibliography

Coté W A 1986 Wood ultrastructure. In: Bever M B (ed.) 1986 *Encyclopedia of Materials Science and Engineering*. Pergamon, Oxford, Vol. 7, p. 5456

Kelly A, Macmillan, N H 1986 *Strong Solids*, 3rd edn. Clarendon Press, Oxford

McLaughlin E C, Tait R A 1980 Failure mechanism of plant fibers. *J. Mater. Sci.* 15: 89–95

Prasad S V, Pavitran C Rohatgi P K 1982 Alkali treatment of coir fibers for coir-polyester composites. *J. Mater. Sci.* 18: 1443–54

Roe P J, Ansell M P 1985 Jute-reinforced polyester composites. *J. Mater. Sci.* 20: 4015–20

Ross P 1986 Materials of biological origin. In: Bever M B (ed.) 1986 *Encyclopedia of Materials Science and Engineering*. Pergamon, Oxford, Vol. 4, p. 2862

Sanadi A R, Prasad S V, Rohatgi P K 1985 Natural fibers and agro-wastes as fillers and reinforcements in polymer composites. *J. Sci. Ind. Res. (India)* 44: 437–42

Sanadi A R, Prasad S V, Rohatgi P K 1986a Sunhemp fibre reinforced polyester, Part I: Analysis of tensile and impact results. *J. Mater. Sci.* 21: 4299–304

Sanadi A R, Prasad S V, Rohatgi P K 1986b SEM observations on the origins of toughness of natural fiber-polymer composites. *J. Mater. Sci. Lett.* 5: 562–64

Starbird I R, Lawler J V 1986 Natural fibers in the world economy. In: Bever M B (ed.) 1986 *Encyclopedia of Materials Science and Engineering*. Pergamon, Oxford, Vol. 4, pp. 3131–47

S. V. Prasad
[Council of Scientific and Industrial Research, Regional Research Laboratory, Bhopal, India]

Nondestructive Evaluation of Composites

The complexity of composite components can vary very widely, from a simple glass-reinforced plastic (GRP) storage tank to a sophisticated aerospace component fabricated from a variety of materials and exhibiting a very complicated layup. Equally there can be an enormous variation in the thickness and overall size of the components, from small carbon-fiber composite (CFC) panels on satellite structures, which might be only 0.25 mm thick, to the 30 m long, 25 mm thick, GRP hull of a naval minehunter. The requirements for nondestructive evaluation (NDE) will therefore also vary widely and it is only possible to give here a short guide to the principles involved and an introduction to the techniques available. For simplicity the discussion will largely be limited to materials having organic matrices. Metal-matrix composites are of growing importance and a significant amount of NDE has been employed on them, but the NDE requirements are as yet far from clear (Reynolds 1986).

1. Nondestructive Evaluation Policy

The principal objective of nondestructive evaluation is to provide assurance on the quality and structural integrity of a particular component. This may be achieved either directly by using nondestructive inspection techniques to detect specific defects, or indirectly by monitoring or controlling the fabrication process. The latter approach is, however, really only an extension of the usual process control procedures which are necessary in order to achieve a consistent product. Additional NDE might be provided to assist in optimizing the fabrication procedure by, for example, the addition of a device to monitor the local state of cure. The majority of NDE is, however, concerned with the detection and characterization of defects or anomalies.

In order to illustrate the types of defects that might be sought, consider, for example, a carbon-fiber composite aerospace component fabricated in an autoclave. The following defects are likely to be of importance: delaminations (separations between plies), disbonds, porosity in the laminates or bond line, fiber misorientation, missing plies, variations in fiber volume fraction (resin-rich or resin-starved areas), foreign inclusions and honeycomb core misfit (gaps and splices).

Clearly, in order to decide which defects must be detected, and to set defect-acceptance standards, it is necessary to understand the effect that a specific defect has on the performance of a given component. Inspection procedures can be both expensive and time-consuming and many composite components have proved to be surprisingly tolerant to defects, so care should be taken to ensure that the requirements for inspection are not unnecessarily rigorous (Stone and Clarke 1987).

As a general guide it may be said that there are two

main approaches to the use of NDE to assess quality. The first approach aims to provide a qualitative assessment and is typified by the way that x-radiography is used to assess whether the level of porosity is acceptable. Radiographs of production components are compared with those of development components. Physical tests of these components against the design criteria enable examples of acceptable and unacceptable quality to be identified. The second approach is less subjective and employs quantitative acceptance standards, such as a rejection of any component containing an area of delamination greater than, say, 500 mm^2.

2. Visual Inspection

The value of simple visual inspection should not be underestimated. It is, for example, possible to obtain useful information about the quality of even quite thick GRP components by examining them visually against a brightly lit background. Even with opaque materials, such as CFC, it is often informative to perform a preliminary visual examination of the surfaces. Poor laminates are sometimes revealed by the presence of porosity or resin-starved areas on the surface, and light polishing of the edges of laminates is often sufficient to reveal edge delaminations. Visual inspection may be enhanced by the use of conventional dye penetrants. Such penetrants may, however, be difficult to remove and should be used with caution if a subsequent repair is intended.

3. Ultrasonic Inspection

This is by far the most widely used inspection method for laminated composite materials, being particularly suited to the detection of defects normal to the interrogating beam. A large range of techniques is available using frequencies of 1–50 MHz to meet requirements ranging from coarse defect (e.g., disbond) detection, through void content determination, to detailed defect characterization. For example, at one extreme there is manual contact scanning using a digital-readout ultrasonic thickness gauge, while at the other there could be a computer-controlled multiaxis immersion scanning system.

A common feature of all ultrasonic techniques is the requirement to couple the transducer to the specimen in such a way that the transfer of energy into and out of the specimen is maximized. The efficiency of this coupling depends on the acoustic properties of the various materials involved. If you take, for example, a water–CFC (or CFC–water) interface then some 73% of the incident energy is transmitted. It is often convenient, therefore, to immerse both the specimen and the transducer in a water bath. When using manual scanning, however, it is usually more practical to use a grease or gel couplant; this is quite satisfactory although the coupling is more variable. Consider now a CFC–air interface. This time only 0.03% of the energy is transmitted and 99.97% is reflected. Because of this, air-filled delaminations, porosity and disbonds give rise to strong echoes. Furthermore the fact that there is a strong echo from the rear face of an air-backed laminate means that it can readily be inspected from one side.

3.1 Immersion Testing

In some cases the use of a gel-coupled contact transducer is unavoidable but immersion systems do have a number of advantages. Not only is the coupling constant, but the fact that the transducer may be located some distance away from the surface of the specimen permits the use of transducers producing focused or collimated beams. The transducer is held in a manipulator so that its orientation and distance from the specimen can be adjusted. The transducer can then be used to scan in a prescribed plane relative to the specimen surface. Information can be obtained about the condition of the specimen and presented in a number of ways, the most usual being the so-called A-, B- and C-scan displays. These presentations are also applicable to nonimmersion systems, but are more difficult to achieve in practice.

3.2 Ultrasonic A-Scan

The simplest presentation, the A-scan, displays the ultrasonic information acquired at a single point in a specimen. It is usually displayed on an oscilloscope, as in Fig. 1, and consists of a series of echoes whose position along the horizontal axis can be calibrated in terms of depth in the sample under test so that the position of a reflector can be measured. The amplitude of each echo will give some indication of the size and nature of the reflector.

It is important to note that, if a defect is present and is of a type that reflects a large fraction of the incident

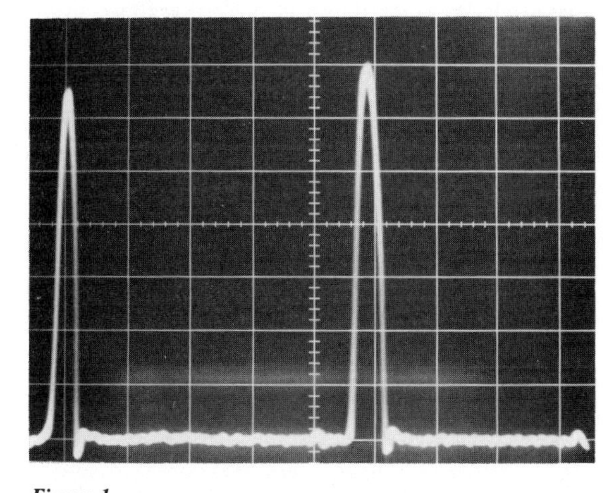

Figure 1
A-scan showing front and back surface reflections from a 10-ply composite

soundwave, then only a very small proportion will propagate beyond the defect. Any additional reflections will be small and will usually not be detected. The result is that the material in the shadow region behind the first defect will not be inspected and additional defects will not be found. This effect is especially noticeable with impact damage which causes multiple delaminations.

3.3 Ultrasonic B-Scan

If the same transducer is moved linearly (usually by a motor drive) in a plane parallel to the surface of the specimen then a number of A-scans can be recorded, corresponding to different positions across the specimen. It is, however, more convenient to extract the essential features of the A-scans, namely the through-thickness position and amplitude of the major peaks, and to combine them in a single display which corresponds to a slice taken through the sample, normal to the surface. Using this type of presentation, known as a B-scan, it is possible to estimate both the depth of reflectors and their lateral extent along the axis of transducer movement. This can be particularly helpful in presenting the distribution of impact damage. An example is shown in Fig. 2 of impact damage in a CFC test specimen where the thickness tapers from 4 to 3 mm.

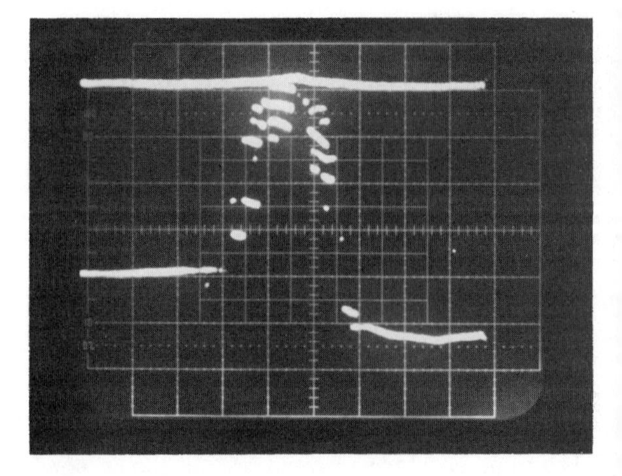

Figure 2
B-scan showing depth of impact damage in a CFC test specimen where thickness tapers from 4 to 3 mm

3.4 Ultrasonic C-Scan

In this type of presentation a more complex scanning system is used such that the transducer is scanned in a plane parallel to the surface of the specimen in a rectilinear raster pattern. A recorder pen is coupled to the transducer manipulator so that the pen movement accurately reproduces the scanning pattern. In this way a plan view of the attenuation distribution is obtained, allowing, for example, an assessment of the lateral extent of damage. It is usual to use an immersion system in which the transducer is scanned in the water tank above the specimen. If a single transducer is employed in a "pulse-echo" configuration, a glass reflector plate placed behind the specimen is often used. The received signal echo from the glass plate is monitored using a time-gate module, which converts the amplitude of the largest reflection within a preset time interval to a voltage. In simple systems this is used to produce a gray-scale presentation on current-sensitive paper showing the variation in attenuation. Figure 3 shows a C-scan of the specimen used as an example of B-scanning in Fig. 2.

Alternatively this data may be stored in a computer, allowing the use of various color presentations. Simple displays require the arbitrary selection of a limited range of colors and are not always easy to interpret, but more sophisticated systems can offer a significant increase in the dynamic range displayed. The big advantage of computer storage, however, lies in its ability to produce, from a single scan, presentations corresponding to different attenuation thresholds. Whether or not computer storage is used it is important to note that the threshold settings are arbitrary and comparatively small changes can result in quite different C-scans being produced. The C-scan will only be meaningful if the relationship between the settings and the parameter plotted is known by calibration.

There are some disadvantages in using immersion testing equipment (Stone and Clarke 1987); alternative techniques are therefore sometimes used.

3.5 Ultrasonic Jet Probes

A commonly used technique, which still uses water as a couplant, is the jet (or squirter) probe technique. Here the component is not immersed; instead, the coupling water impinges onto the surface of the component in a jet along which the ultrasound travels. The water jet acts as a waveguide to the sound. The main difficulty is to make sure, by correct design of the

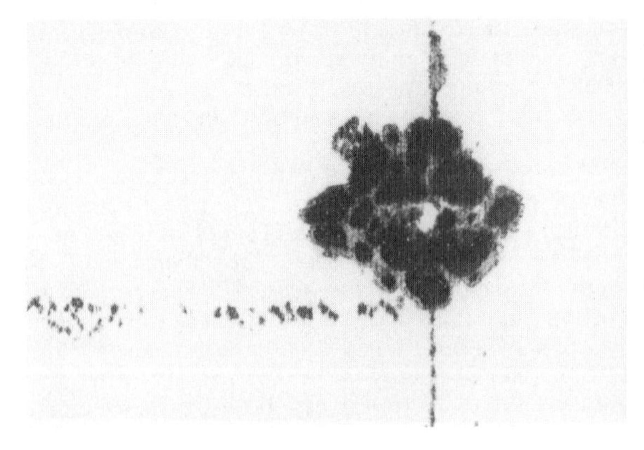

Figure 3
C-scan showing area of impact damage of specimen in Fig. 2

probe assembly, that the water flow is laminar. Because the sound is guided down the water jet, awkward structural geometries can be inspected by employing suitably shaped guide tubes. This technique is generally used in through transmission and at frequencies below about 10 MHz. Although a jet-probe system looks quite different from an immersion system, the basic considerations are the same. In fact it is currently the most widely used technique in the aerospace industry for producing through-transmission C-scans of large components.

3.6 Roller Probes

In applications where the composite must not come into contact with water, roller probes can be used. The ultrasonic transducer element is held inside a wheel and the sound propagated into the specimen through a soft rubber tire. This technique is generally used at low frequencies and can be used for the rapid detection of in-service damage but is not really suitable for its detailed characterization.

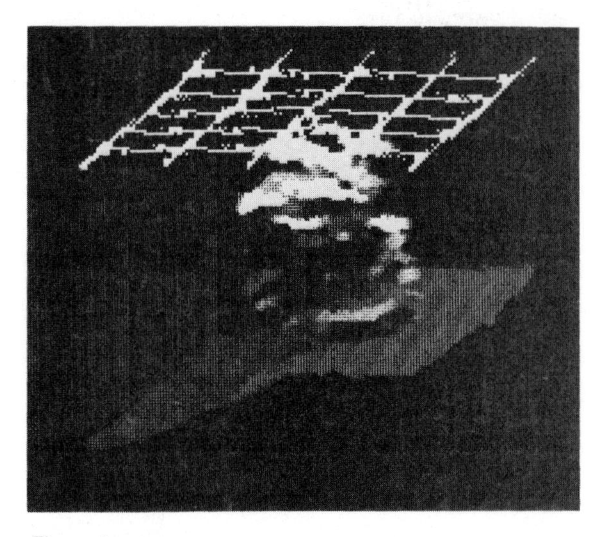

Figure 4
Ultrasonic image of impact damage of specimen in Fig. 2

3.7 Imaging Systems

A considerable amount of work is being carried out in the field of ultrasonic imaging systems, some of which is aimed at imaging damage in composites. Delaminations are of prime importance when considering damage mechanisms in CFC aerospace structures, and an ultrasonic imaging system has recently been developed (Lloyd and Wright 1986) specifically aimed at providing a pictorial presentation of the distribution of delaminations. Essentially such a system uses a combination of the B- and C-scan methods to generate isometrically projected images of defects as pseudo-three-dimensional displays. The software package within this system enables the viewer to select viewing angle and rotational position of the object as well as an option to magnify. Both the area and through-thickness position of the damage can therefore be viewed with a clarity not possible by conventional ultrasonic techniques (see Fig. 4).

4. Radiographic Methods

4.1 X-Radiography

Radiography is based on the differential absorption of penetrating radiation. Because of variations in thickness or differences in absorption characteristics caused by variations in composition, different parts of a component absorb different amounts of radiation. Unabsorbed radiation passing through the component may be recorded on film, viewed on a fluorescent screen or monitored by a radiation detector. Access to the rear surface is therefore necessary, and this may limit the applications for which radiography is suitable. It is generally used to reveal hidden changes in thickness or areas having a different density from the surrounding material. It is therefore widely used to reveal significant areas of porosity, the presence of foreign bodies and the alignment of substructure. Composite materials are, however, relatively poor absorbers of x-radiation so it is necessary to use low-voltage levels (say 20 kV) in order to obtain sufficient contrast. Furthermore, it can only detect those features that have an appreciable dimension parallel to the radiation beam so that its ability to detect planar defects, such as cracks, is strongly dependent on the orientation of the defect. For this reason delaminations are very difficult to detect by conventional radiography.

It is, however, possible to produce a marked increase in contrast by introducing a radio-opaque penetrant into the defective area. This method, often called penetrant enhanced x-radiography (PEXR), is particularly valuable in determining the extent and nature of impact damage. Figure 5 is the enhanced radiograph of the impact damage previously examined ultrasonically.

The damage must, however, break the surface at some point in order to allow access for the penetrant. Translaminar cracks, which are difficult to detect by other methods, are readily revealed by PEXR. Selection of the most suitable penetrant depends on a number of factors (Stone and Clarke 1987), including the degree of opacity required. If the damage is gross then a penetrant exhibiting low attenuation should be used in order to reveal fine detail without saturating the picture. Lesser degrees of damage will require a more opaque penetrant.

In order to take full advantage of the ability of PEXR to give a detailed picture of damage a stereo technique should be employed. This requires that two separate radiographs are taken of a single object, one

Figure 5
Penetrant enhanced x-ray of impact damage

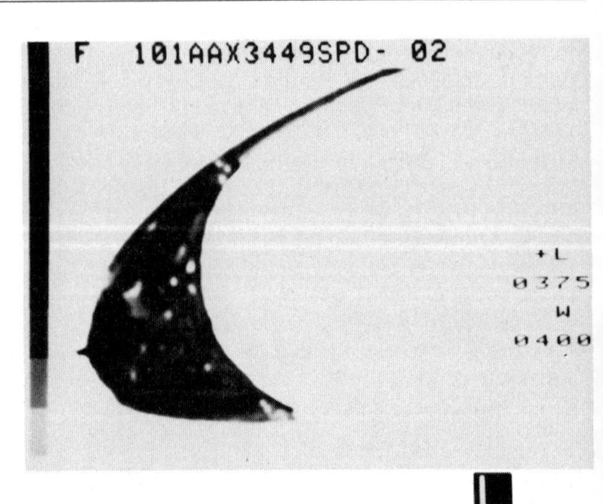

Figure 6
X-ray CAT showing voids in the leading edge of a composite blade

normal to the x-ray beam and another inclined at an angle of about 15°. For small components this is most easily achieved by keeping the film in contact with the component and rotating them together relative to the x-ray beam.

4.2 Computer-Aided Tomography (CAT)

A conventional radiograph is a two-dimensional presentation of a three-dimensional object, the density of the image at each point in the radiograph being governed by the total attenuation experienced by that part of the x-ray beam. There is no information on the contribution that each element of the object makes to the total attenuation. CAT on the other hand measures the intensity of x-ray transmission along many paths in a single plane, and uses a computer to calculate the attenuation contributed by each element in the plane (Gilboy and Foster 1982). Because industrial versions of such systems are expensive, applications to composites have been somewhat limited but they are now being used extensively for the inspection of rocket motors and cases. Even units designed for medical use are proving valuable for components such as helicopter rotor blades, an example of which is shown in Fig. 6.

4.3 Neutron Radiography

Neutron radiography is often regarded as a complementary technique to x-radiography; in particular this is because neutrons are strongly absorbed by the lightest elements (including hydrogen), whereas heavy elements such as lead are relatively transparent to neutrons. Nonreactor sources of neutrons are available but they are at present both expensive and somewhat cumbersome; furthermore, the source size is much greater than that available in x-radiography, thus reducing the resolution attainable. For these and other reasons there have been relatively few applications of neutron radiography to composites. Only for specialized problems, such as determination of the quality of a composite hidden inside a thick metal casing, has its use proved worthwhile.

5. *Thermographic Methods*

These may be conveniently divided into two main types; passive—where there is an externally applied heating or cooling source—and active—where there are internally generated heating or cooling effects. In both cases the surface temperature distribution is monitored and examined for anomalies that indicate the presence of defects. The active method, however, has only really found application in the monitoring of fatigue tests where heat is generated by fretting effects at sites of internal damage.

In the passive method, which has been much more widely employed, the surface of a component is subjected to a rapid temperature change and the subsequent heat flow monitored. A variety of heat sources have been investigated, varying from simple hot-air guns to banks of flash tubes. The way in which the heat is dissipated depends on the thermal conductivity of the composite and on the nature and location of any defects. If there is a delamination in a laminate, for example, then less heat will be transmitted through the defective area resulting in a hot spot on the heated surface, and a cold spot on the back surface. It is possible to monitor either the front or the back surface (provided that this is accessible). By analogy with ultrasonics terminology the former is often called a pulse-echo technique and the latter, through-transmission. It is usual for some kind of infrared-sensitive camera to be used and for the output to be stored on video tape. Not only does this permit subsequent viewing in real time, or at some other rate, but it also allows image processing techniques to be applied if required.

The performance of thermographic systems is, of course, strongly affected by the thermal conductivity of the particular composite. In CFC, for example, the conductivity in the plane of the laminate is about nine times that in the through-thickness direction. This transverse diffusion tends to obscure defects that are not close to the surface. However, the usual through-thickness distribution of impact damage is such that a damaged area that is large enough to be structurally significant will contain limited but measurable damage close to the surface. Thermography is capable of detecting the presence of impact damage in CFC material and disbonds in bonded composite structures (McLaughlin et al. 1987).

6. *Vibration and Mechanical Impedance Methods*

Although vibration techniques formed the basis of some of the most ancient methods of NDT it is only comparatively recently that they have gained a scientific basis (Adams and Cawley 1985). For convenience these techniques may be divided into global and local methods. In the former a whole component is excited by a pulse or by continuous excitation, and natural frequencies and damping characteristics are measured. This approach has been used to provide a rapid method of production inspection for components of simple geometry such as small missile-launcher tubes. With local methods only a small area is excited and its mechanical response measured. Such methods have the advantage of only requiring access to one surface and do not need a couplant. They are particularly suited to the detection of disbonds between thin composite skins and a foam or honeycomb substructure.

7. *Optical Methods*

A number of optical methods have been developed to measure the absolute three-dimensional shape of a component, and hence to detect changes in this shape or to measure in-plane or normal displacements. The presence of anomalies in the shape or displacement patterns may be used to reveal the presence of defects (Walker and McKelvie 1987). Some methods may require Moiré grids or photoelastic coatings to be applied to the surface being inspected. Most methods, however, do not require special surface preparation, although a thin coat of whitener is necessary in order to get sufficient reflection of light from CFC. Methods such as holography and electronic speckle pattern interferometry (ESPI) detect very small displacements and so require a vibration-free environment.

A problem common to all of the optical methods lies in the fact that the component has to be lightly stressed in some way in order to generate the background displacement patterns against which the anomalies are revealed. The way in which these displacements are generated may govern whether or not a particular defect is revealed. Very simple methods are, however, often sufficient. For example, gentle heating with a hot-air gun has proved a very effective way of revealing disbonds between a composite skin and a honeycomb core.

8. *Eddy-Current Inspection*

Eddy-current methods are widely used to inspect metallic materials and they can also be applied to organic-matrix composites containing electrically conducting fibers. However, because the resistivity of even a carbon-fiber composite is very high relative to that of most metals, the techniques employed do differ.

In the eddy-current method, if a current-carrying coil is placed adjacent to a conducting material, eddy currents are induced into the material affecting the impedance of the driving coil. Local changes in resistivity or interruption of the eddy-current field by defects will further alter the impedance. This response is usually characterized in terms of both amplitude and phase angle. A critical factor is the selection of frequency; high frequencies result in better resolution but a decreased depth of penetration. For CFC, however, the resistivity is very high and penetration depth is not usually a problem.

Eddy-current methods have not been used very widely on composites despite the fact that they are comparatively simple to use and the probe does not need to make contact with the surface. The reason for this lies in the ambiguity of the response. Although it provides a simple and effective way of detecting impact damage (Sproat et al. 1986) it is also sensitive to other factors such as volume fraction and laminate thickness.

9. *Acoustic Emission*

In the acoustic emission (AE) technique, a stress of a representative pattern, but of a modest level, is applied to a component, and sensitive transducers are used to monitor the occurrence of stress waves generated by small local failure events. Depending on the degree of sophistication of the instrumentation employed, information can be obtained on the location and severity of these events, and sometimes on their nature. The use of AE to distinguish between various types of failure event is, however, far from easy, even on simple laboratory specimens, and is extremely difficult to achieve on most structural components. There is a wide choice of equipment and signal processing procedures, and care must be taken to select those most suitable for a specific application (Arrington 1987). Some of the most successful applications have been when proof tests have been performed on a significant number of similar components allowing patterns of emission behavior to be established. However the stage at which emission restarts during a repeated loading (the Felicity effect) has also been shown to provide valuable information on individual components.

An interesting alternative is provided by the vibroacoustic emission technique in which the component is excited by a vibration and similar equipment is used to monitor fretting noise emanating from damage.

10. Adhesive-Bonded Joints

Composite components are frequently fabricated by adhesively bonding together pre-cured elements. Poor bonding can easily result in planes of weakness so the inspection of bond lines is often necessary. However, although the detection of bond-line defects is perfectly feasible, nondestructive measurement of bond strength is extremely difficult. Also, a clear distinction must be drawn between the adhesive strength of the interface between the adherend and the adhesive layer, and the cohesive strength of the adhesive itself. Discrete defects, such as disbonds or areas of porosity, are detectable in much the same way as they are in the composite itself. The presence of release film can, however, cause a problem. Small isolated areas of film are usually found without difficulty, but a complete sheet of thin (0.07 mm) polyester film is much harder to find. This is, in fact, very similar to the situation encountered with a "stuck bond," where the two surfaces are in intimate contact but the bond has no strength. Contamination of the adherend surfaces prior to bonding can be highly detrimental and difficult to detect (Stone 1986).

11. Conclusion

The multiplicity of possible defect types in composites and the current uncertainty as to their effect on the physical properties and structural performance means that the NDE of composites is a challenging field. Although the present generation of composites is proving to be surprisingly defect-tolerant, and currently available NDE methods can do much of what is required, the cost of inspection is unacceptably high. Also some areas, such as repair, have so far received little attention. Equally the introduction of new materials will probably bring with it new problems for NDE. However, quantitative NDE is a rapidly advancing field and developments in areas such as automation and signal processing are likely to lead to marked increases in capability.

Bibliography

Adams R D, Cawley P 1985 Vibration techniques in nondestructive testing. In: Sharpe R S (ed.) 1985 *Research Techniques in Nondestructive Testing*, Vol. 7. Elsevier, London

Arrington M 1987 Acoustic emission. In: Summerscales J (ed.) 1987 *Non-destructive Testing of Fibre Reinforced Plastics Composites*, Vol. 1. Elsevier, London

Gilboy W B, Foster J 1982 Industrial applications of computerised tomography with x and gamma radiation. In: Sharpe R S (ed.) 1982 *Research Techniques in Nondestructive Testing*, Vol. 6. Elsevier, London

Lloyd P A, Wright M A 1986 An introductory investigation into the performance of an ultrasonic imaging system. Royal Aircraft Establishment Report TR 86063. RAE, Farnborough, UK

McLaughlin P V, Mirchand M G, Ciekurs P V 1987 Infrared thermographic flaw detection in composite laminates. *J. Eng. Mater. Technol.* 109: 146–50

Reynolds W N 1986 Nondestructive testing techniques for metal matrix composites. Harwell Report AERE R 13040. Harwell Laboratory, Didcot, UK

Sproat W H, Lewis W H 1986 Barely visible impact damage (BVID) detection in aircraft composites. *Proc. AIAA/SOLE 2nd Aerospace Maintenance Conf.* American Institute of Aeronautics and Astronautics, New York

Stone D E W 1986 Non-destructive methods of characterising the strength of adhesive-bonded joints—a review. Royal Aircraft Establishment Report TR 86058. RAE, Farnborough, UK

Stone D E W, Clarke B 1987 Non-destructive evaluation of composites—an overview. *Proc. 6th Int. Conf. on Composite Materials, 2nd European Conf. on Composite Materials.* Elsevier, London

Walker C A, McKelvie J 1987 Optical methods. In: Summerscales J (ed.) 1987 *Non-destructive Testing of Fibre Reinforced Plastics Composites*, Vol. 1. Elsevier, London

D. E. W. Stone and B. Clarke
[Defence Research Agency, Farnborough, UK]

Nonmechanical Properties of Composites

Composite materials have found a number of structural applications in the aircraft and automobile industries, but their use in electronics is also surprisingly widespread. Multiphase composite devices are used as multilayer capacitors, piezoelectric transducers, packages for integrated circuits, high-voltage insulators, magnetic tape, varistors for lightning arrestors, chemical sensors and printed circuit boards.

A number of interesting concepts are involved in the application of composite materials to electronics: sum and product properties; connectivity patterns (which determine field and force concentration); the importance of periodicity and scale in resonant structures; the symmetry of composite materials and its influence on physical properties; polychromatic percolation and coupled conduction paths in composites; varistor action and other interfacial effects; coupled phase-transformation phenomena in composites; and the important role that porosity and inner surface area play in many composite sensors.

1. Classification of Properties

For convenience, the physical and chemical properties of composites can be classified as sum properties, combination properties and product properties. The basic ideas underlying sum and product properties were introduced by van Suchtelen (1972). For a sum property, the composite property coefficient depends on the corresponding coefficients of its constituent phases. Thus the stiffness of a composite is governed by the elastic stiffnesses of its component phases and the mixing rule appropriate to its geometry. In general, the property coefficient of the composite will be between those of its constituent phases.

This is not always true for combination properties, which involve two or more different coefficients. Poisson's ratio is a good example of a combination property since it is equal to the ratio of two compliance coefficients. Some composite materials such as wood have extremely small Poisson ratios, far smaller than its constituent phases.

Product properties are more complex and more interesting. The product properties of a composite involve different properties in its constituent phases; the interactions between the phases often cause unexpected results.

1.1 Sum Properties

The dielectric constant will be used to illustrate a simple sum property. Series and parallel mixing rules delimit the bounding conditions for the dielectric constant K of a diphasic composite:

$$K^n = V_1 K_1^n + V_2 K_2^n + \ldots$$

where K_1 and K_2 are the dielectric constants of the constituent phases, and V_1 and V_2 are their volume fractions. The exponent n is $+1$ for parallel mixing and -1 for series mixing. For many composites, the geometric arrangement is partly series and partly parallel, in which case K can often be described by a logarithmic mixing rule for which the exponent $n \sim 0$.

There are, of course, many other mixing rules in addition to the series and parallel models. Theoretical models have been derived to describe the properties of dispersed spheres, disks, lamellae and needles (Hale 1976).

Ferroelectric ceramics are used as multilayer capacitors because of their high dielectric constants. Depressors are added to a high K capacitor formulation to depress the peak at the Curie point, thus producing a flatter temperature-dependence curve. Bismuth stannate and magnesium zirconate are often used as depressors for barium titanate multilayer ceramics. These additives form a second phase in the grain-boundary regions of the ceramic. The grain-boundary phase has a much lower dielectric constant than $BaTiO_3$ and depresses the dielectric constant of the ceramic, largely through the series mixing rule. Low K boundary phases in series with high K grains have a much greater effect on the permittivity than do those in parallel. The brick-wall model gives a good description of diphasic ceramic dielectrics.

Composite ceramics are also useful in high-voltage applications. The dielectric constant of $BaTiO_3$ multilayer capacitors decreases substantially under high-voltage fields, often by 100% or more. This is normal behavior for ferroelectric materials in which the polarization saturates, but antiferroelectric substances such as $NaNbO_3$ behave differently. The dielectric constant of sodium niobate is nearly independent of bias field as its metastable ferroelectric structure begins to influence the permittivity under high fields.

Capacitor compositions with enhanced permittivity at high fields have been manufactured from composites made from $BaTiO_3$ and $NaNbO_3$. Fast-firing a mixture of $BaTiO_3$ and $NaNbO_3$ causes the $NaNbO_3$ to melt and coat the grains of $BaTiO_3$, producing a composite structure with ferroelectric grains embedded in an antiferroelectric matrix. By adjusting the composition and firing schedule, a capacitor with field-independent permittivity can be produced.

Insulators with low dielectric constants are required for microwave lead-through seals and for high speed computers. Decreasing the dielectric constant by introducing porosity into the ceramic increases the speed of electromagnetic waves travelling along conducting wires embedded in the composite. The speed is doubled by replacing alumina insulation ($K \sim 9$) with porous glass ($K \sim 2$).

1.2 Combination Properties

For simple mixing rules the properties of a composite lie between those of its constituent phases, but combination properties involve two or more coefficients which may average in a different way.

For example, acoustic wave velocity determines the resonance frequency of piezoelectric devices. The velocity of waves propagating along the length of a long, thin rod is $v = (E/\rho)^{1/2}$, where E is Young's modulus and ρ is the rod density. Fiber-reinforced composites often have very anisotropic wave velocities. Consider a compliant matrix material reinforced with parallel fibers. Long, thin rods fashioned from the composite have different properties when the fibers are oriented parallel or perpendicular to the length of the rod. The wave velocities are much faster for rods with longitudinally-oriented fibers (v_L) than for those with transversely oriented fibers (v_T).

Experimental measurements for composites made from steel filaments embedded in epoxy conform closely to the equations for v_L and v_T. It is interesting

Table 1
Examples of product properties (after van Suchtelen 1972)

Property of phase 1	Property of phase 2	Composite product property
Thermal expansion	Electrical conductivity	Thermistor
Magnetostriction	Piezoelectricity	Magnetoelectricity
Hall effect	Electrical conductivity	Magnetoresistance
Photoconductivity	Electrostriction	Photostriction
Superconductivity	Adiabatic demagnetization	Electrothermal effect
Piezoelectricity	Thermal expansion	Pyroelectricity

to note that v_T, the wave velocity for waves travelling transverse to the fibers, is less than the velocity of both epoxy and steel, the two phases that make up the composite. The slowness of this wave results from the different dependencies of density and stiffness on volume fraction. This difference in mixing rules for E and ρ cause the combination property v_T to lie outside the range of the end members. The longitudinal wave v_L behaves more normally; E and ρ follow the same mixing rule and the values for v_L lie between those of the end members.

1.3 Product Properties

The interaction of different properties in the two phases of a composite result in yet a third property, a product property. The combination of different properties of two or more constituents sometimes yields surprisingly large product properties. Indeed, in a few cases, product properties are found in composites that were entirely absent in the phases that make up the composite. Table 1 lists a few of the hundreds of possible product properties (van Suchtelen 1972), including several described in this article.

Several of the most sensitive magnetic field sensors are composite materials that utilize product properties. In the magnetoresistive field plate a composite of InSb and NiSb is directionally solidified to form parallel NiSb needles in an InSb matrix. A long rectangular segment of the composite is electroded across the ends with the NiSb fibers parallel to the electrodes and transverse to the length of the composite. InSb is a semiconductor with a large Hall effect and NiSb is metallic and highly electrically conductive.

When an electric current flows along the length of the bar and a magnetic field is applied perpendicular to the current and perpendicular to the NiSb needles the current is deflected because of the Hall effect. Normally this would result in an electrical field transverse to the current and the magnetic field, but the NiSb needles short out the field. Electric current continues to be deflected as long as the magnetic field is present. The resulting product property is a large magnetoresistance effect.

A different type of magnetic field probe can be made from magnetoelectric composites of ferroelectric $BaTiO_3$ and ferrimagnetic cobalt titanium ferrite. A dense eutectic mixture of the perovskite and spinel-structure phases was obtained by directional solidification, and then was electrically poled to make the $BaTiO_3$ phase piezoelectric. When a magnetic field is applied to the composite the ferrite grains change shape because of magnetostriction, and the strain is passed along to the piezoelectric grains, resulting in an electrical polarization. Magnetoelectric effects a hundred times larger than those in Cr_2O_3 are obtained in this way.

2. Symmetry of Composite Materials

A wide variety of symmetries are found in composite materials, including crystallographic groups, Curie groups, black-and-white groups and color groups. In describing the symmetry of composite materials, the basic idea is Curie's principle of symmetry superposition: *A composite material will exhibit only those symmetry elements that are common to its constituent phases and their geometrical arrangement.* The practical importance of Curie's principle rests upon the resulting influence on physical properties. Generalizing Neumann's law from crystal physics: *The symmetry elements of any physical property of a composite must include the symmetry elements of the point group of the composite.*

2.1 Crystallographic Groups

Laminated composites are good examples of composite materials that conform to crystallographic symmetry. In the unidirectional laminates used as printed circuit boards the glass fibers are aligned parallel to one another, such that the laminate has orthorhombic symmetry (crystallographic point group *mmm*). Mirror planes are oriented perpendicular to the laminate normal and perpendicular to an axis formed by the intersection of the other two mirrors. The physical properties of a unidirectional laminate must therefore include the symmetry elements of point group *mmm*. If the laminate is heated it will change shape because of

Table 2
Symmetry groups of representative composites

Composite type	Symmetry group
Unidirectional laminate	mmm
Cross-ply laminate	$42m$
Angle-ply laminate	222
Tetragonal honeycomb extrusion	
unpoled	$4/mmm$
longitudinally poled	$4mm$
transversely poled	$mm2$
Glass-ceramic	$\infty\infty m$
Polar glass-ceramic	∞m
Ferroelectric–ferromagnetic composite	
unpoled, unmagnetized	$\infty\infty m$
poled, unmagnetized	∞m
unpoled, magnetized	∞/mm'
parallel poled and magnetized	$\infty m'$
transverse poled and magnetized	$2'mm'$

thermal expansion. Less expansion will take place parallel to the fiber axis because glass has a lower thermal expansion and greater stiffness than does the polymer. The laminate will therefore expand anisotropically but will not change its symmetry, i.e., the heated laminate continues to conform to point group mmm.

A cross-ply laminate is made up of two unidirectional laminates bonded together with the fiber axes at 90°. Such a laminate belongs to tetragonal point group $\bar{4}2m$, as indicated in Table 2. Laminated composites with $\pm\theta$ angle-ply alignment exhibit orthorhombic symmetry, which is consistent with point group 222 characteristics.

Other types of symmetry elements can also be introduced during processing. The extruded honeycomb ceramics used as catalytic substrates and as positive-temperature coefficient (PTC) thermistors are an interesting example. By suitably altering the die used in extruding the ceramic slip, a large number of different symmetries can be incorporated into the composite body when the extruded form is filled with a second phase.

Lead zirconate titanate (PZT) honeycomb ceramics have been transformed into piezoelectric transducers by electroding and poling. The symmetry of the honeycomb transducers depends on the symmetry of the extruded honeycomb and also on the poling direction. For a square honeycomb pattern, the symmetry of the unpoled ceramic is tetragonal ($4/mmm$) with a fourfold axis parallel to the extrusion direction. When poled parallel to the same direction, the symmetry changes to $4mm$. Transversely poled composites filled with epoxy are especially sensitive to hydrostatic pressure waves, and in this case the symmetry belongs to orthorhombic point group $mm2$ (Newnham 1986).

2.2 Curie Groups and Magnetic Symmetry

The piezoelectric properties and symmetry of natural composites, such as wood and bone, conform to their texture symmetries (Zheludev 1974). Some texture symmetry groups belong to the 32 crystallographic point groups, but others do not. Composite bodies with texture may also belong to one of the Curie groups: $\infty\infty m$, $\infty\infty$, ∞/mm, ∞m, ∞/m, $\infty 2$ and ∞. Polar glass ceramics with conical symmetry can be used to illustrate this idea. A glass can be crystallized under a strong temperature gradient such that polar crystals grow like icicles into the interior from the surface. Certain glass-ceramic systems, such as $Ba_2TiSi_2O_8$ and $Li_2Si_2O_5$, show sizable pyroelectric and piezoelectric effects when prepared in this manner. Polar glass ceramics belong to the Curie point group ∞m, the point group of a polar vector. As the glass is crystallized in a temperature gradient, its symmetry changes from spherical ($\infty\infty m$) to conical (∞m), the same symmetry found in a poled ferroelectric ceramic.

To describe magnetic fields and magnetic properties it is necessary to introduce black-and-white Curie groups. Magnetic fields are represented by axial vectors with symmetry ∞/mm'. The symbol m' indicates that the mirror planes parallel to the magnetic field are accompanied by time reversal.

The magnetoelectric composite described previously is an excellent illustration of the importance of symmetry in composite materials. In combining a magnetized ceramic (symmetry group ∞/mm') with a poled ferroelectric (symmetry group $\infty m1'$), the symmetry of the composite is obtained by retaining the symmetry of elements common to both groups: $\infty m'$.

An interesting consequence of this symmetry description is its effect on physical properties. According to Neumann's law, the symmetry of a physical property of a material must include the symmetry elements of the point group. The symmetry of a magnetized ceramic and a poled ferroelectric both forbid the occurrence of magnetoelectricity, but their combined symmetry ($\infty m'$) allows it. By incorporating materials of suitable symmetry in a composite, new and interesting product properties can be expected to occur.

The symmetry of a magnetoresistive field plate with current flowing perpendicular to the fibers is $2'mm'$, the same symmetry group found in crossed electric and magnetic fields.

3. Magnetic Composites

Flexible magnets are made by embedding ferrimagnetic ceramic grains in a polymer matrix. The processing is carried out by rolling, extrusion or injection molding. Barium ferrite fillers in nylon or polyphenylene sulfide have sufficient mechanical strength to withstand normal load-bearing environments. To obtain maximum alignment, the ferrite particles are physically rotated in a magnetic field during the molding process. BH energy products equivalent to those of a cobalt steel are obtained in this way. In addition to simple mechanical clamps and latches, molded plastic magnets are used as bearing sleeves, timing-motor rotors, beam-focusing devices for television receivers, and magnetic sensors.

3.1 Magnetic Tape and Disks

Composite magnetic recording media consist of submicroscopic, single-domain particles of magnetic oxides or metals immersed in a polymeric binder that separates the particles and binds them to the substrate. The substrates used in tapes, magnetic cards and floppy disks are generally made from poly(ethylene terephthalate), while rigid disks are fabricated from an Al–Mg alloy. Among the advantages of the particulate composites are low cost, high yields, high roll-coating speeds and independent control of the magnetic, mechanical and thermal properties of the recording media.

Particles of γ–Fe_2O_3 have been used in tapes for more than 50 years, but as the bit length of recorded signals becomes shorter, further improvements in coercivity are required. Coercive fields have been raised from 100 to 500 $A\,m^{-1}$ by impregnating the surface of the iron oxide particles with cobalt. The market for magnetic tape now exceeds eight billion US dollars a year.

4. Transport Properties of Composites

Electric and thermal transport in conductor-filled composites is controlled by percolation. Thick-film resistors, PTC thermistors and multiphase varistors based on percolation are fabricated commercially. Most composite conductors are made up of conducting metal particles suspended in an insulating polymer matrix. Particle contact and percolation require a larger volume fraction when the metal and polymer grains are comparable in size. When the conducting particles are small they are forced into interstitial regions between the insulating particles; this forces the conducting particles into contact with one another, which results in a low percolation limit.

These ideas were borne out by experiments on copper particles embedded in a matrix of poly(vinyl chloride) (Bhattacharya and Chaklader 1982). The critical volume fraction decreased markedly when the Cu particles were far smaller than the polymer particles. When the size ratio was 35:1 the critical volume percent was only 4% Cu. This highly segregated mixing establishes contact between conducting copper particles at a very low ratio of conductor to insulator.

4.1 Thick-Film Resistors

Commercial thick-film resistors are prepared from an ink consisting of coarse ($\sim 5\,\mu m$) glass particles and fine ($< 1\,\mu m$) metallic particles suspended in an organic liquid. Lead borosilicate glasses and ruthenium oxides (RuO_2 or $Bi_2Ru_2O_7$) are commonly employed. After screen printing on an alumina substrate, the ink is dried and slowly fired at temperatures up to 850 °C. Micrographs of the fired resistors reveal a glassy matrix with a network of connected metal particles.

Conduction takes place through the metallic chains in thick-film resistors. For commercial systems the electrical conductivity changes over many orders of magnitude with metallic particle volume fraction v_M. It conforms to a power law $\sigma = K(v_M - v_0)^t$, where K, v_0, and t are constants for a given system. Experimentally it has been observed that the constant v_0 is very small (0.01–0.1), while t is unusually large (2–7). The functional dependence of the conductivity is similar to that predicted by theoretical models for a resistor lattice just above the percolation threshold, but the constants are different. According to theory, v_0 is 0.25 for a simple cubic lattice and t is about 1.6 for many types of resistor lattices.

Pike (1978) has reconciled theory and experiment by proposing a model more consistent with the observed microstructure. The modified percolation model consists of large close-packed glass particles with tiny metal particles connecting the interstitial sites.

4.2 Polychromatic Percolation

Transport by percolation through two or more materials can be visualized in terms of colors. Black and white patterns illustrate percolation in a diphasic solid. Three kinds of percolation are possible: (a) percolation through an all-white path, (b) percolation through an all-black path, and (c) percolation through a combined black and white path. From a composite point of view, the third path is the most interesting because it offers the possibility of discovering phenomena not present in either phase individually. Foremost among these effects are the interfacial phenomena, which arise from the insertion of a thin insulating layer between particles with high electrical conductivity. Varistors, PTC thermistors and boundary-layer capacitors are examples of such materials. In ceramic varistors conducting ZnO grains are surrounded by thin layers of insulating Bi_2O_3. The tunnelling of electrons through this barrier gives rise to the varistor effect.

For three-color systems, a multitude of conduction paths are possible. Connectivity requirements for polychromatic percolation have been discussed from a theoretical viewpoint by Zallen (1977). An example of such a system is an easy-poling piezoelectric composite made up of two kinds of particles mixed in an insulating polymer matrix. The first kind of particulate phase in a piezoelectric composite is PZT (lead zirconate titanate), a ferroelectric ceramic phase that must be poled to make it piezoelectrically active. Poling is difficult because the PZT grains are not in good electrical contact, and when shielded by a polymer, only a small fraction of the poling field penetrates into the ferroelectric PZT particles. A small amount of a second conductive filler material is added to facilitate poling. When a conductor is added and the composite is stressed, electrical contact is established between the ferroelectric grains, making poling possible. Pressure sensors of remarkable sensitivity are obtained in this way.

5. *Composite Thermistors*

A second interesting effect is the dependence of electrical resistance on temperature. PTC thermistors are characterized by a positive temperature coefficient of electrical resistance. Doped barium titanate ($BaTiO_3$) has a useful PTC effect in which the resistance undergoes a sudden increase of four orders of magnitude just above the ferroelectric Curie temperature (130 °C). The PTC effect is caused by insulating Schottky barriers created by oxidizing the grain-boundary regions between conducting grains of rare earth-doped $BaTiO_3$.

Similar PTC effects are observed when polymers are loaded near the percolation limit with a conducting filler. Commercial overload protectors are made from high-density polyethylene with carbon filler. At room temperature the carbon particles are in contact, giving resistivities of only 1 Ω cm, but on heating the polymer expands more rapidly than carbon, pulling the carbon grains apart and raising the resistivity. Polyethylene expands very rapidly near 130 °C, which results in a pronounced PTC effect comparable to that of $BaTiO_3$. A rapid increase in resistivity of six orders of magnitude occurs over a 30 ° temperature rise.

Combined NTC–PTC composites have been fabricated using a vanadium sesquioxide (V_2O_3) filler which has a metal–semiconductor transition near −110 °C with a large increase in conductivity on heating. This material can be incorporated in a composite by mixing V_2O_3 powder in an epoxy matrix. The filler particles are in contact at low temperatures and exhibit an NTC resistance change similar to that observed in V_2O_3 crystals and single-phase ceramics. On heating above room temperature, the polymer matrix expands rapidly, pulling the V_2O_3 grains apart and raising the resistance by many orders of magnitude. This produces a PTC effect similar to the carbon–polyethylene composite. The net result is an NTC–PTC thermistor with a conduction "window" in the range −100 °C to +100 °C. This is a good example of the use of coupled phase transformations in composites.

6. *Connectivity and Tensor Properties*

Connectivity is a key feature in property development in multiphase solids because their physical properties can change by many orders of magnitude depending on the manner in which connections are made.

Each phase in a composite may be self-connected in zero, one, two or three dimensions. It is natural to focus attention on three perpendicular axes because all property tensors are referred to orthogonal systems. If we limit the discussion to diphasic composites, there are ten connectivities: 0–0, 1–0, 2–0, 3–0, 1–1, 2–1, 3–1, 2–2, 2–3 and 3–3. Connectivity patterns for more than two phases are similar to the diphasic patterns, but far more numerous. There are 20 three-phase patterns and 35 four-phase patterns. For n phases the number of connectivity patterns is $(n+3)!/3!n!$.

During the past few years processing techniques have been developed for making piezoelectric composites with different connectivities (Newnham 1986). Extrusion, tape casting, injection molding and fugitive phase methods have been especially successful. The 3–1 connectivity pattern, for instance, is ideally suited to extrusion processing. A ceramic slip is extruded through a die to yield a three-dimensionally connected pattern with one-dimensional holes, which can later be filled with a second phase. Another type of connectivity well suited to processing is the 2–2 pattern, made up of alternating layers of the two phases. The tape casting of multilayer capacitors with alternating layers of metal and ceramic is a way of producing 2–2 connectivity. In this arrangement both phases are self-connected in the lateral directions but not self-connected perpendicular to the layer.

6.1 Stress Concentration

The importance of stress concentration in composite materials is well known from structural studies, but its relevance to electroceramics is not so obvious. Stress concentration is a key feature of many of the piezoelectric composites made from polymers and ferroelectric ceramics. By focusing the stress on the piezoelectric phase, some of the piezoelectric coefficients can be enhanced and others reduced.

For example, consider the hydrostatic piezoelectric coefficient $g_h = (d_{31} + d_{32} + d_{33})/\varepsilon_{33}$ in a 1–2–3–0 composite. This composite is made up of PZT fibers in the poling direction (X_3) and glass fibers in the X_1 and X_2 directions. The fibers are embedded in a foamed polymer matrix. In terms of the 1–2–3–0 symbol, the PZT is self-connected in one dimension, the glass fibers in two dimensions, and the polymer in three dimensions, and the voids in none.

Hydrostatic stress waves are converted to uniaxial stresses inside the composite. Stress components in the X_1 and X_2 directions are carried by the glass fibers, while stresses along X_3 act upon the PZT. Because of its greater compliance, the polymer matrix transfers stress to the fibers. Foaming the polymer reduces the Poisson ratio of the composite, preventing transfer of stress between the X_3 (poling) direction and the orthogonal X_1 and X_2 directions. As a result d_{33} is kept large while d_{31} and d_{32} are reduced. This improves g_h because normally d_{31} and d_{32} are opposite in sign from d_{33}. As an added benefit, the dielectric permittivity ε_{33}

is reduced by eliminating much of the ferroelectric PZT from the transducer. Improvements in g_h of two orders of magnitude have been demonstrated.

Advantageous internal stress transfer can also be utilized in pyroelectric coefficients. If the two phases have different thermal expansion coefficients, there is stress transfer between the phases, which generates electrical polarization through the piezoelectric effect. In this way it is possible to make a composite pyroelectric that is not piezoelectric.

Stress redistribution also occurs in the piezoelectric bimorphs and unimorphs used as fans, printers and speakers. Unimorphs are made by bonding a thin piezoelectric ceramic to a steel shim and driving it in resonance. The steel shim restricts extension of the piezoelectric, causing a large bending motion. Force is traded off for displacement in the unimorph and bimorph.

6.2 Electric Field Concentration

The multilayer design used for ceramic capacitors is an effective configuration for concentrating electric fields. By interleaving metal electrodes and ceramic dielectrics in a 2–2 connectivity pattern, relatively modest voltages can produce high electric fields.

Multilayer piezoelectric transducers are made in the same way as multilayer capacitors. The oxide powder is mixed with an organic binder and tape-cast using a doctor blade configuration. After drying the tape is stripped from the substrate and electrodes are applied with a screen printer and electrode ink. A number of pieces of tape are then stacked, pressed and fired to produce a ceramic with internal electrodes. After attaching leads, the multilayer transducer is packaged and poled. When compared to a simple piezoelectric transducer, the multilayer transducer offers a number of advantages.

(a) The internal electrodes make it possible to generate larger fields for smaller voltages, eliminating the need for transformers for high-power transmitters. Ten volts across a tape-case layer 100 μm thick produces an electric field of 10^5 V m^{-1}, a value not far from the depoling field of PZT.

(b) The higher capacitance inherent in a multilayer design often improves acoustic impedance matching.

(c) Many different electrode designs can be incorporated in the transducer to shape poling patterns, which in turn control the mode of vibration and the ultrasonic beam pattern.

(d) Additional design flexibility can be achieved by interleaving layers of different composition. One can alternate ferroelectric and antiferroelectric layers, for instance, thereby increasing the depoling field.

(e) Grain-oriented piezoelectric ceramics can also be tape-cast into multilayer transducers. Enhanced piezoelectric properties are obtained by aligning the crystallites parallel to the internal electrodes.

(f) Another advantage of the thin dielectric layers in a multilayer transducer is improved electric breakdown strength. The dc breakdown field for ceramics 1 cm thick was less than that of samples 1 mm thick. It is likely that the trend toward thinner specimens will continue, leading to improved poling and more reliable transducers.

See also: Percolation Theory

Bibliography

Bhattacharya S K, Chaklader A C D 1982 Review on metal-filled plastics. *Polym-Plast. Technol. Eng.* 19: 21–51

Hale D K 1976 The physical properties of composite materials. *J. Mater. Sci.* 11: 2105–41

Newnham R E 1986 Composite electroceramics. *Ann. Rev. Mater. Sci.* 16: 47–68

Pike G E 1978 Conductivity of thick film (cermet) resistors. *AIP Conf. Proc.* 40: 366–71

van Suchtelen J 1972 Product properties: A new application of composite materials. *Philips Res. Rept.* 27: 28–37

Zallen R 1977 Polychromatic percolation. *Phys. Rev. B* 16: 1426–35

Zheludev I S 1974 Piezoelectricity in textured media. *Solid State Phys.* 29: 315–59

R. E. Newnham

[Pennsylvania State University, University Park, Pennsylvania, USA]

O

Oxide Inorganic Fibers

This article covers refractory inorganic-oxide fibers, the majority of which comprise combinations of aluminum oxide (alumina, Al_2O_3) and silicon dioxide (silica, SiO_2) in various proportions. There is interest also in fibers of zirconium oxide (zirconia, ZrO_2). Fibers containing a high proportion of SiO_2 (40–55 wt%) with the rest made up of alumina are called aluminosilicate fibers. Such fibers are glassy, find considerable use in high-temperature insulation and comprise by far the greatest tonnage of refractory fibers on the market. Alumina fibers containing lower proportions of SiO_2 (e.g., ~5 wt%) are polycrystalline. They are able to withstand higher use temperatures than glassy fibers and, having a high stiffness, are of interest as a reinforcement, for example in metal-matrix composites. They are used in insulation at temperatures beyond the capability of glassy aluminosilicate fibers.

1. Aluminosilicate Fibers

The range of composition of melt-spinnable aluminosilicates is about 45–60 wt% Al_2O_3 with SiO_2 as the other major component and with minor amounts of Fe_2O_3, TiO_2, CaO and other oxides. The fibers are usually manufactured from kaolin or a related clay mineral, or from blends of alumina and silica (allowing closer control of composition than with clay minerals). The raw material is melted and made into fibers in one of two ways: the melt may be poured into a stream of compressed gas (air or steam) which breaks the melt into fibers, or the melt may be fed to a rapidly rotating disk from which fibers are thrown by centrifugal force. The latter method produces longer and silkier fibers, but in both processes a wide range of fiber diameters (<1 to >10 μm) are formed together with a fraction of nonfibrous material termed shot. The resulting fibers vary from equidimensional fragments to several centimeters in length.

For glassy aluminosilicate fibers, the usual limit to temperature resistance is the devitrification of the glass with, for example, the nucleation and growth of mullite ($3Al_2O_3.2SiO_2$), which reduces mechanical strength drastically. Temperature resistance increases with increasing alumina concentration, fibers containing up to about 52% Al_2O_3 being suitable for use at temperatures up to 1250 °C, and those containing higher levels (up to about 65%) being capable of use at up to 1400 °C. Small additions of chromia (CrO_2) improve temperature resistance. Aluminosilicate fibers are processed into a number of forms such as loose wool, blanket, felt, paper and board. These can be made by dewatering suspensions of the chopped fiber.

The major application of aluminosilicate fibers is in the insulation of furnaces in the metallurgical, ceramic and chemical industries, with a resulting considerable saving in weight and fuel usage. There may be up to a tenfold difference between the mass of a fiber-lined furnace and the mass of one constructed conventionally from refractory brick. This brings very considerable advantages in construction and maintenance; the low thermal mass allows rapid heating and cooling in cyclic operations and a reduction in fuel usage of as much as 40% is not uncommon. Not surprisingly, the use of refractory fibers in high-temperature technology has grown rapidly.

2. Manufacture of High-Alumina Fibers

The incentive to produce inorganic fibers higher in alumina content than can conveniently be manufactured by melt spinning is twofold. The success of melt-spun fibers in thermal insulation at temperatures up to 1400 °C led to a demand for fibers capable of use at higher temperatures, and the potential high stiffness of alumina-rich fibers was attractive for composite manufacture. The high modulus of alumina (α-alumina, 530 GPa), its relatively low density (α-alumina, 4.0 g cm^{-3}) and its nonreactivity make it an attractive candidate for metal reinforcement. The low viscosity of molten alumina and its high melting point (2070 °C) preclude melt spinning, so that processes had to be developed avoiding the melting step. Suspensions of particulate alumina or an alumina hydrate may be extruded as fibers, and then dried and sintered to produce a polycrystalline filament. Similarly, viscous solutions of aluminum salts may be extruded or drawn into fibers, the gel being dried and fired. Other methods investigated include soaking an organic textile fiber (usually cellulosic) in an inorganic salt solution, heating to burn out the organic material, and then firing to sinter and densify the oxide relic. This relic process is in use for the manufacture of ZrO_2 fibers.

2.1 The Slurry Process

The slurry process has been adopted for the production of continuous alumina fibers. The process consists essentially of three stages—slurry preparation, yarn spinning and firing. An aqueous suspension of particulate alumina is prepared, the aqueous phase of which may contain dissolved organic polymers to increase viscosity and to stabilize the suspension, together with additives to control grain size at the sintering stage. The aqueous phase may also contain a dissolved alumina precursor (e.g., an aluminum salt) to aid densification at the sintering stage. The slurry is then extruded into air, collected, dried and fired in two steps—an initial firing at low temperature to control shrinkage, and a final flame-firing to complete conversion to the high-temperature stable form (α-alumina) and to eliminate porosity. The fired fiber may be coated with a thin layer of silica to increase fiber strength by about 50% as a result of the heating of surface flaws. The major fiber made by this process is Fiber FP (manufactured by Du Pont), the diameter of individual filaments of which is about 20 μm in tows of 200 filaments. The grain size of the α-alumina in these fibers is about 0.5 μm.

The properties of FP fibers have recently been improved by modifying the composition with 15–25 wt% of tetragonal zirconia. The new fiber (designated PRD-166) shows a 50% improvement in strength over the α-Al_2O_3 FP fiber but otherwise the morphology is the same—20 μm diameter filaments in 200 filament tows.

2.2 Solution or Sol–Gel Spinning

An alternative process to slurry spinning is the spinning of viscous, concentrated solutions of aluminum compounds that are precursors to the oxide, followed by the drying and heat treatment of the gel fibers so produced. The hexahydrated aluminum cation, $[Al(H_2O)_6]^{3+}$, is stable only in highly acidic solution and as the solution becomes more basic, hydrolysis of the ion occurs and polymeric species are formed in which aluminum atoms are linked by oxygen bridges. Eventually, complex polynuclear species such as $[AlO_4Al_{12}(OH)_{24}(H_2O)_{12}]^{7+}$ are formed. The viscosity of solutions containing such species rises rapidly with the equivalent Al_2O_3 content and, at an equivalent Al_2O_3 concentration of about 45 wt%, a gel or glass-like solid is produced. In the preparation of alumina fibers, a syrupy solution of a basic aluminum salt (usually the chloride) is prepared, to which may be added soluble organic polymers to control rheology (spinning aids), and sources of oxides such as SiO_2, MgO, B_2O_3 may be added as grain-growth inhibitors. The viscosity of the solution used for spinning will depend on the type of spinning process to be used and the type of fiber it is desired to produce—staple fiber or continuous filament. For continuous filament, a solution having a viscosity between 10 and 100 Pa s is desirable. This is extruded through spinneret holes between 100 and 200 μm diameter into dry air ($<65\%$ RH) and the fiber drawn down in diameter during collection. When discontinuous fiber is to be produced, centrifugal spinning may be used, in which droplets of solution are projected into dry air from a rapidly rotating disk; or gas-attenuation spinning may be employed, in which fibers are produced by drag in a stream of gas. For this type of spinning, solutions having a viscosity between 0.5 and 2.0 Pa s will be used, in which the equivalent Al_2O_3 concentration is 25–30 wt%. Drying of the fibers in flight produces solid-gel fibers containing the equivalent of 45 wt% Al_2O_3. These are then subjected to a heat-treatment sequence to complete the drying of the fiber without distortion, to dehydrate hydrous material, to remove the anion (chloride as HCl) and to burn out organic matter. A final stage at a higher temperature is used to convert the alumina to an appropriate ceramic phase, for example, δ-alumina or α-alumina.

Continuous fibers produced by sol–gel processing have diameters between 10 and 20 μm, whereas staple fibers made by centrifugal or gas-attenuation spinning have a mean diameter of about 3 μm.

A particular advantage of sol–gel spinning is the ability to control fiber diameter so that it is in the range 1–7 μm. Such close control is important if a product free from health hazard is required.

In the heat-treatment sequence, the aqueous-gel fiber first decomposes to form an amorphous or poorly crystalline alumina. On further heat treatment, this poorly defined alumina passes through a series of transition alumina-phases (η-, γ-, θ- and δ-aluminas) before α-alumina appears. The process may be stopped at an intermediate stage so that alumina fibers having a variety of properties may be prepared. The η-phase fiber, for example, has a high surface area of $200\ m^2\,g^{-1}$ due to high internal porosity, which makes this form of fiber particularly useful as a catalyst support. To produce fibers for high-temperature use, the heat treatment is continued until the major phases present are δ- and α-alumina. The presence of a few percent of silica in the fiber is essential for the control of grain growth since grain size must be a fraction of fiber diameter if strength is to be preserved. The modulus increases with the α-alumina content. Short staple fiber produced by sol–gel spinning is converted to a variety of forms such as loose wool, blanket, paper and board.

Zirconium oxide (ZrO_2) fibers have also been made by sol–gel spinning using zirconium acetate as a precursor and Y_2O_3 to inhibit grain size and to stabilize the tetragonal phase at room temperature.

3. Properties and Applications of Oxide Inorganic Fibers

The mechanical properties of the major commercially available oxide fibers are given in Table 1. Other important properties of oxide fibers include the preservation of room-temperature mechanical properties at elevated temperatures, resistance to aggressive chemical environments and stability in molten metals.

The earliest application for oxide fibers was for thermal insulation at elevated temperature, and this remains the largest usage of staple fiber in the form of blanket and board. Polycrystalline alumina fibers containing 5% SiO_2 can be used as thermal insulation at temperatures up to 1600 °C. The staple fiber, alone

Table 1
Mechanical properties of major oxide fibers (SD, standard density; LD, low density)

Fiber	Manufacturer	Fiber diameter (μm)	Density ($g\,cm^{-3}$)	Modulus (GPa)	Tensile strength (MPa)	Specific strength (MPa)	Specific modulus (GPa)
Fiber FP							
α-Al_2O_3 yarn	Du Pont	20	3.9	380	>1400	>360	97
PRD-166							
Al_2O_3–ZrO_2 yarn	Du Pont	20	4.2	380	2070	492	90
Saffil RF							
5% SiO_2/Al_2O_3 staple	ICI	1–5	3.3	300	2000	600	90
Saffil HA	ICI	1–5	3.4	>300	1500	440	>90
5% SiO_2/Al_2O_3 staple							
Safimax	SD	3.0	3.3	300	2000	606	90
4% SiO_2/Al_2O_3 semicontinuous	ICI						
	LD	3.5	2.0	200	2000	1000	100
15% SiO_2/Al_2O_3 yarn	Sumitomo	17	3.2	200	1500	470	62
Fiberfrax	Sohio/						
50% SiO_2/Al_2O_3 staple	Carborundum	1–7	2.73	105	1000	360	38
Nextel 312	3M	11	2.7	152	1720	640	56
24% SiO_2/14% B_2O_3 Al_2O_3							
Nextel 440	3M	11	3.1	220	1720	550	71
28% SiO_2/ 2% B_2O_3/Al_2O_3							

or blended with aluminosilicate fibers, is converted to a variety of forms such as boards, blankets and blocks, the thermal conductivity of which can be as low as 0.07 W mK^{-1} at 200 °C and 0.4 W mK^{-1} at 1600 °C. These, used in the lining and construction of a wide variety of high-temperature furnaces, can result in fuel savings of up to 40%. Oxide fibers are also used in the filtration of hot gases and in the removal of impurities from molten metals.

3.1 Metal-Matrix Composites

The use of oxide fibers to reinforce metals such as aluminum alloys, magnesium and titanium has emerged as an important technology and, for example, staple fibers are used to enhance the hot tensile strength, creep resistance, hardness and wear resistance of aluminum pistons for diesel engines. Composites are produced from staple fiber by powder-metallurgical techniques or by squeeze casting, in which a fiber preform is infiltrated with molten alloy under pressure. Typically, in an Al–9Si–3Cu alloy, a volume fraction of 0.24 of random fiber will increase room-temperature modulus from 70 GPa for the unreinforced alloy to 115 GPa, with little elevation of the ultimate tensile strength. However, at 300 °C the ultimate tensile strength of the unreinforced alloy has fallen to 70 MPa while that of the composite remains high at 280 MPa. The modulus of the composite remains high with temperature, being 90 GPa at 300 °C. The incorporation of staple fiber increases hardness, a volume fraction of 0.24 elevating room-temperature hardness from Vickers hardness 131 to 212. Generally, the incorporation of staple fiber raises the working temperature of aluminum alloy pistons by 100–150 °C.

Considerably enhanced properties are obtained when continuous oxide-fibers are used to prepare uniaxial composites. Unidirectional FP-Al_2O_3 fiber/Al composites containing a fiber volume fraction of 0.6 have an axial modulus of 262 GPa, a tensile strength of 690 MPa and a compressive strength of 3.4 GPa, retaining these properties to 316 °C.

The use of metal-matrix composites using oxide fibers is being extended from pistons to other components such as gudgeon pins, connecting rods, valve and valve seats and cylinder liners, and includes the aluminum alloys, magnesium and titanium. In the fabrication of such composites, wetting of the fiber by the metal is important. The presence of a thin silica coating on the fiber appears to be beneficial. Metal-matrix composites, and hence oxide fibers, are set to become increasingly important in automotive engines and in aerospace applications.

4. Toxicity of Oxide Fibers

It is known that inorganic fibers that are durable in tissue can promote a carcinogenic reaction irrespective of chemical composition if they are long (>10 μm) and thin (<1 μm). Continuous oxide fibers present no problem, nor do staple fibers that are greater than 1 μm in diameter. Staple fibers finer than 1 μm should be treated as an inhalation hazard.

Bibliography

Birchall J D 1983 The preparation and properties of polycrystalline aluminum oxide fibres. *J. Br. Ceram. Soc.* 83: 143–45

Birchall J D, Bradbury J A A, Dinwoodie J 1985 Alumina fibres: preparation, properties and applications. In: Watt W, Perov B V (eds.) 1985 *Handbook of Composites*, Vol. 1: *Strong Fibres*. Elsevier, Amsterdam, pp. 115–53

Clyne T W, Bader M G, Cappleman G R, Hubert P A 1985 The use of a δ-alumina fibre for metal matrix composites. *J. Mater. Sci.* 20: 85–96

Dhingra A K 1980a Alumina fibre FP. *Philos. Trans. R. Soc. London, Ser. A* 294: 411–17

Dhingra A K 1980b Metal matrix composites reinforced with FP fibre. *Philos. Trans. R. Soc. London. Ser. A* 294: 559–64

Milieko S T 1985 Fabrication of metal matrix composites. In: Watt W, Perov B V (eds.) 1985 *Handbook of Composites*, Vol. 1: *Strong Fibres*. Elsevier, Amsterdam, pp. 221–94

J. D. Birchall
[Keele University, Keele, UK]

P

Paper and Paperboard

Paper and paperboard production is an environmentally sound, integrated approach to converting natural resources (wood, water, energy, chemicals) into a myriad of useful packaging, communication and specialty products. This article relates the raw materials and processing steps in the manufacture of paper; discusses major paper properties; summarizes applications, production levels and geographic consumption; and indicates environmental, energy and recycling considerations.

1. Raw Materials

Paper is an interlocking network of cellulose fibers enhanced with filler materials and binders. Cellulose fibers, which are the dominant component in paper or paperboard products, have a strong influence on the practical range of physical and mechanical properties of the product. The provision of fiber represents the highest raw material cost for a typical paper mill.

The average "whole" fiber in a paper product is hollow (10–50% void) and ribbon-like with dimensions of 1 mm × 25 μm × 5 μm (length × width × thickness) and weighs 2×10^{-7} g (Dodson 1976). The native cellulose polymer is composed of anhydroglucose monomer units (molecular weight 162) with an overall original degree of polymerization of 2500–8000. Fiber-source and papermaking-process variables combine to produce a wide range of fiber lengths in a paper product. The major effect of wood species (deciduous and coniferous) upon fiber dimensions is shown in Fig. 1. A typical length for hardwood fiber is 0.7 mm and for softwood 2.5 mm.

In contrast to man-made fibers, which are uniform in composition, the wood fiber structure contains four distinct concentric layers (Fig. 2). Each layer has its own orientation of small fibrous elements, fibrils and chemical makeup. The primary wall, which is the first to be removed in either chemical pulping or mechanical treatment, is very thin with random fibrils enmeshed in lignin and intracellular adhesives. There are three sublayers in the adjoining secondary wall. The first, or outer layer (S1) is a moderately thick layer composed of fibrils in laminae, wound at precise large angles relative to the fiber axis and crossing one another for balanced structural strength. The S1 and the primary layers contain most of the lignin in a wood fiber. The secondary wall middle layer (S2) is very thick, lamellar and primary cellulose and hemicellulose. Fibrils in S2 are wound at a single small angle (about 5–20°) relative to the fiber axis. The S2 layer provides most of the characteristic mechanical properties (especially tensile strength) associated with wood fibers and paper. The innermost layer of the secondary wall (S3) resembles the S1 layer in thickness but has fibrils only wound in a single large angle relative to the axis.

Fiber sources are chosen to obtain the desired blend of strength and physical properties for the product. In North America, long, thin softwood fibers are included for tear resistance, while short, stubby hardwood fibers enhance bulk and surface smoothness. In developing countries, where labor costs are relatively low, native vegetable fibers such as bamboo and annual crops (sugar cane bagasse and straw) may be the fiber of choice. These fibers can have exceptional fiber length and strength, but their large-scale use in paper awaits economic production techniques for processing and dispersing them (Clark 1969, McGovern 1982). In industrialized countries, very long man-made polymer fibers such as polypropylene or glass may be blended with cellulose to confer tear resistance to wallpapers and envelopes.

Fiber substitution for traditional hardwoods and softwoods has become increasingly common in certain industrialized regions such as Europe. Eucalyptus pulp has captured more than 40% of the European market for paper-grade hardwood (Gallep 1987). It is exported worldwide mainly from Portugal, Spain and Brazil at levels of 0.5–1 Mt per year from each country. The short thin eucalyptus fibers offer softness and printability benefits in tissue and paper grades (McGrath 1987).

An increasing source of paper fiber is secondary or recycled post-consumer fiber. Major sources of recycled fiber are corrugated cartons, newspapers and shredded business papers. Recycled fiber must be freed of extraneous matter such as inks, coatings and polymeric films. Generally, papers made from recycled fibers have low tensile strength but have increased opacity and better formation than papers made from virgin fibers, because the repeated processing shortens and mechanically disrupts the fibers (Altieri and Wendell 1967).

Waste-paper recycling has risen from a 1968 level of 20% to the 25–30% level in the 1980s, and is projected to approach 35% by the year 2000. Major applications for recycled fiber include newspapers, corrugated paperboard and tissue (Franklin 1986).

A wide range of organic and inorganic additives are used to modify the strength, aesthetics and physical characteristics of paper. Except for dry or wet strength agents, additives generally reduce overall mechanical strength of paper by interfering with hydrogen bonding.

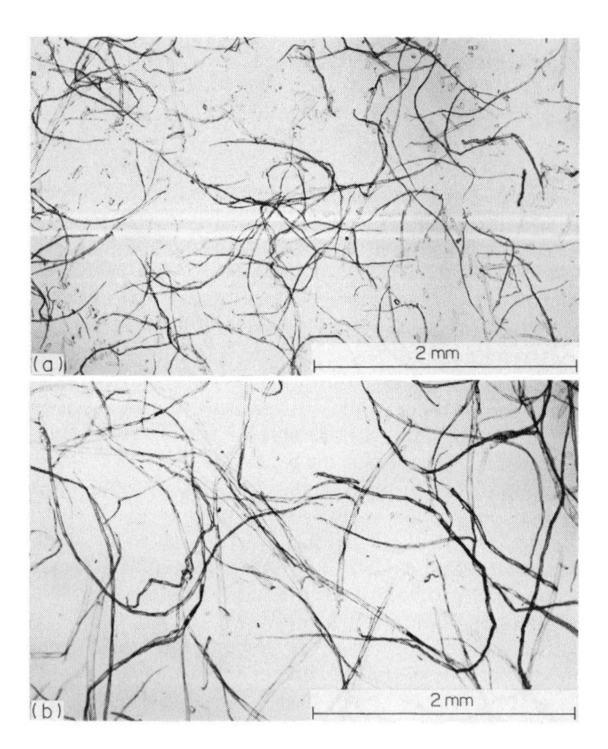

Figure 1
(a) Hardwood (deciduous) fiber for papermaking, and
(b) softwood (coniferous) fiber for papermaking

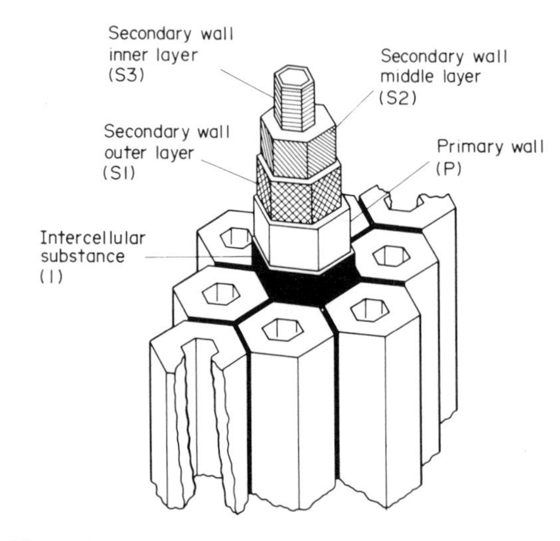

Figure 2
Schematic of cell wall structure of wood fiber

Three major additive categories are fillers, internal size and surface size. Both inorganic (clay, calcium carbonate and titanium dioxide) and organic (urea formaldehyde) fillers are added to paper in order to improve optical, surface and bulk characteristics. Rosin or synthetic materials (alkyl succinic anhydride and alkyl ketene dimer) are added as internal sizing agents to reduce the sensitivity of the fiber web to moisture during printing. Surface sizing with a binder such as starch enhances both surface strength for printing and bulk strength properties for conversion and end-use requirements.

To enhance appearance and printability of paperboard, a very thin layer of pigment and binder is applied. Fine-particle-size clay, titanium dioxide and calcium carbonate pigments are major inorganic coating ingredients. Polystyrene coating pigment enhances gloss. Primary coating binders are starch, styrene butadiene and poly(vinyl acetate).

Large quantities of water and fuel are required to produce a tonne of paper. Water is essential in transporting and treating pulp slurries, and in developing strength in paper products (Campbell effect). Chemical recovery and paper-drying processes are the major consumers of energy.

2. Production of Pulp

2.1 Pulping

Pulping of wood can be done mechanically with or without steam and/or chemicals to obtain fibers in high yield (80–95+%). High-yield fibers have considerable lignin and hemicellulose associated with the cellulose, which causes the fibers to be stiff, to bond relatively poorly and to produce a bulky fiber web. Papers from mechanical pulps have low tensile and bursting strengths.

The pressurized groundwood (PGW) process developed by Finnish papermakers has found application in lightweight coated (LWC) and super calendered (SC) grades, while Swedish-developed processes for chemithermomechanical pulp (CTMP) and chemimechanical pulp (CMP) have proven attractive for making fluff pulp, tissue and bleached paperboard (Breck and Styan 1985).

The "kraft" or sulfate process, illustrated in Fig. 3, is the dominant commercial pulping method (40–70% yield). The pulp has higher cellulose content than mechanical pulp and provides better hydrogen bonding among fibers. The high-pH kraft process typically uses sodium hydroxide (caustic) and sodium sulfide to delignify wood chips under high pressure and temperature in a digester. Unbleached products using kraft pulp include sacks, bags, corrugated board, and saturating paper for laminates. Bleached kraft products include fine papers and printed folding cartons.

The third pulping method, the sulfite process, uses sulfur dioxide combined as a salt (sodium, calcium, magnesium or ammonium). Normal operating ranges during pulping are 120–145 °C and pH 1–9. Sulfite pulps range from greaseproof with relatively high lignin (4%) and high hemicellulose (14%), to dissolv-

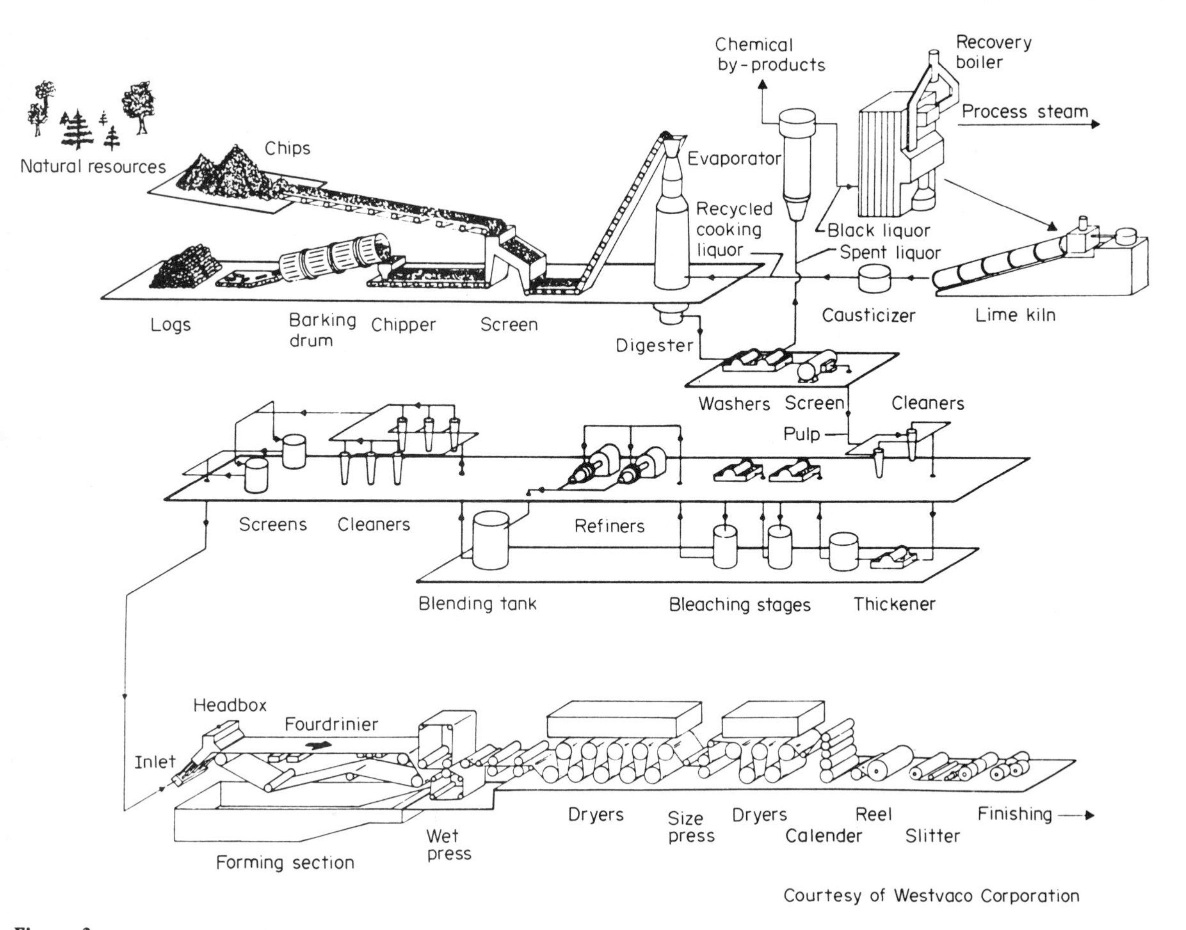

Figure 3
The papermaking process (after Wardrop and Bland 1959)

ing pulp with high alpha-cellulose, low lignin (1%) and low hemicellulose (6%) for cellulose derivatives.

Modern woodyards stress greater pulp uniformity, beginning with screening wood chips by thickness rather than by length, which typically reduces digester rejects by a factor of 1.5 to 3.0 (Christie 1987). A growing trend is pulping in the presence of anthraquinone to increase pulping rate and interfiber bonding potential (Tay et al. 1985).

Nonpaper pulps (Gross 1982) account for about 5% of all the pulp produced and appear chiefly as: (a) chemical cellulose or dissolving pulp, (b) fluff pulp for diapers and other high-bulk products, and (c) speciality pulps used in products ranging from battery separators to plastic moldings.

Chemical coproducts from the papermaking process include speciality chemicals used in pollution control, agriculture, dyestuffs, cement, rubber, protective coatings and plastics. A wide range of activated carbon, lignin and tall (pine) oil products derived from wood are commercially available.

2.2 *Bleaching*

The common bleaching method (Fig. 3) involves removal of lignin and hemicellulose from naturally white cellulose in kraft and sulfite pulps. The process can involve combinations of chlorine, chlorine dioxide, hypochlorite, and less frequently oxygen, in a series of two to six stages. An alkaline wash or extraction stage between bleaching stages removes chlorinated byproducts. Excess bleaching can degrade pulp strength.

Whitening of high-yield (chemimechanical) pulps employs hydrogen peroxide or sodium hydrosulfite which bleach lignin. These agents do not significantly weaken the fiber, but their "bleaching" effect can be reversed by ultraviolet light (newspaper yellowing in sunlight).

The application of chlorine compounds has been reduced in new plants because kraft and sulfite pulp mills have shown increasing interest in first-stage treatment by oxygen as a means of reducing effluent loads and operating costs (Ducey 1986). Expansion in the production of CTMP/CMP utilizes more sodium hydrosulfite and hydrogen peroxide stages to bleach, rather than remove, lignin for these high-yield pulps (Carmichael 1986).

3. Production of Paper and Paperboard.

Cellulose fiber is the major ingredient in paper products, but water is the predominant material used in its manufacture. Water is essential in developing the characteristic strength of the fiber network through hydrogen bonding. The furnish for paper is supplied at a low consistency (0.1 to 1.0%) to equipment which forms, consolidates and dries the web at up to 1700 m min^{-1}.

3.1 Stock Preparation

Prior to forming the wet fiber web, the furnish for the process is subjected to mechanical refining which can fibrillate, cut and/or modify the fibers to increase fiber/fiber contact and thereby enhance strength and hydrogen bonding in the product. Recycled paper is redispersed prior to refining. The dispersion and refining operations occur at 3–12% consistency.

Following refining, the furnish is cleaned and screened to remove foreign materials which detract from quality and runnability. Filler materials added to the low consistency furnish for smoothness and optical properties include clay, titanium dioxide and calcium carbonate. Internal additives for styling and strength (both wet and dry) can be added during stock preparation.

Alkaline papermaking has not advanced in North America as rapidly as projected (Wuerl 1986).

3.2. Forming

There are three categories of equipment for forming a wet paper web: fourdrinier, cylinder and twin-wire. The most common, the fourdrinier system, is described below and illustrated in Fig. 3.

Formation and consolidation of the fiber network are major areas of technical development in the paper industry. During forming and initial consolidation on a fourdrinier machine, 95% of the water is removed to yield a self-supporting web which still contains 4 kg of water per kg of solid material. Fiber orientation and filler and fines distribution can be adjusted through operating conditions during forming.

The first steps in forming a fiber network involve uniform delivery of furnish to headbox, adequately dispersing the furnish, and extruding the dilute slurry from the "slice. " The stock jet impinges onto a fast-moving continuous fabric to begin dewatering. Fiber orientation in the plane of the web is affected by the relative speeds of the stock jet and the forming fabric. Direction (in twin-wire systems) and rate of water removal are used to control filler and fiber fines distribution through the web thickness. Drainage elements which enhance water removal can also promote turbulence and improve stock uniformity.

Relatively recent technical developments allow high-speed formation of the web in simultaneous or sequential multilayer structures. Formation achieved between two conventional wires can minimize web surface differences typical of paper made on a fourdrinier machine, and thereby produce a more balanced product. These developments originated in key papermaking countries including Finland, Germany, Sweden and the USA, and to a lesser extent in Australia, Canada and Japan (Wahlström 1981).

Initial growth in multilayer paper and board production primarily involved tissue and linerboard grades. It is spreading to the mid-range of basis weights to include printing and writing grades. Cited as reasons for interest in multiply products are flexibility in furnish utilization (for example, high-grade chemical pulps in surface layers with mechanical or recycled pulp in core layers) as well as flexibility in product design (O'Brian 1987).

3.3 Pressing

Drainage on a fourdrinier increases web consistency to approximately 20%. A web at this moisture level can be lifted from the supporting fabric and transported to the press section for additional water removal. The brief residence in the press nips doubles the web solids level to 40%. Vacuum, press loading, temperature and felts are parameters used to enhance water removal. A multiple-nip press section is shown in Fig. 3. Laboratory studies of extended nip pressing indicate solids levels as high as 55% are possible where high density papers are acceptable. This level of dewatering could greatly enhance production capacity (Wahlström 1981). In the press section, the wet web can experience a significant draw or machine-direction stress which accentuates fiber alignment and promotes greater strength in this direction at the expense of cross-machine direction strength.

The achievement of extended or wide-nip wet-pressing has resulted from the combined efforts of research institutes and suppliers of equipment and materials. In addition to the overall press design, advances in materials for press rolls and new fabric designs have been implemented. The benefits claimed from wide-nip pressing include higher machine speeds attained by increasing dewatering, as well as potential for improved machine direction and cross-machine direction paper strength properties (Helm 1987). Feltmakers are improving the uniformity of pressure distribution and dewatering capability in the press nip by redesign of the fabric (Coan 1984).

A relatively recent development is press-drying, a pilot process for drying a web while it is restrained in the thickness direction. A key to effectiveness of the process is the use of unbleached kraft. A hardwood furnish can be made to behave like the more expensive softwood furnish in the liner or medium of corrugated boxes, giving: (a) reduced compression creep, (b) higher burst strength, and (c) greater tensile energy absorption (Setterholm 1979).

3.4 Drying

Excessive moisture in the web after pressing is thermally removed. This approach is expensive compared to the preceding mechanical procedures. Available drying methods depend on web porosity and basis weight. Most paper and board grades are dried on steam-heated cylinders. The two sides of the web alternately contact a series of dryers as the moisture level is reduced to approximately 5%. Felts promote web contact with drums and increase drying rate. Figure 3 shows a multicylinder dryer section.

Lightweight tissues and towels can be pressed and dried on a single "Yankee" dryer drum from which the web can be released by a doctor blade. Doctoring also provides creping which increases sheet extensibility, bulk and absorbency. Lightweight paper can also be processed in a "through-dryer" or air float dryer.

3.5 Finishing

Many alternatives are available to enhance base stock surface. It is generally calendered to improve smoothness. Aqueous, solvent or extruded polymeric coatings can be applied to meet end-use requirements. In fine papers and folding cartons, the coating and finishing steps smooth the relatively rough-surfaced base stock to meet high print-quality requirements. Calcium carbonate is the major coating pigment in European paperboard applications because of its availability, low cost and reported benefits of improved smoothness from higher solids content (Wintgen 1987).

Coating and calendering operations can either be performed "on" or "off-machine." Extruding a polymeric film over the paper, or metallizing the paper surface (Carter and Beardow 1983) are done off the paper machine to accommodate differences in speeds and reel size, and special handling requirements.

3.6 Converting

Mechanical compaction is used for extensible paper applications such as sacks for agricultural and chemical products which require stretch and energy absorption. "Clupak" extensible paper was developed in 1958 from an invention by Sanford Cluett, who also discovered the "Sanforized" process for textiles.

Paper and board for packaging is converted by such steps as printing, cutting, folding and glueing to yield folded cartons or corrugated boxes. Fiber drums are formed in wrapped, glued multiple layers, and fastened with lids. Other formed products are pressed—for example, egg cartons and special casings (see *Molded Fiber Products*). Saturating kraft sheets are impregnated with resin before laminating.

4. Properties of Paper and Paperboard

Paper and board properties derive from the raw materials properties and papermaking processes. This section presents: (a) theories on paper mechanical properties, (b) mechanical properties of paper relative to other composite materials, (c) mechanical requirements for paper cartons and boxes, (d) on-line measurement of paper and board properties, and (e) optical, surface, aesthetic and specialty properties.

4.1 Theoretical Models for Paper Mechanical Properties

Paper is generally considered to be a viscoelastic, anisotropic nonlinear composite material. Theoretical models describing mechanical properties of paper include those based on: (a) molecular bonds between fibers, (b) fiber networks, and (c) an orthotropic continuum.

Nissan applied hydrogen-bonding theory to explain changes in Young's modulus for paper in the presence of different solvents, and the effects on paper properties of chemical substitution in cellulose fibers. As noted by Nissan (Mark and Murakami 1983), there is at present no model covering the intermediate scale of microfibril interactions which are believed to be important in bond development and disruption during paper drying and calendering.

A number of models for paper are based on the fiber network scale interactions; examples include models by Cox (1952), Page and Seth (1980) and Van den Akker (1970).

Fiber network models consider polymeric and physical properties of fibers, bonding areas, geometrical network structure, and fracture mechanics. Network models omit consideration of network non-homogeneity, nonplanar fiber orientation, and deformation of the web in the thickness direction. Despite the limitations, these models can predict certain mechanical behavior, for example edgewise compressive strength (Perkins and McEvoy 1981).

Another level of mechanical model considers paper an orthotropic plate or continuum. Because Baum used ultrasonic waves with an appropriate wavelength to paper thickness relationship, he could describe paper in a plate form (Baum et al. 1981). The ultrasonically determined mechanical property values are higher than the corresponding values determined destructively at lower strain rates, but the two approaches provide similar mechanical descriptions. Ultrasonically measured fundamental properties have been correlated with traditional paper tests such as mullen (burst) and edgewise compression strength (Fleischman et al. 1982).

4.2 Fundamental Mechanical Measurements

The theory of composites (Cox 1952) predicts properties such as Young's modulus for paper from fiber characteristics:

$$E_p = 1/3\ E_f\{[1 - W/(L\,RBA)]\,[E_f/(2G_f)]^{1/2}\}$$

where E_p is the sheet Young's modulus, E_f is the fiber modulus in the axial direction, W is the fiber width, L

is the fiber length, RBA is the relative bonded area of sheet, and G_f is the fiber shear modulus of deformation. For well-bonded paper sheets (highly refined and pressed well-delignified pulp) of straight fibers, the in-plane elastic modulus of paper is one-third of the elastic modulus of the component fibers (Page and Seth 1980).

Paper properties related to fracture have been quantified (Dodson 1976). Levels of fracture energy per unit new area are of the order:

(a) zero span tensile: 5×10^7 J cm^{-2} (from area under load-extension curve)

(b) tearing: 0.5 J cm^{-2}

(c) splitting: 5×10^{-3} J cm^{-2}

Paper zero-span tensile strength is directly dependent upon individual fiber strength. A change in zero-span tensile strength with paper orientation indicates a change in fiber orientation. Fiber-to-fiber bonds and frictional forces contribute to tear. The relatively low value for splitting reflects only fiber-to-fiber bonding since fiber orientation in the thickness direction is low and frictional forces are minimal. The energy consumed in propagating a fracture across a paper strip is believed to be of the order of 10^5 J cm^{-1}, with a typical Griffith creeping crack velocity (in a sample of tracing paper) estimated at 6×10^{-4} cm s^{-1} (Dodson 1976).

Edgewise compressive strength of paperboard, which is 30–40% of tensile strength, is important in corrugated boxes and folding cartons. For process conditions producing low fiber bonding (high-yield fiber, low refining and wet pressing), edgewise compressive strength is proportional to bonding, but as the degree of bonding is increased, fiber compressive strength becomes controlling (Seth et al. 1979). These workers found the intrinsic compressive strength of laboratory-made paper to be 3600–7200 N m^{-1}, depending directly on the level of wet pressing over the range 10^2–10^4 kPa.

4.3 Comparison of Paper with Other Composites

Table 1 reveals that paper fibers and papers as composites have quite respectable mechanical properties in comparison to other natural materials (cotton, wood), and when used in an acceptable environment are in some ways superior to man-made "structural materials" such as plastics and reinforced plastics. Like other natural fibers, paper fibers are moderately low in density. They excel in stiffness—on a weight basis (specific stiffness) they are comparable to nylon, or in some cases glass fiber. Paper fiber strength declines when wet, as does the strength of nylon and glass, but its specific dry strength exceeds that of polyethylene, nylon and steel fibers. In the composite called paper, the low density is retained and the specific stiffness is comparable to glass-reinforced

Table 1
Comparative properties of fibers and composites[a] (values are upper bounds)

Material	Density (kg m^{-3})	Initial stiffness (Gpa)	Specific stiffness (km)	Strength Dry (Mpa)	Strength Wet (Mpa)	Specific strength (km)	Breaking strain (%)
Fibers							
Cotton	1360	11	825	860	1070	64	10
Wool	1175	4	347	190	180	17	45
Paper							
unbleached pine kraft	1500	20	1360	1570	1030	107	35
spruce kraft, 30° fibril	1450	30	2110	1000	660	70	
spruce kraft, 5° fibril	1450	70	4926	1800	1200	127	
High-density polyethylene	860	10	1186	640	640	76	20
Nylon	1030	11	1090	420	360	42	19
Glass	2260	70	3160	3000	600	135	4.5
Carbon (graphite)	1720	500	29600	3200	3200	190	1.5
Steel	7150	210	3000	4000	4000	57	10
Composites							
Oak wood (axial/radial)	550	10/2	1850/370	90		17	
Glass-reinforced polyethylene	1630	30	1880	600	300	38	
Paper							
newsprint (MD/CD)[b]	600			22/11	3.2/1.6	3.7/1.8	1
kraft linerboard (MD/CD)[b]	721	8.7/4.1	1230/580	51/26		7.2/3.7	1.1/1.4

[a] Data provided by Dr Gary Baum, Institute of Paper Chemistry [b] MD, machine direction; CD, cross direction

polyethylene. Like some other composite materials, the specific strength of paper is lower than for individual fibers. On a specific (dry) strength basis, paper has approximately one-tenth to one-third the strength of wood, and one-twentieth to one-sixth that of glass-reinforced polyethylene.

Cellulose fibers and paper, when utilized in a manner consistent with their sensitivity to moisture, are therefore remarkably good structural materials for use in low-density applications where stiffness rather than strength is important.

4.4 Overall Carton and Box Mechanical Properties

Numerous "quality control" tests have been devised to predict paper and paperboard performance when it is subsequently formed into folding cartons, or when used in corrugated boxes. In the folding carton "block compression" test, paperboard strips are fitted vertically into slots in top and bottom metal plates and compressed to determine crush resistance. A direct correlation between the cross-machine direction block compression test on paperboard and the top-to-bottom carton compression performance has been shown (Cope 1961).

Koning (1975, 1978) developed a theoretical model and subsequently confirmed its validity for predicting the compressive properties of linerboard as related to the compression strength of corrugated containers. The load at which the container would fail was found to be related to in-plane moduli of elasticity and dimensions of liner and medium.

4.5 On-Line Determination of Properties

An exciting developing area in paper technology is the application of sensing and computing systems to determine on-line the physical uniformity of paper (using lasers), moisture content (radionuclide beams) and elastic moduli (ultrasonic velocities). The potential for rapid feedback or even feedforward control is very appealing.

Ultrasonically determined elastic constants of paper, such as extensional stiffness (elastic modulus × caliper) and shear stiffness (shear modulus × caliper), and out-of-plane properties have been measured in the laboratory, with estimates of tensile strength (Baum and Habeger 1980, Baum et al. 1981).

The paper industry is moving to increase both its process effectiveness and product uniformity by the application of statistical process control (SPC) which relies on conclusions from computed model parameters rather than from original raw data from sensors and instrumentation (Mendel 1987).

4.6 Surface Properties

Because paper and paperboard are frequently used for communicating the written word and/or colorful images, the visual esthetic properties of paper—especially brightness, color and gloss—are often as important to end-users as the mechanical characteristics.

Paper used in household and personal products must have acceptable levels of absorbency, softness, bulk and stiffness, surface roughness and "handle" or feel. For applications where paper or paperboard is a barrier, for example, to light, sound or thermal energy, or where paper's electrical or dielectrical properties are employed, specialty papers have been developed.

Among the most frequently considered non-mechanical end-use properties are optical properties. Corte (1976) notes that, strictly speaking, optical properties are related to "appearance," which is beyond the scope of physical measurement because it involves physiological and psychological factors. Nevertheless, measurement of optical properties is done using sophisticated instrumentation and optical reference materials to indicate opacity, brightness and whiteness, color, gloss and even formation, which is related to local optical contrasts in a sheet.

5. Applications of Paper and Paperboard

Paper has expanded into many markets beyond its initial use in writing and printing, and now competes with cloth, plastic, and insulating products. Paper fiber, combined with man-made fibers and resins, can be molded, pressed, or dry formed (see *Molded Fiber Products*). Familiar applications are laminated counter tops, asphaltic roofing shingles and felts, flooring and hardboard, wallpapers, battery separators, and interior autopanelling. In these applications, paper is an inexpensive, structurally important reinforcing fiber. Whenever paper fibers are associated with wax, asphalt or other polymers, the product will be less amenable to recycling, and some mechanical properties (e.g., stiffness and tear) and aesthetics will be altered.

Important large structural paper products are: (a) corrugated paper and (b) boxes, fiberboard, spirally-wound drums, and (c) molded fiber products such as luggage cases, egg cartons, and combustible cartridge cases. These products may be printed or coated. They are designed for load-bearing under highly adverse conditions related to the material enclosed and the external environment. The ability to make these products inexpensively for shipping or transmitting such diverse materials as powdered chemicals, eggs and refrigerators is a challenge for papermakers and converters.

Opportunities for paper and paperboard products have been changing. Paper is now often used with other materials to achieve an improved combination of properties and economy.

Plastic in the form of bags, cup stock, plates and containers is challenging paper products in markets where extended resistance to moisture or chemicals is required. In the entertainment, communication and business fields, electronic and video systems are being developed which can bypass "hard-copy" paper products. It has been projected that by 1990 about 7% of homes in the USA will rely on videotext in place of printed catalogs (Goodstein 1982).

In other markets, paper fibers are finding increasing roles. In convenience packs for microwave cooking, ovenable paperboard is challenging a market formerly dominated by glass, foil and plastic containers. Forecasts predict a 9% annual growth in food containers, and in paper containers for foods in particular (Technology Forecasts 1981). In multiwall bags for corrosive, granular chemicals, and in electrical and decorative paper-based laminates, paper plies contribute significantly to strength and/or bulk at low cost.

A trend is combining paper with man-made plastic and/or metal films, particularly for food applications. Paperboard provides a significant fraction of the stiffness and bulk for sterilized laminate used in aseptic packaging which can keep unrefrigerated juice, milk, and specialty food products fresh for up to six months (Allan 1982, Mies 1982).

Metallized paper is making inroads in the label market as well as for personal-care products, gift wrap, cigarette inner liners, food cans and dairy wrappers. The product requires an exceptionally high gloss paper, a lacquer which assures smoothness and adhesion for the metallic finish, and metal foil, generally vacuum deposited aluminum. Conventional laminate has a 2:1 paper-to-foil weight ratio, but metallized paper has over 200 parts paper to 1 part foil, considerably reducing raw material cost for the same thickness of packaging material (Carter and Beardow 1983).

6. Production and Consumption of Paper Products

In 1986, world paper and paperboard production exceeded 200 Mt for the first time, giving a per capita value of 42 kg. Certain papermaking countries such as the Nordic group had an annual increase in production in 1986 as low as 1%, while developing areas such as Latin America and Asia reached 7–8% growth levels (Sutton et al. 1987).

Based on the reported 57.8 Mt of paper and paperboard produced in the USA in 1980, it can be computed that for a population of just over 220 million persons (US Department of Commerce 1980), the USA in 1980 used about 260 kg of paper and paperboard products for every man, woman and child in the country. Similar high consumption levels exist in other industrialized countries, and an increase in paper use levels is expected in developing market countries.

The world distribution of paper and paperboard production capacity is as follows: 76% in regions of developed market economies, 14% in centrally planned economies and 10% in developing market economies. The 1987 United Nations Food and Agriculture Organization (FAO) predictions to 1995 indicate between 2.6 and 2.9% growth in both demand and production of paper and board worldwide (FAO 1987). World trade in paper and board will continue to represent 1.5% of the world's total merchandise export (Brusslan 1987).

Table 2 classifies the world pulp and paperboard capacities in five-year increments from 1976 through to the projections for 1991. The disparity between the total wood pulp produced and total paper and paperboard manufactured reflects such items as nonfibrous components used in paper and boards and the use of recycled fibers.

The predominant pulping method will remain chemical pulping. Chemical pulp and pulps made by various mechanical means will both contribute up to a third more tons per year by 1991 than they did in 1976 (FAO 1987).

Table 2.
Total world pulp and paper/paperboard capacity 1976–1991

	Capacity (Mt)			
	1976	1981	1986	1991
Total wood pulp, paper grades	130	142	155	166
mechanical, thermomechanical	31	34	38	43
semichemical (includes chemiground wood)				
groundwood	11	16	10	10
chemical	88	98	107	113
Other fiber pulp (nonwood)	10	12	14	17
Dissolving pulp	6	6	6	6
Total paper and paperboard	180	203	226	248
newsprint	25	29	32	34
printing and writing	40	50	60	70
other paper and paperboards	115	124	134	144

Source: FAO (1987).

Table 3
Total paper and paperboard consumption, 1980–2000

	Consumption (Mt)			
Region	1980	1990	2000	%/Year
World	176.0	265.3	396.9	4.2
North America	67.9	91.6	119.3	2.9
Western Europe	42.9	63.1	89.6	3.8
Japan	18.6	31.8	49.6	5.0
Other	46.5	78.8	137.4	5.6

Source: Hagemeyer and Holt (1982)

Table 4
Regional paper and paperboard consumption, 1980–2000

	Consumption (% of total)		
Region	1980	1990	2000
North America	38.6	34.5	30.0
Western Europe	24.4	23.8	22.6
Japan	10.6	12.0	12.5
Other	26.4	29.7	34.9

Source: Hagemeyer and Holt (1982)

Newsprint accounts for about 15% of the total paper and paperboard production, while printing and writing papers will grow to take up about 30% by 1991. The largest production category is composed mainly of wrapping and packaging papers and boards.

The Technical Association of the Pulp and Paper Industry (TAPPI) has predicted paper and paper board consumption for the years 1980, 1990 and 2000 (Tables 3 and 4). Predictions through the year 2000 estimate a shift from North America and Western Europe to Japan and the developing market regions. This is expected partially through market saturation in North America.

7. Energy, Recycling and Environmental Considerations in Papermaking

Three environmentally attractive characteristics of the paper industry are: (a) it uses a renewable polymer resource, i.e., wood fiber, (b) about half its energy use comes from "waste" sources such as the lignin removed during cooking of chips and bark removed from logs, and (c) its end-products are generally recyclable.

In 1981, the US pulp and paper industry consumed about 2.3×10^{18} J of which roughly 50% was generated in the paper mill. It came from cooking liquor (37.6% of total), hogged fuel (6.7%) and bark (5%). The industry has been converting from using purchased to self-generated energy sources. The self-generated level increased from 40.7 to 50.2% during the period 1972 to 1981 (Grant and Slinn 1982).

The energy per tonne of product varies. A range is from about $3.9–4.9 \times 10^{10}$ J t^{-1} for printing and writing paper (which require bleaching, sizing and, in some cases, coating) to 3.26×10^4 J t^{-1} for linerboard (which is neither bleached nor typically sized and coated) (Hersh 1981). In the paper mill, the highest single energy consumer is the drying process. Steam-heated cylinders use $6.7 \times 10^9–1.3 \times 10^{10}$ J t^{-1}. This can correspond to one-fourth to one-third of the total energy requirement for an integrated mill (Chiogioji 1979).

Economics of the fiber market require that post-consumer materials and "waste" fiber be recycled with a minimum of transportation. Recycling mills face an environmental problem not encountered in mills using virgin fiber, namely, disposition of contaminant, non-fibrous materials, including heavy metals from inks (Wrist 1982).

The greatest commercial use of recycled paper is in corrugated paper for boxes, containers, packing and low-density structural core. In general, recycled products are lower in some properties such as tensile strength.

Environmental interests in papermaking encompass all unit processes from forestry to control of stack gas and mill effluent quality. The balance between environmental and economic issues must be thoroughly evaluated. Wood harvesters optimize fiber quality, harvesting procedures and product mix (the proportion of lumber, pulp mill chips and fuel residuals) to achieve maximum yield of biomass per hectare. Alternatives for higher yield and/or quality are being evaluated also. High-yield pulping, such as thermomechanical, chemimechanical and neutral sulfite, offers a way to obtain more fiber per tonne of wood where end-use requirements allow.

The recovery cycle of the paper mill recycles the major portion of chemicals which pulp the wood while burning lignin for fuel. Efforts are underway to further improve energy recovery through better understanding of black liquor and recovery boiler operation (see Fig. 3). Reductions in water and energy use are being pursued through a more closed process. Developments include: (a) reduction of water requirements through medium-to-high consistency pumping, cleaning and refining: (b) innovative efficient dewatering of paper webs; and (c) advances in drying.

Bibliography

Allan D R 1982 U.S. bleached board supply/demand. *Pulp Pap.* 56(10): 164–68

Altieri A M, Wendell J W Jr 1967 *TAPPI Monograph No. 31: Deinking of Waste Paper.* Technical Association of the Paper and Pulp Industry, Atlanta, Georgia

Anon. 1981 World capacity: FAO sees modest growth in capacity. *Pulp Pap.* 55(11): 23

Baum G A, Habeger C C, 1980 On-line measurement of paper mechanical properties. *Tappi J.* 63(7): 63–66

Baum G A, Brennan D C, Habeger C C, 1981 Orthotropic elastic constants of paper. *Tappi J.* 64(8): 97–101

Bolam F (ed.) 1976 *The Fundamental Properties of Paper Related to its Uses*, Vols. 1 and 2. Technical Division of the British Paper and Board Industries Federation, London

Breck D H, Styan G E 1985 Explaining the increased use of mechanical pulps in high-value papers. *Tappi J.* 68(7): 40–44

Bristow J A, Kolseth P (eds.) 1986 *Paper Structure and Properties*, International Fiber Science and Technology Series, Volume 8. Dekker, New York

Brusslan C 1987 World paper supply will meet 1995's demand. *Am. Papermaker* 50(7): 24–26

Carmichael D L 1986 Uses for high-brightness CMP expand thanks to new bleaching methods. *Pulp Pap.* 60(7): 66–70

Carter J H, Beardow T 1983 A lesson in making metallized paper. *Pap., Film Foil Converter* 57(3): 45–47

Chiogioji M H 1979 *Industrial Energy Conservation*. Dekker, New York

Christie D 1987 Chip screening for pulping uniformity. *Tappi J.* 70(4): 113–117

Clark T F 1969 Annual crop fibers and the bamboos. In: Macdonald R G, Franklin J N (eds.) 1969 *Pulp and Paper Manufacture*, 2nd edn., Vol. 2. McGraw-Hill, New York, pp. 1–74

Coan B 1984 Paper machine clothing: strategies for maximum performance, profit. *Pulp Pap.* 58(4): 55–68

Cope P 1961 Measuring and specifying bulge and crush resistance in cartons and carton board. *Tappi J.* 44(9): 633–36

Corte H 1976 Perception of the optical properties of paper. In: Bolam F (ed.) 1976 *The Fundamental Properties of Paper Related to its Uses*, Vol. 2. Technical Division of the British Paper and Board Industries Federation, London, pp. 626–61

Cox H L 1952 Elasticity and strength of paper and other fibrous materials. *Br. J. Appl. Phys.* 3(3): 72–9

Dodson C T J 1976 A survey of paper mechanics in fundamental terms. In: Bolam F (ed.) 1976 *The Fundamental Properties of Paper Related to its Uses*, Vol. 1. Technical Division of the British Paper and Board Industry Federation, London, pp. 202–26

Ducey M J 1986 Efforts in chemical pulp bleaching technology emphasize cutting costs. *Pulp Pap.* 60(7): 47–50

Fleischman E H, Baum G A, Habeger C C 1982 A study of the elastic and dielectric anisotropy of paper. *Tappi J.* 65(10): 115–8

Food and Agriculture Organization of the United Nations 1987 *Pulp and Paper Capacities, Survey* 1986–1991. FAO, Rome

Franklin W E 1986 Trends in recovery and utilization of waste paper in recycling mills, and other uses of waste paper, 1970–2000. *Tappi J.* 69(2): 28–31

Gallep G 1987 Eucalyptus trade signals increased competition for paper companies. *PIMA* 69(4): 4–5

Goodstein D H 1982 Electronics in the catalog arena: what will be their impact. *Am. Printer Lithogr.* 190(3): 48–49

Grant T J, Slinn R J 1982 *Patterns of Fuel and Energy Consumption in the U.S. Pulp and Paper Industry 1972–1981*. American Paper Institute, New York

Gross R M 1982 Prospects for nonpaper pulp. *Pulp Pap.* 57(9): 45

Hagemeyer R W, Holt S G 1982 A prediction of world printing and writing paper consumption. *Tappi J.* 65(11): 37–40

Helm D J 1987 New tandem-ENP ups speed, output of O-I corrugating medium machine. *Pulp Pap.* 61(6): 128

Hersh H N 1981 *Energy and material flows in the production of pulp and paper*, ANL/CNSV-16. Argonne National Laboratory, Argonne, Illinois

Hunter D 1943 *Papermaking: The History and Technique of an Ancient Craft*. Knopf, New York

Kalish J (ed.) 1982 *Publication Papers: An Appraisal of the Future of Paper, Printing and Publishing at the Start of the Electronic Era*. Miller-Freeman, Brussels

Kline J E 1987 *Paper and Paperboard: Manufacturing and Converting Fundamentals* 3rd edn. Miller Freeman, San Francisco

Koning J W Jr 1975 Compressive properties of linerboard as related to corrugated fiberboard containers: a theoretical model. *Tappi J.* 58(12): 105–08

Koning J W Jr 1978 Compressive properties of linerboard as related to corrugated fiberboard containers: theoretical model verification. *Tappi J.* 61(8): 69–71

MacDonald R G, Franklin J N (eds.) 1970 *Pulp and Paper Manufacture*, 2nd edn., Vol. 3, McGraw-Hill, New York

Mark R E, Murakami K 1983 *Handbook of Physical and Mechanical Testing of Paper and Paperboard*. Dekker, New York

Mark R E, Murakami K 1984 *Handbook of Physical and Mechanical Testing of Paper and Paperboard*, Vol. II. Dekker, New York

McGovern J N 1982 Fibers used in early writing papers. *Tappi J.* 65(12): 57–58

McGrath R 1987 US companies looking for ways to jump on the eucalyptus bandwagon. *Pulp Pap.* 61(7): 92–3

Mendel J M 1987 Statistical process control—basis principles and techniques. *Tappi J.* 70(3): 83–87

Mies W 1982 Aseptic cartons off to a fast start in U.S. *Pulp Pap.* 56(10): 168–69

Nissan A H 1977 *Lectures in Fiber Science in Paper*. Technical Association of the Paper and Pulp Industry, Atlanta, Georgia

O'Brian H 1987 Three-layers give many possibilities. *PPI* 29(4): 73–74

Page D H, Seth R S 1980 Elastic modulus of paper II—the importance of fiber modulus, bonding and fiber length. *Tappi J.* 63(6): 113–16

Paper Trade Journal Statt 1981 Europe's pulp and paper industry travels uncharted economic seas. *Pap. Trade J.* 165(24): 24–25

Perkins R W, McEvoy R P Jr 1981 The mechanics of the edgewise compressive strength of paper. *Tappi J.* 64(2): 99–102

Pulp and Paper 1984 Review of Fifteen Years of Machine Installations—New Paperboard Machines at U.S. and Canadian Mills, 1970–84. Miller-Freeman, San Francisco, California

Seth R S, Soszynski R M, Page D H 1979 Intrinsic edgewise compressive strength of paper—effect of some papermaking variables. *Tappi J.* 62(3): 45–46

Setterholm V C 1979 An overview of press drying. *Tappi J.* 62(3): 45–46

Smith K E 1982 Silvichemical review: wood-based chemical production in U.S., Canada, *Pulp Pap.* 56(11): 73–77

Sutton P, Pearson J, O'Brian H 1987 Paper production and consumption records broken in '86; '87 looks good. *Pulp. Pap.* 61(8): 47–55

Tay C H, Fairchild R S, Imada S E 1985 A neutral-sulfite/SAQ chemimechanical pulp for newsprint. *Tappi J.* 68(8): 98–103
Technology Forecasts and Technology Surveys December 1981 *Packaged Food and Consumer Spending Projections.* PWG Publications, Beverly Hills, California, pp. 5–9
Thorp B A 1982 Development of hydrofoils, hydraulic headboxes, and twin-wire formers. *Pulp Pap.* 58(9): 141–43
US Department of Commerce, Bureau of the Census 1980 *Statistical Abstract of the United States.* US Government Printing Office, Washington, DC
US Department of Commerce 1981 *U.S. Industrial Outlook for 2000—Industries with Projections for 1985.* US Government Printing Office, Washington, DC
Van den Akker J A 1970 Structure and tensile characteristics of paper. *Tappi J.* 53(3): 388–400
Wahlström B 1981 Developments in paper technology in a global perspective. *Sven. Papperstian.* 84(18): 32–39
Wardrop A B, Bland D E 1959 The process of lignification in woody plants. *Biochemistry of Wood, Proc. 4th Int. Congress of Biochemistry.* Pergamon, New York, pp. 76–81
Wintgen M 1987 Board coating with natural ground calcium carbonate in Europe: technology today. *Tappi J.* 70(5): 79–83
Wrist P E 1982 The direction of [paper] production technology development in the 1980s. *Tappi J.* 65(11): 41–45
Wuerl P 1986 Alkaline papermaking dominates papermaking and coating chemicals scene. *Pap. Trade J.* 170(7): 41

J. W. Glomb
[Westvaco, New York, USA]

D. D. Mulligan
[Westvaco, Covington, Virginia, USA]

Particulate Composites

Particulate composites are defined in terms of the size, shape and concentration of the constituents and their roles in load bearing. A large number of composites can be considered as belonging to this class if due account is taken of scale effects. In this article, two categories are considered: metal-bonded composites and resin-bonded composites. They are classified by composition, method of preparation and use.

Mechanical properties, thermal and electrical conductivity, thermal expansion, and dielectric constant, are discussed here and, wherever possible, the theory for the property, which is applicable to the entire class of composites, is given.

1. Definitions

Particulate composites may be defined as follows.

(a) They have components which are not noticeably one or two dimensional (as fibers or lamellae would be) but have similar dimensions in all directions (and in some cases are spheres).
(b) All phases carry a proportion of the load.

Particulate composites therefore form an intermediate class in between dispersion-hardened materials and fiber-reinforced ones, for in the former the particles are very small, their proportion is a few percent, and they act by blocking dislocation movement and therefore yield in the matrix. In fiber reinforcement, by contrast, the proportion of fibers is high and they are the main load bearers. Particulate composites have proportions of the hard phase from a few percent up to 70%.

In complete generality we may include in the class of particulate composites materials such as rock, concrete, ceramics such as porcelain and alumina–silicate (fire-brick), hard metals such as tungsten carbide–cobalt, and filled resins such as reinforced phenolics, epoxies and thermoplastics.

Although we shall not treat the subject in such generality, it is as well to point out that many of the physical properties of particulate composites (elastic moduli, yield and fracture properties, thermal and electrical conductivity, thermal expansion) may be described by theories which are completely general and depend only on the physical properties of the components and their concentrations. The only problem then to be considered in each case is the size of the "representative volume element," that is, the unit of scale above which the material can be considered as a continuum but below which the microstructure must be taken into account. Hill (1963) has defined the representative volume element in precise terms in a paper dealing with theoretical principles which apply to all reinforced solids. Holliday (1966) included the concept of the "representative cell" in a comprehensive discussion of geometrical considerations and phase relationships in composite materials.

2. Metal- and Resin-Bonded Composites

We shall be concerned in this article mainly with metal- and resin-bonded composites rather than with the broader class outlined above. Metal-bonded composites include structural parts made by powder metallurgy, electrical contact materials, metal-cutting and rock-drilling tools, and magnet materials. Several of these have been in use for 50 years or more, but a thorough theoretical understanding of their behavior has only been achieved since the 1970s following the upsurge of work on composites in general and fiber composites in particular. Details of some metal-bonded composites are given in Table 1.

In parallel with the metal-bonded composites is the class of resin-bonded ones of which the earliest examples were the filled phenolics (e.g., bakelite and formica) in which the filler particles served primarily as an adulterant to reduce the cost of the finished molding but were later found to play a significant part in reinforcement and enhancement of other properties.

In resin-bonded composites a great range of different filler types, concentrations and sizes is used depending upon the properties desired and the mode of preparation. The size may vary from submicrometer,

Table 1
Metal-bonded composites

Type	Particle size and concentration	Method of preparation	Uses	Reference
Cemented carbides	0.8–5 μm particles of WC, TiC or TaC in cobalt or other metal of iron group; up to 94% carbide	powder metallurgy and liquid-phase sintering	cutting tools, dies, weights, rock-drilling bits	Jones (1960)
Reinforced electrical conductors	30–400 μm particles of W or Mo in Ag or Cu matrix 30–80% refractory	powder metallurgy and impregnation with conducting matrix	heavy-duty electrical contacts, electrodes for spark erosion machining	Jones (1960)
Electrically conducting composites	50–150 μm particles of graphite as soft phase dispersed in copper, bronze or silver; 5–70% graphite	powder metallurgy	electrical contacts (brushes on motors and other moving parts carrying electric current)	Jones (1960)
Magnetic composites (e.g., Alnico 5)	0.1 μm particles of FeCo precipitate in Al–Ni matrix phase; 75% Fe–Co phase	casting or powder metallurgy	permanent magnets of high coercivity	de Vos (1969)
Bearing alloys				
(a) White metal	10–100 μm particles of hard intermetallic (Cu_6Sn_5 or SbSn) in soft Sn- or Pb-rich matrix; up to 20% hard phase	casting	plain bearings; particularly superior as regards accommodation of foreign bodies	Neale (1973)
(b) Copper base	10–100 μm particles of soft (Sn/Pb) phase in Cu-base matrix; up to 30% soft phase	casting or sintering; plating	plain bearings; inferior to (a) as regards accommodation, but better high-temperature fatigue properties	Neale (1973)
(c) Aluminum base	10–100 μm "islands" of Sn in Al; 20% of soft (Sn) phase	casting and cold work	plain bearings; similar to (b) but better corrosion resistance	Neale (1973)
(d) PTFE-filled porous metal	Up to 100 μm particles of soft PTFE impregnated into porous metal matrix, typically bronze; up to 30% PTFE	sintering	plain bearings	Neale (1973)
(e) Graphite-filled metal	Similar to PTFE impregnation, 8–15% graphite	sintering	plain bearings	Neale (1973)

in the case of silica flour, to several millimeters (e.g., glass spheres or sand), and the concentration from a few percent up to 70%. Typical fillers are silica flour, wood flour, mica (for insulators), china clay, carbon, cellulose, glass spheres, chalk, kaolin, talc, ground slate, cork and metal powders.

The resins used as the matrix fall into three groups.

(a) Phenol, urea or melamine formaldehydes—these are molded by pressure bonding of powders and the curing involves a condensation reaction. Typical uses are for electrical insulators, durable kitchenware and general-purpose moldings.

(b) Epoxy or polyester resins—these are cured with a hardener in a two-part (usually liquid) mixture in an addition reaction which does not involve high pressures in the mold. Typical uses are for electrical insulation at moderate temperatures, molded goods, the building and construction industry and tooling.

(c) Thermoplastic (heat deformable) resins—these include poly(methyl methacrylate), poly(acrylic acid), polypropylene, polyimide, nylon and PTFE. Such resins with filler included may be extruded into rods, tubes or films as well as being injection molded into intricate shapes. Block copolymers, which are two-phase polymer composites, may be included within this class. For a comprehensive account of resin-bonded composites, see Lubin (1982).

3. *Discussion of Technology*

The technology used in each of the cases listed above depends upon the materials employed, the ease of mixing to ensure a uniform dispersion, and the end use. In some cases solid particles of one component may be successfully dispersed in a liquid phase of the other, whereas in other cases both components are or need to be in powder form before they can be mixed and combined to form a composite. In some cases (e.g., with magnet alloys) both sintering and casting can be used and the microstructure of the resulting material differs accordingly, with appropriate end uses for each material. In the case of many polymer-based composites, the polymer is a viscous liquid and powder methods are not possible. With phenolics, however, powders are again available and mixing of resin and filler powders is carried out before pressure molding and curing. The technology may allow control of a critical grain size; for example, in magnet materials the final grain size needs to be of the order of the critical domain size. This is best achieved by powder techniques.

Impregnation of a powder compact by molten metal is used for tungsten- or molybdenum-reinforced copper, whereas solid- or liquid-phase sintering (a situation where the molten phase is formed by reaction) is used for tungsten carbide–cobalt composites.

In deciding upon the appropriate technology many factors other than the physical properties of the composite need to be taken into account (Jones 1960, Lenel 1980, Davidge 1979, Spriggs 1974, Broutman and Krock 1967, Lubin 1982).

4. *Mechanical Properties*

4.1 Elastic Properties

The mechanical properties of particulate composites in the elastic region are well described by the Hashin–Shtrikman bounds as follows. Let G_c, G_1, G_2 and K_c, K_1, K_2 denote the shear and bulk moduli, respectively, where c, 1 and 2 indicate the composite, phase 1 and phase 2, then the following inequalities hold:

$$\frac{c_1}{1+(K_1-K_2)c_2/(K_2+K_1^*)} \leqslant \frac{K_c-K_2}{K_1-K_2} \leqslant \frac{c_1}{1+(K_1-K_2)c_2/(K_2+K_g^*)}$$

and

$$\frac{c_1}{1+(G_1-G_2)c_2/(G_2+G_1^*)} \leqslant \frac{G_c-G_2}{G_1-G_2} \leqslant \frac{c_1}{1+(G_1-G_2)c_2/(G_2+G_g^*)}$$

In these expressions, c_1 is the concentration by volume of phase 1, c_2 that of phase 2, and $c_1+c_2=1$. In addition,

$$K_g^* = \tfrac{1}{3}G_g, \quad K_l^* = \tfrac{1}{3}G_l$$

$$G_l^* = \frac{3}{2}\left(\frac{1}{G_l}+\frac{10}{9K_l+8G_l}\right)^{-1}$$

$$G_g^* = \frac{3}{2}\left(\frac{1}{G_g}+\frac{10}{9K_g+8G_g}\right)^{-1}$$

and G_g, K_g are the greater of G_i, K_i, and G_l, K_l are the lesser of G_i, K_i in the mixture. These bounds were proved by Walpole (1966) and are a more general form of those found by Hashin and Shtrikman (1963).

Figure 1, from the latter reference, illustrates the bounds in the case of a tungsten carbide–cobalt composite, while Fig. 2, from Crowson and Arridge (1977), shows their relevance to experimental results on glass-bond-reinforced epoxy resin.

The bounds are completely general and apply whatever the phase geometry provided each phase is isotropic. Modified bounds for anisotropic phases are to be found in Walpole (1969), and Kantor and Bergman (1984).

For many purposes, however, the simpler "rule-of-mixture" formulae may be used though they lead to bounds which lie outside the Hashin–Shtrikman bounds. The rule-of-mixture bounds assuming uniform strain throughout the composite give M_c

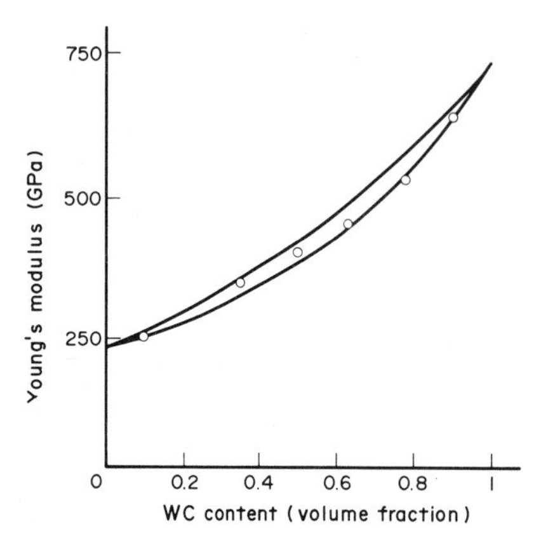

Figure 1
Hashin–Shtrikman bounds for Young's modulus of a WC–Co alloy (after Hashin and Shtrikman 1963)

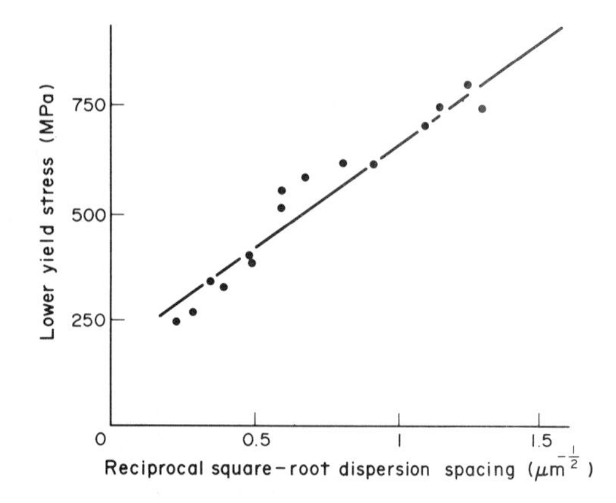

Figure 3
Lower yield points of several hypoeutectoid, eutectoid and hypereutectoid steels as a function of reciprocal square root of dispersion spacing (after Broutman and Krock 1967)

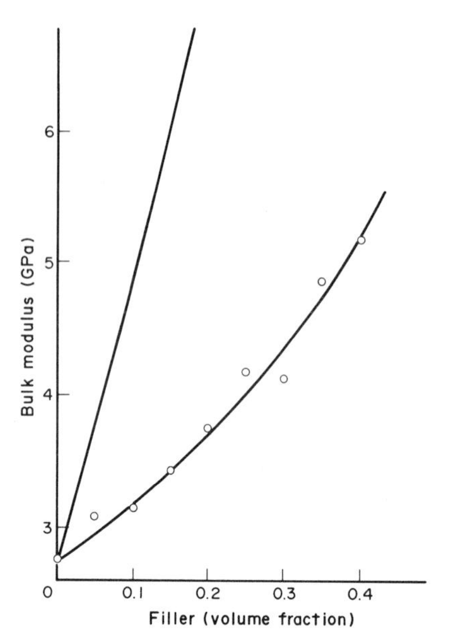

Figure 2
Hashin–Shtrikman bounds for bulk modulus of glass-filled epoxy resin (after Crowson and Arridge 1977)

$= c_1 M_1 + c_2 M_2$, where M is any modulus (Young's, shear or bulk), whereas the bound assuming uniform stress involves the compliances M^{-1}; thus $M_c^{-1} = c_1 M_1^{-1} + c_2 M_2^{-1}$. For comprehensive reviews see Hashin (1983) and Hale (1976).

4.2 Yield and Hardness

When the particles are not deformable by the matrix, theory suggests that the yield strength should be proportional to (mean interparticle spacing)$^{1/2}$, and this is observed for steels (considered as particle-reinforced composites) and for cemented carbides (Fig. 3) at least until the interparticle spacing falls below 0.5 μm (Broutman and Krock 1967). By contrast, when the particles are deformable by the matrix the yield stress is independent of interparticle spacing, and this is observed to be the case in W–Ni–Fe particulate composites. The deformability or otherwise of the dispersed particles thus determines the nature of the composite. If they are deformable the entire flow characteristics of the composite are determined by those of the dispersed phase (i.e., flow stress, work-hardening rate and ultimate elongation are independent of the concentration of the reinforcement). By contrast, where the dispersed phase is nondeformable, the composite is brittle and of low impact strength although the hardness may be very high, leading to uses such as cutting tools. Yield strength, tensile strength and hardness are then all directly proportional to the concentration of the hard phase.

4.3 Fracture

It is found that the fracture toughness K_{Ic} depends upon the concentration of the reinforcing elements in a particulate composite in a manner consistent with a linear variation of the fracture energy $\mathcal{G}$ with volume percent reinforcement. But $K_{Ic} = (E\mathcal{G})^{1/2}$, where E is Young's modulus, which increases linearly with volume percent reinforcement. This would then explain the linear dependence of K_{Ic} upon concentration as is found for example for WC–Co (Chermant and Osterstock 1976) and for TiC-strengthened alumina (Wahi and Ilschner 1980).

5. *Thermal Expansion*

The rule-of-mixtures law, $\alpha_c = c_1\alpha_1 + c_2\alpha_2$, for the expansion coefficient α is close to Kerner's formula

$\alpha = c_1\alpha_1 + c_2\alpha_2 - c_1c_2(\alpha_1 - \alpha_2)$, and the experiments of Crowson and Arridge (1977) on glass-reinforced epoxy resin confirm the rule-of-mixtures law as a reasonable approximation at least for this system. Fahmy and Ragai (1970) and Hartwig et al. (1976), using a composite sphere model, give the formula

$$\alpha_c = \alpha_m - \frac{3(\alpha_m - \alpha_i)(1 - \nu_m)c_i}{2E_m/E_i(1 - 2\nu_i)(1 - c_i) + 2c_i(1 - 2\nu_m) + 1 + \nu_m}$$

where E is Young's modulus and ν Poisson's ratio, and the subscripts i and m refer, respectively, to inclusion and matrix. c_i is the volume concentration of inclusion. The relation has been tested to concentrations of about 40% with good agreement between theory and experiment (see also Hashin 1983). Ishibashi et al. (1979) found that combinations of different fillers and different resin-curing agents enabled the thermal contraction of composites to be matched to that of any metal. This property is also exploited in dental cements composed of poly(acrylic acid) with added alumina.

6. Other Properties

The formulae for dielectric properties of a particulate composite are the same as those for the elastic moduli (Hashin and Shtrikman 1963) if dielectric constants are exchanged for the elastic constants.

For the electrical conductivity σ of binary metallic mixtures, Landauer (1952) found that the relation

$$c_1\frac{\sigma_1 - \sigma_c}{\sigma_1 + 2\sigma_c} + c_2\frac{\sigma_2 - \sigma_c}{\sigma_2 + 2\sigma_c} = 0$$

gave a good fit to experiment.

See also: Dental Composites; Metal-Matrix Composites; Nonmechanical Properties of Composites

Bibliography

Broutman L J, Krock R H 1967 *Modern Composite Materials.* Addison-Wesley, Reading, Massachusetts

Chermant J L, Osterstock F 1976 Fracture toughness and fracture of WC–Co composites. *J. Mater. Sci.* 11: 1939–51

Crowson R J, Arridge R G C 1977 The elastic properties in bulk and shear of a glass bead reinforced epoxy resin composite. *J. Mater. Sci.* 12: 2154–64

Davidge R W 1979 *Mechanical Behaviour of Ceramics.* Cambridge University Press, Cambridge

de Vos K J 1969 Alnico permanent magnet alloys. In: Berkovitz A E, Kneller E (eds.) 1969 *Magnetism and Metallurgy.* Academic Press, New York, pp. 473–512

Fahmy A A, Ragai A N 1970 Thermal expansion behavior of two-phase solids. *J. Appl. Phys.* 41: 5108

Hale D K 1976 The physical properties of composite materials. *J. Mater. Sci.* 11: 2105–41

Hartwig G, Weiss W, Puck A 1976 Thermal expansion of powder filled epoxy resin. *Mater. Sci. Eng.* 22: 261–64

Hashin Z 1983 Analysis of composite materials—a survey. *J. Appl. Mech.* 50: 481–505

Hashin Z, Shtrikman S 1963 A variational approach to the theory of the elastic behaviour of multiphase materials. *J. Mech. Phys. Solids* 11: 127–40

Hill R 1963 Elastic properties of reinforced solids: Some theoretical principles. *J. Mech. Phys. Solids* 11: 357–72

Holliday L 1966 Geometrical considerations and phase relationships. In: Holliday L (ed.) 1966 *Composite Materials.* Elsevier. Amsterdam, pp. 1–27

Ishibashi K, Wake M, Kobayashi M, Katase A 1979 Powder-filled epoxy resin composites of adjustable thermal contraction. In: Clark A F, Reed R P, Hartwig G (eds.) 1979 *Non-Metallic Materials and Composites at Low Temperatures.* Plenum, New York

Jones W D 1960 *Fundamental Principles of Powder Metallurgy.* Arnold, London

Kantor Y, Bergman D J 1984 Improved rigorous bounds on the effective elastic moduli of a composite material. *J. Mech. Phys. Solids* 32: 41–62

Landauer R 1952 The electrical resistance of binary metallic mixtures. *J. Appl. Phys.* 23: 779–84

Lenel F V 1980 *Powder Metallurgy: Principles and Applications.* Metal Powder Industries, Princeton, New Jersey

Lubin G (ed.) 1982 *Handbook of Composites.* Van Nostrand Rheinhold, New York

Neale M J (ed.) 1973 *Tribology Handbook.* Butterworth, London

Spriggs G E 1974 Cemented carbides: Why do they work? *Practical Metallic Composites.* Institution of Metallurgists, London

Wahi R P, Ilschner B 1980 Fracture behaviour of composites based on Al_2O_3–TiC. *J. Mater. Sci.* 15: 875–85

Walpole L J 1966 On bounds for the overall elastic moduli of inhomogeneous systems. *J. Mech. Phys. Solids* 14: 151–62; 289–301

Walpole L J 1969 On the overall elastic moduli of composite materials. *J. Mech. Phys. Solids* 17: 235–51

R. G. C. Arridge
[University of Bristol, Bristol, UK]

Percolation Theory

Percolation theory is applied to quantitative investigations of whether an impervious medium containing holes and passages would allow the passage of a gas or fluid, or whether an electrical insulator containing electrically conducting particles or fibers would allow the passage of an electrical current. If, in the latter example, a composite consisting of an electrically insulating plastic containing a prescribed volume fraction of short thin rods of a conducting metal were considered, percolation theory would provide an estimate of the probability of the existence of a conducting path through the medium as a function of fiber volume fraction, the length of the rods and the range of orientation of the rods with respect to the direction in which the conductivity was measured.

Percolation theory requires the introduction of a network which can be analyzed using statistical techniques. For example, in two dimensions, a net can be introduced which is thought of as a regular assembly of sites joined by bonds so as to generate a square mesh. Other regular geometries are also allowed, such as triangular or hexagonal nets; three-dimensional nets can also be considered. The conductivity of the bonds is regarded as a statistical quantity. For example, let p be the probability that the bond is conducting (having conductivity σ_0) so that $1-p$ is the probability that a bond is an insulator. Percolation theory provides a method of calculating the conductivity in a particular direction of an infinite net as a function of p. Such calculations for finite nets (sometimes with periodic boundary conditions) have been carried out on powerful computers, yielding a statistical variation in estimates of the conductivity which diminishes as the size of the net increases. For low values of p, finite clusters are formed which are separated from each other by nonconducting bonds. Thus in this case the system has zero conductivity. As p is increased, a threshold p_c is reached where there is at least one conducting path through the infinite sample. As p is increased beyond p_c the conductivity increases to its maximum value at $p=1$. Percolation theory suggests that in the neighborhood of the threshold the conductivity is given by the following power-law relation:

$$\sigma \propto \sigma_0(p-p_c)^t$$

where t is a dimensionless index. For two-dimensional square nets it has been shown that $p_c=0.5$ and it is believed that $t \approx 1.35$. A similar relation can be found expressing the conductivity as a function of fiber volume fraction.

An important requirement of percolation theory is that the dimensions of the two-dimensional systems are infinite in two orthogonal directions. Three-dimensional systems must be infinite in three orthogonal directions. This rules out systems which have the shape of tapes or wires. It is worthwhile to mention systems which are infinite in only one direction. Consider an infinitely long wire of uniform cross-section. Divide the system into an infinite number of elements each having the same finite length. A net is then allocated to every element and a conductivity applied to each bond in the nets with probability p so that a fraction p of bonds has conductivity σ_0 while the remaining fraction $1-p$ has zero conductivity. The conductivity of the elements is therefore statistically distributed and dependent on the length of the element. When the conductivities of an infinite number of elements are sampled, a small number will be found to have zero conductivity. Thus the conductivity of the infinitely long wire must be zero, as one expects from weakest-link theory. However, it is tempting to conclude from percolation theory that an infinite wire would have a finite conductivity if $p>p_c$, contrasting sharply with the conclusion from weakest-link theory. The reason for the difference is that percolation theory demands an infinite domain in at least two orthogonal directions in order to maintain a finite conductivity. Researchers should therefore be very careful when applying percolation theory to the study of materials having finite dimensions.

Bibliography

Stauffer D. 1985 *Introduction to Percolation Theory*. Taylor and Francis, Basingstoke, UK

L. N. McCartney
[National Physical Laboratory, Teddington, UK]

Plywood

Plywood comprises a number of thin layers of wood called veneers which are bonded together by an adhesive. Each layer is placed so that its grain direction is at right angles to that of the adjacent layer (Fig. 1). This cross-lamination gives plywood its characteristics and makes it a versatile building material. Plywood has been used for centuries. It has been found in Egyptian tombs and was in use during the height of the Greek and Roman civilizations.

The modern plywood industry is divided into the so-called hardwood and softwood plywood industries. While the names are misnomers, they are in common use, the former referring to industry manufacturing decorative panelling and the latter referring to that which serves building construction and industrial uses. This article will be concerned primarily with the plywood manufactured for construction and for industrial uses in North America.

1. Manufacture

The fundamentals of plywood manufacture are the same for decorative and for construction grades, the primary differences being in the visual quality of the veneer faces.

1.1 Log Preparation

Logs received at the plywood plant are either stored in water or stacked in tiers in the mill yard. From there they are transported to a deck where they are debarked and cut into suitable lengths. The cut sections of logs are referred to as blocks. Debarking a log is most often accomplished by conveying it through an enclosure that contains a power-driven "ring" of scraper knives. This ring has the flexibility to adjust to varying log diameters and scrapes away the bark.

Before transforming the blocks into veneer, many manufacturers find it helpful to soften the wood fibers by steaming or soaking the blocks in hot-water vats. The block conditioning process is

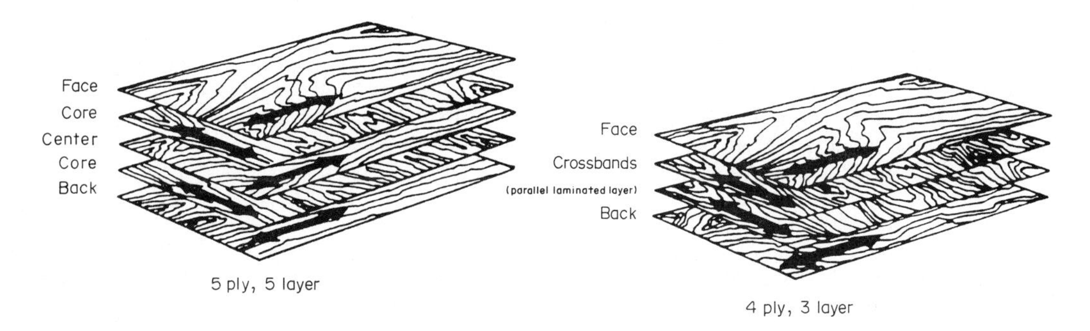

Figure 1
Ply-layer construction of plywood (courtesy of the American Plywood Association)

usually performed after debarking. The time and temperature requirements for softening the wood fibers will vary according to wood species and desired heat penetration. Block conditioning usually results in smoother and higher-quality veneer.

1.2 Conversion into Veneer

The most common method of producing veneer is by the rotary-cutting of the blocks on a lathe. Rotary lathes are equipped with chucks attached to spindles and are capable of revolving the blocks against a knife which is bolted to a movable carriage. New "spindleless" lathes are being introduced which drive the block against a fixed knife with exterior power-driven rollers. Once in place, the chucks revolve the block against the knife and the process of "peeling" veneer begins. The first contact with the lathe knife produces veneer which is not full length. This is due to the fact that the blocks are never perfectly round and are usually tapered to some extent. It is therefore necessary to "round up" the block before full veneer yield is realized. Once this step is complete, the veneer comes away in a continuous sheet, in much the same manner as paper is unwound from a roll. Veneer of the highest grade and quality is usually obtained at the early cutting, knots becoming more frequent when the block diameter is reduced. A knot is a branch base embedded in the trunk of a tree and is usually considered undesirable.

All veneer has a tight and a loose side. The side which is opposite to that of the knife is the tight side because of pressure from the machinery holding it against the knife. The knife side is where the lathe checks occur and is called the loose side. The loose side is always checked to some degree and the depths of the lathe checks will depend upon the pressure exerted by the pressure bar opposite the knife. Various thicknesses of veneer are peeled according to the type and grade of plywood to be produced.

Wood blocks are cut at high speed, and for this reason veneer is fed into a system of conveyors. At the opposite end of the conveyor is a machine called the clipper. Veneer coming from the lathe through the system is passed under a clipper where it is cut into prescribed widths. Defective or unusable veneer can be cut away. The clipper usually has a long knife attached to air cylinders. The knife action is a rapid up-and-down movement, cutting through the veneer quickly. Undried veneer must be cut oversized to allow for shrinkage and eventual trimming. Up to this point the wood has retained a great deal of its natural moisture and is referred to as green. Accordingly, this whole process before drying is called the green end of the mill.

1.3 Veneer Drying and Grading

The function of a veneer dryer is to reduce the moisture content of the stock to a predetermined percentage and to produce flat and pliable veneer. Most veneer dryers carry the veneer stock through the dryer by a series of rollers. The rollers operate in pairs, one above the other with the veneer between, in contact with each roll. The amount of moisture in the dryers is controlled by dampers and venting stacks. Moisture inside the dryer is essential to uniform drying, since this keeps the surface pores of the wood open. The air mixture is kept in constant circulation by powerful fans. The temperature and speed of travel through the dryer is controlled by the operator and is dependent upon the thickness of the veneer species and other important factors.

Dry veneer must be graded, then stacked according to width and grade. Veneer is visually graded by individuals who have been trained to gauge the size of defects, the number of defects and the grain characteristics of various veneer pieces. Veneer grades depend upon the standard under which they are graded, but in North America follow a letter designation A, B, C and D, where A has the fewest growth characteristics and D the most.

1.4 Veneer Joining and Repair

For some grades of plywood, and especially where large panels may be necessary, pieces are cut with a specialized machine, glue is applied and the veneer strips are edge-glued together. The machine usually utilizes a series of chains which crowd the edges of the veneer tightly together, thus creating a long continuous sheet. Other methods of joining veneer may also be employed, such as splicing, stringing and stitching. Splicing is an edge-gluing process in which only two pieces of veneer are joined in a single operation. Stringing is similar to the edge-gluing process except the crowder applies string coated with a hot-melt adhesive to hold the veneers temporarily together before pressing. Stitching is a variation of stringing in which industrial-type sewing machines are used to stitch rows of string through the veneer. This system has the advantage of being suitable for either green or dry stock.

The appearance of various grades of veneer may be improved by eliminating knots, pitch pockets and other defects by replacing them with sound veneer of similar color and texture. Machines can cut a patch hole in the veneer and simultaneously replace it with a sound piece of veneer. While the size and shape vary, boat-shaped and dog-bone-shaped veneer repairs are most common.

1.5 Adhesives

The most common adhesives in the plywood industry are based on phenolic resins, blood or soybeans. Other adhesives such as urea-based resorcinol, polyvinyl and melamine are used to a lesser degree for operations such as edge-gluing, panel-patching and scarfing.

Phenolic resin is synthetically produced from phenol and formaldehyde. It hardens or cures under heat and must be hot pressed. When curing, phenolic resin goes through chemical changes which make it waterproof and impervious to attack by microorganisms. Phenolic resin is used in the production of interior plywood, but is capable in certain mixtures of being subjected to permanent exterior exposure. Over 90% of all softwood plywood produced in the USA is manufactured with this type of adhesive. Another class of phenolics is called extended resins, which have been extended with various substances to reduce the resin solids content. This procedure reduces the cost of the glueline and results in a gluebond suitable for interior or protected-use plywood.

Soybean glue is a protein type made from soybean meal. It is often blended with blood and used in both cold pressing and hot pressing. Blood glue is another protein type and made from animal blood collected at slaughter houses. The blood is spray dried and applied to the plywood and supplied in powder form. Blood and blood–soybean blends may be cold pressed or hot pressed. Protein-based glues (blood or soybean) are not waterproof and therefore are used in the production of interior-use plywood. Protein glues are not currently in common use.

The glue is applied to the veneers by a variety of techniques including roller spreaders, spraylines, curtain coaters and a more recent development, the foam glue extruder. Each of these techniques has its advantages and disadvantages, depending upon the type of manufacturing operation under consideration.

1.6 Assembly, Pressing and Panel Construction

Assembly of veneers into plywood panels takes place immediately after adhesive application. Workmanship and panel assembly must be rapid, yet careful. Speed in assembly is necessary because adhesives must be placed under pressure with the wood within certain time limits or they will dry out and become ineffective. Careful workmanship also avoids excessive gaps between veneers and veneers which lap and cause a ridge in the panel. Prior to hot pressing, many mills prepress assembled panels. This is performed in a cold press which consists of a stationary platen and one connected to hydraulic rams. The load is held under pressure for several minutes to develop consolidation of the veneers. The purpose of prepressing is to allow the wet adhesive to tack the veneers together, which provides easier press loading and eliminates breaking and shifting of veneers when loaded into the hot press. The hot press comprises heated platens with spaces between them known as openings. The number of openings is the guide to press capacity, with 20 or 30 openings being the most common, although presses with as many as 50 openings are in use. When the press is loaded, hydraulic rams push the platens together, exerting a pressure of 1.2–1.4 MPa. The temperature of the platens is set at a certain level, usually in the range 100–165 °C.

2. *Standards*

Construction plywood in the USA is manufactured under US Product Standard PS 1 for Construction and Industrial Plywood (US Department of Commerce 1983) or a Performance Standard such as those promulgated by the American Plywood Association. The most current edition of the Product Standard for plywood includes provisions for performance rating as well. The standard for construction and industrial plywood includes over 70 species of wood which are separated into five groups based on their mechanical properties. This grouping results in fewer grades and a simpler procedure.

Certain panel grades are usually classified as sheathing and include C-D, C-C and Structural I

C-D and C-C. These panels are sold and used in the unsanded or "rough" state, and are typically used in framed construction, rough carpentry and many industrial applications. The two-letter system refers to the face and back veneer grade. A classification system provides for application as roof sheathing or subflooring. A fraction denotes the maximum allowable support spacing for rafters (in inches) and the maximum joist spacing for subfloors (in inches) under standard conditions. Thus, a 32/16 panel would be suitable over roof rafter supports at a maximum spacing of 32 in (800 mm) or as a subfloor panel over joist spaced 16 in (400 mm) on center. Common "span ratings" include 24/0, 24/16, 32/16, 40/20 and 48/24.

Many grades of panels must be sanded on the face and/or back to fulfil the requirements of their end use. Sanding most often takes place on plywood panels with A- or B-grade faces and some C "plugged" faces are sanded as well. Sanded panels are graded according to face veneer grade, such as A-B, A-C or B-C. Panels intended for concrete forming are most often sanded as well and are usually B-B panels. Panel durability for plywood is most often designated as Exterior or Exposure 1. Exterior plywood consists of a higher solids-content glueline and a minimum veneer grade of C in any ply of the panel. Such panels can be permanently exposed to exterior exposure without fear of deterioration of the glueline. Exposure 1 panels differ in the fact that D-grade veneers are allowed. Exposure 1 panels are most often made with an extended phenolic glueline and are suitable for protected exposure and short exterior exposure during construction delays.

3. *Physical Properties*

Plywood described here is assumed to be manufactured in accordance with US Product Standard PS 1, Construction and Industrial Plywood (US Department of Commerce 1983). The physical property data were collected over a period of years (O'Halloran 1975).

3.1 Effects of Moisture Content

Many of the physical properties of plywood are affected by the amount of moisture present in the wood. Wood is a hygroscopic material which nearly always contains a certain amount of water. When plywood is exposed to a constant relative humidity, it will eventually reach an equilibrium moisture content (EMC). The EMC of plywood is highly dependent on relative humidity, but is essentially independent of temperature between 0 °C and 85 °C. Examples of values for plywood at 25 °C include 6% EMC at 40% relative humidity (RH), 10% EMC at 70% RH and 28% EMC at 100% RH.

Plywood exhibits greater dimensional stability than most other wood-based building products. Shrinkage of solid wood along the grain with changes in moisture content is about 2.5–5% of that across the grain. The tendency of individual veneers to shrink or swell crosswise is restricted by the relative longitudinal stability of the adjacent plies, aided also by the much greater stiffness of wood parallel to, as opposed to perpendicular to, the grain. The average coefficient of hygroscopic expansion or contraction in length and width for plywood panels with about the same amount of wood in parallel and perpendicular plies is about 0.002 mm mm^{-1} for each 10% change in RH. The total change from the dry state to the fiber saturation point averages about 0.2%. Thickness swelling is independent of panel size and thickness of veneers. The average coefficient of hygroscopic expansion in thickness is about 0.003 mm mm^{-1} for each 1% change in moisture content below the fiber saturation point.

The dimensional stability of panels exposed to liquid water also varies. Tests were conducted with panels exposed to wetting on one side as would be typical of a rain-delayed construction site. Such exposure to continuous wetting on one side of the panel for 14 days resulted in about 0.13% expansion across the face grain direction and 0.07% along the panel. The worst-case situation is reflected by testing from oven-dry to soaking in water under vacuum and pressure conditions. Results for a set of plywood similar to those tested on one side showed approximately 0.3% expansion across the panel face grain and 0.15% expansion along the panel. These can be considered the theoretical maximum that any panel could experience.

3.2 Thermal Properties

Heat has a number of important effects on plywood. Temperature affects the equilibrium moisture content and the rate of absorption and desorption of water. Heat below 90 °C has a limited long-term effect on the mechanical properties of wood. Very high temperatures, on the other hand, will weaken the wood.

The thermal expansion of wood is smaller than swelling due to absorption of water. Because of this, thermal expansion can be neglected in cases where wood is subject to considerable swelling and shrinking. It may be of importance only in assemblies with other materials where moisture content is maintained at a relative constant level. The effect of temperature on plywood dimensions is related to the percentage of panel thickness in plies having grain perpendicular to the direction of expansion or contraction. The average coefficient of linear thermal expansion is about $6.1 \times 10^{-6}\,\mathrm{m\,m^{-1}\,K^{-1}}$ for a plywood panel with 60% of the plies or less running perpendicular to the direction of expansion. The coefficient of thermal expansion in panel thickness is approximately $28.8 \times 10^{-6}\,\mathrm{m\,m^{-1}\,K^{-1}}$.

The thermal conductivity k of plywood is about 0.11–0.15 $\mathrm{W\,m^{-1}\,K^{-1}}$, depending on species. This compares to values (in $\mathrm{W\,m^{-1}\,K^{-1}}$) of 391 for copper (heat conductor), 60 for window glass and 0.04 for glass wool (heat insulator).

From an appearance and structural standpoint, unprotected plywood should not be used in temperatures exceeding 100 °C. Exposure to sustained temperatures higher than 100 °C will result in charring, weight loss and permanent strength loss.

3.3 Permeability

The permeability of plywood is different from solid wood in several ways. The veneers from which plywood is made generally contain lathe checks from the manufacturing process. These small cracks provide pathways for fluids to pass by entering through the panel edge. Typical values for untreated or uncoated plywood of 3/8 in (9.5 mm) thickness lie in the range 0.014–0.038 $\mathrm{g\,m^{-2}\,h^{-1}\,mm\,Hg^{-1}}$.

Exterior-type plywood is a relatively efficient gas barrier. Gas transmission ($\mathrm{cm^3\,s^{-1}\,cm^{-2}\,cm^{-1}}$) for 3/8 in exterior-type plywood is as follows:

oxygen	0.000 029
carbon dioxide	0.000 026
nitrogen	0.000 021

4. Mechanical Properties

The current design document for plywood is published by the American Plywood Association (1986), and entitled *Plywood Design Specification*. The woods used to manufacture plywood under US Product Standard PS 1 are classified into five groups based on elastic modulus, and bending and other important strength properties. The 70 species are grouped according to procedures set forth in ASTM D2555 (*Establishing Clearwood Strength Values*, American Society for Testing and Materials 1987). Design stresses are presently published only for groups 1–4 since group 5 is a provisional group with little or no actual production. Currently the *Plywood Design Specification* provides for development of plywood sectional properties based on the geometry of the layup and the species, and combining those with the design stresses for the appropriate species group.

4.1 Section Properties

Plywood section properties are computed according to the concept of transformed sections to account for the difference in stiffness parallel and perpendicular to the grain of any given ply. Published data take into account all possible manufacturing options under the appropriate standard, and consequently the resulting published value tends to reflect the minimum configuration. These "effective" section properties computed by the transformed section technique take into account the orthotropic nature of wood, the species group used in the outer and inner plies, and the manufacturing variables provided for each grade. The section properties presented are generally the minimums that can be expected.

Because of the philosophy of using minimums, section properties perpendicular to the face grain direction are usually based on a different configuration than those along the face grain direction. This compounding of minimum sections typically results in conservative designs. Information is available for optimum designs where required.

4.2 Design Stresses

Design stresses include values for each of four species groups and one of three grade stress levels. Grade stress levels are based upon the fact that bending, tension and compression design stresses depend upon the grade of the veneers. Since veneer grades A and natural C are the strongest, panels composed entirely of these grades are allowed higher design stresses than those of veneer grades B, C-plugged or D. Although grades B and C-plugged are superior in appearance to C, they rate a lower stress level because the plugs and patches which improve their appearance reduce their strength somewhat. Panel type (interior or exterior) can be important for bending, tension and compression stresses, since panel type determines the grade of the inner plies.

Stiffness and bearing strengths do not depend on either glue or veneer grade but on species group alone. Shear stresses, on the other hand, do not depend on grade, but vary with the type of glue. In addition to grade stress level, service moisture conditions are also typically presented—for dry conditions, typically moisture contents less than 16%, and for wet conditions at higher moisture contents.

Allowable stresses for plywood typically fall in the same range as for common softwood lumber, and when combined with the appropriate section property, result in an effective section capacity. Some comments on the major mechanical properties with special consideration for the nature of plywood are given below.

Bending modulus of elasticity values include an allowance for an average shear deflection of about 10%. Values for plywood bending stress assume flat panel bending as opposed to bending on edge which may be considered in a different manner. For tension or compression parallel or perpendicular to the face grain, section properties are usually adjusted so that allowable stress for the species group may be applied to the given cross-sectional area. Adjustments must be made in tension or compression when the stress is applied at an angle to the face grain.

Shear-through-the-thickness stresses are based on common structural applications such as plywood mechanically fastened to framing. Additional options include plywood panels used as the webs of I-beams. Another unique shear property is that termed rolling shear (see *Wood Strength*). Since all of the plies in plywood are at right angles to their neighbors, certain types of loads subject them to stresses which tend to make them roll, as a rolling shear stress is induced. For instance, a three-layer panel with framing glued on both faces could cause a cross-ply to roll across the lathe checks. This property must be taken into account with such applications as stressed-skin panels.

Bibliography

American Plywood Association 1986 *Plywood Design Specification*, Form Y510. APA, Tacoma, WA

American Plywood Association 1987a *Grades and Specifications*, Form J20. APA, Tacoma, WA

American Plywood Association 1987b *303 Plywood Siding*, Form E300. APA, Tacoma, WA

American Society for Testing and Materials 1987 *Annual Book of ASTM Standards: Wood*, Vol. 4(09). ASTM, Philadelphia, PA

O'Halloran M R 1975 *Plywood in Hostile Environments*, Form Z820G. APA, Tacoma, WA

Sellers T Jr 1985 *Plywood and Adhesive Technology*. Dekker, New York

US Department of Commerce 1983 *Product Standard for Construction and Industrial Plywood*, PS 1. USDC, Washington, DC (available from the American Plywood Association, Tacoma, WA)

Wood A D, Johnston W, Johnston A K, Bacon G W 1963 *Plywoods of the World: Their Development, Manufacture and Application*. Morrison and Gibb, London

M. R. O'Halloran
[American Plywood Association, Tacoma, Washington, USA]

Polymer–Polymer Composites

The term polymer–polymer composite (PPC) refers to a material in which rigid, rod-like polymer molecules (reinforcement) are dispersed at a molecular level in a flexible coil-like polymer (matrix) of similar chemical composition. With dispersion at this level the materials are sometimes also known as molecular composites.

In the search for higher-strength polymer composites, PPCs offer three main advantages over polymer-fiber-reinforced polymers. Firstly, because of flaws and imperfect alignment of chains within fibers, the strength of an isolated polymer molecule exceeds, by an order of magnitude or more, the strength of fibers produced from the same polymer. Secondly, fiber-reinforced composites can present adhesion problems at the fiber–matrix interface leading to loss of strength. Chemical similarity between the matrix polymer and reinforcement in a PPC at least ensures good "wetting" at the interface. Thirdly, due to the stress transfer region at the fiber ends, it is only when the axial ratio (ratio of the length of the fiber to its diameter) is high enough that the full reinforcement of the fiber is realized. Clearly, with reinforcement being at a molecular level, the diameter is very small and this is unlikely to be such a problem.

1. Methodology of Preparation

Unfortunately, the production of a PPC is not straightforward. Because of the high melting point of rod-like molecules, melt processing is unfavorable. Use of ternary solutions provides the best prospect of obtaining the optimum single phase consisting of a homogeneous dispersion of rigid rods. Statistical thermodynamics of ternary systems comprising a solvent, a rod-like solute and a random-coil chain were predicted by Flory (1978). It has been shown that: (a) aggregates of rigid molecules reject the coiled molecules with a selectivity approaching that of a pure crystal, and (b) for a rigid rod molecule with an axial ratio high enough to achieve reinforcement, a high degree of selectivity occurs even at low concentrations. Since this theory, when closely examined, also provides guidelines for overcoming this problem, it represents a very effective tool in present research on polymer–polymer composites.

Of particular significance is the prediction that, to obtain a homogeneous dispersion, a high degree of molecular orientation of the rod-like molecules in the random coil matrix will be required. This may be achieved by recognizing the liquid-crystalline properties of the rod-like molecules. If the composites are processed from polymer solutions at or near the critical concentration C_{cr} of rod-like molecules necessary to form anisotropic domains, then a high dispersion of the rod-like molecules can be obtained and, because the viscosity of the solution is at a maximum, the rod-like polymer can be readily oriented by an external shear stress.

Alternatives to the physical blending of polymers to form PPCs are being investigated. One approach appears particularly promising: the synthesis of block copolymers composed of flexible-coil segments and rigid-rod block segments.

Table 1
Chemical structure of PPC systems

	Chemical structure	Acronym
System 1 Rod-like polymers	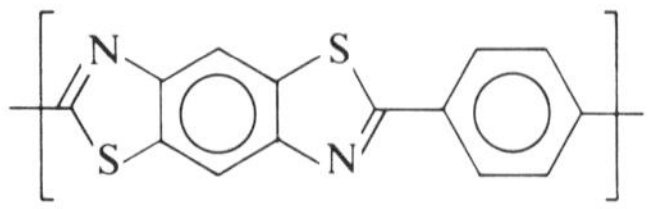	PBT
	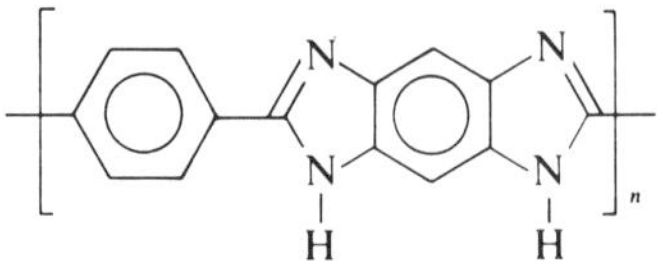	PDIAB
Coil-like polymers	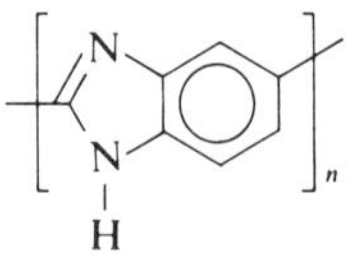	ABPBI
System 2 Rod-like polymers	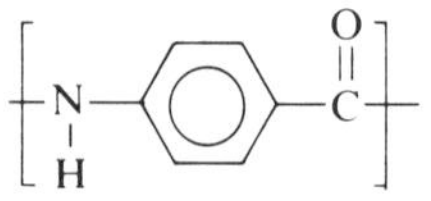	PPTA
Coil-like polymers	$\left[NH-(CH_2)_6-\overset{O}{\overset{\parallel}{C}} \right]$	nylon 6
	$\left[NH-(CH_2)_6NH-\overset{O}{\overset{\parallel}{C}}-(CH_2)_6-\overset{O}{\overset{\parallel}{C}} \right]$	nylon 66

2. Properties

Two PPC systems have received particular attention (Table 1). One is based on rod-like polymers such as poly(p-phenylene benzobisthiazole) (PBT) and poly-(2-phenylene benzobisimidazole) (PDIAB) and a matrix of the more flexible amorphous poly(2-5(6)-benzimidizole (ABPBI). The other involves rigid wholly aromatic polymers such as poly(p-phenylene terephthalamide) (PPTA), which is produced by Du Pont under the trade name Kevlar, and a crystallizable matrix of nylon 6 or nylon 66. The remainder of this article describes some of the characteristics and properties of these two systems.

2.1 Morphology

The morphology of composite films of PBT and ABPBI has been investigated by Hwang et al. (1983) and Krause et al. (1986). Scanning electron microscopy (SEM) reveals that, for a content of rod-like polymer up to 50 wt%, composites vacuum-cast from solution consist of two distinct phases. The rod-like polymers tend to form ellipsoids that are typically 3 μm long. These ellipsoids are chiefly composed of 10 nm PBT crystallites moderately well aligned with the long axis of the ellipsoid. The matrix is ductile and composed primarily of ABPBI. A similar morphology is observed in PDIAB/PBT blends (Husman et al. 1979). The removal of solvent in the casting process clearly causes the polymer concentration to rise above C_{cr} and induce phase separation. With separation occurring at this level, the material cannot really be classed as a PPC. However, when PBT/ABPBI solutions at a concentration less than C_{cr} are dry-jet/wet-spun, no phase separation can be observed using SEM. Wide-angle x-ray scattering and transmission electron microscopy show that fibers heat-treated under tension in air contain well-oriented molecules and crystallites of both ABPBI and PBT which are less than 3 nm in width. Triblock copolymers of ABPBI/PBT/ABPBI show a similar morphology to the physical blend when vacuum cast. However, the size of the ellipsoids is generally smaller and hence the reinforcement more efficient, as the size is controlled by the length of the PBT block. Spun solution of the copolymer does, however, produce fiber in which once more no features greater than 3 nm in size are present.

Composites consisting of aromatic and aliphatic polyamides may be prepared by extruding sulfuric acid solutions of the polymers into a large amount of water and methanol. Takayanagi et al. (1980) have investigated the morphology of such composites prepared under conditions which, because of mixing and turbulence, may be far from equilibrium. For the samples prepared from 7% PPTA, of viscosity average molecular weight $M_v = 34\,000$, and 93% nylon 6, a uniform dispersion of PPTA microfibrils is produced in an isotropic matrix of nylon. A typical diameter of a microfibril is 30 nm. When the viscosity average molecular weight of the PPTA is reduced to 4500 the morphology is retained but the microfibrils are less perfect, with the surface of the fibrils being rougher. The average diameter is reduced to 15 nm with a concomitant increase in the population of microfibrils. Although the diameters of the microfibrils are much smaller than the diameters of fibers used in fiber-reinforced composites, a homogeneous molecular composite has still not been obtained.

The blending of PPTA with aliphatic polyamides induces the crystallization of the latter. Analysis of the crystallization isotherm of nylon 6 shows an Avrami constant equal to 2, which can be interpreted as being due to heterogeneous nucleation followed by two-dimensional disk-shaped growth. It therefore appears that the surface of the well-developed fibrils of PPTA (i.e., those produced from high-molecular-weight molecules) may well act as nucleation sites for crystallization of the nylon. As the surface energy of crystallites in the nylon is greater than that of the amorphous polymer, such a crystallization mechanism will result in optimum adhesion between the microfibrils and the matrix. Examination of the fracture surfaces of specimens broken in liquid nitrogen show no fiber pullout, that is, long sections of the PPTA fibrils protruding from the nylon. This is indicative of good adhesion.

A schematic of the morphology of composites containing high and low molecular weight PPTA fibrils is shown in Fig. 1. The morphology can be likened to the "shish-kebab" morphology observed when crystallization is undertaken in an oriented melt.

2.2 Mechanical Properties

As listed in Table 2, vacuum-cast films of 30% PBT/70% ABPBI have a modulus and tensile strength roughly one order of magnitude less than ABPBI fiber and two orders of magnitude less than PBT. These low-value properties (compared to the homopolymer fibers) are due to the low axial ratio of the PBT phase. The modulus, strength and elongation to fracture of the copolymer is markedly higher. This reflects the reduced scale of phase separation in the copolymer and the increased interfacial strength. In both types of fiber the more efficient reinforcement and orientation of ABPBI and PBT result in high values of the mechanical properties. Examination of the fracture surfaces of the fibers is also revealing. PBT fiber fibrillates extensively upon fracture because of its high axial strength and low lateral strength. The physical blend fibrillates moderately upon fracture, but the copolymer shows no evidence of fibrillation. It therefore has greater ductility and lateral strength than the physical-blend fiber.

For PPTA composites, the effective strength of the microfibrils can be estimated by stretching of specimens to produce uniaxial orientation of the fibrils which then allows a simple property analysis. Data for nylon 6 and a blend of PPTA/nylon 6 having a 3:97 weight ratio, in which both were stretched to an extension ratio of 3:1, is shown in Table 3. Orientation

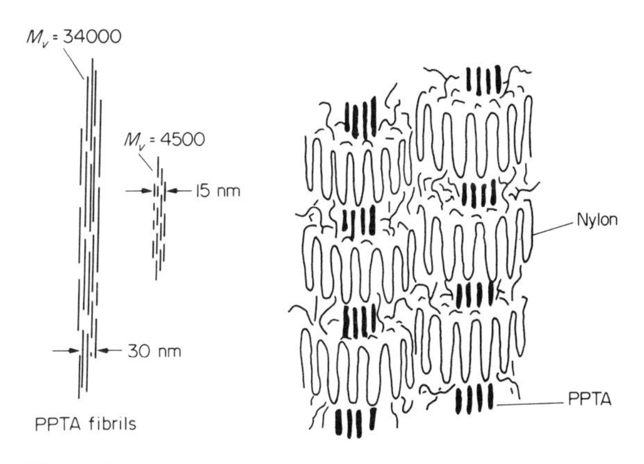

Figure 1
Schematic of the morphology of PPTA-reinforced nylon 6

Table 2
Mechanical properties of PBT, ABPI and 30% PBT/70% ABPI fibers and films

Sample	Modulus (GPa)	Tensile strength (MPa)	Elongation to fracture (%)
PBT Fiber	320	3100	1.1
ABPBI Fiber	36	1100	5.2
30% PBT/70% ABPBI blend film	1.1	35	5.6
30% PBT/70% ABPBI blend fiber	120	1300	1.4
30% PBT/70% ABPBI copolymer film	2.4	220	43
30% PBT/70% ABPBI copolymer fiber	100	1700	2.4

Table 3
Tensile properties of oriented PPTA/nylon 6 of 3/97 weight ratio

	Modulus (GPa)	Elongation to fracture (%)	Ultimate tensile strength (MPa)
Oriented nylon 6	1.18	20	220
Oriented blend	3.35	28	340

Table 4
Tensile properties of PPTA/nylon 6 of 5/95 weight ratio

	Modulus (GPa)	Yield stress (MPa)	Elongation to fracture (%)	Ultimate tensile strength (MPa)
PPTA				
$M_v = 980$	1.45	40	2.7	52
$M_v = 4500$	1.59	46	1.6	54
$M_v = 12300$	1.67	58	0.6	59
Nylon 6	0.91	24	5.3	51

is seen to be clearly effective in reinforcement when comparison is made with the unoriented sample.

If the simple assumption of the volume additivity of the ultimate strengths of PPTA and nylon 6 is adopted, the strength of a PPTA microfibril is estimated to be 4 GPa. This is comparable to or higher than the highest tenacity, 3.46 GPa, of commercial PPTA fiber (du Pont Kevlar 49). Thus even when discontinuity of the PPTA microfibrils in the specimen is neglected, the full strength of PPTA may be realized in a polymer composite. The same observation applies to the effective microfibril modulus, 75 GPa, which falls between the modulus of Kevlar 29 ($E = 61$ GPa) and that of Kevlar 49 ($E = 120$ GPa).

The molecular weight of PPTA has a marked effect on the mechanical properties. Data for a composition ratio of PPTA to nylon 6 of 5:95 is shown in Table 4. On increasing the viscosity average molecular weight from 980 to 12 300, there is a steady increase in modulus, yield stress and breaking stress. Although the content of PPTA is small it is interesting to reflect upon the fact that the stress–strain behavior of the composite approaches that of glass-reinforced nylon containing as much as 30% chopped glass fiber. The elongation to break does, however, vary, with the higher molecular weight, higher modulus and stronger PPTA fibrils restricting the ductility of the matrix more than their low-molecular-weight counterparts.

Microvoids located at the ends of microfibrils or at the interface between the fibrils and the matrix would appear to be the cause of the decrease in ductility. Interestingly, the extensibility of the composites can be improved without loss in strength if a block copolymer of PBTA with nylon 6 or nylon 66 is used. This leads to a morphology of finer microfibrils and a stronger interface between the microfibrils and matrix.

Thus the concept of a polymer–polymer composite has been demonstrated. Although the studies to date are very preliminary, the concept appears to be very promising. The emphasis of future work will be on improving the processing techniques used in order to better control the morphology, dispersion and orientation of the rod-like molecules.

Bibliography

Flory P J 1978 Statistical thermodynamics of mixtures of rodlike particles. *Macromolecules* 11: 1119–44.

Husman G, Helminiak T, Adams W, Wiff D, Benner C 1979 Molecular composites: rodlike polymer reinforcing an amorphous polymer matrix. *Org. Coat. Plast. Chem.* 40: 797–802

Hwang W-F, Wiff D R, Benner C L, Helminiak T E 1983 Composites on a molecular level: phase relationships processing and properties. *J. Macromol. Sci., Phys.* B22: 231–49

Krause S J, Haddock T B, Price G E, Lenhert P G, O'Brien J F, Helminiak T E, Adams W W 1986 Morphology of a phase separated and a molecular composite PBT/ABPBI polymer blend. *J. Polym. Sci., Polym. Phys. Ed.* 24: 1991–2016

Krause S J, Haddock T B, Price G E, Adams W W 1988 Morphology and mechanical properties of a phase separated and a molecular composite 30% PBT/70% ABPBI triblock copolymer. *Polymer* 29: 195–206

Prevorsek D C 1982 Recent advances in high-strength fibers and composites. In: Ciferri A, Krigbaum W R, Meyer R B (eds.) 1982 *Polymer Liquid Crystals.* Academic Press, New York, pp. 329–76

Takayanagi M, Ogata T, Morikawa M, Kai T 1980 Polymer composites of rigid and flexible molecules: system of wholly aromatic and aliphatic polyamides. *J. Macromol. Sci., Phys.* B17 (4): 591–615

D. C. Prevorsek
[Allied-Signal Inc., Morristown, New Jersey, USA]

P. J. Mills
[ICI, Wilton, UK]

R

Recycling of Polymer-Matrix Composites

Recycling is one of the biggest issues facing the composites industry, particularly for large-volume applications. Increasingly stringent environmental regulations are likely to restrict the use of composites in favor of materials that can be recycled cost effectively. Indeed, in order to counter the arguments which may lead to legislation that would limit the nature of composites products for sale, the composites industry as a whole will have to encompass viable primary, secondary and tertiary recycling capabilities. The principal categories of scrap polymer-matrix composites and the principal routes for their recycling are summarized in Sects. 1 and 2, respectively. The matrices used in polymer composites treated in this article are not ones that can be commercially depolymerized to monomer.

1. *Categories of Scrap Composites*

Scrap composites can be conveniently divided into three categories.

(a) Scrap in the form of offcuts, rejects, sprues, and so on arises in the manufacture of composite products. Increasingly this waste material is used in primary recycling by blending it as filler or reinforcement with virgin plastic often of the same chemical origin. This route is subject to the careful control of the levels of contamination in the comminuted composite and to the deterioration in physical properties which may be caused by repeated thermal and mechanical processing. Some thermosets have been successfully recycled since the mid-1980s without adverse effect on quality.

(b) Single grades of contaminated plastic collected from consumers or processors may provide feedstock for primary or secondary recycling, subject to the feasibility of contaminant removal.

(c) Mixtures of two or more grades of composite compounds arise as industrial or consumer scrap. This category of scrap poses a substantial problem in composites recycling, because of the problems associated with automatic identification, separation, and determination and control of composition. Tertiary recycling is expected to be the more appropriate route for recycling.

2. *Recycling Methods*

2.1 Thermoplastic-Matrix Composites

Thermoplastic-matrix composites scrap arising in categories (a) and (b) provides for more straightforward recycling than thermoset-matrix composite scrap, principally because the thermoplastic can be remelted. Fiber attrition and, to a lesser extent, degradation of the matrix polymer, lead to reuse applications with less-demanding physical property requirements (Reinforced Plastics 1991). A principal example of large-tonnage thermoplastic-matrix composite recycling is long-glass-fiber-mat-reinforced polypropylene (GMT). The offcuts from GMT sheet used for thermoforming of products, particularly in the manufacture of automotive parts, can be subsequently used after comminution and used over again as raw material for semifinished sheet. Additionally, the offcuts or reject parts can be ground and used for extrusion or for injection molding.

The consequences of incorporating mechanically reconstituted offcuts of a range of thermoplastic- and thermoset-matrix composites into virgin feedstock as well as reuse directly have been reviewed by Farrisey (1992), and specifically for long-fiber thermoplastic composites by Clegg et al. (1990). The alternative routes of separating the composite into fiber and matrix components by acid dissolution and by solvent swelling have also been reported (Buggy et al. 1993).

2.2 Thermoset-Matrix Composites

The recycling of thermoset-matrix composites presents much more significant difficulties, principally centered on the irreversibility of cross-linking, the fiber attrition associated with comminution, and a polymer content that may be less than 30% of the total weight. The bulk of the material is often glass-fiber reinforcement or a filler, including fire retardants and resin diluents. In tertiary recycling the problem is not just one of recycling the polymer.

Large-volume applications of thermoset-matrix composites include pipes, containers, covers for machinery, housings, switch gear cabinets and highly stressed automotive components such as chassis parts and bumpers.

Greater stability in the supply of scrap is associated with large sources of standardized composite scrap arising from co-operative industry ventures (Rowlands 1991, Weaver 1992). For example, ERCOM, a consortium of four large European

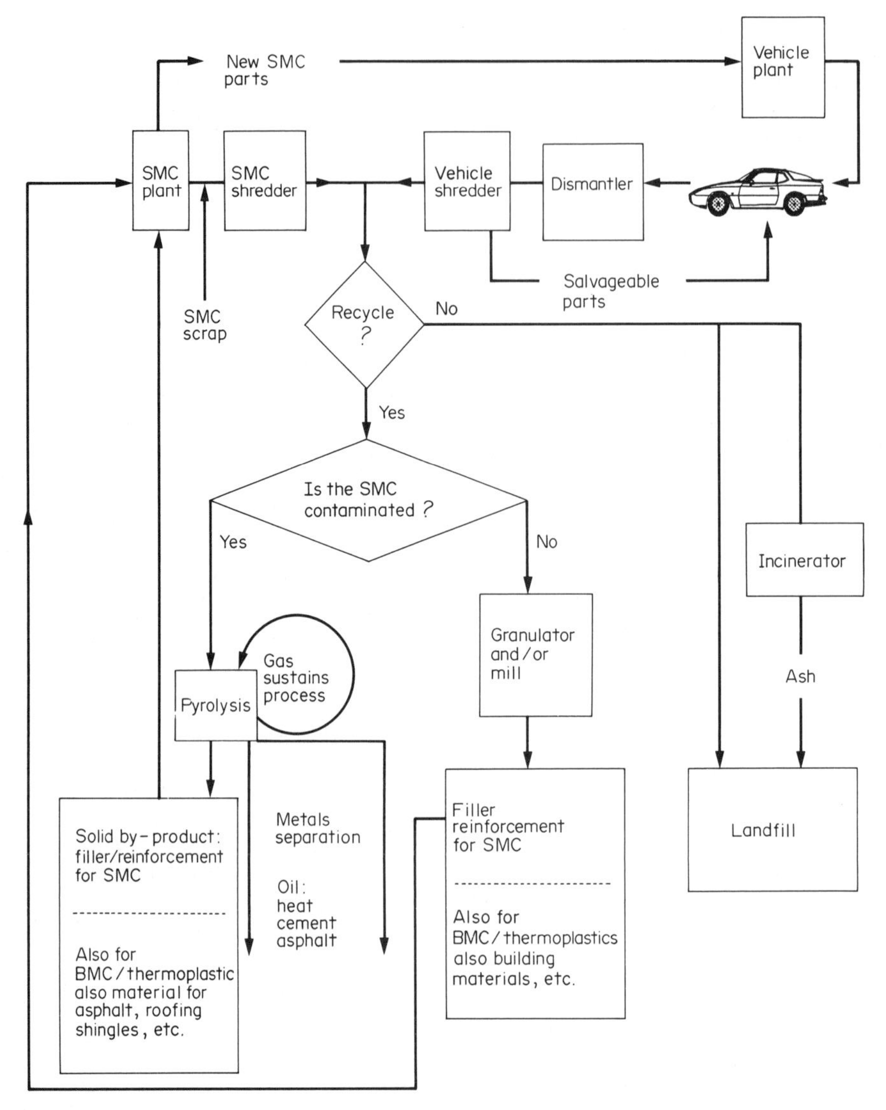

Figure 1
The options for SMC recycling and disposal according to the SMC Automotive Alliance

composite manufacturers in partnership with a number of leading raw materials suppliers, shred components manufactured from polyester and vinylester-based sheet and bulk-molding compounds to a range of well-defined particle sizes. The resultant fiber and powder fractions can be used in the production of new bulk-molding compound (SMC/BMC) components, and can also be used as reinforcing material for thermoplastics and other materials.

Addition of reground process scrap back into uncured resin formulations has been reviewed for a wide range of thermoset-matrix composites (Farrisey 1992), including polyurethanes, phenolics and epoxies in addition to unsaturated polyester SMC. This comprehensive review of thermoset recycling also encompasses the use of hydrolysis, glycolysis and pyrolysis as effective routes for recycling of selected thermosets.

The ERCOM type of initiative is providing for large-scale utilization of composites scrap, and the incentive for the development of more comprehensive secondary and tertiary recycling operations. Considerable imagination is evident in the research and development work (Farrisey 1992, Dennison 1993) which is in progress to identify profitable routes for the recycling of composites materials. These include recycling machinery development, with attention to communitive procedures, and new product development that has to overcome the technical, cost and aesthetic advantages of competing virgin materials. Recent research relates to proposed radical and potentially very substantial disposal options (Pickering and Benson 1991) for composites scrap. Combustion with heat recovery is proposed as a route for utilizing the energy content of the matrix polymer. The behavior of a range of composites during combustion, the form of the ash product and the emissions during combustion have been investigated systematically.

Two industrial processes that utilize the energy content and the inert materials arising in the ash have been proposed. In cement manufacture thermoset-matrix composites may be burned in a cement kiln to utilize their energy content and the mineral materials utilized in the cement klinker. Alternatively, polymeric materials filled with calcium carbonate may be of use as a fuel substitute and sulfur oxide removing agent in coal-fired fluidized bed combustion.

Large tonnage applications for selected recycled composite materials are being developed as shot-blasting media (Farrisey 1992) for selective paint removal, and for use as soil conditioners.

Thermosetting polyester-matrix composites represent a large percentage of composite manufacturing, particularly in the automotive sector, and SMCs have been the subject of considerable research and success in recycling. Comprehensive assessments of the mechanical properties of SMCs containing large proportions of comminuted recyclate have been reported (Bledski et al. 1992, Konig et al. 1992). These confirm that incorporation for new product manufacture is viable, and justify major recovery and reuse initiatives of the form summarized in Fig. 1. The diagram summarizes the options that are available for recovery and reuse of SMCs and, within limitations, can be applied to other thermosetting matrices, including epoxies, phenolics, polyurethane and urea formaldehyde.

2.3 Concluding Remarks

Plastic-matrix composites are amenable to all waste management options. They are capable of being recycled into the chemical and oil feedstocks used in their original manufacture, and especially comminution of the composite to particulate provides for reuse either mixed with virgin material or remolded. Finally, composites can be used as a source of energy in place of fossil fuels, and this option has a major future role if a significant move away from landfill is to be achieved (Dennison 1993).

Bibliography

Bledski A K, Kurek K, Barth Ch 1992 Properties of SMCs with regrind—The effect of particle recycling on the behaviour of the material. *Kunstoffe* 82: 1093–6

Buggy M, Farragher L, Madden W 1993 Recycling of composite materials. In: Hashmi M S J (ed.) 1993 *Proc. Int. Conf. Advances in Materials and Processing Technologies*. Dublin City University, Dublin, pp. 293–304

Clegg D W, McGrath G, Morris M 1990 Properties and economics of reclaimed long fibre thermoplastic composites. *Compos. Manuf.* 1: 85–9

Dennison M T 1993 Plastics recycling: product feedstock or energy?—A future view. *Maach Conf.—Recycle '93* Paper 23/4. Maach Business Services, Davos, Switzerland

Farrisey W J 1992 Thermosets. In: Ehlig R H (ed.) 1992 *Plastics Recycling, Products and Processes*. Hanser, Munich, Germany, pp. 233–59

Konig W, Buhl D, Mobius K H 1992 Particle recycling guarantees the future of SMCs. *Kunstoffe* 82: 667–70

Pickering S J, Benson M 1991 Disposal of thermosetting plastics—Paper 22. *Proc. 2nd Int. Conf. Plastics Recycling*, London. Plastics and Rubber Institute, London

Recycling 1991 The key to future success. *Reinf. Plast.* (July/August): 20–3

Rowlands H E 1991 Thermoset technologies for the automotive industry. *Proc. Auto-tech. '91*. Paper C427/3 Institute of Mechanical Engineers, London p. 105

Weaver A 1992 Recycling in action. *Reinf. Plast.* (February): 32

M. J. Bevis, P. R. Hornsby, W. H. Lee and K. Tarverdi

[Brunel University, Uxbridge, UK]

S

Silicon Carbide Fibers

Silicon carbide fibers have a high structural stability even at high temperatures. This is indicated by an extreme resistance to oxidation combined with good high-temperature strength, which makes them useful as fiber reinforcement in high-temperature composite materials.

Two different types of silicon carbide fibers exist: substrate-based fibers and fine ceramic fibers. Substrate-based fibers generally have a tungsten filament (SiC/W fiber) or a carbon filament (SiC/C fiber) as the substrate. The thickness of the filament lies in the range 100–150 μm. Fine ceramic fibers are based on silicon carbide, have a diameter of around 15 μm and are produced by the pyrolysis of a polycarbosilane precursor.

1. Preparation, Microstructure and Morphology of SiC on a Substrate

The substrate-based fiber is produced in a chemical vapor deposition process in the same type of reaction chamber as that used in boron fiber production (see *Boron Fibers*) but with multiple injection points for the reactant gases. Various carbon-containing silanes have been used as reactants. In a typical process, with CH_3SiCl_3 as the reactant, SiC is deposited on a tungsten core:

$$CH_3SiCl_3(g) \rightarrow SiC(s) + 3HCl(g) \qquad (1)$$

The substrate-based SiC fiber consists of a nearly unreacted core surrounded by a mantle of β-SiC microcrystallites in a preferred orientation where the (111) planes are parallel to the fiber axis. The earliest fibers made in this way had surfaces which were under slight tensile stress, which made the fibers relatively sensitive to surface defects. This was overcome by depositing a thin layer of carbon onto the finished fiber, which considerably increased abrasion resistance. Prolonged use at high temperatures (above 1000 °C) produces an increasing reaction between the SiC sheath and the tungsten core, giving rise to α-W_2C and W_5Si_3, and this eventually limits the use of the fiber. The SiC/W fiber exhibits nodules which are smaller than the amorphous boron nodules in boron fibers.

Silicon carbide fibers are now produced on a carbon-filament substrate which has a diameter of 33 μm. The carbon filament is potentially cheaper than tungsten wire and it has been found that faster SiC filament production is possible using this method. The factors which prevent the use of carbon-filament substrates in boron fiber production do not apply to SiC fiber manufacture.

The SiC/C fiber has a relatively smooth surface (compared with that of the B/W fiber). The density of a 100 μm-thick SiC/W fiber grown on a 12.5 μm-thick tungsten filament is 3.35 g cm^{-3}, while the density of a SiC/C fiber is somewhat lower ($\sim$3.2 g cm^{-3}).

As mentioned above, sensitivity to surface abrasion was overcome by depositing a layer of pyrolytic carbon on the SiC fiber surface. This has the disadvantage, however, of reducing interfacial bonding, particularly in light alloys. To overcome this difficulty SiC fibers on a carbon fiber core are produced with a surface layer of pyrolytic carbon which itself is coated with silicon carbide. These fibers are given the designation SCS and basically there are three types: SCS-2, SCS-8 and SCS-6. SCS-6 has a thicker final SiC layer which makes it suitable for reinforcing titanium, and shows no degradation after 5 hours at 900°C when embedded in a Ti (6Al4V) matrix, as demonstrated by Whatley and Wawner (1985).

2. Properties and Chemical Compatibility

The room-temperature axial tensile fracture-stress distribution of SiC/W fibers contains a broad maximum with a mean value in the range 3–4 GN m^{-2}. Low fracture stresses are caused by surface flaws. At high fracture stresses the fracture is initiated in the core–mantle interface or in the core itself.

The fracture stress of SiC fibers is reduced at higher temperatures. Figure 1 and Table 1 show the reduction of the fracture stress of SiC fibers at relatively low temperatures and short-time exposure conditions and at high temperatures and long-time exposure conditions, respectively. In air, for instance, the SiC/W fiber loses only 20% of its strength after nine minutes exposure at 700°C, while a SiC-coated boron fiber (Borsic) loses 25% of its strength and a B/W fiber is completely degraded. The reduction in tensile strength at higher temperatures is caused by a chemical reaction in the fiber core–mantle interface which results in the formation of α-W_2C and W_5Si_3.

SiC fiber-reinforced composites are mostly used at temperatures above 350°C. This means that strength and chemical compatibility with metal matrices at high temperatures are of the highest importance. At 400°C and 800°C the tensile fracture stress of SiC/W fibers is 90% and 75% of the room-temperature value, respectively. The axial Young's modulus of the SiC/W fibers decreases linearly with increasing temperature from 420 GN m^{-2} at room temperature to 390 GN m^{-2} at 600°C. Above 600°C the modulus decreases somewhat more rapidly.

Table 1
Room-temperature fracture stresses of substrate-based SiC fibers after heating in air and argon at 1200°C for 48 h (after Lindley and Jones 1975)

Fiber	Fiber diameter (μm)	Substrate diameter (μm)	Tensile fracture stress ($GN\,m^{-2}$): Initially	After exposure to air	After exposure to argon
SiC/W	100	12.5	2.99	0.66	0.73
SiC/W	150	12.5	4.00	0.92	0.74
SiC/C	100	33	3.99	1.48	0.81

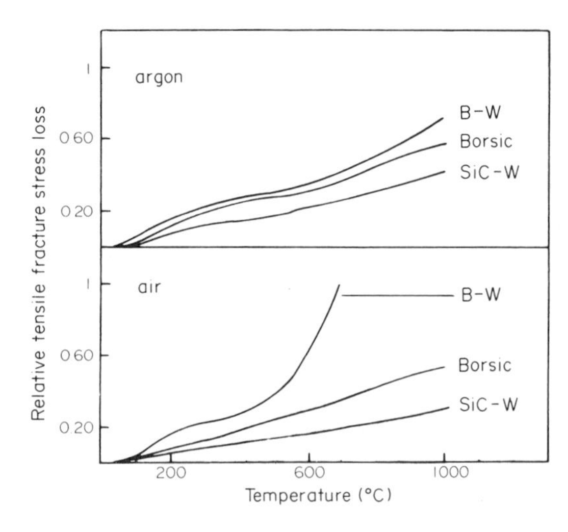

Figure 1
Relative fracture-stress loss at room temperature of different fiber types after heating in air and argon for nine minutes

Below 350°C, SiC-coated boron fibers are superior to SiC fibers as reinforcement for titanium. Above 350°C, titanium-SiC fiber composites have favorable properties in comparison with other composite materials. At high temperature, however, an interfacial reaction between titanium and the SiC fiber occurs: in the reaction zone, Ti_3Si, Ti_5Si_3 and TiC_{1-x} are formed.

3. *Preparation of Fine SiC Fibers*

The manufacture of silicon carbide fibers using a polycarbosilane precursor fiber was first described by Yajima et al. (1976). Polydimethylsilane is made by reacting sodium with dichlorodimethyl silane:

$$n\mathrm{SiCl_2(CH_3)_2} \xrightarrow{\mathrm{Na}} \mathrm{Si{+}(CH_3)_2{+}_n}$$

This is then heated in an autoclave at a pressure of about 10 MPa, resulting in a reorganization of the polymer and the introduction of Si–C into the chain, giving

$$-\!(\mathrm{Si(CH_3)})_n\!- \longrightarrow \left(\begin{array}{c}\mathrm{CH_3}\\ |\\ \mathrm{Si\text{-}CH_2}\\ |\\ \mathrm{H}\end{array}\right)_n$$

The structure of the polycarbosilane is not linear but consists of cycles of six atoms arranged in a similar manner to the cubic structure of β-SiC:

```
Si\
|   C-Si
|       \
C — Si—C
```

The molecular weight of the polymer is, however, low ($M \simeq 1500$) when compared to values of the order of hundreds of thousands for polymers drawn into textile fibers. The polymer is therefore more like a paste and is thus extremely difficult to draw into filaments. In addition the methyl (CH_3) groups are not included in the Si–C–Si chain so that during pyrolysis the hydrogen is driven off leaving a residue of carbon.

After synthesis and drawing, the fibers are subjected to heat treatment in air at 200°C in order to achieve cross-linking of the structure. During this stage some of the silicon bonds to the oxygen giving Si–O–Si although the alternative Si–O–C can also be formed. This oxidation makes the fibers infusible but has the drawback of introducing oxygen into the polymer which remains after pyrolysis. Ceramic fiber is obtained by a slow increase in temperature, in an inert atmosphere, up to 1300°C. The fiber which is obtained by this method contains mostly SiC but also significant amounts of free carbon and excess silicon and oxygen probably combined as SiO_2. This route has been adopted by Nippon Carbon to produce a fiber called Nicalon which contains approximately 65 wt% of microcrystalline silicon carbide and which has a diameter of around 15 μm.

A similar process used by Ube Chemicals leads to an amorphous fiber, produced commercially under the name Tyranno, which contains silicon, titanium, carbon and oxygen. The Tyranno fiber is made by first producing a cross-linking organometallic polymer, polytitanocarbosilane. This is synthesized, as described by Yamamura et al. (1987), by dechlorination of dimethyldichlorosilane mixed with titanium alkoxide, heated to 340°C in N_2 gas and polymerized. Again the molecular weight is low, around 1500. The general structure of the precursor fiber produced is shown below:

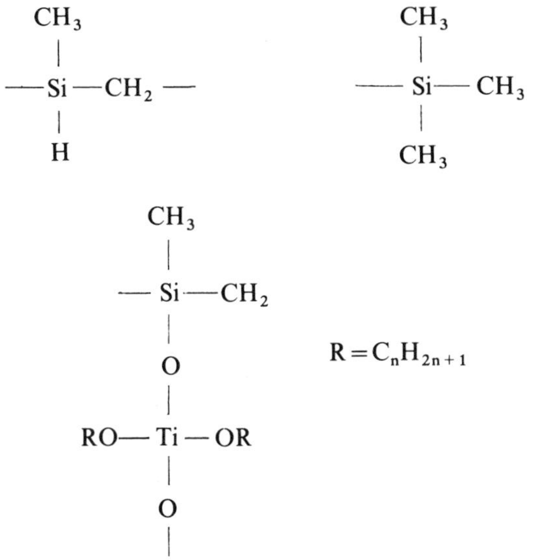

$$-Si(CH_3)(H)-CH_2- \qquad -Si(CH_3)_2-CH_3$$

$$-Si(CH_3)(-CH_2)-O-Ti(OR)_2-O- \qquad R=C_nH_{2n+1}$$

Traces of other compounds are also found in the fiber.

A ceramic fiber is obtained by heating in N_2 gas up to around 1300 °C. The Tyranno fiber has a diameter of between 8 and 12 μm.

Both the Nicalon and Tyranno fiber belong to the new family of fine ceramic fibers and offer the possibility of reinforcing materials for use at high temperatures. The former fiber has been available in relatively small quantities since about 1982 whereas the latter is still at a small pilot plant stage.

4. *Microstructure of Fine Ceramic Fibers*

Of the two fibers mentioned above, only Nicalon has been available sufficiently long enough for detailed studies of its structure to be made.

The only elements detected in Nicalon fibers are silicon, carbon and oxygen. Electron microprobe measurements of the intensity of x-ray emission characteristics of the different elements has revealed the distribution of the elements across the fiber diameter. The resolution of this technique is about 1 μm^3 and has shown a uniform distribution of silicon, carbon and oxygen across the fiber. The presence of oxygen across the diameter shows that it was introduced during the oxidation stage, although a fine layer of SiO_2 could also exist on the surface but would not be detectable by the technique employed.

X-ray diffraction studies of the Nicalon fiber reveal only one diffraction peak corresponding to microcrystalline particles of β-SiC having a size of approximately 1.7 nm. Such small crystals have been observed by Simon and Bunsell (1984a) using dark-field transmission electron microscopy. The structure of the excess silicon, oxygen and carbon in the fiber identified by electron microprobe analysis is therefore assumed not to be crystalline.

A study by x-ray central-beam scattering has allowed segregations of carbon to be identified, as the electron density of carbon is different from that of SiC and SiO_2, and reveals an average size of carbon segregate of about 2 nm. The only paramagnetic phase that can be present in the fiber is free carbon and the detection of a signal by ESR (electron spin resonance) is proof of its existence. The presence of free-carbon segregates in the Nicalon fibers has important consequences in controlling creep at high temperature.

The structure of Nicalon fibers is found to evolve on heating to high temperatures. X-ray studies reveal that the diffraction peaks narrow on heating above 1200°C, indicating an improved organization of the structure. Nicalon fibers show an increase in SiC grain size on being heated to this temperature. Grain growth is found to be slower when the fiber is under load and, under no load, heating can lead to shrinkage. Grain growth stabilizes with a mean grain size of 3 nm. The rate of recrystallization is not affected by the environment in which the fiber is heated; however, an oxide surface layer has been seen to develop during heating in air above 1100 °C.

Under heating conditions which lead to a stabilized grain structure, the mechanical properties of Nicalon fibers are still observed to deteriorate after stabilization. This has been shown by Simon and Bunsell (1984b) to be a consequence of internal reactions between the excess carbon and oxygen in the structure.

Changes in the carbon content of Nicalon fiber were followed during heating between 1000°C and 1500°C, using the ESR technique. Figure 2 shows that the intensity falls after heating for one hour and the fall is more rapid in argon than in air. The quantity of free carbon in the fiber falls from a value of about 30 mol% to a minimum calculated as about 15 mol% when the fiber is heated in air and only 5% when it is heated in argon.

It is clear from the above results that two independent processes are occurring in Nicalon SiC fiber when it is heated above 1000°C. A reorganization of the structure occurs above 1100 °C, leading to shrinkage under low load and stabilizing at a grain size of 3 nm. The free carbon, however, combines with the oxygen in the fiber. This process consumes the carbon and degrades the fiber. Degradation is slowed in air as the

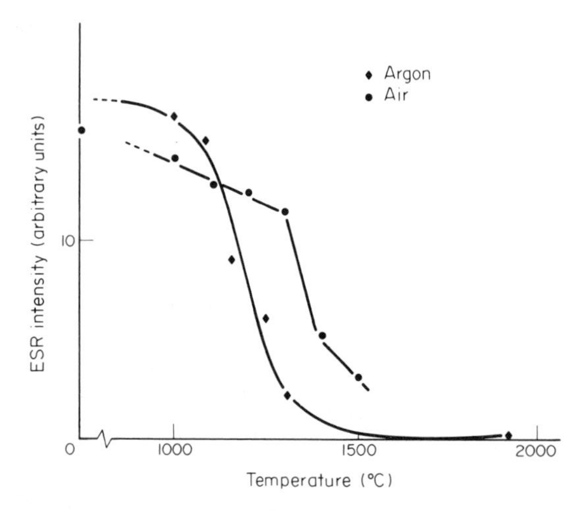

Figure 2
The reduction in the paramagnetic ESR signal obtained from free carbon in Nicalon fibers, after one hour's heat treatment in air and in argon, due to the carbon combining with the oxygen in the structure.

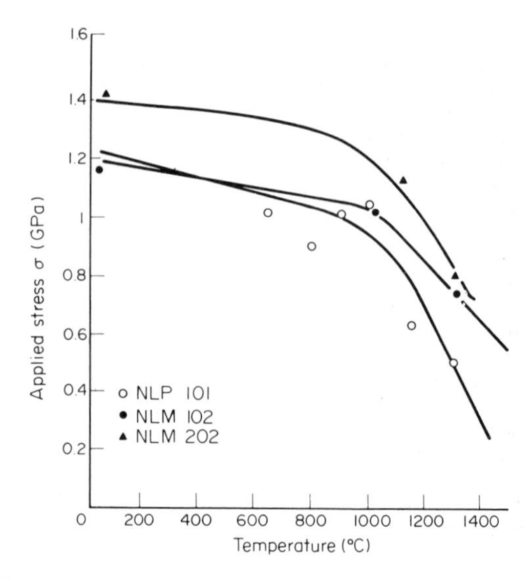

Figure 3
The strengths of three commercially available Nicalon fibers as a function of temperature

oxide surface-coating prevents outgassing of the oxides of carbon formed.

Tyranno fibers are amorphous when received from the manufacturer; however, heating above 1000°C results in a crystallization of the structure. The presence of titanium, up to about 5 wt%, is said to retard crystallization and reduce reactions between the fiber and an aluminum matrix.

5. *Properties of Fine SiC Fibers*

The tensile strength of fibers always shows considerable scatter which can be described in statistical terms by a parameter, usually called the Weibull modulus. The higher the value of this modulus the less the scatter, so that steels can have a Weibull modulus of 60 whereas the corresponding values for ceramics are rarely above 20. Fine ceramic fibers have Weibull moduli of around 4, showing that considerable scatter exists in their tensile properties. Nicalon fiber has a Young's modulus of around 190 GPa and a strength of 1.5 GPa with a density of 2.55 $\mathrm{g\,cm^{-3}}$. Its tensile behavior is linearly elastic.

Tensile tests on Nicalon fibers in both air and argon have shown that the fibers maintain their strengths and Young's moduli up to 1000°C. This can be seen from Fig. 3. Each point represents an average of at least 30 tests.

The fracture surface of the fibers as seen by a scanning electron microscope does not alter in appearance over the temperature range studied (up to 1400°C) and is of a brittle nature. A smooth mirror zone of stable crack-growth is followed by the irregular fracture surface characteristic of rapid failure.

Nicalon SiC fibers have been observed to creep at temperatures above 1000°C. At 1000°C, creep is not observed if the applied stress is below 0.4 GPa and the time-dependent strain is less than $10^{-8}\,\mathrm{s^{-1}}$. Creep curves obtained are of a classical form obeying the equation

$$d\varepsilon/d\tau = A(\sigma - \sigma_o)^n \exp(-\Delta H/RT)$$

where $d\varepsilon/d\tau$ is the creep rate, σ is the applied stress, σ_o is the creep threshold stress below which creep is not observed, ΔH is the activation energy of the mechanisms controlling creep, T is the temperature, and n and A are adjustable parameters.

The creep threshold level is found to decrease as the temperature is increased above 1000°C so that it is practically nonexistent at 1300°C. This threshold stress level is thought to be due to the particles of free carbon in the structure inhibiting movement. A typical activation energy at 1100–1300°C is 490 $\mathrm{kJ\,mol^{-1}}$—in air this value becomes rather lower.

Nicalon fibers have been reported by Favry and Bunsell (1987) to react at relatively low temperatures when in contact with molten metal so that the use of these fibers depends very much on the matrix environment in which they are embedded.

Tyranno fibers have a density of 2.4 $\mathrm{g\,cm^{-3}}$, a room-temperature strength of 2.9 GPa and a Young's modulus of 140 GPa. The strength of Tyranno fibers falls at temperatures above 1000°C although the presence of the titanium in its structure may retard its degradation (Okamura 1987).

See also: Boron Fibers

Bibliography

Crane R L, Krukonis V J 1975 Strength and fracture properties of silicon carbide filament. *Am. Ceram. Soc. Bull.* 54: 184–88
Favry Y, Bunsell A R 1987 Characterisation of Nicalon (SiC) reinforced aluminium wire as a function of temperature. *Compos. Sci. Technol.* 30: 85
Lindley M W, Jones B F N 1975 Thermal stability of silicon carbide fibres. *Nature (London)* 255: 474–75
Mah T I, Mendiratta M G, Katz A P, Mazdiyasni K S 1987 Recent developments in fibre-reinforced high-temperature ceramic composites. *Am. Ceram. Soc. Bull.* 66: 304
Okamura K 1987 Ceramic fibres from polymer precursors. *Composites* 18(2): 107
Simon G, Bunsell AR 1984a Mechanical and structural characterisation of the Nicalon SiC fibre. *J. Mater. Sci.* 19: 3649
Simon G, Bunsell A R 1984b Creep behaviour and structural characterisation at high temperatures of Nicalon SiC fibres. *J. Mater. Sci.* 19: 3658
Whatley W, Wawner F E 1985 Kinetics of the reaction between SiC (SCS-6) filaments and Ti-(6Al-4V) matrix. *J. Mater. Sci. Lett.* 4: 173–75
Yajima S, Hasegawa Y, Hayashi J, Limura M 1978 Synthesis of continuous silicon fibre with high tensile strength and high Young's modulus. *J. Mater. Sci.* 13: 2569–76
Yajima S, Okamura K, Hayashi J, Omori M 1976 Synthesis of continuous SiC fibres with high tensile strength. *J. Am. Ceram. Soc.* 59: 324
Yamamura T, Hurushima H, Kimoto M, Ishikawa T, Shibuya M, Iawa T 1987 Development of new continuous Si–Ti–C–O fiber with high mechanical strength and heat resistance. In: Vincenzini P (ed.) 1987 *High Tech. Ceramics.* Elsevier, Amsterdam, p. 737

A. R. Bunsell
[Ecole Nationale Supérieure des Mines de Paris, Evry, France]

J.-O. Carlsson
[University of Uppsala, Uppsala, Sweden]

Silicon Nitride Fibers

The formation of silicon nitride (Si_3N_4) by the pyrolysis of organometallic polymers containing silicon has been of interest for some years as a potentially convenient route to the production of components of complex shape. Because many precursor organometallic polymers can be readily spun into continuous filament fibers using conventional textile processes, a route also exists to the production of silicon nitride fibers. If these fibers can be made with an adequate degree of control of structure and surface finish, it is expected that, like carbon fiber, they will have high modulus and high strength. Such fibers are of considerable interest as possible reinforcing materials in polymer matrices and for high-temperature applications in ceramic matrix systems where better resistance to oxidation than that shown by carbon fibers may be expected.

1. Precursor Systems

The primary requirement for the polymeric precursor is that the main chain contains silicon. A number of systems have been examined:

(a) polysilazanes +Si–N+_n, with a Si–N chain;

(b) polycarbosilanes $\text{+Si–CH}_2\text{+}_n$, in which CH_2 groups separate the silicon atoms; and

(c) polysilanes +Si–Si+_n.

Of these, the first two types have received the most attention. The polysilazanes contain the necessary nitrogen in the chain conveniently already bonded to the silicon. The polycarbosilanes can be nitrided at temperatures between 500 °C and 800 °C in a nitrogen-containing atmosphere (e.g., ammonia) to effect nitrogen insertion with development of silicon–nitrogen bonds. An important factor for the production and subsequent heat-treatment steps of the organopolymer is the nature of the remaining side groups required to complete the silicon and nitrogen valencies of four and three, respectively. Hydrogen would, theoretically, be ideal; in practice, alkyl (CH_3), aryl (C_6H_5) and halogen (Cl) groups may also be linked with an increasing probability of producing stable silicon–carbon bonds leading to a mixed nitride–carbide fiber, or to the presence of residual chlorine. The polysilazanes and polymers containing silicon–hydrogen bonds suffer from the disadvantage of marked reactivity towards water, and the incorporation of oxygen and the presence of silicon–oxygen bonds in the ceramic fiber; for this reason, dry-box handling is standard practice.

The use of the organofibers as a starting point to ceramic fiber production places further restrictions on the precursor polymer properties; this normally results in a number of compromises. The ideal fiber is based on a linear, symmetrical polymer chain, with strong intermolecular bonding resulting from the presence of polar groups, and of moderately high (5000–20 000) molar mass to give good spinning characteristics.

2. Production

A linear chain polymer softens and deforms on heating so that, after spinning, it is necessary to carry out a preliminary crosslinking ("curing") step in order to bridge polymer chains and to prevent melting and excessive depolymerization and evaporation of the fiber during the later high-temperature pyrolysis stage. Crosslinking requires active sites within the polymer chain. Identification of the best crosslinking method will depend on the polymer chemistry but techniques explored include controlled intermediate temperature (~50–200 °C) oxidation, exposure to ultraviolet (<300 nm) light, to high-energy (2 MeV) electrons and to γ rays. Crosslinking using reactive molecules (e.g., $HSiCl_3$) has been useful in special cases. A properly crosslinked fiber is both insoluble and infusible.

After crosslinking, a high-temperature heat treatment of the fiber can be carried out to bring about the structural rearrangement required to generate a three-dimensional network of silicon–nitrogen bonds and to eliminate unrequired atomic species. Temperatures in the range 1000–1300 °C are normally sufficient to produce a silicon nitride fiber of reasonable purity. Polysilazanes can by pyrolyzed under dry argon; nitrogen or ammonia atmospheres are normally used for fibers based on the polycarbosilanes. An important factor in the pyrolysis process is the rate of temperature rise. This must be slow and, typically, a rate of around 3 °C min^{-1} is used to allow time for ejection of pyrolysis products without excessive damage to the fiber. Mass loss starts at around 300 °C and is completed by about 800 °C.

As with other forms of ceramic, good mechanical properties can only be obtained with materials of good microstructural quality. Defects of any kind lead to strength degradation. Pores, microcracks and surface damage are particularly undesirable; this applies equally to the precursor fiber because of the virtual impossibility of annealing out such defects during pyrolysis. For this reason, precursor chemicals of high purity, homogeneity and freedom from particulate contamination are also required.

During the pyrolysis stage there is a large mass loss (up to 50% or more) as volatile species are eliminated. There is an associated large volume shrinkage of 70% or more (corresponding to an overall linear shrinkage of the order of 25%; typically a 20 μm diameter fiber will shrink to about 15 μm). The density increases from around 1000 kg m^{-3} to around 2500 kg m^{-3}.

3. *Fiber Properties*

The product at the completion of the pyrolysis stage is normally 100% amorphous by x-ray diffraction, colorless and transparent. The composition is normally homogeneous with respect to element distribution, although not usually exactly of the Si_3N_4 stoichiometry. A slight excess of silicon is often found, giving an apparent formula of $Si_3N_{3.7-3.8}$. Spectroscopic examination shows that silicon atoms can be bonded simultaneously to nitrogen, carbon and oxygen atoms. If there is a large excess of residual carbon, this may be present as microcrystalline graphite. The measured densities of about 2500 kg m^{-3} correspond to 80–90% of the expected value for amorphous silicon nitride (2700–2800 kg m^{-3}); the absence of measurable porosity suggests that there is a large void volume within the (Si_3N_4) structure.

Table 1
Properties of fiber produced by pyrolysis at 1200–1300 °C

Diameter (μm)	Tensile strength (GPa)	Modulus (GPa)
4.5	3.36	245
10–15	3.1	260
12	1.75	196
13	1.3	120

The mechanical properties of high-quality amorphous silicon nitride fiber can be very good indeed, although it is clear that the fiber quality is very sensitive to factors such as the nature of the precursor polymer, the steps of the production process and the quality of the surface finish. Fiber diameters are generally in the range 3–20 μm (smaller diameters are associated with health hazards and larger diameters become inconveniently inflexible). There is a strong trend for mechanical property values to deteriorate with increasing fiber dimension and with increasing temperature of pyrolysis. Table 1 shows the values reported for fiber produced by pyrolysis at temperatures in the region of 1200 °C to 1300 °C.

Pyrolysis at temperatures above 1350 °C leads to crystallization of the amorphous silicon nitride to the α-Si_3N_4 crystalline phase and to total loss of strength.

It is clear that optimization of fiber quality has not yet been achieved and work is in progress in all areas. A major question concerns the high-temperature stability of the fiber. The tendency for amorphous silicon nitride to crystallize at temperatures above 1300 °C is well known from earlier studies on amorphous chemical-vapor-deposited silicon nitride. Whether or not silicon nitride fiber can find long-term application at temperatures above 1350 °C is likely to depend on a satisfactory degree of inhibition of the crystallization process (possibly by incorporation of other elements into the structure, as in the case of the amorphous silicon dioxide system).

Silicon nitride fibers and composites containing them are still in the development stage and actual applications have not yet been reported. Research in this area forms part of extended programs of work into low-density, high stiffness-to-weight ratio polymer systems and into ceramic systems suitable for use at very high temperature, possibly where a limited life is acceptable. Much of this work is commercially or militarily sensitive.

Bibliography

Baney R H 1984 Some organo-metallic routes to ceramics. In Hench L L, Ulrich D R (eds.) 1984 *Ultrastructure Processing of Ceramics, Glasses and Polymers*. Wiley, New York, pp. 245–55

Katz A P, Kerans R J 1988 Structural ceramics program at AFWAC materials lab. *Ceram. Bull.* 67(8): 1360–6

Legrow G E, Lim T F, Lipowitz J, Reaoch R S 1987 Ceramics from hydridopolysilazane. *Ceram. Bull.* 66(2): 363–7

Okamura K, Sato M, Hasegawa Y 1987 Silicon nitride fibre and silicon oxynitride fibre obtained by the nitridation of polycarbosilane. *Ceram. Int.* 13: 55–61

Seyferth D, Wiseman G H 1984 Silazane precursors to silicon nitride. In: Hench L L, Ulrich D R (eds.) 1984 *Ultrastructure Processing of Ceramics, Glasses and Polymers*. Wiley, New York, pp. 265–71

F. L. Riley, J.-C. Ko and G. C. East
[University of Leeds, Leeds, UK]

Smart Composite Materials Systems

There is no universally accepted definition of the term "smart material" (also described as intelligent, sense-able, multifunctional or adaptive materials); however, they can be thought of as material systems which manifest their own functions intelligently depending on sensed environmental changes. They are modelled on biological systems with:

(a) sensors acting as a nervous system (right-hand full circle in Fig. 1),

(b) actuators acting like muscles (left-hand full circle in Fig. 1), and

(c) real-time processors acting as a brain to control the system (lowest full circle in Fig. 1).

The different classes of structures relating to the technology are given in Fig. 1. Of particular interest are the following:

(a) *Type I: passive/sensory structures*, which possess a structually integrated microsensor system for determining the state of the structure and possibly the environment in which it is operating;

(b) *Type II: reactive smart structures*, which have a nervous system and an actuator control loop to effect a change in some aspects (stiffness, shape, position, orientation or velocity); and

(c) *Type III: intelligent structures*, which are capable of adaptive learning.

While specialized materials incorporating different aspects of intelligence are being actively researched, most notably in Japan (Takagi 1990), smart materials concepts are most likely to be applied in a more primitive form within host composite materials, where appropriate sensors and actuators can be embedded into the structure during manufacture. Embedded or surface-bonded sensors represent the first stage of development, and it is in this area that most work has been carried out. The second stage concerns the incorporation of embedded or surface-mounted actuators, and the third stage is the control system to interpret the sensor data to decide on appropriate action.

1. Sensor Systems

It is possible to embed sensors into composite components during manufacture to allow internal interrogation of the material. Sensors can be based on acoustic waveguide wires, piezoelectric or optical fibers. Fiber-optic based sensors offer the greatest scope and are receiving the most attention. They are compatible with the fabrication process and are capable of withstanding strains of the same magnitude as the composite itself. The dielectric nature of the material is maintained while the optical fibers provide their own signal paths for sensing the interior of the composite. They offer the prospect of continuously monitoring the composite structure at all stages of its life through

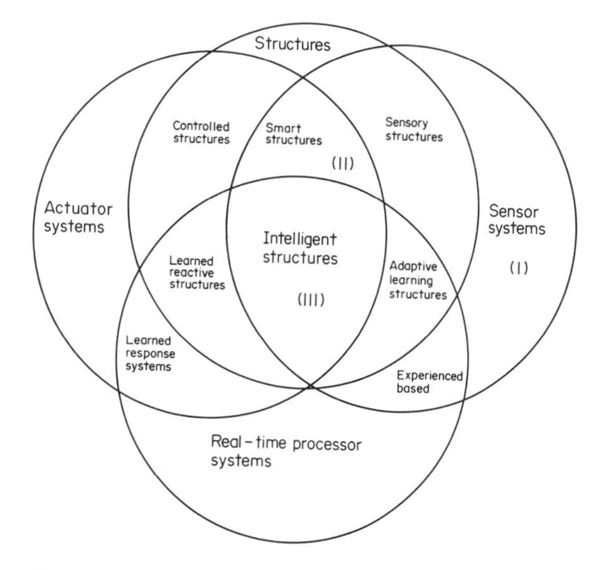

Figure 1
Smart materials classification

fabrication, test qualification and service. The successful development of a passive smart structure is reliant upon devising and adapting appropriate fiber-optic sensors, sensing techniques and multiplexing techniques; and upon the establishment of appropriate fabrication schemes to include the embedded sensors. Fiber-optic sensors rely on the interaction of a physical parameter in the vicinity of a fiber to produce modulation of the transmitted signal. This may take the form of modulation of either amplitude, phase, polarization or mode interactions. Optical fiber sensors can be designed to detect a whole range of physical parameters including temperature, strain, pressure, electric fields and magnetic fields. Distributed and quasidistributed measurements have been demonstrated using optical time domain reflectometry (OTDR) techniques by monitoring reflected signals (Rogers 1988). In sensor applications the system must demodulate the fiber sensor signal and relate the demodulated output to the physical parameters under investigation. The basic sensor types researched for use in composites are described in Sects. 1.1–1.4.

1.1 Microbend Intensity Sensors

The variation in transmitted power through a monomode or multimode fiber optic is related to the periodic microbending effects produced by the composite itself. The characteristics of such a sensor for temperature and strain have relatively low sensitivity and poor predictability. Microbend sensors for state of cure monitoring may have a role to play (Davidson et al. 1992).

1.2 Interferometric Phase Sensors

A range of interferometric sensor configurations exists which all detect the influence of physical perturbations on the phase of coherent laser light propagating in a single mode fiber. Phase shifts can be caused by either a change in the refractive index of the core or a change in the state of strain in the fiber. In simple interferometers the phase change is detected as fringe shifts when the outputs of the sensor and a reference arm are recombined. The effects of temperature and strain are not separable and the output represents an integrated effect over the whole length of fiber.

1.3 Modal Interferometers

A single fiber may be used as an interferometer if it is operated at a wavelength significantly below the cutoff wavelength so as to propagate the first two modes (LP_{01} and LP_{11}). The modal interference is a function of the strain or temperature changes seen by the fiber optic. Detection is either by observing intensity changes in a small part of the signal in the far field or using a detector array to analyze the intensity distributions (Bennett et al. 1988). Such systems are less sensitive than phase sensors but will be adequate for most applications and do not suffer problems associated with fiber couplers on the isolation of the reference arm.

1.4 Polarimetric Sensors

Polarization-maintaining fibers are required and these sensors also use only a single optical fiber. The fibers have axes of stress built in using either preferential doping of the cladding or through shape birefringence using elliptical cores, so that the otherwise orthogonal but degenerate polarization modes become separated and propagate at different velocities. The phase difference or the power distribution induced in the two orthogonal polarization eigenmodes are monitored to assess strain or temperature changes (Dunphy et al. 1987). Localization of the sensing region can be achieved with the use of 45° splice as these ensure that the sensor has temperature-insensitive lead-in and lead-out optical fibers. A differential measurement is made which is ~100 times less sensitive than interferometric systems.

1.5 Distributed Fiber-Optic Sensing Systems

Distributed sensors exploit the one-dimensional nature of optical fibers to make continuous measurements as a function of position along the fiber length. Most designs rely on backscattered light which is usually of very low intensity compared with the forward transmission. Distributed measurements can take place anywhere along a single fiber length where a perturbation takes place. The light backscattered from optical inhomogeneities is used to obtain information on the attenuation properties of the optical fiber along its length. In practice, however, the spatial resolution is limited to ~1 m along the fiber for real-time operation. Consequently a substantial reassessment of "conventional" distributed sensing techniques is required. Quasidistributed measurements look more promising. The basic requirement is to sensitize the fiber at discrete points along the length to create significantly larger signals from each sensing element and obtain a high signal-to-noise ratio. Several approaches have been examined including partially reflective splices, Bragg reflections from laser-etched photorefractive regions of periodic variation in the core refractive index (Measures et al. 1992) and pressure-induced coupling in birefringent fiber (Gusmeroli et al. 1989), though none is ideal for embedded systems. Recent European work (Davidson and Culshaw 1989) has demonstrated a quasidistributed measurement system using up to seven sensed regions of 25 cm length. Thermal effects at discrete points in a Hi-Bi (highly birefringent) fiber are used to produce cross-coupling points to allow demultiplexing of the signal using a "white light" interferometric system.

2. Actuation Mechanisms

To produce controlled or reactive smart structures, actuators are necessary. These can be either retrofitted onto any material by surface bonding or embedded within a composite material. The ideal actuator does not exist. A range of candidate systems is being studied, particularly in the USA where industrial and military interest in reactive smart structures is high. The main systems under investigation are described in Sects. 2.1–2.5.

2.1 Shape Memory Alloys

Shape memory alloy (SMA) based on NiTi (53–57 wt% Ni) Nitinol, when plastically deformed in the low-temperature martensitic phase, can be restored to its original shape or configuration by heating above a characteristic temperature. Plastic strains of 8% can be completely recovered by heat transformation from the deformed martensitic phase to the austenitic phase. However, for cyclic applications a level of <2% is recommended. If the material is restrained from regaining its memory shape, high stresses of up to 700 MPa can be induced. This compares with a yield strength of the martensitic phase of 80 MPa. On transformation to the austenitic form the Young's modulus increases by a factor of 4 and the yield strength by a factor of 10. Shape memory alloys have the ability of changing material properties almost reversibly; typical properties are given in Table 1. SMAs used as embedded actuators take the form of 200–400 μm plastically elongated wires constrained from recovering their normal memorized length during fabrication. The plastically deformed fibers become an integrated part of the composite material structure. When the fibers are resistively heated, they are restrained from recovering to their memorized length by the composite and generate a uniformly distributed shear load along the entire length of the fiber. If the fibers are offset from the neutral axis the structure will deform in a predictable manner.

Table 1
Typical properties of Nitinol

Property	Value
Expansion coefficient (°C^{-1})	10 ppm
Density (kg m^{-3})	6500
Melting temperature (°C)	1270
Transition temperature (°C)	tailorable (0 to 100 °C)
Anneal temperature for permanent damage (°C)	320
Shear modulus: martensite (GPa)	7.5
Shear modulus: austenite (GPa)	22.0
Yield stress: martensite (MPa)	80
Yield stress: austenite (MPa)	620

Such systems have been proposed to control vibrations in large flexible composite structures (Rogers et al. 1988). An alternative method of use is to place the SMA fibers in or on the structure in such a way that when actuated there are no resulting large deformations but instead the structure is placed in a residual state of strain, and the modal response of the structure is changed. Using such techniques, the ability to change the effective stiffness, natural frequencies and mode shapes of composite plates has been demonstrated (Rogers 1992). SMAs are high-force, high-stroke, low-frequency actuators, not suitable for damping high-frequency vibrations since cooling is governed by conduction and radiation loss.

2.2 Piezoelectric Actuators

When stress is applied to piezoelectric materials they develop an electric moment whose magnitude is proportional to the applied stress (Nye 1957). This is a consequence of the direct piezoelectric effect, and the constant of proportionality d_{ijk} (a third-rank tensor) defines the piezoelectric moduli. When an electric field is applied, the dimensions of a piezoelectric material change slightly as a result of the converse piezoelectric effect. The same d-tensor relates the resultant strain to the electric field. Materials in film form can be bonded externally or embedded internally into composite structures to act as actuators. The crystalline subdomains in the films are first aligned (poled) by the application of a large coercive field of ~2 kV mm^{-1} across their thickness. This causes the piezoelectric to grow in the field direction and shrink laterally. The domains remain stable at temperatures less than the Curie temperature. Subsequent application of a field in the poling direction also causes growth in that direction and generates longitudinal and transverse contraction strains. Application of reverse field causes a shrinkage in thickness until a negative coercive field level is reached, after which the poling reverses and the thickness expands again. Piezoelectrics are available in ceramic or polymer form, and the properties of several systems are given in Table 2. The effectiveness of the piezoelectric in the bending of a substrate beam is given by:

$$\text{Effectiveness} = E_{max} d_{31}\{6/[6 + (E_b t_b / E_c t_c)]\}$$

where E_b and E_c are the elastic moduli of the beam and the piezoelectric ceramic, respectively, and t_b and t_c are the respective thicknesses.

The effectiveness data in Table 1 are obtained assuming $t_b/t_c = 10$ and $E_b = 70$ GPa. If the voltage available is limited then the effectiveness per field becomes important and ceramics are much more efficient than polymers. Lead zirconate–titanate (PZT) piezoelectric ceramics are solid solutions of lead zirconate and lead titanate in a perovskite structure, with the ratio determining the

Table 2
Comparison of piezoelectric materials

Material	PZT G1195	PZT G1278	PVDF
Curie temperature (°C)	360	190	100
Electric field E_{max} (kV m^{-1})	600	600	40000
d_{31}[a] (pm V^{-1})	190	250	23
Modulus (GPa)	63	60	2
Effectiveness ($\times 10^{-6}$)	40	51	16
Effectiveness/field (pm V^{-1})	67	85	0.39

a $d_{33} \approx 2.5\, d_{13}$ (units: m V^{-1} = C N^{-1})

phase transition boundaries and resulting properties. For embedding applications the actuator thickness must either be less than the ply thickness or plies must be cut to accommodate the sensor; also the Curie temperature must be higher than the processing temperature of the composite. It is clear that ceramics offer a wider operating temperature range and higher effectiveness per field than polymeric poly(vinylidene fluoride) (PVDF)-based films. Piezoelectrics cannot be inserted directly into conducting carbon-fiber-reinforced plastic (CFRP); a hard insulating layer is required to insulate the actuator and electrical leads. Such factors may have serious consequences on the structural integrity of strength-critical composites. Experiments at Massachusetts Institute of Technology have investigated various aspects of distributed surface-bonded and embedded piezoelectric actuators. Controlled bending and twisting of model aerodynamic surfaces has been achieved using distributed surface-bonded piezoelectrics. The piezoelectric actuators also have the potential to act as dynamic strain sensors, since a dynamic load applied to the piezoelectric results in an electric charge which may be monitored. Hence, if the voltages generated are sufficiently high to be monitored and if the loads generated can be made large enough to cause displacements, piezoelectrics may act as both actuator and sensor; recent work at TRW (Bronowicki 1991) has demonstrated this in tubular composite structures.

2.3 Electrostrictive Materials

One disadvantage of piezoelectric actuators is that the material response is both nonlinear and hysteretic, particularly at high ac voltages where electrostrictive contributions are significant. All materials are electrostrictive but few have a large coefficient. The electrostrictive properties result from a quadratic dependence of permittivity with electric field, in contrast to the true converse piezoelectric effect caused by a linear dependence. Pure electrostrictive materials typified by lead magnesium niobate (PMN) show no piezoelectric effect; they are nonhysteretic but are nonlinear. The perovskite-type crystal expands longitudinally and shrinks laterally on the application of an electric field. The strain effect is quadratic in voltage and decreases with temperature. The low hysteresis makes the effector suitable for open-loop shape control. No poling is necessary but for vibration damping a bias voltage is required. Little work is available in the open literature on the use of electrostrictors in smart composites though there is increasing interest.

2.4 Magnetostrictive Materials

Recent advances in the production of Terfenol-D magnetostrictive alloys of terbium, dysprosium and iron ($\sim Tb_{0.3}Dy_{0.7}Fe_{1.9}$) offer high-strain ($\sim$0.1%) and high-energy-density ($\sim$20 kJ m^{-3}) materials, suitable for magnetic actuation of structures (Hathaway and Clark 1993). Work relating to active space struts using this material is ongoing.

2.5 Electrorheological Fluids

Some initial proof-of-concept work has been carried out to demonstrate that dynamically tuneable smart composites can be produced using electrorheological fluid-based actuators (Gandhi and Thompson 1989). These fluids undergo significant instantaneous reversible changes in materials characteristics, most notably in their bulk viscosity, when subjected to electrostatic potentials. A variety of electrorheological (ER) fluids exist based on micrometer-sized hydrophilic particles suspended in a suitable hydrophobic carrier liquid. Dry field effect fluids have been developed featuring a nonconducting liquid in which is suspended particles of crystalline aluminum silicate zeolites. These fluids are completely dry, with good suspension stability, and are capable of operating at up to 250 °C. It is this type of fluid which has the most potential to modify the damping properties of composite materials. On the application of electric fields of 2–4 kV mm^{-1} across the thickness of an ER fluid layer, the rheological characteristics are dramatically changed and the inherent molecular structure of the ER fluid creates solidlike characteristics, as the particles in the suspension orient themselves in relatively regular chainlike columns. These columnar structures increase the energy dissipation characteristics of the suspension and cause a redistribution of mass in the suspension. On removal of the electrostatic field the particles return to a state of random orientation with fluid viscosities. The transient responses of several cantilever CFRP laminated beams containing ER fluids have been studied. The natural frequencies can be actively changed throughout the frequency spectrum by controlling the voltage imposed on the fluid domains and in this way resonances can be avoided. The problems with this technology relate to the

Table 3
A comparison of actuator materials

	PZT G1195	PVDF	PMN-BA	Terfanol	Nitinol
Actuation mechanism	Piezoceramic	Piezofilm	Electrostrictor	Magnetostrictor	SMA
$\varepsilon_{max}{}^{a}$ (μ strain)	1300	230 dc 690 ac	1300	>2000	80000 dc 20000 ac
E (GPa)	63	2	121	48	30(m), 89(a)[b]
$T_{max}{}^{c}$ (°C)	360	80–120	>500	380	~50
Linearity	good	good	fair	fair	poor
Hysteresis (%)	10	>10	<1	2	5
Temperature sensitivity (% °C^{-1})	0.05	0.8	0.9	high	
Bandwidth	high	high	high	moderate	low

a Maximum strain capability b m = martensitic phase, a = austenitic phase c maximum operating temperature.

means of adequately incorporating and confining the ER fluid into composites and being able to activate the fluid with high voltages. The fluid adds weight and does not enhance the structural performance of the composite.

Relative properties of a variety of actuator materials are given in Table 3.

3. Control Systems

For type III structures containing both embedded sensors and actuators, signals from the sensors will be received and interpreted by microprocessor-based controllers which will communicate with and trigger the actuators to alter the material response. Parallel processing using neural networks is receiving much attention for use in real-time controllers. The concepts are inspired by biological systems where a large number of nerve cells that individually function rather slowly and imperfectly learn collectively to perform complex tasks. Neural networks are made of layers of relatively simple nonlinear elements and are capable of adaptive learning and rapid processing and decision making. If and when sensors are developed to give strain and temperature information from critical areas of say an aircraft, then if the structural integrity is to be assessed, at the very least some comparative reference values will be necessary. Probably the best way to get a baseline would be to fly the aircraft and learn. Neural networks with self-adaptive processing are potentially well suited to handle such high volumes of data (Grossman and Thursby 1989). Although neural network-based controllers do not yet exist in suitable forms to act as the "brain" in a fully intelligent material, it is an area where rapid advances are expected.

4. Materials Issues Relating to Smart Materials

Composite materials with embedded sensors and actuators will only gain acceptance if the structural integrity of the composite is not significantly reduced by the presence of the inclusions, which will always be significantly larger in diameter than the carbon, aramid or glass-reinforcing fibers which are typically ~8–10 μm in diameter. When optical-fiber sensors, typically 100–300 μm in diameter, are embedded in composite laminates there is an inevitable disruption of the reinforcing fibers in the vicinity of the fiber optic. The nature of this disruption is dependent on both the diameter of the embedded sensor and the relative orientation of the fiber optic with respect to neighboring reinforcing plies. For example, sensing fibers lying parallel to the local reinforcement cause a minimum disruption provided the diameter is less than half the ply thickness. Reinforcing fibers laying orthogonally to the sensors are locally deformed, creating a resin-rich region around the sensor. In order to be acceptable the fiber sensor must:

(a) produce a minimum perturbation in the distribution of reinforcing fibers;

(b) not significantly reduce the mechanical properties of the composite;

(c) not suffer from excessive attenuation or damage from the embedding process, such that the sensing technique cannot be applied; and

(d) include a suitable means to input and output the laser light into the system, through pigtails or connectors. Such systems must be robust and compatible with the fabrication process.

4.1 Optical Fiber Embedding

Modifications to the composite manufacturing process are required in order to lay fiber-optic sensors at the required positions and to ensure that signals can be input and output through fiber pigtails or couplers. Successful manufacture has been achieved using autoclave, press and vacuum bag molding of laminates and filament winding of tubes. In the molded laminates the fiber pigtails are usually taken out of the edges of the laminate, whereas in the filament-wound tubes it has been possible to exit the fiber pigtails through the tube surface away from the ends. Extra protection around the emergent parts in the form of small-diameter polytetrafluoroethylene (PTFE) tubing has proved effective. In order to assess the influence of embedded sensors on the mechanical integrity of the composite, transverse tensile specimens were used, with the fiber optic laying parallel to the reinforcement. Various fiber sensors with polymeric, metallic and other hermetic coatings as well as uncoated fibers were tested. Strengths ranging from 56% to 98% of the transverse strength of the base composite were obtained and fiber dimensions and coatings giving the least detrimental effects were identified (Roberts and Davidson 1991). Compression and shear strengths of CFRP were also shown to be unaffected if appropriate choices of sensor and coating combinations are made. Coatings based on polyimides were found to offer the best overall performance and were not thermally or mechanically degraded at the fabrication temperatures of up to 200 °C used in epoxide matrix composites. Such fiber sensors have also been successfully used in high-temperature thermoplastic matrix composites based on polyether ether ketone (PEEK), processed at 390 °C. The coating on the optical fiber is also important in transferring strain from the composite into the fiber optic and so modulates the laser light propagating in the core.

4.2 Important Factors

Many factors remain to be fully investigated before the sensing technology can be applied in real structures (e.g., naval, aircraft or space structures).

(a) Materials considerations include how the fiber affects the composite strength characteristics, and whether the sensors are resistant to the composite environment.

(b) Fabrication aspects include determining the best deposition of the fiber sensors in the composite; how to retain accurate positioning of the fiber optic during manufacture; how to monitor and automate the fabrication process and deal with connection problems; and how to make cost-effective structures and make repairs.

(c) Optical sensing/multiplexing techniques include the kind of fiber sensor to use; how to differentiate important variables; how to make distributed measurements to the required resolutions; and how to design reliable laser connect schemes and miniaturized systems.

(d) System aspects include how the strain sensing relates to scheduled maintenance; which necessary developments in artificial intelligence techniques are required to interpret the data; and how redundancy can be built into the system.

Many of the above factors are equally important as far as actuation is concerned. The actuator developments are at a less advanced stage than for sensors. However, working demonstrators are beginning to be made and tested (Wilson et al. 1990, Bronowicki 1991).

5. Application Areas for Smart Composite Materials

Much of the work relating to smart composites is still at the research stage with the aim of demonstrating the concepts and proving the feasibility of workable schemes in real structures. The first use of embedded fiber-optic sensors will be most likely in process monitoring and control of the fabrication stage of composites. This is particularly important in thick-walled structures based on thermosetting resin matrices where exothermic reactions can cause high temperatures in the center of the structure producing nonuniform cure throughout the thickness and high thermal stresses which can lead to cracking of the structure.

Potential applications in structures include the control of aerodynamic, hydrodynamic and optical surfaces, stealth, robotics, damping/tuning of structures and vibration suppression in transportation (in particular in satellites, helicopters and submarines). In the field of civil engineering, applications are expected in the development of intelligent buildings and in bridge and road monitoring.

Advanced aircraft of the future may use embedded sensors to examine nondestructively the flight worthiness during power up and monitor the dynamic structural response during taxiing (Sendeckyi and Paul 1989). Flight loads and damage growth during flight will be monitored and the pilot advised of any flight restrictions. Maintenance and repair will be performed as needed by automatic analysis of flight sensor data. Smart skins also offer the prospect of improved aircraft performance and avionics capability by integrating sensors, signal processors, power distribution networks and control functions within a composite load-bearing skin structure to form an active interface with the flight environment (Curtis 1989). Such a system could be thin and conformal allowing external aerodynamic contours to be maintained. Further developments may enable the smart skins to monitor hostile threats and initiate countermeasures. Clearly the "smart" aircraft is some years away and significant improvements in current technology will be necessary to bring these ambitious concepts into reality.

Space-based applications for smart materials include space-based platforms and space stations where there are serious vibration and control problems such that it will be necessary to have active control of the vibration of the structure. Piezoelectric sensors and actuators look to be the most promising for this application (De Hart 1989). Health monitoring systems capable of sensing any degradation in large space structures caused by, for example, micrometeorite damage, will be required; here fiber optics are likely to be of most use. It will be necessary to determine the position and extent of any damage in order to initiate repair. The whole range of advanced composite material types is being considered as possible hosts for smart materials technologies. Polymeric matrices (both thermosetting and thermoplastic) based on carbon, glass and aramid reinforcement present the most benign host materials. Reinforced metals, ceramics, and carbon are much more difficult systems to incorporate smart concepts in view of the high temperatures involved in both processing and operation. No suitable embedded sensor systems exist as yet for such host materials though recent work has considered sapphire-fiber sensors as a possible solution (Ayestia et al. 1989).

6. Concluding Comments

The development of smart composite materials systems requires a truly multidisciplinary approach. Innovations relate to combining materials, sensors and actuators along with control systems. New design methodologies will be required to fully utilize and take full advantage of these advanced materials systems.

See also: Nonmechanical Properties of Composites

Bibliography

Ayestia R, Seshu B, Claus R O 1989 High temperature refractory coatings for sapphire waveguides. *Proc. Soc. Photo-Opt. Instrum. Eng.* 1170: 513–20

Bennett K D, McKeeman J C, May R G 1988 Full field analysis of modal domain sensor signals for structural control. *Proc. Soc. Photo-Opt. Instrum. Eng.* 986: 85–9

Bronowicki A J 1991 Controlled structures for space. (Paper presented at Active Materials and Structures Conference, Alexandria, VA, November 1991)

Curtis D D 1989 Fibre optic beam forming of smart skin arrays. *Proc. Soc. Photo-Opt. Instrum. Eng.* 1170: 48–57

Davidson R, Culshaw B 1989 Intelligent composites containing measuring fibre optic networks for continuous self diagnosis. *Proc. Soc. Photo-Opt. Instrum. Eng.* 1170: 211–23

Davidson R, Roberts S S J 1991 Do embedded sensor systems degrade mechanical performance of host composites? In: Knowles G J (ed.) 1991 *Proc. Active Materials and Structures Conf.* Alexandria, VA. Institute of Physics, Bristol, UK, pp. 109–14

Davidson R, Roberts S S J, Zahlan N 1992 Fabrication monitoring of APC-2 using embedded fibre optic sensors. In: Bunsell A R, Jamet J F, Massiah A (eds.) 1992 *Proc. 5th European Conf. Composite Materials*, Bordeaux. European Asociation for Composite Materials, pp. 441–6

De Hart D W 1989 Airforce astronautics laboratory smart structures and skins overview. *Proc. Soc. Photo-Opt. Instrum. Eng.* 1170: 11–18

Dunphy J R, Meltz G, Elkow R M 1987 Distributed strain sensing with a twin core fibre optic sensor. *ISA Trans.* 26(4): 7–10

Gandhi M V, Thompson B S 1989 Dynamically-tunable smart composites featuring electrorheological fluids. *Proc. Soc. Photo-Opt. Instrum. Eng.* 1170: 294–304

Grossman B G, Thursby M H 1989 Smart structures incorporating artificial neural networks, fibre optic sensors and solid state actuators. *Proc. Soc. Photo-Opt. Instrum. Eng.* 1170: 316–25

Gusmeroli V, Vavassori P, Martinelli M 1989 A coherence multiplexed quasi-distributed polarimetric sensor for structural monitoring. *Proc. IEE Optical Fibre Sensors Conf.*, Paris. Springer Proceedings in Physics, Springer, Berlin, pp. 513–18

Hathaway K B, Clark A E 1993 Magnetostrictive materials. *MRS Bull.* (April): 34–41

Measures R M, Melle S, Liu K 1992 Wavelength demodulated Bragg grating fibre optic sensing systems for addressing smart structure critical issues. *Smart Mater. Struct.* 1(1): 36–44

Roberts S S J, Davidson R 1991 Mechanical properties of composite materials containing embedded fibre optic sensors. *Proc. Soc. Photo-Opt. Instrum. Eng.* 1588: 326–41

Rogers A J 1988 Distributed optical fibre sensors for the measurement of pressure, strain and temperature. *Phys. Rep.* 169(2): 99–143

Rogers C A 1992 Mechanics issues of reduced strain actuation. *Proc. 1st European Conf. Smart Structures and Materials*, Glasgow. Institute of Physics, Bristol, UK, pp. 163–75

Rogers C A, Barker D K, Bennett K D, Wynn P 1988 Demonstration of a smart material with embedded actuators for active control. *Proc. Soc. Photo-Opt. Instrum. Eng.* 986: 90–105

Sendeckyi G P, Paul C A 1989 Some smart structure concepts. *Proc. Soc. Photo-Opt. Instrum. Eng.* 1170: 2–10

Takagi T 1990 A concept of intelligent materials. *J. Intell. Mater. Syst. Struct.* 1: 149–56

Wilson G W, Anderson R D, Ikegami R 1990 Shape memory alloys and fibre optics for flexible structure control. *Proc. Soc. Photo-Opt. Instrum. Eng.* 1370: 286–95

R. Davidson
[AEA Technology, Didcot, UK]

Solid Fiber Composites as Biomedical Materials

The use of fiber composites as biomaterials is relatively new; the idea of composite materials, however, dates back to ancient Egypt, where it was common practice to use chopped straw in bricks to prevent the bricks from cracking. Moreover, many of the naturally occurring biomaterials such as bone and bamboo are in fact fiber composites. Bone is a composite of mineral phase (brittle, high-modulus apatite crystals) existing in a ductile organic phase of collagen fibers. Similarly, bamboo is composed of cellulose reinforced by silica, which makes it a remarkably strong material with a high impact strength.

Most of the recent developments in composite materials technology have occurred during the last 20–30 years, mainly in the aircraft industry. Solid fiber composites have only recently been used as biomedical materials. Some of these applications are: fiber-reinforced bone cement; carbon-fiber-reinforced bone plates; and carbon–polyethylene composite prostheses.

1. Mechanical Properties

The main attractiveness of a fiber-reinforced composite derives from the fact that fibers with high strength and stiffness but low fracture toughness can be combined with a low-modulus epoxy resin so that the resulting material has the advantage of complementary properties of both of their constituents. Such material is generally strong as well as light.

When a composite with aligned fibers is subjected to tension, almost the same strains are generated in the fiber and the composite. However, the stress in the fiber is much larger, due to its high modulus. Therefore the breaking strength of the composite mainly depends on the strength of the fibers. With increasing load some fibers may break, but crack propagation is hindered by the ductile matrix and the crack front may be deflected by the weak fiber–matrix interface or it may even stop. Failure of the whole specimen requires fracture of all the fibers in one plane or pull-out of the fibers from the matrix, thus absorbing additional energy in the creation of the large fracture surface. When the composite is subjected to compression parallel to the fibers, the fibers often fail due to buckling and shear. Fibers with higher stiffness resist buckling and may contribute to increased load-carrying capacity.

The strength of fiber-reinforced composites is highly directional, depending on the predominant direction of the fibers, just as bone is stronger in the longitudinal direction than in the transverse direction. Structures which are subjected to a complex system of loads can therefore be made of composites with randomly dispersed fibers so that the strength characteristics are uniform in all directions. This is often the case for artificial total joint parts made of carbon-fiber-reinforced polyethylene.

Simple laws of mixtures for composites provide the following sets of equations under two alternative assumptions. When the strain is uniform throughout the composite,

$$E_c = V_f E_f + V_m E_m \tag{1}$$

when stress is uniform throughout the composite,

$$\frac{1}{E_c} = \frac{V_f}{E_f} + \frac{V_m}{E_m} \tag{2}$$

where E is the elastic modulus and V the volume percent of each constituent; the subscripts c, f and m denote composite, fiber and matrix, respectively. According to Kelly and Davies (1965) the equation for the fracture strength of a composite consisting of a plastic matrix and very short elastic fibers is given by

$$\sigma_c = \sigma_f V_f + (1 - V_f)\sigma_m \tag{3}$$

where σ is the fracture strength. This can be modified for randomly oriented fibers whose length is longer than the critical length. Similarly,

$$E_c = n(1 - 1/2l_c) V_f \sigma_f + \sigma_m (1 - V_f) \tag{4}$$

where the critical fiber length $l_c = \sigma_f d_0 / 2\tau$, d_0 is the fiber diameter, τ is the matrix shear strength at the interface and n is a factor depending on the orientation of fibers; for randomly oriented fibers, $n = 0.33$.

To verify the validity of the above equations, the ultimate tensile strengths and elastic moduli of bone cement reinforced with various volume percentages of carbon fiber can be calculated and the values compared with the experimental data of Litchman et al. (1978), who reported the following properties for Thornel-300 carbon fibers and a bone cement matrix: $\sigma_f = 2.241$ GPa, $\sigma_m = 34.48$ MPa, $E_f = 248.3$ GPa, $l = 1.27 \times 10^{-2}$ m, $l_c = 2.54 \times 10^{-4}$ m, aspect ratio (length–diameter ratio of fibers) = 2500. The results of such a comparison are shown in Table 1, from which it can be concluded that the theoretical prediction is in good agreement (within 10%) with the experimental

Table 1
Theoretical and experimental tensile properties of carbon-fiber-reinforced PMMA

Fiber content (vol%)	Strength (MPa)		Elastic modulus (GPa)	
	Theor.	Expt.	Theor.	Expt.
0.53	37.9	33.5	3.30	3.31
1.05	41.3	38.3	4.51	3.86
1.57	44.8	42.8	5.72	4.83
2.08	48.1	44.5	6.90	5.52

Table 2
Mechanical properties of carbon-reinforced ultrahigh-molecular-weight polyethylene (UHMWPE)

Carbon fiber (vol%)	Density ($g cm^{-3}$)	Strength (MPa)				Flexural Modulus (MPa)
		Tensile	Compressive[a]	Shear	Flexural	
0	0.939	23.1	12.7	21.4	14.4	713
10	0.990		14.2		20.1	1017
15	0.998		15.7		22.6	1405
20	1.026	41.9	16.6	24.9	24.7	1488
25	1.029		17.0		26.6	1520
30					28.0	1607
40		54.0		28.6		

[a] Yield strength

data for handmade samples. From extrapolation of these data, it appears that ~4.9 vol% (3.8 vol% theoretically) of fiber reinforcement will give a modulus of elasticity approximately matching that of bone. However, this much fiber is difficult to mix uniformly and it may not produce the desired effect, due to the formation of voids and bundling of the fibers in the cement.

2. Surgical Applications

2.1 Carbon–Polyethylene Composites

Most of the artificial total joint prostheses used today consist of an ultrahigh-molecular-weight polyethylene (UHMWPE) component articulating against a polished metal part. This type of hip joint prosthesis has been tested for adequate wear performance in use for about 10 years. However, the total wear of UHMWPE over the 30–40 year period required for implants in younger patients may be excessive. Moreover, when a similar prosthesis is designed for joints other than the hip where it may be subjected to more severe stresses and strains, the resistance to creep of the polymeric material employed assumes increasing importance. Even under relatively low loads, polymeric materials can experience gradual flow which may result in permanent deformation. These considerations led to the development of a carbon-fiber-reinforced, ultrahigh-molecular-weight polyethylene composite material (Poly Two, made by Zimmer), which offers improved mechanical, creep, fatigue and wear properties in total joint replacements.

The UHMW polyethylene is reinforced with carbon fibers which are produced by pyrolyzing acrylic fibers in such a way as to obtain an oriented, graphitic structure. The carbon fibers thus produced have high tensile strength and a high modulus of elasticity. The mean fiber length in the composite material is ~3 mm and the fiber distribution and orientation within the matrix are random. The mechanical properties of this material for various percentages of carbon fiber reinforcement are shown in Table 2. It is evident that both mechanical strength and stiffness increase with increasing carbon fiber content. Similarly, tests have demonstrated that UHMWPE with 10–15 vol% carbon fibers has a significantly lower wear rate and increased fatigue life compared with plain UHMW polyethylene. Figure 1 shows an example of a total knee prosthesis in which the tibial component is made of carbon-fiber-reinforced UHMWPE.

2.2 Wire-Reinforced Bone Cement

Self-curing poly(methyl methacrylate) (PMMA) is used extensively as a bone cement in orthopedic surgery for fixation of endoprostheses and the repair of bone defects. However, such bone cement is weak in tension and is also significantly weaker than compact bone. This has prompted several researchers to attempt to improve the mechanical properties of bone cement by reinforcing it with metal wires and graphite fibers.

PMMA incorporating metal wires has been used clinically in the stabilization of the cervical spine. Investigations of the effect of metal wire on the tensile, bending and shear properties of PMMA have shown that wire reinforcement significantly increased the failure stress of PMMA in all these modes. Figure 2 shows an example of the increase in the tensile strength of wire-reinforced bone cement compared with an unreinforced control. As the use of metal wire is not practicable in total joint replacement, due to the narrow annular space between the prosthesis and the bone, carbon or other fiber reinforcement may be used in such cases.

2.3 Fiber-Reinforced Bone Cement

Several authors have shown that graphite or carbon fiber reinforcement increases the tensile, compressive and shear strengths of bone cement (Fig. 3) and increases its fatigue life (Fig. 4). The modulus of elasticity of fiber-reinforced PMMA also increases with increasing fiber content when the fiber volume

Figure 1
Total knee prosthesis with the tibial component and patellar button of carbon-fiber-reinforced UHMWPE (Poly Two, black) and normal UHMWPE (white, upper right hand) (courtesy Zimmer, Inc.)

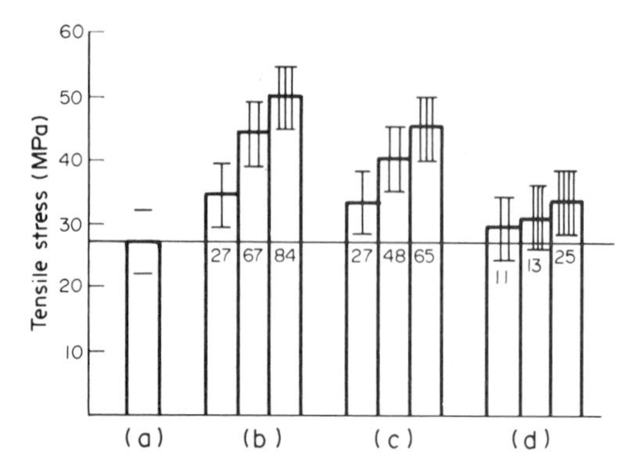

Figure 2
Effect of wire reinforcement on tensile stress supported by PMMA: (a) control; (b) with 1 mm Vitallium wire; (c) with 1 mm stainless steel wire; (d) with 0.5 mm stainless steel wire. Figures on bars indicate percentage increase over control, horizontal lines at top of bars indicate range of variation about mean stress, and vertical lines indicate number of wires

percentage is kept small. A compilation of data on the mechanical properties of carbon-fiber-reinforced bone cement reported by various investigators is given in Table 3. Carbon fiber reinforcement of bone cement can also reduce the deformation due to creep (Fig. 5). Recently several authors have also experimented with aramid and Kevlar fiber reinforcement of bone cement. They demonstrated that addition of these fibers can also improve the mechanical properties of PMMA. The improved mechanical properties of graphite-fiber-reinforced PMMA would allow the clinical use of this material in more diverse applications than is presently possible with normal bone cement.

In order to examine the mechanism of strength improvement, the fracture surfaces of PMMA reinforced with carbon fiber were examined in a scanning electron microscope. Figure 6 shows an electron micrograph of the fracture surface of a specimen which has failed by interfacial shear of the fiber and cement. This shows that the energy absorption capacity of the material was increased due to the fiber pull-outs.

A large rise in temperature during the setting of bone cement may cause tissue necrosis, so reduction of peak temperature is highly desirable in the clinical use of bone cement. Measurements have been made of the

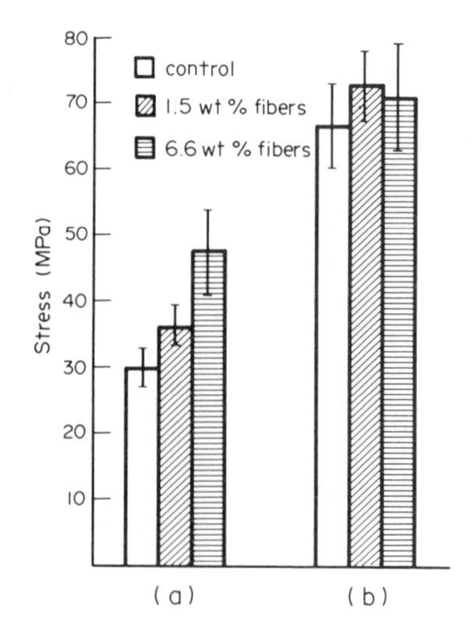

Figure 3
Effect of graphite fiber reinforcement on (a) shear and (b) compressive strengths of bone cement

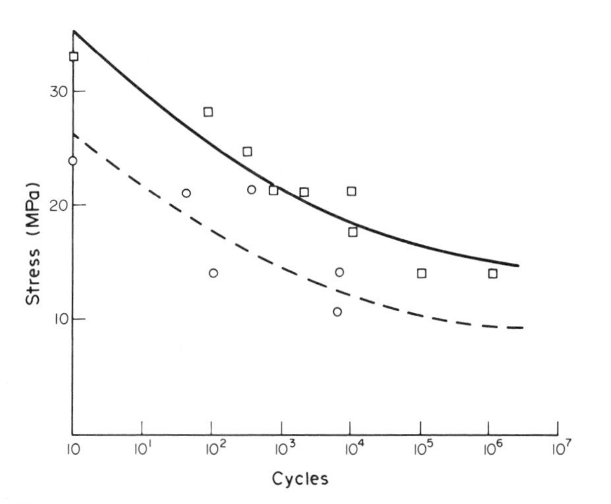

Figure 4
Results of fatigue tests on bone cements: – – –, ○ unreinforced; ———, □ reinforced with carbon fiber (Pilliar and Blackwell 1976)

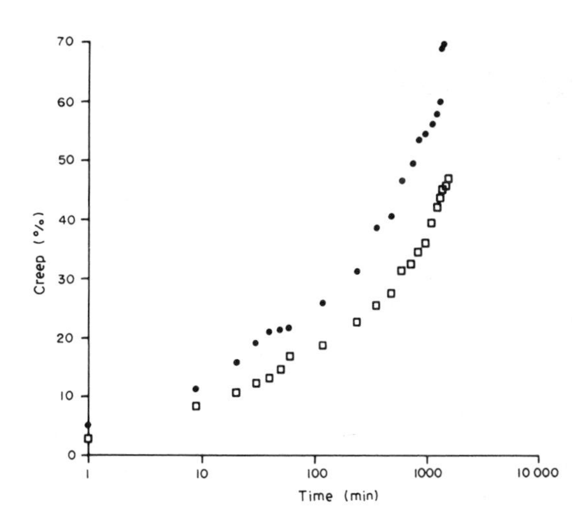

Figure 5
Results of creep tests on PMMA at 10.5 MPa and 25 °C: ●, unreinforced; □, reinforced with carbon fiber

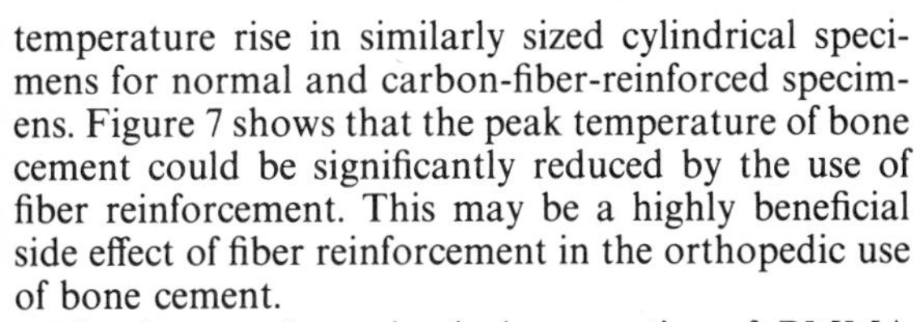

temperature rise in similarly sized cylindrical specimens for normal and carbon-fiber-reinforced specimens. Figure 7 shows that the peak temperature of bone cement could be significantly reduced by the use of fiber reinforcement. This may be a highly beneficial side effect of fiber reinforcement in the orthopedic use of bone cement.

The improved mechanical properties of PMMA reinforced with ultrahigh-strength graphite fiber, Thornel-300 (Union Carbide) or aramid fiber, Kevlar-29, are compared with those of human compact bone in Fig. 8. Incorporation of ~2% Thornel-300 fibers improves the tensile strength of PMMA to ~50% and its modulus of elasticity to ~40% of that of compact human bone.

Results suggest that although significant improvement in the mechanical properties of bone cement can be achieved by fiber reinforcement, it may not be possible to match the properties of compact bone by this means. Further improvement in the mechanical properties of bone cement may be achieved by the use of different types of fibers and by a more uniform mixing of the fiber and cement. A machine-mixed carbon-fiber-reinforced bone cement (Zimmer) in

Table 3
Mechanical properties of carbon-fiber-reinforced PMMA

	Fiber					
Cement	Type	Property	% used	Strength[g] (MPa)	Modulus of Elasticity[g]	Comment
Osteobond[a]	graphite	$l=6$ mm chopped	1,2,3,10[e]	1.7(c), 4.7(f)	4600(c)	
CMW[b]	carbon	$l=6$ mm $d=7$ μm	2[f]		5560(t)	60% increase in σ_t, 100% increase in E
CMW-Porous[b]	carbon	$l=6$ mm $d=10$–15 μm $E=380$–460 GPa	1[f] 2[f]	38(t)	static: 3700 5800 cyclic: 4000(t) 3750(t)	at $\varepsilon=5\times10^{-6}\ s^{-1}$ at $\varepsilon=2\times10^{-6}\ s^{-1}$
Simplex P[c]	carbon (Hercules As-type)	$l=6$ mm $d=8$ μm chopped	1.5[e] 6.6[e]	74(c), 35 (s) 72(c), 47(s)		τ increased 20% and 59%, σ_c 9% for 1.5% fiber
Surgical Simplex (RO)[d]	graphite Thornel-300 (Union Carbide)	$l=12.5$ mm $d=5.0$ μm	0.53[f] 1.05[f] 1.57[f]	34.1(t) 37.8(t) 42.2(t)	3241(t) 4008(t) 5730(t)	crosshead speed = 2.5 mm min^{-1}

[a] Knoell et al. (1975) [b] Pilliar and Blackwell (1976) [c] Saha and Kraay (1979) [d] Litchman et al. (1978) [e] Value in wt% [f] Value in vol% [g] Tensile (t), compressive (c), shear (s), flexural (f)

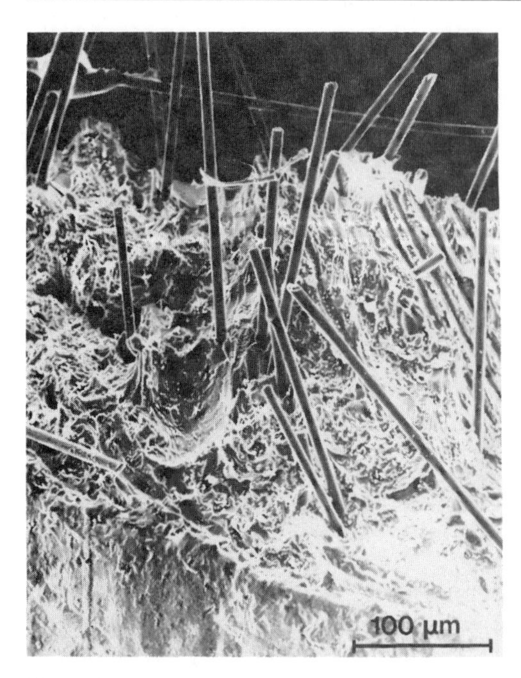

Figure 6
Scanning electron micrograph of the fracture surface of a carbon-fiber-reinforced bone cement specimen tested in shear

which the fiber cross sections are dogbone-shaped for better bonding has recently been used for limited clinical trials.

2.4 Composite Fracture Fixation Plates

At present most internal fixation plates are made of metals, selected primarily for their strength and inertness. However, their high stiffness (the elastic modulus of metals being an order of magnitude higher than that of bone) does not allow the underlying bone to carry a normal stress level. This stress protection induces osteopenia or osteoporosis of the bone, which in severe

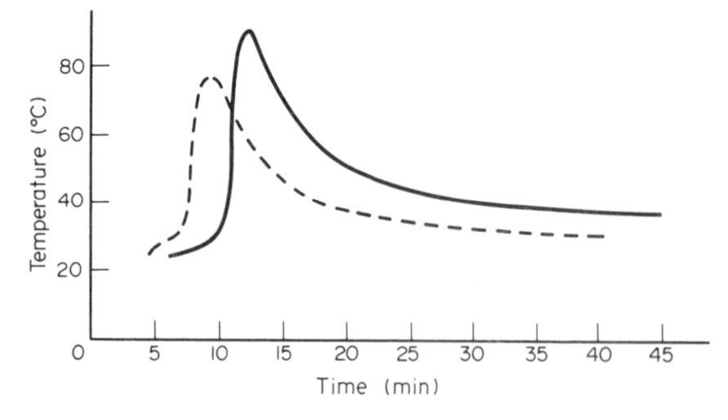

Figure 7
Variation in temperature during rapid polymerization of PMMA specimens 18 mm in diameter: ——, unreinforced; - - - , reinforced with 2% graphite fiber

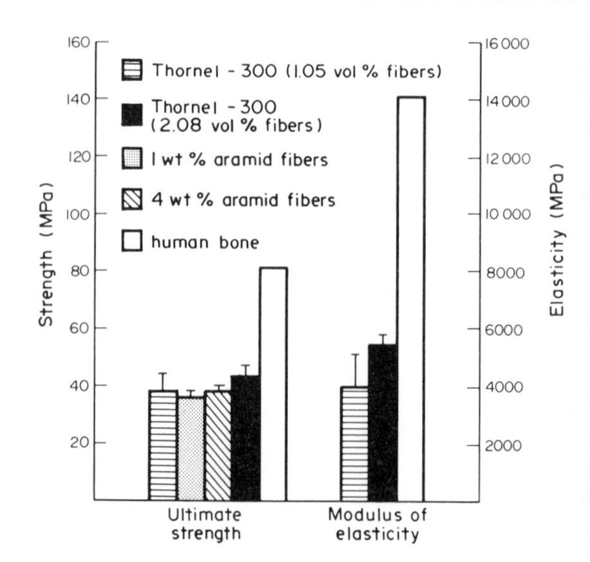

Figure 8
Comparison of tensile properties of fiber-reinforced PMMA and human compact bone

cases may produce bone fracture when the plate is removed. A plate with a modulus of elasticity similar to that of bone, or a plate which decreases in stiffness due to biodegradation, would be ideal. With this objective, several investigators (Woo et al. 1974) have conducted animal studies using fracture plates made of graphite-fiber-reinforced epoxy resin, bone cement and polysulfone. Graphite fibers improved the mechanical properties of epoxy resin such that the properties were very similar to those of human bone (Bradley et al. 1980), but the creep resistance was poor. Animal studies by Woo et al. (1976) showed that such composite plates produced significantly less plate-induced osteopenia than rigid metal plates.

2.5 Sintered-Fiber Metal Composites

Sintered-fiber metal composites have been developed for potential application in skeletal fixation for internal prosthetic devices. Metal fibers are cut into short lengths, compressed in dies and sintered under vacuum to form composite plugs. These composites have shown adequate strength and stiffness. When implanted in the femur of dog and rabbit, bone ingrowth occurred within 2–3 weeks and the bond strength increased initially (up to 2–4 weeks) and then remained relatively constant.

2.6. Composite Materials in Dentistry

Fiber composites have also been used in dentistry (see *Dental Composites*) where the heat-cured acrylic used as a denture base material is reinforced with glass fibers to achieve greater fatigue and impact resistance. Particulate-filled polymer composites are also used as dental restorative materials but have limited durability in the oral environment. These materials do not behave as traditional fiber-reinforced composites, since particulates cannot develop the high stresses that fibers do.

Bibliography

Bradley J S, Hastings G W, Johnson-Nurse C 1980 Carbon fiber reinforced epoxy as a high strength, low modulus material for internal fixation plates. *Biomaterials* 1: 38–40

Evaskus D S, Rostoker W, Laskin D M 1980 Evaluation of sintered titanium fiber composite as a subperiosteal implant. *J. Oral. Surg.* 38: 490–98

Fishbane B M, Pond R B 1977 Stainless steel fiber reinforcement of polymethylmethacrylate. *Clin. Orthop. Relat. Res.* 128: 194–99

Galante J, Rostoker W, Lueck R, Ray R D 1971 Sintered fiber metal composites as a basis for attachment of implants to bone. *J. Bone Jt. Surg.* 53A: 101–14

Kelly A, Davies G J 1965 The principles of fiber reinforcement of metals. *Met. Rev.* 10: 1

Knoell A, Maxwell H, Bechtol C 1975 Graphite fiber reinforced bone cement. *Ann. Biomed. Eng.* 3: 225–29

Litchman H M, Richman M H, Warman M, Mitchell J 1978 Improvement of the mechanical properties of polymethylmethacrylate by graphite fiber reinforcement. *Proc. Orthop. Res. Soc.* 2: 86

McKenna G B, Dunn H K, Statton W D 1979 Degradation resistance of some candidate composite biomaterials. *J. Biomed. Mater. Res.* 13: 783–98

Pal S, Saha S 1982 Stress relaxation and creep behavior of normal and carbon fibre acrylic bone cement. *Biomaterials* 3: 93–96

Pilliar R M. Blackwell R 1976 Carbon fiber reinforced bone cement in orthopedic surgery. *J. Biomed. Mater. Res.* 10: 893–906

Saha S, Kraay M J 1979 Improved strength characteristics of polymethylmethacrylate beam specimens reinforced with metal wires. *J. Biomed. Mater. Res.* 13: 443–57

Saha S, Pal S 1984 Improvement of mechanical properties of acrylic bone cement by fiber reinforcement. *J. Biomech.* 17: 467–78

Saha S, Pal S, Albright J A 1984 Mechanical properties of bone cement: A review. *J. Biomed. Mater. Res.* 18: 435–62

Sclippa E, Pierkerski K 1973 Carbon fiber reinforced polyethylene for possible orthopaedic usage. *J. Biomed. Mater. Res.* 7: 59–70

Taitsman J P, Saha S 1977 Tensile strength of wire reinforced bone cement and twisted stainless steel wire. *J. Bone Jt. Surg.* 59A: 419–25

Tewary V K 1978 *Mechanics of Fibre Composites*. Wiley, New York

Woo Savio L-Y, Akeson W H, Coutts R D, Rutherford L, Doty D, Jemmott G F, Amid D 1976 A comparison of cortical bone atrophy secondary to fixation with plates with large differences in bending stiffness. *J. Bone Jt. Surg.* 58A: 190–95

Woo S, Akeson W H, Leventz B, Coutts R D, Matthews J V, Amiel D 1974 Potential application of graphite fiber and methyl methacrylate resin composites as internal fixation plates. *J. Biomed. Mater. Res.* 8: 321–38

Wright T M, Fukubayashi T, Burstein A H 1981 The effect of carbon fiber reinforcement on contact area, contact pressure, and time-dependent deformation in polyethylene tibial components. *J. Biomed. Mater. Res.* 15: 719–30

S. Saha
[Louisiana State University, Shreveport, Louisiana, USA]

S. Pal
[Jadavpur University, Calcutta, India]

Strength of Composites

Composite materials made from aligned fibers embedded in a plastic matrix are orthotropic, that is, their properties vary symmetrically about orthogonal axes. The characterization of strength therefore requires the measurement of a minimum of five independent strength parameters. Assuming transverse isotropy, the principal parameters are the tensile and compressive strengths measured parallel to, and at right angles to, the fibers and the shear strength (measured parallel to the fibers). The longitudinal tensile strength (that is, the tensile strength parallel to the fibers) is determined largely by the strength properties of the fibers themselves whereas the longitudinal compressive strength is generally governed by fiber stability. The transverse and shear strengths are limited by the relatively low strength properties of the matrix and fiber–matrix bond so they are usually an order of magnitude lower than the longitudinal properties.

In practical applications, the relatively poor transverse and in-plane shear properties can be overcome by the use of multidirectional laminated composites. The strength properties of multidirectional laminates can be determined from those of the unidirectional plies using models based on elastic lamination theory. Relatively simple models will often give results which are sufficiently accurate for initial design, but for high accuracy, effects such as material nonlinearity, the development of noncatastrophic damage and the presence of hygrothermal strains must be included. In practice therefore, it is often more convenient to make direct measurements of the strength of multidirectional composites.

1. The Effect on Strength of Fiber Orientation

The tensile strength of a unidirectional composite loaded at an angle to the fiber direction may be related to the longitudinal and transverse tensile strengths (σ_{1T} and σ_{2T}) and in-plane shear strength (τ_{12}) as illustrated in Fig. 1. The failure mode, and hence the controlling strength parameter, depends on the angle between the fiber and loading axes. For compressive loading the effect of fiber orientation follows a similar pattern, the strength and failure mode being governed by the longitudinal and transverse compressive strengths and the shear strength.

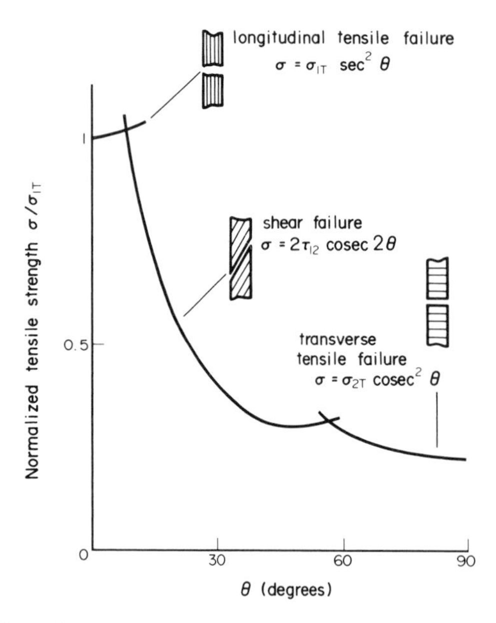

Figure 1
The effect on tensile strength of fiber orientation

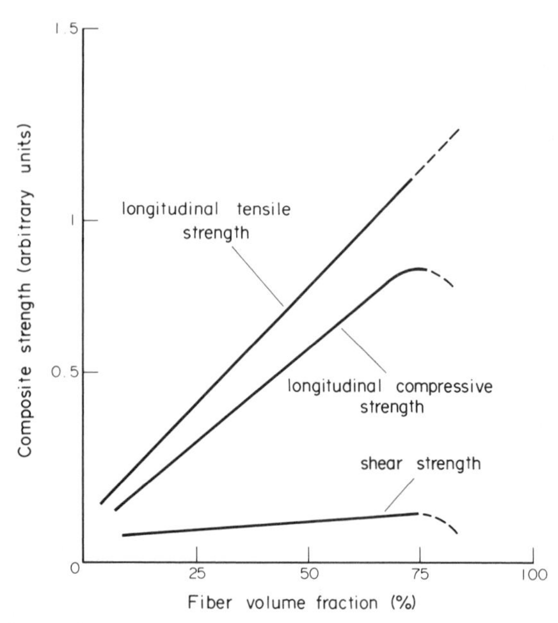

Figure 2
The variation of strength with fiber volume fraction

This description of the effect of fiber orientation is based on the theory of maximum principal stresses which makes the reasonable assumption that failure in any one mode is not influenced by the magnitude of the stress in any other mode. The corresponding theory of maximum principal strains, which is based on similar assumptions concerning the independence of principal strains, results in slightly different curves due to Poisson's ratio effects. Both of these theories have the advantage of simplicity, but they indicate abrupt changes in the relationship between strength and fiber orientation at the points of transition between failure modes. Failed specimens corresponding to the transition points exhibit features of both modes suggesting that they are not entirely independent. Moreover, since a single continuous curve is generally preferred for analysis and computation, a variety of interaction theories have been proposed. Most of these theories are modifications of the various yield criteria developed for homogeneous materials and are therefore used purely empirically. The differences between the various strength predictions are often relatively small compared to the scatter in experimental data so that many of these theories have been used successfully and none is universally preferred.

2. *Variation of Strength with Fiber Volume Fraction*

The variation with fiber volume fraction of the longitudinal tensile and compressive strength and the in-plane shear strength of a typical fiber reinforced plastic is illustrated in Fig 2. Over much of the range (including the great majority of practical composites) the strength parameters vary in direct proportion to the fiber volume fraction. Thus the longitudinal tensile strength, for example, may be defined by the equation

$$\sigma_{1T} = \sigma_f V_f + \sigma_m (1 - V_f)$$

where V_f is the fiber volume fraction, σ_f is the effective fiber strength and σ_m is the stress on the matrix at the time of failure of the fibers. The value of σ_f depends on the complex relationships between the properties of both fibers and matrix and, for engineering purposes, it is usually deduced from the measured strength of the composite at known fiber volume fraction. The failure strains of the fibers and matrix are rarely equal so that, for brittle fibers in an elastic matrix, the value of σ_m is given by $\sigma_f E_m / E_f$, where E_m and E_f are the matrix and fiber moduli, respectively. Alternatively, the yield strength may be more appropriate for a matrix capable of yielding. In many practical composites the contribution of the matrix is comparatively small so that the expression $\sigma_{1T} \simeq \sigma_f V_f$ may be used with little error.

The upper limit to composite tensile strength is determined largely by the maximum fiber volume fraction which may be achieved without loss of integrity due to inadequate continuity of the matrix phase. This will depend on the cross-sectional shape and degree of alignment of the fibers but, in general, fiber volume fractions in excess of about 70% are rarely achieved without excessive fiber damage.

In longitudinal compression, failure can occur by shear across both fibers and matrix but, in general, it occurs by instability of the fibers. In either case, compressive strength varies approximately in proportion to fiber volume fraction and may be described by equations similar to those for tensile strength above. However, the loss of composite integrity at high fiber volume fraction has a more profound effect on compressive strength, since it permits premature instability of the fibers. The transverse and shear strength properties depend primarily on the strength of the matrix and fiber–matrix interface and exhibit only minor increases in strength with fiber volume fraction. These properties are also limited by the loss of composite integrity at higher fiber volume fractions.

3. *Other Factors Affecting Composite Strength*

In composites containing aligned discontinuous fibers, the stress in each fiber is not uniform along its length but builds up from either end. The tensile strength of such composites depends on the relative magnitudes of the mean fiber length and the critical fiber length, which may be defined as the minimum length which will allow the development of sufficient stress to fail the fiber at its midpoint. For advanced fiber-reinforced plastics the critical fiber length could be as much as 50 fiber diameters, and only composites containing fibers very much longer than this will exhibit strengths approaching that of the corresponding continuous-fiber composite. As fiber length is reduced, the reinforcement efficiency decreases until, when fiber length is less than the critical length, failure will no longer involve fracture of the fibers and will be governed by fiber pull-out.

Poor fiber alignment also substantially reduces reinforcement efficiency in discontinuous fiber composites. By examination of the effects of fiber orientation for continuous fibers, illustrated in Fig. 1, it is evident that the contribution to composite strength from fibers more than a few degrees off-axis is greatly reduced. This effect together with the effect of finite fiber length means that, although reinforcement efficiencies in excess of 90% have been observed, values of 50% to 70% are perhaps more typical of aligned discontinuous fiber composites.

The effects on compressive strength of both fiber length and fiber alignment are broadly similar to the effects on tensile strength. However, while fiber length will have virtually no effect on transverse or shear properties, poor fiber alignment will tend to increase the transverse tensile and compressive strengths and also the shear strength.

A particularly important factor affecting the strength of both continuous- and discontinuous-fiber composites is the occurrence of microstructural defects and, in particular, the occurrence of voids. In fiber-reinforced plastics, a small fraction of voids can have a quite disproportionate effect on those strength properties which depend on the integrity of the matrix phase. As an example, Fig. 3 shows the effect of matrix void content on the compressive strength and shear strength of a typical glass-fiber-reinforced plastic (GRP). In practice, void content in fiber-reinforced plastics may be limited to less than about 5% by quite simple manufacturing techniques, and more sophisticated methods, such as those used in production of aerospace components, enable void content to be reduced to 0.5–1%. However, even at these low levels, the effects of void content are not negligible and can contribute significantly to material strength variability.

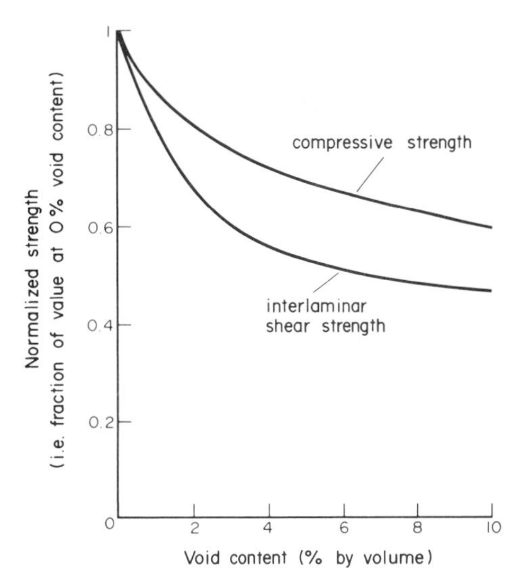

Figure 3
The effect of voids on composite strength

Finally, it is important to note that environmental effects, particularly temperature and moisture absorption in fiber-reinforced plastics, can substantially affect composite strength. Detailed discussion of these effects is beyond the scope of this article but since both moisture and temperature tend to soften plastic matrices and may, in some systems, degrade the fiber–matrix bond the effects are most marked on compressive and shear strength properties.

4. *Strength of Multidirectional Laminates*

Unidirectional composites are often too anisotropic for practical applications, but this may be overcome by the use of laminated materials in which the fiber plies are oriented in two or more prescribed directions. Such laminates can be made most efficient in terms of strength-to-weight ratio if the applied loads are carried by direct stresses in the fibers. Thus, a proportion of the plies should lie parallel to each of the principal tensile and compressive loads whereas the remainder should be oriented at $\pm 45°$ to carry the shear loads.

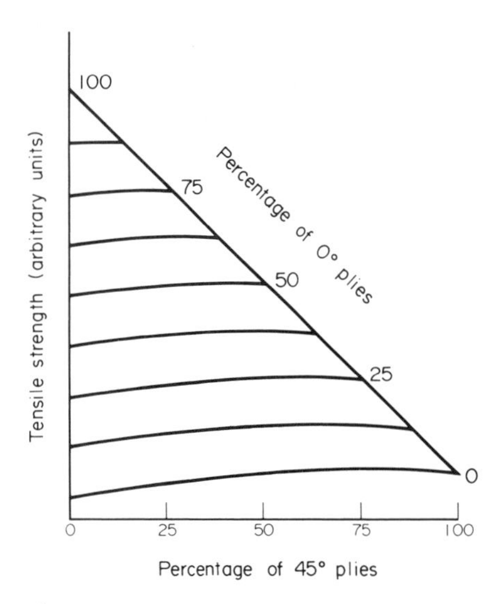

Figure 4
Tensile strength of 0°/±45°/90° family of laminates

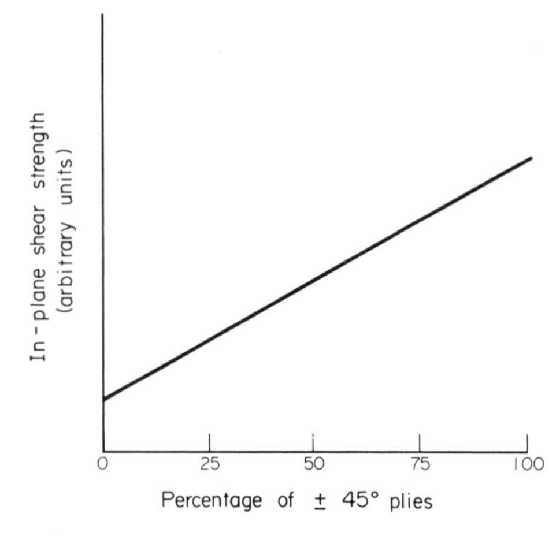

Figure 5
In-plane shear strength of 0°/±45°/90° family of laminates

Varying the orientation and stacking sequence of the plies gives an infinite family of laminates. Many of these laminates will exhibit phenomena not observed in isotropic materials including, for example, coupling between bending, stretching and twisting deformations. However, there is rarely any particular advantage in using more than four fiber directions and, except for a few specific applications in which coupling effects can be exploited to advantage, structural laminates will generally contain only 0°, 90° and ±45° plies assembled in sequences which eliminate coupling effects and exhibit orthotropic behavior, that is, the in-plane properties are symmetric about axes at right angles to each other The strength properties of such laminates clearly depend on the proportions of fiber in each direction and Fig. 4 shows a typical carpet plot of tensile strength in the 0° fiber direction for a carbon-fiber composite. It may be seen that the most important factor is the proportion of fiber lying parallel to the applied load. Except for very low proportions of 0° fiber, the relative proportions of ±45° and 90° fiber have very little effect on tensile strength. Compressive strength varies with layup in a manner which is generally similar but, because the ply-stacking sequence can affect the stability of the load-bearing 0° plies, the use of carpet plots such as that shown above for tensile strength should be treated with caution. The in-plane shear strength of the 0°, 90°, ±45° family of laminates is unaffected by the relative proportions of 0° and 90° plies but increases linearly with the proportion of ±45° plies as shown in Fig. 5.

The most simple model for the prediction of strength properties of multidirectional laminates makes the assumption that the transverse and shear load-carrying capacity of a unidirectional ply is negligible. Thus, for example, the tensile strength along a given axis is given by the product of the longitudinal tensile strength of the unidirectional material and the fraction of fiber plies which are parallel to that axis. By resolving in-plane shear stress into tensile and compressive stresses in the ±45° plies, the shear strength may also be expressed as a simple function of the unidirectional strength properties. By comparison with Figs. 4 and 5, it may be seen that this approach would give conservative strength values of varying accuracy. Nevertheless, the extreme simplicity of such a model makes it particularly attractive for initial design and optimization of composite laminates for specific applications.

More accurate analytical models of the strength of multidirectional laminates are based on the constitutive relationships used to describe the elastic behavior of a unidirectional ply, and to relate that behavior to the elastic properties of multidirectional laminates on the basis of strain compatibility between the plies (see, for example, Ashton et al. 1969). To determine a particular strength property, the first step is to calculate the laminate strains for unit applied load and thence to calculate the longitudinal, transverse and shear stresses and strains in each of the plies. The margin of strength for each mode of loading in each ply may then be calculated, whence it is possible to determine the minimum level of applied load at which one of the plies will fail.

At this point the remaining plies must absorb some additional load and this may be modelled by modification of the stiffness matrix. Recalculation of the ply stresses and strains will indicate whether any further failures will occur immediately or whether further load may be sustained. By an interative procedure, a point will eventually be reached at which the laminate can carry no additional load.

When using models such as that described above, it is generally assumed that the strength properties of the unidirectional material may be applied to each of the plies within the laminate. This assumption is not always valid since, for example, the strain at which cracks form in 90° plies has been shown to depend on ply thickness. However, for most purposes this assumption is not unreasonable. The major differences between the various models of this type lie in the assumptions concerning the load-carrying capability of a failed ply and hence the manner in which the stiffness matrix must be modified.

For accurate strength prediction it will often be necessary to include thermal- and moisture-induced strains into the calculations, and in some cases it may also be necessary to include material nonlinearity. This latter effect may be modelled most accurately by an iterative incremental procedure in which the stiffness properties are adjusted according to the nonlinear stress–strain relationships. Even relatively simple models of the strength of multidirectional laminates require considerable computation so that the great majority are computer based and iterative procedures do not present any particular difficulty.

5. *Measurement of Composite Strength*

The high degree of anisotropy and the wide range of failure mechanisms exhibited by fiber composites give rise to a number of novel problems in test specimen design. Moreover, as a result of differences in the relative magnitudes of the various properties, it is not possible to specify a unique set of specimens which will give satisfactory results for all fiber composites. Various standards have been proposed by, for example, the American Society for Testing and Materials (ASTM) and the British Standards Institution (BSI), and the use of such standards is to be encouraged. However, it is important to understand the principles governing the design of such specimens in order to ensure that the specimen and test method selected for a particular purpose will result in failure by a mechanism appropriate to the intended use of the data. The principles of design and the consequent form of typical specimens are outlined in the following paragraphs.

For unidirectional fiber composites, the very high ratio of longitudinal tensile or compressive strength to shear strength leads to two major problems in the measurement of longitudinal properties. First, since load must be transferred into the body of the specimen predominantly by shear, large shear areas are required at the end fittings to generate adequate longitudinal load. Moreover, any waisting necessary to ensure failure in a region of uniform stress must be very gradual to avoid unwanted shear failures. A further problem arises from the very high stiffness and generally linear-elastic behavior of these materials. Any slight bending of the specimen or misalignment of the load can give rise to significant bending stresses and lead to premature failure. The transverse failing strain is often even lower than the longitudinal failing strain so that sensitivity to bending effects is even greater in the measurement of transverse strength properties.

The problems mentioned above may be largely overcome using specimens of the general form shown in Fig. 6(a), (b) which are based on specimens designed for carbon-fiber-reinforced plastics. For tensile specimens, load is applied by friction grips bearing on thin aluminum alloy or glass-fiber-reinforced plastic tabs bonded to the specimen. The specimen is waisted in thickness which allows a significant reduction in cross-sectional area to be achieved within a reasonable length while meeting the requirement for a shallow waisting curve. This also reduces the bending stiffness in the plane in which test machine misalignment is most likely.

For compression testing, Euler instability is a major problem and the need to shorten the free length conflicts with the need for a shallow waisting curve. For low-modulus materials, fixtures to prevent buckling and to provide rotational restraint of the

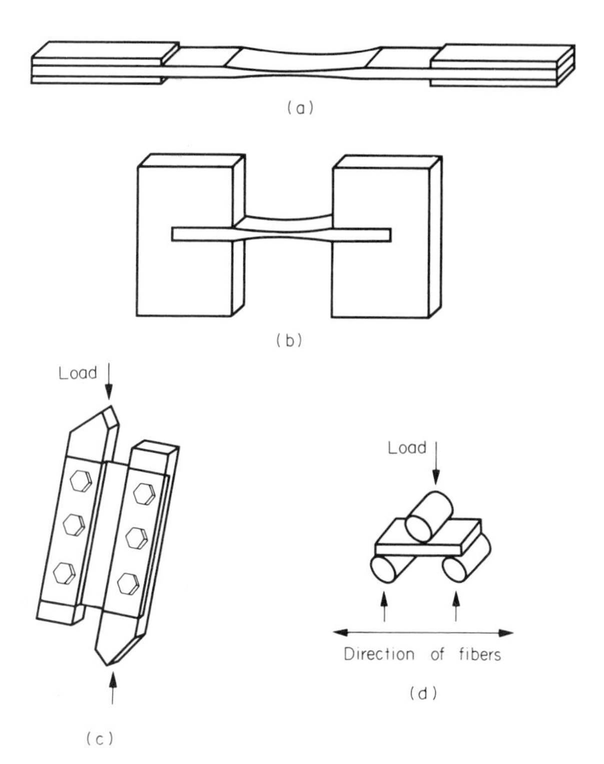

Figure 6
General configuration of test specimens for unidirectional-fiber reinforced plastics: (a) tensile test specimen; (b) compressive test specimen; (c) rail shear test specimen; (d) short beam shear test specimen

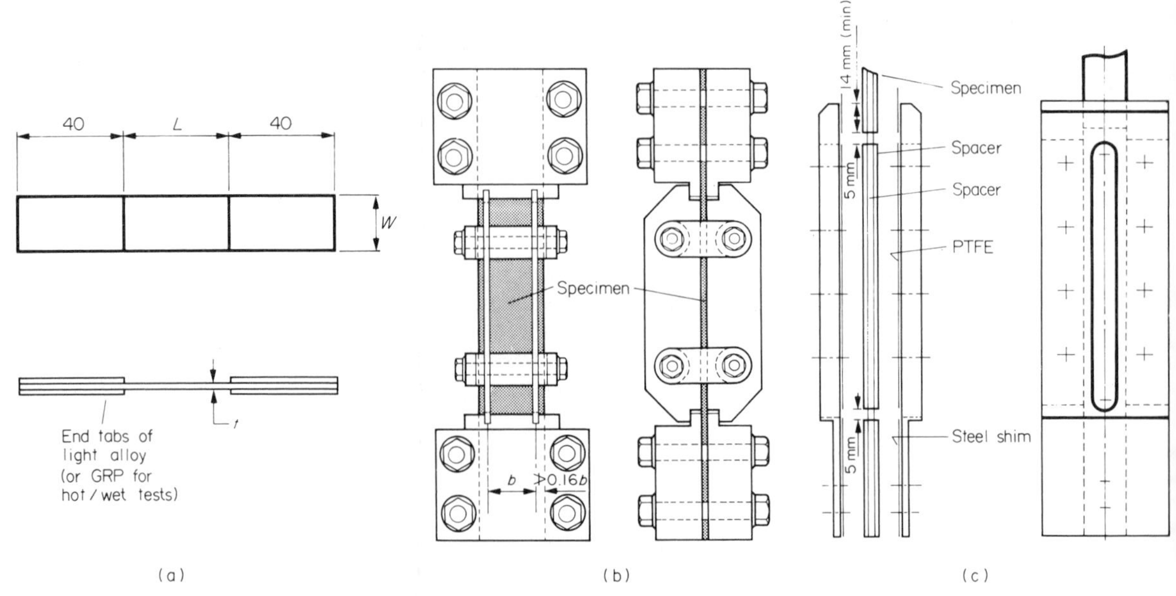

Figure 7
Multidirectional test specimen (a), and typical anti-buckling guides (b) and (c)

ends are necessary. However, for high-modulus fibers, such as carbon, it may be possible to manage without antibuckling guides if the end tabs are replaced by slotted aluminum blocks (Fig. 6(b)) which provide an encastré end condition. A further difficulty results from the fact that any restraint of Poisson expansion by the end fittings can lead to artificially high results for very short specimens. Transverse compression specimens require less waisting since transverse isotropy leads to lower stress concentrations at the end fittings. Thus, in spite of the low transverse stiffness, stability may often be achieved using a short specimen with slotted end blocks.

The main problem associated with the measurement of in-plane shear strength of unidirectional composites is that of ensuring failure in a region of pure shear. Applying torsional load to a thin-walled tube is the most effective technique, but for flat laminates some other technique is required. The rail shear test (Fig. 6(c)) is generally preferred for unidirectional fiber composites. Although not entirely free from stress concentrations, it provides a reasonable measure of shear strength if the fibers are at right angles to the loading rails and can therefore sustain the direct loads occurring along the short free edges. Alternatively, the shear properties of the unidirectional material can be derived from a tensile test on a $\pm 45°$ laminate using a specimen similar to that described for multidirectional laminates below. It can easily be shown that the shear strength is equal to half the tensile strength of the $\pm 45°$ specimen.

The most simple method of evaluating shear strength is the short beam shear test (Fig. 6(d)). This test actually measures interlaminar shear strength near the midplane of the laminate and since there are also direct stresses due to the bending of the beam and stress concentrations due to the loading rollers this test is far from ideal. Nevertheless, its great simplicity makes it an attractive test method which finds widespread use, particularly for quality control purposes.

The measurement of tensile and compressive strength of multidirectional laminates is usually effected using unwaisted specimens as shown in Fig. 7(a). The fact that some of the off-axis fibers run into the end fitting tends to have a local reinforcing effect and the majority of failures occur near the center of the specimen. For specimens containing plies at $\pm\theta°$, the length-to-width ratio (L/W) should be chosen to ensure that no $\theta°$ fibers run from one end fitting to the other. This may not be possible for very small values of θ and results from such specimens should be treated with caution. For compressive loading, a wide range of antibuckling guides has been successfully used and typical examples are illustrated in Fig. 7(b), (c). Measurement of in-plane shear strength should be carried out by torsion of a thin-walled tube. However since multidirectional tubular specimens are expensive and difficult to manufacture, the rail shear test has been used even though the stress concentrations in the vicinity of the short free edges vary with fiber layup.

In summary, it must be recognized that fiber composite materials exhibit a much wider range of failure mechanisms than isotropic materials. The specimen design principles outlined above are based on the need to initiate particular failure mechanisms in regions of uniform stress. It is clearly important to examine all failed specimens to ensure that failure occurred at the correct location and through a mechanism appropriate to the intended use of the data.

6. Acknowledgement

This article is Crown Copyright and is reproduced with the permission of the Controller of Her Majesty's Stationery Office.

See also: Laminates: Elastic Properties; Long-Term Degradation of Polymer-Matrix Composites; Multiaxial Stress Failure

Bibliography

American Society for Testing and Materials 1988 *Annual Book of ASTM Standards.* ASTM, Philadelphia, Pennsylvania

Ashton J E, Halpin J C, Petit P H 1969 *Primer on Composite Materials: Analysis.* Technomic, Stamford, Connecticut

British Standards Institution 1987 *British Standards Institute Catalogue.* BSI, Milton Keynes, UK

Ewins P D 1974 Techniques for measuring the mechanical properties of composite materials. In: *Composites Standards Testing and Design Conf. Proc.* IPC Science and Technology Press, Guildford

Holister G S, Thomas C 1966 *Fibre Reinforced Materials.* Elsevier, London

Jones R M 1975 *Mechanics of Composite Materials.* McGraw-Hill, New York

Kelly A, Macmillan N H 1986 *Strong Solids*, 3rd edn. Clarendon, Oxford

Sih G C, Skudra A M 1986 Failure mechanics of composites. In: Kelly A, Rabotnov Y N (eds.) 1986 *Handbook of Composites*, Vol 3. Elsevier, Amsterdam

R. T. Potter
[Defence Research Agency, Farnborough, UK]

Strength of Composites: Statistical Theories

The traditional ductile structural materials (steel, aluminum, titanium, etc.) generally exhibit quite well-defined strengths. With these materials, use of safety factors dictated mainly by uncertainties as regards the stresses to be encountered in service has proved satisfactory.

There are circumstances, however, under which the use of brittle materials in load-carrying structural components is advantageous or even unavoidable. (Here the term brittle is used to denote materials that sustain negligible deformation or damage during loading until a critical stress is reached, at which they fail catastrophically.) Brittle materials characteristically exhibit a large dispersion in fracture stress. The uncertainty in fracture stress may be comparable to or in some cases may even exceed the uncertainty in service stresses, and must therefore be taken into account when adopting safety margins in design. This need has led to the development of statistical fracture theory.

Owing mainly to the monumental efforts of Weibull in Sweden, statistical fracture theory became a practical engineering tool in the early 1940s. His theory (summarized in Sect. 1) was formulated for brittle isotropic materials such as glass, ceramics, and low ductility metals.

A decade or two later, fibers of extraordinary strength and stiffness for their weight were being developed and incorporated into composites. Since the fibers were brittle there was a widespread tendency to assume that the dispersion in composite failure stress would obey Weibull's theory, but it turned out that it did not. One reason for this is that, although they contain brittle components, composites are not completely brittle. As the applied load increases, many individual fibers typically rupture with consequent composite damage before the composite as a whole fails. The flaw population thus increases with the applied load, in contrast to Weibull theory where it remains unchanged until the moment of failure.

Although Weibull theory does not apply to composites, it applies quite well to fibers that are used as reinforcement. Moreover, the theory for a composite as a whole bears a pleasingly simple relation to Weibull theory. For these reasons an account of statistical fracture theory for brittle materials is presented first.

1. Weibull Theory

Consider first the case of a single fiber of length L. Such a fiber can be conceptually divided into a series of very short fibers connected end to end. It is clear that when subjected to a tensile load such a fiber will fail when the weakest constituent element fails, just as the strength of a chain is that of its weakest link. Because of this analogy, the analysis developed below is known as weakest link theory (WLT).

Weibull (1939) assumed that a fiber contains many flaws of varying strength randomly distributed over its length. Under these circumstances the probability of rupture of an element of length ΔL_i at stress σ is equal to the probability that the element contains a flaw weaker than σ. Thus

$$(\Delta P_{\mathrm{f}})_i = n(\sigma)\Delta L_i \tag{1}$$

where $n(\sigma)$ is the number of flaws per unit length of fiber having a strength less than σ. (More precisely, it is the average number; this wording eliminates the problem of dealing with the fluctuation of n from one unit of length to the next because of the random positioning of the flaws.)

Since the sum of the probabilities of failure and

survival must be unity, the probability of survival of the ith element is

$$(P_s)_i = 1 - (\Delta P_f)_i \quad (2)$$

The probability of survival of the entire fiber is the product of the probabilities of survival of the constituent elements, i.e.,

$$P_s = \prod_{i=1}^{N} (P_s)_i = \prod_{i=1}^{N} [1 - (\Delta P_f)] \quad (3)$$

where N is the total number of elements.

Now if all ΔL values are infinitesimal, then all probabilities of failure ΔP_f are also infinitesimal and except for infinitesimal terms of second or higher order,

$$P_s = \prod_{i=1}^{N} \exp[-(\Delta P_f)_i] = \exp[-\sum_{i=1}^{N} (\Delta P_f)_i] \quad (4)$$

Eqn. (4) is a general result that applies to any structure failing according to the rules of weakest link theory. Weibull called the sum appearing at the end of Eqn. (4) the risk of rupture. Combining Eqn. (4) and Eqn. (1) we obtain

$$P_s = \exp[-Ln(\sigma)] \quad (5)$$

The cumulative probability of failure $P_f(\sigma)$ (i.e., the probability that the fiber will have ruptured by the time the stress reaches σ) is thus given by

$$P_f = 1 - P_s = 1 - \exp[-Ln(\sigma)] \quad (6)$$

All WLTs for brittle materials must be expressible in this general form.

In applying WLT it is convenient to express $n(\sigma)$ in parametric form. Weibull found that in most cases experimental data on fracture can be represented with satisfactory accuracy when it is assumed that $n(\sigma)$ obeys the equation

$$n(\sigma) = (\sigma/\sigma_0)^m \quad (7)$$

The two parameters m and σ_0 are commonly referred to as Weibull's shape and scale parameter, respectively. Materials that obey WLT and conform (in good approximation) to the distribution given in Eqn. (7) are known as Weibull materials.

The expected number of flaws of strength less than σ in a Weibull fiber of length L, which we shall call Q, is given by

$$Q = L(\sigma/\sigma_0)^m \quad (8)$$

This result is shown graphically in Fig. 1. In what is probably the simplest approximation to apply, failure is assumed to occur at the stress at which $Q = 1$. This leads to

$$\sigma_f = \sigma_0 L^{-1/m} \quad (9)$$

where σ_f is the failure stress. Equation (9) leads to the Weibull size effect, which states that when the failure stress of a fiber of length L_1 is known, the failure stress of a fiber of length L_2 can be determined by use of the equation

$$(\sigma_f)_2 = (\sigma_f)_1 (L_1/L_2)^{1/m} \quad (10)$$

The size effect is shown graphically in Fig. 2.

The probability that fiber failure will occur at a stress lying between σ and $\sigma + d\sigma$ is given by

$$dP_f = (dP_f/d\sigma)d\sigma \quad (11)$$

Figure 3 shows the failure probability density distribution function $dP_f/d\sigma$ for several values of m when the defect strength distribution obeys Eqn. (7). It is seen that the resulting curves are similar to the Gaussian distribution familiar from probability theory. In fact they can be regarded as somewhat skewed Gaussians that have the desirable feature (not shared by the Gaussian itself) that failure is predicted not to occur at zero stress. (A second objection to the Gaussian distribution is that it cannot be put in the form of Eqn. (6), and it is therefore not a WLT.) The coefficient of

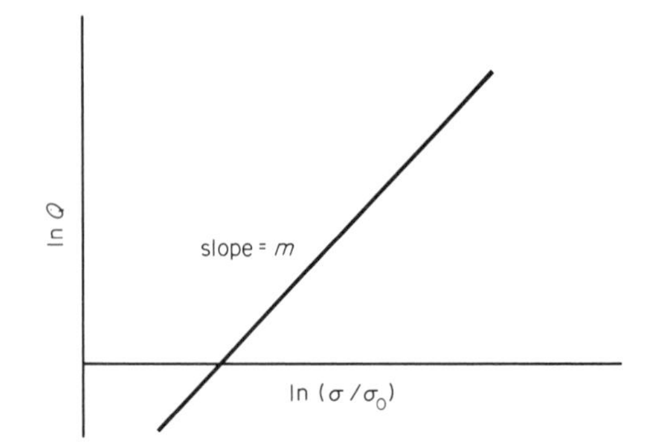

Figure 1
Average number of ruptures in a fiber vs. applied stress (Note that for Weibull material $\ln Q = \ln[-\ln(1-P)]$

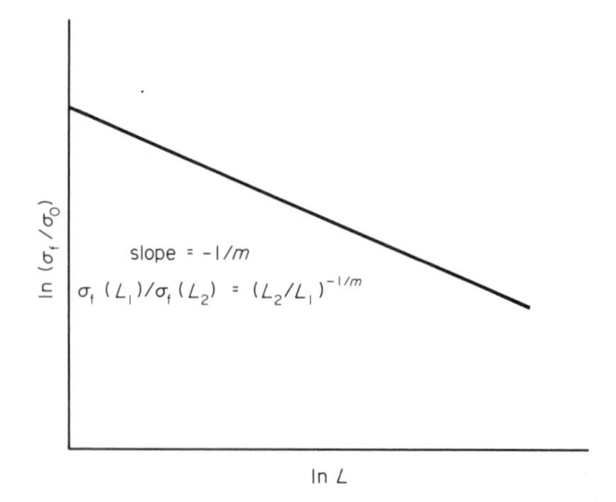

Figure 2
Weibull size effect

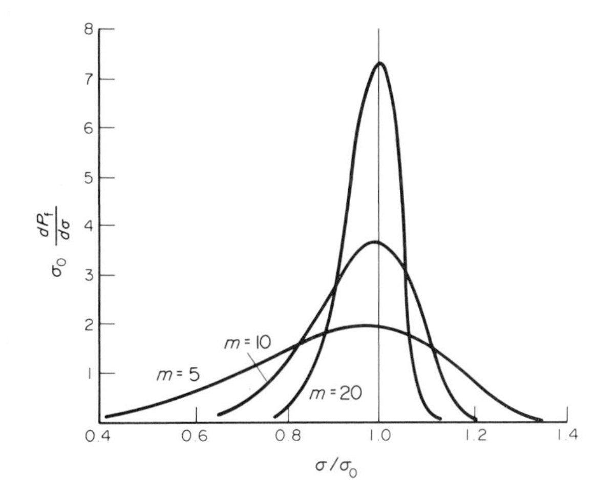

Figure 3
Failure probability density distribution for several values of Weibull's shape parameter m

variation (CV) of the Weibull failure distribution, defined as the standard deviation divided by the mean, is given approximately by

$$\mathrm{CV} = 1.2/m \qquad (12)$$

This relation is often utilized for the determination of m. The Weibull shape parameter m is thus very important; it not only describes the flaw distribution, but also determines both the coefficient of variation and the magnitude of the size effect.

The scale parameter σ_0, as noted earlier, is one measure of the failure stress of a unit length of fiber. Other measures that might be used include the mode, the median and the mean of the probability density distribution. These various measures generally differ from each other by less than half of a standard deviation. For $m > 20$ (a situation applying to most composites) this means the various measures differ by less than 3%. For simplicity we will therefore use σ_0 to represent the failure stress of a unit length of fiber; it is slightly higher than the mode of the distribution.

The preceding discussion assumed uniform stress over the length of the fiber. If the stress level varies with location, one must use the more general relation

$$P_f(\sigma) = 1 - \exp\left[-\int_0^L (\sigma/\sigma_0)^m dL\right] \qquad (13)$$

Weibull's theory for isotropic solids is essentially the same as that just described for fibers. All equations continue to apply but with volume V replacing length L and with $n(\sigma)$ redefined to be the average number of flaws per unit volume with strength less than σ. For example, the failure probability of a solid of volume V in uniform tension σ is

$$P_f(\sigma) = 1 - \exp[-V(\sigma/\sigma_0)^m] \qquad (14)$$

and the size effect is given by

$$(\sigma_f)_2 = (\sigma_f)_1 (V_1/V_2)^{1/m} \qquad (15)$$

The failure probability of a body in nonuniform uniaxial tension is given by

$$P_f(\sigma) = 1 - \exp\left[-\int_0^V (\sigma/\sigma_0)^m dV\right] \qquad (16)$$

In the case of a rectangular bar in pure bending this leads to

$$P_f(\sigma) = \frac{V}{2(m+1)} (\sigma/\sigma_0)^m \qquad (17)$$

where σ is the maximum tensile stress in the specimen. Here we have assumed with Weibull that compressive stresses have no effect. The factor $V/[2(m+1)]$ is sometimes referred to as the effective volume, because it is the volume that would have the same fracture probability when loaded in uniform tension at peak bending stress as the actual bending specimen has under its bending loads.

Statistical fracture theory for polyaxial stress states is much more complicated, and will not be discussed here because it is not needed for an understanding of the following treatment of composite failure. However, interested readers may find a survey article by Batdorf (1978) on statistical fracture theory for brittle materials useful in this and other contexts.

2. Fracture Behavior of Uniaxially Reinforced Composites

A typical uniaxially reinforced composite is constructed of many parallel fibers having high strength and high modulus held together by a relatively weak low-modulus matrix material. In such a composite the load is carried mainly by the fibers. The following analysis, which is based on a conceptual model of composite failure pioneered by Rosen (1964) and Zweben and Rosen (1970) assumes that the individual fibers behave according to Weibull's theory. We will find that the composite as a whole, however, does not.

As the tensile load on such a fiber increases, isolated fiber ruptures (singlets) occur at the locations of particularly weak spots (Fig. 4). Such isolated breaks do not normally cause failure because the matrix transfers the load from the broken fiber to its neighbors in the break plane, and back again into the broken fiber some distance from the break. The overloaded portion of a neighboring fiber is very short and therefore strong and generally able to support the excess load. The length δ over which the broken fiber carries a reduced load is known as the ineffective length. It is the same as the length over which each of the neighboring fibers is subjected to an overload.

As the load increases, the number of singlets increases, and eventually the overloaded segment of a

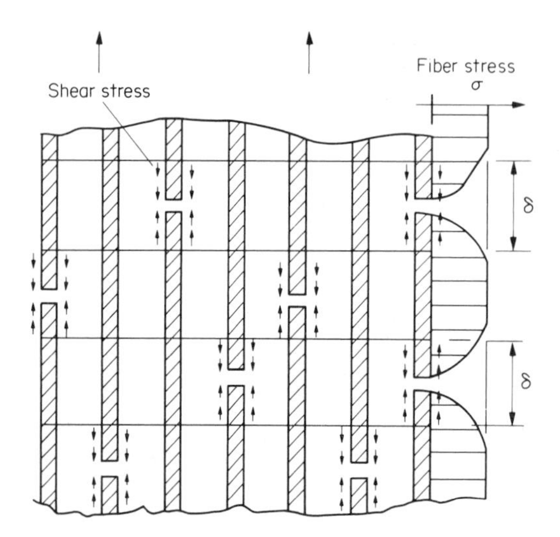

Figure 4
Zweben–Rosen model of damage produced by tension in uniaxially reinforced composite

fiber adjacent to a singlet will be abnormally weak and therefore will rupture, resulting in a double break (doublet). This process continues, generating more—and higher order—multiplets until one multiplet reaches a critical size and rupture of the composite as a whole ensues.

Consider now a composite consisting of N fibers each of length L. Normally $\delta \ll L$, so that the presence of even a considerable number of fiber fractures does not significantly reduce the total length of fiber subjected to stress σ. Under these circumstances, Q_1, the number of singlets that are formed in the course of increasing fiber stress from 0 to σ, is the total length of fiber multiplied by the number of flaws per unit length weaker than σ, i.e.,

$$Q_1 = NL(\sigma/\sigma_0)^m \tag{18}$$

Actually Q_1 is a random variable which follows a Poisson distribution with the mean given by Eqn. (18). Here and in the following analysis it is assumed that $N \gg 1$; in this way the differing environment of the relatively rare edge flaws will not significantly influence the result.

Let us assume that each singlet has n_1 nearest neighbors each subjected to an overload over a segment of length δ_1. The load on any neighboring fiber is a maximum in the plane of the break, and this maximum will be designated $c_1\sigma$, where $c_1 > 1$. For convenience we define a conceptual effective length λ_1 of an overloaded segment such that when subjected to a uniform stress $c_1\sigma$ it will have the same failure probability as the actual nonuniformly loaded segment of length δ_1 has. The total effective length of the overloaded segments adjacent to the Q_1 singlets is $Q_1\lambda_1 n_1$. The number of ruptures expected in this group of overloaded segments, which is the same as the number of singlets that are converted to doublets during the loading process, is given by

$$Q_2 = Q_1\lambda_1 n_1(c_1\sigma/\sigma_0)^m \tag{19}$$

Generalizing to higher order multiplets,

$$Q_{i+1} = Q_i\lambda_i n_i(c_1\sigma/\sigma_0)^m \tag{20}$$

Note that the Q's are not in general the numbers of multiplets in existence at a particular stress because in the loading process some will have been converted into higher order multiplets. Instead the number of i-plets in existence at stress σ is given by

$$g_i = Q_i - Q_{i+1} \tag{21}$$

It is evident that to calculate the number of multiplets as a function of applied stress, much detailed information such as the number of nearest neighbors, the stress concentration factors, the effective length of overload and so on, is necessary. Before discussing this, however, it is instructive to see how the results are to be interpreted.

Figure 5 is a logarithmic plot of the number of multiplets of various orders as a function of σ. The plot of Q_1 is the same as that for Q in Fig. 1, except that Q is the number of ruptures that would occur in loading a single fiber of length L to stress σ, whereas Q_1 is the number that would occur in N fibers. In both cases the slope of the logarithmic plot is m.

The doublet line has slope $2m$. At low stress the number of doublets is very much less than the number of singlets, but as stress rises the number of doublets increases much faster than the number of singlets, and at a certain stress the two are equal. Since all doublets are assumed to come from singlets the Q_2 line becomes the same as the Q_1 line above that stress. What this means, of course, is that above that stress the singlet is unstable and is converted into a doublet. Generalizing this argument, an i-plet is unstable at or above the stress at which it joins the $(i+1)$-plet line.

Consider now what happens as load is applied. At the stress designated σ_1 in Fig. 5 the first singlet appears, but the specimen does not fail because this stress is far below the singlet instability stress, designated σ_{1u} in the figure. At stress σ_2 the first doublet appears, but it is also stable because $\sigma_2 < \sigma_{2u}$. Extending this argument, failure of the composite as a whole occurs at the σ-intercept of the thicker line representing the envelope of the Q lines.

From Eqns. (18)–(20) it follows that the number of multiplets of any order is proportional to NL, so that a change of composite volume from N_1L_1 to N_2L_2 simply translates the lines of Fig. 5 vertically through a distance $\ln(N_2L_2/N_1L_1)$. Such a vertical translation changes the value of the σ intercept; this allows us to find the failure stress as a function of volume NL and to construct the composite analog to Weibull's size effect as shown in Fig. 2. This is done in Fig. 6. Again the failure line is shown as a thicker line, and the

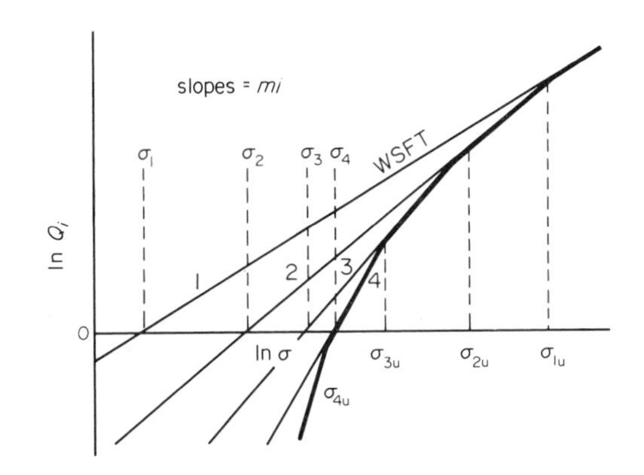

Figure 5
Number of multiplets in uniaxially reinforced composite vs. applied stress (schematic) (WSFT, Weibull's single fiber theory). Lines 1, 2, 3, 4 have $i=1$, 2, 3, 4, respectively

various segments have slopes that are the negative reciprocals of the corresponding segments in Fig. 5: $-1/m$, $-1/2m$, $-1/3m$, etc. If an increasing stress is applied to a composite of the size indicated by the vertical dashed line in the figure, the first singlet occurs at stress σ_s, the first doublet at stress σ_d and so on. Failure does not occur, however, until the dashed line reaches the failure line, which for the case illustrated corresponds to quadruplet instability. A smaller composite would fail by instability of a lower-order multiplet, and a larger composite would fail by instability of a higher order multiplet. Note that the failure lines in Fig. 6 do not end as shown but continue indefinitely with increasing NL.

Many simplifications have been made in the preceding analysis in order to focus attention on the most important features of the fracture process. For instance, while singlets and doublets are all alike, triplets can take two forms (linear array or vertices of a 45° right triangle) in a square array of fibers, or any of three forms in a hexagonal close-packed array. The number of possible multiplet shapes increases rapidly with multiplet order. The stress concentration factor is not the same for all nearest neighbors in multiplets of order two or more; nor is the effective length. Moreover, stress concentration factors have been calculated for only a few nearest neighbors in only a few selected higher order multiplets (Hedgepeth and Van Dyke 1967). The statistical treatment itself was also simplified to highlight the way the effective Weibull shape parameter depends on composite size.

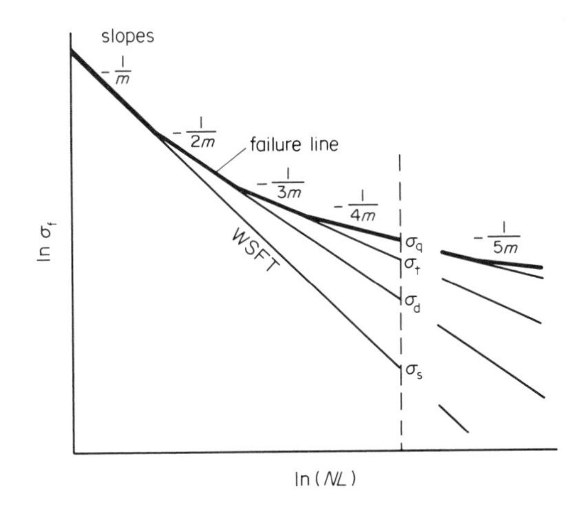

Figure 6
Damage accumulation and failure in uniaxially reinforced composite (schematic)

3. *Relation to Work Done Elsewhere*

In their treatment of tensile failure of composites, Zweben and Rosen (1970) modelled the composite as a chain of microbundles each of length δ as illustrated in Fig. 4. Because of the difficulty of coping with complexities such as those mentioned in the preceding section, they carried out their parametric analysis of three-dimensional-composite failure only to the first appearance of doublets, and proposed using the corresponding stress as a conservative estimate of the strength of the composite (which of course it is). They also showed that this criterion leads to a size effect in which the effective Weibull modulus is $2m$. Tamuzs (1979), employing a similar approach, carried out numerical analyses to very high multiplet order for a number of composites, and reported good agreement between theory and experiment.

Harlow and Phoenix (1978a, b) were the first to obtain exact results for the chain-of-microbundles model. Unfortunately, although ingenious, their method was applicable only to a tape that was one fiber thick (two-dimensional case) with $N \leqslant 14$, this limit being imposed by available computer capacity. Also the method assumed that the entire load rejected by an i-plet would be taken up by the two immediately adjacent neighbors. Actually a substantial part is transferred to more distant neighbors and the fraction transferred increases with increasing multiplet order (Hedgepeth and Van Dyke 1967). Nevertheless the work is valuable in providing a standard that can be used to assess the degree of inaccuracy resulting from the mathematical approximations in more versatile methods. Later, Harlow and Phoenix devised a very accurate approximate method based on a recurrence analysis that is free of most of the limitations on the exact solution.

The theories discussed above are laborious to apply. Asymptotic approximations that are much easier to use have been developed by Smith (1982). These approximations lead to failure lines quite similar to those presented here.

The failure lines in the exact solution are actually

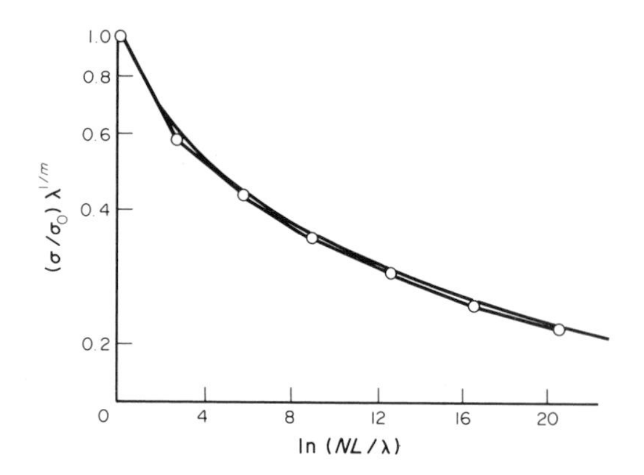

Figure 7
Comparison of failure stress predicted by approximate theory of Sect. 2 and that of exact theory of Harlow and Phoenix for $m=5$

smooth curves. The shape of the lines in the present solution (Figs. 5 and 6) is a consequence of the assumption that the numbers of defects and of multiplets of any order are always the expected numbers, and that all of them have the expected strength. Such theories should perhaps be referred to as quasi-probabilistic. In an exact treatment the fluctuations from the expected values serve to smooth out the failure line and to eliminate the vertices.

To illustrate the use of the exact solution to measure the mathematical inaccuracies in an approximate method, Fig. 7 was prepared. The smooth curve is the exact solution. Since the same curve applies for all values of λ_i, it is convenient to choose $\lambda_i=1$. This choice converts the coordinates to the more familiar ones, σ/σ_0 and NL. The results of the present theory are then readily found using the following values of the needed parameters: $n_i=2$, $\lambda_i=1$, and $c_i=1+0.5i$. Setting $Q_i=1$ to find the failure stresses of the various multiplets, we obtain

$$(\sigma_f)_i/\sigma_0=\left[NL\,2^{i-1}\left\{\prod_{k=1}^{i-1}(1+0.5k)\right\}^m\right]^{-1/im} \tag{22}$$

A logarithmic plot of the above relation yields the line shown in Fig. 7 (for clarity, only the envelope of the Q-lines is given in the figure). The circles are the vertices in the approximate solution. The difference in predicted failure stress between the two solutions is generally only a few percent. This encourages the view that the simple theory of Sect. 2 is accurate enough for most practical purposes.

In more recent work Phoenix and Smith (1983) and Phoenix and Tierney (1983) have analyzed three-dimensional composites using asymptotic approaches. Usually they have assumed that the entire load rejected by a multiplet is carried by its immediate neighbors. Plausible assumptions are then made regarding the way in which the load is distributed among the neighbors for any given shape of multiplet. Batdorf and Ghaffarian (1984), on the other hand, have assumed that all neighboring fiber elements are equally loaded regardless of multiplet shape, and that the load divided between them is that calculated by Hedgepeth and Van Dyke (1967), with the gaps in their calculated results filled by interpolation. Neither approach is strictly correct, and the differences between their predictions are generally less than the uncertainties imposed by our ignorance of the values of the parameters appearing in both theories (stress concentration factors, ineffective lengths, etc.).

4. Theoretical Predictions

In spite of the difficulties mentioned at the end of Sect. 2 there are a number of important general conclusions that can be drawn from the theory of composite failure. They are explained below.

4.1 Relation to Weibull Theory

There are some striking similarities to the Weibull theory. To aid in identifying them, observe that the recursion relation in Eqn. (20) implies that

$$Q_i=NL(\sigma/\sigma_{0i})^{m_i} \tag{23}$$

where $m_i=mi$ and σ_{0i} is a complicated algebraic function of all the c_k's, λ_k's and n_k's for which $k<i$. First of all, note that as in Weibull theory, the fracture characteristics depend on the total volume of the composite (or NL), not the shape. In both theories, however, this generalization has to be qualified; if the surface area is sufficiently large relative to the volume, surface flaws will influence or may even dominate the failure process, and the statement is then no longer true. Second, since the probability density distribution for strength of multiplets of any given order, as given in Eqn. (23), is a Weibull distribution, many of the results of Weibull theory can be taken over directly. In particular, we can conclude that if the volume is such that a k-plet is of critical size, the probability of failure at stress σ is

$$P_f(\sigma)=1-\exp[-NL(\sigma/\sigma_{0k})^{mk}] \tag{24}$$

and the failure stress will be approximately

$$\sigma_f=\sigma_{0k} \tag{25}$$

In addition the coefficient of variation will be given approximately by

$$\mathrm{CV}=1.2/m_k=1.2/mk \tag{26}$$

4.2 Critical Crack Size

As composite size increases, the critical crack size increases. There seems to be no way of checking the validity of this prediction experimentally (at least for three-dimensional composites), but the critical crack sizes obtained by Barry (1978) using a Monte Carlo approach are in good agreement with those obtained analytically by Batdorf and Ghaffarian (1984).

4.3 Coefficient of Variation

According to theory, the coefficient of variation decreases without limit, but this is because only the effects of random positioning of fiber defects are considered. It is to be expected that when the variance arising from this cause is sufficiently small, minor sources of strength variation (e.g., variations in fiber diameter, fiber volume ratio, matrix strength and stiffness and fiber–matrix bond strength) will ensure that some strength variability is present no matter how large the composite becomes.

4.4 Size Effect

According to theory, failure stress decreases with increasing composite size, but the decrease is less than Weibull theory would predict. The physical explanation of this is that the multiplet population density of a large specimen at its failure stress is lower than that of small specimens at their (higher) failure stresses. It is obviously of great practical importance to know the size above which failure stress does not decrease significantly. This will depend on the parameters appearing in the theory, which in turn depend on the composition of the composite. Some data suggest that for graphite epoxy even small laboratory coupons may approach this critical size (Batdorf and Ghaffarian 1984). In tension experiments by Hitchon and Phillips (1978) on graphite epoxy composites having volumes less than a cubic inch the strength reduction was reported to be only 10% for a twenty-seven-fold volume change. The size effect problem is of course crucial for a rocket manufacturer who tests a number of filament-wound rocket cases 15 cm or so in diameter and uses the results to predict the failure stress of (say) a rocket case 3 m in diameter. Although here the volume ratio is 8000:1, data obtained at Thiokol in Utah, USA, suggest that the strength difference between two such sizes may turn out to be very small.

4.5 Damage Tolerance

Theory implies that composites, especially large ones, may be very damage-tolerant. This is because if the size is such that the critical multiplet order is j, a preexisting i-plet such that $i<j$ will not grow until the stress exceeds the level at which other i-plets are created and start to grow. Above this stress there is no more reason for the preexisting i-plet to be the ancestor of the j-plet responsible for the eventual failure than there is for any of the competing i-plets created during the loading process to be the ancestor.

5. Discussion

One of the most pressing issues from a practical point of view is the identification of the size range over which the theory is important. It is clearly essential in finding the relation between the strength of a composite and that of its constituents, and the relation between strengths of small composite specimens. Above some size, theory suggests that both the size effect and the strength variability become small and therefore relatively unimportant. The validity of such a conclusion and, if found to be valid, the critical size, need clarification.

The theory presented in Sect. 3 may not be the most accurate available, but it has some compensating features. It is probably the easiest to understand and apply, for it deals with physical quantities (number of multiplets) rather than mathematical abstractions (probabilities) and relates them to stress levels with algebraic rather than transcendental equations. In so doing it leads to a better appreciation of the damage sustained by composites prior to the final fracture. Since it is not based on the chain-of-microbundles model it can assign different ineffective lengths to different multiplet orders. Also it can be modified to allow interactions between multiplets in different microbundles, thus permitting extension to such matters as accounting for the commonly observed jagged or broom-type fracture surfaces. In practical applications, the errors introduced by the various approximations employed in the approach are probably smaller than those due to our uncertainties regarding the proper values of the parameters common to all the theories.

Virtually all present theories assume that the fibers form some kind of regular array (e.g., square, hexagonal, etc.). Unfortunately most composites of structural interest are far from regular; interfiber distances vary, and the number of nearest neighbors, even for a singlet, may vary from place to place. An attempt has been made to deal theoretically with the first of these complications (Batdorf and Ghaffarian 1984) but little has been done about the second. More work is clearly needed on the effects of such geometrical imperfections so that the behavior of real composites can be better understood.

The statistical theory for uniaxially reinforced composites has been extended to cover the problem of creep rupture by Phoenix and Tierney (1983). The theory for either short term or creep failure can probably be extended to orthogonally-reinforced composites for stress states in which the shear is zero in the principal axis system. Whether either theory can be extended to more complex composites remains to be seen.

Bibliography

Barry P W 1978 The longitudinal tensile strength of unidirectional fibrous composites. *J. Mater. Sci.* 13:2177–87

Batdorf S B 1978 Fundamentals of the statistical theory of fracture. In: Bradt R C, Hasselman D P H, Lange F M (eds.) 1978 *Fracture Mechanics of Ceramics 3*. Plenum, New York, pp. 1–30

Batdorf S B, Ghaffarian R 1984 Size effect and strength

variability of unidirectional composites. *Int. J. Fract.* 26: 113–23
Harlow D G, Phoenix S L 1978a The chain-of-bundles probability model for the strength of fibrous materials I: Analysis and conjectures. *J. Compos. Mater.* 12: 195–214
Harlow D C, Phoenix S L 1978b The chain-of-bundles probability model for the strength of fibrous materials II: A numerical study of convergence. *J. Compos. Mater.* 12: 314–34
Hedgepeth J, Van Dyke P 1967 Local stress concentrations in imperfect filamentary composite materials. *J. Compos. Mater.* 1: 294–309
Hitchon J W, Phillips D C 1978 The effect of specimen size on the strength of cfrp. *Composites* X: 119–24
Phoenix S L, Smith R L 1983 A comparison of probabilistic techniques for the strength of fibrous materials under local load sharing. *Int. J. Solids Struct.* 19: 479–96
Phoenix S L, Tierney L J 1983 A statistical model for the time dependent failure of unidirectional composite materials under local elastic load sharing among fibers. *Eng. Fract. Mech.* 18: 193–215
Rosen B W 1964 Tensile failure of fibrous composites. *AIAA J.* 2: 1985–91
Smith R L 1982 A probability model for fibrous composites with local load sharing. *Proc. R. Soc. London, Ser. A* 372: 294–309
Tamuzs V P 1979 Some peculiarities of fracture in heterogeneous materials. In: Sih G C, Tamuzs V S (eds.) 1979 *Proc. 1st USA–USSR Symp. Fracture of Composite Materials.* Sijthoff and Nordhoff, The Netherlands
Weibull W 1939 A statistical theory for strength of materials. *Ingeniorsvetenskapsakad. Handl.* 151: 1–29
Zweben C, Rosen B W 1970 A statistical theory of material strength with application to composite materials. *J. Mech. Phys. Solids* 18: 189–206

S. B. Batdorf
[Laguna Hills, California, USA]

Structure–Performance Maps

The purpose of the structure–performance map is to present in a concise manner the effective properties of composites based upon various reinforcement forms and fiber–matrix combinations. The maps are constructed using the results of extensive analytical studies of the thermoelastic properties of unidirectional laminated composites as well as two-dimensional and three-dimensional textile structural composites with polymeric, metal and ceramic matrices. The effectiveness and uniqueness of the reinforcement forms can be evaluated from the maps. Also, these maps can guide engineers in selecting materials for structural design, and researchers in identifying the needs of future work.

1. Material Systems

The primary materials for reinforcement include graphite, boron, glass and aramid fibers in continuous, discontinuous (e.g., short fibers, whiskers and particulates) or textile forms. Other reinforcement materials with a promising potential of development are silicon carbide and alumina fibers. A number of other ceramic fibers and whiskers have also received considerable attention. Table 1 lists the mechanical and physical properties of several fibers. The superior strength and modulus of fibers are also demonstrated in Fig. 1 on the specific basis (strength/density and modulus/density).

In the following discussions, continuous fibers in unidirectional laminae and woven-fabric composites are considered. It should be noted that the term woven-fabric composites is used to encompass composites reinforced with two-dimensional (2-D) and three-dimensional (3-D) textile preforms.

The types of matrix materials along with their mechanical and physical properties are given in Table 2. Polymeric, metal and glass/ceramic matrices for composites are included.

2. Analytical Modelling

The analytical modelling work for constructing the structure–performance map focuses on unidirectional laminae, and 2-D and 3-D woven-fabric composites.

Unidirectional laminae are the building blocks of laminated composites. Various approaches have been successfully developed to predict the thermoelastic properties of unidirectional laminae, such as the mechanics-of-materials approach and elasticity approach. The off-axis thermoelastic properties of a unidirectional lamina inclined at an angle to the principal material direction can be obtained through proper transformation of the tensorial quantities. The cross-ply laminate, which is a special case of the off-axis laminate, is composed of unidirectional laminae with principal material directions oriented alternately at 0° and 90° to the laminate axis. The laminated angle-ply

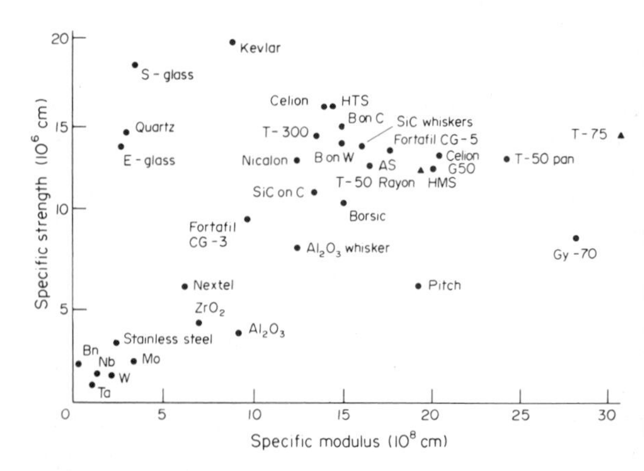

Figure 1
Variation of specific strength with specific volume for several fibrous materials

Table 1
Properties of selected fibers

	Density ρ ($g\,cm^{-3}$)	Longitudinal tensile modulus E_1 (GPa)	Transverse tensile modulus E_2 (GPa)	Poisson's ratio ν_{12}	Shear modulus G_{12} (GPa)	Longitudinal tensile strength σ (MPa)	Longitudinal thermal expansion coefficient α_1 ($10^{-6}\,°C^{-1}$)	Transverse thermal expansion coefficient α_2 ($10^{-6}\,°C^{-1}$)
Glass	2.45	71	71	0.22	30	3500	5	5
Kevlar	1.47	154	4.2	0.35	2.9	2800	−4	54
Graphite (AS)	1.75	224	14	0.20	14	2100	−1	10
Graphite (HMS)	1.94	385	6.3	0.20	7.7	1750	−1	10
Boron	2.64	420	420	0.20	170	4200	5.0	5.0
SiC	3.2	406	406	0.20	169	3395	5.2	5.2
Al_2O_3	3.9	385	385		154	1400	8.5	8.5

Table 2
Properties of selected matrices

	Density ρ ($g\,cm^{-3}$)	Tensile modulus E (GPa)	Poisson's ratio ν	Shear modulus G (GPa)	Tensile strength σ (MPa)	Thermal expansion coefficient α ($10^{-6}\,°C^{-1}$)
Epoxy	1.246	3.5	0.33	1.25	35	57.5
PEEK	1.30	4	0.37	1.4	70	45
Aluminum	2.71	69	0.33	26	74	23.6
Titanium	4.51	113.8	0.33	43.5	238	8.4
Magnesium	1.74	45.5	0.33	7.5	189	26
Borosilicate glass	2.23	63.7	0.21	28	90	3.25

composite is made up of off-axis unidirectional laminae in $+\theta$ and $-\theta$ orientations.

The effective thermoelastic properties of laminated composites can be predicted from the well-established classical laminated-plate theory. The key laminate geometric parameters are fiber orientation, lamina thickness, lamina stacking sequence and thermoelastic properties of the individual laminae. Comprehensive discussion of the analysis for laminated composites can be found in the article *Laminates: Elastic Properties*.

The state of the art in the modelling of 2-D textile structural composites is discussed in the article *Woven-Fabric Composites: Properties*. In the construction of performance maps, a range of fabric materials with increasingly larger size of repeating unit—from plain weave to 8-harness satin weave—are considered.

In the case of 3-D fabric composites, the braided-fiber preform is adopted for modelling purposes. The concepts developed for analysis can be readily applied to other types of 3-D preforms. The "unit cell" approach has been proposed, in which a fiber assumes a position along a diagonal in the unit cell and defines an angle with respect to the braiding axis. Two analytical approaches have been developed: the fiber interlock model and fiber inclination model. These are used for predicting the thermoelastic properties as functions of fiber spatial orientation, fiber volume fraction, and braiding parameters. The fiber inclination model is employed for the performance-map analysis which treats the unit cell of the 3-D braided composite as an assemblage of four inclined unidirectional laminae. The classical laminated-plate theory is employed to derive the effective laminate thermoelastic properties. The validity of these analytical methods has been substantiated by experimental characterizations of 3-D braided reinforcements in both polymeric- and metal-matrix composites.

3. Structure–Performance Maps

Four types of reinforcement forms are under consideration: unidirectional laminated angle-plies with the off-axis angle θ ranging from 0° to 90°; a 0°/90° cross-ply; two-dimensional woven fabric with n_g ranging from 2 (plain weave) to 8 (8-harness satin); and three-dimensional braided composites with the braiding angle between a fiber segment and braiding axis in the range 15°–35°.

The results of the parametric studies have been used to construct maps of the following four types:

(a) longitudinal Young's modulus E_x vs. transverse Young's modulus E_y;

(b) longitudinal Young's modulus E_x vs. in-plane shear modulus G_{xy};

(c) longitudinal Young's modulus E_x vs. Poisson's ratio ν_{xy}; and

(d) longitudinal thermal expansion coefficient α_x vs. transverse thermal expansion coefficient α_y.

The fiber volume fraction of all the composites presented in the maps is assumed to be 60%.

The structure–performance maps for various combinations of fiber and matrix materials and reinforcement forms are shown in Figs. 2–6. The general characteristics of the maps are summarized below.

The in-plane thermoelastic properties of unidirectional laminae depend strongly on fiber orientation. Unidirectional reinforcement provides the highest elastic stiffness along the fiber direction. The 0°/90° cross-ply yields identical thermoelastic properties in 0° and 90° orientations. Its in-plane shear rigidity is poor. The in-plane thermoelastic properties of angle-ply laminates are also dependent upon the fiber orientations. The longitudinal Young's modulus of an angle-ply laminate is lower than that of a unidirectional lamina. Better transverse elastic property and in-plane shear resistance can be achieved through stacking of the unidirectional laminae with different fiber orientations. For ±45° angle-ply, the in-plane stiffness is relatively low, while the shear modulus reaches a maximum.

The 2-D biaxial woven-fabric composites can provide balanced in-plane thermoelastic properties within a single ply. They behave similarly to 0°/90° cross-plies, although the fiber "waviness" tends to reduce the in-plane efficiency of the reinforcements. As the fabric construction changes from plain weave to 8-harness satin, the frequency of crimp due to fiber crossover is reduced, and the fabric structure approaches that of 0°/90° cross-ply.

The thermoelastic properties of three-dimensional braided composites also show a strong dependence on fiber orientation. Three-dimensionally braided composites have demonstrated good in-plane properties, which are comparable to those of unidirectional angle-

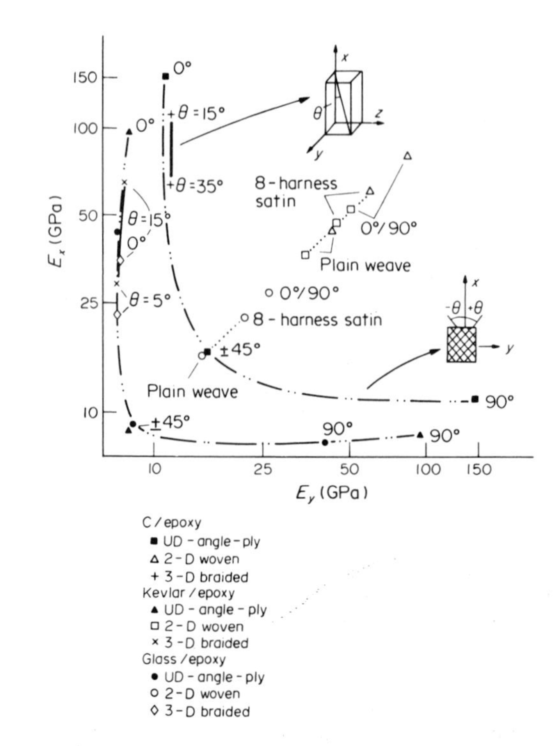

Figure 2
Variation of longitudinal Young's modulus E_x with transverse Young's modulus E_y for C–epoxy, Kevlar–epoxy and glass–epoxy composites

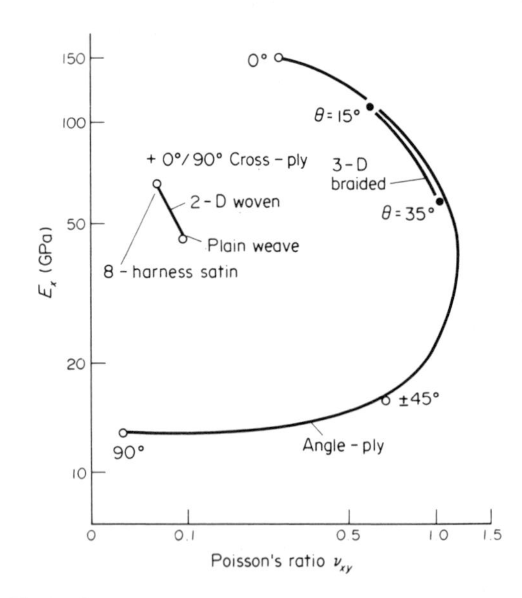

Figure 3
Variation of longitudinal Young's modulus E_x with Poisson's ratio for C–epoxy composites

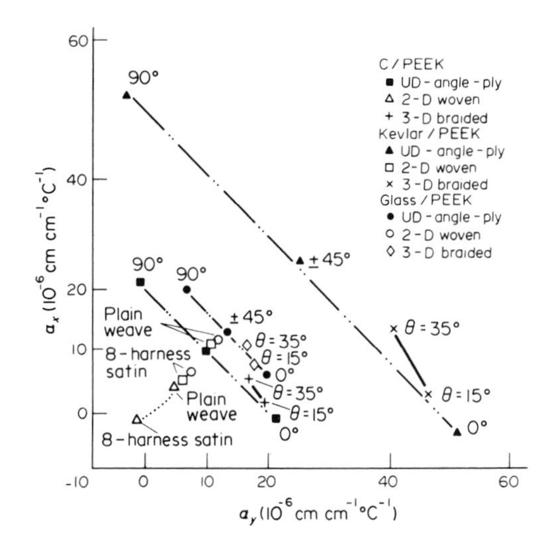

Figure 4
Variation of longitudinal thermal expansion coefficient α_x with transverse thermal expansion coefficient α_y for C–PEEK, Kevlar–PEEK and glass–PEEK composites

plies with the same range of fiber orientation. The longitudinal Young's moduli and in-plane shear rigidities of 3-D braided composites with braiding angles ranging from 15° to 35° are better than those of 2-D woven-fabric composites. However, the transverse Young's moduli are lower and the major Poisson's ratios higher than those of 2-D woven-fabric composites. It is a unique capability of 3-D braided composites that they can provide both stiffness and shear rigidity along the thickness direction. Also,

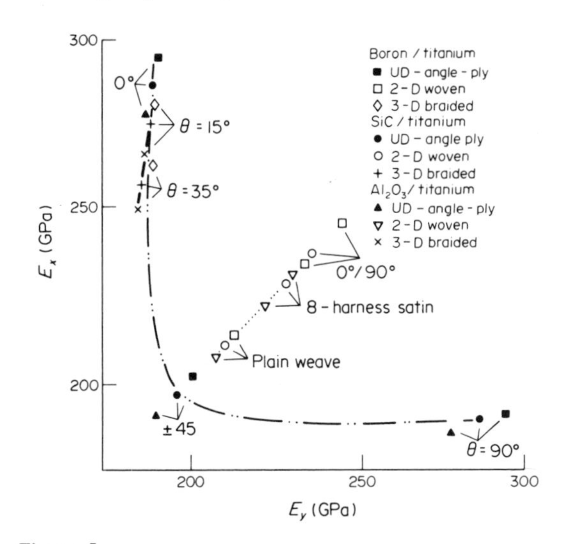

Figure 5
Variation of longitudinal Young's modulus E_x with transverse Young's modulus E_y for B–Ti, SiC–Ti and Al_2O_3– Ti composites

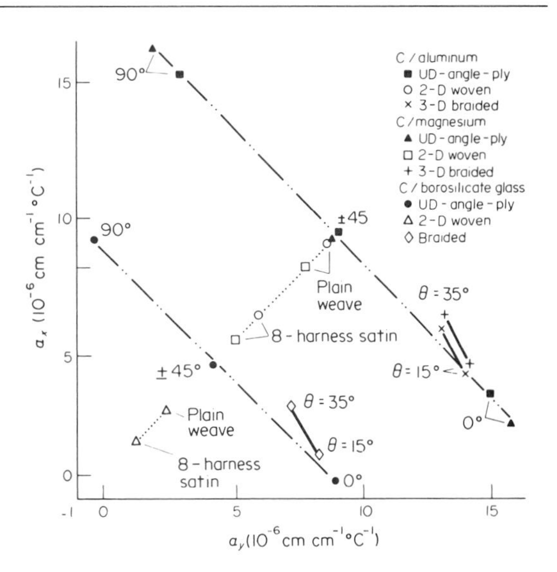

Figure 6
Variation of longitudinal thermal expansion coefficient α_x with transverse thermal expansion coefficient α_y for C–aluminum, C–magnesium and C–borosilicate-glass composites

because of the integrated nature of fiber arrangement, there are no interlaminar surfaces in braided composites, such as would be found in unidirectional and 2-D fabric laminate composites. The elimination of interlaminar failure is responsible for the enhanced damage tolerance and impact resistance in 3-D braided composites.

Figure 2 shows the variation of E_x with E_y for composites of carbon, Kevlar and glass-fibers in epoxy matrix. The variation of the major Poisson's ratio with various forms of reinforcement is demonstrated in Fig. 3 for carbon/epoxy composites. The lateral contractions under axial loading of 3-D braided composites are relatively higher. A unique characteristic of Kevlar and graphite fibers is their negative coefficients of thermal expansion along the axial direction. The thermal expansion coefficient map for polyether ether ketone (PEEK) matrix composites is shown in Fig. 4. Composites with zero thermal expansion and hence dimensional stability can be achieved by suitable material selection, geometric design and fiber hybridization.

A wide variety of metal matrix materials has been adopted, ranging from low-melting-point, lightweight metal alloys such as those of aluminum and magnesium to various superalloys for application at elevated temperature. The variation of E_x with E_y for titanium matrix composite is shown in Fig. 5. The high specific strength and high resistance to corrosion, as well as low thermal expansion coefficient, make titanium-based composites attractive for aerospace applications.

Figure 6 compares the thermal expansion coefficients α_x and α_y for carbon-fiber-reinforced aluminum, magnesium and borosilicate glass matrices. The low thermal expansion characteristic of borosilicate glass composite is evident.

The advances in analytical modelling have laid the foundation for the understanding of structure–performance relationships of laminated and fabric-based composites. Maps of other mechanical and physical properties of various advanced composites can be constructed.

See also: Laminates: Elastic Properties; Three-Dimensional Fabrics for Composites; Woven-Fabric Composites: Properties

Bibliography

Ashby M F, Gandi C, Taplin D M R 1979 Fracture-mechanism maps and their construction for FCC metals and alloys. *Acta Metall.* 27: 699

Chou T-W, Kelly A, Okura A 1985 Fiber-reinforced metal-matrix composites. *Composites* 16: 187

Chou T-W, Yang J M 1986 Structure–performance maps of polymeric, metal and ceramic matrix composites. *Metall. Trans.* 17A: 1547-59

Frost H J, Ashby M F 1982 *Deformation-Mechanism Maps.* Pergamon, Oxford

Gandi C, Ashby M F 1979 Fracture-mechanism maps for materials which cleave: FCC, BCC and HCP metals and ceramics. *Acta Metall.* 27: 1565

Ishikawa T, Chou T-W 1982a Stiffness and strength behavior of woven-fabric composites. *J. Mater. Sci.* 17: 3211-20

Ishikawa T, Chou T-W 1982b Elastic behavior of woven fabric composites. *J. Compos. Mater.* 16: 2–19

Ishikawa T, Chou T-W 1983a One-dimensional micro-mechanical analysis of woven fabric composites. *AIAA J.* 21: 1714–21

Ishikawa T, Chou T-W 1983b In-plane thermal expansion and thermal bending coefficients of fabric composites. *J. Compos. Mater.* 17: 92–104

Ishikawa T, Chou T-W 1983c Thermoelastic analysis of hybrid fabric composites. *J. Mater. Sci.* 18: 2260–68

Ishikawa T, Chou T-W 1983d Non-linear behavior of woven fabric composites. *J. Compos. Mater.* 17: 399–413

Kelly A 1985 Composites in context. *Compos. Sci. Technol.* 23: 171

Ma C L, Yang J M, Chou T-W 1986 Elastic stiffness of three-dimensional woven fabric composites, ASTM-STP 893. American Society for Testing and Materials, Philadelphia, Pennsylvania

Majidi A P, Yang J M, Chou T-W 1986 *Toughness Characteristics of Three-dimensionally Braided Al_2O_3/Al–Li Composites.* The Metallurgical Society, Warrendale, Pennsylvania

Vinson J R, Chou T-W 1975 *Composite Materials and Their Use in Structures.* Wiley, New York

Yang J M, Ma C L, Chou T-W 1986 Fiber inclination model of 3-dimensional textile structural composites. *J. Compos. Mater.* 20: 472–84

T-W. Chou
[University of Delaware, Newark, Delaware, USA]

T

Thermosetting Resin Matrices

This article discusses the chemistry of major thermosetting resin matrices used in composites.

Unlike thermoplastics which, by definition, are reversibly melt-processable and usually of high molecular weight and high viscosity, thermoset resins are irreversibly formed from low molecular-weight precursors of low viscosity. Compared to thermoplastic composites, the initial low viscosity enables higher concentrations of both fibers and fillers to be incorporated into thermosetting matrices while still retaining good fiber wet-out and filler dispersion.

After compounding with either fibers, fillers or both, the resin is cured to give a three-dimensional cross-linked polymeric matrix of essentially infinite molecular weight. Once cross-linked, thermosets cannot be reversibly processed at elevated temperatures. The inherent three-dimensional network of thermosets results in significant advantages over thermoplastics, including greater dimensional stability, less flow under stress, greater resistance to solvents and, in general, a lower coefficient of thermal expansion.

The curing of thermoset resins has been extensively studied. In particular, Gillham (1979) has developed a generalized model to account for the effect of curing temperature on the two macroscopic processes of gelation and vitrification which occur during cure. On the molecular level, gelation corresponds to the incipient formation of branched molecules of mathematically infinite molecular weight and occurs at a calculable degree of reaction for a particular reactive system. Vitrification usually follows gelation, and is the point at which the thermoset becomes glassy as a consequence of the network becoming tighter through further chemical reaction. Subsequent chemical reaction is retarded once vitrification has been achieved. Gillham's model facilitates an understanding of the properties of thermosets in terms of the gelation and vitrification processes.

Fibrous reinforcements used in thermoset composites are usually coated with a sizing solution prior to compounding. The nature of the size is dependent on the chemistry of the resin matrix, and is necessary in order to achieve optimum physical properties from the composite. The two most important components of the size are a film former and a coupling agent. The film former holds the fiber filaments together as a strand, and is generally selected to be soluble in the matrix resin. Thus, during compounding the film former allows the breakup of fiber bundles and the uniform distribution of fibers throughout the resin. Poly(vinyl acetate) is a typically used film former for unsaturated polyester and vinyl ester matrices.

The principal role of the coupling agent is to promote wetting of the fibers and to ensure good stress transfer between the weak matrix polymer and the stronger fibers by acting as a bridge between the two. This is achieved by the presence of both fiber and resin functional groups within the coupling agent. Typical coupling agents are chromium complexes, titanates or silanes, containing resin-reactive groups such as epoxy, vinyl, amino or methacryl. The choice of reactive group depends on the resin used in the composite. For example, coupling agents having reactive vinyl groups would normally be used in unsaturated polyesters or vinyl esters, whereas nylon composites would use coupling agents with amino-reactive groups. In some instances, coupling agents also tend to reduce the absorption of water by a composite, thereby improving its electrical properties and reducing the degradation of other physical properties by water. Some coupling agents, notably titanates, are reported to produce significant decreases in the viscosity of filled resins, allowing much higher loadings of fillers to be achieved.

A consequence of poor fiber wet-out can be the presence of voids within the composite. Voids are particularly detrimental to physical properties, since these act as stress-concentration sites and promote premature failure of the composite. They also result in unsupported regions of fibers, which both reduce stress transfer between fiber and matrix and increase the susceptibility of the fiber to chemical attack. Voids may be introduced into a composite by poor mixing techniques entrapping air between the fibers or filler particles. Alternatively, they may arise from the presence of volatiles or by chemical reaction occurring within the composite. Voids may be minimized by the use of proper fabricating techniques and by matching the size on the fiber with the resin matrix.

The remainder of this article discusses in depth the chemistry of thermoset resins derived from unsaturated polyesters and vinyl esters; epoxy resins containing aliphatic, aromatic and cycloaliphatic groups; polyurethanes derived from polyfunctional isocyanates and polyols; polyimides and certain formaldehyde-based resins. While this list is not exhaustive, it represents several billion kilograms per year of material in a wide variety of applications.

1. Unsaturated Thermoset Resins

Any system containing polyfunctional vinyl groups dissolved in a vinyl monomer can, in principle, be cured to form a thermoset resin. In practice, the vinyl groups present in the system must be capable of reasonably rapid copolymerization, preferably by free-radical initiators. For this reason, and for economic considerations, the unsaturated resin matrices that are of greatest interest in composites are polyesters containing maleate/fumarate unsaturation and vinyl esters containing methacrylate/acrylate groups. Styrene is the most widely used monomer, because it is a solvent for a variety of unsaturated systems, it is relatively inexpensive, and it is capable of rapid copolymerization with both maleate/fumarate and methacrylate/acrylate unsaturation.

The general reaction scheme for producing an unsaturated polyester resin is given in Scheme 1. The resulting unsaturated polyester moiety is dissolved in a polymerizable solvent, usually styrene. During cure, styrene copolymerizes with the polyester through the unsaturation inherent in the R_3 portion of the molecule, to give a cross-linked resin. End groups X and Y depend on the mole ratio $a/(b+c)$ of diol to diacid used in the polymerization. The degree of polymerization, n,is small and usually between 3 and 15. A wide variety of polyesters is available, based on the choice of the diacid, which is generally derived from phthalic, isophthalic, terephthalic or adipic acid in combination with an unsaturated diacid, which is almost always maleic or fumaric. The diol which is copolymerized with the diacid is generally based on ethylene glycol, propylene glycol, ethoxylated or propoxylated bisphenol A. (More specifically, so-called general-purpose unsaturated polyester resins are made using ethylene glycol as the diol, either orthophthalic or isophthalic acid as the saturated diacid, and fumaric acid—usually generated in situ from maleic anhydride—as the unsaturated diacid.) An example of a bisphenol A based unsaturated polyester resin, which after curing has superior corrosion resistance, is shown in Scheme 2. This molecule is also dissolved in styrene and copolymerizes in an analogous manner to Scheme 1.

Polyesters based on unsaturated diols have also been prepared, but such systems are generally more expensive, and are not as commercially important as maleate/fumarate-based unsaturated polyesters. The flexibility of polyesters may be controlled by judicious choice of diacids and diols, such that relatively flexible polyesters are produced from highly aliphatic precursors, whereas high-modulus, brittle materials with increasing glass-transition temperatures may be derived from combinations containing large amounts of aromatic diacids and/or aromatic diols.

The range of applicability of polyesters may be further extended by replacing some of the styrene with other monomers, such as methyl methacrylate to improve weathering; diallyl phthalate, divinyl benzene, di- and triacrylates to increase the heat distortion temperature; or highly brominated or chlorinated monomers to enhance fire retardance.

While classical maleate-based polyesters date back several decades, over the last ten years a number of methacrylate- and acrylate-based polyfunctional systems have emerged. These are generally classified as vinyl esters, but some variations may contain urethane as well as ester bridging groups. Scheme 3 shows the reaction scheme and structure of a typical vinyl ester, which has a relatively low viscosity in styrene solution, and which after cure has good physical properties at elevated temperatures, and good corrosion resistance.

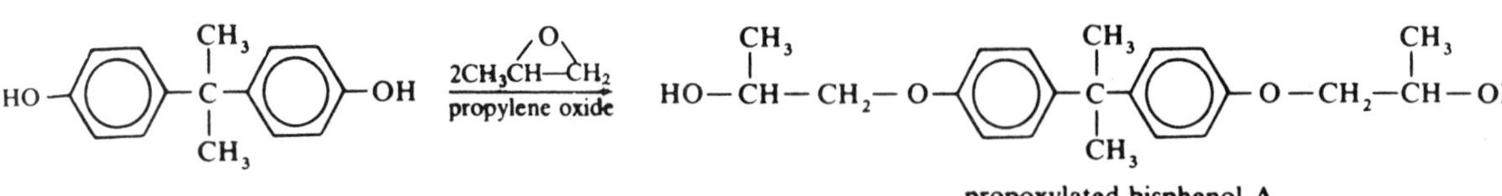

$$a\,HO{-}R_1{-}OH + b\,HO{-}\overset{O}{\overset{\|}{C}}R_2\overset{O}{\overset{\|}{C}}{-}OH + c\,HO{-}\overset{O}{\overset{\|}{C}}R_3\overset{O}{\overset{\|}{C}}{-}OH \longrightarrow X{-}\left(O{-}R_1{-}O{-}\overset{O}{\overset{\|}{C}}R_2\overset{O}{\overset{\|}{C}}{-}O{-}R_1{-}O{-}\overset{O}{\overset{\|}{C}}R_3\overset{O}{\overset{\|}{C}}\right)_n{-}Y$$

diol; saturated acid (or anhydride); unsaturated acid (or anhydride); unsaturated polyester

Scheme 1

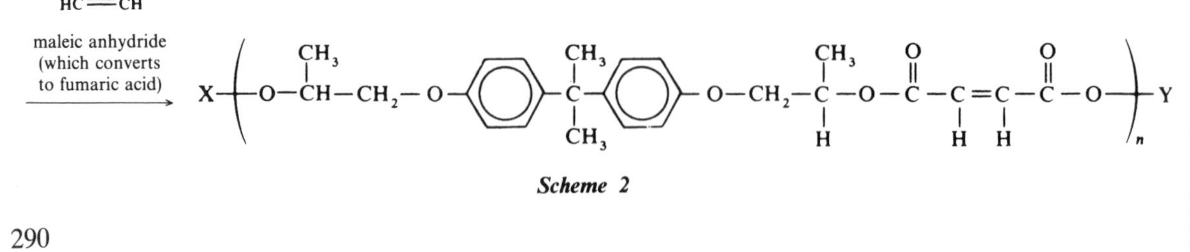

$$HO{-}C_6H_4{-}C(CH_3)_2{-}C_6H_4{-}OH \xrightarrow[\text{propylene oxide}]{2CH_3CH{-}CH_2\ (\text{epoxide, }O)} HO{-}\underset{CH_3}{CH}{-}CH_2{-}O{-}C_6H_4{-}C(CH_3)_2{-}C_6H_4{-}O{-}CH_2{-}\underset{CH_3}{CH}{-}OH$$

propoxylated bisphenol A

maleic anhydride (which converts to fumaric acid)

$$\longrightarrow X{-}\left(O{-}\underset{CH_3}{CH}{-}CH_2{-}O{-}C_6H_4{-}C(CH_3)_2{-}C_6H_4{-}O{-}CH_2{-}\underset{H}{\overset{CH_3}{C}}{-}O{-}\overset{O}{\overset{\|}{C}}{-}\underset{H}{C}{=}\underset{H}{C}{-}\overset{O}{\overset{\|}{C}}{-}O\right)_n{-}Y$$

Scheme 2

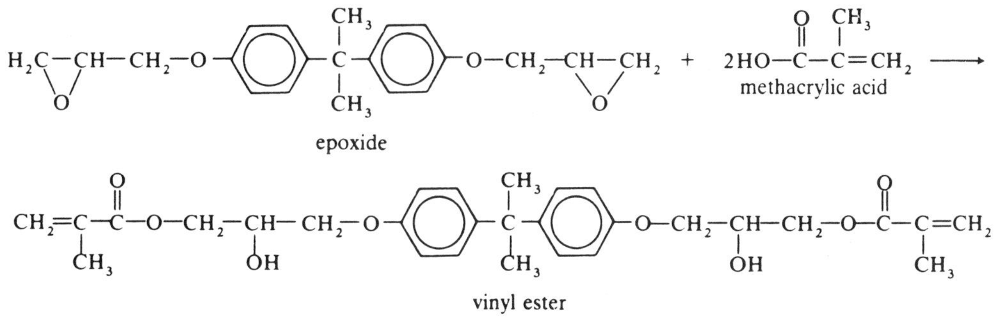

$$H_2\overset{O}{\overbrace{C-CH}}-CH_2-O-C_6H_4-C(CH_3)_2-C_6H_4-O-CH_2-\overset{O}{\overbrace{CH-CH_2}} + 2HO-\overset{O}{\overset{\|}{C}}-\underset{}{C}(CH_3)=CH_2 \longrightarrow$$

epoxide   methacrylic acid

$$CH_2=C(CH_3)-\overset{O}{\overset{\|}{C}}-O-CH_2-\underset{OH}{CH}-CH_2-O-C_6H_4-C(CH_3)_2-C_6H_4-O-CH_2-\underset{OH}{CH}-CH_2-O-\overset{O}{\overset{\|}{C}}-C(CH_3)=CH_2$$

vinyl ester

Scheme 3

The vinyl ester is dissolved in a polymerizable solvent which is frequently styrene. However, unlike unsaturated polyesters, vinyl esters containing either methacrylate or acrylate unsaturation will copolymerize with a wider range of unsaturated solvents. Copolymerization occurs through the α, ω (beginning and end) unsaturation of the vinyl ester, to give a cross-linked resin. Vinyl esters generally contain α, ω unsaturation instead of mostly internal double bonds characteristic of classical maleate polyesters. At room temperature, the physical properties of castings of cured polyesters and cured vinyl esters are fairly similar. However, important differences are seen at elevated temperatures, where retention of structural properties is generally better for the vinyl esters. Since vinyl esters are generally lower in molecular weight and viscosity than many polyesters, physical properties of composites that are influenced by wet-out of the filler and fiber may also be improved.

Thermoset resins derived from vinyl esters, vinyl urethanes and bisphenol A propylene oxide-based polyesters show improved corrosion resistance over general-purpose resins. Corrosion resistance is generally improved by increasing the aromatic character, the symmetry of the aromatic diacid or diol, the cross-link density and the steric hindrance at the ester link. These factors also tend to reduce the absorption of water and other polar molecules. This is important in the selection of a resin matrix, because interaction of water or acid at the fiber-resin interface can affect fatigue resistance, as well as the nature of crack propagation.

The substitution of methacrylate/acrylate unsaturation for maleate/fumarate double bonds in polyester systems affects not only the physical properties, but also the rate of cure and the choice of polymerizable solvent that may be utilized. Apart from styrene, maleate/fumarate double bonds copolymerize with only a very limited number of other monomers. Thus, in maleate-based polyesters, it is not possible to completely replace styrene with, say, methyl methacrylate, although fully curable polyesters are achievable using blends of styrene and methyl methacrylate. Methacrylate-based vinyl esters have the advantages that they may be copolymerized with a variety of vinyl monomers, including methyl methacrylate, and that complete substitution of styrene in these systems is not only possible but in some instances desirable, since it extends the range of physical properties achievable and could play an imporatant role in the development of specialty composites.

Although there are a variety of polyunsaturated vinyl monomer systems that may be polymerized by ionic catalysts, these have not become important in polyester and vinyl ester chemistry, because of the nature of the specific unsaturated groups and because cationic and anionic catalysts are more difficult to handle, being sensitive to moisture and other impurities. Free-radical-initiated curing is therefore used in virtually all commercially important systems. Since the free-radical moiety adds to the vinyl group and becomes part of the polymer chain, it is not a true catalyst but an initiator. Initiators include a variety of organic and inorganic systems capable of decomposition into free radicals over the temperature range of about $0-175$C. Organic peroxides, hydroperoxides and certain azo nitriles and peroxysulfates with or without amines or transition metal accelerators are typical of initiators often used. However, there are numerous more exotic initiators, and the use of mixed free-radical initiators is often necessary to achieve the appropriate degree of cure, as well as acceptable rates of curing. Furthermore, since curing is an exothermic process, the choice of initiator affects not only the rate of cure, but also the peak temperature achieved during cure. This can reach 200° C or higher, and may affect the volatilization of styrene, the stresses stored in the system and the practical thickness that may be used. Shrinkage on cure is very pronounced for polyunsaturated systems, and this can cause surface and structural inhomogeneities. Thermoplastic low-profile agents may be added to counter the problem, although at the expense of other properties.

Since unsaturated polyesters, and to a lesser degree vinyl esters, are relatively inexpensive, they are usually used to form composities with glass fibers. Traditionally, these have been prepared by labor-intensive hand lay-up or spray-up processes. More recently, there has been considerable interest in closed molding, either the compression molding of sheet-molding compound, or the injection molding of bulk-molding compound. Although polyesters cannot be B-staged (see Sect. 2) like epoxies, these molding compounds may be regarded as analogous to epoxy prepregs. Such molding compounds usually contain three major components. These are a polyester or vinyl ester resin matrix, short, randomly dispersed or continuous glass-fiber reinforcement, and a particulate filler such as calcium carbonate. After mixing the components together, the molding compound is allowed to maturate or thicken, until it acquires a suitable viscosity for optimum molding. Conventional thickening processes involve the reaction of alkaline earth oxides or hydroxides with the polyester molecule to form ionic cross-links. A novel approach has recently been described, whereby the thickening process is accomplished by the in situ formation of a polyurethane network. This system allows for more rapid maturation, closer control of molding viscosity and a much longer shelf life as compared to conventionally maturated molding compound.

2. *Epoxies*

The generic term epoxy resins describes a class of thermosetting resins prepared by the ring-opening polymerization of compounds containing an average of more than one epoxy group per molecule. Commercially important epoxides are generally formed by the reaction of epichlorohydrin with hydrogen-reactive molecules, or via controlled olefin oxidation using peracetic acid. Key building blocks of many epoxides are based on bisphenol A and epoxy novolac resins, such as those derived from phenol-formaldehyde or cresol-formaldehyde. Chemical structures typical of these important classes of epoxides are given in Scheme 4 and Formula (1). Scheme 4 shows an example of a simple epoxide based on bisphenol A and epichlorohydrin. The curing agent ring-opens the epoxide groups. Since each molecule of curing agent can react with several epoxide groups, a cross-linked structure is formed by the curing agent acting as a multifunctional bridge between the epoxide moieties. Formula (1) shows the unit structure of an epoxy

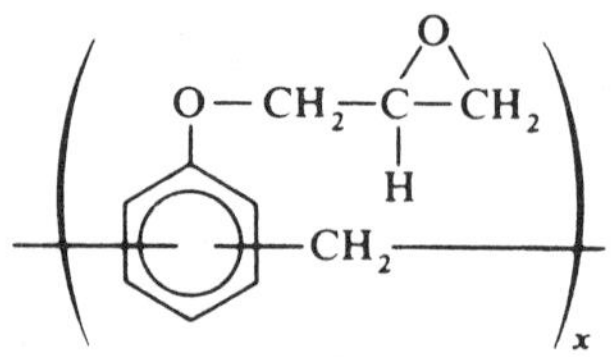

O—CH_2—C(H)—CH_2 (epoxide, O bridging)

—[C_6H_3(O—…)]—CH_2—]$_x$

(1)

derivative of a novolac (phenol-formaldehyde prepolymer).

Epoxides may be polymerized either cationically or anionically, using a wide variety of catalysts including bases, acids, anhydrides and hydroxy compounds. Depending on the required physical properties of the cross-linked resin, the backbone of the epoxide may be either aliphatic, cycloaliphatic, heterocyclic or aromatic. For example, epoxies with aliphatic backbones have lower glass-transition temperatures than those with aromatic backbones, and the substitution of aliphatic groups with aromatic moieties results in a significant increase in the retention of properties at elevated temperatures. Cycloaliphatic backbones are used to prepare epoxy resins having improved stability, especially to light and outdoor weathering.

Not only do the physical and chemical properties of epoxy resins depend on the nature of the backbone, but they also depend on the choice of epoxide polymerization initiator. For example, the use of aromatic anhydrides as initiators usually results in bisphenol A based epoxides with superior heat resistance and chain stiffness. Initiators are sometimes used in conjunction with accelerators to increase the curing rate or to better control the pot life. The rate of polymerization depends on both the nature of the epoxide and the curing agent; the use of curing agents that are too reactive may increase the viscosity of the system so rapidly that layout becomes difficult, and internal stresses in the system adversely affect physical properties.

A balance of properties is often achieved by the addition of reactive diluents to the system. There are many such additives, but C_4–C_{12} aliphatic glycidyl ethers are frequently added to provide increased chain flexibility with increasing chain length, whereas the

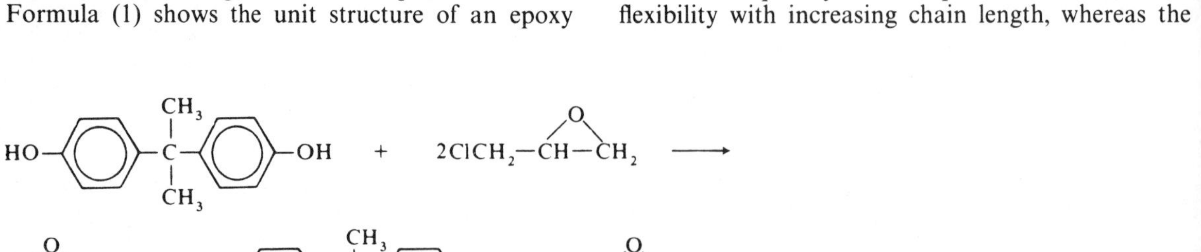

HO—C_6H_4—C(CH_3)$_2$—C_6H_4—OH + 2ClCH$_2$—CH—CH$_2$ (epoxide, O bridging) ⟶

CH$_2$—CH—CH$_2$—O—C_6H_4—C(CH_3)$_2$—C_6H_4—O—CH$_2$—CH—CH$_2$ (epoxide rings) $\xrightarrow[\text{(e.g., amines or anhydrides)}]{\text{epoxide curing agent}}$ cross-linked epoxy resin

Scheme 4

addition of aromatic glycidyl ethers generally improves chemical and solvent resistance. Appropriate reactive diluents may also be used to control viscosity and pot life.

High-performance epoxies have been prepared with a variety of phenolics and aromatic amines. Those resins derived from novolacs are especially noteworthy for their improved chemical resistance and retention of physical properties at higher temperatures compared to bisphenol A epoxies. A high cross-link density, and the large number of rigid phenyl groups in the structure, generally improves corrosion resistance over aliphatic and mixed bisphenol A types. Epoxy derivatives of novolacs are extremely viscous liquids at room temperature. However, on heating, the viscosity rapidly decreases and solventless systems may be formulated, thus making them easier to process in many applications.

Many specialty epoxy resins are known based on bisphenol F and bisphenol S, as well as on other bi- and polyphenols. Also of interest are the tetrafunctional epoxides derived from epichlorohydrin and aromatic diamines such as methylene dianiline. In conjunction with graphite and polyaramide fibers, these epoxides may be used to produce high-performance composites having excellent structural and thermal properties. A further group of specialty epoxies are those heterocyclic epoxides based on diazine or triazine five- and six-membered ring structures. Important members of this class include epoxies derived from isocyanurates and hydantoin. These epoxies are especially noteworthy for excellent wet-out of a variety of fillers and reinforcing fibers, and for low smoke generation during burning.

Epoxy resins have good mechanical properties and are less brittle and tougher than either unsaturated polyesters or vinyl esters. These properties, together with the low shrinkage on cure and their good bondability to other materials, make epoxy resins desirable composite-matrix materials. Most advanced composites based on epoxy resins contain carbon fibers, since the higher cost of epoxy resins together with their superior properties justify the use of more expensive fibers than E-glass. Epoxy-carbon composites have excellent physical properties in terms of modulus and strength and have many applications in the aerospace industry. Epoxy resins have an advantage over unsaturated polyesters, in that they can be partially cured or B staged. Thus, the reinforcement is preimpregnated with liquid resin and partially cured to give a prepreg. Prepreg material can then be subsequently molded by a fabricator, without the fabricator requiring a knowledge of resin chemistry and detailed information on resin handling. The overall purpose of prepreg is generally to produce a quality product at a low cost.

Since the mid-1970s there has been a steady progression of new high-performance epoxies that have emerged in response to the physical- and chemical-property requirements for resins in advanced composite applications. A broad spectrum of properties is required in both commercial and military aerospace applications; the latter is dominated by especially demanding specifications for hot wet properties, damage tolerance, higher T_g (glass transition temperature) and, in general, greater toughness in combination with environmental resistance. Several Japanese companies along with Dow, Shell and Ciba–Geigy worldwide represent some of the major participants in the research into, and development of, new families of high-performance epoxies. Of special interest are the epoxy resins based on tetraglycidyl 4,4′ methylenedianiline (MDA), and another family of epoxies based on tris-hydroxyphenyl derivatives. Both families are being developed with the aim of improving key physical properties and processing in advanced composite systems. Some of these new products can, with modification, be adapted to applications in electronics including printed wiring boards and encapsulants. Epoxies based on tetramethyl biphenol (produced by oxidative coupling of 2,6 xylenol) are emerging as a promising family of resins from Japan. Because of the high purity of this biphenol monomer and the melt characteristics of the resulting epoxy, applications in powder coatings and electronics are of special interest. These epoxies represent only a few examples of the many newer epoxy developments described in detail in both US and non-US (especially Japanese) patent literature.

3. Polyurethanes

Scheme 5 shows an example of polyurethane formation via a *t*-amine catalyzed reaction of diisocyanate and a diol containing an aliphatic group R. The reaction of a multifunctional alcohol with a polyfunctional isocyanate produces polyurethanes which are generally thermoset resins, but which may be thermoplastic if the functionality of the reacting groups is maintained below 2.0, and the process is operated free from side reactions. Functionality may be defined as the number of groups per molecule capable of undergoing the polymerization reaction. Although polyurethanes have been made with a variety of polyisocyanates, the majority of commercial production is based on either toluene diisocyanate (TDI) (2) or methylene diisocyanate (MDI) (3). Most of the latter is actually used as polymeric MDI (4),

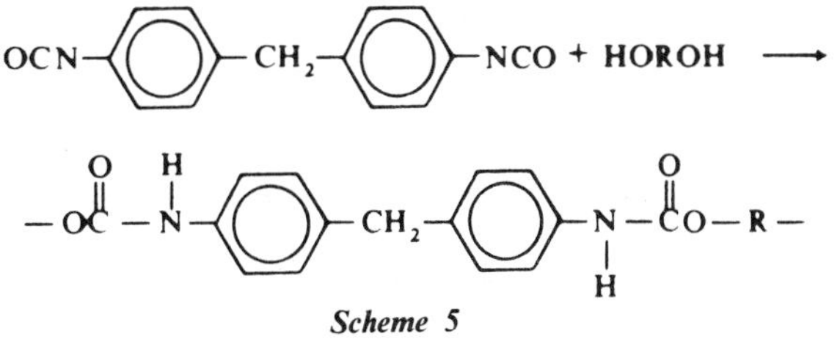

$$\mathrm{OCN{-}C_6H_4{-}CH_2{-}C_6H_4{-}NCO + HOROH \longrightarrow}$$

$$\mathrm{-OC(=O){-}NH{-}C_6H_4{-}CH_2{-}C_6H_4{-}NH{-}C(=O){-}O{-}R{-}}$$

Scheme 5

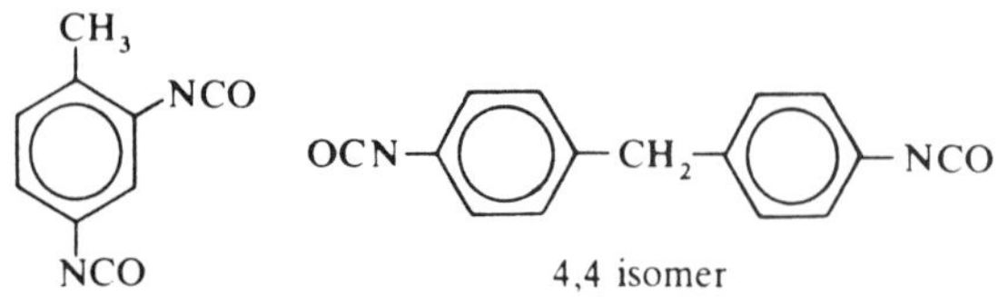

2,4 isomer

4,4 isomer

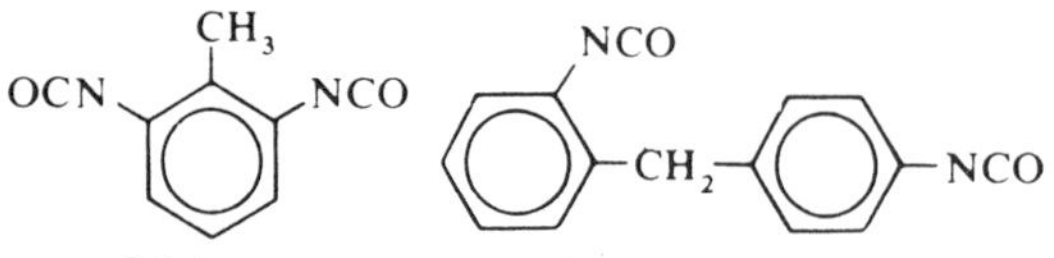

2,6 isomer

(**2**)

2,4′ isomer

(**3**)

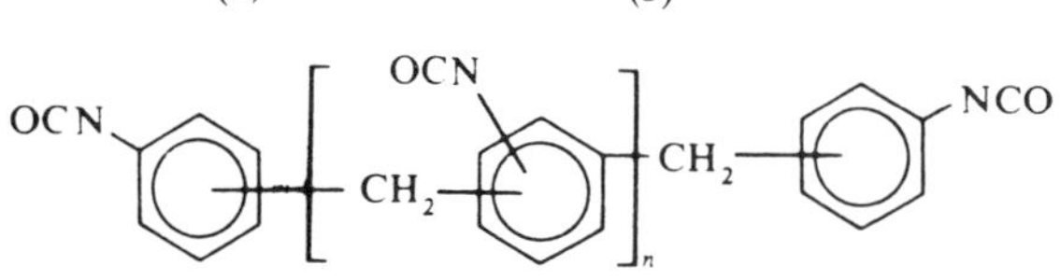

(**4**)

with an isocyanate functionality of greater than 2 and usually less than 3.

The polyols most commonly used are ethers derived from complex alcohols and sugars. Examples include glycols, glycerine, sorbitol and sucrose ethers derived from ethylene oxide or propylene oxide. Polyols derived from bisphenol A and polyesters are also used. Because polyester diols are generally less stable to hydrolysis and are generally more expensive, polyether polyols dominate as the coreactant of choice, especially for rigid and flexible foams that represent in excess of 80% of a multibillion kilogram market. However, the market for branched polyester polyols, although relatively small as a percentage of the total, is still significant.

As in all thermoset resins, the physical properties achievable in polyurethanes depend on the degree of cross-linking and the nature and length of the chain segments (aliphatic, cycloaliphatic, aromatic) between polyol and polyisocyanate functional groups. Polyurethanes based on TDI and triols are generally used to produce highly flexible systems, whereas polymeric MDI and polyols with functionality of 4–6 are preferred for rigid systems. Blowing agents such as CO_2 or low-boiling-point halocarbons are used in preparing thermoset elastomeric or rigid foams. Polyurethanes with a wide variety of properties have been prepared by utilizing mixtures of polyols or mixtures of isocyanates. Since the market is very large and highly competitive, economics often dictate the particular choice.

Many compounds will initiate the polymerization of isocyanates and alcohols, including acids, bases, inorganic and organic salts, amines and alkoxides. Commercially important initiators are amines and alkyl tin salts, especially tin octoate and laurate. In general, the stronger the base strength of amines, the greater their reactivity. Furthermore, the reactivity of the system can be slowed substantially by increasing the steric hindrance on either the amine group, the alcohol or the isocyanate.

Under certain conditions, isocyanates will react with neighboring isocyanate groups to produce dimers or trimers. These trimers are much more stable than the parent isocyanate, and they have achieved commercial importance as highly cross-linked thermoset resins, known as polyisocyanurates. Scheme 6 illustrates the structure of the isocyanurate ring and the formation of the polyisocyanurate network. The trimerization reaction is catalyzed by inorganic salts, especially the acetate or formate salt of calcium, sodium and potassium, trialkyl phosphine or trialkyl amines.

There is probably no class of resins more versatile and complex than those derived from isocyanate chemistry, since elastomers, plastics and fibers have all been commercially prepared. Further modifications and extension of the chemistry into composites are under active study. The fact that certain MDI and polyol mixtures can be maintained in the liquid state and then polymerized in seconds by highly reactive catalysts is the basis for reaction injection molding (RIM),which promises further expansion of polyurethane science and technology. RIM is based on the rapid injection of separate streams of isocyanate, polyol and catalyst, or other compounds, into a mold in which polymerization and molding take place simultaneously.

In order to increase modulus and heat-distortion temperature and to reduce the coefficient of thermal expansion of RIM polyurethanes, reinforced RIM systems (RRIM) are being developed, in which nonparticulate fillers, such as wollastonite and mica, and short-aspect-ratio fibers such as 0.15 cm milled-glass

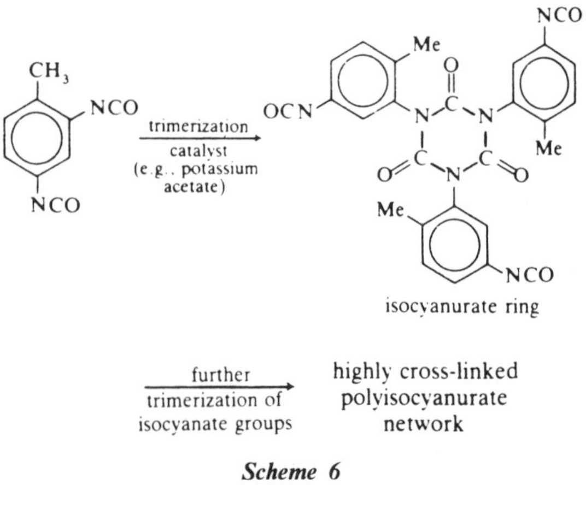

trimerization catalyst (e.g., potassium acetate)

isocyanurate ring

further trimerization of isocyanate groups

highly cross-linked polyisocyanurate network

Scheme 6

fibers, are added to the liquid components and injected into the mold. This technique will have a significant impact on the transportation industry, especially automotive, because of the lower cost and increased speed of molding in comparison to other techniques such as the compression molding of sheet-molding compound.

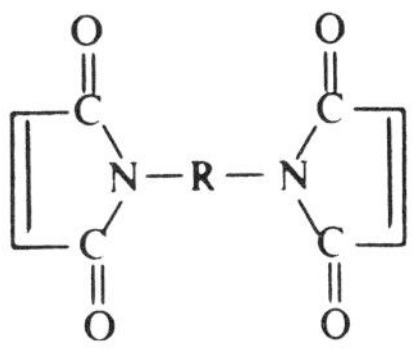

(5)

4. Polyimides

The reaction of a primary amine with an anhydride produces an amide and a vicinal carboxyl group, which on further reaction converts to an imide and water. Scheme 7 illustrates the polyimide structure, and shows one of the important members of the polyimide family—namely, the highly aromatic, high-performance reaction product of pyromellitic dianhydride with 4,4′-diaminodiphenyl ether. Each reaction sequence can be controlled, and products are available commercially as the intermediate polyamic acid or as the finished polyimide fabricated in custom parts or film. Formula (5) is a representation of less aromatic polyimides with aliphatic/aromatic groups (R) derived from a diamine or diisocyanate: for example, those derived from 4,4′-diaminodiphenyl ether with 2 moles of maleic anhydride to produce α,ω unsaturated bismaleimides. This stable intermediate can then be reacted with diamines or free-radical initiators to produce high-molecular-weight polymers by addition polymerization, without the elimination of by-product water. Although less thermally stable than the fully aromatic polyimides, the absence of water as a by-product from the reaction makes composite processing and curing easier.

The early success of Rhone Poulenc in commercializing bismaleimide/MDA copolymers in demanding composites for electronic applications has triggered considerable interest in other bismaleimide copolymers. However, these early resins were too brittle to achieve the toughness necessary for advanced composites. Important considerations in reducing brittleness in bismaleimide systems include: the reduction of cross-link density on cure by reducing the level of bismaleimide unsaturation in the prepolymer; the substitution of coreactant side groups with alkyl groups; copolymerization with comonomers that contain alkyl groups (or other flexibilizing groups); and modification of the chemistry to permit polyblends with rubbers or other polymers, including high-performance thermoplastics, to toughen the system and increase fracture toughness. In recent years copolymers of 2,2′diallyl bisphenol A- and MDA-derived bismaleimide have been introduced which are reported to be tougher than Michael-addition copolymers of bismaleimides and MDA. Other bismaleimide formulations based on copolymerization with bis-allylphenyl compounds are available which can be melt processed and thermally cured to produce dense, tough, high-temperature-resistant thermoset resins. The chemistry is versatile, allowing polyblends with rubbers and other polymeric compositions which can provide a range of physical and chemical properties of interest in filament winding, injection molding and prepregging.

Since some polyimides may be dissolved (albeit in sulfuric acid, fuming nitric acid or antimony trichloride), they are not truly speaking thermoset resins. They are included here because of their superior structural and physical stability compared to either polyester or epoxy thermosets, especially at elevated temperatures. Truly thermoset polyimides have been made via addition of polyfunctional coreactants, or by

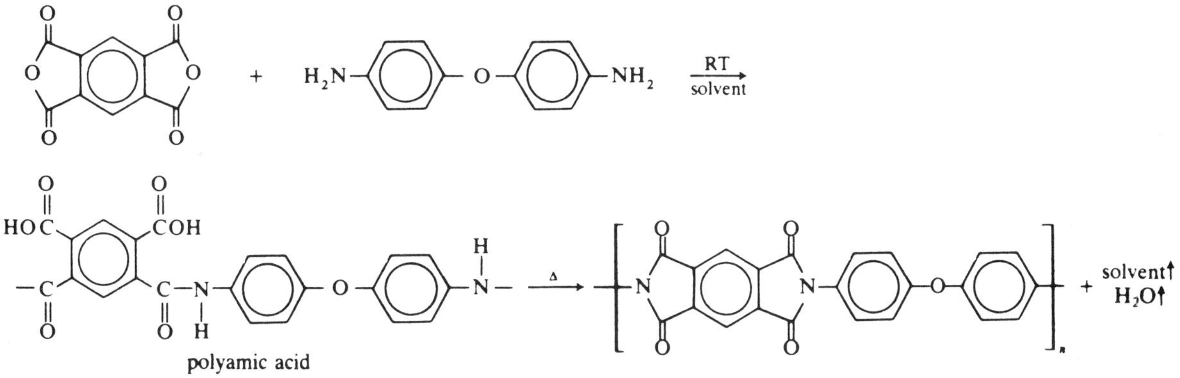

polyamic acid

Scheme 7

further cross-linking of the structure via the ketone moiety at elevated temperatures. Substitution of aliphatic for aromatic groups in the diamine detracts from the thermal stability of the polyimides, and when the number of aliphatic methylene groups is less than about six, decomposition of the polymer occurs around the melt temperature.

Polymerization is generally carried out in solution, by reacting the dianhydride and diamine in a polar solvent such as dimethyl formamide at a temperature below 50°C. As solvent is removed, the intermediate amide/acid (polyamic acid) condenses with the elimination of water, to form the polyimide. Polyimides based on 4,4′ diamino diphenyl ether have greater toughness and oxidative stability than those derived from the cheaper 4,4′ diamino diphenyl methane derivative. In spite of the high cost and relative difficulty in fabrication, a few polyimides have grown substantially in usage. Properties of most interest include outstanding thermal stability at 300°C and above, excellent strength and modulus compared to other organic polymers, excellent fire retardance by virtue of the reduced level of hydrogen, and high dielectric strength. The intractable nature and rigidity of the polyimide, and the need to remove solvent and water, have triggered interest in variations based on amide/imide, ester/imide, epoxy/imide, and bismaleimide, which cures without elimination of water. However, in these cases, there is a compromise in the retention of physical properties at elevated temperature. Essentially all of the polyimides and modified polyimides used in composites are directed to high-performance applications, where cost and fabrication are less important than the properties afforded in combination with superfiber reinforcement such as polyaramide, carbon and specialty inorganic materials. However, the outstanding properties of composites derived from highly aromatic polyimides are tempered by the additional requirement to remove the by-product water or volatiles that form during polymerization. Water is not only detrimental to the physical properties of the polymer, but also can give rise to voids and internal stresses, which cause further deterioration of composite properties. Consequently, extra care is required in working with polyimide-based composites to remove water from all components in the system.

5. *Formaldehyde-Based Resins*

Resins based on formaldehyde may be either thermoplastic or thermoset. The most important are those derived from phenol or substituted phenols, urea and melamine. All of these products became important commercially in the period from about 1910 to 1940, with the phenol-formaldehyde resins being the oldest, and the melamine-based systems the most recent. The markets for all are still substantial. Phenol, urea and melamine may be made to react with formaldehyde to form the methylol derivative, which upon further condensation forms products that are initially thermoplastic. However, these may be cured by excess formaldehyde or polyfunctional curing agents and heat, to yield cross-linked thermoset resins. Control of the curing reaction was the key to the successful development of the phenol-formaldehyde resins, and paved the way for subsequent formaldehyde condensation polymers. These resins may be polymerized by acid or base catalysts, and their polymerization chemistry is sensitive to formaldehyde ratio, temperature and pH. The chemical structures for the phenol, urea and melamine formaldehyde resins are outlined in Scheme 8, which shows the reaction of bifunctional formaldehyde with trifunctional phenol (6), tetrafunctional urea (7), and hexafunctional melamine (8). These reactions produce the methylol derivatives of each $-R-CH_2OH$ group, which through a series of condensations produce a methylene bridge and water. The degree of substitution $-R_x$, and the fraction of methylol groups condensed, determines whether the product is a stable liquid of moderate viscosity (novolac), a solid or a cured thermoset resin.

Formaldehyde-based resins may be appropriately modified and blended with fillers, pigments or reinforcing materials while they are thermoplastic, and subsequently cured to produce composites. However, none of the resins has achieved significance in high-performance reinforced-composite systems.

6. *Trends in Thermoset Resins*

The previous sections summarize the chemistry of the more important thermoset resins useful in composites containing fillers, fibers or both. Since World War II, there has been a developing interest in high-performance fiber-reinforced composites. The 1970s has seen a revitalization of this interest, particularly to

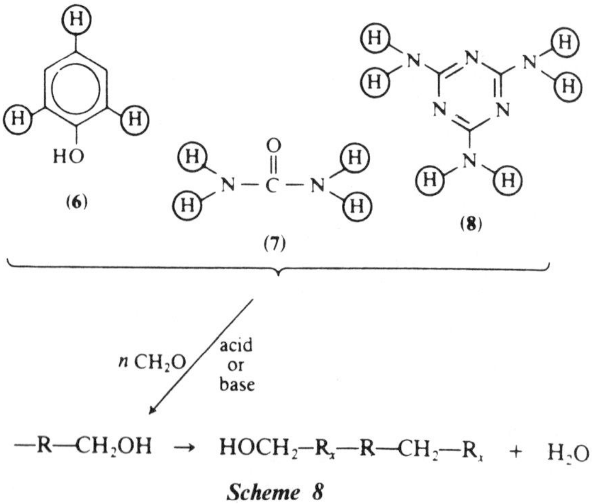

(6) (7) (8)

$n\,CH_2O$, acid or base

$$-R-CH_2OH \rightarrow HOCH_2-R_x-R-CH_2-R_x + H_2O$$

Scheme 8

produce more energy-efficient materials to replace metals and ceramics. The number of new organic polymers of commercial significance has been declining in recent years. The thrust in the 1980s appears to be moving in the direction of optimizing existing systems, by balancing physical properties through composite formulation and through a greater appreciation of the chemistry and physics of these complex heterogeneous systems. More versatile polymer systems based on polyblends will extend the range and balance of properties achievable in composites. By the same token, polymer systems based on handling or generating toxic materials during fabrication will limit use, except in highly controlled, automated systems. Systems based on styrene, formaldehyde and some of the more volatile aromatic isocyanates are a few examples. Increased interest is developing in hybrid resins based on a marriage of unsaturated polyesters with urethanes, and also in curable systems based on coreacting two or more different resin systems to produce interpenetrating thermosets. An example of the latter system is a polyblend of unsaturated polyester or vinyl ester resin in a prepolymerized polyurethane network.

Fast-reaction chemistry to polymerize and mold simultaneously, such as with reaction injection molding (RIM) will become even more important in the future, and practical methods for reinforcing such systems will be further developed. This market, now dominated by polyurethanes, will be increasingly vulnerable to fast reactions based on modified epoxies, as well as derivatives containing amides, imides or acrylate unsaturation that are liquid around room temperature. Composites based on combinations of thermoplastic resins in thermoset systems are receiving increased attention in an effort to further improve properties, extend the range of applicability, or aid in processing economics.

See also: Carbon-Fiber-Reinforced Plastics; Glass-Reinforced Plastics: Thermosetting Resins; Hybrid Fiber–Resin Composites

Bibliography

Batzer H, Lohse F 1979 *Introduction to Macromolecular Chemistry*, 2nd edn. Wiley, New York

Bauer R S (ed.) 1979 *Epoxy Resin Chemistry*, ACS Symposium Series 114, American Chemical Society, Washington, DC

Bauer R S (ed.) 1983 *Epoxy Resin Chemistry* II, ACS Symposium Series 221. American Chemical Society, Washington, DC

Bruins P F (ed.) 1976 *Unsaturated Polyester Technology*. Gordon and Breach, New York

Gillham J K 1979 Formation and properties of network polymeric materials. *Polym. Eng. Sci.* 19:676–82

Knop A, Scheib W 1979 *Chemistry and Application of Phenolic Resins*. Springer, New York

Lenz R W 1967 *Organic Chemistry of Synthetic High Polymers*. Interscience, New York

Magrans J J, Ferarini L J 1979 Unique electrical and mechanical properties of ITP™ SMC. *Proc. 34th SPI Conf., New Orleans*. Society for Plastics Industry, New York, pp. 1–6

Monteiro H A 1986 Matrix systems for advanced composites. *Pop. Plast.* 31(4) :20–27

Nir Z, Gilwee W J, Kourtides D A, Parker J A 1985a Polyfunctional epoxies 1: rubber-toughened brominated and nonbrominated formulations for graphitic composites. *Polym. Comp.* 6(2): 65–71

Nir Z, Gilwee W J Kourtides D A, Parker J A 1985b Polyfunctional epoxies 2: nonrubber versus rubber-toughened brominated formulations for graphite composites. *Polym. Comp* 6(2) :71–81

Potter W G 1976 *Uses of Epoxy Resins*. Chemical publishing, New York

Sandler S R, Karo W 1974, 1977 *Polymer Synthesis*, Vols. 1,2. Academic Press, New York

Saunders K J 1973 *Organic Polymer Chemistry*. Chapman and Hall, London

Stevens M P 1975 *Polymer Chemistry — An Introduction*. Addison–Wesley, Reading, Massachusetts

A. J. Restaino and D. B. James
[ICI Americas, Wilmington, Delaware, USA]

Three-Dimensional Fabrics for Composites

Three-dimensional (3-D) fabrics for structural composites are fully integrated continuous fiber assemblies having multiaxial in-plane and out-of-plane fiber orientation. More specifically, a three-dimensional fabric is one that is fabricated by a textile process, resulting in three or more yarn diameters in the thickness direction with fibers oriented in three orthogonal planes. The engineering applications of three-dimensional fabrics for composites date back to the 1960s, responding to the need in the emerging aerospace industry for parts and structures to be capable of withstanding multidirectional mechanical and thermal stresses. Since most of these early applications were for high-temperature and ablative environments, carbon–carbon composites were the principal materials. As indicated in a review article by McAllister and Lachman (1983), the early carbon–carbon composites were reinforced by biaxial (2-D) fabrics. Beginning in the early 1960s, it took almost a whole decade and the trial of numerous reinforcement concepts including needled felts, pile fabrics and stitched fabrics to recognize the necessity of three-dimensional fabric reinforcements in addressing the problem of poor interlaminar strength in carbon–carbon composites (Laurie 1970, Adsit et al. 1972, Schmidt 1972). Although the performance of a composite depends a great deal on the type of matrix and the nature of the fiber–matrix interface, it appears that much can be learned from the experience of the role of fiber architecture in the processing and performance of carbon–carbon composites.

The recent expansion of interest in three-dimensional fabrics for resin, metal and ceramic matrix composites is a direct result of the current trend in the expansion of the use of composites from secondary to primary load-bearing applications in automobiles, surgical implants, aircraft and space structures. This requires a substantial improvement in the damage tolerance and reliability of composites. In addition, it is also desirable to reduce the cost and broaden the usage of composites from aerospace to automotive applications. This calls for the development of a capability for quantity production and the direct formation of structural shapes. In order to improve the damage tolerance of composites, a high level of through-thickness and interlaminar strength is required. The reliability of a composite depends on the uniform distribution of the materials and consistency of interfacial properties. The structural integrity and handleability of the reinforcing material for the composite is critical for large-scale, automated production. A method for the direct formation of the structural shapes would therefore greatly simplify the laborious hand lay-up composite formation process. With the experience gained in the three-dimensional carbon–carbon composites and the recent progress in fiber technology, computer-aided textile design and liquid molding technology, three-dimensional fabric structures are increasingly being recognized as serious candidates for structural composites.

The importance of three-dimensional fabric-reinforced composites in the family of textile structural composites is reflected in several books on the subject (Chou and Ko 1989, Tarnopol'skii et al. 1992). This article is intended to provide a concise introduction to three-dimensional textile fabrics for composites. The discussion will focus on the preforming process and structural geometry of the three basic classes of integrated fiber architecture: woven, knit and braid. A fourth class, the orthogonal nonwoven three-dimensional structure, is not treated here. More information on the subject can be found in the work of Herrick (1977), Fukuta et al. (1982), McAllister and Lachman (1983), Bruno et al. (1986), Fukuta and Aoki (1986) and Geoghegan (1988).

1. Three-Dimensional Textile Fabrication Processes

The conversion of fiber to textile preform can be accomplished via the "fiber-to-fabric" (FTF) process, the "yarn-to-fabric" (YTF) process and combinations of the two. An example of the FTF process is the Noveltex method developed by P. Olry at SEP (Société Européen de Propulsion, Bordeaux, France). The Noveltex concept is based on the entanglement of fiber webs by needle punching. A similar process has been developed in Japan by Fukuta using fluid jets in place of the needles to create through-thickness fiber entanglement.

YTF processes are popular means for preform fabrication wherein the linear fiber assemblies (continuous filament) or twisted short-fiber (staple) assemblies are interlaced, interlooped or intertwined to form two- or three-dimensional fabrics. Examples of preforms created by the YTF processes are shown in Fig. 1. A comparison of the basic YTF processes is given in Table 1.

In addition to the FTF and YTF processes, textile preforms can be fabricated by combining structure and process. For example, the FTF webs can be incorporated into a YTF preform by needle or fluid jet entanglement to provide through-thickness reinforcement. Sewing is another process that can combine or strategically join FTF and/or YTF fabrics together to create a preform having multidirectional fiber reinforcement (Palmer 1989) (Fig. 2).

1.1 Three-Dimensional Woven Fabrics

Three-dimensional woven fabrics are produced principally by the multiple-warp weaving method which has long been used for the manufacturing of double cloth and triple cloths for bags, webbing and carpets. By the weaving method, various forms of fiber architecture can be produced including solid orthogonal panels (Fig. 3(a)), variable thickness

Table 1
A comparison of three-dimensional yarn-to-fabric formation techniques

Preforming process	Basic direction of yarn introduction	Basic formation techniques
Weaving	two (0°/90°)	Interlacing (by selective warp and fill) insertion of 90° yarns into 0° yarn system
Braiding	one (machine direction)	Intertwining (position displacement)
Knitting	one (0° or 90°)	Interlooping (by drawing warp or fill) loops of yarns over previous loops
Nonwoven	three or more (orthogonal)	Mutual fiber placement

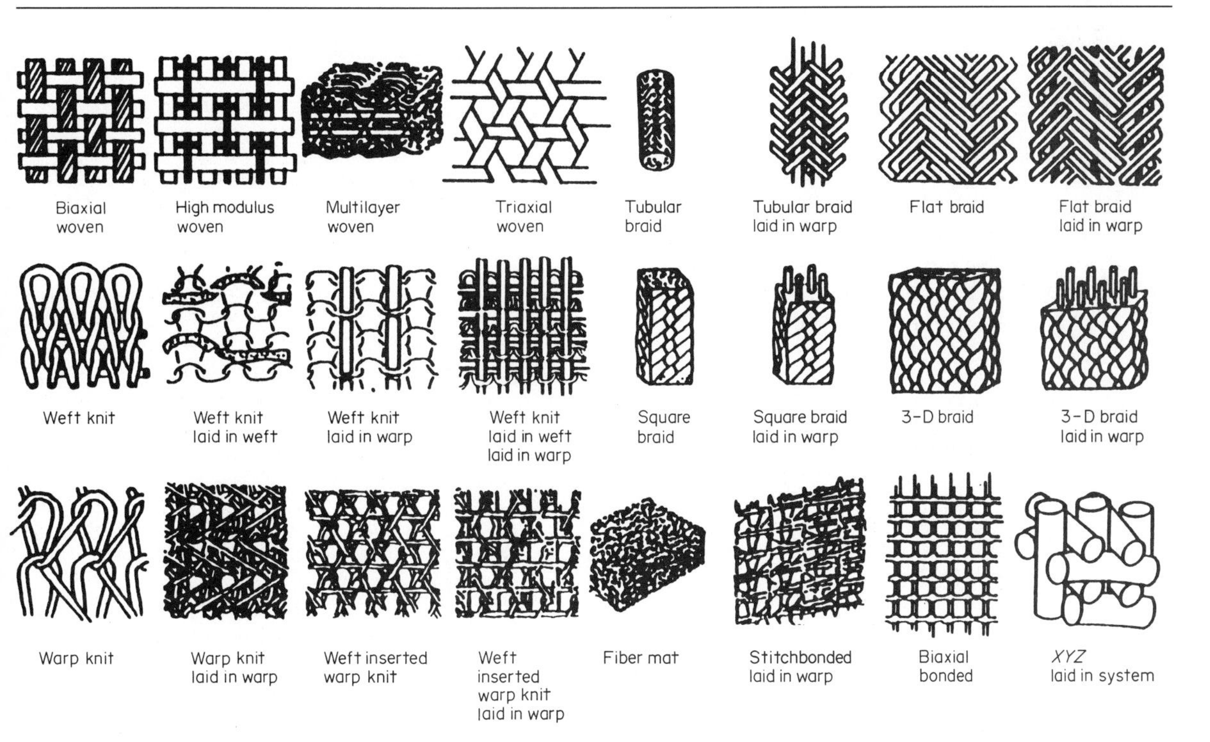

Figure 1
Examples of yarn-to-fabric preforms

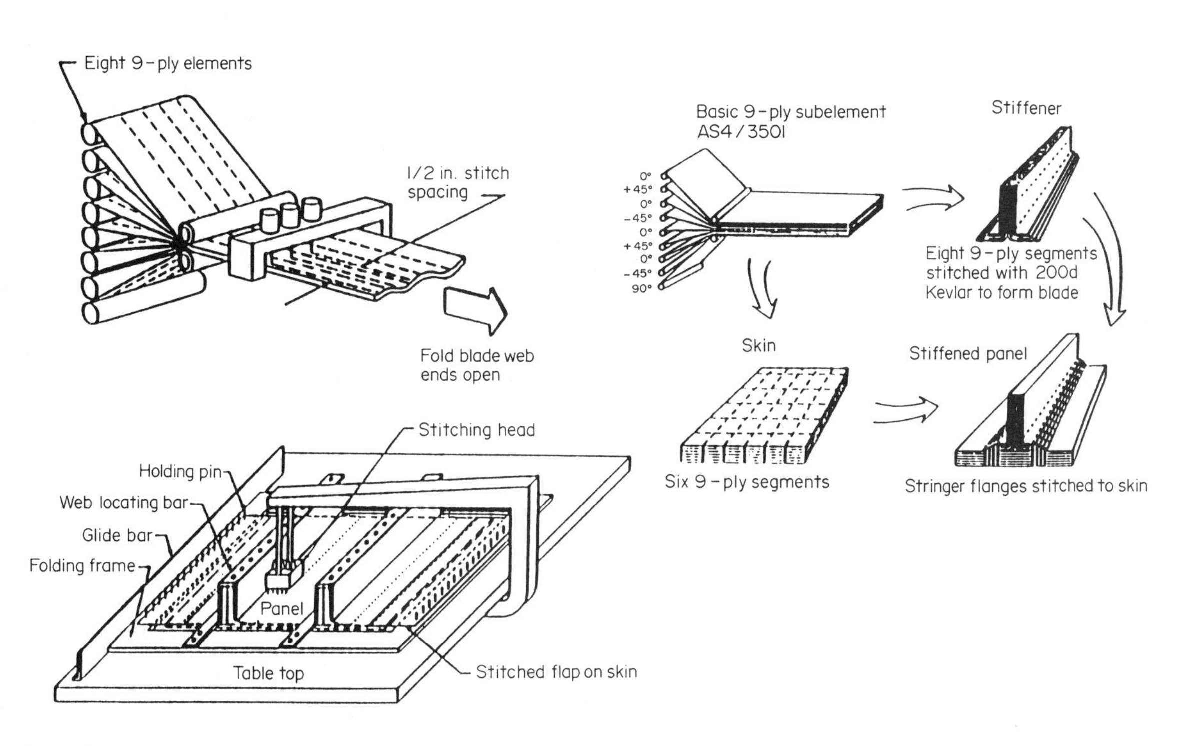

Figure 2
Combination of FTF and YTF processes

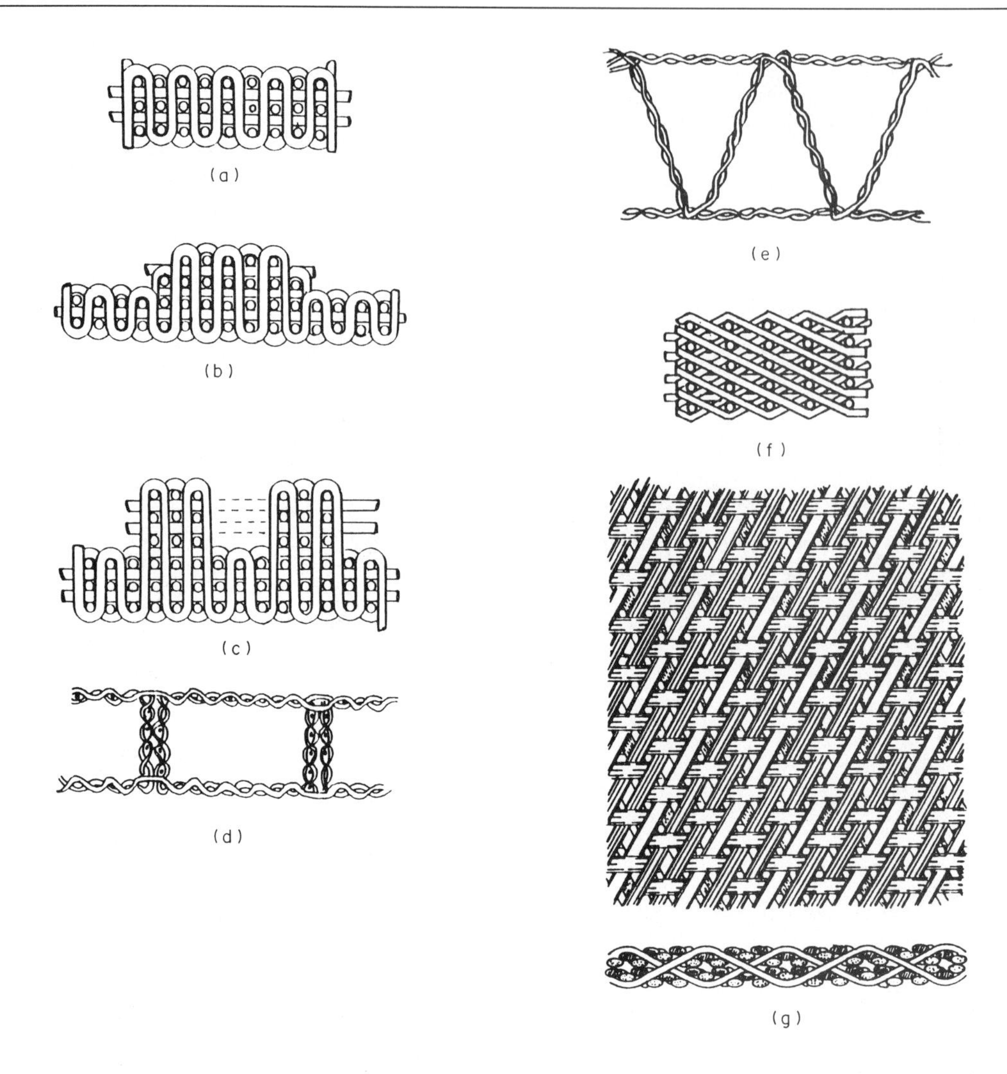

Figure 3
Three-dimensional woven fabrics: (a) solid orthogonal panel; (b), (c) variable thickness solid panels; (d) box beam structure; (e) trusslike structure; (f) angle interlock structure; (g) multilayer triaxial structure

solid panels (Figs. 3(b), (c)), and core structures simulating a box beam (Fig. 3(d)), or trusslike structure (Fig. 3(e)). Furthermore, by proper manipulation of the warp yarns, as exemplified by the angle interlock structure (Fig. 3(f)), the through-thickness yarns can be organized into a diagonal pattern. To address the inherent lack of in-plane reinforcement in the bias direction, new progress is being made in triaxial weaving technology by Dow (1989) to produce multilayer triaxial fabrics as shown in Fig. 3(g).

1.2 Knitted Three-Dimensional Fabrics

Knitted three-dimensional fabrics are produced either by the weft knitting or warp knitting processes. An example of a weft knit is the near net shape structure knitted under computer control by the Pressure Foot process (Williams 1978) (Fig. 4(a)). In a collapsed form this preform has been used for carbon–carbon aircraft brakes. The unique feature of the weft knit structures is their conformability (Hickman and Williams 1988). While the weft knitted structures have applications in limited

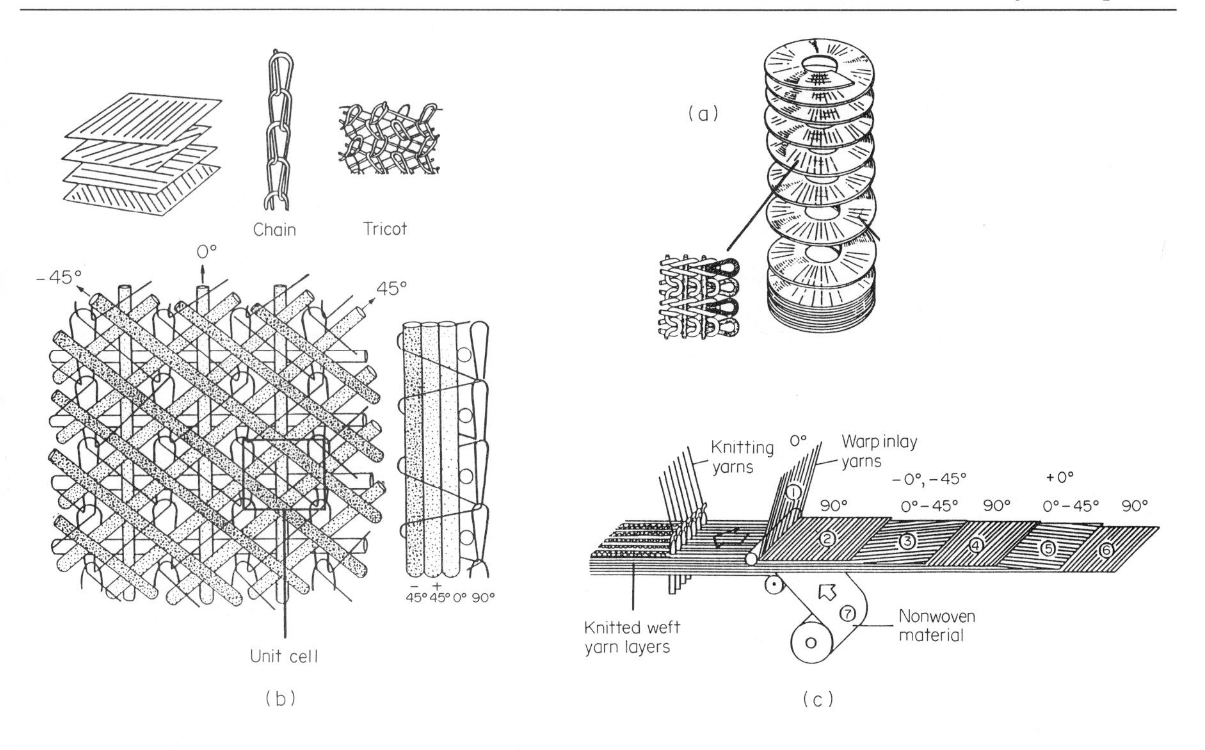

Figure 4
Three-dimensional knitted fabrics: (a) near net shape structure produced by the Pressure Foot process; (b) MWK structure; (c) impaled MWK structure

areas, the multiaxial warp knit (MWK) three-dimensional structures are more promising and have undergone a great deal more development (Ko et al. 1986, Ko and Kutz 1988). From the structural geometry point of view, the MWK fabric systems consist of warp (0°), weft (90°) and bias ($\pm\theta$) yarns held together by a chain or tricot stitch through the thickness of the fabric, as illustrated in Fig. 4(b). The major distinctions between these fabrics are the linearity of the bias yarns, the number of axes and the precision of the stitching process. The latest commercial nonimpaled MWK fabric is produced by the Mayer Textile Corporation, utilizing a multiaxial magazine weft insertion mechanism. The attractive feature of this system is the precision of yarn placement, with four layers of linear or nonlinear bias yarns plus a short fiber mat arranged in a wide range of orientations. Furthermore, the formation of stitches is done without piercing through the reinforcement yarns at a production rate of $100\,\mathrm{m\,h^{-1}}$.

An example of impaled MWK is the LIBA or Hexcel system, as shown in Fig. 4(c). Six layers of linear yarns can be assembled in various stacking sequences and stitched together by knitting needles piercing through the yarn layers. While this piercing action unavoidably damages the reinforcing fiber, it also permits the incorporation of a fiber mat as a surface layer for the composite. The latest addition to this class of structure is the multiaxial fabric produced by the Malimo process which permits stitching through a greater number of layers of fabric due to the unique needling action associated with the process (Pönitz 1993).

1.3 Three-Dimensional Braided Fabrics

Three-dimensional braiding technology is an extension of the well-established two-dimensional braiding technology wherein the fabric is constructed by the intertwining of two or more yarn systems to form an integral structure. Three-dimensional braiding is one of the textile processes wherein a wide variety of solid complex structural shapes (Fig. 5(a)) can be produced in an integral manner resulting in a highly damage-resistant structural preform. Figure 5(b) shows two basic loom setups in circular and rectangular configurations (Ko 1989a, b). The three-dimensional braids are produced by a number of processes including the track and column method (Brown and Ashton 1989) (Fig. 5(c)) and the two-step method (Popper and McConnell 1987) (Fig. 5(d)), as well as a variety of displacement braiding techniques. The basic braiding motion includes the alternate *X* and *Y* displace-

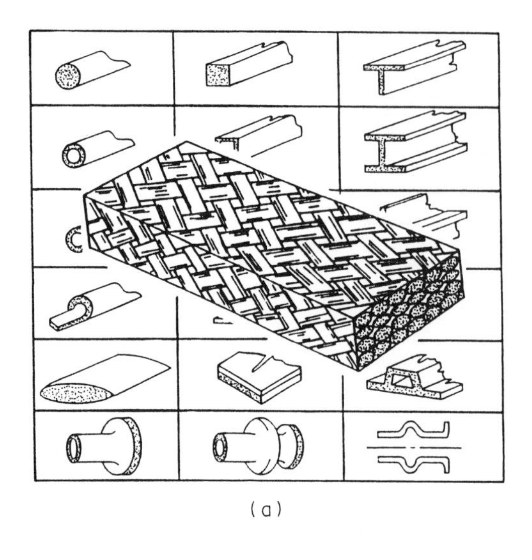

(a)

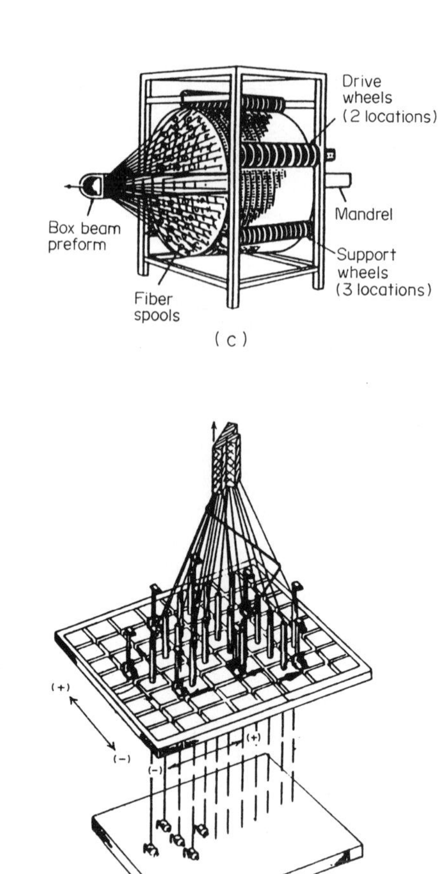

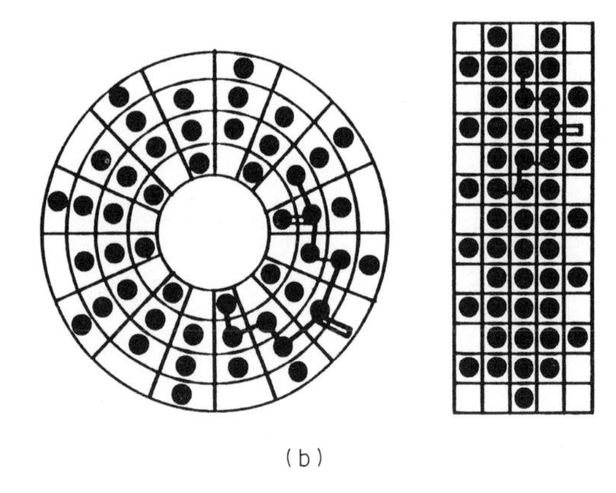

(b)

Figure 5
Three-dimensional braided fabrics: (a) examples of solid complex structural shapes; (b) circular and rectangular loom setups; (c) track and column method; (d) two-step method

ment of yarn carriers followed by a compacting motion. The formation of shapes is accomplished by the proper positioning of the carriers and the joining of various rectangular groups through selected carrier movements.

2. *Selection of Three-Dimensional Fabrics for Composites*

The key criteria for the selection of three-dimensional fabrics for composites are: the range of multiaxial, in-plane reinforcement ($0/90/\pm\theta$); amount of through-thickness reinforcement; and capability for net shape manufacturing. Depending on the application, all or some of these features are required. Specifically, one should recognize the geometric limits of three-dimensional fabrics which are functions of the fiber packing density in the component yarns, yarn bundle size and fabric weave tightness in terms of interlacing density.

Expressed in terms of fiber volume fraction, V_f, and fiber orientation, the geometric and processing kinematics information will be particularly useful for gas and liquid infiltration of matrix material into the three-dimensional fiber network. Likewise, this information is equally important in predicting the mechanical properties of the three-dimensional composites.

In a systematic study of the process–structure

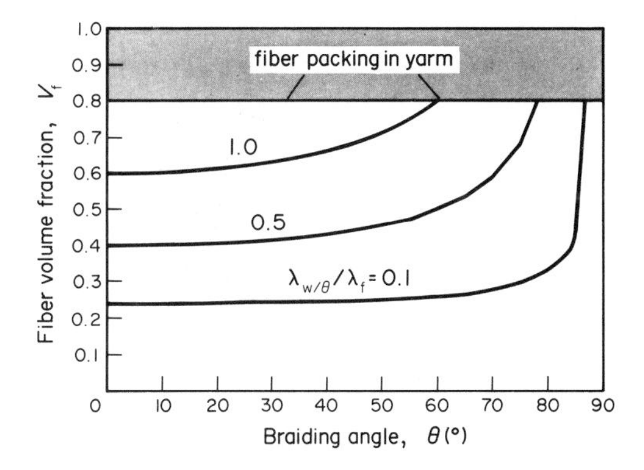

Figure 6
Process window of fiber volume fraction for three-dimensional wovens, where $\lambda_{w/\theta}$ is the linear density of the warp or web yarn, λ_f is the linear density of the fill yarn and η is the fabric tightness factor (representing the proportion of total fiber cross section to unit cell cross section)

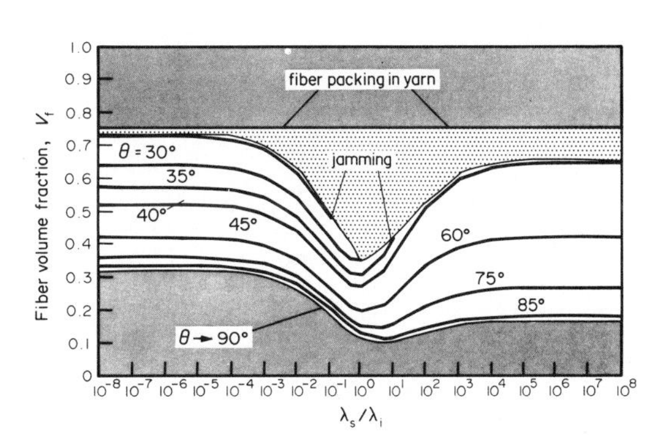

Figure 7
Fiber volume fraction versus ratio of stitch-to-insertion yarn linear density (tricot stitch, fiber packing fraction $\kappa = 0.75$, fiber density $\rho = 2.5\ \mathrm{kg\ m^{-3}}$, $f_i = 5$, and $\eta = 0.5$). Key process variables include the orientation angle of bias yarns (θ), the ratio of stitch-to-insertion yarn linear density (λ_s/λ_i), and the fabric tightness factor (η)

relationship of textile preforms carried out by Ko and Du (1992), it was found that the maximum fiber volume fraction is limited by the yarn packing density as well as the fabric geometric tightness factor. For example, Figs. 6, 7 and 8 show the dynamic interaction of fiber volume fraction, fiber orientation and fabric geometric parameters for three-dimensional woven angle interlock, multiaxial warp knit, and three-dimensional braided structures. From Fig. 6 it can be seen that for a given fiber volume fraction the orientation of the through-thickness yarn varies with the warp and filling yarn linear density ratio. Similarly, as shown in Fig. 7, the relative size of the stitch yarn and insertion yarn has a significant effect on the fiber volume fraction–fiber orientation relationship. A

Table 2
Engineering and processing parameters for textile preforms

Preform	Fiber orientation, θ (°)	V_f	*Processing parameter*
Linear assembly	θ, yarn surface helix angle		Bundle tension, transverse compression, fiber diameter, number of fibers, twist level
roving	$\theta = 0$	0.6–0.8	
yarn	$\theta = 5$–10	0.7–0.9	
Woven	θ_f, yarn orientation in fabric plane		Fiber packing in yarn, fabric tightness factor, yarn linear density ratios, pitch count, weaving pattern
	θ_c, yarn crimp angle		
2-D biaxial	$\theta_f = 0/90$, $\theta_c = 30$–60	~0.5	
2-D triaxial	$\theta_f = 0/90/\pm30$–60, $\theta_c = 30$–60	~0.5	
3-D woven	$\theta_f = 0/90$, $\theta_c = 30$–60	~0.6	
Nonwoven	θ_x, fiber/yarn orientation along X axis		2-D nonwoven: fiber packing in fabric, fiber distribution; 3-D orthogonal: fiber packing in yarn, yarn cross-section, yarn linear density ratios
	θ_y, fiber/yarn orientation along Y axis		
	θ_z, fiber/yarn orientation along Z axis		
	θ_{xy}, fiber distribution on fabric plane		
2-D nonwoven	θ_{xy} = uniform distribution, θ_z	0.2–0.4	
3-D orthogonal	θ_x, θ_y, θ_z	0.4–0.6	
Knit	θ_s, stitch yarn orientation		Fiber packing in yarn, fabric tightness factor, yarn linear density ratios, pitch count, stitch pattern
	θ_i, insertion yarn orientation		
2-D weft knit	$\theta_s = 30$–60	0.2–0.3	
3-D MWK	$\theta_s = 30$–60, $\theta_i = 0/90/\pm30$–60	0.3–0.6	
Braid	θ, braiding angle		Fiber packing in yarn, fabric tightness factor, braid diameter, pitch length, braiding pattern, carrier number
2-D braid	$\theta = 10$–80	0.5–0.7	
3-D braid	$\theta = 10$–45	0.4–0.6	

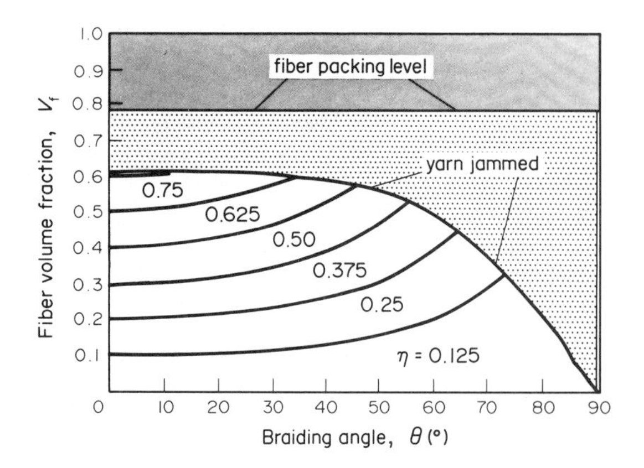

Figure 8
Relationship of fiber volume fraction to braiding angle for various tightness factors (η)

finer stitch yarn tends to yield fabrics with lower bias yarn angle at higher fiber volume fraction. For three-dimensional braids, the maximum fiber volume fraction is 60% which decreases as the tightness factor decreases. However, the interaction of tightness factor and braiding angle is clearly illustrated by the increase of fiber volume fraction with braiding angle for the same braid tightness.

Finally, as a general reference, the practical range of the engineering and processing parameters for various textile preforms, including three-dimensional fabrics, are summarized in Table 2.

Acknowledgements

Much of the work on textile preforms reported herein and specifically two- and three-dimensional fiber-reinforced glass composites has been supported by the Office of Naval Research and the Army Research Office. The assistance provided by Mitchell Marmel of the Fibrous Materials Research Laboratory in the preparation of this manuscript is greatly appreciated.

See also: Woven Fabric Composites: Properties

Bibliography

Adsit N R, Carnahan K R, Green J E 1972. In: Corten H T (ed.) 1972 *Composite Materials: Testing and Design (Second Conference)*. STP 497, American Society for Testing and Materials, Philadelphia, PA, pp. 107–20

Brown R T, Ashton C H 1989 Automation of 3-D braiding machines. *Proc. 4th Textile Structural Composites Symp*, Drexel University, Philadelphia, PA

Bruno P S, Keith D O, Vicario A A Jr 1986 Automatically woven three dimensional composite structures. *SAMPE Q*. 17(4): 10–16

Chou T W, Ko F K 1989 *Textile Structural Composites*. Elsevier, Amsterdam

Dow R M 1989 New concept for multiple directional fabric formation. *Proc. 21st Int. SAMPE Tech. Conf.* Society of Aerospace Material and Process Engineers, Anaheim, CA, pp. 558–69

Fukuta K, Aoki E 1986 3-D fabrics for structural composites. *Proc. 15th Textile Res. Symp.*

Fukuta K, Aoki E, Onooka R, Magatsuka Y 1982 Application of latticed structural composite materials with three dimensional fabrics to artificial bones. *Bull. Res. Inst. Polym. Text.* 131: 159

Geoghegan P J 1988 DuPont ceramics for structural applications—The SEP Noveltex technology. *3rd Textile Struct Composites Symp.*, Drexel University, Philadelphia, PA

Herrick J W 1977 Multidimensional advanced composites for improved impact resistance. *10th National SAMPE Tech. Conf.* Society of Aerospace Material and Process Engineers, Anaheim, CA

Hickman G T, Williams D J 1988 3-D knitted preforms for structural reactive injection molding (SRIM). *Proc. 4th Annu. Conf. Advanced Composites*. ASM International, Materials Park, OH, pp. 367–70

Ko F K 1989a Preform fiber architecture for composites. *Ceram. Bull.* 68(2): 401–14

Ko F K 1989b Three-dimensional fabrics for structural composites. In: Chou T W, Ko F K (eds.) 1989 *Textile Structural Composites*. Elsevier, Amsterdam, pp. 129–71

Ko F K, Du G W 1992 Processing of textile preforms. *Science and Innovation in Polymer Composites Processing*. MIT Press, Cambridge, MA

Ko F K, Kutz J 1988 Multiaxial warp knit for advanced composites. *Proc. 4th Annu. Conf. Advanced Composites*. ASM International, Materials Park, OH, pp. 377–84

Ko F K, Pastore C M 1989 *Atkins and Pearce Industrial Handbook on Braiding*. Atkins and Pearce, Covington, NY

Ko F K, Pastore C M, Yang J M, Chou T W 1986 Structure and properties of multidirectional warp knit fabric reinforced composites. In: Kawata K, Umekawa S, Kobayashi A (eds.) 1986 *Composites '86: Recent Advances in Japan and the United States*. Japan Society for Composite Materials, pp. 21–8

Laurie R M 1970 Polyblends and composites. Appl. Polym. Symp. 15: 103–11

McAllister L E, Lachman W L 1983 In: Kelly A, Mileiko S T (eds.) 1983 *Handbook of Composites*, Vol. 4. North-Holland, Amsterdam, p. 109

O'Shea J 1988 Autoweave: A unique automated 3-D weaving technology. *3rd Textile Structural Composites Symp*. Drexel University, Philadelphia, PA

Palmer R 1989 Composite preforms by stitching. *4th Textile Structural Composites Symp*. Drexel University, Philadelphia, PA

Pastenbaugh J 1988 Aerospatiale technology. *Proc. 3rd Textile Structural Composites Symp*. Drexel University, Philadelphia, PA

Pönitz W 1993 technical literature, Mayer Textile Machinery Company, Oberhausen, Germany

Popper P, McConnell R 1987 A new 3-D braid for integrated parts manufacturing and improved delamination resistance—The 2-step method. 32nd *Int. SAMPE Symp. Exhibition*. Society of Aerospace Material and Process Engineers, Anaheim, CA, pp. 92–102
Scardino F L 1989 Introduction to textile structures. In: Chou T W, Ko F K (eds.) 1989 *Textile Structural Composites*. Elsevier, Amsterdam
Schmidt D L 1972 Carbon–carbon composites. *SAMPE J.* 8: 9
Stover E R, Mark W C, Marfowitz I, Mueller W 1979 Preparation of an omniweave-reinforced carbon–carbon cylinder as a candidate for evaluation in the advanced heat shield screening program. AFML TR-70283. US Air Force
Tarnopol'skii Y M, Zhigun I G, Polyakov V A 1992 *Spatially Reinforced Composites*. Technomic, Lancaster, PA
Williams D J 1978 New knitting methods offer continuous structures. *Adv. Compos. Eng.* (Summer): 12–13

F. K. Ko
[Drexel University, Philadelphia, PA, USA]

Toughness of Fibrous Composites

A polymer containing long strong fibers can fail by any one or a combination of often competing mechanisms of fracture. The failure mechanisms in a damage zone include debonding of the fiber–matrix interface, breakage of the fiber at points of weakness, matrix microcracking between debonded fibers, and fiber pullout in the crack wake (Beaumont and Harris 1972, Wells and Beaumont 1985a, b). If the composite is in the form of a laminate, then crack extension from the root of a notch is generally preceded by splitting and delamination in the damage zone (see *Failure of Composites: Stress Concentrations, Cracks and Notches*).

These failure processes have characteristics that depend on the microstructure and properties of the matrix, fiber and interface (or interphase). An understanding of these mechanisms, their dominance and the conditions under which they operate is important in engineering design because, for a particular application, they determine the ductility of the component or structure, the toughness of the composite and the crack-advance rate that can lead to premature fracture in a cyclic loading situation.

Models have been developed which describe these mechanisms and enable estimates to be made of the various potential contributions to the toughness of the composite (see *Failure of Composites: Stress Concentrations, Cracks and Notches*, Beaumont 1986, Wells and Beaumont 1985a).

1. Modelling Toughness

The aim is to derive for each mechanism an energy absorption equation to predict toughness, based on the physically sound microscopic process.

Our starting point assumes a matrix crack spanned by an unbroken brittle fiber that has partially debonded (Fig. 1) (Wells and Beaumont 1985a, b, c). We select an equation which describes the buildup of a nonuniform stress along the length x of the debonded fiber:

$$\sigma(x)=\sigma_p-(\sigma_p-\sigma_d)\exp(-\beta x) \qquad (1)$$

In Eqn. (1), σ_p is the maximum stress carried by the fiber due to friction at the interface, σ_d is the fiber stress at the tip of the cylindrical interfacial crack and β is an elastic coefficient that takes into account the Poisson contraction of fiber under load. The terms σ_p, β and σ_d can be calculated by inserting values of the appropriate material parameters (Wells 1984) into the following equations (Wells and Beaumont 1985a, b):

$$\sigma_p=\frac{\varepsilon_0 E_f}{v_f}$$

$$\beta=\frac{4\mu v_f E_m}{E_f d(1+v_m)}$$

$$\sigma_d=\left[\frac{8E_f\Gamma_i}{d}\right]^{1/2}$$

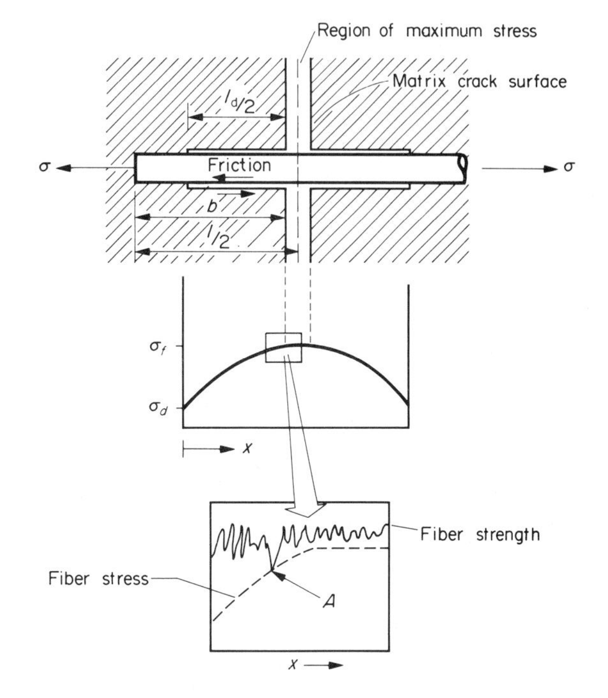

Figure 1
A matrix crack spanned by a brittle fiber. Fiber fracture occurs at the point marked A

Here, E_f, ν_f and E_m, ν_m are the tensile moduli and Poisson's ratios of the fiber and matrix, respectively, ε_0 is the misfit strain, due to the differences in thermal expansion coefficients between fiber and matrix, set up during fabrication, μ is the coefficient of friction between sliding fiber and matrix socket, d is the fiber diameter and Γ_i is the interfacial fracture energy.

As the load on the fiber increases, its diameter decreases and the interfacial (debond) crack spreads until the fiber eventually snaps at σ_f. Rewriting Eqn. (1):

$$\sigma_f = \sigma_p - (\sigma_p - \sigma_d)\exp(-\beta l_d/2) \quad (2)$$

Hence the final debond length of fiber l_d which extends on both sides of the matrix crack is

$$l_d = \frac{2}{\beta}\ln\left(\frac{\sigma_p - \sigma_d}{\sigma_p - \sigma_f}\right) \quad (3)$$

If $\sigma_d > \sigma_f$, no debonding occurs; if $\sigma_f > \sigma_p$ and $\sigma_f > \sigma_d$, debonding extends along the entire length of fiber (Wells and Beaumont 1985a).

The likelihood is that the fiber break will not occur adjacent to the surface of the matrix crack even though the position of maximum fiber stress coincides with the matrix crack plane (Fig. 1). This is because of the variability of strength of a brittle fiber like carbon or glass (Wells and Beaumont 1985a). A statistical model based on Weibull analysis, which predicts the distribution of fiber breaks and hence the average fiber pullout length $\langle l_p \rangle$, has been proposed for long fibers (Wells and Beaumont 1985a) and for short fibers (Wells and Beaumont 1988) but is beyond the scope of this article. Simply, it turns out that

$$\langle l_p \rangle \approx l_d/7 \quad (4a)$$

for glass fibers and Kevlar fibers and

$$\langle l_p \rangle \approx l_d/35 \quad (4b)$$

for carbon fibers.

2. Toughening Mechanisms

2.1 Toughening by Interfacial Debonding

While the fracture energy of an interface is small, the total surface area of cylindrical cracks along fiber–matrix interfaces can be extremely large. It follows that the contribution ΔJ_i this mechanism makes to the enhancement of toughness of the composite is in proportion to the total surface area (Wells and Beaumont 1985a, b):

$$\Delta J_i = 8 l_d \Gamma_i f/d \quad (b > l_d/2) \quad (5a)$$

where f is the fiber volume fraction and l_d can be determined using Eqn. (3) and the appropriate material properties. In a composite containing short fibers where one end of a fiber is close to the matrix crack, the maximum length of debond crack, l_d, given by Eqn. (3), will be unattainable if the fiber pulls out (Fig. 1). The model therefore has to be modified for the case where $b < l_d/2$, where b is defined in Fig. 1:

$$\Delta J_i = 8 b \Gamma_i f/d \quad (b < l_d/2) \quad (5b)$$

2.2 Toughening by Fiber Fracture

A loaded debonded fiber dissipates energy when it snaps. An estimation of this energy can be made from a knowledge of the states of stress in the fiber and matrix immediately before and after fiber fracture (Wells and Beaumont 1985a, b).

The enhancement of composite toughness per unit fracture area due to fiber fracture is

$$\Delta J_f = \frac{f}{E_f}\left[\frac{\sigma_p^2 l_d}{2} - \frac{(\sigma_p - \sigma_d)^2(\exp(-\beta l_d) - 1)}{2\beta} + \frac{2\sigma_p(\sigma_p - \sigma_d)(\exp(-\beta l_d/2) - 1)}{\beta}\right] \quad (b > l_d/2) \quad (6)$$

Likewise, the model has to be modified for short-fiber composites when $b < l_d/2$ by replacing l_d with b in Eqn. (6).

2.3 Toughening by Fiber Pullout

The work done to pull out a fiber whose broken end is a distance l from the matrix crack plane is

$$U = \int_0^l \frac{\pi d^2}{4}\sigma(x)\,dx \quad (7)$$

where $\sigma(x)$ is given by Eqn. (1). Combining Eqns. (1) and (7), the toughness enhancement per unit area of fracture surface is

$$\Delta J_p = f\sigma_p\left[l + \frac{(\exp(-\beta l) - 1)}{\beta}\right] \quad (8)$$

If the average pullout length is $\langle l_p \rangle$ (Wells and Beaumont 1985a) then to a good approximation

$$\Delta\langle J_p \rangle = f\sigma_p\left[\langle l_p \rangle + \frac{(\exp(-\beta\langle l_p \rangle) - 1)}{\beta}\right] \quad (9)$$

3. Toughness Maps

The failure mechanism which dominates (that is, the mechanism that makes the greatest contribution to the total work done to fracture the composite) will depend on the property values which appear in the above equations. First, data on these properties, such as fiber strength and modulus, fiber diameter, matrix modulus and toughness, and interfacial bond strength, are gathered. Second, the fiber debond length and average fiber pull-out length are predicted (Wells and Beaumont 1985a, b); also the frictional stress distribution parameters σ_p and β are estimated from these material properties. Values of these parameters, together with the appropriate equations, are then used to predict the maximum energy dissipated for each mechanism (Eqns. (5), (6) and (9)). Furthermore, a

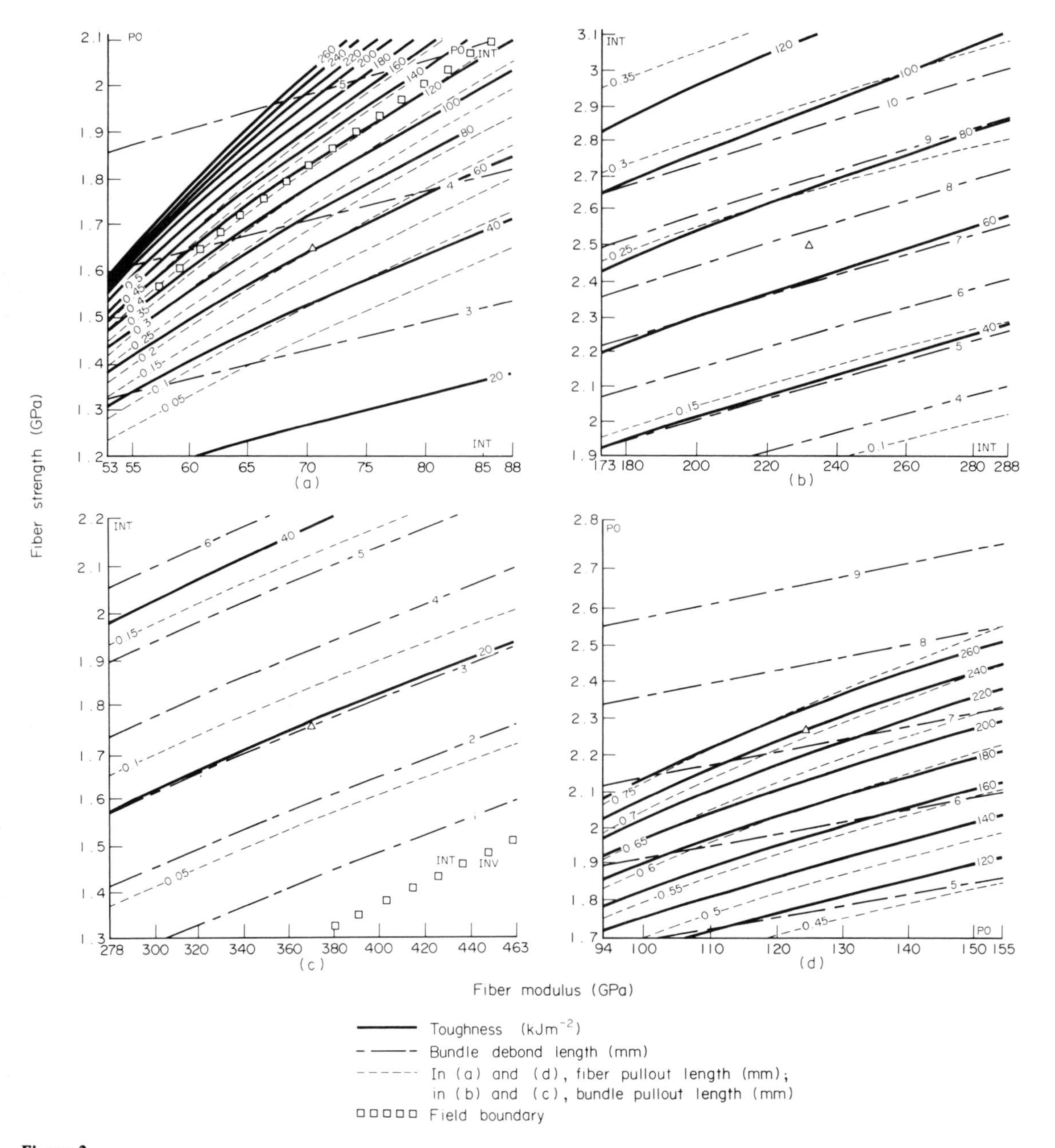

Figure 2
Variation of properties with fiber modulus and fiber strength for (a) E-glass–epoxy composite, (b) high-strength-carbon–epoxy composite, (c) high-modulus-carbon–epoxy composite and (d) Kevlar 49–epoxy composite EL = fiber fracture, INT = debonding, INV = invalid field, PO = pullout

computer is used to construct a diagram or map by allowing any two of the material properties to vary in turn, the remaining parameters being held constant, and for the two variables to be plotted against one another. These two properties form the two axes of the map. Contours of constant total toughness are superimposed on the map. Finally, predictions of l_d and $\langle l_p \rangle$ can be displayed as contours on the map.

Consider the map shown in Fig. 2(a). The map is divided into fields which show the region of fiber strength and fiber modulus for which a particular mechanism is dominant, i.e., where it contributes more than 50% of the total toughness. The field boundaries, shown by small squares, are the loci of points at which the two mechanisms make an equal contribution to the total toughness of the composite. Examples of other maps are shown in Figs. 2(b)–(d) and 3–5 (Wells and Beaumont 1985b).

The maps all show toughness increasing rapidly with fiber strength, and decreasing slowly with increasing fiber stiffness. Likewise, toughness increases with increasing fiber (and fiber-bundle) diameter as well as with decreasing matrix modulus and decreasing fiber–matrix bond strength. Toughness of composites increases with increasing matrix toughness also, but the dependence is a weak one. The gradient of the toughness contours and their spacing indicates the sensitivity of the toughness of the composite to a particular material property.

The models yield reasonable estimates of toughness of the composite, ranking the Kevlar-fiber composite the toughest and the high-modulus carbon-fiber composite the least tough. Interfacial debonding is the principal mechanism of toughening of the glass- and carbon-fiber composites, while fiber pullout is the origin of toughening of the Kevlar-fiber composite.

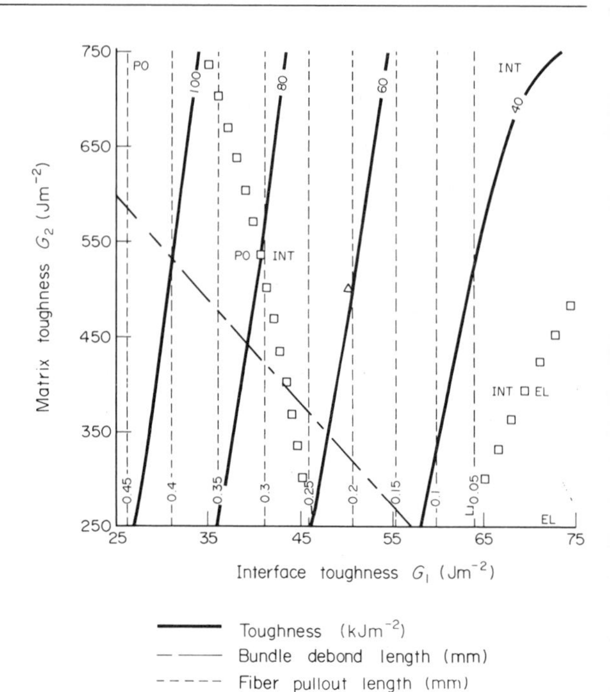

Figure 3
Variation of properties with matrix toughness and interface toughness of an E-glass–epoxy composite

4. Toughness of Short-Fiber Composites

The maximum toughness of a composite containing aligned short fibers dispersed in the matrix can be predicted by summing Eqns. (5), (6) and (9).

Figure 6 shows the variation of predicted toughness with fiber length for fibers with a Weibull modulus of $m=100$ and $m=7$, respectively. The toughness increases to 95% of its maximum when $l \sim 6l_d$. For fibers of uniform strength ($m=100$), behavior is similar except when $l > l_d$, where the toughness rises more slowly because of a reduction in fiber pullout energy. A substantial proportion of the energy absorbed is due to the mechanism involving interfacial debonding J_i and fiber fracture J_f. Consequently, a peak in average fiber pullout length $\langle l_p \rangle$ does not appear in the toughness prediction. The predicted maximum toughness of an aligned 2-D short glass-fiber-epoxy mat is therefore about 11 kJ m^{-2} (Fig. 6), about a factor of 6 smaller than the predicted value of 60 kJ m^{-2} for a unidirectional continuous-fiber composite (Fig. 2(a)) (Wells and Beaumont 1985c).

A crude allowance for the random orientation of short fibers can be made by halving the maximum toughness of a similar composite containing dispersed aligned fibers. The assumption made is that the fibers produce a 2-D random mat and therefore only half of them can be considered as being aligned parallel to each of the two perpendicular directions.

Friedrich (1980) has measured the fracture toughness K_c of an E-glass-fiber–PET (poly(ethylene terephthalate)) composite containing fibers 200 μm in

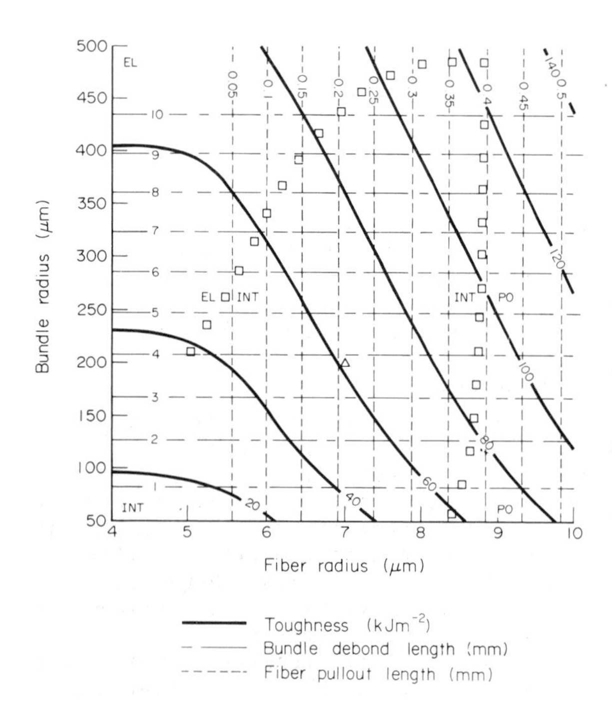

Figure 4
Variation of properties with bundle radius and fiber radius of an E-glass–epoxy composite

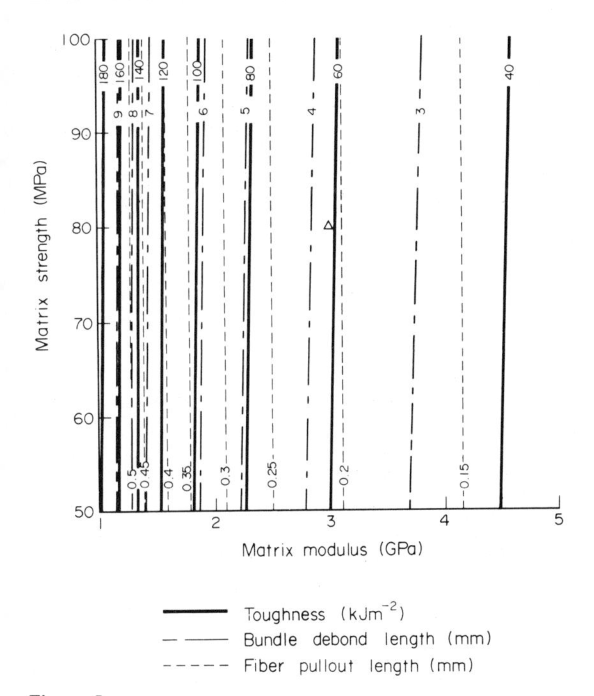

Figure 5
Variation of properties with matrix strength and matrix modulus of an E-glass–epoxy composite

length ($f = 0.5$), dispersed randomly. Using the linear elastic fracture mechanics relation

$$J_c = \frac{K_c^2}{E} \tag{10}$$

and Friedrich's measured values (20 GPa for the tensile modulus E and 8 MPa m$^{1/2}$ for the fracture toughness) a toughness J_c of about 3.2 kJ m^{-2} is calculated. This is in reasonable agreement with the predicted value and supports the validity of this approach.

5. Implications

The theoretical models reviewed in this article provide a basis for assessing trends in the toughness of polymer composites. Although comparison between theory and experiment is sometimes difficult since many of the microstructural parameters which appear in the equations have not been measured, the analyses do provide insight into the origins of toughness. Moreover, they indicate ways of improving the resistance to cracking of this class of material.

There is an idea that toughening mechanisms can combine to provide multiplicative (synergistic) effects, e.g., a combination of a crack bridging mechanism (fiber pullout) with a crack process zone mechanism (interfacial debonding) can generate a toughness much

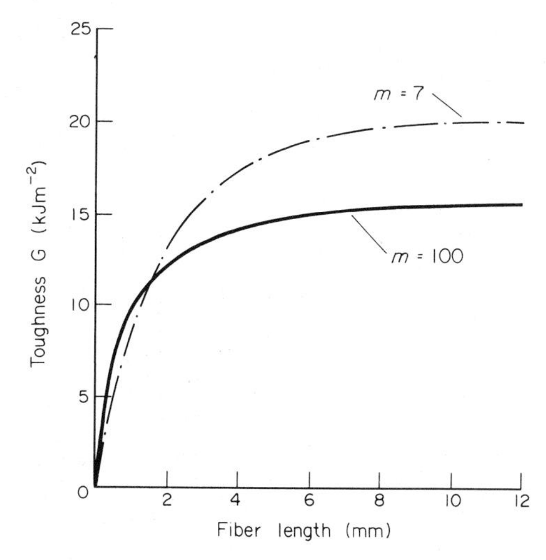

Figure 6
Variation of predicted toughness with fiber length for fibers with differing Weibull moduli

higher than that created if the two forms of contribution were added together. Physically, interaction between the two kinds of toughening mechanism results in an increase in both crack-process-zone size and crack-wake-bridging-zone size. Such behavior needs to be investigated.

See also: Failure of Composites: Stress Concentrations, Cracks and Notches

Bibliography

Beaumont P W R 1986 Toughness of composites. In: Bever M B (ed.) 1986 *Encylopedia of Materials Science and Engineering*, Vol. 7. Pergamon, Oxford, pp. 5120–22

Beaumont P W R, Harris B 1972 The energy of crack propagation in carbon-fiber reinforced resin systems. *J. Mater. Sci.* 7:1265

Friedrich K 1980 Report No. CCM-80-17. Center for Composite Materials, University of Delaware, Newark, Delaware, USA

Wells J K 1984 Micromechanisms of fracture of fibrous composites. Ph.D. Thesis, University of Cambridge Engineering Department

Wells J K, Beaumont P W R 1985a Debonding and pull-out processes in fibrous composites. *J. Mater. Sci.* 20:1275

Wells J K, Beaumont P W R 1985b Crack tip energy absorption processes in fibrous composites. *J. Mater. Sci.* 20:2735

Wells J K, Beaumont P W R 1988 The toughness of a composite containing short brittle fibers. *J. Mater. Sci.* 23:1274

P. W. R. Beaumont
[University of Cambridge, Cambridge, UK]
J. K. Wells
[BP Research Centre, Sunbury, UK]

W

Whiskers

Whiskers are fibers which have been grown under controlled conditions that lead to the formation of high-purity single crystals. The resultant highly ordered structure produces not only unusually high strengths but also significant changes in electrical, optical, magnetic, ferromagnetic, dielectric, conductive and even superconductive properties.

The tensile strength properties of whiskers are far above those of the current high-volume reinforcements, and represent the potential for producing the highest attainable composite strengths.

1. Methods of Whisker Growth

More than 100 materials, including metals, oxides, carbides, halides, nitrides, graphite and organic compounds, have been prepared as whiskers. Whiskers may be grown from supersaturated gas phases, from melts, from solutions by chemical decomposition or electrolysis or from solids.

Whiskers grow by two different mechanisms: basal growth and tip growth. In basal growth, the atoms of growth migrate to the base of the whisker and extrude the whisker from the substrate. This type of growth has been observed on tin-plated structures subject to stress. On the other hand, Al_2O_3 (sapphire), SiC, Si_3N_4 and AlN whiskers are formed by tip growth at temperatures high enough that the vapor pressure of the whisker or whisker-forming material becomes significant. In tip growth, the atoms attach themselves to the tip of a growing whisker by various mechanisms.

Some metallic whiskers have been produced by sublimation or evaporation of the metal, which is transported in the vapor phase to a lower-temperature growth site where, under high supersaturation conditions, condensation produces whiskers. Zinc, cadmium and other metallic whiskers have been grown in a simple chamber with a thermal gradient between the source and the growth site. An inert-gas carrier may be used to help control the rate of whisker growth.

The hydrogen reduction of metal salts has been a frequently used method for the production of metal whiskers such as nickel, iron, copper, silicon and gold. The best temperature is usually near or slightly above the melting point of the source material (Brenner 1956, 1958).

Among oxides prepared as whiskers are Al_2O_3, MgO, MgO–Al_2O_3, Fe_2O_3, BeO, MoO_3, NiO, Cr_2O_3 and ZnO. A simple vapor-transport method of growth consists of heating the metal in a suitable atmosphere (e.g., wet hydrogen, a moist inert gas or air). For example, when a stream of moist hydrogen is passed over aluminum powder which has been heated to 1300–1500°C, a mass of acicular sapphire is deposited in a cooler part of the furnace (Webb and Forgeng 1957).

Since about 1969, the major interest in whiskers has shifted from Al_2O_3 to SiC. A main reason for this shift was the initial indication that SiC whiskers are easier to wet and bond in low-temperature-metal matrices and in polymer matrices. Also, SiC whisker growth is less susceptible to branching growth and thermal degradation. Therefore, there was a greater potential for producing a uniform product and for scale-up of production.

A process for producing α-SiC uses bulk SiC vaporized by heating under reduced pressure, with the whiskers forming on nucleation sites containing lanthanum or another catalyst. β-SiC whiskers have been grown by hydrogen reduction of methyltrichlorosilane onto carbon substrates at 1500°C (Ryan et al. 1967). Another process uses combinations of chlorosilanes, CO and CH_4 as source gases for the production of β-SiC. This process has produced relatively large quantities of high-quality whiskers.

A number of different whiskers have been grown by the VLS (vapor–liquid–solid) technique, including α-Al_2O_3, B, GaAs, GaP, Ge, MgO, $NiBr_2$, NiO, Se, Si and SiC. Figure 1 illustrates the sequence of VLS growth. At a preselected elevated temperature the solid catalyst particle melts and forms the liquid catalyst ball. Carbon and silicon atoms in the vapor feed are extracted by the liquid catalyst, which soon becomes a supersaturated solution of carbon and silicon that precipitates the solid SiC on the substrates. As precipitation continues, the whisker grows, lifting the catalyst ball.

The advantage of VLS growth is that it can be controlled and limited by the location and type of catalyst. A uniform-size catalyst produces a uniform-size whisker (Shyne et al. 1968).

2. Basic Properties of Whiskers

Whiskers are a generic class of materials having mechanical strengths equivalent to the binding forces of adjacent atoms. For example, SiC whiskers have a tensile strength greater than 27 GPa and a Young's modulus greater than 500 GPa. Thus, in contrast to conventional materials—which contain a multiplicity of grain boundaries, voids, dislocations and imperfections—the single-crystal whisker approaches strutural perfection and has eliminated almost all such

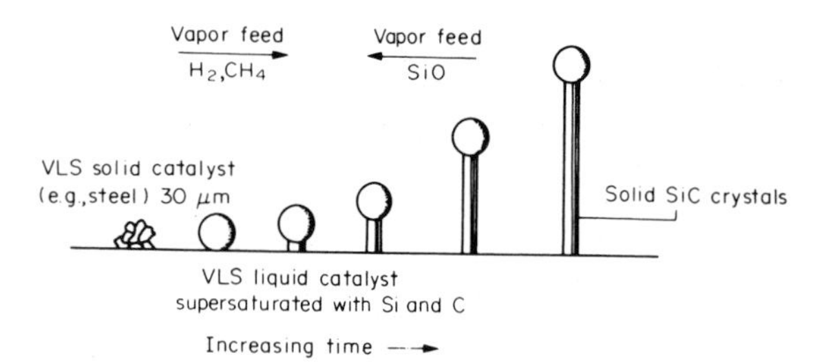

Figure 1
VLS growth sequence of SiC whisker at 1400 °C

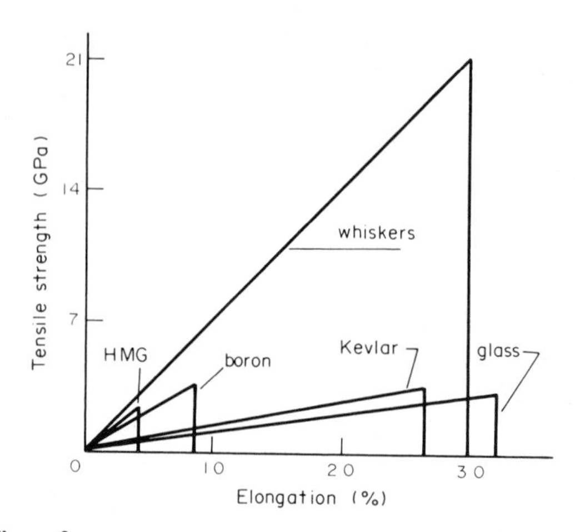

Figure 2
Tensile strength and elongation of reinforcing fibers (HMG= high-modulus graphite)

defects. Because of this internal perfection, the strength of whiskers is not strictly limited by their surface perfection. This is very significant in that it gives whiskers an unusual toughness and nonfriability in handling as compared with polycrystalline fibers and fiberglass.

If the whisker's strength is plotted against its diameter, it is found that as the whisker becomes smaller (and therefore more nearly perfect) its strength increases rapidly. The tensile strength, elongation and modulus of whiskers (see Fig.2) compare favorably with those of other reinforcing fibers such as glass, high-modulus graphite. Kevlar and boron. Whiskers have a five- to six-fold higher basic fiber strength, the high elongation of Kevlar (about 3%), and a modulus greater than that of boron. The disadvantage of whiskers is that an improved technology for preparing discontinuous fiber composites must be developed in order to utilize all their reinforcing potentials.

3. Physical Characteristics

Some of the most common whiskers and their properties are given in Table 1. Ceramic whiskers are unique in that they can be strained elastically as much as 3% without permanent deformation, compared with less than 0.1% for bulk ceramic crystals. In addition, whiskers consistently exhibit much less strength deterioration with increasing temperature than the best conventional high-strength alloys. This would be expected because of the absence of imperfections, which are prone to permit slip. Sapphire whiskers have been found to retain tensile strengths as high as 700 MPa at 2000°C. No appreciable fatigue effects have been observed in whiskers. They can be handled roughly and can be milled and chopped or otherwise worked without having their strength impaired.

Table 1
Physical properties of whiskers

Material	Density (10^3 kg m^{-3})	Melting point (°C)	Tensile Strength (GPa)	Young's modulus (GPa)
Aluminum oxide	3.9	2082	14–28	550
Aluminum nitride	3.3	2198	14–21	335
Beryllium oxide	1.8	2549	14–21	700
Boron oxide	2.5	2449	7	450
Graphite	2.25	3593	21	980
Magnesium oxide	3.6	2799	7–14	310
Silicon carbide (α)	3.15	2316	7–35	485
Silicon carbide (β)	3.15	2316	7–35	620
Silicon nitride	3.2	1899	3–11	380

4. Types of Whiskers and Whisker Products

Whiskers have been produced in a range of fiber sizes and in three forms: grown wool, loose fiber and felted paper. The wool has a fiber cross section of 1–30μm and aspect ratios of 500–5000:1. The bulk density is approximately 0.028 g cm^{-3}, which is less than 1% solids. This is a very open structure, suitable for a vapor-deposition coating on the whiskers or direct use as an insulation sheet. Loose fibers are produced by processing the larger diameter fiber in a blender to yield lightly interlocked clusters of fibers with aspect ratios of 10–200:1. In felt or paper, the whiskers are randomly oriented in the plane of the felt and have fiber aspect ratios ranging from 250–2500:1. The paper as felted has approximately 97% void volume and a density of 0.06–0.130 g cm^{-3}.

4.1 Cobweb Whiskers—Ultrasmall and Ultrastrong

The future of fine-fiber technology may be foreshadowed by new experimental fibers called cobweb whiskers. These fibers are so small that they are seen only as a blue cloud formed by light refraction similar to that which causes the sky to appear blue. Electron micrographs at 200 000 × show the fibers to be 180 Å in diameter. Their strength can only be estimated.

5. Whisker-Reinforced Ceramic-Matrix Composite Development

Ceramic-matrix composites (CMCs) have been around for centuries, as demonstrated by the straw-filled bricks fabricated by the Israelites for the Egyptians in the time of Moses. Ceramic-whisker-reinforced CMCs did not begin to emerge until the late 1960s and early 1970s. In 1967 DeBoskey and Hahn reported that the addition of sapphire whiskers to a dense alumina body "increases the strength of the body and changes the mode of fracture of the ceramic." The development of ceramic-whisker-reinforced CMCs was then renewed with intense effort when SiC whiskers were incorporated in alumina and mullite (Wei and Becher 1985), molybdenum disilicide, glass and glass ceramics (Gac et al. 1985, 1986), magnesium-aluminate spinel (Panda and Seydel 1986), transformation-toughened zirconia (Claussen et al. 1986) and hot-pressed Si_3N_4 (Gac and Petrovic 1985, Shalek et al. 1986). In all of these cases the strength and/or fracture toughness of the composites was improved over that of the monolithic matrix material.

6. Strengthening and Toughening Mechanisms in Ceramics

Mechanisms of strengthening and toughening in ceramic-matrix composites have been reviewed by Rice (1985). Potential toughening mechanisms include fiber pullout, crack deflection, matrix microcracking, crack bowing and crack bridging.

From a practical materials engineering standpoint, three universal features are most important in fiber toughening. First, a high fiber volume fraction is required, preferably 20 vol% or more. Second, a larger-diameter fiber is more desirable than a smaller-diameter fiber. This argues for using 3–10 μm diameter VLS SiC whiskers rather than 0.05–0.5 μm diameter SiC whiskers produced by the vapor–solid (VS) process (Gac et al. 1986). Finally, the higher the fiber tensile strength the better. Thus a high-strength single-crystal SiC whisker would be preferred over a chopped polycrystalline SiC fiber. Fiber aspect ratio and interfacial shear strength are important, but both are governed by the materials system being dealt with and the specific toughening mechanism(s) which may be capitalized on.

Two other concepts merit discussion: prestressing of the fibers and matrix, and modulus transfer of load from matrix to fibers. When developing a composite, the thermal expansion–contraction mismatch between the fiber and matrix should be considered. This mismatch results in residual stress which can directly influence the load-bearing and toughening characteristics of the composite.

7. Silicon Carbide Particulate- or Whisker-Reinforced Metal-Matrix Composites

Silicon carbide particulate composites or whisker-reinforced metal-matrix composites (MMC) exhibit a blend of reinforcement and matrix properties, and properties resulting from their interaction. A typical reinforcement has a diameter of 0.1–6 μm and an aspect ratio of 100–200. Fabrication of these materials is carried out either by casting or compaction from metal powders after mixing with a required amount of reinforcement. The former approach is cheaper but powder-metallurgy processing is the most common technique employed. In powder-metallurgy processing the aluminum powder particle oxide skins must be broken up by metalworking before the highest composite properties can be attained. These MMCs have the advantage of being formable by standard metalworking practices: extrusion, forging, rolling, machining, drilling, grinding and welding. The composite strength is retained to a temperature of approximately 300°C compared to 200 °C for the unreinforced alloy. At all temperatures, the composite elastic modulus is increased substantially over that of the matrix: from 70 GPa without reinforcement to 110 GPa for a fiber volume fraction of 20 vol% SiC whiskers. The tensile yield strength is also increased substantially, from 210 to 300 MPa at 20 vol% SiC (whiskers), and the hardness (Vickers HV10) increases from 130 to 210 for 24 vol% SiC (whiskers). Similar results are obtained for SiC particle–reinforced 6061 aluminum alloy: yield strength increases from 280 to 430 MPa and modulus increases from 70 to 115 GPa from 25 vol% SiC (particles). The ductility decreases from 12% to about 4.5% from 25 vol% SiC (particles)–6061 aluminum alloy. The shear strength for the same material is 280 MPa. The coefficient of thermal expansion decreases monotonically by a factor of 2 for an increase in particle or whisker volume fraction from 0 to 0.50. Fracture toughness values are lower than for the alloy matrix, typically 0.5–0.7 times those of the matrix. The strength properties given above are about doubled for a 7090 aluminum alloy. The fatigue limit at 10 megacycles is improved from 80 MPa for the alloy to 130 MPa for the SiC (whiskers)–Al composite.

These composites respond to heat treatment similarly to the matrix alloy but the aging time needed to achieve peak strength is reduced. Differences between particulate and whisker strengthening are small, and anisotropy in strength and modulus properties for SiC (whiskers)–Al is small. Overall mechanical property improvements in SiC–Mg are of about the same magnitude as for SiC–Al.

Bibliography

Brenner S S 1956 Growth of whiskers by the reduction of metals salts. *Acta Metall.* 4:62–74

Brenner S S 1958 Growth and properties of "whiskers." *Science* 128: 569–75
Brenner S S, Sears G W 1956 Mechanism of whisker growth—III: Nature of growth sites. *Acta Metall.* 4:268–70
Claussen N, Weisskopf K L, Ruhle M 1986 Tetragonal zirconia polycrystals reinforced with SiC whiskers. *J. Am. Ceram. Soc.* 69: 288-302
DeBoskey W R, Hahn H 1967 Opaque lightweight armor—Final Report. Defense Technical Information Center report AD822526
Gac F D, Milewski J V, Petrovic J J, Shalek P D 1985 Silicon carbide whisker reinforced glass and ceramics, *Metal Matrix, Carbon, and Ceramic Matrix Composites 1985*, National Aeronautics and Space Administration Conference Publication 2406, National Technical Information Service, Springfield Virginia, pp. 53–72
Gac F D, Petrovic J J 1985 Feasibility of a composite of SiC whiskers in an $MoSi_2$ matrix. *J. Am. Ceram. Soc., C* 68 (8): 200–201
Gac F D, Petrovic J J, Milewski J V, Shalek P D 1986 Performance of commercial and research grade SiC whiskers in a borosilicate glass matrix. *Ceram. Eng. Sci. Proc.* 7:978–82
Griffith A A 1920 The phenomena of rupture and flow in solids. *Philos. Trans. R. Soc. London, Ser A* 221:163–98
Katz H S, Milewski J V 1978 Whiskers. In: Katz H S, Milewski J V (eds.) 1978 *Handbook of Fillers and Reinforcements for Plastics*. Van Nostrand Reinhold, New York, pp. 446–64
Levitt A P (ed.) 1970 *Whisker Technology*. Wiley–Interscience, London
Milewski J V 1978 Whiskers and microfibers. In: Seymour R B (ed.) 1978 *Additives for Plastics*, Vol. 1, *State of the Art*. Academic Press, London, pp. 79–122
Milewski J V 1979 Short-fiber reinforcements: Where the action is. *Plast. Compd.* 2(6): 17–37
Panda P C, Seydel E R 1986 Neat-net-shape forming of magnesia–alumina spinel/silicon carbide fiber composites. *Am. Ceram. Soc. Bull.* 65 (2): 338–41
Parratt N J 1972 *Fibre-Reinforced Materials Technology*. Van Nostrand Reinhold, London
Rice R W 1985 Ceramic matrix composite toughening mechanisms: an update. *Ceram. Eng. Sci. Proc.* 6(7–8) 589–607
Ryan C E, Marshall R C, Hawley J J, Berman I, Considine D 1967 The Conversion of Cubic to Hexagonal Silicon Carbide as a Function of Temperature and Pressure. Report AF-CRL-67 0436. USAF Cambridge Research Laboratory, Hanscomfield, Massachusetts
Shalek P D, Petrovic J J, Hurley G F, Gac F D 1986 Hot-pressed SiC whisker/Si_3N_4 matrix composites. *Am. Ceram. Soc. Bull.* 65 (2) 351–56
Shyne J J, Milewski J V, Shaver R G, Cummingham A L 1968 Development of processes for the production of high quality long length whiskers. Technical Report AFML-TR 67-402. USAF Materials Laboratory, Dayton, Ohio
Webb W W, Forgeng W D 1957 Growth and defect structure of sapphire microcrystals. *J. Appl. Phys.* 28:1449–54
Wei G C, Becher P F 1985 Development of SiC-whisker-reinforced ceramics. *Am. Ceram. Soc. Bull.* 64 (202): 298–304

J. V. Milewski
[Los Alamos National Laboratory, Los Alamos, New Mexico, USA]

Wood–Polymer Composites

Over the years many improvements in wood properties have been achieved by various treatments and modifications. A relatively recent development is the production of wood–polymer composites. These composites offer desirable aesthetic appearance and high compression strength, hardness and abrasion resistance; they also have improved dimensional stability.

1. Wood Modification

Traditional wood treatments include the use of tars, pitches, creosote, resins and salts to coat the surface or fill its porous structure. Table 1 illustrates the range of new treatments introduced during the period 1930 to 1960. Some of the monomers are of the condensation type and react with the hydroxyl groups in the wood, whereas other chemicals react with hydroxyl groups to form cross-links. Another group of compounds simply bulk the wood by replacing the moisture in the cell walls.

During the 1960s treatment with vinyl-type monomers that could be polymerized into the solid polymer by means of free radicals was introduced (Siau et al. 1965). This vinyl polymerization is an improvement over the condensation reaction because the free-radical catalyst does not degrade the cellulose as do the acid and base catalysts used with other treatments. Vinyl polymers vary in their properties, ranging from soft rubber to hard brittle solids depending upon the groups attached to the carbon–carbon backbone. Examples of vinyl monomers used in wood–polymer composites include styrene, methyl methacrylate, vinyl acetate and acrylonitrile. In general, such vinyl polymers simply bulk the wood structures by filling the void spaces. The free radicals used for the polymerization reaction are usually produced via the use of temperature-sensitive catalysts or ^{60}Co γ radiation: in each case the vinyl polymerization is the same.

2. Chemistry of the Vinyl Polymerization Process

"Vazo" (Du Pont 1967) or 2.2′ -azobisisobutyronitrile is preferred over peroxide catalysts because of its low decomposition temperature and its nonoxidizing nature. Vazo will not bleach dyes dissolved in the monomer during polymerization. Vazo decomposes as follows:

$$(CH_3)_3C{-}N{=}N{-}C(CH_3)_3 \rightarrow 2(CH_3)_3C\cdot + N_2 \quad (1)$$

The rapid decomposition of Vazo with increasing temperature means that the reaction can be initiated at the moderate temperature of 60 °C. Since the decomposition rate of Vazo is negligible at 0 °C, the catalyzed monomer can be stored safely for months at this temperature.

The use of radiation as a source of free radicals has many inherent complications, but it does have advantages. Since the monomer is not catalyzed, it can be

Table 1
Wood modification processes (after Meyer and Loos 1969)

Modification	Details
Acetylation	Hydroxyl groups reacted with acetic anhydride and pyridine catalyst to form esters. Capillaries empty. Antishrink efficiency (ASE) about 70%
Ammonia treatment	Evacuated wood exposed to anhydrous ammonia vapor or liquid at 1 MPa. Bends in 1.25 cm stock up to 90°
Compreg process	Compressed wood–phenolic–formaldehyde composite. Dried treated wood compressed during curing to collapse cell structure. Relative density 1.3–1.4. ASE ~ 95%. Usually thin veneers for cutlery handles
Cross-linking	Catalyst 2% $ZnCl_2$ in wood then exposed to paraformaldehyde heated to 120 °C for 20 min. ASE ~ 85%. Drastic loss of toughness and abrasion resistance
Cyanoethylation	Reaction with acrylonitrile (ACN) with NaOH catalyst at 80 °C. Fungi resistant, impact strength loss
Ethylene-oxide treatment	High-pressure gas treatment, amine catalyst. ASE to 65%
Impreg process	Noncompressed wood–phenolic–formaldehyde composite. Thin veneers, soaked, dried and cured under mild pressure. Swells cell wall, capillaries filled. ASE ~ 75%. Used in modelling cars
Irradiation	Exposure to 10^6 rad of γ radiation gives slight increase in mechanical properties. Above this level cellulose is degraded and mechanical properties decrease rapidly. Low exposure used to temporarily inhibit growth of fungi
Ozone treatment	Gas-phase treatment degrades cellulose and lignin, pulping action
β-propiolactone treatment	β-propiolactone diluted with acetone, wood loaded and heated. Grafted polyester side chains on swollen cell wall cellulose. Carboxyl end groups reacted with copper or zinc to decrease fungi attack. Compression strength increased
Staybwood	Heat-stabilized wood. Wood heated to 150-300 °C
Staypak	Heat-stabilized compressed wood. Wood heated to 320 °C, then compressed, at 2.75–27.5 MPa, then cooled and pressure released. Used for handles and desk legs

stored at ambient temperature as long as the proper amount of inhibitor is maintained. The rate of free-radical generation is constant for a given amount of γ radiation and does not increase with temperature as with the heat-sensitive catalysts. When γ radiation passes through wood and vinyl monomer, it leaves behind a trail of ions and excited states, which in turn produce free radicals. These free radicals, like the catalyst free radicals, initiate the vinyl-monomer polymerization reaction (Meyer 1965).

3. *Impregnation Process*

Figure 1 represents the essential components of a system for the vinyl-monomer impregnation of wood. In this process, air is first evacuated from the wood vessels and cell lumens by means of a vacuum pump. The pump is then isolated from the system and the catalyzed monomer, containing cross-linking agents and on occasion dyes, is introduced into the evacuated chamber through a reservoir at atmospheric pressure. The wood must be weighted so that it does not float in the monomer solution. In the radiation process the catalyst is omitted from the monomer. After the wood is covered with the monomer solution, air at atmospheric pressure is admitted, or dry nitrogen in the case of the radiation process. The monomer solution immediately flows into the evacuated wood structure to fill the void spaces. The time to fill the wood depends upon the structure of the wood and the viscosity of the monomer solution.

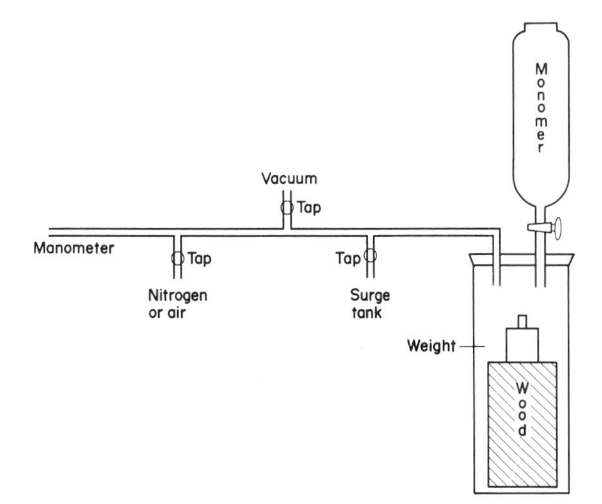

Figure 1
Essential components of a system for the vinyl-monomer impregnation of wood

After the impregnation is complete, the wood–monomer is removed and placed in an explosion-proof oven, or in the γ-radiation source, for curing. On a laboratory scale or in a small production unit the wood–monomer is wrapped in aluminum foil before placing in the curing oven at 60°C. In larger production units the wood–monomer is placed directly into the curing oven, usually in the basket which held the wood during impregnation. In the radiation-cure procedure the thin metal can, in which the wood was impregnated, is flushed with nitrogen and is lowered into a water pool next to the ^{60}Co source (Witt and Morrissey 1972). With high-vapor-pressure monomers, the wood surface is depleted to some extent by surface evaporation, but this depleted zone is usually removed by machining. Methyl methacrylate has a vapor pressure of 40 mm Hg at room temperature, whereas *t*-butyl styrene has a vapor pressure of only 0.8 mm Hg.

4. Monomers for Wood–Polymer Composites

Many different vinyl monomers have been used to make wood–polymers during the past ten years (Langwig et al. 1969), but methyl methacrylate appears to be the preferred monomer for both the catalyst–heat and radiation processes. In fact, methyl methacrylate is the only monomer that can be economically polymerized using γ radiation. On the other hand, all types of liquid vinyl monomers can be polymerized with Vazo or peroxide catalysts.

All vinyl monomers contain inhibitors to prevent premature polymerization during transport and storage. If these inhibitors are not removed before polymerization, the catalyst or radiation must generate enough free radicals to use up the inhibitor before polymerization can initiate. Wood also contains natural inhibitors, depending upon the species of wood. Monomers extract the soluble fractions from the wood structure and these extractives can inhibit the polymerization as well as causing excessive foaming under vacuum.

The polymerization of vinyl monomers is an exothermic reaction and a considerable amount of heat is released, above $75\,kJ\,mol^{-1}$. Since wood is an insulator due to its cellular structure, heat flow into and out of the wood–monomer is restricted. The temperature of the wood–monomer–polymer composite increases rapidly once the reaction is started, and can reach temperatures as high as 250 °C in thick pieces (Duran and Meyer 1972). This increases the vapor pressure of the moisture in the cell walls and drives it out of the wood, causing shrinkage and distortion of the original shape. Wood–polymer composites cured by the catalyst–heat process must be machined to final shape after treatment. Soluble dyes can be added to the vinyl-monomer solution—these produce a three-dimensional depth of color not present in surface-finished wood.

5. Physical Properties and Commercial Applications

Improvements of the physical properties of wood–polymer composites are related to polymer loading. This, in turn, not only depends upon the permeability of the wood species, but also on the particular piece of wood being treated (Young and Meyer 1968). Sapwood is filled to a much greater extent than heartwood for most species. Table 2 gives a few examples of loading and physical-property improvement. Compressive strength, hardness and abrasion resistance are the three most improved properties. Bending, impact resistance, toughness, rupture modulus, work to the proportional limit and other properties are improved to some extent, again depending upon the species involved (Langwig et al. 1968). Along with improved strength wood–polymer composites also have a desirable aesthetic appearance and improved dimensional stability, depending on polymer loading.

Table 2
Selected physical property data on wood–polymer composites

Species		Polymer loading (%)	Density increase (%)	Compressive-Strength increase (%)	Tangential-hardness increase (%)
Sugar maple	S	40	65	160	229
	H	38	58	125	200
Basswood	S	63	168	425	626
	H	62	160	288	505
Yellow birch	S	37	58	146	215
	H	31	43	56	120
Beech	S	36	53	201	261
	H	24	30	30	112
Red pine	S	51	100	636	523
	H	8	7	1	1

S = sapwood H = heartwood

Commercial production of wood–polymer composites for flooring began in the mid-1960s using the radiation process. One company, Perma-Grain Products, is still producing parquet flooring and other items using γ radiation. Commercial catalyst–heat wood–polymer production began in 1967 when the American Machine and Foundry Company produced novel billiard cues. Since 1967 the catalyst–heat system has found its way into many small specialty operations making wood–polymer-composite items, such as archery bows, golf clubs, bagpipes, flutes, guitar fret boards, drum sticks, buttons, jewelry, office-desk items and parquet flooring.

Bibliography

Du Pont 1967 *VAZO Vinyl Polymerization Catalyst*, Du Pont Product Bulletin. Du Pont, Wilmington, Delaware

Duran J A, Meyer J A 1972 Exothermic heat released during catalytic polymerization of basswood–methyl methacrylate composites. *Prod. J.* 6: 59–66

Langwig J E, Meyer J A, Davidson R W 1968 Influence of polymer impregnation on mechanical properties of basswood. *For. Prod. J.* 18: 33–36

Langwig J E, Meyer J A, Davidson R W 1969 New monomers used in making wood-plastics. *For. Prod. J.* 19: 57–61
Meyer J A 1965 Treatment of wood-polymer systems using catalyst–heat techniques. *For. Prod. J.* 15: 362–64
Meyer J A 1977 Wood–polymer composites and their industrial applications. In: Goldstein I S (ed.) 1977 *Wood Technology: Chemical Aspects*, American Chemical Society Symposium Series No. 43. American Chemical Society, Washington, DC, pp. 301–25
Meyer J A, Loos W E 1969 Processes of, and products from, treating southern pine wood for modification of properties. *For. Prod. J.* 19: 32–38
Siau J F, Meyer J A, Skaar C 1965 Dimensional stabilization of wood. *For. Prod. J.* 15: 162–66
Witt A E, Morrissey J A 1972 Economics of making irradiated wood–plastic products. *Mod. Plast.* 49: 78–82
Young R A, Meyer J A 1968 Heartwood and sapwood impregnation with vinyl monomers. *For. Prod. J.* 18(4): 66–68

J. A. Meyer
[State University of New York, Syracuse, New York, USA]

Wood Strength

Wood is the most commonly used natural composite material. It also serves as the raw material for wood-based composites such as plywood or particleboard. Composite action in wood occurs on several levels. Individual cell-wall layers of wood are like filament-wound composites, with the cellulose microfibrils as the filaments embedded in a matrix of lignin and hemicellulose. Cell-wall layers with different angles of filament orientation are next assembled into the cell wall as a layered composite. On a yet larger scale, woods with a pronounced growth-ring structure may form a layered composite which alternates earlywood bands containing thin-walled cells with latewood bands containing thick-walled cells. Perpendicular to the growth rings, and touching every longitudinally aligned cell at least once, are radially oriented, ribbon-like structures called rays which provide a radial stiffening and reinforcement.

In the living tree, wood has several functions, one of which is structural support. The wood of the tree trunk has to withstand the compression loads from the weight of the crown above, and it has to resist bending moments resulting from wind forces acting on the entire tree. The ability of wood to resist loads (i.e., its strength) depends on a host of factors. These factors include the type of load (tension, compression, shear), its direction, and wood species. Ambient conditions of moisture content and temperature are important factors, as are past histories of load and temperature. Strength of a wood sample also depends on whether it is a small, clear piece free of defects or a piece of lumber with knots, splits and the like. This article will be concerned with the strength of clear wood and the major factors that influence it.

1. The Anisotropic Nature of Wood Strength

The cellular structure of wood and the physical organization of the cellulose-chain molecules within the cell wall make wood highly anisotropic in its strength properties. On the basis of the structural organization of wood, three orthogonal directions of symmetry can be distinguished. The longitudinal (L) direction, also referred to as the direction parallel to grain, is parallel to the cylindrical axis of the tree trunk and also to the long axis of the majority of the constituent cells. This direction has the highest proportion of primary bonds resisting applied loads, and is therefore the direction of greatest strength. The other two directions are the radial (R) and tangential (T) directions, which are perpendicular and parallel, respectively, to the circumference of the tree trunk. The latter two directions are also known collectively as directions perpendicular to grain. The strength perpendicular to grain is low because loads are resisted predominantly by secondary bonds.

The tensile strength of air-dry (~12% moisture content) softwoods parallel to grain is of the order of 70–140 MPa. For hardwoods it is often greater but may be less, depending on the particular species. The tensile strengths in the tangential and radial directions are about 3–5% and 5–8%, respectively, of the strength parallel to grain. The tensile strength perpendicular to grain of wood is thus only a small fraction of what it is parallel to grain, and in practical applications tensile stresses perpendicular to grain are avoided to the largest extent possible.

The compression strength (strength as a short column with a slenderness ratio of 11 or less) of air-dry softwoods parallel to grain is of the order of 30–60 MPa, which is much less than the tensile strength. The degree of anisotropy is also less, the compression strength perpendicular to grain being about 8–25% of the value parallel to grain. In contrast to tensile strength, which is always higher radially than tangentially, the compression strength may be either higher or lower, depending on species. Furthermore, compressive strength perpendicular to grain is often a minimum at 45° to the growth rings, that is, intermediate to the radial and tangential directions. The degree of anisotropy in compression depends on species, and in general is less in species of higher density. This is because compression failure of wood as a porous material occurs principally by instability at several levels. In compression parallel to grain, failure of cellulose microfibrils by folding can occur at stresses as low as one half of the ultimate failure stress. At higher stresses, a similar pattern of folding takes place on the cell-wall level, and these eventually aggregate into massive compression failures. In compression perpendicular to grain, failure occurs by collapse and flattening of the cells. Species with higher density have less porosity and thicker cell walls, which makes the cells much more resistant to collapse.

The shear strength of wood is characterized by six principal modes of shear failure, as illustrated in Fig. 1. The shear plane or failure plane may be in one of the three principal planes of wood (the LT, LR and TR planes), and failure in each shear plane may occur by sliding in one of two principal directions. These six modes may be divided into three groups: shear perpendicular to grain, shear parallel to grain, and rolling shear. In shear parallel to grain, both the shear plane and the sliding direction are parallel to grain. In air-dry softwoods, the shear strength parallel to grain is of the order of 5–12 MPa. In shear perpendicular to grain, where the shear plane and the sliding direction are both perpendicular to grain, wood is most resistant, but because of its high degree of anisotropy often fails first in some other mode such as compression perpendicular to grain. Limited data indicate that the shear strength perpendicular to grain is of the order of 2.5–3 times the shear strength parallel to grain. In rolling shear, so-called because the wood fibers can be thought of as rolling over each other as failure occurs in this mode, the strength is least, amounting to 10–30% of the shear strength parallel to grain. There are no consistent or major differences between modes within each of the three groups of shear modes.

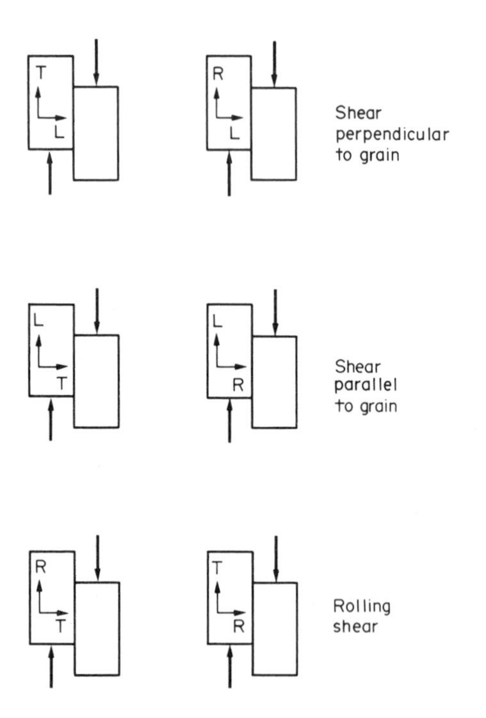

Figure 1
Schematic representation of the six principal modes of shear failure in wood, divided into three groups as shown

Certain problems relating to the strength of wood members can be dealt with by using linear elastic fracture mechanics. The required material parameters, the values of the critical stress intensity factor K_c, also vary because of the anisotropic nature of wood. In the opening mode of crack loading (mode I), there are six principal systems which are somewhat analogous to the six modes of shear failure. The crack plane may be located in one of three principal planes of wood structure, and in each of these the crack may propagate in one of two principal directions. Similarly, six systems are found in the other two modes of crack loading, the forward shear mode (mode II) and the transverse shear mode (mode III), and thus 18 fracture parameters are required to completely characterize wood. Since mode I is the most important, most of the available data are for K_{Ic}. Particular systems are denoted by two subscripts, the first representing the direction normal to the crack plane, and the second the direction of crack propagation. The RT, TR, RL and TL systems (i.e., the four systems where the crack plane is parallel to grain) have very similar K_{Ic} values. Those for the remaining systems, namely LT and LR, where the crack plane is perpendicular to grain, are again similar to each other but almost an order of magnitude greater than the other four. Thus, "weak" and "strong" systems can be distinguished. K_{Ic} values in the weak systems of air-dry softwoods are of the order of 150–500 kPa m$^{1/2}$, and those in the strong systems are 6–7 times greater. Weak system K_{IIc} values are of the order of 1000–2500 kPa m$^{1/2}$.

2. *Methods of Determining Wood Strength*

Many countries have established national standards for testing the mechanical properties of wood. By and large, the objectives and methods are similar. In the USA the methods are given in ASTM Designation D143–82, *Standard Methods of Testing Small Clear Specimens of Timber*. This standard has provisions both for sampling and for the tests themselves. The strength tests included are static bending, two types of impact bending, compression parallel and perpendicular to grain, tension parallel and perpendicular to grain, shear parallel to grain, hardness, and cleavage. All tests must be made at controlled (room) temperature and moisture content, which is usually either green (moisture content above the fiber saturation point) or 12%, and at prescribed speeds of testing. Test specimens are cut from sticks 50 × 50 mm in cross section with the grain parallel to the long dimension. European systems are usually based on sticks 20 × 20 mm in cross section.

The static-bending test is the most important single test. The test specimen is 50 × 50 × 760 mm and is center loaded as a simply supported beam over a span of 710 mm. Load-deflection data are taken so that an elastic modulus of bending may be calculated. The maximum bending moment is used to calculate the modulus of rupture, which is the presumed stress at the extreme fiber of the beam assuming that stress varies linearly through the depth of the beam. Actually, since the compression strength of wood is much less than its tensile strength, there is initial yielding on the compression side, followed by development of visible compression failures and enlargement of the compression zone. The neutral surface shifts toward the tension side as tensile stresses continue to increase. Maximum moment is reached when there is failure in tension. The modulus of rupture is the characteristic bending strength of wood, and its value is intermediate between the tensile and compressive strengths.

The dimensions of compression parallel to grain specimens are 50 × 50 × 200 mm. The maximum load is used to calculate the stress at failure, which is usually referred to as maximum crushing strength. For compression perpendicular to grain specimens, the dimensions are 50 × 50 × 150 mm. The specimen is laid flat, and the load applied through a metal loading block to the middle third of its upper surface (Fig. 2d). This is intended to simulate the type of loading as found in a stud or column bearing on a wood sill, so that the edges of the area under load are laterally supported by the adjacent unloaded area. In this test, a clearly defined maximum load cannot usually be obtained and only a fiber stress at proportional limit is computed.

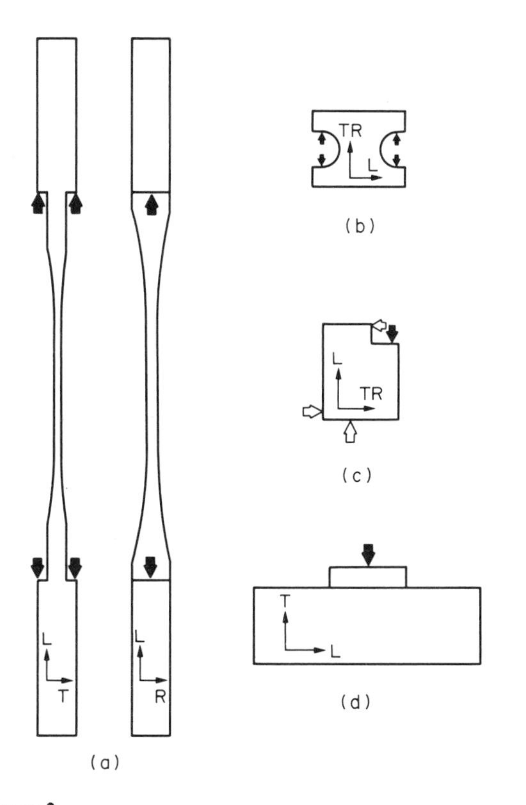

Figure 2
Standard test specimens for testing clear wood: (a) tension parallel to grain: (b) tension perpendicular to grain: (c) shear parallel to grain; and (d) compression perpendicular to grain. Application of loads is indicated schematically

Shear tests are generally made only parallel to grain. The specimen is 50 × 50 × 63 mm, with a 13 × 20 mm notch (Fig. 2c). The notch is cut so that the failure plane is the LT or LR plane in alternate specimens, and the results are averaged.

The four tests described so far are the most significant tests, since allowable stress values for visually graded structural lumber are based on their results. Data obtained from the remaining tests are useful for special purposes and for comparing relative strength of different species for particular applications. This includes tensile tests, because allowable stresses in tension parallel to grain are based on modulus of rupture data and there are no allowable stresses in tension perpendicular to grain. Tensile tests parallel to grain are difficult to make because of the high degree of anisotropy. Specimens are 460 mm long, have a 25 × 25 mm cross section at the ends, and are necked down to a minimum cross section of 4.8 × 9.5 mm (Fig. 2a). Load is applied to the shoulders of notches introduced near the ends. Load transfer from machine to specimen is therefore through shear over an area of 50 cm^2 at each end. Data for tensile strength parallel to grain are not available for most species and are not usually included in tables of strength data. Tensile tests perpendicular to grain are made on specimens with circular notches at each end, and loads are applied (through suitable fixtures) to the inside surface of the notches (Fig. 2b). Because of the geometry of the specimen, the tensile stresses perpendicular to grain are distributed very unevenly over the minimum cross section, so that the tensile strength is underestimated by about one-third.

The two types of impact bending tests are impact bending and toughness. The impact bending test involves dropping a hammer from successively greater heights, the maximum height to cause failure being recorded. This test is now considered to be obsolete. The toughness test is a single-blow impact test using a pendulum to break a 20 × 20 × 280 mm specimen by center-loading over a 240 mm span. The energy required to break the specimen is recorded. Toughness test data are not available for more than a few species.

The cleavage test, which is intended to measure the splitting resistance of wood, is made with a specimen which resembles the tension perpendicular to grain specimen. It has the same circular notch, but only at one end, and is 95 mm long overall. The load per unit of width, as applied at the notch, which causes splitting is recorded.

Hardness in wood is measured by embedding a spherical indentor of 11.3 mm diameter to a depth of 5.6 mm, and recording the load required to do so. Indentations are made on all three principal wood

Table 1
Strength data for four wood species at 12% moisture content

Property	Eastern white pine	Douglas fir	White oak (overcup oak)	Yellow poplar
Static bending: modulus of rupture (MPa)	59	85	87	70
Compression parallel to grain: maximum crushing strength (MPa)	33	50	43	38
Compression perpendicular to grain: fiber stress at proportional limit (MPa)	3.0	5.5	5.6	3.4
Shear parallel to grain: maximum shear strength (MPa)	6.2	7.8	13.8	8.2
Tension parallel to grain: maximum tensile strength (MPa)	78	130	101	154
Tension perpendicular to grain: maximum tensile strength (MPa)	2.1	2.3	6.5	3.7
Impact bending: height of drop causing complete failure (m)	0.46	0.79	0.97	0.61
Toughness: work to complete failure (J)	13	32	37	24
Cleavage: splitting force per unit width (kN m^{-1})	28	32		49
Hardness: force to cause 5.6 mm indentation (kN)				
side grain	1.7	3.2	5.3	2.4
end grain	2.1	4.0	6.3	3.0

planes. Values obtained on the RT plane are referred to as end hardness and those on the LT and LR planes are known collectively as side hardness. Strength values for a few species as obtained from the various tests described are shown in Table 1.

3. Factors Affecting Wood Strength

Various factors can strongly affect the strength of wood. The compression strength parallel to grain, for example, can more than double when drying wood from the green condition to 12% moisture content. The major factors can be divided into three groups as shown below.

3.1 Factors Related to Wood Structure

Although the nature and composition of wood substance is basically the same for all wood, the types of cells, their proportions and arrangements differ greatly from one species to the next. Balsa, a well-known lightweight wood, has an air-dry modulus of rupture of 19.3 MPa, as compared with values of over 200 MPa for a few African and South American tropical species.

Since wood is a biological material, it is subject to environmental and genetic factors that influence its formation. As a result, wood strength is very variable, not only from one species to the next, but also within a species. It will even vary depending on location within a single tree stem. Within a species, values of wood strength follow approximately a normal distribution. The variability of a particular species, as measured by the standard deviation, is generally proportional to the mean value. Therefore, it is convenient to express the standard deviation as a percentage of the mean, which is referred to as the coefficient of variation. Some strength properties tend to be more variable than others, but for most the coefficient of variation is of the order of 20%. This means that 95% of the values will fall in the range from about 60–140% of the mean value.

Much of the variation between and within species can be attributed to differences in wood density. The density of wood substance is constant at 1500 kg m^{-3}, so that wood density is in effect a measure of porosity. As a result, there is a moderate to high degree of correlation between wood density and strength. It can be expressed as

$$S = k\rho^n$$

where S is a strength property, ρ is the wood density, and k and n are constants depending on the particular property. Values of n range from 1.00 to 2.25. If n is larger than unity it suggests that as density increases there is not only an increase in the amount of wood substance (which would imply $n = 1$) but qualitative changes in wood structure as well.

Grain direction has a major effect on strength as already discussed in detail. In structural lumber, which is used as an essentially linear load-bearing element (tension or compression member, or beam), it is desirable that the direction of greatest strength—the direction parallel to grain—should coincide with the long geometric axis. However, in practice the grain direction is often at an angle to the edge of the piece, a condition which is referred to as cross grain. In such a case, uniaxial stress along the geometric axis will have components both parallel and perpendicular to grain. The effective strength could be predicted if a suitable criterion for failure under combined stresses were available. For wood, such a general theory has not yet been found. An empirical relation which is often used is the Hankinson equation:

$$N = PQ/(P \sin^2\theta + Q \cos^2\theta)$$

where N is the strength at an angle θ to the grain, and P and Q are the strengths parallel and perpendicular to the grain, respectively. The variation of strength with grain angle is similar to the variation in Young's modulus, so that even small angles can have a major effect on strength.

In addition to cross-grain, the major strength-reducing characteristics of structural lumber are knots. They are portions of branches which have become part of the stem through normal growth. Although generally harder and denser than the surrounding wood, their grain is more or less perpendicular, so that they contribute little to the strength parallel to the geometric axis of the piece. An additional factor is that the grain of the wood around the knot becomes distorted, representing localized cross-grain which also reduces strength.

3.2 Factors Related to the Environment

Equilibrium moisture content of wood is a function of ambient conditions of relative humidity and temperature. Above the fiber saturation point moisture content has no effect on wood strength, but below it strength increases upon drying for most properties. A notable exception is toughness, which does not change much and may even decrease upon drying. The amount of increase depends on the particular property. Shear strength parallel to grain increases by about 3% for each 1% decrease in moisture content below the fiber saturation point; for maximum crushing strength in compression parallel to grain the increase is 6% per 1% moisture-content decrease. Most tabulated data give strength values when green (S_g) and at 12% moisture content (S_{12}). The strength (S_m) at moisture content M, in the range from 8% moisture content to the fiber saturation point (for strength adjustments commonly assumed to be 25%), can be estimated by the equation

$$S_m = S_{12}(S_g/S_{12})^{[(M-12)/13]}$$

Below 8% moisture content this exponential relation no longer holds as the rate of strength increase becomes less, and in the case of some properties there may even be a maximum at 6 or 7% moisture content. For example, tensile strength both parallel and perpendicular to grain decreases somewhat as the moisture content is decreased below this limit. Since the moisture content of wood in use is rarely much below 6%, this is of little practical significance.

Temperature has both immediate, reversible effects on wood strength and time-dependent effects in the form of thermal degradation. One immediate effect is that strength decreases as temperature is increased. There is an interaction with moisture content, because dry wood is much less sensitive to temperature than green wood. Some properties, such as maximum crushing strength in compression parallel to grain and modulus of rupture, are more sensitive to temperature than others, such as tensile strength parallel to grain. As a general rule, the effect amounts to a decrease in strength of 0.5–1% for each 1 °C increase in temperature. This is applicable in the temperature range of approximately −20 to 65 °C. Temperature effects over a greater range for several properties are illustrated in Fig. 3.

3.3 Agents Causing Deterioration

Deterioration or degradation of wood can be caused by biological agents, by chemicals, and by certain forms of energy. The principal biological agents of concern are decay fungi and wood-destroying insects, although damage may also be sustained because of bacteria, marine borers, and woodpeckers. Damage by wood-destroying insects is by removal of wood substance, so that the effect on strength could be estimated by an assessment of the volume of material removed. In the case of wood decay, such estimates are more difficult because in the early stages of decay wood may appear sound but may have already suffered significant strength loss. To date, there are no satisfactory methods for estimating the residual strength of partially decayed wood members.

Chemicals which do not swell wood may have little or no effect on wood strength. Swelling chemicals, such as alcohols and some other organic solvents, may affect strength the same way water does, in proportion to the amount of swelling. The effect of such chemicals disappears if the chemicals are removed from the wood. Other chemicals may cause degradation of

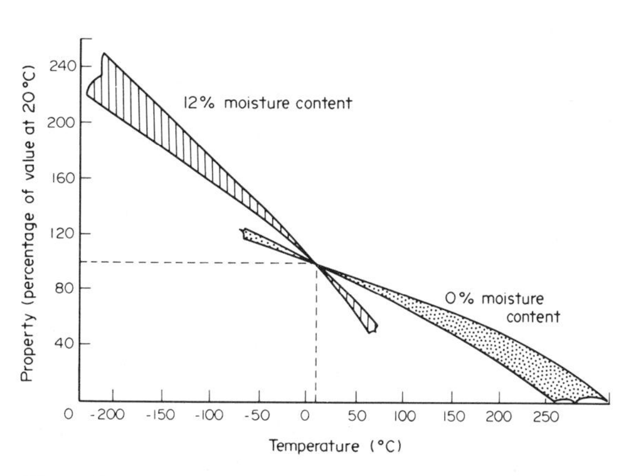

Figure 3
Immediate effect of temperature on bending strength, tensile strength perpendicular to grain, and compression strength parallel to grain. Variability is indicated by the width of the bands (adapted from *Wood Handbook*)

wood by such mechanisms as acid hydrolysis of cellulose. In general, wood is relatively resistant to mild acids but is degraded by strong acids or alkali. Heartwood tends to be more resistant to chemical degradation than sapwood, and softwoods are generally more resistant than hardwoods. Some chemicals used for preservative treatments do affect strength, and also the temperatures used in the treating process can have a permanent effect. Fire-retardant treatments usually result in some loss of strength. Common practice is to take a reduction of 10% in modulus of rupture for design purposes, but available data indicate that the effect on bending strength may be greater.

Wood under sustained loads is subject to static fatigue, commonly referred to as the duration of load factor. Available data show that for clear wood load capacity is linearly related to the logarithm of load duration. Load capacity decreases by 7–8% for each decade of increase on the logarithmic time scale. In structural lumber, the duration of load factor may be less pronounced, but current practice is to make adjustments for duration of loading in design. Dynamic fatigue in wood has not been studied in great detail. Results of tests on clear wood have shown that for fully reversed loading the fatigue strength after 30×10^6 cycles is about 30% of the static strength.

Wood can be degraded with large doses of nuclear radiation. Doses of γ rays in excess of 1 Mrad will cause a measurable effect on tensile strength; at a dose of 300 Mrad the residual tensile strength will be about 10%.

A more common form of degradation is by heat. The amount of degradation is a function not only of temperature but also time of exposure. Measurable effects on strength can be observed at temperatures as low as 65 °C. Degradation is more severe if the heating medium supplies water than if it does not. Softwoods are somewhat more resistant to thermal degradation than hardwoods. As in the case of other forms of degradation, including decay, toughness or shock resistance is affected first, followed by modulus of rupture, while stiffness is affected least. For softwood heated in water, a 10% reduction in modulus of rupture takes 280 days at 65 °C. At 93 °C the same reduction takes 8 days, and at 150 °C it takes only about 3 hours.

See also: Strength of Composites

Bibliography

American Society for Testing and Materials 1987 *Annual Book of ASTM Standards*, Vol. 4. 09: *Wood*. American Society for Testing and Materials, Philadelphia, Pennsylvania

Bodig J, Jayne B A 1982 *Mechanics of Wood and Wood Composites*. Van Nostrand Reinhold, New York

Dinwoodie J M 1981 *Timber, Its Nature and Behavior*. Van Nostrand Reinhold. New York

Forest Products Laboratory 1987 *Wood Handbook: Wood as an Engineering Material*. US Government Printing Office, Washington, DC

Gerhards C C 1982 Effect of moisture content and temperature on the mechanical properties of wood: An analysis of immediate effects. *Wood Fiber* 14: 4–36

Kollmann F F P, Côté W A Jr 1968 *Principles of Wood Science and Technology*, Vol. 1, *Solid Wood*. Springer, Berlin

Patton-Mallory M, Cramer S M 1987 Fracture mechanics: A tool for predicting wood component strength. *For. Prod. J.* 37 (7/8): 39–47

Schniewind A P 1981 Mechanical behavior and properties of wood. In: Wangaard F F (ed.) 1981 *Wood: Its Structure and Properties*. Pennsylvania State University, University Park, Pennsylvania

A. P. Schniewind
[University of California, Berkeley, California, USA]

Woven-Fabric Composites: Properties

The term woven-fabric composites is used here to encompass composite materials reinforced with two-dimensional (2-D) and three-dimensional (3-D) textile preforms. Because of their unique combination of light weight, flexibility, strength and toughness, textile materials have long been recognized as an attractive reinforcement for composites. The recent revival of interest in woven-fabric composites is a result of the need for significant improvements in intra- and interlaminar strength and damage tolerance for structural composite applications. As the needs for composites are being expanded to large-scale structural components, textile reinforcements have increasingly been considered for providing adequate integrity as well as shapeability for near-net-shape manufacturing. This article focuses on the stiffness and stress–strain behavior of 2-D fabric composites. Sources of information on thermal expansion behavior and properties of hybrid materials and 3-D fabric composites can be found in the Bibliography.

Two-dimensional woven fabrics exhibit good stability in the mutually orthogonal warp and fill directions; they provide more balanced properties in the fabric plane than unidirectional laminae. The bidirectional reinforcement in a single layer of fabric enhances impact resistance. The ease of handling and low fabrication cost have made fabrics attractive for structural applications. Triaxially woven fabrics, made from three sets of yarns which interlace at 60° angles, offer improved isotropy and in-plane shear rigidity. There are also other two-dimensional fabrics in the forms of knit fabric and weft-inserted warp-knit constructions.

1. Geometric Characteristics

An orthogonal-woven fabric consists of two sets of interlaced yarns. The length direction of the fabric is known as the warp, and the width direction is referred

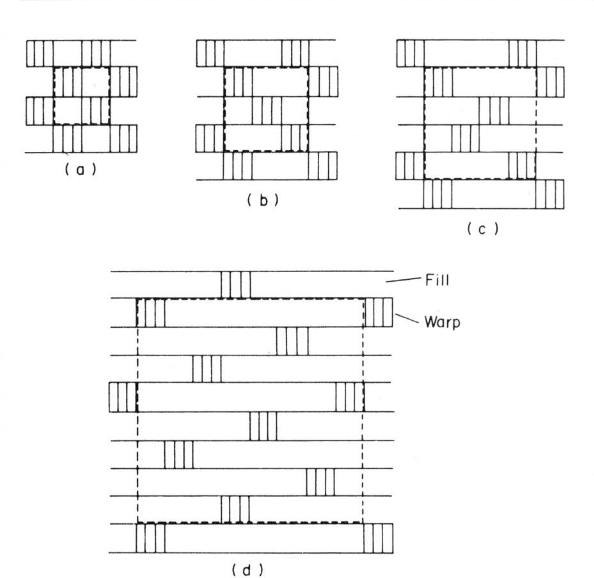

Figure 1
Examples of woven fabric patterns: (a) plain weave ($n_g = 2$); (b) twill weave ($n_g = 3$); (c) 4-harness satin ($n_g = 4$); and (d) 8-harness satin ($n_g = 8$)

to as the fill or weft. The various types of fabrics can be identified by the pattern of repeat of the interlaced regions as shown in Fig. 1. Two basic geometrical parameters can be defined to characterize a fabric; n_{fg} denotes that a warp yarn is interlaced with every n_{fg}th fill yarn and n_{wg} denotes that a fill yarn is interlaced with every n_{wg}th warp yarn. Here, consider only the case of $n_{wg} = n_{fg} = n_g$ for both hybrid and nonhybrid fabrics. Fabrics with $n_g \geqslant 4$ are known as satin weaves. As defined by their n_g values, the fabrics in Fig. 1 are termed plain weave ($n_g = 2$), twill weave ($n_g = 3$), 4-harness satin ($n_g = 4$), and 8-harness satin ($n_g = 8$). The regions in Fig. 1 enclosed by the dotted lines define the "unit cells" or the basic repeating regions for the different weaving patterns. Note also that one side of the fabrics in Fig. 1 is dominated by the fill yarns, whereas the other side is dominated by the warp yarns.

2. Analytical Modelling

Three analytical techniques are available for modelling the thermoelastic behavior of 2-D woven fabric composites. These are known as the mosaic model, fiber crimp model and bridging model. The theoretical basis of the analysis is the classical laminated plate theory. The limitations and applicability of these models are briefly outlined below.

A fabric composite idealized by the mosaic model can be regarded as an assemblage of pieces of asymmetric cross-ply laminates. The key simplification of the model is the omission of the fiber continuity and undulation (crimp) that exist in an actual fabric, and it provides a convenient estimation of the upper and lower bounds of the elastic stiffness and compliance constants.

The crimp (fiber undulation) model is developed to take into account the continuity of fibers in a fabric composite. Figure 2 depicts the unit cell geometry of the model where the boundary surfaces of the fill and warp yarns after matrix consolidation are denoted by $h_1(x)$ and $h_2(x)$, respectively. The width of an ellipsoidally shaped yarn is denoted by a, and a_u indicates the width of the crimp region. Because of the existence of the pure matrix region, the overall fiber volume fraction of the fabric composite can be different from that in the yarn regions. The analysis of the model is based upon the assumption that the laminated plate theory is applicable to each infinitesimal piece (dx) of the model along the fill direction. The orientation angle of the fill yarn is denoted by θ. Having identified the elastic properties of the individual pieces as a function of x, the averaged in-plane stiffness and compliance properties can be obtained for the unit cell. This modelling approach is applicable to the fill and warp directions separately for unbalanced fabrics. Because of its essentially one-dimensional nature, the model is more suitable for fabrics with low n_g values.

The bridging model is developed for satin-weave-based composites where the interlaced regions are not closely connected and are surrounded by regions with straight yarns, or cross-plies (see Fig. 1(d)). Figure 3(a) shows the hexagonal shape of the repeating unit in a satin weave. It is then modified to a square shape (Fig. 3(b)) for simplicity of calculation. A schematic view of the bridging model is shown in Fig. 3(c) for a repeating unit which consists of the interlaced region and its

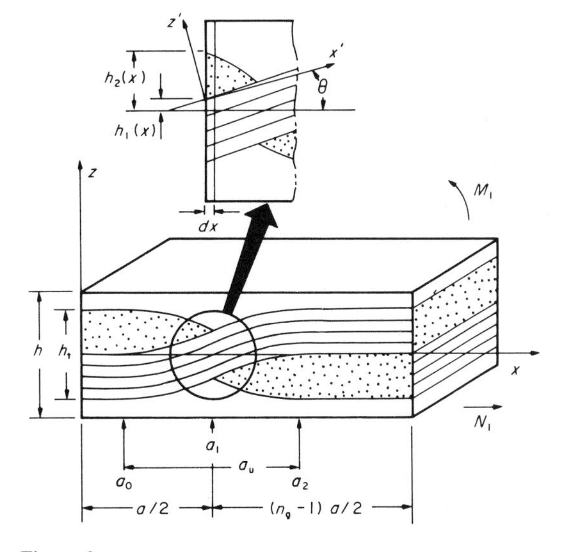

Figure 2
Fiber crimp model

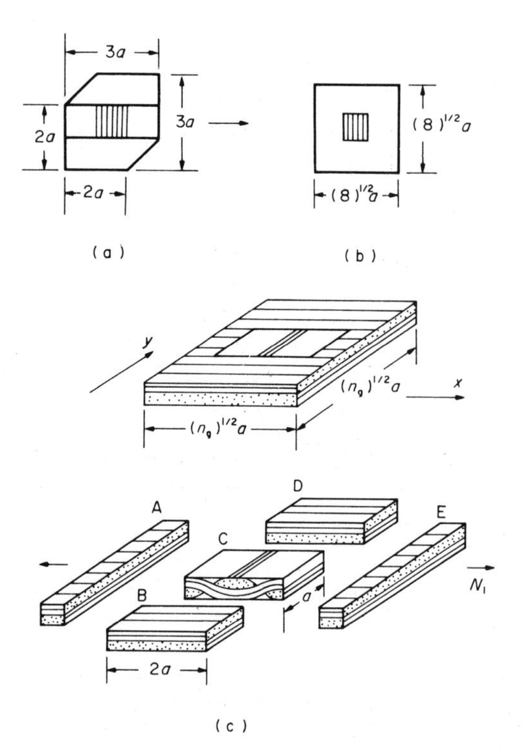

Figure 3
Concept of the bridging model: (a) shape of the repeating unit of 8-harness satin, (b) modified shape for the repeating unit, (c) idealization for the bridging model

surrounding areas. This model is valid only for satin weaves where $n_g \geqslant 4$. The four regions labelled by A, B, D and E consist of straight fill yarns, and hence can be regarded as pieces of cross-ply laminate. Region C has an interlaced structure with an undulated fill yarn. Although the undulation and continuity in the warp yarns are ignored in this model, the effect is expected to be small because the applied load is assumed to be in the fill direction.

The in-plane stiffness in region C, where $n_g = 2$, can be calculated from the crimp model and has been found to be much lower than that of a cross-ply laminate. Therefore, regions B and D carry higher loads than region C and all three regions act as bridges for load transfer between regions A and E. It is also assumed here that regions B, C and D have the same averaged mid-plane strain and curvature. Then the overall elastic properties of the unit cell can be obtained from an averaging technique.

3. *Elastic Properties*

The experimental measurements of the elastic properties of woven-fabric composites are documented for plain and 8-harness carbon-fabric-reinforced epoxy. The plain-weave prepreg is composed of T-300 graphite fibers with a yarn width of 2 mm and prepreg thickness of about 0.2 mm. The number of filaments in the fill and warp yarns is 3000. Ply numbers of 1, 4, 8 and 20 are selected for the plain-weave composites. The 8-harness satin prepreg is composed of graphite fibers and 2-ply laminates are made. The yarn width is 1 mm and the prepreg thickness is about 0.35 mm.

The experimental results of in-plane stiffness A_{11}, nondimensionalized by the corresponding A_{11} of the cross-ply laminate, are presented in Fig. 4 as functions of $1/n_g$ (see *Laminates: Elastic Properties* for the definition of A_{11}). Recall that $n_g = 2$ and 8 for plain-weave and 8-harness satin weave, respectively. The plain-weave data points are based upon 4-ply laminates. The cross-ply values are calculated from the elastic properties of unidirectional carbon/epoxy composites with 65% of fiber volume fraction.

The analytical predictions of composite in-plane stiffness are also presented in Fig. 4. These include the upper bound (UB) and lower bound (LB) predictions of the mosaic model for all n_g values, crimp model (CM) for $4 \geqslant n_g \geqslant 2$, and bridging model (BM) for $n_g \geqslant 4$. It should be noted that in the development of the crimp model and bridging model based upon the classical lamination theory there is no restriction on the local out-of-plane deformation, or warping due to the in-plane force; this corresponds to the local warping allowed (LWA) case in Fig. 4. On the other hand, the local warping constrained (LWC) case arises due to the mutual constraint in out-of-plane deformation in a multilayer fabric composite. It is understood that both LWA and LWC are limiting cases. Also it is assumed in the numerical calculations that $h = h_t$ and $a = a_u$ (Fig. 2).

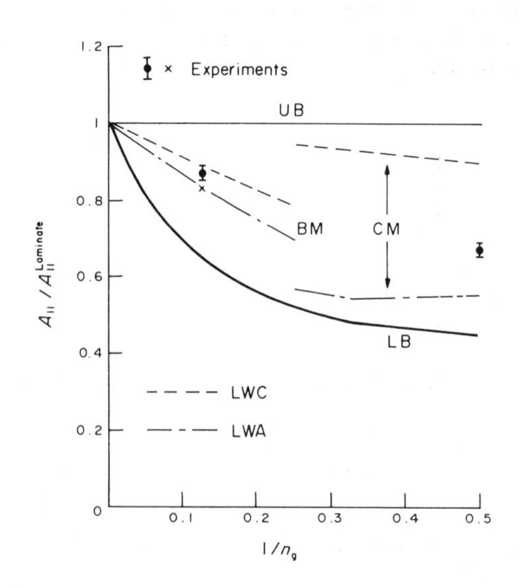

Figure 4
Relationships between nondimensionalized in-plane stiffness and $1/n_g$

A good correlation between theory and experiment has been found for 8-harness satin composites. The experimental data lie between the LWC and LWA predictions where the discrepancy between the two is small. These results suggest good predictability of the bridging model for satin weave composites. For $n_g < 4$, the crimp model, along with the LWC and LWA assumptions, provides somewhat narrower ranges of predictions than those of the upper and lower bounds from the mosaic model. Although the experimental data lie within the range of these predictions, the limitations of the mosaic and crimp models are obvious. The influence of the crimp ratio h/a on in-plane elastic modulus A_{11} also has been examined. The results indicate that plain-weave composites are much more sensitive to the crimp ratio than 8-harness satin.

Figure 5 demonstrates the dependency of elastic moduli of plain weave composites on the number of plies. The in-plane modulus increases from the value for 1-ply, which is slightly higher than the LWA prediction, and reaches values slightly lower than the LWC prediction for 20-ply thick laminate. The levelling off in modulus occurs at 8-ply. The stiffening effect in Fig. 5 is attributed to the suppressing of the warping of the laminate as the number of layers increases.

The correlation between fiber volume fraction V_f of the fabric composite and n_g has been identified based upon the model of Fig. 3. V_f decreases linearly with $1/n_g$, and is approximately 82% of the fiber volume fraction of impregnated yarns for $n_g = 2$. Because of the frequency in fiber crimp, plain-weave fabrics can be closely packed when the "hills" and "valleys" in the fabric profile (Fig. 2) of adjacent layers are allowed to penetrate, a higher fiber volume fraction of approximately 96% of the value of the impregnated yarns can be attained.

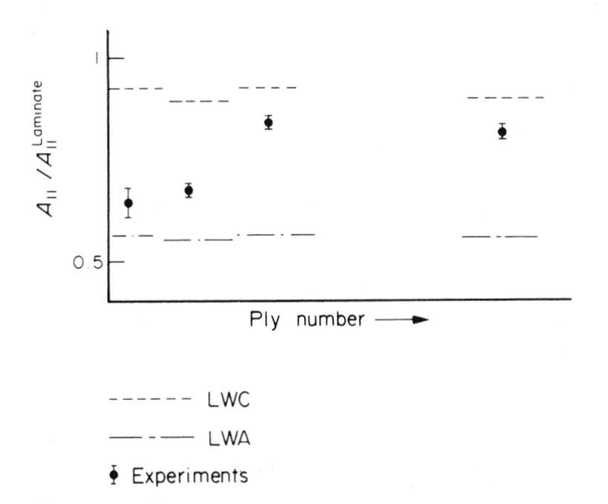

Figure 5
Dependency of in-plane stiffness on ply number in plain-weave composites

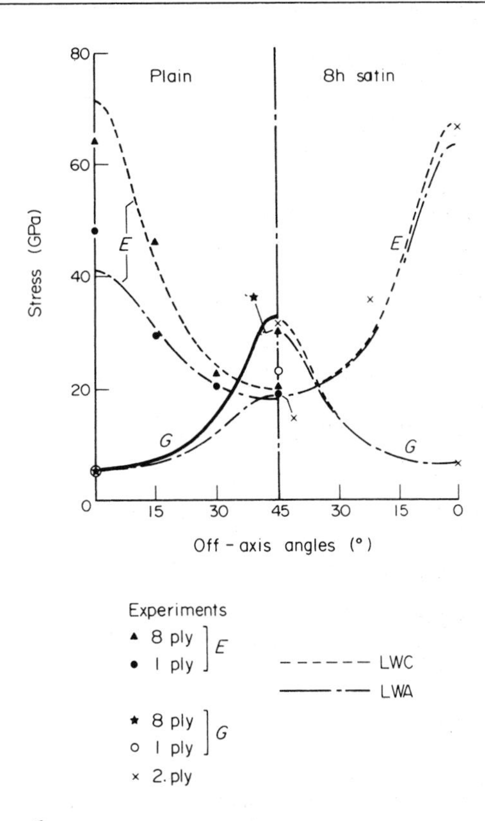

Figure 6
Off-axis elastic moduli of plain weave and 8-harness satin

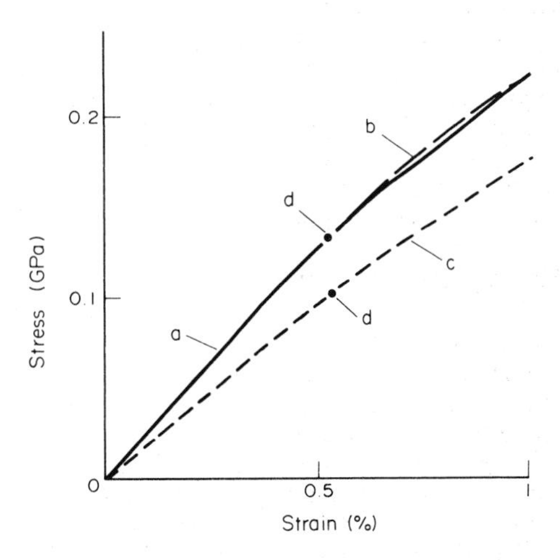

Figure 7
Theoretical and experimental stress–strain curves for a glass/polyimide composite, $V_f = 50\%$ in yarns: a, bridging model solution without bending for 8-harness satin (overall $V_f = 47.7\%$); b, an experimental curve; c, fiber crimp model solution without bending for plain weave (overall $V_f = 40.9\%$); d, knee points

The in-plane shear property, $A_{66}/h = G$, also decreases linearly with $1/n_g$ and hence with V_f. However, A_{66} is much less sensitive to $1/n_g$ than A_{11} is.

The in-plane off-axis elastic modulus E and shear modulus G are presented in Fig. 6. The experimental materials of plain weave and 8-harness satin fabrics are considered to be balanced in the fill and warp directions. Besides the results of A_{11} (Fig. 5), Fig. 6 suggests the dependence of G on ply number for 45° off-axis measurements.

4. Stress–Strain Behavior

Both the crimp and bridging models have been extended to the study of the nonlinear stress–strain behavior of woven fabric composites after initial failure, known as the knee phenomenon. The essential experimental fact for the knee phenomenon is that the breaking strain in the transverse direction (warp yarn) is much smaller than that in the longitudinal direction when the fabric is loaded in the fill direction.

The knee behavior is attributed to the concentration of strain at the center of fiber undulation (Fig. 2) which induces transverse matrix cracking in the warp yarn region and hence, local softening. The concept of successive failure of the warp yarns and the bridging idealization have been combined to study the knee phenomenon and stress–strain behavior of satin composites. Figure 7 compares the numerical and experimental results of stress–strain curves of an 8-harness satin fabric composite of glass/polyimide under bending free condition. The knee point is defined by a deviation of 0.01% in strain from the linear strain. It is also observed from the analysis that the elastic stiffness and knee stress in satin composites are higher than those of in-plane weave composites due to the presence of the bridging regions.

See also: Laminates: Elastic Properties; Structure–Performance Maps; Three-Dimensional Fabrics for Composites

Bibliography

Ishikawa T 1981 Anti-symmetric elastic properties of composite plates of satin weave cloth. *Fibre Sci. Technol.* 15: 127–45

Ishikawa T, Chou T W 1982 Elastic behavior of woven hybrid composites. *J. Compos. Mater.* 16: 2–19

Ishikawa T, Chou T W 1982 Stiffness and strength behavior of woven fabric composites. *J. Mater. Sci.* 17: 3211–20

Ishikawa T, Chou T W 1983 In-plane thermal expansion and thermal bending coefficients of fabric composites. *J. Compos. Mater.* 17: 94–104

Ishikawa T, Chou T W 1983 One-dimensional micromechanical analysis of woven fabric composites. *AIAA J.* 21: 1714–21

Ishikawa T, Chou T W 1983 Nonlinear behavior of woven fabric composites. *J. Compos. Mater.* 17: 399–413

Ishikawa T, Chou T W 1983 Thermoelastic analysis of hybrid fabric composites. *J. Mater. Sci.* 18: 2260–68

Ishikawa T, Matsushima M, Hayashi Y, Chou T W 1985 Experimental confirmation of the theory of elastic moduli of fabric composites. *J. Compos. Mater.* 19: 443–58

Rogers K F, Phillips L N, Kingston-Lee D M, Yates B, Overy M J, Sargent J P, McCalla B A 1977 The thermal expansion of carbon fiber-reinforced plastics, part 1: The influence of fiber type and orientation. *J. Mater. Sci.* 12: 718–34

Yates B, Overy M J, Sargent J P, McCalla B A, Kingston-Lee D M, Phillips L N, Rogers K F 1978 The thermal expansion of carbon fiber-reinforced plastics, part 2: The influence of fiber volume fraction. *J. Mater. Sci.* 13: 433

Zweben C, Norman J C 1976 Kevlar/Thornel 30 hybrid fabric composites for aerospace applications. *SAMPE Q.* 1

T.-W. Chou
[University of Delaware, Newark, Delaware, USA]

LIST OF CONTRIBUTORS

Contributors are listed in alphabetical order, together with their addresses. Titles of articles which they have authored follow in alphabetical order. Where articles are co-authored, this has been indicated by an asterisk preceding the article title.

Arridge, R G C
H H Wills Physics Laboratory
University of Bristol
Royal Fort
Tyndall Avenue
Bristol
BS8 1TL
UK
Particulate Composites

Bader, M G
Department of Materials Science and Engineering
University of Surrey
Guildford
GU2 5XH
UK
Glass-Reinforced Plastics: Thermoplastic Resins

Batdorf, S B
5536B Via La Mesa
Laguna Hills, CA 92653
USA
Strength of Composites: Statistical Theories

Beardmore, P
Engineering and Research Staff
Ford Motor Company
PO Box 2053
Dearborn, MI 48121-2053
USA
Automotive Components: Fabrication

Beaumont, P W R
Department of Engineering
University of Cambridge
Trumpington Street
Cambridge CB2 1PZ
UK
Failure of Composites: Stress Concentrations, Cracks and Notches
Toughness of Fibrous Composites

Bevis, M J
Wolfson Centre for Materials Technology
Brunel University
Uxbridge
UB8 3PH
UK
**Recycling of Polymer-Matrix Composites*

Birchall, J D
Department of Chemistry
Keele University
Keele
Staffordshire
ST5 8BG
UK
Oxide Inorganic Fibers

Bonfield, W
Department of Materials
Queen Mary and Westfield College
University of London
Mile End Road
London
E1 4NS
UK
Artificial Bone

Bowen, D H
Materials Development Division
AEA Technology
Building 47
Harwell Laboratory
Didcot
Oxfordshire
OX11 0RA
UK
Applications of Composites: An Overview
Manufacturing Methods for Composites: An Overview

Bunsell, A R
Centre des Materiaux
École Nationale Supérieure des Mines de Paris
BP 87
F-91003 Evry Cédex
France
Boron Fibers
Long-Term Degradation of Polymer-Matrix Composites
**Silicon Carbide Fibers*

Calvert, P
Department of Materials Science and Engineering
University of Arizona
4715 E Fort Lowell Road
Tucson, AZ 85712
USA
In Situ Polymer Composites

Carlsson, J-O
Institute of Chemistry
University of Uppsala
Box 532
S-751 21 Uppsala 1
Sweden
**Silicon Carbide Fibers*

Chou, T-W
Department of Mechanical and Aerospace Enginering
University of Delaware
107 Evans Hall
Newark, DE 19711
USA
Laminates: Elastic Properties
Structure–Performance Maps
Woven-Fabric Composites: Properties

Clarke, B
MS6 Division
Defence Research Agency
Farnborough
Hampshire
GU14 6TD
UK
**Nondestructive Evaluation of Composites*

Collings, T A
Materials and Structures Department
X32 Building
Defence Research Agency
Farnborough
Hampshire
GU14 6TD
UK
Joining of Composites

Cooke, T F
Textile Research Institute
PO Box 625
Princeton, NJ 08542
USA
Fibers and Textiles: An Overview

Davidson, R
AEA Technology
Harwell Laboratory
Didcot
Oxfordshire
OX11 0RA
UK
Smart Composite Materials Systems

De Luca, P L
Armtek Defense Products Co.
PO Box 848
85-901 Avenue 53
Coachella, CA 92236
USA
**Molded Fiber Products*

East, G C
Division of Ceramics
School of Materials
University of Leeds
Leeds
LS2 9JT
UK
**Silicon Nitride Fibers*

Fishman, S G
United Technologies Research Center
East Hartford, CT 06108
USA
**In Situ Composites: Fabrication*

Gent, A N
Institute of Polymer Science
University of Akron
Akron, OH 44325
USA
**Automobile Tires*

Glomb, J W
Westvaco
290 Park Avenue
New York, NY 10017
USA
**Paper and Paperboard*

Goettler, L A
Monsanto Company
PO Box 5444
260 Springside Drive
Akron, OH 44313-0444
USA
Fiber-Reinforced Polymer Systems: Extrusion

Gray, W
Southwest Research Institute
PO Drawer 28510
6220 Culebra Road
San Antonio, TX 78284
USA
**Composite Armor*

Greer, A L
Department of Materials Science and Metallurgy
University of Cambridge
Pembroke Street
Cambridge
CB2 3QZ
UK
Multilayers, Metallic

Hancox, N L
Materials Development Division
AEA Technology
Harwell Laboratory
Didcot
Oxfordshire
OX11 0RA
UK
High-Performance Composites with Thermoplastic Matrices

Hannant, D J
Department of Civil Engineering
University of Surrey
Guildford
GU2 5XH
UK
Fiber-Reinforced Cements

Haresceugh, R I
Structural Engineering Division
British Aerospace plc
Military Aircraft Division
Warton Aerodrome
Preston
PR4 1AX
UK
Aircraft and Aerospace Applications of Composites

Harris, B
School of Materials Science
University of Bath
Claverton Down
Bath
BA2 7AY
UK
Hybrid Fiber–Resin Composites

Hodgson, A A
65 Wellesley Drive
Wellington Park
Crowthorne
Berkshire
RG11 6AL
UK
Asbestos Fibers

Holt, D
Westland Aerostructures Ltd
East Cowes
Isle of Wight
PO32 6RH
UK
Helicopter Applications of Composites

Hornsby, P R
Wolfson Centre for Materials Technology
Brunel University
Uxbridge
UB8 3PH
UK
**Recycling of Polymer-Matrix Composites*

Jaffe, M
Hoechst Celanese Research Division
86 Morris Avenue
Summit, NJ 07901
USA
High-Modulus High-Strength Organic Fibers

James, D B
221 Aronimink Drive
Newark, DE 19711
USA
**Thermosetting Resin Matrices*

Jeronimidis, G
Department of Engineering
University of Reading
Whiteknights
Reading
RG6 2AY
UK
Natural Composites

Johnson, A F
DFVLR
Pfaffenwaldring 38-40
D-7000 Stuttgart 80
Germany
Glass-Reinforced Plastics: Thermosetting Resins

Kelly, A
Vice-Chancellor
University of Surrey
Guildford
GU2 5XH
UK
Halpin–Tsai Equations
Kirchhoff Assumption (Kirchhoff Hypothesis)
Multiaxial Stress Failure
Multiple Fracture

Ko, F K
Fibrous Materials Research Laboratory
Department of Materials Engineering
Drexel University
Philadelphia, PA 19104
USA
Three-Dimensional Fabrics for Composites

Ko, J-C
Division of Ceramics
School of Materials
University of Leeds

Leeds
LS2 9JT
UK
**Silicon Nitride Fibers*

Lankford, J
Southwest Research Institute
PO Drawer 28510
6220 Culebra Road
San Antonio, TX 78284
USA
**Composite Armor*

Lee, W H
Wolfson Centre for Materials Technology
Brunel University
Uxbridge
UB8 3PH
UK
**Recycling of Polymer-Matrix Composites*

Lemkey, F D
United Technologies Research Center
East Hartford, CT 06108
USA
**In Situ Composites: Fabrication*

Livingston, D I
Livingston Associates
731 Frank Boulevard
Akron, OH 44320
USA
**Automobile Tires*

McAllister, L E
Allied Signal Inc.
Bendix Aircraft Brake &
Strut Division
PO Box 10
3520 West Westmoor Street
South Bend, IN 46624
USA
Carbon–Carbon Composites

McCartney, L N
Division of Materials Applications
National Physical Laboratory
Teddington
Middlesex
TW11 0LW
UK
Percolation Theory

McLean, M
Department of Materials Science
Imperial College of Science, Technology
and Medicine
Prince Consort Road
London
SW7 2BB
UK
Creep of Composites

Meyer, J A
College of Environmental Science
and Forestry
State University of New York
Syracuse, NY 13210
USA
Wood–Polymer Composites

Milewski, J V
Ceramic-Powder Metallurgy Section
Los Alamos National Laboratory
CMB-6 Mail Stop 770
Los Alamos, NM 87545
USA
Whiskers

Mills, P J
Wilton Materials Research Institute
ICI plc
PO Box 90
Wilton
Middlesbrough
Cleveland
TS6 8JE
UK
**Carbon-Fiber-Reinforced Plastics*
**Polymer–Polymer Composites*

Morrell, R
National Physical Laboratory
Teddington
Middlesex
TW11 0LW
UK
Nanocomposites

Mulligan, D D
Westvaco
Covington Research Center
Covington, VA 24426
USA
**Paper and Paperboard*

Newnham, R E
Materials Research Laboratory
Pennsylvania State University
University Park, PA 16802
USA
Nonmechanical Properties of Composites

O'Halloran
American Plywood Association
7011 South 19th Street

Tacoma, WA 98411
USA
Plywood

Pal, S
Department of Mechanical Engineering
Jadavpur University
Calcutta
West Bengal
India
**Solid Fiber Composites as Biomedical Materials*

Phillips, D C
Kobe Steel Ltd
10 Nugent Road
Surrey Research Park
Guildford
GU2 5AF
UK
Fiber-Reinforced Ceramics

Potter, R T
Materials and Structures Department
Defence Research Agency
Farnborough
Hampshire
GU14 6TD
UK
Strength of Composites

Prasad, S V
Materials Division
Regional Research Laboratory
Council of Scientific and Industrial Research
Hoshangabad Road
Habibganj Naka
Bhopal 462 026 (MP)
India
Friction and Wear Applications of Composites
Natural-Fiber-Based Composites

Prevorsek, D C
Allied Signal Inc.
Corporate Technology
PO Box 1021R
Morristown, NJ 07960-1021
USA
**Polymer–Polymer Composites*

Proctor, B A
Fibres and Composites Department
R & D Laboratories
Pilkington Brothers plc
Lathom, Ormskirk
Lancashire
L40 5UF
UK
Continuous-Filament Glass Fibers

Restaino, A J
615 Black Gates Road
Wilmington, DE 19803
USA
**Thermosetting Resin Matrices*

Riley, F L
Division of Ceramics
School of Materials
University of Leeds
Leeds
LS2 9JT
UK
**Silicon Nitride Fibers*

Roberts, T A
Advanced Healthcare Ltd
Duke's Factory
Chiddingstone Causeway
Tonbridge
Kent
TN11 8JU
UK
Dental Composites

Saha, S
Department of Orthopedics
Louisiana State University Medical Center
PO Box 33932
Shreveport, LA 71130
USA
**Solid Fiber Composites as Biomedical Materials*

Schniewind, A P
Forest Products Laboratory
University of California, Berkeley
1301 South 46th Street
Richmond, CA 94804
USA
Wood Strength

Signorelli, R A
NASA Lewis Research Center
Cleveland, OH 44135
USA
Metal-Matrix Composites

Singer, L S
525 Race Street
Berea, OH 44017-2220
USA
Carbon Fibers

Smith, P A
Department of Materials Science and Engineering
University of Surrey
Guildford

GU2 5XH
UK
Carbon-Fiber-Reinforced Plastics

Stone, D E W
MS6 Division
Defence Research Agency
Farnborough
Hampshire
GU14 6TD
UK
Nondestructive Evaluation of Composites

Talreja, R
School of Aerospace Engineering
Geogia Institute of Technology
Atlanta, GA 30332-1050
USA
Fatigue of Composites

Tarverdi, K
Wolfson Centre for Materials Technology
Brunel University
Uxbridge
UB8 3PH
UK
Recycling of Polymer-Matrix Composites

Williams, J C
Technical Consultant
512 Braxton Place
Alexandria, VA 22301
USA
Molded Fiber Products

Zweben, C
Martin Marietta Astro Space Division
Building 100
Valley Forge Space Center
PO Box 8555
Philadelphia, PA 19101-8555
USA
Fibrous Composites: Thermomechanical Properties

SUBJECT INDEX

The Subject Index has been compiled to assist the reader in locating all references to a particular topic in the Encyclopedia. Entries may have up to three levels of heading. Where there is a substantive discussion of the topic, the page numbers appear in ***bold italic*** type. As a further aid to the reader, cross-references have also been given to terms of related interest. These can be found at the bottom of the entry for the first-level term to which they apply. Every effort has been made to make the index as comprehensive as possible and to standardize the terms used.